Lecture Notes in Computer Science   2124

Edited by G. Goos, J. Hartmanis, and J. van Leeuwen

Springer
Berlin
Heidelberg
New York
Barcelona
Hong Kong
London
Milan
Paris
Tokyo

Władysław Skarbek (Ed.)

Computer Analysis of Images and Patterns

9th International Conference, CAIP 2001
Warsaw, Poland, September 5-7, 2001
Proceedings

 Springer

Series Editors

Gerhard Goos, Karlsruhe University, Germany
Juris Hartmanis, Cornell University, NY, USA
Jan van Leeuwen, Utrecht University, The Netherlands

Volume Editor

Władysław Skarbek
Warsaw University of Technology
Faculty of Electronics and Information Technology
Institute of Radioelectronics
ul. Nowowiejska 15/19, 00-665 Warsaw, Poland
E-mail: W.Skarbek@ire.pw.edu.pl

Cataloging-in-Publication Data applied for

Die Deutsche Bibliothek - CIP-Einheitsaufnahme

Computer analysis of images and patterns : 9th international conference ;
proceedings / CAIP 2001, Warsaw, Poland, September 5 - 7, 2001. Wladyslaw
Skarbek (ed.). - Berlin ; Heidelberg ; New York ; Barcelona ; Hong Kong ; London ;
Milan ; Paris ; Tokyo : Springer, 2001
    (Lecture notes in computer science ; Vol. 2124)
    ISBN 3-540-42513-6

CR Subject Classification (1998): I.4, I.5, I.3.3, I.3.7, J.2, I.7

ISSN 0302-9743
ISBN 3-540-42513-6 Springer-Verlag Berlin Heidelberg New York

Springer-Verlag Berlin Heidelberg New York
a member of BertelsmannSpringer Science+Business Media GmbH

http://www.springer.de

© Springer-Verlag Berlin Heidelberg 2001
Printed in Germany

Typesetting: Camera-ready by author, data conversion by PTP-Berlin, Stefan Sossna
Printed on acid-free paper   SPIN: 10839930   06/3142   5 4 3 2 1 0

Preface

Computer analysis of images and patterns is a scientific field of longstanding tradition, with roots in the early years of the computer era when *electronic brains* inspired scientists. Moreover, the design of vision machines is a part of humanity's dream of the artificial person.

I remember the 2nd CAIP, held in Wismar in 1987. Lectures were read in German, English and Russian, and proceedings were also only partially written in English. The conference took place under a different political system and proved that ideas are independent of political *walls*. A few years later the Berlin Wall collapsed, and Professors Sommer and Klette proposed a new formula for the CAIP: let it be held in Central and Eastern Europe every second year. There was a sense of solidarity with scientific communities in those countries that found themselves in a state of transition to a new economy. A well-implemented idea resulted in a chain of successful events in Dresden (1991), Budapest (1993), Prague (1995), Kiel (1997), and Ljubljana (1999).

This year the conference was welcomed at Warsaw. There are three invited lectures and about 90 contributions written by more than 200 authors from 27 countries. Besides Poland (60 authors), the largest representation comes from France (23), followed by England (16), Czech Republic (11), Spain (10), Germany (9), and Belarus (9). Regrettably, in spite of free registration fees and free accommodation for authors from former Soviet Union countries, we received only one accepted paper from Russia.

Contributions are organized into sessions corresponding to the scope of the conference: image analysis (20 papers), computer vision (12), pattern recognition (12), medical imaging (10), motion analysis (8), augmented reality (4), image indexing (7), image compression (8), and industrial applications (6). Several brilliant results are presented and in my opinion the average level of quality of the contributions is high. New trends in these disciplines are well represented.

The 9th conference on *Computer Analysis of Images and Patterns* was organized at Warsaw University of Technology, in September 2001, under the auspices of its Rector, Professor Jerzy Woznicki. We appreciate the kind patronage of the International Association for Pattern Recognition (IAPR) and the Polish Association for Image Processing, the Polish Section of Institute of Electrical and Electronics Engineers, and of the Institute of Radioelectronics in the Department of Electronics and Information Technology.

Major sponsorship was received from Altkom Akademia S.A., a private educational institution in Poland. We also thank The Foundation for the Development of Radiocommunication and Multimedia Technologies for support.

Władysław Skarbek
CAIP 2001 chair

Table of Contents

Image Indexing (MPEG-7)

Image Compression

Pattern Recognition

Augmented Reality

Industrial Applications

Image Analysis

Computer Vision

MPEG-7: Evolution or Revolution?

Mirosław Bober

Visual Information Laboratory
Mitsubishi Electric Information Center Europe
20 Frederic Sanger Road, Surrey Research Park
Guildford, GU1 2SE, UK
miroslaw.bober@vil.ite.mee.com

Abstract. The ISO MPEG-7 Standard, also known as a Multimedia Content Description Interface, will be soon finalized. After several years of intensive work on technology development, implementation and testing by almost all major players in the digital multimedia arena, the results of this international project will be assessed by the most cruel and demanding judge: the market. Will it meet all the high expectations of the developers and, above all, future users? Will it result in a revolution, evolution or will it just simply pass unnoticed?

In this invited lecture, I will review the components of the MPEG-7 Standard in the context of some novel applications. I will go beyond the classical image/video retrieval scenarios, and look into more generic image/object recognition framework relying on the MPEG-7 technology. Such a framework is applicable to a wide range of new applications. The benefits of using standardized technology, over other state-of-the art techniques from computer vision, image processing, and database retrieval, will be investigated. Demonstrations of the generic object recognition system will be presented, followed by some other examples of emerging applications made possible by the Standard. In conclusion, I will assess the potential impact of this new standard on emerging services, products and future technology developments.

Keywords: MPEG-7, multimedia database retrieval, multimedia object recognition

W. Skarbek (Ed.): CAIP 2001, LNCS 2124, p. 1, 2001.
© Springer-Verlag Berlin Heidelberg 2001

The MPEG-7 Visual Description Framework –
Concepts, Accuracy, and Applications

Jens-Rainer Ohm

Rheinisch-Westfälische Technische Hochschule Aachen,
Institut für Nachrichtentechnik
Melatener Str. 23, D-52072 Aachen, Germany
ohm@ient.rwth-aachen.de

Abstract. This paper gives a brief introduction into the Visual part of the forth-coming new standard MPEG-7, the "Multimedia Content Description Interface". It then emphasizes on the aspects how the Visual Descriptors of MPEG-7 were optimized for efficiency, compactness and behavior similar to human visual characteristics. The MPEG-7 descriptors were mainly designed for signal identi-fication and recognition in the context of multimedia applications ; however, they are applicable wherever interoperability between distributed systems de-signed for the task of visual information recognition need a standardized inter-face. I this sense, MPEG-7 may become a key element in the process of conver-gence of multimedia related applications with computer vision systems.

Keywords: MPEG-7, visual descriptor, application scenario

1 Introduction

Recently, ISO's Moving Pictures Experts Group (MPEG) has finalized the standardi-zation of the "Multimedia Content Description Interface", called MPEG-7, which shall provide standardized feature description tools for audiovisual data [1][2]. Part 3 of the standard is handling Visual feature description. Even though database retrieval is fore-seen as one of the first and most illustrative applications of MPEG-7, the general ap-plicability of this description framework is much broader, installing a link between image/video coding and visual-signal recognition techniques. MPEG-7 is not a com-petitor with previous MPEG standards (MPEG-1, MPEG-2, MPEG-4), which were mainly designed for encoding of audiovisual signals with the goal of *reconstruction and rendering* with highest quality. Moreover, it is a step towards the next-higher level of signal representation, describing *signal features*, not *signal samples*. This is highly valuable in the context of systems dealing with the *content* of audiovisual signals, such as automatic identification and recognition systems. The need for a standard arises when interoperability between distributed systems on visual information is necessary, or communication between a human user and a machine about visual information is

W. Skarbek (Ed.): CAIP 2001, LNCS 2124, pp. 2–10, 2001.

required. This paper gives a coarse overview about the Visual elements of the MPEG-7 standard (section 2), explain how the MPEG-7 Visual Descriptors were designed taking into account the characteristics of human perception (section 3), and point out possible applications from different areas as indicated above (section 4).

2   The MPEG-7 Standard and Its Visual Part

The meaning of "content features" to be described by MPEG-7 is widespread, and can consist of elements for
–   high-level description (e.g. manually-generated metadata like authoring information, scripting and editing information) ;
–   mid-level description (e.g. rule-based semantic categories of objects or subjects present within a scene) ;
–   low-level description (e.g. basic visual features like color, texture, shape, geometry, motion within a scene or of a camera).

The work reported in this contribution concentrates on the low-level visual feature description aspects, in which case automatic extraction of features from the data is usually possible, and definition of matching criteria for the similarity using a specific feature type is more or less unique. Even though, it is not the intention that MPEG-7 standardizes the feature extraction, nor the search/retrieval algorithms, which may be differently optimized for specific applications. However, for some features at a very low level, the feature extraction process must be more or less unique, and had to be specified as the semantic meaning of a description element. This is of high importance for interoperability between different systems, or an automatic system and a human, where the "common understanding" about what a specific feature description means must be clear.

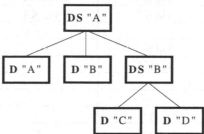

Fig. 1. Example of a simple hierarchical MPEG-7 description scheme structure.

An MPEG-7 description is structured in terms of *Description Schemes* (**DS**) and *Descriptors* (**D**), the latter ones instantiated as *Descriptor Values* (**DV**). A description scheme can contain one or more descriptor(s) and/or subordinate description scheme(s). An example is illustrated in Fig.1, where the **DS** "A" is the top-level DS, containing the **D**s "A" and "B", and **DS** "B", which again is containing **D**s "C" and

"D". A simple hierarchical structure is shown here, but recursive nesting, and linking of DSs is possible as well. Each single descriptor usually characterizes one single feature of the content. The associated descriptor value can be a scalar or a vector value, or some simple structure of different value types, depending on the nature of the descriptor.

Furthermore, MPEG-7 specifies a *Description DefinitionLlanguage* (**DDL**), which is an extension of the XML (eXtensible Markup Language) allowing definition of **DS** and **D** structures (syntactic structure in terms of composition and data types). As XML is a textual format, there exists a complementary *Binary Format fo MPEG-7 data* (**BiM**), which is more efficient in terms of data size of description. In the Visual part of MPEG-7, containing all description elements related to images, video, graphics etc., each **D** is specified syntactically in **DDL** and in a specifically tuned, compact binary representation. Further, semantics is explained, which specifies the exact meaning of **DV**s related to each **D**.

Table 1. Overview MPEG-7 Visual Descriptors.

	MovingRegionDS	*VideoSegmentDS*	*StillRegionDS*
Color			
Scalable Color	via K/F	via K/F	x
Group Of Frames Histogram	x	x	
Dominant Color	via K/F	via K/F	x
Color Layout	via K/F	via K/F	x
Color Structure	via K/F	via K/F	x
Texture			
Homogeneous Texture	via K/F	via K/F	x
Edge Histogram	via K/F	via K/F	x
Texture Browsing	via K/F	via K/F	x
Shape			
Region Shape	via K/F	--	x
Contour Shape	via K/F	--	x
Motion			
Camera Motion	--	x	--
Motion Trajectory	x	--	--
Parametric Motion	x	x	--
Motion Activity	x	x	--
Localization			
Region Locator	--	--	x
Spatio-Temporal	x	--	--

The most important MPEG-7 **DS**s related to Visual **D**s are the MovingRegionDS, the VideoSegmentDS and the StillRegionDS. Table 1 gives an overview about the most important Visual Ds and their relationship with these three DS types. The notion "via

K/F" means that a descriptor originally designed for still image/region can be attached to a video or moving region via a keyframe.

The set of Visual Descriptors covers elementary methods that are frequently employed in tasks like automatic image/video analysis, spatial/temporal segmentation, similarity analysis, as applicable in computer vision, content-based browsing/retrieval, surveillance and many other areas. Even though many systems for these applications have been developed already, proprietary methods have been used in specification of the visual features. As this may be useful for optimized standalone systems, it is not a viable solution in networked environments, where distributed components must interoperate. Moreover, one of the main purposes of the development of MPEG-7 has been *encoding* of signal features, i.e. achieving the best tradeoff between compactness (e.g. number of bits, number of values in a Descriptor) and expressiveness of a feature representation. As a by-product, optimum feature encoding also allows classification based on these features with minimum complexity : This is due to the fact that the comparison, e.g. based on a feature vector distance calculation, is directly proportional to the dimensionality, data type and bit precision of the underlying values. Moreover, there is some evidence that there exists a close relationship between information-theoretic concepts and expressiveness of a feature descriptor. As an example, Fig. 2 shows some retrieval results that were achieved in experimentation based on the Scalable Color Descriptor, which is a multi-resolution representation of an HSV color histogram [3]. These are parameterized against the entropy per histogram line, which for a discrete number J of histogram value levels with probabilities $P(j)$ is given as

$$H = - \sum_{j=1}^{J} P(j) \, \log_2 P(j) \tag{1}$$

and the number of histogram lines (varied between 32-256). Lower value of "ANMRR" (see next section) means better retrieval, i.e. better expressiveness of the descriptor.

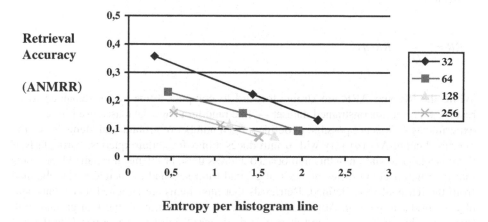

Fig. 2. Compactness versus retrieval accuracy of a color histogram (HSV color space).

3   Criteria for Visual Descriptor Efficiency

Core experiments were extensively conducted during the MPEG-7 standardization process to compare different proposed, competing technologies and optimize the selected approaches. For Visual **D**s, the retrieval application was found to be the best model : A good retrieval result in response to a visual-feature based query would be a good indicator for the expressiveness of the descriptor. The so-called *query by example* paradigm has been employed as the primary method for evaluations. In query-by-example, the respective descriptor values are extracted from the query image, and then matched to the corresponding descriptors of images contained in a database. In order to be objective in the comparisons, a quantitative measure was needed. This required for each Descriptor specification of a data set, a query set and the corresponding ground-truth data. The ground-truth data is a set of visually similar images for a given query image.

For statistically valuable results, the number of queries should be about 1% of the number of images in the database. For example, in the color experiments, the color image dataset consisted of more than 5000 images [4], and a set of 50 queries, each with specified ground truth images, has been defined. Based on these definitions, queries were run, and the query results had to be weighted based on some numeric measure. A very popular measure for this purpose is the Retrieval Rate (RR), defined for query q as

$$RR(q) = \frac{NF(\alpha, q)}{NG(q)}, \tag{2}$$

where $NG(q)$ is the size of the ground truth set for a query q, and $NF(\alpha, q)$ is the number of ground truth images found within the first $\alpha NG(q)$ retrieved items. $RR(q)$ can take values between 0 and 1, where 0 stands for "no image found", and 1 for "all images found". The factor α shall be 1, where a larger α is more tolerant to occasional misses. If (2) is performed over the whole set of NQ queries, the Average Retrieval Rate (ARR) is given by

$$ARR = \frac{1}{NQ} \sum_{q=1}^{NQ} RR(q). \tag{3}$$

While the RR and ARR are straightforward to compute, some issues remained to be resolved. For an unconstrained dataset — like random image databases used in retrieval experiments — it is not possible to define a fixed number of ground truth items for all the queries. Letting $NG(q)$ vary with q introduces a bias for certain queries, particularly if there is a large variation in this number. To address these problems, normalized measures that take into account different sizes of ground truth sets and the actual ranks obtained from the retrieval were defined. Retrievals that miss items are assigned a constant penalty. Consider a query q. Assume that as a result of the retrieval, the k-th ground truth image for this query q is found at a specific $Rank(k)$. Further, a number K is defined which specifies the "relevant ranks", i.e. the ranks that would still count as feasible in terms of subjective evaluation of retrieval. We therefore define a modified rank, *Rank*(k)* as:

$$Rank * (k) = \begin{matrix} Rank(k) & if & Rank(k) & K(q) \\ 1.25 \ K & if & Rank(k) > K(q) \end{matrix}$$

(4)

$$K = \min\{4 \ NG(q), 2 \ \max[NG(q) \forall q]\}$$

From (4), we get the Average Rank for query q

$$AVR(q) = \frac{1}{NG(q)} \sum_{k=1}^{NG(q)} Rank * (k)$$

(5)

To eliminate influences of different $NG(q)$, the *Modified Retrieval Rank*

$$MRR(q) = AVR(q) - 0.5 - \frac{NG(q)}{2}$$

(6)

is defined, which is always larger than 0, but with upper margin still dependent on *NG*. This finally leads to the *Normalized Modified Retrieval Rank*

$$NMRR(q) = \frac{MRR(q)}{1.25K + 0.5 - \frac{NG(q)}{2}}$$

(7)

which can take values between 0 (indicating whole ground truth found) and 1 (indicating nothing found). From (7), finally the *Average Normalized Modified Retrieval Rank* (ANMRR) is defined :

$$ANMRR = \frac{1}{NQ(q)} \sum_{q=1}^{NQ(q)} NMRR(q)$$

(8)

The Measure of ANMRR has been used in the MPEG-7 Core Experiments wherever the whole set of queries consisted of ground truth sets with variable size, which was the case in the Color Descriptors group and for the Edge Histogram Descriptor. In most other cases, it was possible to define equally-sized ground truth sets deterministically, such that it was possible to employ the ARR measure (2). In any case, observation of human visual characteristics, as expressed by the ground truth definitions, has been an important design goal. There is evidence that the ANMRR measure approximately coincides linearly with the results of subjective evaluation about retrieval accuracy of search engines. Fig. 3 shows the high correlation between this measure and the retrieval quality explored in a subjective experiment [5].

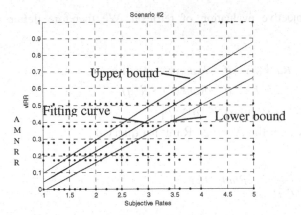

Fig. 3. Least squares fit of ANMRR on subjective rates (color similarity retrieval experiment)

4   Application Scenarios

Most MPEG-7 Visual applications will be based on distance measurements between a specific item's visual feature(s) and some feature reference. A distance measure that was found adequate for similarity measurements with most of the MPEG-7 Visual **Ds** is the L_1 norm. Assume that the feature with index i of two Visual items A and B is described in a vector-valued Descriptor as $\mathbf{vect}^{(i)}(A)$ and $\mathbf{vect}^{(i)}(B)$, respectively. Then, the L_1 distance with regard to this feature is defined as

$$d_i = L_1(A, B) = \left\| \mathbf{vect}^{(i)}(A) - \mathbf{vect}^{(i)}(B) \right\| \tag{9}$$

If there is a set of different descriptors describing a Visual item, a similarity-cost function will be associated with each descriptor, and final ranking is performed by weighted summation of all descriptors' similarity results. If I features with individual matching criteria d_i are compared, normalization is necessary. This can, e.g., be based on the variance of distances

$$s_i = \frac{d_i}{\sigma^2(d_i)}. \tag{10}$$

Further, an individual weighting factor w_i can be associated with each feature, such that highest similarity is achieved for the Visual item, which minimizes the following cost function (cf. Fig. 4a) :

$$s = \sum_{i=1}^{I} w_i \, s_i \quad with \quad \sum_{i=1}^{I} w_i = 1. \tag{11}$$

Depending on the application, this weight factor can be adapted according to the relevance of a feature. This aspect, however, is beyond the scope of this paper.

(11) expresses a linear combination of different feature distances, which signifies an exhaustive comparison over all feature values available. To speed up computation, it is useful to employ a coarse-to-fine approach (Fig. 4b). If specific features can be classified as dominant, a coarse pre-selection is possible, which identifies a subset of visual items with highest similarity. Likewise, for one single feature, a specific part of the description can play a dominant role, as it is the case e.g. for the smaller-size histograms in the Scalable Color Descriptor. Descriptions of additional features or of the same feature at a finer granularity level are then compared in subsequent phases of the search only on the pre-selected items. A proper implementation of a coarse-to-fine search will yield best results with a much faster comparison than the parallel approach.

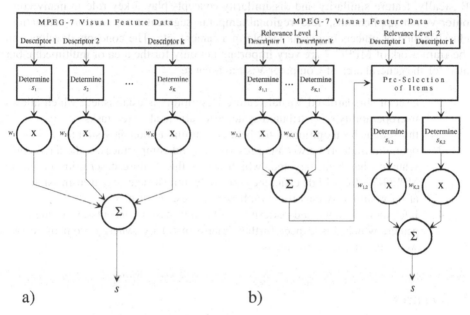

Fig. 4. a Linearly-weighted combination of distances derived from different Visual Ds. **b** Coarse-to-fine matching

The concepts explained so far can either be employed for classification of *similarity* or *dissimilarity*. In the area of multimedia content (images/video), typically similarity would be used for retrieval – e.g. finding most similar images in databases. Fig.5 shows examples for similar images classified based on the color feature (top : indoor scenes with persons) and on the basis of texture feature (bottom), where the query image is the one shown at top/leftmost position. Likewise, maximum similarity criterion can be used if one single representative image shall be selected from a set of given similar images, e.g. selection of a keyframe from a video shot. Based on MPEG-7 **Ds**, it is straightforward to select the image which is most similar to the centroid (mean over all weighted feature values) of the cluster established by the set.
Visual dissimilarity is useful in the case when an unexpected event occurs. This is usually detected if the distance (e.g. (11)) exceeds some pre-defined threshold value.

Dissimilarity measures are useful e.g. in the shot boundary detection (video segmentation), or in surveillance applications (automatic generation of alarms).

During the conference, application examples from the areas of image/video database retrieval, video segmentation and video browsing / keyframe selection will be demonstrated.

5   Conclusions

Basically, feature similarity and dissimilarity concepts play a key role in many computer-vision applications, like regional/temporal segmentation, motion estimation, classification of objects or events visible in a camera, etc. The concepts developed in the framework of MPEG-7 are very important not only for the area of multimedia, but also for the general area of Computer Vision, because

1. a set of standardized Visual feature Descriptors is established, which allows interoperability of distributed systems for artificial vision tasks ;
2. this set has been optimized for behavior similar to human vision ;
3. optimum tradeoff between expressiveness and compactness of the data representation has been targeted, which means that feature description based on MPEG-7 offers high efficiency not only for storage and transmission, but also for similarity classification based on these descriptors ;
4. Information-theory and pattern-classification based concepts are brought together, which lets expect further improvement by synergy from these two well-explored scientific areas.

References

1. ISO/IEC/JTC1/SC29/WG11 : "Text of ISO/IEC 15938-3 Multimedia Content Description Interface – Part 3 : Visual. Final Committee Draft", document no. N4062, Singapore, March 2001.
2. ISO/IEC/JTC1/SC29/WG11 : "MPEG-7 Visual Experimentation Model (XM), Version 10.0", document no. N4063, Singapore, March 2001.
3. Zier, D., Ohm, J. -R. : Common Datasets and Queries in MPEG-7 Color Core Experiments, ISO/IEC JTC1/SC29/WG11 (MPEG) document no. M5060, Melbourne, October 1999.
4. Ohm, J.-R., Makai, B.: Nonlinear representation of histogram bin values, ISO/IEC JTC1/SC29/WG11 (MPEG) document no. M5756, Noordwijkerhout, March 2000.
5. Ndjiki-Nya, P. J., Restat, P., Meiers, T., Ohm, J. -R., Seyferth, A., Sniehotta, R. : Subjective Evaluation of the MPEG-7 Retrieval Accuracy Measure (ANMRR), ISO/IEC JTC1/SC29/WG11 (MPEG) document no. M6029, Geneva, May 2000.

MPEG-7 Color Descriptors and Their Applications

Leszek Cieplinski

Mitsubishi Electric Information Technology Centre Europe,
Visual Information Laboratory,
20 Frederick Sanger Road,
Guildford, Surrey GU2 7YD
United Kingdom
Leszek.Cieplinski@vil.ite.mee.com

Abstract. Color is one of the most important and easily identifiable features for describing visual content. The MPEG standardization group developed a number of descriptors that cover different aspects of this important visual feature. The objective of this paper is to review them and present some applications of the color descriptors by themselves and in combination with other visual descriptors.

Keywords: image indexing, color descriptor, MPEG-7

1 Introduction

Color is one of the most important and easily identifiable features of visual content. A significant amount of research has been conducted on various aspects of color characterization relating to its perception, coherency and spatial distribution. As a result, several approaches to its description have been proposed. MPEG-7 has standardized a subset of these approaches in the form of color descriptors.

This paper provides an overview of color descriptors defined by the MPEG-7 Standard, as well as to explain the process and motivation for their selection. Some applications of the MPEG-7 color technology are also briefly introduced.

Detailed normative information on each descriptor can be found in the Standard document [1] (currently in Final Committee Draft). The non-normative components are specified in the Experimentation Model document [2]. Finally, a general overview of the MPEG-7 standard is given in [3]. All the documents listed above can be found on the official MPEG website: http://www.cselt.it/MPEG/. An overview of the MPEG-7 visual part can be also found in [4,5].

This paper is structured as follows. It starts with a description of the process used for the selection of color descriptors in section 2. Section 3 presents the color descriptors currently defined in MPEG-7 (Final Committee Draft [1]) along with some of their applications. Finally, section 4 concludes and describes possible future work.

W. Skarbek (Ed.): CAIP 2001, LNCS 2124, pp. 11–20, 2001.

2 Core Experiment Process

Following the MPEG tradition, the MPEG-7 group used the Core Experiment (CE) process to select the best techniques for standardization. The procedures differed somewhat between the feature groups, but were the same for all the color descriptors.

Early in the standardization process, the color experts agreed to define a common color dataset (CCD) by selecting 5,466 images from the MPEG-7 content set. Following that, 50 common color queries (CCQ) with associated sets of ground truth images were agreed upon by consensus among the experts [6].

The evaluation process required a retrieval performance measure. This was based on the retrieval rank and normalized to accommodate the differently sized ground truth sets. The resulting measure, called Normalized Modified Retrieval Rank (NMRR) is defined as

$$\text{NMRR}(q) = \frac{\text{AVR}(q) - 0.5[1 + NG(q)]}{\text{Rank}^{\max} - 0.5[1 + NG(q)]} \tag{1}$$

where q is the current query, $NG(q)$ is the ground truth set size, $\text{Rank}^{\max}$ is the upper limit of the retrieval rank, and the Average Retrieval Rank (AVR) is

$$\text{AVR}(q) = \frac{1}{NG(q)} \sum_{k=1}^{NG(q)} \text{Rank}^*(k), \tag{2}$$

where $\text{Rank}^*(k)$ is the retrieval rank capped by the $\text{Rank}^{\max}$ defined above. The Average Normalized Modified Retrieval Rank (ANMRR) is finally defined as the average of the NMRR over the query set. For details and a theoretical explanation, see [7]. An experimental study of the relationship between the ANMRR measure and subjective perception of retrieval performance can be found in [8].

3 Color Descriptors

MPEG-7 defines five color descriptors covering different aspects of color and application areas. The Dominant Color descriptor characterizes an image or image region in terms of a small number of dominant color values and some statistical properties related to those. Scalable Color is a color histogram with efficient encoding based on the Haar transform. Color Structure is an extension of the color histogram that incorporates some associated structural information. Color Layout describes the spatial layout of color within an image. Finally, Group of Frames/Group of Pictures Color is an extension of scalable color to an image sequence/collection.

Besides the "proper" descriptors listed above, the MPEG-7 Standard contains two basic blocks or datatypes: Color Space and Color Quantization. These are not descriptors in their own right – they mainly serve as a reference for the specification of color space and its quantization in other descriptors.

The Color Space datatype is used to specify the color space that a given descriptor refers to. It defines four color spaces: RGB, YCbCr, HVS, and HMMD. To allow extra flexibility, a new color space can be defined by specifying the linear transformation from the RGB color space.

The Color Quantization datatype specifies the quantization of the given color space. It support linear quantization of the color components, with the number of bins for each component specified independently.

It should be noted that only the Dominant Color descriptor actually uses these two datatypes directly. All the other descriptors fix the color space and specify their own quantization, but refer to the Color Space for the definition of transformations.

3.1 Dominant Color

This descriptor provides a compact description of the representative colors of an image or image region. Its main target applications are similarity retrieval in image databases and browsing of image databases based on single or several color values. The representative colors can be indexed in the 3-D color space, which allows for efficient indexing of large databases [9].

In its basic form, the Dominant Color descriptor consists of the number of dominant colors (N), and for each dominant color its value expressed as a vector of color components (c_i) and the percentage of pixels (p_i) in the image or image region in the cluster corresponding to c_i. Two additional fields, spatial coherency (s) and color variance (v_i), provide further characteristics of the color distribution in the spatial and color space domains. Spatial coherency describes the spatial distribution of pixels associated with each representative color, high value means that pixels of similar color are co-located. Color variance describes the variation of the color values of the pixels in a cluster around the corresponding representative color. It can be used to obtain a more precise characterization of the color distribution.

The extraction procedure[1] for the dominant color uses the Generalized Lloyd Algorithm [2,10] to divide the set of pixel values corresponding to a given image region into clusters in the color space. The algorithm minimizes a global distortion measure D, which is defined as

$$D = \sum_{n=1}^{N_I} \|\mathbf{x}_n - c_i(n)\|^2, \tag{3}$$

where N_I is the number of pixels, $\mathbf{x}_n$ is the n-th color vector, and $c_i(n)$ is the cluster center (representative color) for the n-th pixel. The procedure is initialized with one cluster consisting of all pixels and one representative color

---

[1] That is, the extraction method used in the core experiments. MPEG-7 does not standardize the extraction and matching procedures for descriptors, although the recommended methods will be published as part of a technical report based on the Experimentation Model document [2].

computed as the centroid (center of mass) of the cluster. The algorithm then follows a sequence of centroid calculation and clustering steps until a stopping criterion (minimum distortion or maximum number of iterations) is met. The clusters with the highest distortion are divided by adding perturbation vectors to their centroids until the distortion falls below a predefined threshold or the maximum number of clusters has been generated. The resulting percentages are uniformly quantized to 5 bits. The color values are quantized according to the Color Quantization element if it is present, otherwise uniform quantization to 5 bits per color component is used.

The spatial coherency is calculated by selecting each representative color and calculating the per-cluster coherency as the average connectivity of the pixels within the cluster and then computing a weighted average of these values, using percentages p_i as the weights. It is then non-uniformly quantized to 5 bits.

Finally, the color variances are computed as variances of the pixel values within each cluster and non-uniformly quantized to 1 bit per component.

The matching function used depends on the components present in the query and target descriptors. The basic matching function D_{DC} uses only the percentages and color values and is defined as follows

$$D_{DC} = \sum_{i=1}^{N_1} p_{1i}^2 + \sum_{j=1}^{N_2} p_{2j}^2 - \sum_{i=1}^{N_1}\sum_{j=1}^{N_2} 2a_{1i,2j}p_{1i}p_{2j}, \tag{4}$$

where p_1 and p_2 correspond to query and target descriptor, and $a_{i,j}$ is the similarity coefficient between two colors c_i and c_j, specified in [2].

If the spatial coherency field is present, then the matching function D_{DC}^S is defined as

$$D_{DC}^S = w_a|s_1 - s_2|D_{DC} + w_b D_{DC}, \tag{5}$$

where s_1 and s_2 are the spatial coherencies of the query and target descriptors, and w_a and w_b are fixed weights, set to 0.3 and 0.7 respectively.

If the color variance field is present, the matching function is based on modeling of the color distribution as a mixture of 3D Gaussian distributions with parameters defined as color values and variances [11]. Calculation of the squared difference between the query and target distributions then leads to the following formula for the matching function

$$D_{DC}^V = \sum_{i=1}^{N_1}\sum_{j=1}^{N_1} p_{1i}p_{1j}f_{1i1j} + \sum_{i=1}^{N_2}\sum_{j=1}^{N_2} p_{2i}p_{2j}f_{2i2j} + \sum_{i=1}^{N_1}\sum_{j=1}^{N_2} p_{1i}p_{2j}f_{1i2j}, \tag{6}$$

where

$$f_{xiyj} = \frac{1}{2\pi\sqrt{v_{xiyjl}v_{xiyju}v_{xiyjv}}} \exp\left[-\left(\frac{c_{xiyj}^{(l)}}{v_{xiyj}^{(l)}} + \frac{c_{xiyj}^{(u)}}{v_{xiyj}^{(u)}} + \frac{c_{xiyj}^{(v)}}{v_{xiyj}^{(v)}}\right)\right] \tag{7}$$

and

$$c_{xiyj}^{(l)} = (c_{xi}^{(l)} - c_{yi}^{(l)})^2, \quad v_{xiyj}^{(l)} = (v_{xi}^{(l)} - v_{yi}^{(l)})^2. \tag{8}$$

In the equations above, $c_{xi}^{(l)}$ and $v_{xi}^{(l)}$ are the dominant color values and variances, x, y index the query and target descriptors, i, j the descriptor components, and l, u, v the components of the color space.

Table 1 shows a comparison of ANMRR results for two descriptor sizes using different matching functions obtained on the CCD/CCQ dataset using the MPEG-7 XM software. As the extraction results in different numbers of colors for different images, the average over the dataset ($N_{colors}^{(avg)}$) is given. The resulting descriptor size can be easily calculated from the bit allocations specified above. It can be seen that reasonable results are obtained even for the basic version of the descriptor and a significant improvement can be achieved by using the optional fields.

Table 1. ANMRR results for Dominant Color.

$N_{colors}^{(avg)}$	ANMRR(D_{DC})	ANMRR(D_{DC}^S)	ANMRR(D_{DC}^V)
3	0.31	0.30	0.25
5	0.25	0.21	0.16

The Dominant Color descriptor has been applied in combination with the Contour Shape descriptor in a Web-based video retrieval system [12]. A real-time object recognition system using MPEG-7 descriptors (Dominant Color, Contour Shape and Multiple View), which has been developed by Mitsubishi Electric ITE-VIL, was presented during the 2nd MPEG-7 Awareness Event in Singapore in March 2001 (see http://www.mpeg-7.com for details).

3.2 Scalable Color

The Scalable Color descriptor is a color histogram extracted in HSV color space, and encoded for storage efficiency.

The descriptor extraction starts with the computation of the color histogram with 256 bins in the HSV color space with hue (H) component quantized to 16 bins, and saturation (S) and value (V) quantized to 4 bins each. This initial version is then passed through a series of 1-D Haar transforms, starting with H axis, followed by S, V and H. The result is a set of 16 low-pass coefficients and up 240 high-pass coefficients. Due to the redundancy of the original histogram, the high-pass coefficients tend to have low (positive and negative) values. This property can be exploited in two ways.

Firstly, some, or even all, of the high-pass coefficients can be discarded, leading to descriptors with lower number of coefficients, starting with the original 256 and going down through 128, 64, 32 to 16. In the last case only the low-pass coefficients are preserved.

Secondly, the number of bits used can be reduced by scaling of the coefficients to different numbers of bits. In the most extreme case, only the signs of the high-pass coefficients are encoded.

The default matching function for Scalable Color is based on the L_1 metric and is given by,

$$D_{SC} = \sum_{i=1}^{N} |\mathbf{H}_A[i] - \mathbf{H}_B[i]|. \tag{9}$$

The matching can be performed either in the Haar transform space or in the original color histogram space after reverse a Haar transform. The retrieval performance is very similar for both cases, generally slightly higher if the histogram is reconstructed. In principle, any other matching method suitable for histograms can be used, although it was found that L_1 metric gave very good retrieval performance in the MPEG-7 core experiments.

Table 2 shows a comparison of ANMRR results for five settings of the number of coefficients (N_{coeffs}) and varying number of bitplanes discarded, which is given in parentheses [13]. The total descriptor size varies between 54 bits for 16 coefficients and 6 bitplanes discarded, and 2310 for 256 coefficients with no discarded bitplanes.

Table 2. ANMRR results for Scalable Color.

N_{coeffs}	ANMRR(6)	ANMRR(3)	ANMRR(0)
16	0.45	0.20	0.19
32	0.34	0.16	0.13
64	0.21	0.14	0.09
128	0.14	0.10	0.06
256	0.11	0.08	0.05

Typical applications of the descriptor include similarity search in multimedia databases and browsing of large databases. Scalable Color also forms the basis of the Group of Frames/Pictures Color descriptor.

3.3 Color Layout

The Color Layout descriptor compactly characterizes the spatial distribution of color within an image.

The Color Layout uses an array of *representative* colors for the image, expressed in the YCbCr color space, as the starting point for the descriptor definition. The size of the array is fixed to 8x8 elements to ensure scale invariance of the descriptor. The representative colors can be selected in several ways, the simplest one being the average color of the corresponding image block. The array obtained in this way is then transformed using the Discrete Cosine Transform (DCT), which is followed by zig-zag re-ordering [14]. The low-frequency DCT coefficients are preserved and quantized to form the descriptor. By default, 6 coefficients are used for the luminance component and 3 for both chrominance components but different numbers can be specified by means of descriptor attributes (numOfYCoeff and numOfCCoeff for luminance and chrominance components, respectively).

The default matching function is essentially a weighted sum of squared differences between the corresponding descriptor components, specifically,

$$D_{\text{CL}} = \sqrt{\sum_i w_i^y (Y_i - Y_i')^2} + \sqrt{\sum_i w_i^b (Cb_i - Cb_i')^2} + \sqrt{\sum_i w_i^r (Cr_i - Cr_i')^2},$$

(10)

where Y, Cb and Cr are the DCT coefficients of the respective color components, w_i^y, w_i^b, w_i^r are weights chosen to reflect the perceptual importance of the coefficients and the summation is over the number of coefficients.

Table 3 shows a comparison of ANMRR results for different numbers of coefficients [15]. Note that these results were obtained on a modified data set that was defined to exercise the layout of the color in the image [16]. The ANMRR on the standard CCQ dataset with 6+3+3 combination is 0.09 [17].

Table 3. ANMRR results for Color Layout.

number of coefficients	ANMRR	size (bits)
6+3+3	0.50	64
6+6+6	0.36	94

The Color Layout descriptor can be used for fast searching of databases as well as filtering in broadcasting applications. Another interesting application that has been investigated within the MPEG-7 core experiment process is description of video clips, where Color Layout is combined with the Time Series structure [18]. A demonstration of a realtime broadcast video monitoring system using the Color Layout descriptor was presented by NEC at the 2nd MPEG-7 Awareness Event in Singapore in March 2001 (see http://www.mpeg-7.com for details).

3.4   Color Structure

The Color Structure descriptor is a generalization of the color histogram that captures some spatial characteristics of the color distribution in an image. It is defined in the HMMD color space using non-uniform color quantization, specific to Color Structure [1], to between 32–256 colors.

The descriptor is defined by means of a *structuring element*, which is a contiguous set of pixels, typically of rectangular shape. In the process of extracting the Color Structure descriptor, this element is moved across the whole image, one or more pixels at a time, depending on the image size. At each position, the color of each pixel covered by the structuring element is determined. For each of these colors, the histogram bin containing its count is incremented by one.

The actual descriptor is obtained by normalization and non-linear quantization of the final histogram.

As in the case of Scalable Color, Color Structure has the property of scalability in the sense that a descriptor with a larger number of bins can be reduced to one with a smaller number of bins. Due to the non-linear quantization of the color space, this *bin unification* process, explained in more detail in [1] and [2], is relatively complex. For reasons related to interoperability, this process is also applied to obtain the descriptor if the number of colors is smaller than 256.

Since the structure of the descriptor is the same as for color histogram, the same matching functions can be used. The default matching function is the L_1 metric (see Eq. 9), but as in the case of Scalable Color, any histogram matching method can be used.

Table 4 shows a comparison of ANMRR results for four bin number settings at four different bin quantizations [19]. It is interesting to note that the difference in performance between 120 to 184 bins is smaller than the accuracy with which the results are presented here, which shows that the 120-bin quantization of the color space is sufficient for the subset of colors characteristic of typical images. It can be seen that the Color Structure descriptor has very good retrieval

Table 4. ANMRR results for Color Structure.

N_{coeffs}	ANMRR(1 bit)	ANMRR(2 bits)	ANMRR(8 bits)
32	0.81	0.34	0.11
64	0.76	0.27	0.07
120	0.73	0.23	0.05
184	0.71	0.23	0.05

performance. The main application envisaged for it is image/video retrieval in multimedia databases, particularly when high accuracy is required.

3.5   Group of Frames/Group of Pictures Color

This descriptor is an extension of the Scalable Color to a group of frames/pictures. As such it is particularly applicable to video database searching and browsing and filtering of broadcast video material, particularly in combination with hierarchical clustering of the video frames [20].

The structure of the GoF/GoP Color descriptor is identical to that of Scalable Color with the exception of one additional field, *aggregation*, which specifies how the color pixels from the different pictures/frames were combined before the extraction of the color histogram. The possible values are *average, median* and *intersection* [21].

The *average* is obtained simply as the average of the color histograms for all the frames/pictures in the group:

$$\mathbf{GoFP}^{\text{average}}[i] = \frac{1}{N} \sum_{k=1}^{N} \mathbf{H}_k[i], \tag{11}$$

where k indexes the frames/pictures and i indexes the histogram bins (the same convention is followed in the two following equations). The $\mathbf{GoFP}^{\text{average}}$ generally gives retrieval performance for typical video clips [21].

The *median* histogram is obtained by calculating the median value of each histogram bin over the frames/pictures and assigning this value to the resulting histogram bin:

$$\mathbf{GoFP}^{\text{median}}[i] = \text{median}\,(\mathbf{H}_0[i], \ldots, \mathbf{H}_{N-1}[i])\,. \tag{12}$$

The use of the median helps to eliminate the effect of outliers in the frame/picture groups. Therefore, $\mathbf{GoFP}^{\text{median}}$ improves retrieval performance in cases of occlusion, lighting changes, etc.

The *intersection* histogram is obtained by calculating the minimum value of each histogram bin over the frames/pictures and assigning this value to the resulting histogram bin:

$$\mathbf{GoFP}^{\text{intersection}}[i] = \min_k\,(\mathbf{H}_k[l])\,. \tag{13}$$

The intersection finds the minimum common colors in the frames/pictures and can therefore be used in applications that require the detection of a high level of correlation in the color.

4   Conclusions

We have presented MPEG-7 color descriptors and described their main features and performance, We have seen that they cover a large spectrum of applications. Nevertheless some issues related to color characterization are still being investigated and discussed, and may be incorporated in future versions of MPEG-7. Some of them, in particular color "temperature", i.e. a subjective feeling of warmth or cold associated with different color are undergoing tests in core experiments. Other issues, such as color calibration and the colorimetric color spaces and their usefulness in the context of the MPEG-7 Standard are still being discussed.

References

1. Cieplinski, L., Kim, M., Ohm, J.-R., Pickering, M., Yamada, A. (eds.): Text of ISO/IEC 15938-3/FCD Information Technology–Multimedia Content Description Interface–Part 3 Visual. ISO/IEC JTC1/SC29/WG11 (MPEG) document no. N4062, March (2001)
2. Cieplinski, L., Kim, M., Ohm, J.-R., Pickering, M., Yamada, A. (eds.): MPEG-7 Visual Experimentation Model. ISO/IEC JTC1/SC29/WG11 (MPEG) document no. N4063, March (2001)
3. Martinez, J.M. (ed.): Overview of the MPEG-7 Standard. ISO/IEC JTC1/SC29/WG11 (MPEG) document no. N4031, March (2001)
4. Manjunath, B.S., Sikora, T., Salembier, P. (eds.): Introduction to MPEG-7: Multimedia Content Description Language. John Wiley (2001) (to be published)

5. Manjunath, B.S., Ohm, J.-R., Vasudevan, V.V., Yamada, A.: Color and Texture Descriptors. IEEE Transactions on Circuits and Systems for Video Technology (2001) (to be published)
6. Zier, D., Ohm, J.-R.: Common Datasets and Queries in MPEG-7 Core Experiments. ISO/IEC JTC1/SC29/WG11 (MPEG) document no. M5060, October (1999)
7. Vinod, V.V. (ed.): Description of Core Experiments for MPEG-7 Color/Texture Descriptors. ISO/IEC JTC1/SC29/WG11 (MPEG) document no. N2929, October (1999)
8. Ndjiki-Nya, P., Restat, J., Meiers, T., Ohm, J.-R., Seyferth, A., Sniehotta, R.: Subjective Evaluation of the MPEG-7 Retrieval Accuracy Measure (ANMRR). ISO/IEC JTC1/SC29/WG11 (MPEG) document no. M6029, May (2000)
9. Deng, Y., Manjunath, B.S., Kenney, C., Moore, M.S., Shin, H.: An Efficient Color Representation for Image Retrieval. IEEE Transactions on Image Processing 10, January (2001) 140–147
10. Gersho, A., Gray, R.M.: Vector Quantization and Signal Compression. Kluwer Academic Publishers, Boston/Dordrecht/London (1993)
11. Cieplinski, L.: Results of Core Experiment CT4 on Dominant Color Extension. ISO/IEC JTC1/SC29/WG11 (MPEG) document no. M5775, March (2000)
12. Bober, M., Asai, K., Divakaran, A.: A MPEG-4/7 Based Internet Video and Still Image Browsing System. Proceedings of SPIE 4209 (Multimedia Systems and Applications III) (2001) 33–38
13. Ohm, J.-R., Makai, B., Smolic, A.: Results of CE5 on Scalable Representation of Color Histograms. ISO/IEC JTC1/SC29/WG11 (MPEG) document no. M6285, July (2000)
14. Pennebaker, W.B., Mitchell, J.L.: JPEG Still Image Data Compression Standard. Van Nostrand Reinhold, New York, USA (1993)
15. Yamada, A.: Results of Core Experiment VCE-2. ISO/IEC JTC1/SC29/WG11 (MPEG) document no. M6445, November (2000)
16. Yamada, A.: A Proposal of Extended Color Dataset, Queries and Ground Truths for Corel Photos. ISO/IEC JTC1/SC29/WG11 (MPEG) document no. M6446, November (2000)
17. Yamada, A.: Results of Core Experiment on Compact Color Layout Descriptor. ISO/IEC JTC1/SC29/WG11 (MPEG) document no. M6446, October (1999)
18. Fujita, T., Shibata, Y.: Results of Core Experiment VCE-1. ISO/IEC JTC1/SC29/WG11 (MPEG) document no. M6534, November (2000)
19. Messing, D., van Beek, P., Sezan, I.: Results of CE CT1: Color Structure Histogram + HMMD – Improved Performance and Interoperability. ISO/IEC JTC1/SC29/WG11 (MPEG) document no. M6218, July (2000)
20. Krishnamachari, S., Abdel-Mottaleb, M.: Image Browsing Using Hierarchical Clustering. Proceedings of the Fourth IEEE Symposium on Computers and Communications, ISCC (1999)
21. Ferman, A.M., Krishnamachari, S., Abdel-Mottaleb, M., Tekalp, A.M., Mehrotra, R.: Core Experiment on Group-of-Frames/Pictures Histogram Descriptors (CT7). ISO/IEC JTC1/SC29/WG11 (MPEG) document no. M5124, October (1999)

Texture Descriptors in MPEG-7

Peng Wu[1], Yong Man Ro[2], Chee Sun Won[3], and Yanglim Choi[4]

[1]Dept. of Electrical and Computer Engineering, UC Santa Barbara, CA 93106-9560, USA
peng@ece.ucsb.edu
[2]Multimedia Group, ICU. Yusung-gu POBOX 77, Taejon, S. Korea
yro@icu.ac.kr
[3]Dept. of Electronics Eng., Dongguk Univ., Seoul, 100-715, S. Korea
cswon@dongguk.edu
[4]M/M Lab. SAIT, KiHeung, YongIn, KyungKi-Do 449-712 S. Korea
yanglimc@samsung.com

Abstract. We present three descriptors of texture feature of a region. Namely, the homogeneous texture descriptor (HTD), the edge histogram descriptor (EHD), and the perceptual browsing descriptor (PBD). They are currently included in the Committee Draft of the MPEG-7 Visual (ISO/IEC 15938-3). Each descriptor has a unique functionality and application domain. HTD and EHD describe statistical distribution of the texture and are useful for image retrieval application, while HTD is for homogeneously textured region and EHD is for multi-textured natural image or sketch. PBD is a compact descriptor suitable for quick browsing application.

Keywords: texture descriptor, MPEG-7

1 Introduction

MPEG-7 establishes a universal M/M description interface and describing content-based feature is one of primary objectives. Image texture is one of important visual features and applications using texture includes the retrieval, browsing, and indexing. Three texture have been recommended in the MPEG-7 Committee Draft (CD) Part3: Visual (ISO/IEC 15938-3). We will brief each of them below.

The homogeneous texture descriptor (HTD) describes a precise statistical distribution of the image texture. It's a vector of 62 integers coming from the Gabor filter response of 30 frequency layout channels. It enables to classify images with high precision. HTD is to be used for similarity retrieval applications.

The edge histogram descriptor (EHD) is an 80-bin histogram representing the local edge distribution of an image. It is to be used for an image retrieval application where the data images are not necessarily homogeneously textured, e.g., natural images, sketch, or clip art images etc. Also, it supports a query based on sub image blocks.

The perceptual browsing descriptor (PBD) is designed for an application where features with perceptual meaning are needed to browse the database. It is very compact and describes a high level perceptual semantics of an image texture; texture regularity, directionality and coarseness.

W. Skarbek (Ed.): CAIP 2001, LNCS 2124, pp. 21–28, 2001.
© Springer-Verlag Berlin Heidelberg 2001

In the following, we present the definition, semantics, extraction, and test results. The readers are advised to read the references to get an in-depth understanding.

2   Homogeneous Texture Descriptor (HTD)

2.1   Definition and Semantics

HTD is composed of 62 numbers. The first two are the mean and the standard deviation of the image. The rest are the energy and the energy deviation of the Gabor filtered responses of the "channel", in the subdivision layout of the frequency domain in Figure1. This design is based on the fact that response of the visual cortex is band-limited and brain decomposes the spectra into bands in spatial frequency [1,2,3,4,5].

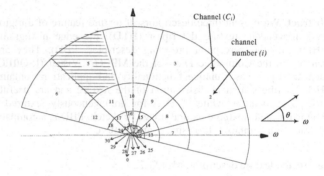

Fig. 1. Frequency Domain Division Layout

The center frequencies of the channels in the angular and radial directions are such that $\theta_r = 30$ r 0 r 5, $\omega_s = \omega_0$ 2^{-s} 0 s 4, $\omega_0 = 3/4$. The equation for the Gabor wavelet filters is the following.

$$G_{P_{s,r}}(\omega,\theta) = \exp\frac{-(\omega-\omega_s)^2}{2\sigma_{P_s}^2}\quad \exp\frac{-(\theta-\theta_r)^2}{2\sigma_{\theta_r}^2} \tag{1}$$

Note that σ_{ω_r} and σ_{θ_r} is the standard deviation in the radial and the angular direction, where the neighboring filters meet at the half of the maximum

2.2   Extraction

The extraction of the mean and the standard deviation is straightforward. For the rest, the Radon transform followed by 1-D Fourier transform is applied. Let $F(\omega,\theta)$ be the result. Then the energy e_i and the energy deviation d_i of the ith channel is

$$e_i = \log[1+p_i], d_i = \log[1+q_i], \text{ where } p_i = \sum_{\substack{\omega=0+ \ \theta=0^\circ+}}^{1} \sum^{360^\circ} I_{sr}^2,$$ (2)

$$q_i = \sqrt{\sum_{\substack{\omega=0+ \ \theta=0^\circ+}}^{1}\sum^{360^\circ}\{I_{sr}^2 - p_i\}^2}, \ I_{sr} = G_{P_{s,r}}(\omega,\theta) \ |\omega| \ F(\omega,\theta)$$

$|\omega|$ is the Jacobian between the Cartesian and the Polar. Then the HTD is written as

$$HTD = [f_{DC}, f_{SD}, e_1, e_2, ..., e_{30}, d_1, d_2, ..., d_{30}]$$ (3)

2.3 Experimental Results

Let HTD_i and HTD_j be the HTD of image i and j, then their similarity is,

$$d(i,j) = \text{distance}(HTD_i, HTD_j) = \sum_k \left| \frac{w(k)[HTD_i(k) - HTD_j(k)]}{\alpha(k)} \right|$$ (4)

Where $w(k)$ is the weight and $\alpha(k)$ is the standard deviations of kth descriptor values in the database. We performed tests on MPEG-7 dataset, which consists of the Brodatz images (T1), ICU images (T2), and their scale and orientation variations [7]. The results for T1 and T2 are shown in Table 1.

Table 1. Average Retrieval Rates (ARR) for the HTD

Data set	ARR (Average Retrieval Rate %)
T1	77.32
T2	90.67

3 Edge Histogram Descriptor

3.1 Definition and Semantics

The edge histogram descriptor (EHD) represents local edge distribution in the image. It describes edges in each 'sub-image', which is obtained by dividing the image using 4x4 grid as in Fig. 2. Edges in the sub-image are classified into five types; vertical, horizontal, 45-degree, 135-degree, and non-directional. Occurrence of each type becomes a histogram bin, producing 80 histogram bins overall. The order of these bins is shown in Table 2.

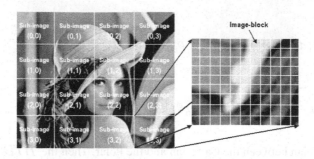

Fig. 2. The sub-image and Image-block.

Table 2. Semantics of local edge bins

Bins	Semantics	Bins	Semantics
Bin[0]	Vertical edge at (0,0)	Bin[5]	Vertical edge at (0,1)
Bin[1]	Horizontal edge at (0,0)	:	:
Bin[2]	45degree edge at (0,0)	:	:
Bin[3]	135 degree edge at (0,0)	Bin[78]	135 degree edge at (3,3)
Bin[4]	Non-direc. edge at (0,0)	Bin[79]	Non-direc. edge at (3,3)

The histogram bin values are normalized by the total number of the image-blocks. The bin values are then non-linearly quantized to keep the size of the histogram as small as possible [9]. We assigned 3 bits/bin and 240 bits are needed in total [8].

3.2 Edge Extraction

Since the EHD describes non-directional edge and no-edge cases, an edge extraction scheme based on image-block (instead of pixel) is needed, as in Fig. 2. For the block-based edge extraction schemes, we refer the reader to methods in [10][11][12].

3.3 Experimental Results

For a good performance, we need the global edge distribution for the whole image and semi-global, horizontal and vertical edge distributions. The global is obtained by accumulating 5 EHD bins for all the sub-images. For the semi-global, four connected sub-images are clustered as in Fig. 3 [13]. In total, we have 150 bins (80 local + 5 global + 65 semi-global). The 11639 MPEG-7 image dataset [7] is used for the test. Using an absolute distance measure [12][13], the ANMRR was as low as 0.2962. EHD showed good results both in query by example and by sketch, especially for natural images with non-uniform textures and clipart images.

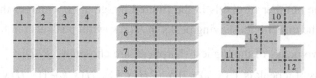

Fig. 3. Clusters of sub-images for semi-global histograms

4 Perceptual Browsing Descriptor

4.1 Definition and Semantics

The syntax of the Perceptual Browsing Descriptor (PBD) is the following.

$$PBD = \begin{bmatrix} v_1 & v_2 & v_3 & v_4 & v_5 \end{bmatrix} \tag{5}$$

v_1 represents the regularity. v_2, v_3 represent the two directions that best capture the directionality. v_4, v_5 represent the two scales that best capture the coarseness.

4.2 Extraction

As in [17], the image is decomposed into a set of filtered images, $W_{mn}(x, y)$, $m = 1, ..., S$ $n = 1, ..., K$. The overall extraction process is depicted in Fig. 4.

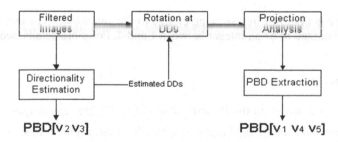

Fig. 4. Extraction of Perceptual Browsing Descriptor

Dominant Direction can be estimated in spatial domain [17] or in frequency domain [16], [18]. Due to the aperture effect [18], the spatial domain approach seems to be more adequate [14]. We used S directional histograms of the S K images.

$$H(s, k) = \frac{N(s, k)}{\sum_{1}^{K} N(s, k)}, \quad s = 1, ..., S \text{ and } k = 1, ..., K \ (0.1) \tag{6}$$

$N(s,k)$ is the number of pixels in the filtered image at (s, k) larger than a threshold t_s [1]. Among the direction(s) having peaks in $H(s,k)$ with peak also in the neighboring scale, with two peaks of the highest sharpness, are chosen as $PBD[v_2]$ and $PBD[v_3]$. The sharpness of a peak is $C(s,k) = 0.5 \ (2H(s,k) - H(s,k-1) - H(s,k+1))$ [14].

For the coarseness, two projections, $P_H^{(mn)}$ and $P_V^{(mn)}$ are computed as follows.

$$P_H^{(mn)}(l) = \iint W_{mn}(x,y)\delta(x\cos\theta_{DO1} + y\sin\theta_{DO1} - l)dxdy \quad P_V^{(mn)}(l) = \iint W_{mn}(x,y)\delta(x\cos\theta_{DO2} + y\sin\theta_{DO2} - l)dxdy \quad (7)$$

θ_{DO1}, θ_{DO2} are the two dominant directions. Let $P(l)$ be the projection corresponding to θ_{DO1} (similar for θ_{DO2}). The Normalized Auto-correlation Function (NAC) is computed as in (8). Let $p\_posi(i)$, $p\_magn(i)$ and $v\_posi(j), v\_magn(j)$ be the positions and magnitudes of the peaks and valleys in $NAC(k)$. And let dis and std be the mean and the standard deviation of the distances of successive peaks. The projections with std/dis less than a threshold become candidates. Let $m^*(H)$ and $m^*(V)$ be the scale index of candidate $P_H^{(mn)}$ and $P_V^{(mn)}$ with the maximum contrast (defined in (8)). Then $PBD[v_4] = m^*(H)$ and $PBD\ v_5 = m^*\ V$.

$$NAC(k) = \frac{\sum_{m=k}^{N-1} P(m-k)P(m)}{\sqrt{\sum_{m=k}^{N-1} P^2(m-k) \sum_{m=k}^{N-1} P^2(m)}} \qquad contrast = \frac{1}{M}\sum_{i=1}^{M} p\_magn(i) - \frac{1}{N}\sum_{j=1}^{N} v\_magn(i) \qquad (8)$$

The regularity is obtained by assigning credits to the candidates. The credits are then added and quantized to be an integer between 1 and 4. For more details, see [19].

4.3 Experiments

For each 512×512 image in the Brodatz album [15], 19 256×256 images are obtained by partitioning and rotating the large image by $30°$ steps. Figure 5(a) shows dominant direction estimation of the T069. Note also that the estimated orientations in Figure 5(b) match reasonably well with the perceived directionality of the texture.

Subjective tests are also conducted with 25 texture groups. The ground truth was formed by human observation. The comparison result is shown in Table 3

---

[1] At each scale s, the threshold is set to $t_s = \mu_s + \sigma_s$, where μ_s and σ_s are the mean and the standard deviation over the all K filtered images at that scale s.

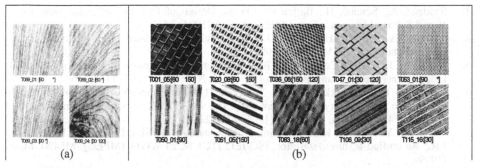

Fig. 5. (a) Detection of the gradual directionality change   (b) Detected dominant direction(s)

Table 3. Subjective Evaluation results

	Texture Pattern	% of matches	
		First	Second
S1	T001, T003, T020, T021, T022, T032, T036, T046, T047, T052, T053, T056, T064, T065, T077, T082, T084, T085, T094, T095	98	88
S2	T050, T051, T083, T106, T115	100	

The regularity and coarseness are also evaluated using images in Table 4. Five people were asked to quantify the regularity and coarseness. For regularity, the two values were within ±1 deviation for 29 out of 30 images. For the scales, assuming that the scales match if the two values are within ±1, they were in agreement for 26/30 (87%).

Table 4. Images used in subjective evaluation

#	T001,T002,T006,T007,T009,T012,T014,T018,T020,T021,T023,T025,T026,T037,T039, T053,T055,T056,T064,T065,T067,T071,T075,T088,T094,T096,T097,T103,T107, T111

5 Summary and Conclusion

Texture is one of salient features in content-based representation of images. Due to variety of the applications, MPEG-7 recommends three texture descriptors. Their efficiency and robustness are carefully checked through the CE process by the MPEG members. The descriptors are more or less complimentary to each other. It is reasonable to speculate that a combination scheme would give richer functionality. Future research leads to investigation of this aspect.

References

1. Manjunath, B.S., Ma, W.Y.: Texture Features for Browsing and Retrieval of Image Data. IEEE Transactions on PAMI, Vol. 18, No. 8, August (1996)
2. Chellappa, R.: Two-dimensional discrete Gaussian Markov random field models for image processing. Pattern Recognition, vol. 2 (1985) 79-112

3. Saadane, A., Senane, H., Barba, D.: On the Design of Psychovisual Quantizers for a Visual Subband Image Coding. SPIE, Vol. 2308 (1994) 1446
4. Saadane, A., Senane, H., Barba, D.: An Entirely Psychovisual based Subband Image Coding Scheme. SPIE, Vol. 2501 (1995) 1702
5. Daugman, J.G.: High Confidence Visual Recongnition of Persons by a Test of Statistical Independence. IEEE Trans. PAMI, vol.15, no.11, November (1993) 1148-1161
6. Lambrecht, C.J.: A Working Spatio - Temporal Model of Human Vision System for Image Restoration and Quality Assessment Applications. IEEE International Conference on ASSP vol. 4. New York, NY, USA (1996) 2291-2294
7. Ro, Y.M., Yoo, K.W., Kim, M., Kim, J., Manjunath, B.S., Sim, D.G., Kim, H.K., Ohm, J.R.: An unified texture descriptor. ISO/IEC JTC1 SC29 WG11 (MPEG), M5490. Maui (1999)
8. ISO/IEC/JTC1/SC29/WG11: Core Experiment Results for Edge Histogram Descriptor (CT4). MPEG Document M6174. Beijing (2000)
9. ISO/IEC/JTC1/SC29/WG11: CD 15938-3 MPEG-7 Multimedia Content Description Interface-Part 3. MPEG Document W3703. La Baule (2000)
10. Vaisey, J., Gersho, A.: Image compression with variable block size segmentation. IEEE Tr. Signal Process., vol 40, no 8 (1992) 2040-2060
11. Won C.S., Park, D.K.: Image block classification and variable block size segmentation using a model-fitting criterion. Optical Eng., vol 36, no 8 (1997) 2204-2209
12. ISO/IEC/JTC1/SC29/WG11: MPEG-7 Visual XM 8.0. W3673. La Baule (2000)
13. Park, D.K., Jeon, Y.S., Won, C.S., Park, S.-J.: Efficient use of local edge histogram descriptor. Proc. of ACM Multimedia 2000 Workshops. Marina del Rey (2000) 51-54
14. Brodatz, P.: Textures: A photographic album for artists & designers. Dover, NY (1966)
15. Liu, F., Picard, R.W.: Periodicity, directionality, and randomness: Wold features for image modeling and retrieval. MIT Media Lab Technical Report No. 320, March (1995)
16. Manjunath, B.S., Ma, W.Y.: Texture features for browsing and retrieval of image data. IEEE Transactions on PAMI 18(8) (1996) 837-842
17. Tamura, H., Mori, S., Yamawaki, T.: Texture features corresponding to visual perception. IEEE Trans. On Sys. Man, and Cyb, SMC 8(6) (1978)
18. Weszka, J.S., Dyer, C.R., Rosenfeld, A.: A comparative study of texture measures for terrain classification. IEEE Trans., Sys., Man, Cyber., SMC-6, Apr (1976) 269-285
19. Wu, P., Manjunath, B.S., Newsam, S., Shin, H.D.: A texture descriptor for browsing and similarity retrieval. Journal of Signal Processing: Image Communication, Volume 16, Issue 1-2, September (2000) 33-43

An Overview of MPEG-7 Motion Descriptors and Their Applications

Ajay Divakaran

Mitsubishi Electric Research Laboratories, 571, Central Avenue Suite 115,
New Providence, NJ 07974, USA
ajayd@merl.com

Abstract. We present an overview of the MPEG-7 motion descriptors viz. motion trajectory, camera motion, parametric motion and motion activity. These descriptors cover a wide range of functionality and hence enable several applications. We present the salient parts of the syntax, the semantics and the associated extraction of each of these descriptors. We discuss the possible applications for these descriptors and associated complexity trade-offs. We then describe a case study of a low complexity video browsing and indexing system that capitalizes on the simple extraction, compactness and effectiveness of the motion activity descriptor. This system relies on feature extraction in the compressed domain, which makes dynamic feature extraction possible. It combines the MPEG-7 motion activity descriptor and a simple color histogram to achieve both video summarization (top-down traversal) and indexing (bottom-up traversal) and thus enables a user-friendly video-browsing interface.

Keywords: MPEG-7, motion-based video indexing and summarization, motion activity

1 Introduction

The motion features of a video sequence constitute an integral part of its spatio-temporal characteristics, and are thus indispensable for video indexing. Past work on video indexing using motion characteristics has largely relied on camera motion estimation techniques, trajectory-matching techniques and aggregated motion vector histogram based techniques [5-13,15,17,26,27,28].

Expressing the motion field, coarse or fine, of a typical video sequence requires a huge volume of information. Hence, the principal objective of motion-based indexing is concisely and effectively encapsulating essential characteristics of the motion field. Such indexing techniques span a wide range of computational complexity, since the motion fields can be sparse or dense, and the processing of the motion fields ranges from simple to complex. Motion feature extraction in the MPEG-1/2 compressed domain has been popular because of the ease of extraction of the motion vectors from the compressed bitstream. However, since compressed domain motion vectors consti

W. Skarbek (Ed.): CAIP 2001, LNCS 2124, pp. 29–40, 2001.

tute a sparse motion field, they cannot be used for computing descriptors that require dense motion fields.

Motion-based indexing is useful by itself since it enables motion-based queries, which is useful in domains such as sports and surveillance in which motion is the dominant feature. For example, the motion activity descriptor can help detect exciting moments from a soccer game, or the motion trajectory descriptor can help distinguish a lob from a volley in a tennis game. Motion-based indexing has also been shown to significantly improve the performance of similarity based video retrieval systems [15], when combined with other fundamental features such as color, texture and shape. Motion descriptions can also be the basis for complementary functionalities, such as video hyperlinking based on trajectories [18,20], refined browsing based on motion [32], or refinement of table of content [19,20].

The rest of the paper is organized as follows. Section 2 provides an overview of the MPEG-7 motion descriptors. Section 3 discusses possible applications for these descriptors and related practical issues. Section 4 presents a video browsing system based on the motion activity feature. Section 5 concludes the paper and presents possibilities for future work.

2   Overview of MPEG-7 Motion Descriptors

The MPEG-7 motion descriptors cover the range of complexity and functionality mentioned in the introduction, enabling MPEG-7 to support a broad range of applications. The common guiding principle is to maintain simple extraction, simple matching, concise expression and effective characterization of motion characteristics.

They are organized as illustrated in fig 1.

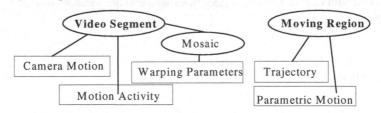

Fig. 1. MPEG-7 motion descriptions

2.1   Camera Motion Descriptor

2.1.1   Salient Syntax and Semantics

This descriptor characterizes 3-D camera motion parameters. It is based on 3-D camera motion parameter information, which can be automatically extracted or generated by capture devices. The camera motion descriptor supports the following well-known basic camera operations: fixed, panning (horizontal rotation), tracking (horizontal transverse movement, also called traveling in the film industry), tilting (vertical rotation), booming (vertical transverse movement), zooming (change of the focal length), dollying (translation along the optical axis), and rolling (rotation around the optical axis).

2.1.2 Highlights of Extraction

Note that this descriptor can be instantiated by the capture device during capture. If that is not possible, the extraction relies on getting the camera motion parameters from the motion field. The camera motion parameters consist of three camera translation parameters T_x, T_y, T_z, three camera rotation parameters R_x, R_y, R_z, and one camera zoom parameter R_{zoom}. The camera motion according to these 7 parameters induces an image motion described as follows:

Equation 1:

$$u_x = -\frac{f}{Z}(T_x - xT_z) + \frac{x.y}{f}\ R_x - f.\left(1 + \frac{x^2}{f^2}\right)\ R_y + y\ R_z + f.\tan^{-1}\frac{x}{f}.\left(1 + \frac{x^2}{f^2}\right)\ R_{zoom}$$

Equation 2:

$$u_y = -\frac{f}{Z}(T_y - yT_z) - \frac{x.y}{f}\ R_y + f.\left(1 + \frac{y^2}{f^2}\right)\ R_x - x\ R_z + f.\tan^{-1}\frac{y}{f}.\left(1 + \frac{y^2}{f^2}\right)\ R_{zoom},$$

These equations are obtained by using perspective transformation of a physical point P(X,Y,Z) on a rigid object in 3-D to a point p(x,y) in the retinal plane. u_x and u_y are the x and y components of the image velocity at a given image position (x,y), Z is 3-D depth and f is the camera focal length.

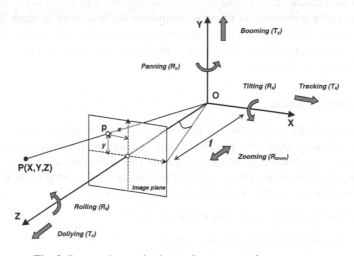

Fig. 2. Perspective projection and camera motion parameters

2.2 Motion Trajectory

Motion trajectory is a high-level feature associated with a moving region, defined as a spatio-temporal localization of one of its representative points (e.g. the centroid).

2.2.1 Salient Syntax and Semantics

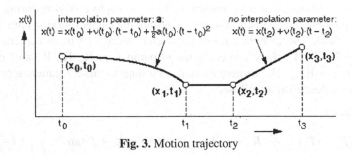

Fig. 3. Motion trajectory

The trajectory model is a first or second order piecewise approximation along time, for each spatial dimension. The core of the description is a set of *keypoints*, representing the successive spatio-temporal positions of the described object (positions of one representative point of the object, such as its center of mass). They are defined by their coordinates in space (2D or 3D) and time. Additionally, interpolating parameters can be added to specify non-linear interpolations between keypoints.

2.2.2 Highlights of Extraction

The extraction in this case relies on the availability of segmented spatio-temporal regions in the content. This would be either through the alpha channel of MPEG-4 or by carrying out a segmentation prior to computing the trajectory to mark the regions. Once the regions are segmented, the motion trajectory is defined as the locus of the center of mass of the region. Note that if the segmentation is already done, the descriptor is quite easy to extract.

2.3 Parametric Motion

This descriptor addresses the motion of objects in video sequences, as well as global motion. If it is associated with a region, it can be used to specify the relationship between two or more feature point motion trajectories according to the underlying motion model. The descriptor characterizes the evolution of arbitrarily shaped regions over time in terms of a 2-D geometric transform.

The parametric model the descriptor expresses are:

Translational model: $v_x(x, y) = a_1,\ v_y(x, y) = a_2$

Rotation/scaling model: $v_x(x, y) = a_1 + a_3x + a_4y$
$v_y(x, y) = a_2 - a_4x + a_3y$

Affine model: $v_x(x, y) = a_1 + a_3x + a_4y$, $v_y(x, y) = a_2 + a_5x + a_6y$

Planar perspective model: $v_x(x, y) = (a_1 + a_3 x + a_4 y) / (1 + a_7 x + a_8 y)$
$v_y(x, y) = (a_2 + a_5 x + a_6 y) / (1 + a_7 x + a_8 y)$

Parabolic model: $v_x(x, y) = a_1 + a_3 x + a_4 y + a_7 xy + a_9 x^2 + a_{10} y^2$
$v_y(x, y) = a_2 + a_5 x + a_6 y + a_8 xy + a_{11} x^2 + a_{12} y^2$

where $v_x(x, y)$ and $v_x(x, y)$ represent the x and y displacement components of the pixel at coordinates (x, y). The descriptor should be associated with a spatio-temporal region. Therefore, along with the motion parameters, spatial and temporal information is provided.

2.3.1 Highlights of Extraction

This descriptor relies on computationally intense operations on the motion field to fit the various models. It is the most computationally complex of the MPEG-7 motion descriptors.

2.4 Motion Activity

A human watching a video or animation sequence perceives it as being a slow sequence, fast paced sequence, action sequence etc. The activity descriptor [3] captures this intuitive notion of 'intensity of action' or 'pace of action' in a video segment. Examples of high 'activity' include scenes such as 'goal scoring in a soccer match', 'scoring in a basketball game', 'a high speed car chase' etc. On the other hand scenes such as 'news reader shot', 'an interview scene', 'a still shot' etc. are perceived as low action shots. Video content in general spans the gamut from high to low activity. This descriptor enables us to accurately express the activity of a given video sequence/shot and comprehensively covers the aforementioned gamut. It can be used in diverse applications such as content re-purposing, surveillance, fast browsing, video abstracting, video editing, content based querying etc.

To more efficiently enable the applications mentioned above, we need additional attributes of motion activity. Thus our motion activity descriptor includes the following attributes.

2.4.1 Salient Semantics and Syntax Intensity

This attribute is expressed as a 3-bit integer lying in the range [1,5]. The value of 1 specifies the lowest intensity, whereas the value of 5 specifies the highest intensity. Intensity is defined as the variance of motion vector magnitudes, first appropriately normalized by the frame resolution and then appropriately quantized as per frame resolution (see [3]). In other words, the thresholds for the 3-bit quantizer are normalized with respect with respect to the frame resolution and rate.

DominantDirection

This attribute expresses the dominant direction and can be expressed as an angle between 0 and 360 degrees

Spatial Distribution Parameters: Nsr, Nmr, Nlr

Short, medium and long runs of zeros are elements of the motion activity descriptor that provide information about the number and size of active objects in the scene. Their values are extracted from the thresholded motion vector magnitude matrix, which has elements for each block indexed by (i,j). Each run is obtained by recording the length of zero runs in a raster scan order over this matrix. The thresholded motion vector magnitude matrix is given by:

For each object or frame the "thresholded activity matrix" C^{thresh}_{mv} is defined as:

$$C^{thresh}_{mv}(i, j) = \begin{array}{l} C_{mv}(i, j), \text{if } C_{mv}(i, j) \quad C^{avg}_{mv} \\ 0, \text{otherwise} \end{array}$$

where the motion vector matrix and average motion vector magnitude for an MxN macro-block frame are defined as:

$$C_{mv} = \{R(i, j)\} \text{ where } R(i, j) = \sqrt{x^2_{i,j} + y^2_{i,j}} \text{ for inter blocks and } R(i, j) = 0 \text{ for intra}$$
$$\text{blocks}$$

$$C^{avg}_{mv} = \frac{1}{MN} \sum_{i=0}^{M-1} \sum_{i=0}^{N-1} C_{mv}(i, j)$$

From the thresholded motion vector magnitude matrix, the zero run-lengths are classified into three categories, short, medium and long, which are normalized with respect to the frame width (see [3] for details).

SpatialLocalizationParameter(s)

A 3-bit integer expressing the quantized percentage duration of activities of each location (i.e. division block).

TemporalParameters

This is a histogram consisting of 5 bins. The histogram expresses the relative duration of different levels of activity in the sequence. Each value is the percentage of each quantized intensity level segments duration compared to the whole sequence duration.

2.4.2 Highlights of Extraction

It is evident from the semantics that the motion activity descriptor relies on simple operations on the motion vector magnitudes. Most often these are MPEG-1/2/4 motion vectors. It lends itself therefore to compressed domain extraction and is also easy to compute since it does not rely on any pre-processing step such as segmentation. The intention of the descriptor is to capture the gross motion characteristics of the video segment. It is the least computationally complex of the MPEG-7 motion descriptors.

3 Applications of Motion Descriptors

Possible applications of MPEG-7 motion descriptors are:
1. **Content-based querying and retrieval from video databases** – Since all the descriptors are compact, they would lend themselves well to this application.
2. **Video Browsing** – The motion activity descriptor can help find the most active parts of a soccer game for example. The camera motion descriptor would also be helpful, as would the motion trajectory descriptor.
3. **Surveillance** – The basic aim in this case would be event detection in stored or live video. The motion activity descriptor is the easiest to apply, but the other descriptors could also be used to detect specific actions.
4. **Video Summarization** – In [30] we have shown that the intensity of motion activity of a video sequence is in fact a direct indication of its summarizability. In the next section, we describe an application that uses the summarizability notion to dynamically generate video summaries.
5. **Video Re-Purposing** - The motion descriptors can be used to control the presentation format of content e.g. by dropping more frames when the motion activity intensity is low and fewer when it is high, or by slowing down the replay frame rate when the camera motion is a fast pan etc.

While there are perhaps other possible applications for the MPEG-7 motion descriptors, some issues emerge from considering the above applications:
a. Extraction Complexity – Low extraction complexity, i.e., fast generation of content description, is obviously desirable for all applications, but for applications such as consumer video browsing, it is essential. To this end, descriptors that lend themselves to computation with motion vectors extracted from MPEG-1/2/4 or other compressed video are the most desirable. The parametric motion descriptor and the motion trajectory descriptor require a dense motion field for accurate computation. The camera motion and motion activity descriptors can be successfully extracted from compressed domain motion vectors. Thus they are the most suitable for video browsing applications. The wider scope of the motion activity descriptor makes it our first choice for incorporation into a video browsing system.
b. Inferring Higher level features from low-level features – All four described descriptors are low-level descriptors. However, in the right context, they can provide hints to higher level features. For instance, the most interesting moments in a soccer game are often marked by high motion activity and camera motion. In a specific domain, a certain trajectory would immediately identify an event e.g. in a game of tennis. Therefore, all four descriptors are highly applicable to systems that use lower-level features to infer higher level features. Such systems typically have less stringent requirements on computational complexity since they are seldom real-time. Note that for a short video program, the motion activity and the camera motion descriptors already provide a simple way to infer higher-level features like interesting play etc. However, such inferences are still broad. In all likelihood, a domain-specific knowledge-based in-

ference superstructure would be required to detect semantic events using the MPEG-7 motion descriptors.

c. Content-Based Lower Level Operations – The MPEG-7 motion descriptors would be extremely useful in providing hints for lower level operations such as transcoding from one bitrate to another, content enhancement, content presentation etc

4   A Video Browsing System Based on Motion Activity

As mentioned earlier, the compactness and ease of extraction of the motion activity descriptor make it the best candidate for a video browsing system. In our application, we use the intensity and the spatial distribution attributes of the motion activity descriptor. In Figure 4, we illustrate finding the most active shots in a video program

Fig 4. Video browsing – Extracting the 10 most "active" video segments in a news program

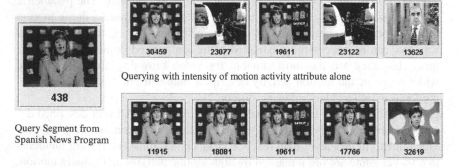

Querying with intensity of motion activity attribute alone

438

Query Segment from
Spanish News Program

Querying with combination of spatial and intensity of motion activity attributes

Fig 5. Illustration of video browsing with and without descriptors of spatial distribution of motion activity

using the intensity of motion activity. Note that the results are a mixture of sports and other shots. In Figure 5, we illustrate the use of the spatial distribution attribute of the motion activity distributor. As can be seen, the spatial distribution goes beyond the intensity of motion activity and helps match similar activities such as in this case, talking. In Figure 6, we again illustrate the use of the spatial attribute in matching shots with similar activities within a news video program in the first row, since the query and the retrieval shots are all "head and shoulders" shots. In the next two rows, we show that we can use the motion activity descriptor as a first line of attack to prune the search space, before using a bulky (64 bin) color histogram to get the final matches in the subsequent stage. Note that the results are similar are to those got by using color alone in row 2. Note that the color histogram is also extracted in the compressed domain.

Fig 6. Video indexing using motion activity combined with color

Fig 7. Illustration of adaptive sub-sampling approach to video summarization. The top row shows a uniform sub-sampling of the surveillance video, while the bottom row shows an adaptive sub-sampling of the surveillance video. Note that the adaptive approach captures the interesting events while the uniform sub-sampling mostly captures the highway when it is empty. The measure of motion activity is the average motion vector magnitude in this case which is similar to the MPEG-7 intensity of motion activity descriptor in performance.

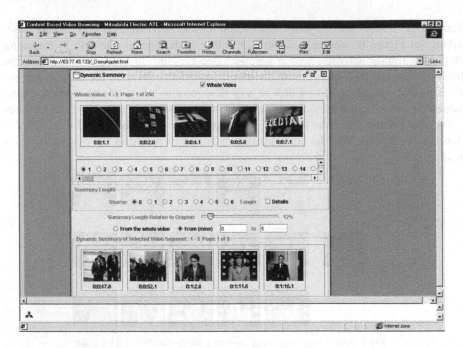

Fig 8. WWW-based interface for dynamic video summarization
(http://63.77.49.133/_DemoApplet.html) or (http://63.77.49.131/_DemoApplet.html)

Figures 4, 5 and 6 illustrate the video indexing capabilities of the motion activity descriptor. We have recently shown that the intensity of motion activity is an indication of the summarizability of the video sequence[30]. This gives rise to a sampling approach to video summarization in which video sequences are sub-sampled heavily if the activity is low and not so heavily if the activity is high [4,33]. We can thus vary the length of the summary of a shot by varying the sub-sampling rate. We illustrate this approach in Figure 7.

We illustrate our system in Figure 8. We uniformly sample the video program roughly every 1.5 seconds and extract features in the MPEG-1 compressed domain for each of those 1.5 second shots, as well as thumbnails. We do so to circumvent shot detection and to maintain one-pass feature extraction. We extract the four-element motion feature vector as mentioned earlier, as well as a color histogram taken from the dc values of the first intra-coded frame of the shot. The interface displays all the thumbnails, and each video shot can be played by clicking on the corresponding thumbnail. A query can be launched by right clicking on the desired query thumbnail and choosing the desired combination of features. So far we have described the indexing or bottom-up traversal capabilities of the system.

In addition to the above, our system provides a dynamic summarization framework that relies on the intensity of motion activity. As can be seen from figure 8, the user can specify any portion of the video program by entering the starting and stopping times, and immediately receive a summary of the video sequence. The summary is displayed in the form of an array of key frames, but can also be played as a concatenated video sequence. In this manner, the user can traverse the content top-down, until he gets the desired portion. The thumbnails received as a result of the summarization , also lend themselves to the indexing described earlier. Thus our system provides the user with extremely flexible traversal of the content using top-down (video summarization) or bottom-up (video indexing) traversal.

Note that our system is Web-based, and can thus serve clients of varying capabilities and bandwidths. We have already demonstrated this system to the Japanese press and to the MPEG-7 Awareness event in Singapore in 2001.

5 Conclusion

We presented an overview of the MPEG-7 motion descriptors and discussed their applications. We then motivated and presented a video browsing system based on motion activity. Our system is easy to use and computationally simple. In future work, we plan to incorporate the camera motion descriptor into the video-browsing framework, as well as refine the existing summarization by adding audio features.

References

1. Jeannin S., Divakaran, A.: MPEG-7 visual motion descriptors. Invited paper To appear in: IEEE Trans. Circuits and Systems for Video Technology special issue on MPEG-7, **11(6)** (2001)
2. Divakaran A., Sun, H.: A Descriptor for spatial distribution of motion activity. In: Proc. SPIE Conf. on Storage and Retrieval from Image and Video Databases, San Jose, CA 24-28 Jan. (2000)
3. MPEG-7 Visual part of the XM 4.0. In: ISO/IEC MPEG99/W3068, Maui, USA, Dec. (1999)
4. Kadir A. Peker: Video indexing and summarization using motion activity. Ph.D. Dissertation, New Jersey Institute of Technology, Electrical Engineering Dept. (2001)
5. Akutsu & al.: Video indexing using motion vectors. In: Proc. Visual Communications and Image Processing, SPIE **1818** (1992) 1522-1530
6. Ardizzone E. & al.: Video indexing using MPEG motion compensation vectors. In: Proc. IEEE International Conference on Multimedia Computing and Systems (1999)
7. Davis, J.W.: Recognizing movement using motion histograms, Technical Report No. 487, MIT Media Laboratory Perceptual Computing Section (1998)
8. Dimitrova N., Golshani, F.: Motion recovery for video content analysis. ACM Trans. Information Systems, **13** (1995) 408-439
9. Kobla V., Doermann D., Lin, K.-I., Faloutsous, C.: Compressed domain video indexing techniques using DCT and motion vector information in MPEG video. In: Proc. SPIE Conference on Storage and Retrieval for Image and Video Databases V, SPIE Vol. 3022, (1997) 200-211
10. Sahouria, E.: Video indexing based on object motion. M.S. thesis, Dept. of EECS, UC Berkeley

11. Tan, Y.P., Kulkarni, S.R., Ramadge, P.J.: Rapid estimation of camera motion from compressed video with application to video annotation. IEEE Trans. on Circuits and Systems for Video Technology, **10** (2000) 133-146.
12. Tse, Y.T., Baker, R.L.: Camera zoom/pan estimation and compensation fore video compression. In: Proc. SPIE Conf. On Image Processing Algorithms and Techniques II, Boston, MA (1991) 468-479
13. Wolf, W.: Key frame selection by motion analysis. In: Proc. ICASSP 96, Vol. II (1996) 1228-1231
14. Jeannin, S., Jasinschi R. & al.: Motion Descriptors for Content-Based Video Representation. Signal Processing: Image Communication Journal, August 2000 (to appear)
15. Chang, S.F., Chen, W., Meng, H.J., Sundaram, H., Zhong, D.: A Fully Automated Content-Based Video Search Engine Supporting Multi-Objects Spatio-Temporal Queries. IEEE Transactions on Circuit and Systems for Video Technology, **8** (1998) 602-615
16. Marr, D.: Vision. W. H. Freeman, San Francisco (1982)
17. Niblack, W., Ponceleon, D.B., Petkovic D. & al.: Updates to the QBIC System. In: Proc. IS&T SPIE, Storage and Retrieval for Image and Video Databases VI, San Jose, Vol. 3312 (1998) 150-161
18. Hori, O., Kaneko, T.: Results of Spatio-Temporal Region DS Core/Validation Experiment. In: ISO/IEC JTC1/SC29/WG11/MPEG99/M5414, Dec. 1999, Maui.
19. Llach, J., Salembier, P.: Analysis of video sequences: table of contents and index creation. In: Proc. Int'l Workshop on very low bit-rate video coding (VLBV'99), Kyoto, Japan (1999) 52-56
20. Jeannin, S., Mory, B.: Video Motion Representation for Improved Content Access. IEEE Transactions on Consumers Electronics, August (2000)
21. Information technology - Coding of Audio-visual Objects. Standard MPEG4, Version 1. ISO/IEC 14496
22. Lee, K.W., You, W.S., Kim, J.: Quantitative analysis for the motion trajectory descriptor in MPEG-7. In: ISO/IEC JTC1/SC29/WG11/MPEG99/M5400, Dec. 1999, Maui.
23. Szeliski, R.: Image Mosaicing for Tele-reality Applications. In: IEEE Workshop on Applications of Computer Vision, IEEE Computer Society (1994) 44-53
24. Irani, M., Anandan P. & al.: Efficient Representation of Video Sequences and their Application. Signal Processing: Image Communications, **8** (1996) 327-351
25. URL: http://www.cselt.it/mpeg/ official MPEG site
26. URL: http://www.virage.com, Virage system
27. URL: http://www.qbic.almaden.ibm.com, QBIC system
28. URL: http://www.ctr.columbia.edu/videoq/, VideoQ system
29. Peker K., Divakaran, A.: Automatic Measurement of Intensity of Motion Activity of Video Segments. In: Proc. SPIE Conference on Storage and Retrieval from Multimedia Databases, San Jose, CA (2001)
30. Divakaran A., Peker, K.: Video Summarization Using Motion Descriptors. In: Proc. SPIE Conference on Storage and Retrieval from Multimedia Databases, San Jose, CA (2001)
31. Bober, M., Asai K., Divakaran, A.: MPEG-4/7 based internet video and still image browsing system. In: Proc. SPIE Conference on Multimedia Systems and Applications, Boston, MA (2000)
32. Divakaran A., Vetro A., Asai K., Nishikawa, H.: Video browsing system based on compressed domain feature extraction. IEEE Trans. Consumer Electronics, Aug. 2000
33. Peker, K., Divakaran A., Sun, H.: Constant Pace Skimming and Temporal Sub-sampling of Video Using Motion Activity. To appear in: ICIP 2001

MPEG-7 MDS Content Description Tools and Applications

Ana B. Benitez[1], Di Zhong[1], Shih-Fu Chang[1], and John R. Smith[2]

[1] Department of Electrical Engineering, Columbia University,
1312 Mudd, 500 W 120th Street, MC 4712, New York, NY 10027, USA
{ana, dzhong, sfchang}@ee.columbia.edu
[2] IBM T. J. Watson Research Center,
30 Saw Mill River Road, Hawthorne, NY 10532, USA
jrsmith@watson.ibm.com

Abstract. In this paper, we present the tools specified by the MDS part of the MPEG-7 standard for describing multimedia data such as images and video. In particular, we focus on the description tools that represent the structure and semantics of multimedia data to whose development we have actively contributed. We also describe some of our research prototype systems dealing with the extraction and application of MPEG-7 structural and semantic descriptions. These systems are AMOS, a video object segmentation and retrieval system, and IMKA, an intelligent multimedia knowledge application using the MediaNet knowledge representation framework.

Keywords: MPEG-7, multimedia content descriptor, retrieval system

1 Introduction

In recent years, there has been an important increase of available multimedia data in both the scientific and consumer domains and facilities to access the multimedia data. However, the extraction of useful information from the multimedia data and the application of this information in practical systems such as multimedia search engines are still open problems. The most important barrier has been the lack of a comprehensive, simple, and flexible representation of multimedia data that enables flexible, scalable, and efficient multimedia applications.

MPEG-7 aims at standardizing tools for describing multimedia data to tear this barrier. The MPEG-7 framework consists of Descriptors (Ds), Description Schemes (DSs), a Description Definition Language (DDL), and coding schemes. Descriptors represent features or attributes of multimedia data such as color, texture, textual annotations, and media format. Description schemes specify more complex structures and semantics grouping descriptors and other description schemes such as segments of, and objects depicted in, multimedia data. The description definition language allows defining and extending descriptors and description schemes. Finally, coding

W. Skarbek (Ed.): CAIP 2001, LNCS 2124, pp. 41–52, 2001.
© Springer-Verlag Berlin Heidelberg 2001

schemes to compress descriptions and tools are needed to satisfy the storage and transmission requirements. MPEG-7 is currently in Final Committee Draft and scheduled to become an international standard in October 2001.

The specification of the MPEG-7 standard is divided into several parts, which correspond to the different MPEG groups working on it [5] [8]. The parts of the MPEG-7 standard are systems - transmission and encoding of MPEG-7 descriptions and tools -, DDL - description definition language based on XML-Schema -, video - Ds and DSs for video data -, audio - Ds and DSs for audio data -, Multimedia Description Scheme (MDS) - generic Ds and DSs for any media-, reference software - reference implementation of the standard -, and conformance - guidelines and procedures for testing conformance to MPEG-7.

This paper presents the MPEG-7 MDS tools [4], which are organized on the basis of their functionality into basic elements, content description, content management, content organization, navigation and access, and user interaction. We have actively contributed to the development of the MDS tools, especially to the content description tools on which we focus on this paper. Content description tools describe the perceivable content of multimedia data including its structure and semantics.

The creation and application of MPEG-7 descriptions are outside the scope of the MPEG-7 standard. However, MPEG-7 is becoming a significant driver of new research for multimedia analysis, storage, searching, and filtering [6], among others. In this paper, we also present two of our research prototypes systems, AMOS and IMKA, which demonstrate the extraction and application of MPEG-7 structure and semantic descriptions, respectively. AMOS is a video object segmentation and retrieval system [7]. IMKA is an intelligent multimedia knowledge application using the MediaNet knowledge representation framework [2] [3].

This paper is organized as follows. In section 2, we provide an overview of the MPEG-7 MDS tools. We describe the content description tools in detail providing examples in section 3. Section 4 presents the research prototype systems AMOS and IMKA. Finally, we conclude with a summary in section 5.

2   Overview of MPEG-7 MDS Tools

The MPEG-7 MDS tools [4] are organized on the basis of their functionality into basic elements, content description, content management, content organization, navigation and access, and user interaction, as shown in Figure 1. We shall provide an overview of the MPEG-7 MDS tools in this section.

The *basic elements* are the schema tools, basic datatypes, media localization tools, and basic tools repeatedly used to define other MPEG-7 MDS tools. The schema tools define the root and top-level elements for MPEG-7 descriptions, and the base types and packaging of MPEG-7 tools. The root element is the starting element of complete or partial MPEG-7 descriptions to allow both the complete and incremental transmission of MPEG-7 descriptions. MPEG-7 descriptions can be associated metadata such as such as version, creator, and rights. Top-level elements are the elements that immediately follow the root element in a description such as multimedia content entity (e.g., image, video, and multimedia collection.), abstraction (e.g., world, model, and summary), and management (e.g., user and creation) elements. The base types are the tools from which the MPEG-7 tools (e.g., Ds and DSs) are derived. Finally, package tools describe the packaging of MPEG-7 tools into hierarchical folders.

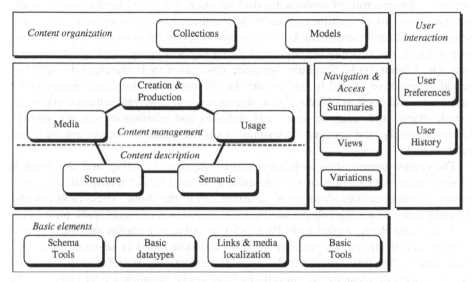

Fig. 1: Organization of the MPEG-7 MDS tools.

Basic datatypes represent constrained integer and real values, vectors and matrices, country and region codes, references, unique identifiers, time points and duration, etc. Media localization tools represent links to multimedia data using media URIs, inline media, and spatio-temporal locators. Some basic tools are text annotation tools, classification schemes and controlled term tools. Text annotation tools represent unstructured and structured textual annotations. Textual annotations can be structured by questions such as who, where, and when, by keywords, or by syntactic dependency relations. Classification scheme and controlled term tools describe hierarchical classifications of textual and graphical terms in subject-like areas, relations among the terms, and references to terms in classification schemes. Other basic tools describe persons, group of persons, organizations, geographical locations either real or fictional, relations and graph structures, criteria for ordering descriptions, audience's affective response to multimedia data, and the pronunciation of a set of words.

The *content management* tools relate to the media, creation and usage of the multimedia data. This information is not usually perceived in the media. Media tools describe media information of the multimedia data as a unique identifier and a set of media profiles. A media profile describes a coded version of the multimedia data in terms of format, quality, and transcoding hints, among others. Creation tools describe the creation and production of the multimedia data including the creation process (e.g., title, abstract, creator, creation place, time, and tools), the classification (e.g., target audience, genre, and rating), and related material (e.g., trailers and web pages for movies). Usage tools describe the usage process of the multimedia data including the rights, the publication (e.g., emission and edition), audience, and financial results.

The *content description* tools describe the perceivable content of multimedia data in terms of the structure and semantics of multimedia data. Structure description tools represent the structure of multimedia data in space, time, and media source by describing segments of multimedia data such as still regions, video segments, and audio segments, and attributes, hierarchical decompositions, and relations of segments. Correspondingly, the structure description tools are segment entity, attribute, decomposition, and relation tools. Similarly, semantic description tools represent the narrative world depicted, or related to multimedia data by describing semantic entities in the narrative world such as objects, agent objects, events, concepts, semantic states, semantic places, and semantic times, and attributes and relations of semantic entities. Correspondingly, the semantic description tools are semantic entity, attribute, and relation tools.

The *content organization* tools address the collection, modeling, and classification of multimedia data. Collection tools describe unordered sets of multimedia data, segments, semantic entities, descriptors, or mixed sets of the above. These tools can also describe relations among collections. There are three types of model tools: probability, analytical, and cluster model tools. Probability model tools associate statistics or probabilities to collections; analytical model tools associate labels or semantics to collection; finally, the cluster model tools can associate not only statistics or probabilities but also labels or semantics to collections. Finally, classification model tools describe known collections in order to classify unknown ones.

The *navigation and access* tools describe summaries, partitions and decompositions, and variations of multimedia data. Hierarchical and sequential summary tools describe hierarchical or sequential summaries of multimedia data. Hierarchy summaries are composed of highlight segments of the multimedia data that are organized in hierarchies. Sequential summaries are composed of images and audio clips that can be synchronized and presented to the user at different speeds. Partitions and decomposition tools describe decomposition of images, video, or audio-visual data in space and/or frequency partitions. Variation tools describe relations between different variations of multimedia data (e.g., summary, revision, and modality conversion).

Finally, the *user interaction* tools describe user preferences and history in the consumption of multimedia data. User preference tools describe preferences of users pertaining to the filtering, search, and browsing of multimedia data (e.g., creation and classification preferences). The user history tools describe history of users in consuming the multimedia data as set of user actions (e.g., record, pause, and play) with associated temporal information.

3   Content Description Tools

We have actively contributed to the development of the MDS descriptors and description schemes, in particular, to the content description tools, which are described in more detail in this section. As mentioned in the previous section, content description tools describe the structure and semantics of multimedia data. Figure 2 shows the structure and semantic descriptions of an image; the corresponding XML description is provided in the Appendix.

3.1 Structure Description Tools

The structure description tools represent the structure of multimedia data in space, time, and media source by describing general and application-specific *segments* of multimedia data together with their attributes, hierarchical decompositions, and relations.

The most basic visual segment is the *still region*, which is a group of pixels in a 2D image or a video frame. Figure 2 shows three examples of still regions corresponding to the full image and the two regions depicting the persons in the image (SR1, SR2, and SR3). Other strictly visual segments are the *video segment*, defined as a group of pixels in a video, and the *moving region*, defined as a set of pixels in a group of frames in a video. The audio segment is defined as a group of samples in an audio sequence. Both *audio-visual segment* and the *audio-visual region* contain audio and visual data in an audio-visual sequence corresponding to a group of frames and audio samples synchronized in time, and an arbitrary group of pixels and audio samples, accordingly.

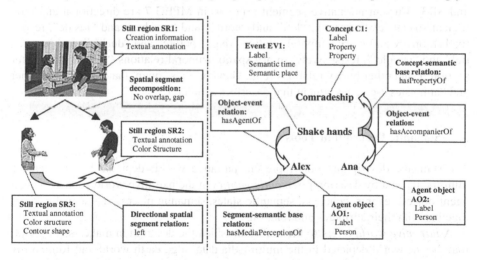

Fig. 2: Structure description (left) and semantic description (right) of an image.

The application-specific segments are *mosaics* (a panoramic view of a video segment), *3D still regions* (3D spatial region of a 3D image), *image* and *video texts* (still region and moving region that correspond to text, respectively), *ink segments* (temporal segment of electronic ink data), *multimedia segments* (composite of segments forming a multimedia presentation such as an MPEG-4 presentation or a web page), and edited video segments (video segments resulting from editing work).

Visual and audio features, and media, creation and usage information of the multimedia data represented by segments can be described using the visual and audio Ds and DSs, and the media, creation and usage tools, respectively. *Segment attribute* description tools describe other attributes of segments related to spatio-temporal, media, and graph masks; importance for matching and point of view; creation and media of ink segments; and hand writing recognition (e.g., lexicon and results).

The *segment decomposition* tools describe the decomposition of segments in hierarchies of segments to form, for example, tables of contents of multimedia data. MPEG-7 has defined four types of segment decompositions: spatial, temporal, spatio-temporal, and media source decompositions. An image can be decomposed spatially into a set of still regions corresponding to objects in the image, which, at the same time, can be decomposed into other still regions. Figure 2 exemplifies the spatial decomposition of a still region into two still regions. Similar decompositions can be generated in time and/or space for video and other multimedia data. A media source decomposition divides segments into its media constituents such as audio and video tracks. The sub-segments resulting from a decomposition may overlap in time, space, and/or media; furthermore, their union may not cover the full time, space, and media extents of the parent segment, thus leaving gaps. The spatial decomposition shown in Figure 2 has gaps but no overlaps.

The *segment relation* description tools describe general relations among segments. Figure 2 shows an example of the spatial relation *left* between two still regions (SR2 and SR3). Current normative segment relations in MPEG-7 are directional and topological spatial relations (e.g., "left" and "north", and "touches" and "inside", respectively), binary and n-ary temporal relations (e.g., "before" and "during", and "sequential" and "parallel", respectively), n-ary spatio-temporal relations ("union" and "intersection"), and other binary relations (e.g., "keyFor" -a segment that is key for another- and "annotates" -e.g., a segment that describes another-).

3.2   Semantic Description Tools

The semantic description tools represent narrative worlds depicted by or related to multimedia data by describing *semantic entities* in the narrative world such as objects, agent objects, events, concepts, semantic states, semantic places, and semantic times, together with their attributes and relations.

A *narrative world* refers to the reality in which the description makes sense, which may be the world depicted in the multimedia data (e.g., earth world and Roman mythology world). *Objects* and *events* are perceivable entities that exist or take place in time and space in the narrative world, respectively. *Agent objects* are objects that are persons, group of persons, or organizations. As example, the semantic description in Figure 2 shows an event and two agent objects corresponding to "Shake hands" (EV1), and the two persons "Alex" and "Ana" (AO1 and AO2). *Concepts* are entities that cannot be perceived in the narrative world or described as the generalization of perceivable semantic entities. "Comradeship" (C1) in Figure 2 is a concept. *Semantic states* are parametric attributes of semantic entities and semantic relations at a specific point in time and space (e.g., weight and height of person). Finally, *semantic places and times* are locations and times in the narrative world, respectively. The event "Shake hands" have associated a semantic place and a semantic time.

Semantics entities can be described in terms of labels, a textual definition, properties, and features of the segments where they appear. *Semantic attribute* description tools describe other attributes of semantic entities related to abstraction levels and semantic measurements in time and space. Abstraction refers to the process of taking a concrete semantic description of multimedia data (e.g., "Alex is shaking hands with Ana" in the image shown in Figure 2) and generalizing it to a set of multimedia data (media abstraction, e.g., "Alex is shaking hands with Ana" in any picture) or a set of concrete semantic descriptions (formal abstraction, e.g., "Any man is shaking hands with any woman").

The *semantic relation* description tools describe general relations among semantic entities and other entities. Current normative semantic relations in MPEG-7 are relations among semantic entities (e.g., "generalizationOf"–a semantic entity has a more general meaning than another, "hasAgentOf"-object that initiates the action of an event-, "hasAccompanierOf"-object that is a join agent in an event-, and "hasPropertyOf"-concept that represents properties of a semantic entity-), relations among semantic entities and semantic relations ("membershipFunctionFor"-semantic state that parameterizes the degree of a semantic relation-), relations among semantic entities and segment (e.g., "hasMediaPerceptionOf"-segment that depicts a semantic entity- and "hasMediaSymbolOf"-segment that symbolizes a semantic entity-), and relations among semantic entities and models (e.g., "exemplifiedBy"-model that exemplifies a semantic entity- and "hasQualityOf"-analytical model that qualifies attributes of a semantic entity). Figure 2 includes examples of some semantic relations.

4 Research Prototype Systems

The creation and application of MPEG-7 descriptions are outside the scope of the MPEG-7 standard. However, MPEG-7 is becoming a significant driver of new research for multimedia analysis, storage, searching, and filtering, [6], among others. In this section, we present two of our research prototypes systems, AMOS and IMKA, which demonstrate the generation and application of MPEG-7 structure and semantic descriptions, respectively, in a retrieval application.

4.1 AMOS: Video Object Segmentation and Search System

AMOS is a video object segmentation and retrieval system [7]. In this framework, a video object (e.g. person, car) is modeled and tracked as a set of regions with corresponding visual features and spatio-temporal relations (see Figure 3.a). The region-based model also provides an effective base for similarity retrieval of video objects.

AMOS effectively combines user input and automatic region segmentation for defining and tracking video objects at a semantic level. First, the user roughly outlines the contour of an object at the starting frame, which is used to create a video object with underlying homogeneous regions. This process is based on a region segmentation method that involves color and edge features and a region aggregation method that classifies regions into foreground and background. Then, the object and the homogeneous regions are tracked through successive frames. This process uses affine motion models to project regions from frame to frame and a color-based region growing to determine the final projected regions. Users can stop the segmentation at any time to correct the contour of video objects. Extensive experimental results have demonstrated excellent results. Most tracking errors are caused by uncovered regions and can be corrected with a few user inputs.

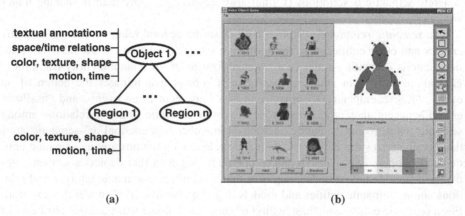

(a)   (b)

Fig. 3. (a) Video object representation in AMOS. (b) Query interface of AMOS: query results (left), query canvas (top right), and feature weights (bottom right).

AMOS also extracts salient regions within video objects that users can interactively create and manipulate. Visual features and spatio-temporal relations are computed for video objects and salient regions and stored in a database for similarity matching. The features include motion trajectory, dominant color, texture, shape, and time descriptors. Currently three types of relations among the regions of a video object are supported: orientation spatial (angle between two regions), topological spatial (contains, does not contain, or inside), and directional temporal (start before, at the same time, or after). Users can enter textual annotations for the objects.

AMOS accepts queries in the form of sketches or examples and returns similar video objects based on different features and relations. Figure 3.b shows the query interface of AMOS. The query process of finding candidate video objects for a query uses a filtering together with a joining scheme. The first step is to find a list candidate regions from the database for each query region based on the visual features. Then, the region lists are joined to obtain candidate objects and their total distance to the query is computed by matching the spatio-temporal relations.

The mapping of the video object representation of AMOS to MPEG-7 Ds and DSs in straightforward. In fact, the AMOS system has been used to generate descriptions to evaluate parts of MPEG-7 [1]. A video can be described as a video segment that is spatio-temporally decomposed into moving regions corresponding to the segmented video objects. In the same way, these moving regions can be decomposed into other moving regions corresponding to the salient regions of the objects. The visual features, textual annotations, and spatio-temporal relations can be described using the normative visual, textual annotation, and segment relation tools.

4.2   IMKA: Intelligent Multimedia Knowledge Application

IMKA, is an intelligent multimedia knowledge application using the MediaNet knowledge representation framework [2] [3]. MediaNet uses multimedia information for representing semantic and perceptual information about the world. MediaNet knowledge bases can be built from partially annotated collections of multimedia data and used to enhance the retrieval of multimedia data.

MediaNet represents the world using concepts and relationships between the concepts that are defined and exemplified by multimedia information such as text, images, video sequences, and audio-visual descriptors. In MediaNet, concepts can represent either semantically meaningful objects or perceptual patterns in the world. MediaNet models the traditional semantic relationship types such as generalization and aggregation but adds additional functionality by modeling perceptual relationships based on feature descriptor similarity and constraints. Figure 4 shows an example of a MediaNet knowledge base illustrating the concepts Human and Hominid.

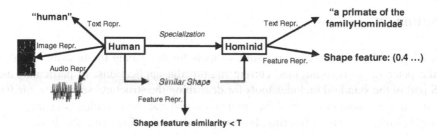

Fig. 4. MediaNet knowledge base illustrating the concepts Human and Hominid.

We construct a MediaNet knowledge base semi-automatically from a partially annotated image collection using the electronic lexical system WordNet and an image network built from multiple example images, extracted color and texture descriptors. First, stop words are removed from the textual annotations. Then, the words in the textual annotations are inputted to WordNet to obtain the list of relevant concepts and relationships between them with human supervision. In this process, a human supervisor removes the meanings associated with each word returned by WordNet that do not apply to the images. A concept is created for each remaining sense. Only a subset of the semantic relationships from WordNet is considered: opposite, specialization/generalization, and part/compose of. Finally, automatic visual feature extraction tools are used to extract features from the images and feature centroids for concepts.

An intelligent image retrieval system for images has been implemented by extending a typical image retrieval system with a MediaNet knowledge base and a query processor that translates and expands queries across multiple media modalities. First, the query processor classifies each incoming query into a set of relevant concepts based on the media representations of the concepts. The initial set of relevant concepts is then extended with other semantically similar concepts. A visual query is issued to the descriptor similarity-based search engine for the incoming query and each relevant concept (using centroids). Finally, the results of all the queries are merged into a unique list based on how similar the concepts that generated those results were to the initial user query. We have found that MediaNet can improve the performance of image retrieval applications for semantic queries (e.g., find Tapirs and other animals); however, additional experiments are needed to further demonstrate the performance gain of using MediaNet in an image retrieval system.

MediaNet knowledge bases can be encoded using MPEG-7 semantic and model tools, which could greatly benefit the exchange and re-use of knowledge and intelligence among multimedia applications. Each concept in MediaNet can be a semantic entity. The textual representations of a concept can be textual labels of the semantic entity. Other media representations of concepts can be described using probability models (e.g., centroids of concepts) and audio-visual examples of semantic entities. Relationships among concepts in MediaNet can be encoded as relationships among the corresponding semantic entities.

5 Summary

MPEG-7 provides a comprehensive suite of tools for describing multimedia data that has the potential to revolutionize current multimedia applications. In particular, the MDS part of the standard includes tools for describing the structure, semantics, media, creation, usage, collections, models, summaries and views of multimedia data that allow efficiently searching, filtering, browsing, and access of multimedia. In addition, MDS tools can also describe user preferences and history in consuming multimedia data that allow personalized multimedia services and devices. However, MPEG-7 has also introduced many new research problems related to the extraction and application of MPEG-7 descriptions. This paper has provided an overview of the MPEG-7 MDS tools focussing on the tools for describing the structure and semantics of multimedia data. It has also presented two research prototypes systems that can extract and use semantic and structure descriptions in a retrieval scenario.

References

1. Benitez, A. B., Chang, S.-F.: Validation Experiments on Structural, Conceptual, Collection, and Access Description Schemes for MPEG-7, Digest of the IEEE 2000 International Conference on Consumer Electronics (ICCE-2000), Los Angeles, CA, June 13-15 (2000)

2. Benitez, A. B., Smith, J. R.: New Frontiers for Intelligent Content-Based Retrieval, Proceedings of the IS&T/SPIE 2001 Conference on Storage and Retrieval for Media Databases, Vol. 4315, San Jose, CA, Jan. 24-26 (2001)
3. Benitez, A. B., Smith, J. R., Chang, S.-F.: MediaNet: A Multimedia Information Network for Knowledge Representation, Proceedings of IS&T/SPIE 2000 Conference on Internet Multimedia Management Systems, Vol. 4210, Boston, MA, Nov. 6-8 (2000)
4. MPEG MDS Group, "Text of ISO/IEC 15938-5 FCD Information Technology - Multimedia Content Description Interface - Part 5 Multimedia Description Schemes", ISO/IEC JTC1/SC29/WG11 MPEG01/M7009, Singapore, March (2001)
5. MPEG, Working Documents for MPEG-7 Standard, http://www.cselt.it/mpeg/working_documents.htm.
6. Chang, Y. C., Lo, M. L., Smith, J. R.: Issues and solutions for storage, retrieval, and search of MPEG-7 documents, Proceedings of IS&T/SPIE 2000 Conference on Internet Multimedia Management Systems, Vol. 4210, Boston, MA, Nov. 6-8 (2000)
7. Zhong, D., Chang, S.-F.: An Integrated System for Content-Based Video Object Segmentation and Retrieval, IEEE Transactions on Circuits and Systems for Video Technology, Vol. 9, No. 8, Dec. (1999) 1259-1268
8. IEEE Transactions on Circuits and Systems for Video Technology, Special Issue on MPEG-7, June (2001)

Appendix: XML for Example Structure and Semantic Descriptions

The XML for the structure description in Figure 2 follows.

```
<StillRegion id="SR1">
  <CreationInformation>
    <Creation>
      <Creator>
        <Role><Name>Photographer</Name></Role>
        <Person>
          <Name>
            <GivenName>Seungyup</GivenName>
            <FamilyName>Paek</FamilyName>
          </Name>
        </Person>
      </Creator>
    </Creation>
  </CreationInformation>
  <TextualInformation>
    <FreeTextAnnotation>
      Alex shakes hands with Ana
    </FreeTextAnnotation>
  </TextualInformation>
  <SpatialSegmentation overlap="false" gap="true">
    <StillRegion id="SR2">
      <TextualInformation>
        <FreeTextAnnotation> Alex </FreeTextAnnotation>
      </TextualInformation>
      <VisualDescriptor xsi:type="ColorStructureType">
        ...
      </VisualDescriptor>
    </StillRegion>
    <StillRegion id="SR3">
```

```
                <TextualInformation>
                  <FreeTextAnnotation> Ana </FreeTextAnnotation>
                </TextualInformation>
                <Relation xsi:type="DirectionalSpatialSegmentRelationType"
                          name="left" target="#SR2"/>
                <VisualDescriptor xsi:type="ColorStructureType">
                  ...
                </VisualDescriptor>
                <VisualDescriptor xsi:type="ContourShapeType">
                  ...
                </VisualDescriptor>
              </StillRegion>
            </SpatialSegmentation>
          </StillRegion>
```

The XML for the semantic description in Figure 2 follows. We assume the XML for the structure and semantic descriptions are in the same XML file.

```
    <Semantic>
      <Label><Name>Alex shakes hands with Ana </Name></Label>
      <SemanticBase xsi:type="EventType" id="EV1">
        <Label><Name>Shake hands</Name></Label>
        <Relation xsi:type="ObjectEventRelationType"
                  name="hasAgentOf" target="#AO1"/>
        <Relation xsi:type="ObjectEventRelationType"
                  name="hasAccompanierOf" target="#AO2"/>
        <Relation xsi:type="ConceptSemanticBaseRelationType"
                  name="hasPropertyOf" target="#C1"/>
        <Relation xsi:type="SegmentSemanticBaseRelationType"
                  name="hasMediaPerceptionOf" target="#SR1"/>
        <SemanticPlace>
          <Label><Name>Columbia University</Name></Label>
        </SemanticPlace>
        <SemanticTime>
          <Label><Name>9:45am, May 27, 1998</Name></Label>
          <Time><TimePoint>1998-05-27T09:45+01:00</TimePoint></Time>
        </SemanticTime>
      </SemanticBase>
      <SemanticBase xsi:type="AgentObjectType" id="AO1">
        <Label><Name>Alex</Name></Label>
        <Relation xsi:type="SegmentSemanticBaseRelationType"
                  name="hasMediaPerceptionOf" target="#SR2"/>
        <Agent xsi:type="PersonType">
          <Name><GivenName>Alejandro</GivenName></Name>
        </Agent>
      </SemanticBase>
      <SemanticBase xsi:type="AgentObjectType" id="AO2">
        <Label><Name>Ana</Name></Label>
        <Agent xsi:type="PersonType">
          <Name><GivenName>Ana</GivenName></Name>
        </Agent>
      </SemanticBase>
      <SemanticBase xsi:type="ConceptType" id="C1">
        <Label><Name>Comradeship</Name></Label>
        <Property>Associate</Property> <Property>Friend</Property>
      </SemanticBase>
    </Semantic>
```

Image Retrieval Using Spatial Color Information

Krzysztof Walczak

Institute of Computer Science Warsaw University of Technology
Nowowiejska 15/19, 00-665 Warsaw, Poland
kwl@ii.pw.edu.pl

Abstract. This paper presents a very efficient and accurate method
for retrieving images based on spatial color information. The method
is based on a regular subblock approach with a large number of blocks
and minimal color information for each block. Binary Thresholded
Histogram and Extended Binary Thresholded Histogram are defined.
Only 40 numbers are used to describe an image. Computing the distance
is done by a very fast bitewise sum mod 2 operation.

Keywords: content based image retrieval, color matching, precision and
recall

1 Introduction

A key aspect of image databases is the creation of robust and efficient indices
which are used for content based retrieval of image. To achieve this aim several
features such as color, shape, texture etc. are employed. In particular, color
remains the most important low-level feature which is used to build indices for
database images. The color histogram is the most popular index, due primarily
to its simplicity. Unfortunately, it tends to give many false positivies, especially
when image database is large. Therefore using color layout is a better solution
to image retrieval. A natural approach is to divide the whole image into regular
parts and extract color features from each of the parts [2,4]. Another similar
approach is the quadtree color layout [8]. In this approach the entire image is split
into a quadtree structure and each tree branch has a histogram to describe its
color. Unfortunately, this regular subblock-based approach is computation and
storage expensive. Moreover, it cannot provide accurate spatial color information
for a small number of parts. Increasing the number of parts increases spatial
color information but it also increases the memory required to store histograms
and the time required to compare them. For example, in the work [4] an image
was partitioned into a 3*3 array of 9 subblocks. Thus 10 histograms need to be
calculated, one for the entire image and one for each of the 9 subblocks. This
method requires large computation but does not provide accurate spatial color
information.

Quite a different approach is to segment the image into regions with salient
color features [12]. The position and color set features of each region are re-
membered for later queries. The disadvantage of this approach is the generally
a difficult problem of good image segmentation.

W. Skarbek (Ed.): CAIP 2001, LNCS 2124, pp. 53–60, 2001.

In the work [10] each pixel of a particular color is classified as either coherent or incoherent. Then scattered pixels are distinguished from clustered pixels. It permits improving the accurate local color information.

In the next method [13] the authors used five partially overlapping fuzzy regions from which they extract the average color and the covariance matrix of the color distribution.

Among other methods the work [5] is where a color co-occurrence matrix is constructed and the correlogram, as the similarity measures, is used.

This work is based on a regular subblock approach. But in comparison to other methods the technique is quite different. The idea is as follows: large subblocks but the minimum color information for each block. This idea results from the following fact: If a human is observing even a very complicated image, he is able to distinguish only few colors for a small part of the image. In our work we will show that for each parts the information about the number of pixels of given color is not essential. Binary information about occuring coarsed color is quite sufficient. Thanks to this assumption a very fast image retrieval is obtained.

2   HSV Color Space

We would like to take into consideration human color perception and recall. To achieve this aim using the RGB space has several inconveniences. Above all there is not any correlation among human color perception and RGB values. If we are to describe the color content of an image, we would use terms such as bright blue, yellow or brown, not RGB values. Furthermore it is very difficult to divide histograms into small number of bins in such a way that any bin describes a color discerned by humans. Therefore, HSV space is much better for our work. We would like to create a histogram with a small number of bins. For HSV space it is very natural.

For images using the RGB color space it is necessary to make the conversion from RGB to HSV space. The method of this conversion for example is given in the works [1,3].

However the main problem is how to select histogram bins.

For instance in [9] only HS coordinates to form a two-dimensional histogram are used. The H and S dimensions were divided into N and M bins, respectively, for a total N*M bins. Authors of the work [9] chose N=8, M=4 so the number of all bins is 32.

In the other works authors distinguished bright chromatic colors, chromatic colors and black and white colors. In work [1] it is shown experimentally that bright chromatic colors are with *value* > 75% and *saturation* >= 20%. Colors with *value* < 25% are classified as black, and colors with *saturation* < 20% and *value* > 75% can be classified as white. All remaining colors are chromatic.

In our method we create bins according to the above partition and we distinguish white, black, bright chromatic and chromatic colors. In HSV space H coordinate is particularly important, since it represents color in a manner which is proven to imitate human color recognition and recall. We divide bright chromatic colors into 9 bins by divide H into 9 nonregular parts. However, we divide the chromatic colors into two parts according to the division S for 2 parts and further each of them into 9 bins by dividing H into 9 nonregular parts. It results from the fact, that many researchers claim that the saturation histogram has a multimodal nature. To sum up we have 29 bins: 2 bins for white and black colors, 9 bins for bright chromatic colors and 18 (2*9) bins for chromatic colors.

It should be underlined that we divide H dimension into nonregular way according to table 1.

Table 1. Division of H dimension

$h < 25$ or $h > 335$	red
$25 <= h < 48$	
$48 <= h < 66$	
$66 <= h < 160$	green
$160 <= h < 185$	
$185 <= h < 200$	
$200 <= h < 270$	blue
$270 <= h < 295$	
$295 <= h < 335$	

It results from the observation how colors are located in color space with changing the H value. Red, green and blue colors occupy more place than other medial colors. We assume that between each main colors: red, green and blue are 2 medial colors, so we have 9 parts of H value.

Each bin contains the percentage of pixels in the image that have a corresponding value of V, H and S for that bin (the sum of all bins equals 100).

3   Binary Thresholded Histogram

In Binary Thresholded Histogram (BTH) only salient colors which can be discerned by human in an image are specified. If we are observing small part of the image, then in general we can distinguish only a few colors.

In order to obtain BTH it is necessary to do the following steps:

Step 1 Creating Thresholded Histogram

In Thresholded Histogram all values less than the given threshold are equal to 0. In this work the threshold is equal to 4%. For the following fragment of the histogram

1, 23, 3, 11, 2, 2, 34, ...

we obtain:

0,23,0,11,0,0,34,...

Step 2 Creating BTH

BTH is obtained from Thresholded Histogram by binarization. If the value in Thresholded Histogram is not equal to 0, we set the value to 1, otherwise we set the value to 0.

For example we have:

0,1,0,1,0,0,1,...

Value 1 in BTH means, that this position of histogram contains essential value. Through experiments we find out that a number of the value 1 in BTH is small and changes between 1 and 8 for small parts of the image (each image was divided into 40 parts). On average it is equal to 3.5 (for our image database).

BTH introduces a big simplification and it can not be used for all images. BTH gives excellent results if the image is divide into small parts (in our case into 40 parts).

4 Retrieval

We base on division of an image into a number of parts. We propose a division into 40 parts (5 rows and 8 columns) for rectangular images. For all parts it is necessary to compute BTH. BTH has 29 bits and therefore, it is possible to locate these bits in one number of type unsigned int (in C language). The number of that type has 32 bits so 3 bits remain free. Each image is described by 40 integer numbers. In these 40 numbers a huge amount of information is contained in the color layout.

The query is described also by 40 BTH keeping in 40 numbers. The distance between two BTH h and g is computed as follows:

$$d = \sum_{i=1}^{29} (h_i \oplus g_i)$$

If BTH h is kept in the number x and BTH g is kept in the number y, to compute the distance it is enough to count positions with value 1 in vector r, where

r = x ^ y

where operation ^ denotes bitewise a sum mod 2 (in C language).

Computing the distance is performed very fast by bitewise operation. For example let us consider the fragments of BTH h and g:

h:

0 1 0 1 0 0 1 ...

g:

0 0 0 1 0 1 1 ...

By doing bitewise the sum mod 2 we have:

0 1 0 0 0 1 0 ...

In this case the distance is 2. It means, that histograms are different in two places. In the first image there is a color which is absent in the second image and vice-versa. The remaining colors are identical.

The distance between two image is simply the sum of the distance between individual parts:

$$D = \sum_{i=1}^{40} d_i$$

According to the given definition the smallest distance value indicates the database image that matches best.

In Fig. 1 there is a query and an example of retrieved images. The distances between the query and images are as follows: 54,64,68,87,91. The distances are not very small but the similarity is very good.

Fig. 1. Retrieval of top left image. Retrieved images are from top left to bottom right in order of best match.

5   Extended Binary Thresholded Histogram

The disadvantage of all histogram techniques is the possibility that the number of relevant images not retrieved can be large. It follows from the fact that a similarity can exist on the basis of not two appropriate bins but two adjacent bins. It is possible if pixels in the image bin are located near of border of the bin and if pixels in adjacent query bin are located near the same border. For example, let us consider the bin in histogram query having the value 3 and the appropriate bin for an image having the value 24. In this case in BTH for query is value 0 and in BTH for image is value 1. This bins are not similar. Yet, the similarity is possible if in the query there are a lot of pixels in adjacent bin located near the border. In our work the resolution of this problem is proposed by using the Extended Binary Thresholded Histogram. Each bin of that histogram has value 1, iff the appropriate bin in Extended Histogram has a value greater than a given threshold. For Extended Histogram the number of pixels in the bins is calculated in extended borders. For example, the bin for bright chromatic red color has borders:

$$v >= 75 \land s >= 20 \land (h < 25 \lor h > 335)$$

The borders for the same bin in Extended Histogram are:

$$v >= 70 \land s >= 15 \land (h < 30 \lor h > 330)$$

The borders of the bin are greater by 5% for v i s and by 5° for h. The borders of bins overlap and each bin contains the part of pixels from adjacent bins. It follows that

$$h_i = 1 => h_i^E = 1$$

where h_i denotes a bin for BTH and h_i^E denotes a bin for Extended BTH.

Therefore by using Extended BTH it is only possible to decrease the distance between the query and the image.

Let us introduce the following notation:

q - variable keeping BTH for query

im - variable keeping BTH for image

qExt - variable keeping Extended BTH for query

If value for im_i is 0, the distance is computed using q_i and if the value for im_i is 1, the distance is computed using $qExt_i$. It results from the fact that only if im_i is 1, it is possible to decrease the distance. Then we have:

```
r = ~im & (q ^ im) | im & (qExt ^ im) = q & ~im | ~qExt & im
```

where ~, ^, |, & denotes bitewise operations. The number of value 1 in vector r is the distance.

Extended BTH can decrease the distance and it can be used as a supplementary method for matching.

For query from Fig. 1 the distances between query and retrieved images by Extended BTH are: 43,54,59,75,76.

6   Discussion of Result

The effectiveness of retrieval performances can be measured by recall and precision [6]:

$$recall = A/(A + C)$$

$$precision = A/(A + B)$$

where A is the number of relevant images retrieved, B is the number of not relevant images retrieved, and C is the number of relevant images not retrieved.

For our method precision is very high because the probability that the distance between two images is small and that the images are not similar, is very small. It follows from the following reasoning: Let us think again what it means that the distance is equal 40 for example (we have 40 parts of the image). It can mean that for each parts the distance is 1. There is 1 additional color for each parts and remaining colors are identical.It is also possible that for 20 parts the distance is 2. It means that for 20 parts only 1 color is different. It is also possible, that few parts are quite dissimilar but the remaining are completely similar. In all this case the entire image is similar to query.

During experiments a query was created also by modifications of an image. These modifications were as follows:

1. small translation of the image
2. small rotation of the image
3. small change of colors for each pixel
4. adding noise to an image

In all these cases the distance between query and image was very small. Therefore our method is better than other methods which base on regular sub-block approach in which spatial information is encoded too strictly i.e. during matching it is hard to take into consideration even for small rotations and translations of an image. In our method a small translation or rotation of an image does not change BTH for each part or perhaps changes it to a very small degree.

Much worse a matter is with recall, because the number of relevant images not retrieved can be large. If the query is created from the image by large translation, then similar images cannot be found.

There exists a simple way of improving the value of recall. It is sufficient to cut from a query the most essential fragment including for example 3*5 parts and then matching this fragment to an image by finding the best position. The number of test amounts to 6 ((5-3) * (8-5) = 6) so it can be accepted.

It is also possible to obtain matching using Extended BTH which can also improve the value of recall.

7   Conclusion

In this paper a very fast and effective method of image retrieval has been presented. The method is based on spatial color information obtained from each part of the image. It should be emphasized that the number of these parts can not be too small. In the work division into 40 parts for rectangular images has been proposed.

In the paper Binary Thresholded Histogram and Extended Thresholded Histogram have been defined. Both have 29 bits which allows to locate each of them in one number. As a result, each image is described by only 40 numbers. Computing the distance is done by a very fast bitewise sum mod 2 operation.

The value of precision for the proposed method is very high. In the work the way of improving the value of recall has been proposed.

References

1. Androutsos, M. et al.: A novel vector-based approach to color image retrieval using a vector angular-based distance measure. Computer Vision and Image Understanding, **75** (1999) 46–58
2. Faloutsos, C. et al.: Efficient and effective querying by image content. Journal of Intelligent Information Systems, **3** (1994)
3. Foley, J.,D. et al.: Using color in computer graphics. IEEE Computer Graphics and Applications, **5** (1988) 25-27
4. Gong, Y. et al.: An image database system with content capturing and fast image indexing abilities. In: Proceedings of the International Conference on Mutimedia Computing and Systems. IEEE, Boston, MA (1994) 121-130
5. Huang, J. et al.: Image indexing using color correlogram. In: Proc. of IEEE Conf. on Computer Vision and Pattern Recognition (1997)
6. Di Lecce, V., Guerriero, A.: An evaluation of the effectiveness of image features for image retrieval. Journal of Visual Communication and Image Representation, **10** (1999) 351–362
7. Li, Z. et al: Illumination invariance and object model in content-based image and video retrieval. Journal of Visual Communication and Image Representation, **10** (1999) 219–244
8. Lu, H., Ooi, B., Tan, K.: Efficient image retrieval by color contents. In: Proc. of the 1994 Int. Conf. on Applications of Databases (1994)
9. Ortega, M. et al.: Supporting ranked boolean similarity queries in MARS. IEEE Trans. on Knowledge and Data Engineering, **10** (1998) 905–925
10. Pass, G. et al.: Comparing images using color coherence vectors. In: Proc. ACM Conf. on Multimedia (1966)
11. Rui, Y., Huang, T., S.: Image retrieval: current techniques, promising directions, and open issues. Journal of Visual Communication and Image Representation **10** (1999) 39–62
12. Smith, J., R., Chang, S.,F.: Single color extraction and image query. In: Proc. IEEE Int. Conf. on Image Proc. (1995)
13. Stricker, M., Dimai, A.: Spectral covariance and fuzzy regions for image indexing. Machine Vision and Applications **10** (1997) 66–73
14. Vinod, V., V., Murase, H.. Image retrieval using efficient local-area matching. Machine Vision and Applications, **11** (1998) 7–14

Lifting-Based Reversible Transforms for Lossy-to-Lossless Wavelet Codecs

Artur Przelaskowski

Institute of Radioelectronics, Warsaw University of Technology,
Nowowiejska 15/19, 00-665 Warszawa, Poland
Tel:+48 22 660-7917, fax: +48 22 825-1363
arturp@ire.pw.edu.pl

Abstract. Reversible transforms applied in wavelet coder to realize lossy-to-lossless compression are considered in this paper. 1-D wavelet transform possible to be customized in part II of nowadays JPEG2000 standard is optimized to increase an efficiency of the first lossy phase of compression process. Different classes of reversible transforms were analyzed, evaluated in experiments, and compared one another in a sense of effectiveness, complexity and possibility of further optimization. Suitable selection of reversible wavelet transform can increase effectiveness of the coder even up to 2.7 dB of PSNR for 0.5 bpp in comparison to standard 5/3 transform. New reversible transform generated with lifting scheme was proposed. It overcomes all other in both phases of lossy-to-lossless compression (up to 0.4 dB of PSNR in comparison to the state-of-art transforms of JPEG2000 standardization process). Therefore, an efficiency of reversible wavelets can be comparable to irreversible wavelets effectiveness in several cases of lossy compression.

Keywords: wavelet coding, integer-to-integer transform, lifting scheme

1 Introduction

Wavelets are useful tool in compression applications because of their great potential of original data modeling. Majority of real life images may be sparsely approximated by transformation base functions with finite support, suitable regularity and enough amount of vanishing moments. Resultant significant decorrelation of input signals assures compactness of information in spatial and frequency domain. Recent wavelet coders take advantage of reversible wavelet transformation to realize lossy-to-lossless progressive compression which is very useful in image data transmission systems. Reversible integer-to-integer wavelet transforms are invertible in finite-precision arithmetic and map input integers to output integers. More efficient irreversible transforms give floating point coefficients and lossless data encoding is not possible because of necessity of additional quantization converting floating point data to integers.

Investigation of new integer transforms is important for practice because of JPEG2000 standard development and expected worldwide applications. Part I of this standard [1] includes two wavelet transforms, the integer 5/3 and float 9/7. Only 5/3 transform provides considerable lossless compression in relation to very low com-

W. Skarbek (Ed.): CAIP 2001, LNCS 2124, pp. 61–70, 2001.

plexity and exhibits a minimum of ringing when quantized. Application of additional reversible and irreversible wavelet transforms is customized in Part II which is developed by JPEG committee (ISO/IEC JTC1/SC29 WG1). They could be fixed or arbitrary defined to increase reduction of redundancy over scales in image domain.

We looked for more effective reversible transforms to make lossy compression comparable-in-efficiency to irreversible wavelet compression reference. Therefore, we investigated the most useful integer-to-integer transforms. State-of-art transforms from literature and JPEG2000 standardization process were tested and new ones were designed. This investigation is presented in section 2, tests and results of our experiments are reported in section 3, followed by conclusions presented in final section 4.

2 Investigation of Effective Integer-to-Integer Transforms

Both lossy and lossless performance of reversible integer transforms in wavelet compression was investigated. We characterised transforms by the number of vanishing moments of the analysing high pass filter $\tilde{N}$ and the number of vanishing moments of the synthesizing high pass filter N in a form $(\tilde{N}, N)$. Additionally, we used a notation $L/\tilde{L}$ to indicate the lengths L and $\tilde{L}$ of low pass and high pass analysis filters, respectively. Computational complexity was concerned as a number of addition and shift operation required in one-level 1-D decomposition per two input samples (even and odd). Divisions are replaced by bit shifts and multiplications by shift and add operations, similarly to [2].

Lifting scheme is a design approach of biorthogonal wavelets with compact support which provides their reversibility [3] and supports enough degrees of freedom to satisfy certain constraints assumptions (such as regularity, vanishing moments, frequency localization and shape) and custom design the wavelet. This scheme could be used as a tool for construction of reversible transforms by rounding off the results of each filtering in dual lifting and lifting steps before correction of proceeded value of odd and even sample, respectively. Lifting starts with lazy wavelet transform given by $s^{(0)}[n] \doteq x[2n]$ and $d^{(0)}[n] \doteq x[2n+1]$ to formulate entree low pass and high pass subbands of input signal $x[n]$. Next, they are replaced in successive (k-indexed) lifting steps of forward transform according to:

$$d^{(k)}[n] = d^{(k-1)}[n] + \left\lfloor \sum_i p^{(k)}[i]s^{(k-1)}[n-i] + \frac{1}{2} \right\rfloor$$

$$s^{(k)}[n] = s^{(k-1)}[n] + \left\lfloor \sum_i u^{(k)}[i]d^{(k)}[n-i] + \frac{1}{2} \right\rfloor ,$$

(1)

where the filter coefficients p (prediction) and u (updating) are computed by factorization of polyphase matrix of any perfect reconstruction filter bank. Rescaling is omitted in our customization process (we assume scaling factor equals to 1).

2.1 State-of-Art Reversible Transforms for Compression

Many effective reversible transforms were applied in lossless wavelet coders. Most of them are defined and tested in [2][4][5]. Simple one called S transform (1,1) is rather ineffective in a case of natural image compression but can be useful in reversible encoding of compound images with a lot of edges and high frequency components [4] and noisy medical ultrasound images. The S transform is simply:

$$d[n] = d^{(1)}[n] = d^{(0)}[n] - s^{(0}[n]$$
$$s[n] = s^{(1)}[n] = s^{(0)}[n] + \tfrac{1}{2} d^{(1)}[n] \; . \tag{2}$$

We selected three classes of the most useful and efficient transforms.

Factorised Effective Irreversible Biorthogonal Transforms

An important example of this is 9/7 transform [6] realised in reversible manner (9/7rev). Irreversible (floating point) version of this transform (9/7irrev) was used as a reference in our tests. By factorisation, this smooth (4,4) kernel was implemented in four lifting steps with rounding giving floating point lifting coefficients [4]. Therefore, arithmetic implementation is not perfect and computational costs are increased.

Generally, the efficiency of reversible 9/7 is poor in both: lossy and lossless manner. Taking into account its higher complexity (even with coefficients approximation by integers divided by powers of 2 from [2] 9/7rev is almost three times more complex than binary 9/7 transform from table 1) one can conclude the inefficiency of such scheme of reversible transform generation. More effective irreversible transforms are more complex than 9/7 and making their reversible forms by factoring poliphase matrix into lifting steps seems to be not profitable.

Interpolating Transforms

They are wavelet transforms built from the interpolating Deslauriers-Dubuc scaling functions [7] derived independently by Reissell [8], Wei et al. [9], Strang and Nguyen [10], Sweldens [3]. These transforms were found as the most useful in lossy-to-lossless wavelet coders because of high transformation efficiency and simplicity of software and hardware applications. The coefficients of interpolating transforms are integers divided by powers of 2, giving perfect arithmetic and fast chip execution. Low pass filter is half band ($h(z) + h(-z) = 2$) in this case. Since, for even samples $h_e(z) = 1$ factorisation can be made only in two lifting steps [11], i.e. one step prediction (dual lifting) is followed by single primal lifting step. Generally, ones can state that in image processing the number of dual vanishing moments is much more important than the number of primal vanishing moments. Hence, the optimisation effort should be mostly directed to initial prediction step in analysis process. It was confirmed in our experiments.

Effective biorthogonal wavelets built from interpolating scaling functions were selected by considering the results presented in [2][4][5][12][13]. Their definitions are collected in table 1. Transforms are grouped into two classes: oriented on satisfying number of vanishing moments and regularity, and oriented on maximising of coding gain with reduced number of vanishing moments and comparable complexity.

Table 1. Effective reversible forward transforms built from interpolating scaling functions. Transform characteristics: taps, vanishing moments, computational complexity, reference.

Transform	Definition
5/3 (2,2) 7 [15]	$d[n] = d^{(0)}[n] - \frac{1}{2}(s^{(0)}[n] + s^{(0)}[n+1])$ $s[n] = s^{(0)}[n] + \frac{1}{4}(d[n-1] + d[n]) + \frac{1}{2}$
9/7 (4,2) 12 [4]	$d[n] = d^{(0)}[n] + \frac{1}{16}(s^{(0)}[n-1] + s^{(0)}[n+2] - 9(s^{(0)}[n] + s^{(0)}[n+1])) + \frac{1}{2}$ $s[n] = s^{(0)}[n] + \frac{1}{4}(d[n-1] + d[n]) + \frac{1}{2}$
13/7 (4,4) 16 [4]	$d[n] = d^{(0)}[n] + \frac{1}{16}(s^{(0)}[n-1] + s^{(0)}[n+2] - 9(s^{(0)}[n] + s^{(0)}[n+1])) + \frac{1}{2}$ $s[n] = s^{(0)}[n] + \frac{1}{32}(-d[n-2] - d[n+1]) + 9(d[n-1] + d[n])) + \frac{1}{2}$
17/7 (4,6) 22 [5]	$d[n] = d^{(0)}[n] + \frac{1}{16}(s^{(0)}[n-1] + s^{(0)}[n+2] - 9(s^{(0)}[n] + s^{(0)}[n+1])) + \frac{1}{2}$ $s[n] = s^{(0)}[n] + \frac{1}{1024}(9(d[n-3] + d[n+2]) - 59(d[n-2] + d[n+1]) +$ $+ 306(d[n-1] + d[n])) + \frac{1}{2}$
17/11 (6,4) 22 [9]	$d[n] = d^{(0)}[n] + \frac{1}{256}(-3(s^{(0)}[n-2] + s^{(0)}[n+3]) + 25(s^{(0)}[n-1] + s^{(0)}[n+2]) -$ $-150(s^{(0)}[n] + s^{(0)}[n+1])) + \frac{1}{2}$ $s[n] = s^{(0)}[n] + \frac{1}{32}(-d[n-2] - d[n+1]) + 9(d[n-1] + d[n])) + \frac{1}{2}$
21/11 (6,6) 28 [5]	$d[n] = d^{(0)}[n] + \frac{1}{256}(-3(s^{(0)}[n-2] + s^{(0)}[n+3]) + 25(s^{(0)}[n-1] + s^{(0)}[n+2]) -$ $-150(s^{(0)}[n] + s^{(0)}[n+1])) + \frac{1}{2}$ $s[n] = s^{(0)}[n] + \frac{1}{512}(3(d[n-3] + d[n+2]) - 25(d[n-2] + d[n+1]) +$ $+ 150(d[n-1] + d[n])) + \frac{1}{2}$
13/7crf (4,2) 16 [13]	$d[n] = d^{(0)}[n] + \frac{1}{16}(s^{(0)}[n-1] + s^{(0)}[n+2] - 9(s^{(0)}[n] + s^{(0)}[n+1])) + \frac{1}{2}$ $s[n] = s^{(0)}[n] + \frac{1}{16}(-d[n-2] - d[n+1]) + 5(d[n-1] + d[n])) + \frac{1}{2}$
13/11crf (6,1) 18 [5]	$d[n] = d^{(0)}[n] + \frac{1}{256}(-3(s^{(0)}[n-2] + s^{(0)}[n+3]) + 25(s^{(0)}[n-1] + s^{(0)}[n+2]) -$ $-150(s^{(0)}[n] + s^{(0)}[n+1])) + \frac{1}{2}$ $s[n] = s^{(0)}[n] + \frac{5}{16}(d[n-1] + d[n]) + \frac{1}{2}$
17/11crf (6,2) 22 [5]	$d[n] = d^{(0)}[n] + \frac{1}{256}(-3(s^{(0)}[n-2] + s^{(0)}[n+3]) + 25(s^{(0)}[n-1] + s^{(0)}[n+2]) -$ $-150(s^{(0)}[n] + s^{(0)}[n+1])) + \frac{1}{2}$ $s[n] = s^{(0)}[n] + \frac{1}{16}(-d[n-2] - d[n+1]) + 5(d[n-1] + d[n])) + \frac{1}{2}$

Low-cost 5/3 transform is very useful in reversible encoding of compound images but rather ineffective for natural images. Lower cost improvement was done by 9/7 transform, and successively higher cost improvement by transforms with increased number of vanishing moments. Three last kernels were elaborated by Canon Research Centre [5] to increase the effectiveness of interpolated wavelets.

Transforms of Extra Lifting Steps

It is a conception to extend the efficient and low cost S or 5/3 transforms. One procedure of reversible transform creation is S+P (S transform plus Prediction) [14]. The forward S+P starts from simple S transform (equations 1) followed by prediction made from a set of low pass and old adjacent high pass coefficients. Effective example of this procedure is SPB.

Another extension procedure given by Calderbank *et. al.* [4] allows to generate a family of (2+2,2) transforms with an extra lifting step extending 5/3 transform as:

$$d[n] = d^{(1)}[n] - \alpha(-1/2s^{(1)}[n-1] + s^{(1)}[n] - 1/2s^{(1)}[n+1]) \tag{3}$$
$$+ \beta(-1/2s^{(1)}[n] + s^{(1)}[n+1] - 1/2s[n+2]) + \gamma(d^{(1)}[n+1] + \tfrac{1}{2}$$

where coefficients •, •, and • can be fitted to optimise encoding efficiency. The most efficient representatives are placed in table 2. First one 5/11a with four vanishing moments of analysis high pass filter improves the effectiveness of 5/3 filter, mostly in a case of lossy compression. Its modification 5/11b done by Adams [2] has got reduced number of vanishing moments and increased coding gain.

Table 2. Effective reversible forward transforms of extra lifting step.

Transform	Definition
SPB (2,1) 11 [14]	$d^{(1)}[n] = d^{(0)}[n] - s^{(0)}[n]$ $s[n] = s^{(0)}[n] + \tfrac{1}{2}d^{(1)}[n]$ $d[n] = d^{(1)}[n] + \tfrac{1}{8}(2s[n-1] + s[n] - 3s[n+1] + 2d^{(1)}[n+1]) + \tfrac{1}{2}$
5/11a (4,2) 13 [4]	$d^{(1)}[n] = d^{(0)}[n] - \tfrac{1}{2}(s^{(0)}[n] + s^{(0)}[n+1])$ $s[n] = s^{(0)}[n] + \tfrac{1}{4}(d^{(1)}[n-1] + d^{(1)}[n]) + \tfrac{1}{2}$ $d[n] = d^{(1)}[n] + \tfrac{1}{16}(s[n-1] - s[n] - s[n+1] + s[n+2])) + \tfrac{1}{2}$
5/11b (2,2) 13 [2]	$d^{(1)}[n] = d^{(0)}[n] - \tfrac{1}{2}(s^{(0)}[n] + s^{(0)}[n+1])$ $s[n] = s^{(0)}[n] + \tfrac{1}{4}(d^{(1)}[n-1] + d^{(1)}[n]) + \tfrac{1}{2}$ $d[n] = d^{(1)}[n] + \tfrac{1}{32}(s[n-1] - s[n] - s[n+1] + s[n+2])) + \tfrac{1}{2}$

We optimised scheme (3) and added fourth lifting step to increase an accuracy of updating. Hence, 19/11 transform extending 5/11a was defined as follows:

$$d^{(1)}[n] = d^{(0)}[n] - \tfrac{1}{2}(s^{(0)}[n] + s^{(0)}[n+1])$$
$$s^{(1)}[n] = s^{(0)}[n] + \tfrac{1}{4}(d^{(1)}[n-1] + d^{(1)}[n]) + \tfrac{1}{2} \tag{4}$$
$$d[n] = d^{(1)}[n] + \tfrac{1}{16}(s^{(1)}[n-1] - s^{(1)}[n] - s^{(1)}[n+1] + s^{(1)}[n+2]) + \tfrac{1}{2}$$
$$s[n] = s^{(1)}[n] + \tfrac{1}{32}(-d[n-2] + d[n-1] + d[n] - d[n+1]) + \tfrac{1}{2} \; .$$

Complexity of resulting 19/11 transform is equal to 19. Generally, adding extra lifting steps to simple two-lifts transforms did not allow us to increase the performance of wavelet decomposition significantly.

2.2 Lifting Scheme-Based Customisation

An analysis of the properties (number of taps, vanishing moments and lifting steps, complexity, coding gain value) of efficient transforms and the results of conducted tests were important in selection of the most useful transforms. Starting with two step lifting scheme, prediction was optimised to increase coding gain of the transform. Primal lifting step was realised in less complex way as a conclusion of introductory considerations and experiments. Next, based on lifting we constructed several transforms of different complexity examining relation between scheme complexity, quality of original signal approximation, and resulting coding gain (with Markov models and simulated mean images) in different decomposition modes and coder conceptions. Finally, the most useful 17/11n (4,2) transform was proposed as follows:

$$d[n]=d^{(0)}[n]+\tfrac{1}{64}(-s^{(0)}[n-2]-s^{(0)}[n+3]+7(s^{(0)}[n-1]+s^{(0)}[n+2])-38(s^{(0)}[n]+s^{(0)}[n+1]))+\tfrac{1}{2}$$
$$s[n]=s^{(0)}[n]+\tfrac{1}{16}(-d[n-2]-d[n+1])+5(d[n-1]+d[n]))+\tfrac{1}{2} \ . \tag{5}$$

Computational complexity of this transform is equal to 20, filter coefficients are presented in table 3, and scaling functions and wavelets are shown in fig. 1. This kernel is a result of interpolating transform optimisation done by more suitable signal approximation in prediction step.

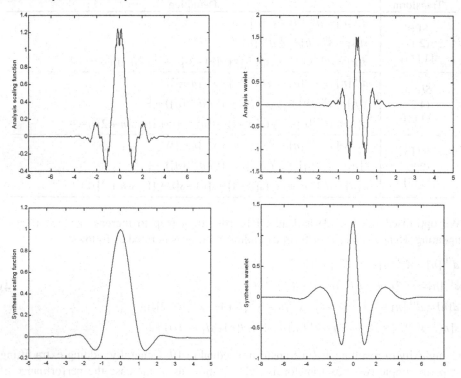

Fig. 1. Scaling functions and wavelets of analysis and synthesis for kernel 17/11n.

Table 3. Coefficients of low pass filters of new transform.

Trans-form	$h(z)/\sqrt{2}$	$\tilde{h}(z)/\sqrt{2}$
17/11n	$2^{-7}(z^{-5}-7z^{-3}+38z^{-1}+64$ $+38z-7z^{3}+z^{5})$	$2^{-10}(z^{-8}-12z^{-6}+68z^{-4}-64z^{-3}-116z^{-2}+320z^{-1}+630$ $+320z-116z^{2}-64z^{3}+68z^{4}-12z^{6}+z^{8})$

3 Tests

Selection of the most effective reversible transforms in wavelet-based compression scheme was verified in many experiments with three natural test images: Lena, Barbara and Goldhill and other JPEG2000 test images (Target and Seismic). SPIHT [16] and JPEG2000 (according to verification model VM8.6) codecs were used in performance tests. The chosen results are presented in tables 4, 5 and 6, and in figure 2.

Table 4. The evaluation of reversible wavelet transforms in lossless compression (with 6 level Mallat decomposition). The values of bit rate (bpp) are presented.

Transform	SPIHT Coder					JPEG2000 Coder				
	Lena	Bar-bara	Gold-hill	Target	Mean	Lena	Bar-bara	Gold-hill	Tar-get	Mean
9/7rev	4.24	4.75	4.81	3.09	4.223	4.33	4.73	4.86	2.50	4.105
5/3	4.22	4.79	4.75	**2.51**	4.068	4.31	4.78	4.83	**2.13**	4.013
9/7	4.19	4.70	4.74	2.64	4.068	4.28	4.69	4.83	2.26	4.015
13/7	4.18	4.68	4.74	2.64	4.060	**4.27**	4.65	4.83	2.25	4.000
17/7	4.18	4.66	4.75	2.63	4.055	**4.27**	4.64	4.83	2.24	3.995
17/11	4.18	4.65	4.75	2.68	4.065	**4.27**	4.64	4.83	2.28	4.005
21/11	**4.17**	4.64	4.75	2.67	4.058	**4.27**	**4.62**	4.83	2.27	3.998
13/7crf	4.18	4.67	4.75	2.63	4.058	4.27	4.66	4.83	2.24	3.998
13/11crf	4.19	4.67	4.76	2.71	4.083	4.28	4.66	4.83	2.30	4.018
17/11crf	4.18	4.64	4.75	2.66	4.058	**4.27**	**4.62**	4.83	2.26	3.995
SPB (S+P)	4.20	4.72	4.78	2.64	4.085	-	-	-	-	-
5/11a	4.19	4.71	4.74	2.66	4.075	4.28	4.70	4.83	2.25	4.015
5/11b	4.20	4.74	4.74	2.63	4.078	4.29	4.73	**4.82**	2.21	4.013
17/11n	4.18	**4.63**	**4.73**	2.63	4.043	**4.27**	4.63	4.83	2.24	3.993

Table 4 presents rather small improvement of lossless compression by suitable reversible transform selection in both tested coders. Generally, more taps and more vanishing moment transforms gave slightly higher performance (mean bit rate improvements is up to 0.6% in comparison to 5/3 transform). We can suggest that that inserting the high pass sample in extra step is not so much profitable in most cases. The 9/7 interpolating transform with similar-to-5/11 complexity gave slightly better results and seems to be more useful.

From table 5 we can suggest that more sophisticated transforms achieved significant improvement of reconstructed image quality in comparison to the 5/3. Considering the efficiency and complexity, the most useful transforms from each class were selected as follows: 21/11, 17/11crf and SPB. They were compared to 17/11n in next test image compression– see figure 2. Effectiveness of 17/11n was more attractive in lossy data encoding. The performance improvement over 5/3 transform was close to 2.5 dB of PSNR for 0.5 bpp, and improvement over any other tested transform was close to 0.4 dB of PSNR for 0.5 bpp in several cases. Additionally, it was shown that in several cases the efficiency of reversible transform is higher than reference efficiency of irreversible transform - see the results of 17/11n and 9/7irrev for Barbara and Target images. Hence, 17/11n transform may be useful for many applications.

The relations and general tendency between transform efficiency were similar in two test coders. Using both of them makes our considerations more reliable.

Table 5. The results of reversible transform selection in lossy compression (with 6 level Mallat decomposition). The results of PSNR are presented.

Codec	Transform	Lena		Barbara		Goldhill		Target	
		0.25	0.5	0.25	0.5	0.25	0.5	0.25	0.5
SPIHT	9/7irrev	*34.11*	*37.21*	*27.58*	*31.39*	*30.56*	*33.13*	*24.27*	*29.90*
	9/7rev	33.53	36.23	26.94	30.57	30.02	32.36	22.57	29.17
	5/3	33.19	36.27	26.44	29.87	30.17	32.65	22.60	28.84
	9/7	33.62	36.55	26.73	30.52	30.20	32.68	23.31	29.73
	13/7	33.76	36.71	27.12	30.94	30.35	32.83	24.15	29.97
	17/7	33.81	36.72	27.28	31.11	**30.37**	32.81	24.51	30.79
	17/11	33.78	36.74	27.22	31.22	30.32	32.78	24.55	30.21
	21/11	33.83	**36.79**	27.35	31.40	30.36	32.80	24.67	30.47
	13/7crf	33.78	36.71	27.28	31.08	30.36	**32.84**	24.61	31.08
	13/11crf	33.72	36.66	27.06	30.96	30.22	32.67	23.77	29.61
	17/11crf	33.84	36.76	**27.41**	31.42	**30.37**	32.81	24.88	31.21
	SPB (S+P)	33.64	36.56	26.87	30.57	30.22	32.60	21.99	29.27
	5/11a	33.54	36.51	26.58	30.40	30.13	32.60	22.93	29.45
	5/11b	33.43	36.46	26.57	30.19	30.19	32.68	22.72	29.09
	17/11n	**33.89**	36.76	**27.41**	**31.47**	30.36	32.80	**25.01**	**31.51**
JPEG2000	5/3	33.28	36.32	27.38	30.92	30.15	32.77	26.71	33.15
	9/7	33.49	36.57	27.56	31.38	30.13	32.69	27.05	33.83
	13/7	33.70	36.70	28.04	31.81	30.32	32.85	27.32	34.18
	17/7	33.69	36.74	28.16	31.98	30.33	**32.89**	28.48	34.78
	17/11	33.72	33.67	28.12	32.05	30.31	32.80	27.91	34.30
	21/11	33.78	36.76	28.30	32.21	30.37	32.87	28.39	34.57
	13/7crf	33.68	36.74	28.22	32.05	30.34	32.88	28.71	35.00
	13/11crf	33.58	36.63	27.82	31.64	30.16	32.68	27.16	33.42
	17/11crf	**33.79**	36.70	**28.34**	32.26	**30.38**	32.83	28.97	34.94
	5/11a	33.44	36.49	27.56	31.37	30.06	32.63	26.72	33.17
	5/11b	33.39	36.45	27.57	31.26	30.12	32.72	26.73	33.28
	17/11n	33.76	**36.79**	28.33	**32.31**	30.35	32.86	**29.07**	**35.09**

Moreover, destructive influence of increased number of lifting steps on efficiency of transformation suggested by Adams and Kosentini [2] was not confirmed in our analysis. We constructed transform with four lifting steps, which simply extends 2 and 3 steps transforms, and outperforms them in most cases of lossless and lossy compression -see the results in table 6. Our understanding is that an impact of better approximation of natural image data (done in additional lifting steps) on final compression efficiency is more significant than losing effectiveness by approximation error of rounding process in successive lifting steps.

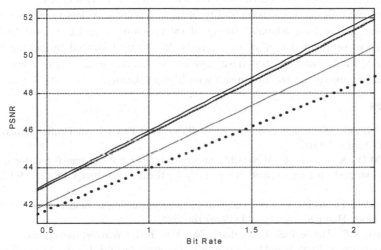

Fig. 2. Compression efficiency evaluation of SPIHT coder with different reversible transforms to be applied. Three kernels as effective representatives of mentioned groups are considered: SPB (dotted), 21/11 (dashdot), 17/11crf (dashed), and proposed 17/11n (solid). As a reference 5/3 transform (point) is used. Approximated curves for Seismic test image are presented.

Table 6. An influence of the number of transform lifting steps on lossless compression performance. The values of bit rate (bpp) are presented.

Transform	Reversible compression (bit rate [bpp])				Lossy compression (PSNR [dB])					
					Lena		Barbara		Goldhill	
	Lena	Bar-bara	Gold hill	Mean	0.25	0.5	0.25	0.5	0.25	0.5
2 lifting steps (5/3)	4.22	4.79	4.75	4.587	33.19	36.27	26.44	29.87	30.17	32.65
3 lifting steps (5/11a)	**4.19**	4.71	**4.74**	4.547	33.54	**36.51**	26.58	30.40	30.13	32.60
4 lifting steps (19/11n)	**4.19**	4.68	**4.74**	4.537	**33.66**	36.50	**27.08**	**30.78**	**30.20**	**32.70**

4 Conclusions

The role of wavelet coders constructed according to JPEG 2000 paradigm seems to be fundamental in a development of new lossy-to-lossless compression techniques for internet, telemedicine and other applications. More complex but however more regular, similar-to-data reversible transforms with greater coding ability indicated by coding gain value could be useful for practice because of considerable compression efficiency improvement. The optimisation of reversible transform improves coder performance by increasing information compactness in the first parts of progressively coded data stream. Proposed integer-to-integer transform gave comparable to irreversible transforms compression efficiency for both coders in several cases.

Because of obvious decomposition ability of interpolating wavelets family we plan a future research based on adaptive fitting of interpolating wavelet support to data characteristics in segmented regions of interests. It will be followed by coding gain optimisation of initial wavelets for these regions. Scale and space adaptive wavelets could be clearly constructed and optimised with lifting scheme.

References

1. ISO/IEC: JPEG 2000 image coding system. JPEG 2000 Part I Final Draft International Standard (August 2000)
2. Adams, M.D., Kossentini, F.: Reversible integer-to-integer wavelet transforms for image compression: performance, evaluation and analysis. IEEE Trans. Image Process. **9** (2000) 1010-1024
3. Sweldens, W.: The lifting scheme: a custom-design construction of biorthogonal wavelets. Appl. Comput. Harmonic. Analysis **3** (1996) 186-200
4. Calderbank, R.C., Daubechies, I., Sweldens, W., Yeo, B.-L.: Wavelet transforms that map integers to integers. Applied and Computational Harmonic Analysis (ACHA) **5** (1998) 332-369
5. Sablatnig, J., Seiler, R., Jung, K.: Report on Transform Core Experiment: Compare the lossy performance of various reversible integer wavelet transforms. In: ISO/IEC JTC 1/SC 29/WG 1 N 915, 1998
6. Antonini, M., Barlaud, M., Mathieu, P., Daubechies, I.: Image coding using wavelet transform. IEEE Trans. Image Process. **1** (1992) 205-220.
7. Deslauriers, G., Dubuc, S.: Interpolation dyadique. In Fractals, dimensions non entieres et applications. Masson, Paris (1987) 44-55.
8. Reissell, L.-M.: Wavelet multiresolution representation of curves and surfaces. CVGIP: Graphical Models and Image Process., **58** (1996) 198-217.
9. Wei, D., Tian, J., Wells, O. Jr., Burrus, C.S.: A new class of biorthogonal wavelet systems for image transform coding. IEEE Trans. Image Process., **7** (1998) 1000-1013.
10. Strang, G., Nguyen, T.: Wavelets and filter banks, Wellesley-Cambridge Press (1996).
11. Daubechies, I., Sweldens, W.: Factoring wavelet transforms into lifting steps. J. Fourier Anal. Appl., **4** (1998) 247-269
12. Bilgin, A., Sementelli, P.J., Sheng, F., Marcellin, W.: Scalable image coding using reversible integer wavelet transforms. IEEE Trans. Image Process., **9** (2000) 1972-1977
13. Adams, M.D., Kharitonenko, I., Kossentini, F.: Report on Core Experiment CodEff4: Performance Evaluation of Several Reversible Integer-to-Integer Wavelet Transforms in the JPEG-2000 Verification Model (Version 2.1). In: ISO/IEC JTC 1/SC 29/WG 1 N 1015, 1998
14. Said, A., Pearlman, W.A.: An image multiresolution representation for lossless and lossy compression. IEEE Trans. Image Process. **5** (1996) 1303-1310
15. Le Gall, D., Tabatabai, A.: Sub-band coding of digital images using symmetric short kernel filters and arithmetic coding techniques. In: Proc. Of IEEE International Conference on Acoustics, Speech, and Signal Process., **2** (1988) 761-764
16. Said, A., Pearlman, W.A.: A new fast and efficient image codec based on set partitioning in hierarchical trees. IEEE Trans. Circ. & Syst. Video. Tech., **6** (1996) 243-250.

Coding of Irregular Image Regions by SA DFT

Ryszard Stasiński

Instytut Elektroniki i Telekomunikacji, Politechnika Poznańska
Piotrowo 3A, 60-965 Poznań, Poland
rstasins@et.put.poznan.pl

Abstract. In the paper the new transform adapting to shapes of irregular image segments is introduced, the shape-adaptive (SA) DFT. Its definition is based on periodic data extension rather than data shifts, hence, in contrast to SA DCT segment reconstruction is possible even if part of contour data is missing. Visually the quality of images reconstructed from the part of SA DFT samples is almost as good as for the SA DCT, especially for high compression ratios.

Keywords: shape adaptive transform, Discrete Cosine Transform, Discrete Fourier Transform

1 Introduction

The introduced in the paper shape adaptive discrete Fourier transform (SA DFT) is an alternative to the shape-adaptive DCT (SA DCT) [1]. SA DCT is a method for coding of irregularly shaped image segments. It is included in the MPEG-4 image compression standard as an auxiliary technique. It consists of defining the two-dimensional DCT in such a way that samples inside an irregular image region are transformed, only, hence, *shape adaptivity* of the transform. This is an obvious advantage of applications in which an image is treated as a collection of *objects*, like multimedia, databases etc. Namely, after segmentation of an image into regions ('objects') we can code them without any excess of information, as the coding blocks need not be rectangular anymore. Moreover, the technique performs much better than simple shape adaptive methods based on DCT for rectangular blocks, probably due to the lack of interference with arbitrary defined samples situated outside the processed region [2], [4].

Nevertheless, when compared with other shape-adaptive coding methods the SA DCT exhibit some drawbacks. Namely, the method is very susceptible to shape errors caused by transmission channel or storage media noise, which is in contrast to extrapolation methods [4]. That is why the use of DFT instead of DCT is considered in the paper, section 2. The DFT of a vector can be interpreted as the Fourier transform of a periodic signal one period of which is processed. We need not to shift the signal as in the SA DCT, it suffices to extend it periodically and to compute the DFT starting from the beginning of a row/column. Secondly, if part of information about the shape border is missing, but we can deduce correct row/column DFT sizes, the image can be correctly reconstructed by periodic extension of inverse DFTs in the decoder, section

W. Skarbek (Ed.): CAIP 2001, LNCS 2124, pp. 71–76, 2001.

3. Then, some device or a human should decide which image samples belong to the region, and which are obtained by shape periodic extension. Similarly, as the SA DCT, SA DFT can be easily extended to process correctly signal DC component, section 4.

2 SA DFT Definition

Let us consider a one-dimensional vector $x(n)$ of size N, in which N_S samples belong to a region, and the remaining ones are background samples. Assume for the moment that the shape is convex, i.e. that shape samples form contiguous clusters inside image rows and columns, i.e. they are not separated by background samples. The one-dimensional SA DFT method is:

- Extend periodically shape samples in the data vector, i.e. if $x(n)$ is a shape sample, then set $x(n+kN_S)=x(n)$, $k=...,-1,0,1,...$, $0 \leq n+kN_S < N$. In fact, only the beginning vector samples $x(0),x(1),...,x(N_S-1)$ should be defined.
- Compute the N_S-point DFT for the first N_S vector samples. Due to the DFT redundancy for real valued data DFT samples for indices greater than $N_S/2$ should be rejected.
- Scale the results. In the paper the scaling factor is $1/\sqrt{N_S}$ for output samples of N_S-point DFT, which makes the 2-D transform orthogonal [2-4].

The method is generalized to two-dimensions in the same way as separable transforms: firstly, the 1-D algorithm above processes each signal row, secondly, the algorithm is applied to each column of the results. Note that the first column is real (it contains samples $X(0)$), in addition for even N_S final row $X(N_S/2)$ samples are real, too. For avoiding column DFT redundancy row DFT samples $X(N_S/2)$ should be removed from the rows and processed by a separate column DFT. Additionally, if the shape is not convex, then before making row/column periodic extensions we should 'shrink the holes' in shape filled with background samples by shape samples shifts in such a way that shape samples make one contiguous block beginning on the position of the first row/column sample. The order of computations can be reversed, column transformations can precede the row ones, which usually leads to different results [4], Fig.1.

3 Experiment

The SA DFT has been realized as a Matlab function and tested on the segment 'face' from image #6 from QCIF sequence 'Carphone' used in experiments in [4]. For making the results comparable to those from [4], the logarithms of the basis restriction errors have been compared:

$$\varepsilon = 10 \log_{10} \frac{\sum_{i=1}^{N_S} x^2(i)}{\sum_{i=1}^{N_S} (x(i) - \tilde{x}(i))^2}$$

$\tilde{x}(i)$ is an approximation of datum $x(i)$. It appears that due to worse data compression properties basis restriction error curves for SA DFT lay below those for the SA DCT and are less steep, Fig.1 , the difference grows from zero when only the greatest transform samples are preserved up to approximately 2 dB when 40% of the greatest transform samples is used for image reconstruction. Nevertheless, when 5-10% greatest transform samples are retained the subjective visual qualities of reconstructed images are quite similar, Fig.2. This is in part due to the fact that errors of images reconstructed from SA DFT concentrate on the shape border, which is not that important visually, Fig.2. Namely, because of implied periodicity of signals processed by the DFT the last data sample is followed by the first one, and if they are very different, then we have an edge, hence, usual problems with edge distortions.

In Figure 3 shape reconstruction property of the SA DFT is illustrated. The first image shows the shape of the convex version of the 'face' segment used for experiments in [4], pixels belonging to the segment are gray, and those of the background are black. To the right the segment reconstructed from the 10% greatest SA DFT coefficients is shown. Then, incomplete shape information is visualized, it is assumed that we know only the sizes of row DFTs, actual positions of segment samples inside rows are missing. Finally, the segment reconstructed from 10% greatest SA DFT coefficients using incomplete shape information is shown, inverse DFT samples have been extended periodically along rows. As it can be seen, shape pixels are identical to those above, only the background is not black anymore.

4 SA DFT Improvements

For the sake of simplicity the compared in the paper SA DFT and SA DCT versions are orthogonal, which means that they are not DC preserving [2-4]. Nevertheless, for such large data segments as the 'face' one the differences between results for SA DCT (and SA DFT) versions are negligible. Moreover, the construction of DC preserving SA DFT version is straightforward. Namely, similarly as for the SA DCT the structure of the DC separated and ΔDC corrected [2] SA DFT is as follows:

- Subtract mean segment value from the data.
- Compute the SA DFT.
- Replace transform $X(0,0)$ sample by the segment mean value (possibly multiplied by a constant).

The $X(0,0)$ sample reconstruction consists in finding such multiplier for this sample eigenfuction that the reconstructed from SA DCT image mean value is nullified. Then, image mean value is added for final image restitution [2].

Basis restriction error curves for the DC corrected SA DFT and SA DCT are almost the same as those in Fig.1.

5 Conclusion

The problem of reducing SA DCT susceptibility to contour reconstruction errors is addressed in the paper. It is proposed to replace shift operations of this transform by periodic data extension, and the DCT by the DFT; in this way the SA DFT is obtained. The transform indeed allows correct reconstruction of irregular segment samples when nothing more than numbers of samples in either rows or columns are provided. Visually partly reconstructed images from the new transform are almost as good as those for the SA DCT.

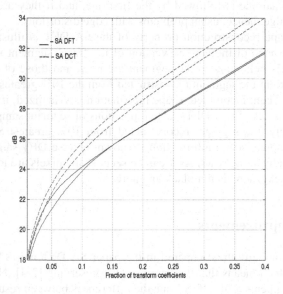

Fig. 1. SA DFT/DCT basis restriction error as a function of used transform samples fraction for segment 'face', upper curves are valid when rows are processed first.

References

1.   Sikora, T., Makai, B.: Shape-adaptive DCT for generic coding of video. IEEE Trans. Circuits Syst. Video Technol., **5** (1995) 59-62
2.   Kauff, Schüür, K.: An extension of shape-adaptive DCT (SA-DCT) towards DC separation and ΔDC correction. In: Proc. 1997 Picture Coding Symposium, Berlin (1997) 647-652
3.   Kaup, Panis, S.: On the performance of the shape adaptive DCT in object-based coding of motion compensated difference images. ibid. (1997) 652-657
4.   Stasinski, R., Konrad, J.: A new class of fast shape-adaptive orthogonal transforms and their application to region-based image compression. IEEE Trans. Circuits Syst. Video Technol., **9** (1999) 16-34

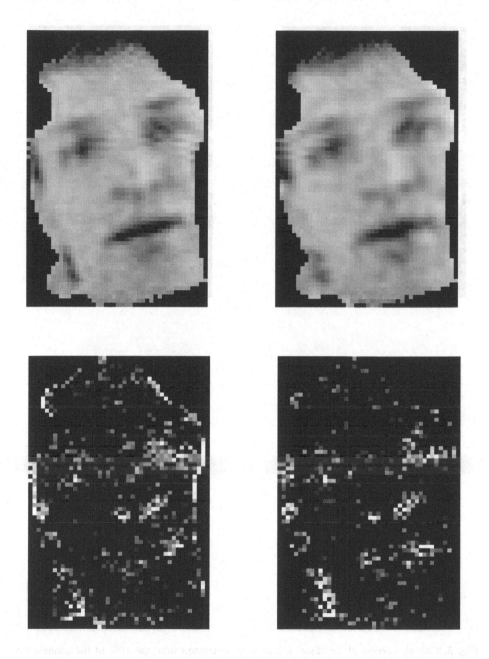

Fig. 2. 'Face' segment reconstructed from 10% of greatest transform samples of the SA DFT (upper left), and SA DCT (upper right), and absolute values of errors between the reconstructions and original image for SA DFT (lower left) and SA DCT (lower right). Note high error values (bright pixels) on the border of the image reconstructed from SA DFT.

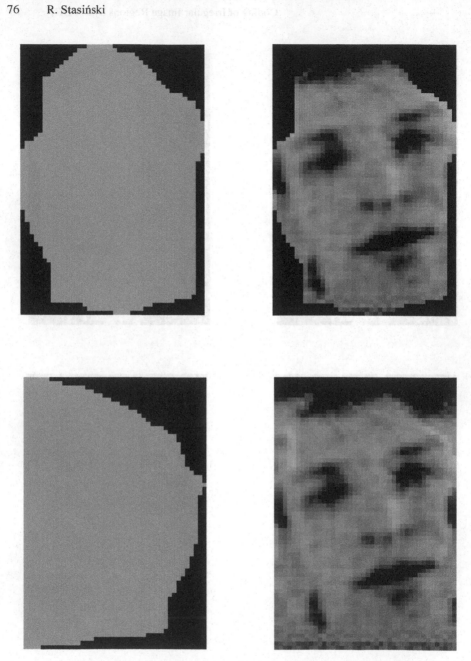

Fig. 3. Convex version of the 'face' segment reconstructed from the 10% of the greatest SA DFT samples (right images) on the basis of full shape information (upper left image), and row DFT sizes (lower left image).

Fast PNN Using Partial Distortion Search

Olli Virmajoki[1], Pasi Fränti[1], and Timo Kaukoranta[2]

[1] Department of Computer Science, University of Joensuu,
P.O. Box 111, FIN-80101 Joensuu, FINLAND
olli.virmajoki@mail.kajak.fi, franti@cs.joensuu.fi
[2] Turku Centre for Computer Science (TUCS), Department of Computer Science,
University of Turku, Lemminkäisenkatu 14A, FIN-20520 Turku, FINLAND
tkaukora@cs.utu.fi

Abstract. Pairwise nearest neighbor method (PNN), in its exact form, provides good quality codebooks for vector quantization but at the cost of high run time. We consider the utilization of the partial distortion search technique in order to reduce the workload caused by the distance calculations in the PNN. By experiments, we show that the simple improvement reduces the run time down to 50-60%

Keywords: Vector quantization, pairwise nearest neighbor method, partial distortion search

1 Introduction

Vector quantization maps a set of N input vectors $\{x_i\}$ in a K-dimensional Euclidean space to a set of M code vectors $\{c_i\}$ [1]. The *pairwise nearest neighbor method* (PNN) [2] is a simple and well-known method for generating the codebook. It starts by initializing a codebook of size N, where each training vector is considered as its own code vector. Two code vectors are merged in each step of the algorithm and the process is repeated until the codebook reduces to the desired size M. The PNN can also be combined with the *generalized Lloyd algorithm* (GLA) [3], as was done in [4], used as a component in the *split-and-merge* algorithm [5], or as the crossover method in *genetic algorithm* [6].

The main drawback of the PNN is its slowness as the original implementation requires $O(N^3K)$ time [7]. An order of magnitude faster algorithm has recently been introduced [8], but the method is still lower bounded by $O(N^2K)$, which is more than the $O(NMK)$ time required by the GLA. Additional improvements are therefore needed in order to make the exact algorithm competitive also in speed.

We propose to use partial distortion search (PDS) technique to be used with the PNN in order to reduce the workload from the distance calculations. The PDS technique was originally proposed by Bei and Gray [9] for fast code vector search under the Euclidean distance metric. The idea of the PDS method is to terminate a single

W. Skarbek (Ed.): CAIP 2001, LNCS 2124, pp. 77–84, 2001.
© Springer-Verlag Berlin Heidelberg 2001

distance calculation immediately when the partially distance exceeds the shortest distance found so far. This idea is independent on the chosen metrics and therefore it can be applied to the PNN.

In the rest of the paper, we recall the exact PNN method and its fast implementations. We then show how the PDS method can be applied both in the initialization stage, and in the main loop of the PNN. By experiments, we show that the method achieves speed-up of about 50-60 % with a very simple implementation and without any additional data structures.

2 Pairwise Nearest Neighbor Method

The aim of the PNN algorithm is to find a codebook C of M code vectors (c_i) by minimizing the average squared Euclidean distance between the training vectors and their representative code vectors:

$$f(C) = \frac{1}{N} \sum_{i=1}^{N} \left\| x_i - c_{p_i} \right\|^2 , \tag{1}$$

where p_i is the cluster (partition) index of the training vector x_i. *Cluster* is defined as the set of training vectors that belong to the same partition a:

$$s_a = \left\{ x_i \middle| p_i = a \right\} . \tag{2}$$

The basic structure of the PNN method is shown in Fig. 1. It starts by initializing each training vector x_i as its own code vector c_i. In each step of the algorithm, two clusters (code vectors) is merged and the centroid of the merged cluster is used as a new code vector. The cost of merging two clusters s_a and s_b ("*distance*" of the clusters) can be calculated as [2]:

$$d_{a,b} = \frac{n_a n_b}{n_a + n_b} \left\| c_a - c_b \right\|^2 , \tag{3}$$

where c_a and c_b are the cluster centroids and n_a and n_b are the cluster sizes.

```
PNN(X, M) → C, P
    s_i ←{x_i} ∀ i∈[1,N];
    m ← N;
    REPEAT
        (s_a, s_b) ← NearestClusters();
        MergeClusters(s_a, s_b);
        m ← m-1;
        UpdateDataStructures();
    UNTIL m=M;
```

Fig. 1. Structure of the exact PNN method.

The exact variant of the PNN applies local optimization strategy: all possible cluster pairs are considered and the one (a, b) increasing the distortion least is chosen. The clusters are then merged and the process is repeated until the codebook reaches the size M. Straightforward implementation of this takes $O(N^3 K)$ time because there are $O(N)$ steps, and in each step there are $O(N^2)$ cost function values to be calculated.

A much faster variant of the PNN can be implemented by maintaining for each cluster a pointer to its nearest neighbor [8]. The nearest neighbor nn_a for a cluster s_a is defined as the cluster minimizing the merge cost. In this way, only few nearest neighbor searches are needed in each iteration. The method is denoted as *fast exact PNN*.

3 Partial Distortion Search

Let s_a be the cluster, for which we seek the nearest neighbor. We use full search, i.e., calculate the distance values $d_{a,j}$ (see Eq. 3) between s_a and all other clusters s_j. Let d_{min} be the distance of the best candidate found so far. The distance is calculated cumulatively by summing up the squared differences in each dimension. The cumulative summation is non-decreasing, as the individual terms are non-negative. The calculation can therefore be terminated and the candidate rejected if the partial distance value exceeds the current minimum d_{min}.

3.1 Simple Variant

The implementation of the partial distance calculation is shown in Fig. 2. The distance function consists of the *squared Euclidean distance* $(e_{a,j})$ of the cluster centroids, and a weighting factor $(w_{a,j})$ that depends on the cluster sizes:

$$e_{a,j} = \sum_{k=1}^{K} \left(c_{ak} - c_{jk} \right)^2 . \tag{4}$$

$$w_{a,j} = \frac{n_a \, n_j}{n_a + n_j} . \tag{5}$$

Here c_{ak} and c_{jk} refer to the k'th component of the corresponding vector. The weighting factor $w_{a,j}$ is calculated first, and the squared Euclidean distance $e_{a,j}$ is then cumulated by summing up the squared differences in each dimension. After each summation, we calculate the partial distortion value $(w_{a,j} \cdot e_{a,j})$ and compare it to the distance of the best candidate (d_{min}):

$$w_{a,j} \, e_{a,j} \quad d_{min} . \tag{6}$$

The distance calculation is terminated if this condition is found to be true. The calculations of the partial distortion require an additional multiplication operation and an extra comparison for checking the termination condition. We refer this as the *simple variant*. Speed-up can be achieved if this extra work does not exceed the time saved by the termination.

3.2 Optimized Variant

The extra multiplication in (6) can be avoided by formulating the termination condition as:

$$e_{a,j} \quad \frac{d_{min}}{w_{a,j}} . \tag{7}$$

The right part of the equation can now be calculated in the beginning of the function, and only the comparison remains inside the summation loop. We refer this as the *optimized variant*. As a drawback, there are one extra division due to (7) and extra multiplication outside the loop.

The computational efficiencies of the two variants are compared to that of the full search in Table 1. The simple variant is faster when the dimensions are very small and in cases when the termination happens earlier. The equation (6) is also less vulnerable to rounding errors than (7). The optimized variant, on the other hand, produces significant improvement when the dimensions are very large.

MergeCost(s_a, s_j, d_{min}) $\rightarrow d$;
  $e \leftarrow 0$;
  $k \leftarrow 0$;
  $w \leftarrow n_a \cdot n_j / (n_a + n_j)$;
  REPEAT
    $k \leftarrow k + 1$;
    $e \leftarrow e + (c_{ak} - c_{jk})^2$;
    $d \leftarrow w \cdot e$;
  UNTIL ($d > d_{min}$) OR ($k = K$);
  RETURN d;

Fig. 2. Pseudo code for the distance calculation in the (simple) PDS method.

Table 1. Summary of the arithmetic operations involved in the distance calculations. The value $q \in [0,1]$ refers as the proportion of the processed dimensions.

Variant:	*	/	+
Full search	$k + 2$	1	$2k + 1$
Simple variant	$2kq + 1$	1	$2kq + 1$
Optimized variant	$kq + 2$	2	$2kq + 1$

3.3   Initial Candidate

Overall, the effectiveness of the method depends on the quality of the current candidate. It is therefore important to have a good initial candidate so that the d_{min} would be small already at the early stages of the calculations. In this way, more distance calculations can be terminated sooner. In the previous iteration, the nearest neighbor for s_a was one of the clusters that were merged. It is expected that the distance to the merged cluster remains relatively small and, therefore, we take this as the first candidate. This minor enhancement turns out to provide significant improvement in the algorithm, see the experiments in Section 4.

4   Test Results

We generated training sets from six images: *Bridge, Camera, Miss America, Table tennis, House* and *Airplane*. The vectors in the first two sets (*Bridge, Camera*) are 4×4 pixel blocks from the gray-scale images. The third and fourth sets (*Miss America, Table Tennis*) have been obtained by subtracting two subsequent image frames of the original video image sequences, and then constructing 4×4 spatial pixel blocks from the residuals. Only the first two frames have been used. The fifth and sixth data sets (*House, Airplane*) consist of color values of the *RGB* images. Applications of this kind of data sets is found in image and video image coding (*Bridge, Camera, Miss America, Table tennis*), and in color image quantization (*House, Airplane*).

The effect of the initial candidate in the PDS is first studied as a function of the vector dimension. For this purpose, we use artificially generated training sets with the parameters $N=1024$, $M=256$, and varying the vector size K from 16 to 1024. The results are shown in Fig. 3, and they clearly demonstrate the importance of the initial guess for training sets with large vector dimension. The improvement is less significant for training sets with $K<16$ but still large enough to be useful. In the following, we assume that the initial guess is always used.

The performance of the two PDS variants (simple and the optimized variant) is summarized in Fig. 4 for the six training sets. The results show that the simple variant is better for the training sets (*House* and *Airplane*) with small vector dimensions ($K=3$). This is because of the extra division operation in the optimized variant. In fact, the optimized variant is even slower than the original PNN. For the other training sets (*Bridge, Camera, Miss America, Table tennis*), the optimized variant works much better and gives always equal to or better result than the simple variant. The significance of the division operation is much smaller in case of vector with higher dimension.

We would also like to note that the cost of the division operation is hardware dependent. The results indicate that the relative performance of the optimized variant increases as a function of K, but one should not draw any final conclusions about the exact value of K, after which the optimized variant outperforms the simple variant.

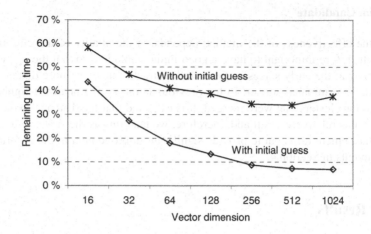

Fig. 3. Remaining run time relative to full search PNN for the optimized PDS with and without initial guess.

Table 2 gives more detailed workload of the different PNN variants for three training sets, one from each category. The overall run times of the PDS is then summarized in Table 3. From the experiments we can see that the results greatly depend on the training set.

The effect of the vector size is demonstrated in Fig. 5 with the artificial training sets described earlier in this section. Obviously, the overall run time increases as a function of the vector size. On the other, the relative improvement of the speed-up methods also increases, which partly compensates the increase in the vector size.

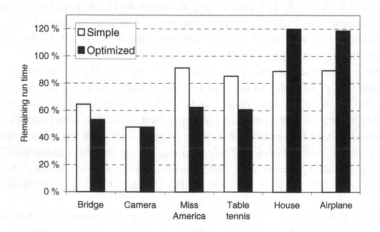

Fig. 4. Remaining run time relative to full search PNN for the simple and optimized variants of the PDS.

Table 2. The average number of distance calculations per nearest neighbor search, the average number of processed vector dimension during the distance calculation, and the total number of processed vector dimensions per search on average.

		Distance calculations / search	Dimensions / distance calculation	Dimensions / search
Bridge, $N=4096$, $M=256$, $K=16$	Full	2208.6	16.0	35338.3
	PDS	2208.7	3.0	6534.2
House, $N=34112$, $M=256$, $K=3$	Full	16514.8	3.0	49544.5
	PDS	16533.6	1.1	17538.3
Miss America, $N=6480$, $M=256$, $K=16$	Full	3404.8	16.0	54476.9
	PDS	3401.5	4.8	16194.3

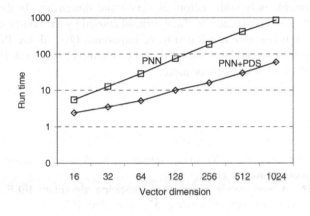

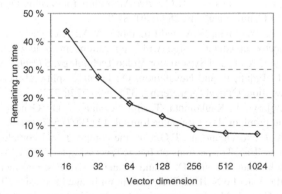

Fig. 5. Run time (in seconds) as a function of the code vector dimension (above), and remaining run time relative to the full search PNN (below).

Table 3. Run times (in seconds) for the six training sets ($M=256$).

	Bridge	Camera	House	Airplane	Miss America	Table tennis
Full search	79	73	1524	8812	229	2895
PDS	42	35	1826	10460	143	1756
Remaining	53%	48%	120%	119%	62%	61%

5 Conclusions

We have proposed the partial distortion search (PDS) for speeding-up the PNN. The effect of the method have studied thoroughly and the results show that the PDS method works well in most cases and achieves similar speed-up than was obtained within the GLA; roughly about 50% with the favorable training sets. The use of the PDS is questionable only with vectors of very small dimension. In this case, the improvement can be overwhelmed by the overhead caused by the additional test. The use of the initial candidate was also found to be important. Overall, the PDS is very efficient with vectors of very large dimension. For example, less than 10% run time is required with vectors of size 256 or more.

References

1. Gersho, A. and Gray, R.M.: Vector Quantization and Signal Compression. Kluwer Academic Publishers, Dordrecht (1992)
2. Equitz, W.H.: A new vector quantization clustering algorithm. IEEE Transactions on Acoustics, Speech, and Signal Processing, **37** (1989) 1568-1575
3. Linde, Y., Buzo, A. and Gray, R.M.: An Algorithm for Vector Quantizer Design. IEEE Transactions on Communications, **28** (1980) 84-95
4. de Garrido, D.P., Pearlman, W.A. and Finamore, W.A.: A clustering algorithm for entropy-constrained vector quantizer design with applications in coding image pyramids. IEEE Transactions on Circuits and Systems for Video Technology, **5** (1995) 83-95
5. Kaukoranta, T., Fränti, P. and Nevalainen, O.: Iterative split-and-merge algorithm for VQ codebook generation. Optical Engineering, **37** (1998) 2726-2732
6. Fränti, P., Kivijärvi, J., Kaukoranta, T. and Nevalainen, O.: Genetic algorithms for large scale clustering problem. The Computer Journal, **40** (1997) 547-554
7. Shanbehzadeh, J. and Ogunbona, P.O.: On the computational complexity of the LBG and PNN algorithms. IEEE Transactions on Image Processing **6** (1997) 614-616
8. Fränti, P., Kaukoranta, T., Shen D.-F. and Chang, K.-S.: Fast and memory efficient implementation of the exact PNN. IEEE Transactions on Image Processing, **9** (2000) 773-777
9. Bei, C.-D. and Gray, R.M.: An improvement of the minimum distortion encoding algorithm for vector quantization. IEEE Transactions on Communications, **33** (1985) 1132-1133

Near-Lossless Color Image Compression with No Error Accumulation in Multiple Coding Cycles

Marek Domański and Krzysztof Rakowski

Politechnika Poznańska, Instytut Elektroniki i Telekomunikacji Piotrowo 3A, 60-965, Poznań, Poland
domanski@et.put.poznan.pl

Abstract. The paper comprises study on accumulation of errors produced by near-lossless JPEG-LS in the consecutive compression-decompression cycles. Paper proves that alternatively, lossless compression can be performed on luminance and chrominance with reduced representation bit numbers. The advantage is that the errors do not accumulate in the consecutive cycles of compression because rounding errors of the RGB $\rightarrow$ YC$_R$C$_B$ $\rightarrow$ RGB transformation do not accumulate in the consecutive transformation cycles. Exemplary experimental data that verify these statements are included in the paper.

Keywords: JPEG-LS, near-lossless image compression

1 Introduction

This paper deals with compression of continuous-tone color natural images, i.e. the images that represent real-world scenes and that these images have been produced by capturing these real scenes using a camera, scanner or other device that convert light stimuli into some electronic signals. The content of the paper is related to lossless (reversible) coding that is used to compress digital image data in such a way that the original data file can be retrieved without any change of bits. The paper deals mostly with near-lossless coding that tries to overcome the principal drawback of lossless techniques related to modest compression ratios achieved. Near-lossless coding trades off between compression ratio and distortion kept very small. Specification of the allowed distortion strongly depends on a specific application. Definitions of „near losslessness" use mostly one of the following criteria:
1. Portion of pixels which are not corrupted during the compression/ decompression process , e.g. 95% [1],
2. Maximum error d of the color component value [2].
The very well known and widely used image compression standard JPEG [3,4] exploits lossy (irreversible) compression modes as well as the lossless (reversible) mode of operation. The lossless mode is based on intraframe prediction and entropy

W. Skarbek (Ed.): CAIP 2001, LNCS 2124, pp. 85–91, 2001.

coding, i.e. on the algorithm that is entirely different from that used by irreversible coding based on the DCT transform. The coming modern JPEG 2000 standard [5] unifies lossless and lossy compression of images into one framework based on the wavelet transform. Despite of its paramount functional properties, for purely lossless operation, JPEG 2000 is inferior to a recently developed lossless compression standard called as JPEG-LS [6]. This new lossless image compression standard is based on the LOCO-I (*low-complexity context-based lossless image compression*) algorithm [7] and usually allows higher compression ratios than those achievable by lossless (reversible) coding in JPEG 2000 [8]. Because of its high compression efficiency, the technique know as CALIC (*context-based adaptive lossless/nearly lossless image coding*) [9] is considered as a reference technique for research results in lossless image compression. This highly sophisticated algorithm exhibits high computational cost (often four times higher than LOCO-I) but only slightly higher compression ratio as compared to LOCO-I (up to 5%).

Applications of image compression are related to digital picture storage and transmission. Even highly sensitive medical images are disseminated via communication networks. Recent rapid development of telemedicine is related to exploitation of communication networks for multimedia transmission. In a networked environment, an individual image can be transmitted several times. Each time it is encoded and decoded. Entirely lossless compression does not change image content but near-lossless compression degrades somewhat image quality in each encoding/decoding cycle. The user who is not able to perceive the degradation caused by a single coding/decoding cycle is often unaware of coding errors accumulated after several cycles of compression. Even small and well acceptable distortion caused by a single cycle of compression can be amplified to a very unacceptable distortion even after few transmissions. Unfortunately, most of near-lossless algorithms do not respect this problem. For example, application of the near-lossless mode of the LOCO-I algorithm, i.e. the JPEG-LS international standard results often in a loss of 10 dB of the peak signal-to-noise ratio after about 10 cycles of encoding and decoding. The experience of authors is that medical histological images exhibited easily perceivable erroneous coloration after already 3 or 4 cycles of encoding/decoding using the JPEG-LS algorithm. Such a situation is absolutely not acceptable in many applications like telepatology where a physician makes vital decisions by considering colors of specific items in a microscopic image. The goal of the paper is to examine the process of accumulation of coding errors in the consecutive cycles of compression and decompression of the near-lossless JPEG-LS. Another goal is to show how the JPEG-LS standard can be easily used for near-lossless compression without accumulation of coding errors.

2 Color Transformations for Image Compression

The images are mostly acquired and displayed in the RGB color space. Unfortunately, the *R, G* and *B* components are highly correlated, therefore their straightforward encoding is not efficient. Decorrelation of components improves the results of further compression. Theoretically optimal is the adaptive Karhunen-Loéve transformation, which is rather sophisticated in application. Employment of the Karhunen-Loéve transformation matched to average statistical properties of a whole image leads to results, which are often even poorer than those obtained by usage of YC_RC_B color coordinates [10]. Therefore rough decorrelation of color components is mostly obtained by a linear transformation to the YC_RC_B color space, opponent color space or another similar one. Unfortunately those linear transformations are mostly represented by matrices with noninteger elements. Therefore application of such transformations prior to actual compression leads to some rounding errors. The whole cycle *color transformation – compression – inverse color transformation* is not exactly lossless, i.e. is not exactly reversible. A very interesting property of the transformation RGB YC_BC_R has been described in [11]. A proposition has been proved that tells us that the rounding errors do not accumulate in the consecutive cycles of the RGB YC_BC_R RGB color transformations. Thus the errors after several cycles of compression and decompression are the same as those after a single cycle. General conditions for a color transformation have been given in [12]. A transformation does not cause error accumulation if it fulfills these conditions. In order to review the problem more carefully, let us assume that an image is represented by R, G and B components. For the sake of brevity let us assume 8-bit representation of component samples. Nevertheless the considerations given below can be adopted for any number of bits of pixel representations. A linear color transformation and the corresponding inverse transformation can be expressed as

$$\begin{matrix} D \\ E \\ F \end{matrix} = \begin{matrix} c_1 \\ c_2 \\ c_3 \end{matrix} + \frac{1}{256} \begin{matrix} t_{11} & t_{12} & t_{13} \\ t_{21} & t_{22} & t_{23} \\ t_{31} & t_{32} & t_{33} \end{matrix} \begin{matrix} R \\ G \\ B \end{matrix} \qquad (1)$$

$$\begin{matrix} R \\ G \\ B \end{matrix} = \frac{1}{256} \begin{matrix} s_{11} & s_{12} & s_{13} \\ s_{21} & s_{22} & s_{23} \\ s_{31} & s_{32} & s_{33} \end{matrix} \begin{matrix} D-c_1 \\ E-c_2 \\ F-c_3 \end{matrix} \qquad (2)$$

where mostly $c_1 = c_2 = c_3 = 0$. Nevertheless for the YC_BC_R color space, there is $D = Y$, $E = C_B$, $F = C_R$ and $c_1 = 16$, $c_2 = c_3 = 128$ in order to match the dynamic range defined by ITU-R. In order to simplify the considerations and without loss of generality, the equations can be simplified to

$$\begin{matrix} D \\ E \\ F \end{matrix} = \begin{matrix} t_{11} & t_{12} & t_{13} \\ t_{21} & t_{22} & t_{23} \\ t_{31} & t_{32} & t_{33} \end{matrix} \begin{matrix} R \\ G \\ B \end{matrix} \qquad (3)$$

$$\begin{matrix} R \\ G \\ B \end{matrix} = \begin{matrix} s_{11} & s_{12} & s_{13} \\ s_{21} & s_{22} & s_{23} \\ s_{31} & s_{32} & s_{33} \end{matrix} \begin{matrix} D \\ E \\ F \end{matrix} \qquad (4)$$

The RGB samples are integers from the interval [0,255]. The exact D, E and F component values obtained according to Eq. 3 have to be rounded in order to obtain integers D^{rd}, E^{rd} and F^{rd} from the interval [0,255]. Then again the recovered R^d, G^d and B^d samples obtained from D^{rd}, E^{rd} and F^{rd} according to Eq. 4 have to be rounded to integers from D^{rd}, E^{rd} and F^{rd} according to Eq. 4 have to be rounded to integers $R^{rd}R^{rd}$,

G^{rd} and B^{rd} from the interval [0,255]. In the second cycle new values $D^{(2)}$, $E^{(2)}$, $F^{(2)}$ are calculated according to Equation 3 that defines a vector transformation corresponding to the second cycle.

The following property has been proved in [12]:

After the first cycle, all consecutive cycles of forward and backward transformations do not influence the values R^{rd}, G^{rd}, B^{rd} obtained in the first cycle. The latter corollary tells us that further transformation cycles do not introduce additional rounding errors. Thus the errors after several cycles of compression and decompression are the same as those after a single cycle. A good example of a transformation which exhibits the following properties is the transformation $RGB \quad YC_BC_R$.

$$\begin{matrix} Y \\ C_B \\ C_R \end{matrix} = \begin{matrix} 16 \\ 128 \\ 128 \end{matrix} + \frac{1}{256} \begin{matrix} 65.738 & 129.057 & 25.064 \\ -37.945 & -74.494 & 112.439 \\ 112.439 & -94.154 & -18.285 \end{matrix} \begin{matrix} R \\ G \\ B \end{matrix} \tag{5}$$

$$\begin{matrix} R \\ G \\ B \end{matrix} = \frac{1}{256} \begin{matrix} 298.082 & 0 & 408.583 \\ 298.082 & -100.291 & -208.120 \\ 298.082 & 516.411 & 0 \end{matrix} \begin{matrix} Y-16 \\ C_B-128 \\ C_R-128 \end{matrix} \tag{6}$$

For the boundary values of 0 and 255, rounding of R, G and B does not fulfill the conditions of the above consideration and slight accumulation of errors occurs which is mostly negligible. For the transformation (5) and (5) the accumulation ends already in the second cycle.

3 Near-Lossless Compression with No Accumulation of Coding Errors

Surprisingly, the following strategy is equivalent to near-lossless compression that does not lead to accumulation of coding errors in the consecutive cycles of coding and decoding.

The steps of coding are:
1. Transform an image from the RGB to YC_BC_R color space.
2. Round the components Y, C_B, C_R.
3. Compress the YC_BC_R representation using a lossless technique, e.g. JPEG-LS.

In the experiments, we prove that this technique is competitive to near-lossless JPEG-LS from the point of rate-distortion characteristic. Nevertheless this proposed but very obvious strategy does not lead to error accumulation in the consecutive coding cycles.

4 Experimental Results

For JPEG-LS near-lossless compression with $d = 1$ and 2, the values of PSNR [dB] and maximum accumulated error are plotted versus cycle number for the well known test image "Lena". For the sake of brevity the experimental results are described for the test image "Lena" only. Nevertheless the results for other test images are similar and they prove the conclusions.

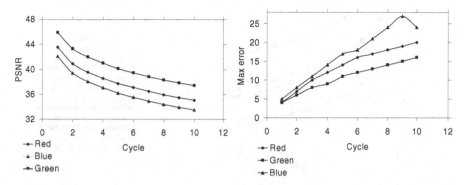

Fig. 1. Near-lossless JPEG-LS with d =1 : PSNR [dB] and maximum accumulated error for test image "Lena".

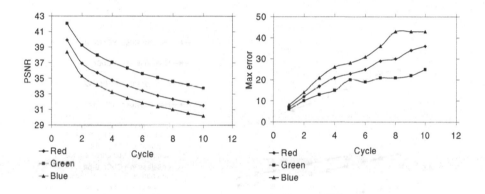

Fig. 2. Near-lossless JPEG-LS with d =2 : PSNR [dB] and maximum accumulated error for test image "Lena".

The plots prove that the accumulated coding error increases after each coding cycle. Similar plots have been obtained for other test images. These plots prove that the accumulated errors are significant. Already second cycle reduces the quality about 2-3 dB as compared to first cycle. The loss of 10 dB of the peak signal-to-noise ratio after about 10 cycles of encoding and decoding is common. Near-lossless mode of JPEG-LS in the RGB and YC_BC_R color spaces is compared with lossless compression in the YC_BC_R space with reduced precision of samples (Figs. 3 and 4). For near-lossless JPEG-LS, accumulated coding errors cause significant decrease of compression ratio K for consecutive cycles. These coding errors are like noise and therefore are difficult do encode.

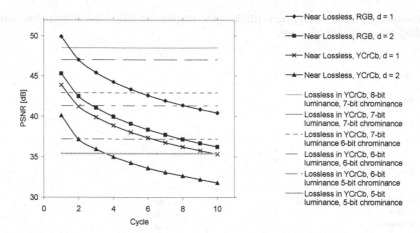

Fig. 3. Near-lossless mode of JPEG-LS in the RGB and YC_BC_R color spaces compared with lossless compression in the YC_BC_R space with reduced precision of samples: PSNR [dB] versus number of coding cycles for test image "Lena".

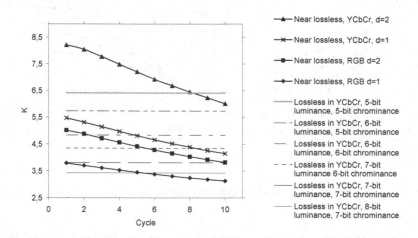

Fig. 4. Near-lossless mode of JPEG-LS in the RGB and YC_BC_R color spaces compared with lossless compression in the YC_BC_R space with reduced precision of samples: Compression ratio K versus number of coding cycles for test image "Lena".

Application of near-lossless JPEG-LS leads to K up to 40 – 60 % greater than in the RGB color space. Unfortunately, image quality is respectively worse. For example, lossless JPEG-LS in the RGB color space with $d=1$ already after the second cycle is inferior to lossless JPEG-LS with 7-bit samples of Y, C_B, C_R. Here, "inferior" means lower compression ratio and lower quality of the reconstructed image. In general, similar conclusions are valid for $d=2$ and for the YC_BC_R color space and for corresponding number of bits of YC_BC_R representation used for lossless compression.

Note that PSNR lower than about 40 dB usually corresponds to visible image degradations. Such compression cannot be described as near-lossless. The degradation caused by rounding of YC_BC_R is related to appearance of false contours while near-lossless compression introduces more noise.

5 Conclusions

The experiments prove that standard near-lossless JPEG-LS compression technique suffers from error accumulation in the consecutive cycles of coding and decoding. The errors accumulate after each cycle and large errors occur after only few cycles. Moreover the cumulated coding noise reduces compression ratio in the consecutive cycles.

The color transformation RGB YC_RC_B introduces errors in the first cycle only. Further cycles do not produce rounding errors. Therefore, for multiple compression cycles, lossless JPEG-LS with reduced sample precision in YC_RC_B is superior to near-lossless JPEG-LS

References

[1]   Karray, L., Rioul, O., Duhamel, P.: L -coding of images: a confidence interval criterion. In: Proc. IEEE Int. Conf. Image Processing, Vol. 2 (1994) 888-892

[2]   Chen, K., Ramabadran, T.V.: Near-lossless compression of medical images through entropy-coded DPCM. IEEE Trans. Medical Imaging, 13 (1994) 538-548

[3]   Information technology – digital compression and coding of continuous-tone still images: requirements and guideline. ISO/IEC IS 10918-1 / ITU-T Rec. T.81

[4]   Pennebaker, W.B., Mitchell, J.L.: JPEG still image compression standard. New York, Van Nostrand Reinhold (1993)

[5]   Information technology — JPEG 2000 image coding system. ISO/IEC FCD 15444

[6]   Lossless and near-lossless compression of continous-tone still images. ISO/IEC DIS 14495

[7]   Weinberger, M.J., Seroussi, G., Sapiro, G.: LOCO-I: a low complexity lossless image compression algorithm. In: Proc. IEEE Data Compression Conf., New York (1996)

[8]   Santa-Cruz, D., Ebrahimi. T.: An analytical study of JPEG 2000 functionalities. In: Proc. IEEE Int. Conf. Image Processing (ICIP), Vancouver, Canada, Vol. 2 (2000) 49-52

[9]   Wu, X., Memon, N.: Context-based adaptive lossless codec. IEEE Trans. Communications, 45 (1997) 437-444

[10] Yovanof, G., Sullivan, J.: Lossless predictive coding of color graphics. Image Processing Algorithms and Techniques, Proc. SPIE, 1657 (1992) 68-82.

[11] Domański M., Rakowski K.: A simple technique for near-lossless coding of color images. In: IEEE Int. Symp. Circuits and Systems, Geneva, Vol. III (2000) 299 – 302

[12] Domański M., Rakowski R.: Color transformations for lossless image compression. In: Proceedings of EUSIPCO 2000, Tampere, Vol. 3 (2000) 1361-1364

Hybrid Lossless Coder of Medical Images with Statistical Data Modelling

Artur Przelaskowski

Institute of Radioelectronics, Warsaw University of Technology,
Nowowiejska 15/19, 00-665 Warszawa, Poland
Tel:+48 22 660-7917, fax: +48 22 825-1363,
arturp@ire.pw.edu.pl

Abstract. Methods of lossless compression of medical image data are considered in this paper. Chosen classes of efficient algorithms were constructed, examined and optimised to conclude the most useful tools for creation of medical image representation. 2-D context-based prediction schemes, and statistical models of entropy coder were fitted to different characteristics of US, MR and CT images. The SSM technique of suitable-to-image characteristics scanning followed by statistical modelling of the context in arithmetic coder was found out as the most effective in most cases. Average bit rate value over test images is equal to 2.54 bpp for SSM coder and significantly overcomes 2.92 bpp achieved for CALIC. Efficient hybrid encoding method (SHEC) was proposed as a complex tool for medical image archiving and transmission. SHEC develops SSM by including CALIC-like coder for archiving the highest quality images and JPEG2000-like wavelet coder for transmission of high and middle quality images in telemedicine systems.

Keywords: medical image compression, reversible hybrid coders, statistical modeling

1 Introduction

Effective compression of medical images is actually useful and desired tool to solve the problems of overloaded networks and data base interfaces, and increasing amount of necessary data store media. More efficient lossy techniques are not acceptable for most medical imaging exams in an opinion of majority of medical specialists' bodies. Most of the newest reversible techniques use heuristic prediction or encoding models, which are efficient only for certain classes of images. The complexity constraints necessitate the use of intuition and heuristics in context modelling instead of universal coding modelling, which has provable optimality based on complex stochastic considerations.

The CALIC [1] is a state-of-art image coder because of applied image-independent order of traversing pixels that can provide adjacent, enclosing (360°) modelling contexts for a maximum number of pixels. Additionally, inventive algorithmic way of

W. Skarbek (Ed.): CAIP 2001, LNCS 2124, pp. 92–101, 2001.

forming and quantizing the modelling contexts to alleviate the problem of context dilution was developed. Alternative coders can fit model to real local data characteristics and capture true information in adaptive way but they are too complex and time consuming, not useful in practise. An example is TMW [2] which contains global optimisation, blended linear prediction and implicit segmentation with splitting the compression process into an analysis step and a coding step. The computational complexity of TMW is several orders of magnitude greater than CALIC but efficiency improvement is between 2 and 10 percent over CALIC for several test images. Less complex adaptive algorithm with weaker effectiveness was proposed by G. Motta *et al.* [3] They used multiple adaptive linear prediction models, locally optimised on a pixel-by-pixel basis and rather simple scheme of entropy coding. In JPEG-LS method, simple nonlinear prediction is followed by more sophisticated entropy encoding scheme with statistical modelling [4]. Four-order context is used. John Robinson's algorithm - BTPC (binary tree predictive coding) [5] applies shape-based predictors and integrated adaptive Huffman encoder. In shape-based prediction the relative values of the four surrounding pixels are interpreted as edge, flat, point, ridge, etc., and a different, simple prediction is formed for each case. The fundamental stage of these coders is entropy encoding based on sophisticated statistical modelling of image data dependencies.

The purpose of our research is to encode medical images in the most effective way. We analysed and tested state-of-art coders. Because of poor effectiveness of the most efficient image coders, e.g. CALIC and JPEG-LS in compression of ultrasonic images, we looked for more adaptive scheme to capture all kinds of information contained in medical image data sets. We presented optimisation process of coding methods in section 2, especially data scanning, modelling, prediction and wavelet encoding. Complimented ideas was merged in hybrid coder scheme in section 3. The experimental results of our research in lossless encoding of US, MR and CT images was reported in section 4 and summarising remarks are concluded in section 5.

2 Optimisation of Coding Methods

A class of medical images is differentiated. A group of high quality images: MR (magnetic resonance), CT (computed tomography) and CR (computed radiography) is characterised by important particulars, low noise level and sharp edges. Lower quality noisy US (ultrasonic) and NM (nuclear medicine) images are the sets of less correlated samples, where estimated sophisticated prediction models to remove linear or nonlinear dependencies give unprofitable results. For example, significant clearly distinguishable edges with high magnitude of gradients are absent in US images. These images contain rather weak edges and structure of contours with lower gradient magnitude, which are placed in noisy areas. A nature of US image noise is very difficult to model without some simplifications. It is characteristic granular or mottled noise called speckle. Many predictive and interpolating models efficient in encoding of natural images do not work properly in compression of lower quality medical images.

Effective schemes of image lossless coder may contain the following components:

- data scanning in an image space fitted to data characteristics to form one-dimensional data stream as an input of entropy coder;
- prediction or interpolation (HINT) where a pixel value is guessed on a base of surrounding pixels (linear and non-linear models are used) formed as finite causal 2-D prediction context; prediction errors are input data of entropy coder;
- transformation-based decorrelation methods: e.g. wavelet transform with selected reversible filters and decomposition schemes; the coefficients are entropy coded;
- entropy encoding (mostly arithmetic), where input data are coded basing on statistical model of 2-D causal context.

We considered and combined these elements into one hybrid coding scheme.

2.1 Image Data Scanning

A priori knowledge and initial estimation of data correlation level to extract the best scanning order for concrete image were applied. Next, arithmetic coder with modelled context was used as entropy encoder. Simple raster scan, raw-column interleaved scan, and other -content or -semantic oriented scanning methods based on a priori knowledge were tested. The examples including data ordering in growing circles curve starting from a middle point of the image were shown in fig. 1.

Moreover, Hilbert differential transformation (HD) of the image samples as more sophisticated scanning technique was tested. Hilbert-type operation is asymptotically optimal in terms of its correlation structure. Lempel and Ziv [6] suggested that the entropy of images can be asymptotically achieved by this type of scanning followed by a particular one-dimensional coding algorithm. An intelligible description of the HD applied for medical image modelling and compression is given by Y. Zhang [7]. Hilbert scan is characterised by 'clustering' or 'locally preserving' capability to be utilised for improving the image compression.

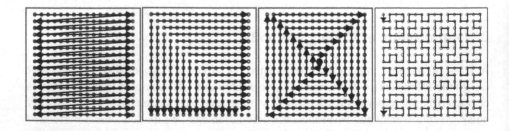

Fig. 1. Data ordering methods in image data scanning process, in sequence from left to right: raw-oriented raster scan, interleaved scan, circle scan and Hilbert scan for 16   16 image block.

Effective scanning method can make context-based data models simpler and less complex. We looked for effective indicators of the most efficient scanning method combined with suitable causal 2D prediction or statistical modelling. A basic measure of the information content or 'uncertainty' is entropy (called unconditional entropy):

$$H(X) = \sum_{x \in A_X} P(x)\log\frac{1}{P(x)}, \tag{1}$$

where ensemble X is a model of source information given by random variable x with a set of possible outcomes called source alphabet: $A_X = \{\alpha_1, \alpha_2, ..., \alpha_K\}$ having probabilities $\{p_1, p_2, ..., p_K\}$ with $P(x = \alpha_k) = p_k$, $p_k \geq 0$ and $\sum_{k=1}^{K} p_k = 1$. Such measure is not suitable as scanning indicator because of context absence. Reliable information content measure called joint entropy is difficult to estimate for finite data sets. Most practical conditional entropy approximates joint entropy and includes context-based data dependency. The conditional entropy of X given context C with alphabet A_C is defined as:

$$H(X \mid C) = \sum_{x,c \in A_X, A_C} P(x,c)\log\frac{1}{P(x \mid c)}. \tag{2}$$

Entropy coders are based on such information characteristics. Causal context C of statistical model is dependent on scanning method. Analytically, the scanning of an image is the data ordering in 1-D sequence of pixel intensities x_i. It is defined by bijection f from the closed interval of integers $[1, ..., N^2]$ which are the pixel indexes in original image $N \times N$ to the set of ordered pairs $\{(i, j) : 1 \leq i, j \leq N\}$ of image location of successive pixels. Thus, the context C support and bijection f should be chosen to minimise $H(X \mid C(f))$. Ordered pixel intensities are as follows: $x_{f(1)}, x_{f(2)}, ..., x_{f(N^2)}$. Utilising of such context in encoder leads to reduction of output bit rate.

Another measure could be used for estimation of 2D prediction efficiency. A difference data stream is created from N^2 ordered input $x_{f(i)}$ data in the following way: $e_{f(i)} = x_{f(i)} - x_{f(i-1)}$, $i = 2, ..., N^2$ and $e_{f(1)} = x_{f(1)}$. Consequently, conditional entropy of a difference stream E (with alphabet A_E) called differential entropy is defined as:

$$H(E \mid C_E) = \sum_{e,c_e \in A_E, A_{CE}} P(e,c_e)\log\frac{1}{P(e \mid c_e)} \tag{3}$$

Memon et al. [8] considered image information source as realisation of an isotropic Gaussian random field with correlations decreasing with distance and pixel values rounded to the nearest integer. Isotropic assumption are not always fulfilled but is helpful when image orientation is not known (none distinct direction of maximum correlation is clearly noticeable). They proved that in this case simple raster scan is more effective than Hilbert scan. These considerations can be widen to Laplacian sources of differential images. In this case raster scan is also more efficient than Hilbert scan. It was really confirmed by the experiments reported in this paper. Also conditional entropy (eq. 2) and differential entropy (eq. 3) as characteristics of the scanning potential for statistical modelling and prediction were tested. A difference between conditional entropy and entropy (eq. 1) is an indicator of statistical modelling potential and differential entropy related to conditional entropy is an indicator of prediction potential. Detection of the cases when 2-D prediction or interpolation of scanned image is not effective for compression allows us to classify medical images into two groups of predictable and unpredictable data sets to increase archiving and transmission potential by selection of proper encoding tools.

2.2 2D Statistical Data Modelling

Entropy encoding of f-scanned data was optimised by context modelling (shape and alphabet), and quantization. Scanning-statistical modelling method (SSM) was the result of our considerations. Context-based probability model for arithmetic coder was realised as linear combination of 12 neighbours surrounding any pixel in a causal way. Alphabet of such 1st order context was uniformly quantized to overcome context dilution problem caused by great complexity of the statistical model used in entropy encoder. Adaptively built model cannot learn the statistics of a source with large number of states fast enough to make a contribution to the compression of the every image being coded. Therefore, quantization of the shape and alphabet of statistical model could be effective in reduction of dilution limitation by decreasing the number of possible model states in the following way:

$$\hat{x} \doteq \frac{x_i \ C_{x(f)}}{q} a_i x_i + \frac{1}{2} , \tag{4}$$

where random variable $\hat{x}$ represents 1st order context with quantized alphabet. Set of coefficients a_i and value q are estimated for each image by linear regression. Typically, $q = 2$ for complexity reduction. Finally, an estimation of the conditional probability model $P(x \mid \hat{x})$ is done adaptively during encoding process.

2.3 Causal 2D Prediction and Interpolation

Performance of SSM technique could be improved in some cases by entering 2-D prediction-interpolation of scanned data and providing suitable 'interface' with scanning and followed statistical modelling. The coefficients of prediction and interpolation models were estimated by linear regression scheme, slightly modified to minimise first order entropy of prediction error stream. Therefore, close-to-optimal prediction models could be fitted automatically for a whole image data set, selected regions or adaptively modified point-by-point in a backward manner. Different techniques of context and prediction function modelling were tested. Context shape, order and alphabet were fitted to encoded stream characteristics to decrease data correlation and dilution of statistical model.

Many image features, such as edge intensity gradient and edge orientation, differences in textures, and generally local data dependencies could be better modelled in a completely enclosing context, characterizing pixel value from each direction. However, because of causal way of model construction it could be the context of pixels distant from modelled one and thus ineffective. It was noticed in HINT [9]. We optimised basic interpolation scheme by fitting number of levels to concrete image characteristics. Noissy, less correlated image data sets were coded efficiently with only five levels of image pyramid, while high quality CT images were interpolated effectivelly with 8 levels pyramid. Also regularity of pixels grid in succesive levels were distorted. The state-of-art realization of mixed prediction-intepolation ideas is CALIC. But a set of fixed heuristic predictors of CALIC were completely unsuitable for several US test images. Another solution was based on multiple predictors with blending its predictions together to give a single predicted pixel value [2]. But the model cost was growing up significantly and made it rather unpractical.

2.4 Wavelet Decomposition and Progressive Coding

By applying reversible transforms lossless wavelet coders were developed [10][11]. Wavelet data decorrelation can be considered as additional step of lossless coding algorithm followed by bitplane or 'whole magnitude', and sign-based statistical modelling and entropy encoding. Data modelling is often based on zerotree structure (SPIHT [11]), and sometimes limited to rectangular blocks which are encoded independently (JPEG2000).

We tried to select the most effective reversible transforms and decomposition procedures. Close to twenty reversible transforms known from state-of-art compression applications [12][13][14] were used to find out the optimal form of medical image decorrelation. The simple S-transform for US images and 17/11 interpolating wavelet [12] for others were selected are the most useful. Besides basic multiresolution dyadic decomposition called mallat, several others like spacl, packet, fbi (examples from JPEG2000 part II draft), uniform, standard, and quincunx schemes (characterised in [15]) were investigated to be useful. Surprisingly, quincunx decomposition occurred the most efficient in most cases of medical image compression.

Additional benefits of wavelet compression according to JPEG2000 paradigm are as follows: natural progressiveness, scalability, multiresolution restoration, ROI selection, easy to achieve embedded data stream, and great efficiency of lossy data packets in lossy-to-lossless scheme.

3 Description of the SHEC

Our concept is that medical images should be encoded in different way depending on medical image characteristics. Useful-for-medical-applications coder should be able to alter encoding scheme relaying on entropy-based criteria or customer a priori knowledge. The SSM is proposed as fundamental groundwork of compression technique. Because prediction-interpolation can slightly improve performance for high-correlated images it may be switched on for that cases. Good automatic indicator is to compare conditional and differential entropy, and both of them to unconditional entropy. Wavelet data decomposition may be applied in several cases with noticeable efficiency and great usefulness. Nevertheless, statistical modelling should be adopted to prediction errors or wavelet data characteristics in that two cases.

We proposed Statistically-driven Hybrid Entropy Coder (SHEC) to encode all classes of medical images. An idea of this coder is shown in figure 2. CALIC as efficient prediction-interpolation scheme may be applied for high-quality CT, CR or MR images in archiving applications. For other images, SSM is the most efficient for archiving, especially for difficult to encode US images. Embedded wavelet coder with selected progression and decomposition schemes is very useful for encoding of MR and CT images in teleradiology. Middle-quality MR images can be efficiently compressed and quickly previewed in data-base applications with wavelet codecs.

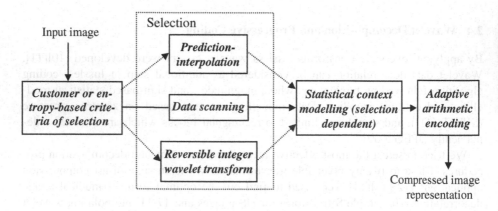

Fig. 2. Hybrid lossless coder with switched decomposition procedure.

As a characteristics of medical images and their susceptibility on prediction-interpolation-statistical modelling compression, the following indicator was used:

$$I_C \quad \frac{H(X \mid C) - H(E \mid C_E)}{H(X)} \tag{5}$$

The entropy values are estimated for reference Hilbert scan.

4   Tests and Results

The following images were used in our experiments: four US images, four MR images and four CT images from different imaging and digitalisation systems. The images US1, MR1 and CT1 are JPEG2000 test images. Bit rate reduction of coded image representation was the most important matter of the optimisation. The results of compression efficiency evaluation tests are presented in tables 1 and 2.

Table 1 Analysis of image data scanning techniques. Two indicators of scanning effectiveness according to equations (2) and (3) were used. An arithmetic coder to encode ordered 1-D data sets was used. The values of bit rate and entropy in bits per symbol are presented.

Input		Scanning method											
		Raster			Interleaved			Circle			Hilbert		
Image	$H(X)$	$H(X\|C)$	$H(E\|C_E)$	BR	$H(X\|C)$	$H(E\|C_E)$	BR	$H(X\|C)$	$H(E\|C_E)$	BR	$H(X\|C)$	$H(E\|C_E)$	BR
US1	6.66	4.29	4.80	**4.81**	4.69	5.26	5.31	4.65	5.16	5.26	4.68	5.25	5.29
US2	4.09	2.00	2.94	**2.12**	2.16	3.10	2.28	2.25	3.20	2.40	2.22	3.17	2.34
US3	2.92	0.97	1.02	**0.89**	1.24	1.38	1.20	1.30	1.42	1.27	1.26	1.46	1.25
US4	4.32	3.09	3.51	**3.17**	3.32	3.84	3.39	3.18	3.64	3.25	3.25	3.74	3.32
CT1	4.56	1.69	1.56	**1.61**	1.87	1.77	1.77	1.74	1.62	1.64	1.80	1.94	1.84
CT2	5.77	3.02	2.87	3.20	3.10	2.97	3.30	2.92	2.74	**3.09**	3.06	3.20	3.33
CT3	4.43	1.65	1.66	1.68	1.73	1.76	1.74	1.57	1.55	**1.60**	1.71	1.90	1.79
CT4	5.30	2.63	2.67	2.78	2.76	2.88	2.96	2.57	2.59	**2.73**	2.73	2.89	2.92
MR1	5.21	3.20	3.36	3.44	3.18	3.36	3.45	3.16	3.37	**3.44**	3.20	3.48	3.51
MR2	5.74	3.57	3.8s1	3.99	3.59	3.81	3.99	3.56	3.77	**3.94**	3.61	3.89	4.03
MR3	5.11	3.18	3.41	**3.42**	3.28	3.54	3.59	3.21	3.48	3.51	3.23	3.52	3.53
MR4	6.37	3.86	3.74	4.33	3.79	3.76	4.28	3.75	3.71	**4.17**	3.86	4.08	4.43
Mean		2.76	2.95	2.95	2.89	3.12	3.10	2.82	3.02	3.03	2.88	3.21	3.13

Significant changes of final coding efficiency are visible in the tests of the different scanning methods (table 1). Simple row-oriented raster scan gave the best results for majority of medical images, especially when image orientation was not clearly identified. The circle scanning can improve effectiveness in the cases with dominant circle correlations, especially for the images of centred round organs (CT and MR images of head and chest). Poor efficiency of Hilbert scan at each case was noticed.

The linear prediction, very efficient in natural image coding is not useful in any case of US image compression – see table 2. Moreover, even low-order predictors are profitable for other medical images (MR, CT) in comparison to the only scanning techniques. Increasing the order of predictor does not influence significantly on encoding performance. HINT is slightly less efficient than 2-D prediction.

Generally, applying of high order models is more profitable in statistical encoding than in prediction step. More important is approximation of enclosed context in interpolation methods. Progressive scan and context is more efficiently formulated in CALIC than in HINT. Closer and wider context modelling makes this state-of-art method optimal in scanning-prediction–statistical modelling class of reversible coders. We noticed that CT images (the highest I_C value) are susceptible to CALIC compression but MR images (with lower I_C) are more efficiently compressed by SSM. Wavelet coders are also more effective in this case.

The group of US images (with the lowest value of I_C) should be encoded only by SSM because of very poor prediction-interpolation and wavelet decomposition efficiency in most cases. Relations between three simple entropy measures are good indicators of prediction and statistical modelling effectiveness. First order conditional entropies are good measures of information for presented entropy coders because of context quantization leading to 1st order causal context dependent on scanning method.

Table 2. Chosen results of compression efficiency evaluation of referenced and optimised coders. The bit rates of compressed data sets are presented. WD is optimised wavelet decomposition, SM means statistical modelling.

Input		Reference coders				Wavelet coders		Statistical coders		
Image	I_c	CALIC	BTPC	JPEG-LS	JPEG-2000	SPIHT	WD+SPIHT	2-D PR +SM	HINT +SM	SSM
US1	-.09	4.79	5.31	4.92	4.89	4.87	4.83	4.75	4.80	**4.43**
US2	-.23	3.35	3.81	3.44	3.81	4.29	3.72	3.66	3.16	**1.94**
US3	-.07	1.88	1.81	2.28	1.95	3.29	1.98	1.65	2.52	**0.89**
US4	-.11	5.16	5.13	5.45	5.20	5.47	5.02	4.61	4.78	**3.00**
US mean	*-.125*	*3.80*	*4.02*	*4.02*	*3.96*	*4.48*	*3.89*	*3.67*	*3.82*	*2.57*
CT1	-.03	**1.19**	1.50	1.32	1.37	1.57	1.43	1.48	1.68	1.25
CT2	-.02	**2.45**	2.91	2.48	2.52	2.65	2.56	2.51	2.9	2.51
CT3	-.04	**1.49**	1.73	1.52	1.93	2.13	1.98	1.87	2.07	1.60
CT4	-.03	**2.29**	2.76	2.35	2.61	2.63	2.60	2.71	2.80	2.55
CT mean	*-.030*	*1.86*	*2.23*	*1.92*	*2.11*	*2.24*	*2.14*	*2.14*	*2.36*	*1.99*
MR1	-.05	2.89	3.37	2.98	2.91	2.92	2.91	2.96	3.02	**2.84**
MR2	-.05	**3.36**	3.94	3.50	3.48	3.48	3.47	3.51	3.68	3.48
MR3	-.06	3.05	3.54	3.12	3.11	3.08	3.07	3.04	3.17	**3.04**
MR4	-.03	3.13	3.80	3.39	3.00	3.28	3.17	3.14	3.62	**2.89**
MR mean	*-.048*	*3.11*	*3.66*	*3.25*	*3.12*	*3.19*	*3.16*	*3.16*	*3.37*	*3.06*
Mean	*-.068*	*2.92*	*3.30*	*3.06*	*3.06*	*3.30*	*3.06*	*2.99*	*3.18*	*2.53*

The results collected in table 2 proved excellent performance of SSM coder. Efficiency improvement is even up to 50% for US images, and middling 13% for all tested images in comparison to CALIC. Wavelet coder efficiency may be significantly increased by proper decomposition scheme selection (see WD+SPIHT contra SPIHT results in table 2, especially for US2 and US3).

5 Conclusions

Entropy coder with statistical modelling is effective in compression of a majority of tested medical images. Entropy-based indicators are useful for evaluation of prediction potential in several cases of CT and MR images. Tested models of predictors and data statistics applied in entropy coder gave significant compression efficiency in comparison to the state-of-art image reversible coders. Complexity of SSM is acceptable for practice.

We proposed SHEC as a complex tool for medical image archiving and transmission. It develops SSM by including CALIC-like coder for archiving the highest quality images and JPEG2000-like wavelet coder for progressive transmission of high and middle quality images.

Better statistical modelling of data dependencies seems to be the most profitable to increase the effectiveness of medical image compression. However, too complex statistical models are not suitable for practice, and CALIC-like predictors are not so effective in archiving and progressive data transmission. Hence, adaptive combination of selected data decomposition modes correlated to locally fitted statistical models seems to be helpful.

References

1. Wu, X.: Lossless Compression of Continuous-tone Images via Context Selection, Quantization, and Modelling. IEEE Trans. Image Process. **6** (1997) 656-664
2. Meyer, B., Tischer, P.: TMW - a New Method for Lossless Image Compression. In: International Picture Coding Symposium PCS97 - Conference Proceedings (1997)
3. Motta, G., Storer, J.A., Carpentieri, B.: Adaptive Linear Prediction Lossless Image Coding. In: Proc. of IEEE Data Compression Conference (1999)
4. Weinberger, M., Seroussi, G., Sapiro, G.: The LOCO-I Lossless Image Compression Algorithm: Principles and Standarization into JPEG-LS. IEEE Trans. Image Process., **9** (2000) 1309-1324
5. Robinson, J.A.: Efficient General-purpose Image Compression with Binary Tree Predictive Coding. IEEE Trans. Image Process. **6** (1997) 601-608
6. Lempel, A., Ziv, J.: Compression of Two-dimensional Data. IEEE Trans. Inform. Theory, **32** (1986) 2-8
7. Zhang, Ya-Qin, Loew, M.H., Pickholtz, R.L.: A Methodology for Modeling the Distributions of Medical Images and Their Stochastic Properties. IEEE Trans. Medical Imaging **9** (1990) 376-383
8. Memon, N., Neuhoff, D.L., Shende, S.: An analysis of some common scanning techniques for lossless image coding. IEEE Trans. Image Process. **9** (2000) 1837-1848
9. Ross, M., Viergever, M.A.: Reversible interframe compression based on HINT (hierarchical interpolation) decorrelation and arithmetic coding. Proc. SPIE, **1444** (1991) 283-290
10. Marcellin, M. W., Gormish, M.J., Bilgin, A., Boliek, M.P.: An Overview of JPEG-2000. In: Proc. of IEEE Data Compression Conference (2000) 523-544
11. Said, A., Pearlman, W.A.: A new fast and efficient image codec based on set partitioning in hierarchical trees. IEEE Trans. Circ. & Syst. Video. Tech. **6** (1996) 243-250
12. Sweldens, W.: The lifting scheme: a custom-design construction of biorthogonal wavelets. Appl. Comput. Harmonic. Analysis, **3** (1996) 186-200
13. Adams, M.D., Kossentini, F.: Reversible Integer-to-Integer Wavelet Transforms for Image Compression: Performance Evaluation and Analysis. IEEE Trans. Image Process. **9** (2000) 1010-1024
14. Calderbank, R.C., Daubechies, I., Sweldens, W., Yeo, Boon-Lock: Wavelet Transforms that Map Integers to Integers. Applied and Computational Harmonic Analysis **5** (1998) 332-369
15. Przelaskowski, A.: Performance evaluation of JPEG2000-like data decomposition schemes in wavelet codec. To be presented at ICIP 2001.

A Simple Algorithm for Ordering and Compression of Vector Codebooks

Maciej Bartkowiak and Adam Łuczak

Poznań University of Technology, Institute of Electronics and Telecommunications
ul. Piotrowo 3A, 60-965 Poznań, Poland, tel. (+4861) 6652171, fax (+4861) 6652572
{mbartkow, aluczak}@et.put.poznan.pl

Abstract. The problem of storage or transmission of codevectors is an essential issue in vector quantization with custom codebook. The proposed technique for compression of codebooks relies on structuring and ordering properties of a binary split algorithm used for codebook design. A simple algorithm is presented for automatic ordering of the codebook entries in order to group similar codevectors. This similarity is exploited in efficient compression of the codebook content by the means of lossless differential coding and lossy DCT-based coding. Experimental results of two compression experiments are reported and show that a small compression gain can be achieved in this way.

Keywords: vector quantization, codebook compression

1 Introduction

Various applications of vector quantization in image data compression are studied for years. Out of two scenarios which are using an universal codebook and using a custom codebook, the latter offers significantly better quality of reconstructed images at the cost of additional storage and/or transmission of the codebook, which is prohibitive in achieving high compression efficiency. Therefore some coding algorithms are being investigated for more efficient representation of the codebook content.

Hereafter, we refer to the codebook $\mathbf{X} = \{\underline{X}_1, \underline{X}_2, \dots \underline{X}_N\}$ as a set of vectors $\underline{X}_i = \left[x_1^i, x_2^i, \dots x_K^i\right]^T$ calculated from image samples, subband/wavelet samples, etc. However following discussion is not limited to any particular data representation, a simple direct vector quantization of the blocks of pixels is used as a benchmark for experimental simulations.

2 Tree-Structured Codebooks

Among several approaches to codebook design, tree-structured techniques which result in tree-structured codebooks exhibit important advantages over non-structured

W. Skarbek (Ed.): CAIP 2001, LNCS 2124, pp. 102–109, 2001.
© Springer-Verlag Berlin Heidelberg 2001

ones, at the cost of slightly reduced performance of both the codebook design and the actual vector quantization stages. First, designing an optimal codebook in highly-dimensional space using an optimization method is very difficult and computationally intractable, while hierarchical methods (such as these employed in tree-structured techniques) offer significant savings in computational complexity. Second, tree-structured codebook allows for a simplified the actual vector quantization stage, whereby, instead of full search, the input vector is hierarchically compared to the successive nodes of the tree.

Codebook design based on tree-structured algorithm is usually an iterative procedure of hierarchical partitioning of the multidimensional data space into disjoint regions. Starting from just one codevector representing the whole data set, the codebook grows in each step. One cluster of vectors in a particular region is split in each iteration, and a respective codevector is replaced by two or more vectors representing the new regions just formed, usually centroids of data clusters. A very popular variant of this approach is a binary split (BS) technique, wherein the region under consideration is always split into two subregions and two new clusters are formed (cf fig. 1), which is associated with codebook growth by 1 codevector:

$$\mathbf{X}^{(n+1)} \stackrel{=}{=} \mathbf{X}^{(n)} \ \{\underline{X}_{i'}, \underline{X}_{i''}\} - \{\underline{X}_i\} \tag{1}$$

where $\mathbf{X}^{(n)}$ denotes the codebook in n-th stage of the binary split algorithm,
$\underline{X}_i$ denotes the codevector being replaced by two new codevectors, $\underline{X}_{i'}$ and $\underline{X}_{i''}$.

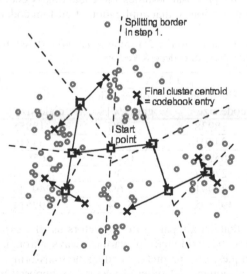

Fig. 1. Hierarchical data bi-partitioning by the binary split algorithm during consecutive steps of the BS algorithm (2D case)

Various rules of region splitting may be applied to the BS algorithm. The most typical approach is to split the region along a hyperplane passing through its centroid in some strictly defined direction related to the data distribution within the region. One of the simplest approaches, proposed by Linde, Buzo and Gray [1] relies on the strengths of their generalized Lloyd algorithm (GLA) known as LBG algorithm (which is similar to the K-means algorithm of MacQueen, [2]) for K=2. It consists of simple iterative LBG-based refinements of randomly initiated split.

An important near-optimal solution (cf. e.g. [3,4,5]) is derived from the principal component analysis (PCA). The cluster in question is being split along the direction of maximum data variation around its centroids which is determined by the dominant eigenvector of the data covariance matrix within the cluster.

3 Codebook Compression: Existing Approaches and the New Proposed Technique

Codebook compression is often considered as a method for effective increasing the efficiency of data coding based on vector quantization [6]. In general, two approaches are practiced: lossless and weakly-lossy coding. Straightforward Huffman coding of the component values of codevectors brings marginal (if any) compression gain (cf Table 1) due to their distribution being almost uniform.

Experiments show, that lossy compression utilizing quantization often distorts the codebook in way that makes it no longer optimal for quantizing the image it has been primarily designed for, and thus the whole process loses the advantage of using a custom codebook. Our proposal assumes the codebook is coded without any loss or with only marginal distortion so that application of custom codebook is still justified.

Table 1. Compression gain (over PCM representation) offered by direct application of Huffman coding to the values of codebook vectors

Codebook size	LENA		BOATS	
	Huffman code size	Compression gain [%]	Huffman code size	Compression gain [%]
4x4x128	18938	0%	18855	0%
4x4x256	34401	0%	34310	0%
4x4x512	65243	0.4%	65042	0.7%
4x4x1024	127097	3%	126964	3.1%
4x4x2048	249991	4.6%	249249	4.9%
4x4x4096	495895	5.4%	492514	6%

Experiments show, that certain pairs of codevectors usually exhibit mutual similarity. This similarity may be exploited in various ways through application of data compression techniques, e.g. predictive coding or transform coding, applied across vectors, i.e. along a sequence of ordered vectors, as opposed to intra-vector coding. The efficiency of such compression strongly depends on the correlation between consecutive members of the sequence, therefore the effort is directed toward defining an ordering rule,

$$\Omega : \{\underline{X}_i\} \quad \{s_i\}, \quad 1 < s_i < N \tag{2}$$

that maximizes this correlation. Such rule is in fact the one that minimizes the total length of a chain of ordered codevectors,

$$\Omega_{opt} = \arg\min_{\Omega} \sum_{i=1}^{N-1} \left\| \underline{X}_{S_i} - \underline{X}_{S_{i+1}} \right\| \tag{3}$$

Such order may be determined through exhaustive search, which however would be unreasonable considering the little gain in compression ratio achieved.

4 Automatic Codebook Ordering

The proposed simple technique relies on structuring and ordering properties of a binary tree. The algorithm is based on a observation that, for reasonably sized tree-structured codebooks, in most cases the aforementioned similarity is mostly observed between pairs of codevectors belonging to the same part of the tree. Therefore the ordering is performed automatically during the codebook design process and is defined by the two inductive rules:

1. Before each step of the binary split algorithm, the codebook in its current stage of growth is assumed to be already ordered (which is trivial for the first and the second step, when codebook consists only of one or two codevectors, respectively),

$$\mathbf{X}^{(n)} = \left\{ \underline{X}_{S_1}, \underline{X}_{S_2}, \ldots \underline{X}_{S_i}, \ldots \underline{X}_{S_N} \right\} \tag{4}$$

2. Each replacement of the codevector $\underline{X}_{S_i}$ with two new codevectors, $\underline{X}_{S_i'}$ and $\underline{X}_{S_i''}$, is associated with inserting two new members into the existing sequence, in the place of the old member. Out of two possibilities, the one is chosen that results in lower the total chain length (cf. Fig. 2):

$$\left\| \underline{X}_{S_{i-1}} - \underline{X}_{S_i'} \right\| + \left\| \underline{X}_{S_i''} - \underline{X}_{S_{i+1}} \right\| < \left\| \underline{X}_{S_{i-1}} - \underline{X}_{S_i''} \right\| + \left\| \underline{X}_{S_i'} - \underline{X}_{S_{i+1}} \right\| \tag{5a}$$
$$\mathbf{X}^{(n+1)} = \left\{ \underline{X}_{S_1}, \ldots \underline{X}_{S_i'}, \underline{X}_{S_i''}, \ldots \underline{X}_{S_N} \right\}$$

or

$$\left\| \underline{X}_{S_{i-1}} - \underline{X}_{S_i'} \right\| + \left\| \underline{X}_{S_i''} - \underline{X}_{S_{i+1}} \right\| > \left\| \underline{X}_{S_{i-1}} - \underline{X}_{S_i''} \right\| + \left\| \underline{X}_{S_i'} - \underline{X}_{S_{i+1}} \right\| \tag{5b}$$
$$\mathbf{X}^{(n+1)} = \left\{ \underline{X}_{S_1}, \ldots \underline{X}_{S_i''}, \underline{X}_{S_i'}, \ldots \underline{X}_{S_N} \right\}.$$

Thanks to hierarchical partitioning scheme, major data clusters are separated in first order into different branches of the binary tree, so are separated within the sequence the codevectors representing them. As the smallest clusters are divided last, the corresponding centroids codevectors most likely result being placed next to each other in the sequence.

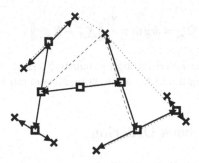

Fig. 2. Example ordering of the tree nodes from Fig. 1 after second step (dashed line) and fourth step (dotted line) of the binary split algorithm

5 Application to Codebook Compression

The proposed coding techniques- lossless differential coding and lossy transform coding both make use of the increased correlation between consecutive codevectors that results from the proposed ordering. For the purpose of coding, K data vectors of length N are formed by scanning the corresponding values x_j^i across all codevectors:

$$\underline{U}_j = \left[x_j^1, x_j^2, x_j^3, \ldots x_j^N \right], \quad j = 1 \ldots K . \tag{6}$$

In case of predictive differential coding, difference vectors are calculated, for each pair of neighboring codevectors in the ordered sequence. For the first vector, its difference with the average value of its coordinates is calculated.

$$\underline{D}_i = \underline{X}_i - \underline{X}_{i-1} , \quad i = 2 \ldots N \tag{7}$$

$$\underline{D}_1 = \underline{X}_1 - \tfrac{1}{K} \sum_{j=1}^{K} x_1^j .$$

Subsequently, the set of differential values is encoded using Huffman code with custom dictionary.

Lossy encoding of the codebook data involves calculating discrete cosine transform of the vectors $\underline{U}_j$, quantization and Huffman coding:

$$\left[F_k \right]_j^T = \mathbf{C}\, \underline{U}_j^T \tag{8}$$

$$Q_k = \begin{cases} F_k & k = 0 \\[2mm] \dfrac{F_k}{q} & k > 0 \end{cases}$$

where $\mathbf{C}$ denotes the DCT matrix, Q_k denotes the quantized DCT coefficient, and q denotes the quantization factor. Due to weak quantization, zero-valued coefficients do not occur as frequently as they do in case of typical image transform coding. Therefore, LRL coding does not offer significant gain in compression ratio.

6 Experimental Results

Three series of experiments have been performed in order to estimate the influence of codebook ordering on the compression gain. Codebooks consisting of 128, 256, 512, 1024, 2048 and 4096 vectors were designed with a PCA-based binary split algorithm using samples of test images partitioned onto 4x4 blocks. The codebooks were ordered using the proposed algorithm and further encoded using both lossless differential coding and lossy DCT-based coding.

Table 2. Compression gain resulting from application of lossless differential coding of codevectors after codebook ordering using the proposed algorithm

Codebook size	LENA		BOATS	
	Huffman code size	Compression gain [%]	Huffman code size	Compression gain [%]
4x4x128	14558	11%	15501	5.4%
4x4x256	27267	16.7%	29420	10%
4x4x512	52356	20%	55677	15%
4x4x1024	100612	23%	109329	16.6%
4x4x2048	199400	23.9%	213939	18.4%
4x4x4096	384797	26.6%	414979	20.8%

As shown in table 2, application of differential coding to codebook compression brings small improvement in terms of positive compression gain (few percent to over 25 percent of the number of bits required to represent the codebook) which is achieved without any decrease of the quality of reconstructed codebook (and hence without any loss in image quality). This compression gain grows especially for larger codebooks.

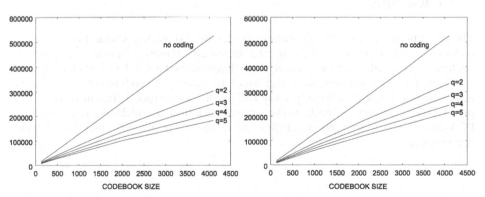

Fig. 3. Experimental results for lossy DCT-based coding of codebooks for test images LENA (left plot) and BOATS (right plot): total number of bits required to store the codebook versus the codebook size (number of codevectors) for various quantization coefficients

Compression gain offered by lossy DCT-based coding scheme is much larger (over 50% and more). The interesting conclusion is that the gain only very slightly grows with the codebook size (cf fig. 3).

The last experiment used codebooks reconstructed after lossy DCT-based coding to vector quantize the original images. Comparison of overall performance of the coding scenarios with and without codebook compression shows that the differences are small. Only for very small values of quantization factor, q=2 , one can observe a consistent improvement in coding efficiency. In case of stronger quantization of the DCT coefficients, the distortions introduced to the codebook make it unfortunatelly no longer optimal and decrease the quality of reconstructed images.

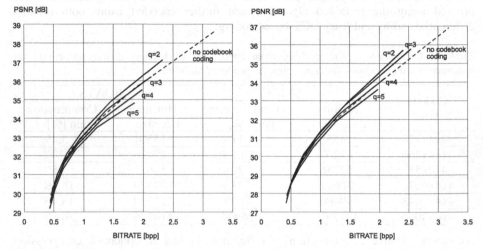

Fig. 4. Comparison of overall vector quantization performance without (dashed line) and with (solid lines) lossy codebook compression for test images LENA (left plot) and BOATS (right plot)

7   Conclusions

The paper proposes a simple algorithm to order the codevectors during tree-structured codebook design, which results in the increased similarity between neighboring vectors. This similarity is exploited by differential and DCT-based coding applied across vectors. Experiments show that small increase of the coding efficiency can be achieved with lossless or weakly lossy coding of codevectors. Efficient compression of codebooks is very difficult, however. We reckon that codebooks obtained form PCA-based binary splitting algorithm are close to optimal and therefore heavily decorrelated.

References

1. Linde, Y., Buzo, A., Gray, R.M.: An Algorithm for Vector Quantizer Design. IEEE Trans. on Communications (1980) 84-89
2. MacQueen, J.B.: Some Methods for Classification and Analysis of Multivariate Observations. In: Proc. 5th Berkeley Symp. Math. Statistics and Probability (1967) 281-297
3. Morgan, J.N., Sonquist, J.A.: Problem in the Analysis of Survey Data, and a Proposal. J. Amer, Static. Assoc. **58** (1963) 415-434
4. Orchard, M., Bouman, C.: Color Quantization of Images. IEEE Trans. on Sig. Proc. **39** (1991) 2677-2690
5. Wu, X., Zhang, K.: A Better Tree-Structured Vector Quantizer. Proc. IEEE Data Compression Conf, IEEE Computer Society Press, LA (1991) 392-401
6. Dionysian, R., Ercegovac, M.: Vector quantization with compressed codebooks. Signal Processing: Image Communication **9** (1996) 79-88

MPEG 2-Based Video Coding with Three-Layer Mixed Scalability

Marek Domański and Sławomir Maćkowiak

Poznań University of Technology,
Institute of Electronics and Telecommunication,
Piotrowo 3A, 60-965, Poznań, Poland
Phone: +48 61 66 52 762
Fax: +48 61 66 52 572
{domanski, smack}@et.put.poznan.pl

Abstract. The paper describes a three-layer video coder based on spatio-temporal scalability and data partitioning. The base layer represents video sequences with reduced spatial and temporal resolution. Decoding of a middle layer gives full resolution images but with lower quality as compared to those obtained from the enhancement layer also. The bitrate overhead measured relative to the single layer MPEG-2 bitstream varies about 5% - 25% for progressive television test sequences. The base layer is fully MPEG-2 compatible and the whole structure exhibits high level of compatibility with individual building blocks of MPEG-2 coders. The paper reports experimental results that prove useful properties of the coder proposed.

Keywords: MPEG-2, spatio-temporal scalability, enhacement layer

1 Introduction

There exists growing demand for multilayer scalable video codecs that are suitable for video transmission over heterogeneous communication networks characterized by various available levels of Quality of Service. On the other hand, the service providers demand that the data are broadcasted once to a group of users accessed via heterogeneous links. For this purpose, the transmitted bitstream has to be partitioned into some layers in such a way that some layers are decodable into video sequences with reduced spatial resolution, temporal resolution or signal-to-noise ratio (SNR). This functionality is called spatial, temporal or SNR scalability, respectively.

For practical reasons, the number of layers is often limited to three or even two layers as in most profiles of MPEG 2 [1,2]. This papers deals with a proposal of a video coding system with three-layer scalability (cf. Fig. 1).

The paper is focused on the functionality of spatial scalability that is already provided by the MPEG-2 [1,2] and MPEG-4 [3] video compression standards. Unfortunately, respective standard implementations are mostly related to unacceptably high bitrate overheads mostly about 50% to 70% of the single-layer bitrate. Among other proposals for spatial scalability, those based on application of subband/wavelet decomposition are the most popular [4-9] but they suffer from the problems with bit allocation and interframe coding.

W. Skarbek (Ed.): CAIP 2001, LNCS 2124, pp. 110–117, 2001.

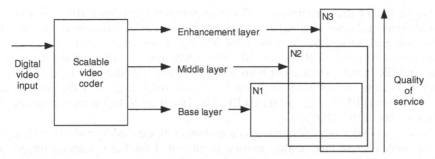

Fig. 1. Three-layer scalable video transmission in heterogeneous network that consists of sub-networks N1, N2 and N3 accessible via links with various Quality of Service

The goal of the work is to achieve possible low bitrate overhead, i.e. the total bitrate related to all three layers of a scalable bitstream should be as close as possible to the bitrate of the respective single-layer coder. The paper deals with scalable video coders that compress digital television signals to bitrates of order of few megabits per second. These bitrates are suitable, e.g. for television broadcasting, the service of video on demand and video transmission using ADSL technology. Moreover it is required that the system produces layers with similar bitrates.

In order to meet the latter requirement for two-layer systems, it has been proposed to combine the spatial scalability with other types of scalability thus reducing the base layer bitrate. The recent proposals are based on a combination of spatial and SNR scalability [10] or spatial and temporal scalability [11, 12]. This paper deals with the latter approach of spatio-temporal scalability that was considered in two versions hitherto. The first one exploited three-dimensional spatio-temporal decomposition [11]. The second version was based on partitioning of data related to B-frames [11, 12]. This approach as well as that based on combination space and SNR scalabilities were quit successful.

The paper deals with a novel version of spatio-temporal scalability based on partitioning of data related to B-frames where subband decomposition is exploited for the I-frames only [13,14]. The paper extends these concepts onto three-layer systems.

The important assumption made for this work is that high level of compatibility with the MPEG video coding standards would be ensured. In the paper, the MPEG-2 video coding standard is used as reference but the results are also applicable to the MPEG-4 systems with minor modifications. In particular, it is assumed that the low-resolution base layer bitstream is fully compatible with the MPEG-2 standard.

2 Spatio-Temporal Scalability with B-Frame Data Partitioning

For the sake of simplicity, spatio-temporal scalability will be reviewed for the simplest case of two-layer systems, i.e. systems that produce the base layer bitstream and the enhancement layer bitstream. Base layer bitstream allows for restoration of video with all resolutions (horizontal, vertical, temporal) halved as compared to the input video sequence. Enhancement layer bitstream can be additionally used for restoration of the full-quality video.

Let us assume that the number of B-frames between two consecutive I- or P-frames is odd. Temporal resolution reduction can be achieved by partitioning of the stream of B-frames: each second frame is included into the enhancement layer only. Therefore there exist two types of B-frames: BE-frames which exist in the enhancement layer only and BR-frames which exist both in the base and enhancement layers. The base layer represents the subband LL from I-, P- and BR-frames, and the enhancement layer represents BE-frames, subbands LH, HL, HH from I-, P-frames and hierarchical enhancement of the BR-frames.

Base layer coder is implemented as a motion-compensated hybrid MPEG-2 coder. In the enhancement layer coder, motion is estimated for full-resolution images and full-frame motion compensation is performed. Motion vectors are transmitted for the base layer. Another motion vectors MV_e are estimated for the enhancement layer. In the enhancement layer, difference values (MV_e-MV_b) are transmitted.

Improved prediction is used for the BR-frames, which are the B-frames represented in both layers. Each macroblock in a full-resolution BR-frame can be predicted from three reference frames: previous reference frame (I- or P-frame), next reference frame (I- or P-frame), current reference frame (BR-frame). The improvement on standard MPEG-2 prediction within a single layer consists in another decision strategy. The best prediction/interpolation is chosen from all three possible reference frames: previous, future and interpolated.

The I-frames are split into four subbands using wavelet decomposition. The LL-subband (the subband of lowest frequencies) is sent with the base layer bitstream while the other three subbands are encoded for middle and enhancement layers.

3 Three-Layer Scalable System

In the previous section a two-layer scalable coder has been described. Experimental data [12-14] obtained by the authors prove that the enhancement layer bitrate is mostly larger than that of the base layer. Therefore a reasonable three-layer system can be designed by use of data partitioning of the enhancement layer, i.e. the enhancement layer data are partitioned into the middle layer data and the new enhancement layer data according to the rule described by Table 1.

The enhancement layer repeats slice headers because each layer has to be error resilient. The repeated slice headers allow the decoder to resynchronize the enhancement layer bitstream after a transmission error.

The middle layer contains N_m first pairs (*run, level*) for individual blocks. The control parameter N_m influences bitrate allocation between the middle and enhancement layers.

The bitrates in the individual layers can be controlled individually to some extent. The experience of the authors shows that all the three bitrates can be set to similar values for a wide range of the overall bitrates of the scalable coder. The coder structure is shown in Fig. 2.

Table 1. Data partitioning in a three-layer system

Middle layer	Enhancement layer of the three-layer system
1. Headers 2. Motion vectors 3. Low frequency DCT coefficients	1. Slice headers 2. Remaining DCT coefficients

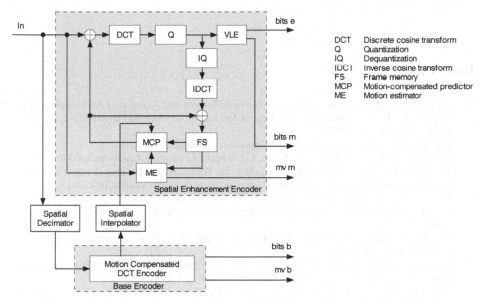

DCT   Discrete cosine transform
Q   Quantization
IQ   Dequantization
IDCT   Inverse cosine transform
FS   Frame memory
MCP   Motion-compensated predictor
ME   Motion estimator

Fig. 2. The general structure of a three-layer coder (*bits b* and *mv b* – base layer, *bits m* and *mv m* – middle layer, *bits e* – enhancement layer)

4 Experimental Results and Conclusions

The verification model has been prepared as software written in C++ language. The coder is implemented for processing of progressive 4:2:0 720 × 576 sequences with 50 frames per second. The base layer coder is standard MPEG 2 coder that processes video in the SIF format but both the middle layer and the enhancement layer are in the full television resolution.

For the sake of brevity, the results are given for two video sequences only. The Table 2 summarizes the results for bitrates in the range of 4 – 7 Mbps. The results for a nonscalable MPEG 2 MP @ ML coder are given for comparison. Figures 3 and 4 include rate-distortion plots for all layers. Fig. 5 shows frames reconstructed from all three layers of quality.

The coder is able to produce three bitstreams with similar bitrates. Such bit allocation is very advantageous for practical applications. With the same bitrate as by

MPEG-2 nonscalable profile, the scalable coder proposed reaches almost the same quality. The bitrate overhead due to scalability is about 5% - 25%. The codec proposed outperforms spatially scalable MPEG-2 [1] or MPEG-4 [3] coders which generate bitrate overheads often exceeding 50% even for two-layer versions.

Acknowledgement. The work has been supported by Polish Committee for Scientific Research under Grant 8 T11D 009 17. Sławomir Maćkowiak is Fellow of the Foundation for Polish Science.

References

1. ISO/IEC IS 13818, Generic Coding of Moving Pictures and Associated Audio Information
2. Haskell, B., Puri, A., Netravali, A.: Digital Video: An Introduction to MPEG-2. Chapman & Hall, New York (1997)
3. ISO/IEC IS 14496, Generic Coding of Audiovisual Objects
4. Tsunashima, T., Stampleman, J., Bove, V.: A Scalable Motion -Compensated Gharavi, H., Ng, W.Y.: H.263 Compatible Video Coding and Transmission. In: Proc. First International Workshop on Wireless Image/Video Communication, Loughborough (1996) 115-120
5. Gharavi, H., Ng, W.Y.: H.263 Compatible Video Coding and Transmission. In: Proc. First International Workshop on Wireless Image/Video Communication, Loughborough (1996) 115-120
6. Bosveld, F.: Hierarchical Video Compression Using SBC, Ph.D. dissertation, Delft Univ. of Technology, Delft (1996)
7. Senbel, S., Abdel-Wahab, H.: Scalable and Robust Image Compression Using Quadtrees. Signal Processing: Image Communication 14 (1999) 425-441
8. Shen, K., Delp, E.: Wavelet Based Rate Scalable Video Compression. IEEE Trans. Circ. Syst. Video Technology 9 (1999) 109-122
9. Chang, P.-C., Lu, T.-T.: A Scalable Video Compression Technique Based on Wavelet and MPEG Coding. In: Proc. Int. Conf. on Consumer Electronics (1999) 372-373
10. Benzler, U.: Spatial Scalable Video Coding Using a Combined Subband-DCT Approach. IEEE Trans. Circ. Syst. Video Technology 10 (2000) 1080-1087
11. Domański, M., Łuczak, A., Maćkowiak, S., Świerczyński, R.: Hybrid Coding of Video with Spatio-Temporal Scalability Using Subband Decomposition. In: Signal Processing IX: Theories and Applications, Rhodes (1998) 53-56
12. Domański, M., Łuczak, A., Maćkowiak, S.: Spatio-Temporal Scalability for MPEG Video Coding. IEEE Trans. Circ. Syst. Video Technology 10 (2000) 1088-1093
13. Łuczak, A., Maćkowiak, S., Domański, M.: Spatio-Temporal Scalability Using Modified MPEG-2 Predictive Video Coding. In: Signal Processing X: Theories and Applications, Tampere (2000) 961-964
14. Domański, M., Łuczak, A., Maćkowiak, S.: On Improving MPEG Spatial Scalability, In: Proc. Int. Conf. Image Proc., Vancouver (2000) 848-851

Table 2. The experimental results for BT.601 progressive test sequences.

Test sequence	Fun Fair				Cheer			
Single layer coder (MPEG-2)								
Bitstream [Mb/s]	4.0	5.18	6.0	7.0	4.0	5.21	6.0	7.0
Average PSNR [dB] for luminance	30.6	32.2	33.0	33.8	30.6	31.9	32.8	33.7
Proposed scalable coder								
Base layer average PSNR [dB] for luminance	30.5	33.0	34.1	34.1	29.5	32.0	33.1	33.1
Average PSNR [dB] for luminance recovered from both base and middle layers	28.5	29.7	29.6	29.7	28.7	29.7	30.2	30.3
Average PSNR [dB] for luminance recovered from all three layers	30.6	32.1	33.0	33.8	30.4	31.9	32.8	33.7
Base layer bitstream [Mbit/s]	1.40	2.16	2.50	2.50	1.40	2.15	2.50	2.50
Middle layer bitstream [Mbit/s]	2.00	2.18	2.42	2.80	1.97	2.03	2.28	2.67
Enhancement layer bitstream [Mbit/s]	1.00	1.40	1.84	2.77	0.92	1.22	1.65	2.64
Base layer bitstream as percent of the total bitstream	31.8	37.6	37.0	31.0	32.6	39.8	38.9	32.0
Middle layer bitstream as percent of the total bitstream	45.5	38.0	35.8	34.7	45.9	37.6	35.5	34.2
Enhancement layer bitstream as percent of the total bitstream	22.7	24.4	27.2	34.3	21.5	22.6	25.6	33.8
Scalability overhead [%] (as compared to single-layer coding)	10.0	10.8	12.6	15.2	7.25	2.6	7.2	11.6

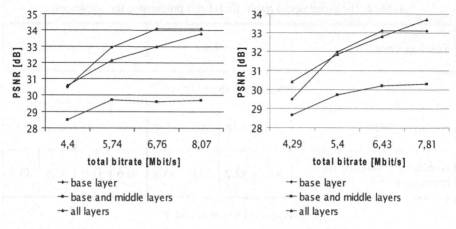

Fig. 3. PSNR for test sequences *Funfair* and *Cheer*

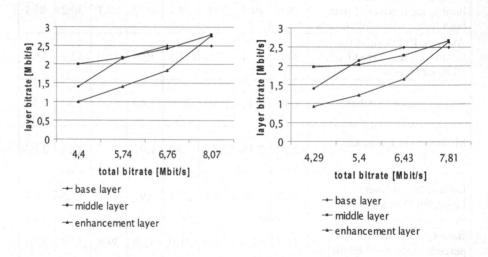

Fig. 4. Layer bitrates for test sequences *Funfair* and *Cheer*

Fig. 5. A frame from the base, middle and enhancement layer of a test sequence *Funfair* (for the coding parameters given in Table 3)

The Coefficient Based Rate Distortion Model for the Low Bit Rate Video Coding

Grzegorz Siemek

Warsaw University of Technology
Ul. Nowowiejska 15/19
00-665 Warszawa, Poland

Abstract. A low bit rate video coding requires strict buffer regulations and low buffer delay. Thus a macroblock level rate control is necessary. However, an MB-level rate control is costly at low bit rates since there is an additional overhead if the quantization parameter is changed frequently within a frame. This paper presents the rate distortion model that selects the number of significant coefficients to code in a macroblock. To do this we derive the model for rate and distortion in terms of the number of encoded coefficients with pre-computed quantization accuracy. Rate-Distortion trade-off is solved then by the Lagrange optimization and the formula is obtained that indicate how many coefficients should be coded.

Keywords. Video coding, rate distortion optimization, H.263, MPEG-4 quantization

1   Introduction

In typical block-based video coders, such as those standard based H.263[1][6] and MPEG[2] and experimental implementations[12][13], the current video frame to be encoded is decomposed into macroblocks of 16x16 pixels per block and the pixel values are transformed into set of coefficients using discrete transform. These coefficients are then quantized and encoded with some type of entropy coding. The number of bits and distortion for a given macroblock depend on the macroblock quantization parameter QP. If the quantization parameter is changed on the macroblock basis, there is additional bit expense. For instance in H.263, two bits are needed to indicate the existence of the differential quantization parameter. Furthermore two additional bits need to be transmitted in order to change QP. For the low bit rate applications this overhead consumes a bandwidth and could cause a loss in compression efficiency. The proposed algorithm pre-computes the frame level quantization parameter QP. The macroblock level quantization is controlled by the number of significant coefficients that should be encoded with QP threshold.

W. Skarbek (Ed.): CAIP 2001, LNCS 2124, pp. 118–124, 2001.

2   Frame Level Quantization

The optimal quantization parameter QP for the whole frame is computed after the motion estimation and compensation (MCP) and coding mode decision. QP can be computed as in TMN8 [1] rate control algorithm based on the analytical model of rate and distortion functions. Alternatively the QP can be found by applying the Lagrangian optimization on the cumulative operational rate – distortion curve that is gathered during the additional scanning of transformed coefficients for each of the macroblock. The gathered statistic defines the number of significant coefficients subject to the quantization threshold $M(QP)$. Knowing the budget R for the current frame and average cost per the significant coefficient r for the previous frame, the number of significant coefficients to code is estimated as $M = R/r$. Thus the frame level quantization parameter QP is chosen from the M(QP) characteristics. In the computation the budget $R_T > R$ can be used to assure that the target budget B is reached in final allocation step.

3   Rate and Distortion Models

The average cost of encoding the significant coefficient is modeled by the following formula[11]:

$$\frac{R_i}{M_i} = C + \log_2 \frac{N}{M_i} \tag{1}$$

where:
C - is an empirical model constant
N - is the block size, e.g. 64, 256.
M_i - is the number of the significant coefficients to code
R_i - is the estimated bit cost per block.
The distortion function is described by the simple formula:

$$D_i = L \frac{D_{max\,i}}{M_i} \tag{2}$$

Where:
$D_{max\,i}$ denotes maximum distortion value determined during MCP and mode decision for the i-th block.
L - is an empirical model constant.
We want to find the expression for the number of the significant coefficients $M_1, \ldots, M_K$ that minimizes the distortion in (2) subject to the constraint that the total number of bits, i.e. the sum of the macroblock Ri must be equal to R:

$$M_1,...,M_K,\lambda = \arg\min_{M_1,...,M_K,\lambda}$$ (3)

$$\sum_{i=1}^{K} L\frac{D_{max\,i}}{M_i} + \lambda \sum_{i=1}^{K} M_i C + M_i \log_2 \frac{N}{M_i} - R$$

By setting partial derivates to zero in (3), we obtain the following expression for the optimized number of significant coefficients to code:

$$M_i$$ (4)
$$\frac{-\lambda N \log_2 e + \sqrt{(\lambda N \log_2 e)^2 + 4\lambda L D_{max\,i} C'}}{2\lambda C'}$$

where $C' = C - 2\log_2 e$ and λ is Lagrange multiplier.

The value of λ can be computed from the following equations:

$$\frac{D_i}{M_i} = -L\frac{D_{max\,i}}{M_i^2} = -\frac{D_i}{M_i}$$ (5)

$$\frac{R_i}{M_i} = C + \log_2 \frac{N}{M_i} - \log_2 e = \frac{R_i}{M_i} - \log_2 e$$ (6)

Thus

$$-\sum_{i=1}^{K} \frac{D_i}{M_i} + \lambda \sum_{i=1}^{K} \frac{R_i}{M_i} - K\log_2 e = 0$$ (7)

$$\lambda = \frac{\sum_{i=1}^{K} d_i}{\sum_{i=1}^{K} r_i - K\log_2 e} \quad \frac{\overline{d}}{\overline{r} - \log_2 e}$$ (8)

$\overline{d}$ - denotes average distortion per coded coefficient in the previous frame
$\overline{r}$ - denotes average bit cost per significant coefficient in the previous frame.

4   Rate Control Algorithm

The available bit budget R for the current frame is computed in frame layer rate control that is realized as in TMN8 [1] rate control algorithm. After the budget B is defined we apply the search for the optimal frame quantization parameter QP. The number of significant coefficients is computed on the macroblock basis by applying expressions (4) and (8) and updating the average distortion and bit cost per encoded coefficient. Below the whole algorithm is outlined:

```
1. Compute current bit budget for the whole frame based
   on frame layer rate control algorithm.
```

2.   Compute the characteristic M(QP).

3. Define the optimal QP for the current frame based on M(QP), average bit cost per significant coefficient and budget R.

4. For each of the block:

Compute the number of significant coefficients to code based on expression (4) and (8).

Set all remaining $N-M_i$ coefficients in the current block to zero.

Code current block with QP accuracy.

Update the model parameters $\overline{d}$ and $\overline{r}$, C and L after encoding the current block

5. Update the model parameters $\overline{d}$ and $\overline{r}$, C and L after encoding the whole frame.

The advantage of the presented algorithm is that it operates with fixed quantization parameter QP and does not use DQUANT parameter to signal quantization changes on the macroblock level. The varying part is the number of significant coefficients that are coded with QP accuracy. Such quantization technique can be understood as non-linear approximation of an original signal f(x) with help of a few basis vectors from a base   that best approximate f(x) in that base[11].

5   Applicability of the Algorithm to DCT and Wavelet Based Video Coders

The presented algorithm can be applied both to standard DCT based video coders and proprietary wavelet-based video coders.
In DCT coding the quantization parameter QP should take values from the quantization range. The macroblock mode decision can be reduced to modes 3 of INTRA, INTER and INTER_4V and excluding the INTRA_Q, INTER_Q, and INTER4V_Q modes. The DQUANT parameter is never used. The bit savings done in syntax can be allocated for better texture encoding.
For the embedded wavelet based coding[13] the quantization parameter shall be the power of 2 $QP=2^n$. For the proper reconstruction a decoder needs to know the maximum threshold 2^{max} of a significant coefficients and the number of bits after

which it should stop. In [13] the idea of an embedded wavelet coder was presented that can be stop at self described End of Coding point (EOC). Because all of the encoded coefficients are represented with the same accuracy there is no need to send to the decoder the information about the maximum threshold.

6   Experimental Results

In figure 1 there is presented a plot of the average cost of the coding of a significant coefficient. It is seen that the bit cost of a significant coefficient varies slowly in function of the quantization parameter QP. The bigger changes are experienced for very large QP values. In that case a lot of macroblocks are quantized to zero, and the coefficient statistics is not very dense. In figure 2 there is a plot that displays the changes of the average cost on the macroblock basis. It is seen that changes are quite big and therefore for the proper bit allocation the macroblock level model update is necessary. Figure 3 displays the comparison of the number of significant coefficients that were computed with the new algorithm and the number of significant coefficients that were coded in TMN8 rate control algorithm in H.263 codec. Generally TMN8 rate control algorithm computes QP parameter. Significant coefficients are those that are above 2QP. Although the exact comparison is not possible, the TMN8 model can serve as very good new model verifier. It is seen that both curves have similar characteristics.

7   Conclusions

In this paper the rate distortion model was presented that depends on the number of significant coefficients instead of the quantization parameter. The model was verified with the state of the art TMN8 rate control algorithm. The results shows that it can be used in rate control algorithm for low bit rate video coding. Currently the rate control is under research and it is expected that the frame by frame PSNR and buffer occupancy characteristics resulted in using the presented model would also be very close to those produced by TMN8 rate control. Because the new model allows bit savings in video syntax we expect that our proposal could outperform TMN8 results for the low bit rate video coding.

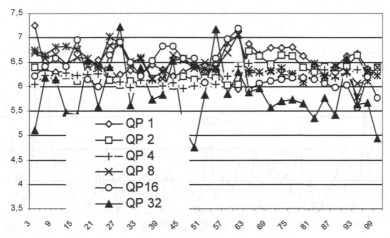

Fig. 1. The average cost per frame of the encoding significant coefficient in foreman.qcif sequence for various quantization parameters. The sequence was coded with variable rate at 10fps.

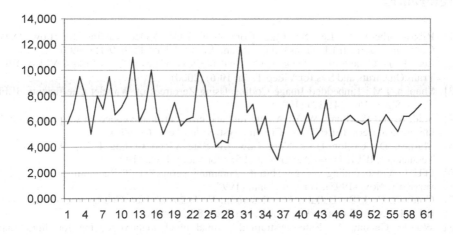

Fig. 2. The average cost per macroblock of the encoding significant coefficient in foreman.qcif sequence for 99 frame. The sequence was coded with variable rate at 10fps, QP = 10.

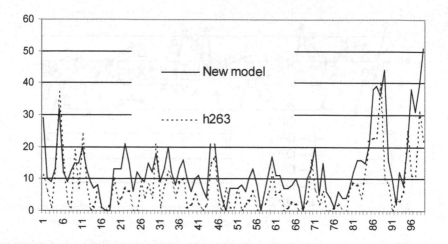

Fig. 3. The verification of the new model. Figure plots the number of encoded coefficients in H.263 coder and the number of predicted significant coefficients in the new model for 99 frame of foreman.qcif. H.263 coder was working TMN8 rate control with 30kbs, 10fps.

References

[1] Ribas-Corbera, J., Lei S.: Rate Control in DCT Video Coding for Low Delay Communications. IEEE Trans.On Ciruts and Sys.for Video Tech. **9(1)** (1999)
[2] Lee, H.-J., Chiang, T., Zhang T.-Q.: Scalable Rate Control for MPEG-4 Video. IEEE Trans.On Ciruts and Sys.for Video Tech. **10(6)** (2000)
[3] Shapiro, J.M.: Embedded Image Coding Using Zerotrees of Wavelet Coefficients. IEEE Trans. Signal Proc. **41(12)** (1993)
[4] Said, A. and Pearlman, W.A.: A new fast and efficient image codec on set partitioning in hierarchical trees. IEEE Trans. Circuits Syst. Video Tech. **6** (1996)
[5] Marphe, D., Cyclon H.L.: Very Low Bit Rate Video Coding Using Wavelet-Based Techniques. IEEE Trans.On Ciruts and Sys.for Video Tech. (1997)
[6] ITU-T, Video coding for low bitrate communication. ITU-T Recommendation H.263, version 1, Nov. (1995), version 2, Jan. (1998)
[7] Cote, G., Erol, B., Gallant, M., Kosentini, F.: H.263+:Video Coding at Low Bit Rates. IEEE Trans.On Ciruts and Sys.for Video Tech. **8(7)** (1998)
[8] Wu, S., Gersho, A.: Rate-constrained optimal block-adaptive coding for digital tape recording of HDTV. IEEE Trans. On Circuts and Sys.for Video Tech. **1** (1991) 100-112
[10] Sullivan, G.J., Wiegand, T.: Rate-Distortion Optimization for Video Compression. IEEE Sig. Proc. Mag., Nov. (1998)
[11] Mallat, S., Falzon, F.: Analysis of Low Bit Rate Image Transform Coding. IEEE Trans.Sig. Proc, April (1998)
[12] Siemek, G.: Rate–distortion optimization of arbitrary set of embedded wavelet coders for low bit rate video compression. Recpad, Portugal, Porto (2000)
[13] Siemek, G.: The SPIHT coder implementation for low bit rate video coding. Accepted to ICASSP2001, Salt Lake City, Utah, USA, 7-11 May (2001)

Shape-Adaptive DCT Algorithm – Hardware Optimized Redesign

Krzysztof Mroczek

Warsaw University of Technology, Institute of Radioelectronics
ul. Nowowiejska 15/19, 00-655 Warsaw, Poland
K.Mroczek@elka.pw.edu.pl

Abstract. This article refers to the shape-adaptive DCT (SA-DCT) algorithm developed by Sikora and Makai in 1995. It is an important tool for encoding texture of arbitrary shaped video objects and can be included in MPEG-4 video codecs. In this paper a modification of normalized version SA-DCT redesigned for intraframe coding is presented. Simulations results show that this solution outperforms standard SA-DCT in rate-distortion sense. Efficiency is close to improved version SA-DCT for intraframe coding, known as ΔDC-SA-DCT. But computational overhead is smaller than for ΔDC-SA-DCT. Therefore, this solution may be attractive for hardware implementations.

Keywords: shape-adaptive DCT, hardware design

1. Introduction

The SA-DCT algorithm, introduced in 1995 [1], is one of the most promising approach for shape adaptive transforms. This method is a relative low complexity solution which outperforms padding methods in rate-distortion sense [2]. For natural video SA-DCT efficiency is close to more complex solutions as Gilge transforms [3]-[4]. Another important property of SA-DCT is backward compatibility with standard DCT for square NxN blocks. Due to these benefits this transform can be used in object and region-based video coding systems. It has been also included in MPEG-4 video verification model (VM)[5].

The basic idea of SA-DCT includes several steps. First pixels in every vertical column are aligned by shifting to the upper border. After the shifting, vertical 1D SA-DCT transform is performed on every column data vector X_j with $L = N_j$ (1) elements, according to:

$$B_j = S_L \ DCT_L \ X_j \tag{2}$$

where

$$DCT_L(i, j) = \gamma \cos \ i(j + 1/2)\frac{\pi}{L} \tag{3}$$

and $\gamma = 1/\sqrt{2}$ for i = 0 and $\gamma = 1$ else.

W. Skarbek (Ed.): CAIP 2001, LNCS 2124, pp. 125–133, 2001.
© Springer-Verlag Berlin Heidelberg 2001

S_L represents scaling factor and for vertical vectors depends on N_j. Next 1D column coefficients $b_{i,j}$ are shifted horizontally toward left border giving $c_{i,j}$. In the last step 1D transforms are performed on $L = M_i$ element raw vectors C_i:

$$D_i = S_L \ DCT_L \ C_i \tag{4}$$

where D_i is i-th raw vector of final 2D SA-DCT coefficients.

For inverse transform respective steps are performed in reverse order and the transform 1D formulas are:

$$C_i^* = \frac{2}{L \ S_L} \ DCT_L^T \ D_i^* \qquad X_j^* = \frac{2}{L \ S_L} \ DCT_L^T \ B_j^* \tag{5,6}$$

for raw ($L=M_i$) and column ($L=N_j$) vectors.

Original algorithm formulation [1] represents dc-preserving transform and has been optimized for intraframe coding, when mean value contains typically most of signal energy. In this case

$$S_L = 4/L \tag{7}$$

This version of transform is called also nonorthonormal SA-DCT (NO-SA-DCT), because 1D transforms are not normalized. It is possible to normalize the basis functions of SA-DCT, by setting

$$S_L = \sqrt{\frac{2}{L}} \tag{8}$$

Unfortunately, this transform, called pseudoorthonormal SA-DCT (PO-SA-DCT), isn't dc-preserving. As shown in [2] PO-SA-DCT is an optimal solution only for interframe coding only, where mean value of motion compensated image segments is typically close to 0. But using PO-SA-DCT to intraframe coding is not recommended, because of mean weighting defect artifact. Kauff and Schüür proposed the extension of PO-SA-DCT suitable for intraframe coding by separating 2D DC value which is applied to the initial data segment before processing forward PO-SA-DCT [3]. For this extension, called ΔDC-SA-DCT, first single mean value of 2D image segment is calculated of all pixels belonging to the segment. Next this mean is substracted from all initial segment data and 2D PO-SA-DCT is applied. Finally, 2D DC value is replaced by scaling mean value. While performing the inverse 2D transform appropriate ΔDC correction procedure is required.

In this paper another modification of PO-SA-DCT is presented to achieve dc-preserving property required for intraframe coding. This modification based on 1D DC separation and appropriate scaling, called 1DDCS-SA-DCT, has performance close to ΔDC-SA-DCT and better than NO-SA-DCT. But computational overhead for 1DDCS-SA-DCT is smaller than for ΔDC-SA-DCT. Therefore, this solution may be attractive for hardware implementations, such as, dedicated transform cores.

The paper is organized as follows: First, the modification idea and the architecture of 2D 1DDCS-SA-DCT are described and the benefit for transform computational complexity is shown. Next efficiency comparison of the considered SA-DCT algorithms and LPE padding method, based on rate-distortion simulation results of several MPEG-4 video sequences, are presented. And finally some discussion about hardware implementation aspects is presented.

2. Extension of PO-SA-DCT toward 1D DC Separation (1DDCS-SA-DCT)

To avoid mean weighting defect of PO-SA-DCT, several additional processing steps are applied as shown in Fig 2. First mean values mc_j for pixels from every column vector X_j are calculated in encoder:

$$mc_j = \frac{\sum\limits_{i=0}^{N_j-1} x_{i,j}}{N_j}, \; j= 0,..,N\text{-}1 \tag{9}$$

In parallel column forward transforms (2) are performed. Next DC coefficients $b_{0,j}$ are replaced with mc_j*f, where f represents positive scaling factor. Finally, standard raw PO-SA-DCT (4) are applied to raw vectors C_i giving $d_{i,j}$ coefficients, which have the same values as in case of standard 2D PO-SA-DCT procedure, except for first raw $d_{0,j}$. It's easy to prove that the results are the same as in case of processing input zero mean data vectors $x'_{i,j} = x_{i,j} - mc_j$. Then every column DC coefficient would be reset:

$$b_{0,j} = 0 = 1 / \sqrt{N_j} \sum\limits_{i=0}^{N_j-1} x_{i,j} - N_j \; mc_j \tag{10}$$

According to $DCT_L(i,j)$ (3) matrix property, remaining coefficients $b_{i,j}$, $i \neq 0$ would have the same values as in case of applying PO-SA-DCT to original X_j vector. It means that it is possible to omit the preparation of zero mean vectors in encoder and perform reverse correction in decoder side.

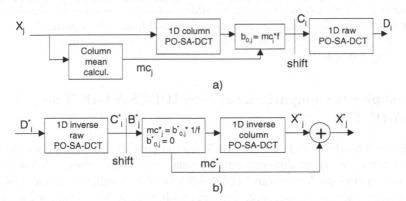

a)

b)

Fig. 1. Block diagram of 1DDCS-SA-DCT. a) Forward transform. b) Inverse transform. Mean correction is performed in the last processing stage, instead of after inverse raw transforms.

The foregoing guarantees that 1DDCS-SA-DCT is a dc-preserving transform, which means that a transform gives only one nonzero coefficient for constant-valued input signal [3]. When an image data segment has nonzero mean value ms, the 2D $d_{i,j}$ coefficients can be presented as sum of two elements:

$$d_{i,j} = dms_{i,j} + ds_{i,j} \qquad (11)$$

where $dms_{i,j}$ represents coefficients after transforming uniform color segment of ms values and ds_i, represents transform results for zero mean data segment $xs_{i,j} = x_{i,j} - ms$. When 1D column transform is applied to constant-valued vectors, the only 1D dc value is set to $ms*f$. Other coefficients are equal to 0. After performing raw transforms, only the first raw vector containing column DC values can give nonzero output. Because all column DC coefficients have the same value the only nonzero coefficient is $dms_{0,0}$. It means that dc- preserving property is accomplished and whole segment mean energy is packed to 2D DC coefficient.

For inverse transform the simple reverse procedure is performed for correction, as shown in Fig. 2b. After every horizontal inverse transform a mean value mc^*_j is extracted by multiplying $b^*_{0,j}$ coefficient by $1/f$ and after that $b^*_{0,j}$ coefficient is reset. Next standard column inverse transform is performed for every data vector. Finally, the previously extracted mean values are added to every output coefficient for appropriate vector.

From VLSI point of view it seems that one dedicated circuit for both standard and shape-adaptive DCT/IDCT procedures could be a good solution for region based coding. If we set scaling factor

$$f = 4/\sqrt{2} \qquad (12)$$

described above correction procedure guarantees compatibility with standard DCT for 8 x 8 blocks. It is possible to apply fixed correction circuit which will perform computations for all values of L without additional control overhead or easily turn it off for full 8x8 blocks. Single fixed constant for multiplication is preferable, because area cost of multiplier circuit would be small. As shown in section 4 it gives also good coding efficiency.

3.   Complexity Comparison between 1DDCS-SA-DCT and ΔDC-SA-DCT

Both 1DDCS-SA-DCT and ΔDC-SA-DCT use the same 1D transform core and there are differences in correction procedures, data and control flows. A number of arithmetic operations for forward 1DDCS-SA-DCT is smaller than for ΔDC-SA-DCT, because there is not preprocessing stage with preparation zero mean data, which need maximum N^2 additions in ΔDC-SA-DCT. Also it is not necessary to compute N inner products for column DC coefficients, which are replaced with scaled mean values. In both cases there is similar number of additions for sum computation. But for this new proposal maximum N additional multiplications are required for computation of N different mean values, while there is one multiplication for ΔDC-SA-DCT. Multiplications by appropriate inverses are useful to avoid division. The inverses of vector or segment length can be stored in LUT or ROM. Because of vector oriented computation flow in this new approach, only N values are needed, while for ΔDC-SA-DCT there are N^2 different constants. The computation of column means may be performed in parallel with column transforms computation for the same data vectors. There is enough time for computation of mean value before using it in correction stage. For this reason it is easy to pipeline operations for high speed architectures and reduce register files length. For ΔDC-SA-DCT, first segment mean calculation have to be performed before applying 2D transform.

For inverse 1DDCS-SA-DCT correction procedure is very simple and for means extraction only N multiplications and N reset operations are needed. Resets may be simply avoided by selecting 0 instead. Extracted mean values can be stored in N registers and used in the last correction step. Finally $N^2 + N$ simple arithmetic operations are needed which do not use any shape signals. Due to multiplication by constant value the area of additional multiplier would be significantly reduced.

Another possibility is to omit N^2 additions in the last stage and to apply only N rescaling multiplications:

$$b_{0,j}^* <= b_{0,j}^* \frac{\sqrt{N_j}}{f} \tag{13}$$

But access to shape parameters $\sqrt{N_j}$ and signal by signal multiplication is required.

For ΔDC correction procedure in inverse ΔDC-SA-DCT there are more arithmetic operations and shape signals have to be processed with access to $\sqrt{N_j}$ values. It is possible to perform ΔDC correction after inverse raw subtransform. Then maximum $3N-2$ additions, N multiplications and 1 division are needed [3]. For final step also N^2 additions are needed. It is also possible to perform correction in post-processing mode after computation of whole 2D transform [5]. In this case maximum $2N^2+2N-1$ additions, N multiplications and 1 division are required. In each case computational overhead for 1DDCS-SA-DCT is smaller than for ΔDC-SA-DCT.

4. Experimental Results

To check 1DDCS-SA-DCT efficiency, several experiments have been made using MoMuSys MPEG-4 Verification Model software. Various segmented image sequences have been coded and rate-distortion characteristics for particular video objects have been analyzed. In addition to the proposed algorithm, in figures named as 1DDCS, three approaches: ΔDC-SA-DCT (DDC), NO-SA-DCT (NO) and LPE padding[1] (LPE) were simulated. Comparison between 1DDCS and PO-SA-DCT (PO) for intra coding was also performed. The following experimental setup was used:

---

[1] after setting SADCT_disable flag to 1

- 100 first frames of every sequence was selected
- QCIF and CIF sequences (AKIYO, BREAM, WEATHER, CHILDREN) were coded at 10 Hz. CCIR sequence NEWS was coded at 30 Hz.
- Five different quantization values were chosen.
- For inter macroblocks PO-SA-DCT was applied. For intra macroblocks one of four SA-DCT based transforms was set.
- For 1DDCS scaling factor $f = 4/\sqrt{2}$

In the main experiment every frame was coded in intra mode. The number of bits for coding luminance component of all boundary VOP blocks was evaluated with achieved PSNR. Also the total number of bits for VOP luminance component with PSNR were computed. Bitrate values were averaged over 100 frames. As a result rate-distortion curves for sequence WEATHER QCIF, CHILDREN CIF and NEWS CCIR are depicted in Fig. 3. For boundary blocks, 1DDCS-SA-DCT outperforms NO-SA-DCT, PO-SA-DCT and LPE method and is close to ΔDC-SA-DCT. The gain compared to NO-SA-DCT is up to 0.6 dB. Also the advantages of this solution are visible in comparison with PO-SA-DCT applied to intra blocks, then the gain is up to 1 dB. The difference between ΔDC-SA-DCT and 1DDCS-SA-DCT is maximum 0.5 dB. When all blocks are considered, the differences between particular transforms are relatively smaller. In this case maximum difference between ΔDC-SA-DCT and 1DDCS-SA-DCT is less than 0.1 dB and depends on ratio of boundary and full object blocks. When this ratio is small, i.e. for large objects, using 1DDCS-SA-DCT gives very close results to ΔDC-SA-DCT. In every case PSNR values for both boundary and all blocks are higher for 1DDCS-SA-DCT comparing with ΔDC-SA-DCT for the same quantization parameters. Moreover, it is still possible to change rate-distortion characteristics for 1DDCS-SA-DCT when any adaptation of factor f is possible. It was checked that setting $f < 4/\sqrt{2}$ decreases bitrate.

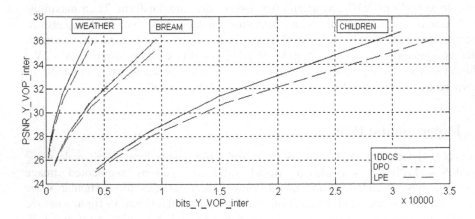

Fig. 2. Rate-distortion curves for interframe coded sequences WEATHER, BREAM (QCIF) and CHILDREN (CIF). Results are averaged over all boundary and object blocks

Similarly small differences are noticed when second experiment was performed and encoder was switched to interframe mode, when only the first frame was encoded

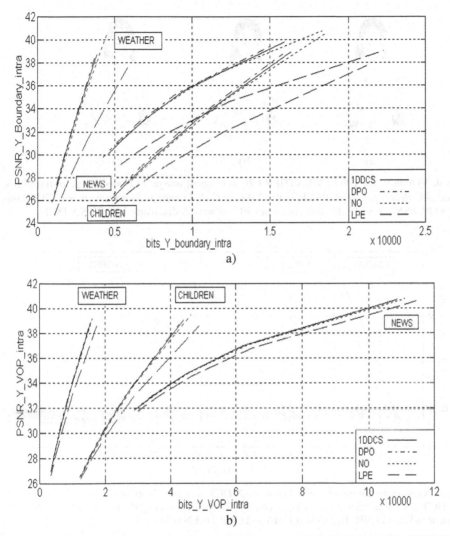

Fig. 3. Rate-distortion curves for intraframe coded sequences WEATHER QCIF, CHILDREN CIF and NEWS CCIR using 1DCCS-SA-DCT, PO-SA-DCT, NO-SA-DCT and LPE. a) Averaged results over boundary blocks only. b) Over all boundary and object blocks

in intra mode whereas others in inter mode (IPPP...). For all these sequences, the differences between using ΔDC-SA-DCT and 1DDCS-SA-DCT are unnoticeable. In Fig. 2 rate distortion curves for inter boundary and all blocks are depicted for WEATHER, BREAM (both QCIF) and CHILDREN CIF sequence. Taking into account typical situation for coding video sequences, when also motion compensation prediction is useful, the efficiency of 1DDCS-SA-DCT is practically the same as for ΔDC-SA-DCT and significantly better than LPE padding.

a)   b)   c)

Fig. 4. VOP in frame 100 of WEATHER CIF a) Original image. b) Coded image in intraframe mode, 48276 bits per VOP, PSNR = 40.17 dB. c) Coded image in interframe mode, 19152 bits per VOP, PSNR = 38.15 dB. Intra boundary blocks were coded using 1DDCS-SA-DCT.

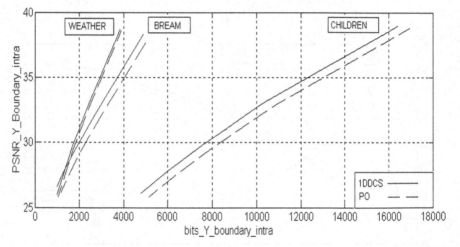

Fig. 5. Gain comparison between 1DDCS-SA-DCT and pseudoorthonormal SA-DCT (PO-SA-DCT) when applied to intraframe coding. Results are averaged over boundary blocks for sequences WEATHER, BREAM (QCIF) and CHILDREN (CIF)

5. A Brief Discussion of Implementation Issues

When VLSI implementations are considered, the efficient solution for both cost-effective and high throughput architectures for DCT/IDCT is fixed point arithmetic usage for computations. Typically, several scalar processing units for inner product computation can work in parallel to achieve desirable throughput. When fixed point arithmetic is used, both multiplier-based (MAC) and distributed arithmetic (DA) based datapath elements can be developed for inner product computations. Architectures either in the regular "direct" forms or in "mixed" forms are possible for SA-DCT/IDCT. For "mixed" forms 1D *LxL* matrices are decomposed to several stages and number of multiplication is decreased. Both ROM-based DA and adder based DA architectures [6]-[7] can be applied.

6. Conclusion

The basic idea of the presented modification is to avoid the mean weighting defect of normalized SA-DCT by appropriate rescaling of column DC coefficients, which may be realized by column mean values separation. It is easy to perform correction procedure in decoder side without access to shape parameters. For this procedure the number of arithmetic operations is significantly smaller than for ΔDC-SA-DCT and there is no division. The computational complexity is close to standard SA-DCT algorithm. The data flow is regular and it's easy to implement simple fixed correction circuit for high throughput architectures. Simulation results show that the efficiency of this method is close to ΔDC-SA-DCT. Therefore this solution may be attractive for hardware realizations, like processor cores, designed for standard and shape-adaptive DCT/IDCT for both intra- and interframe coding, where constant coefficients are the same as for normalized SA-DCT.

References

1. Sikora, T., Makai, B.: Shape-Adaptive DCT for Generic Coding of Video. IEEE Transactions on Circuit and Systems for Video Technology **5** (1995) 59-62
2. Kaup, A.: Object-Based Texture Coding of Moving Video in MPEG-4. IEEE Transactions on Circuit and Systems for Video Technology **9** (1999) 5-15
3. Kauff, P., Schüür, K.: Shape-Adaptive DCT with Block-Based DC Separation and ΔDC Correction. IEEE Transactions on Circuit and Systems for Video Technology **8** (1998) 237-242
4. Gilge, M., Engelhardt, T., Mehlan, R.: Coding of Arbitrarily Shaped Image Segments Based on Generalized Orthonormal Transform. Signal Processing: Image Communications **1** (1989) 153-180
5. MPEG-4 Video Verification Model version 16.0, ISO/IEC JTC1/SC29/WG11
6. Pirsch, P., Demassieux, N., Gehrke, W.: VLSI Architectures for Video Compression – a Survey. Proceedings IEEE, February (1995) 220-246
7. Chang, T.S., Kung, C.S., Jen, C.-W.: A Simple Processor Core Design for DCT/IDCT, IEEE Transactions on Circuit and Systems for Video Technology, April (2000) 439-447

Superquadric-Based Object Recognition

Jaka Krivic and Franc Solina

University of Ljubljana
Faculty of Computer and Information Science
Computer Vision Laboratory
Tržaška 25, 1000 Ljubljana, Slovenia
{jakak, franc}@lrv.fri.uni-lj.si
http://lrv.fri.uni-lj.si

Abstract. This paper proposes a technique for object recognition using superquadric built models. Superquadrics, which are three dimensional models suitable for part-level representation of objects, are reconstructed from range images using the recover-and-select paradigm. Using an interpretation tree, the presence of an object in the scene from the model database can be hypothesized. These hypotheses are verified by projecting and re-fitting the object model to the range image which at the same time enables a better localization of the object in the scene.

Keywords: superquadrics, part-level object modeling, range images

1 Introduction and Motivation

In computer vision, many different models have been used for describing different aspects of objects and scenes. Part-level models are one way of representing 3D objects, when particular entities that they describe, correspond to perceptual equivalents of parts. Several part-level shape descriptions are required to represent an articulated object. One of the more popular types of volumetric part-level descriptions are superquadrics [1,11,13,3,12,5]. They are solid models that represent standard geometrical solids as well as shapes in between.

Pentland [11] was the first who used superquadrics in the context of computer vision. However, Solina and Bajcsy's method for recovery of superquadrics from pre-segmented range images became more widespread [5]. Also, several methods for segmentation with superquadrics have been developed. A tight integration of segmentation and model recovery was achieved [9] by combining the "recover-and-select" paradigm [8] with the superquadric recovery method [13]. The paradigm works by recovering independently superquadric part models everywhere on the image, and selecting a subset which gives a compact description of the underlying data. *Segmentor* is an object-based implementation of the "recover-and-select" segmentation paradigm using superquadrics and other parametric models [5]. Superquadrics, their mathematical properties, recovery from images and its applications are presented in detail in [5].

W. Skarbek (Ed.): CAIP 2001, LNCS 2124, pp. 134–141, 2001.

Segmentor lacks the reliability of segmentation of rough, natural shapes, which are not very close to ideal superquadric shapes. The superquadric models can not expand easily on rough surfaces and complex shapes as easily as on smooth regular objects which results generally in over-segmentation. Nevertheless, our starting hypothesis was that the part-level description obtained by the *Segmentor* system (see Fig. 2a and 2d) is good and stable enough for recognition of part-level models. The configuration of parts and their rough shape should provide enough constraints for successful matching with the models of possible objects. The object hypotheses can be subsequently verified by fitting the object model directly to the range data.

2 Object Recognition Scheme

The results of processing a range image with the *Segmentor* is a set of superquadric models, each with known position, orientation, size and shape. Reconstructed superquadrics represent parts that constitute the scene. An object recognition system that searches for matches between parts of the scene and parts of the modeled object can be used, called "model based matching" [3,10, 2,4]. For identifying the object in the scene we adopted the interpretation tree method [4,14]. Our object recognition system uses the following three steps:

1. range image segmentation and superquadric recovery using the *Segmentor*,
2. search for feasible interpretations of the stored model in the scene using interpretation trees, and
3. hypothesis verification by projecting object models into the scene.

The second and third step can be interleaved, to early eliminate those hypothesis that do not make sense.

2.1 Object Model

We decided to use human figures as generic articulated test objects for our recognition task. We were not interested in the specific problem of modeling human form, although systems using superquadrics for modeling humans do exist [6,7]. Since the workspace of our range scanner is rather small, toy figures representing "Commander Data" from the Star Trek series (Fig. 1 a) were used. Their arms and legs are flexible and the figurines can thus be configured into many different poses.

We built the model of the figurine manually. The model consists of superquadrics (Fig. 1 b). Each superquadric represents one of the major body parts: head, torso, a pair of upper and lower arms, and a pair of upper and lower legs. Due to the limited scale of parts which can be recovered on the selected range image resolution by the *Segmentor*, the model does not include distinct models of hands and feet. Each body part is described by a superquadric of a particular size and shape. The torso is given a central role in the model. The head and upper arms and legs are attached to it via joints (Fig. 1 c)). For each

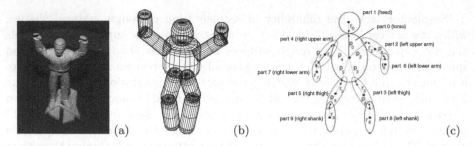

Fig. 1. (a) The object. The object model consists of two levels: (a) superquadric part models define the size and shape of individual parts, (b) the structural level defines how parts are connected to each other.

of those parts a joint position relatively to the center of the part itself (r_i) and to the center of the torso (p_i) is defined. Similar is true for lower extremities. The parameter values for all parts were obtained by measuring the figurine.

In this paper, p_i and r_i stand for the relative positions of the center of the part i in the model, as described above, c_i is the center of superquadric that matches, or should match part i, R is a ZYZ rotation matrix, and ϕ_i, θ_i and ψ_i are rotation angles for part i.

The test figurine is interesting in several ways. It is fairly realistic and therefore cannot be perfectly modeled by superquadrics. Since the surfaces are not smooth, the reconstruction of superquadrics on their range images is less stable. The flexibility of body joints makes the matching problem more complex than if the object part configuration would be rigid.

2.2   Model Matching

The input to model matching is the set of superquadric models resulting from the *Segmentor* and the stored object model. The process of recognizing an object can be viewed as matching scene parts with part models of the stored body model which are arranged in an interpretation tree. The search for correct interpretation begins at the root. The root expands to all possible matches for the first model part. When the search reaches a leaf one gets a consistent interpretation. But because the constraints involved in checking consistency of a match are local in nature, the interpretation does not have to make sense globally. In general, there is no guarantee that a found interpretation makes global sense. These interpretations must therefore be taken only as hypotheses and should be further verified.

Checking the Local Consistency of a Match. Reconstructions of superquadrics on the object in different poses and different viewpoints differ greatly, except in the case of the head. Since superquadric parameters cannot be directly used for the comparison of two superquadrics [5], we decided to base the consistency check merely on the volume of superquadrics. If a volume of a scene part is within the interval given by the model part, the two parts represent a possible match.

Matching the Head. The analysis of superquadric reconstructions of the human body showed that the head was the most consistently reconstructed body part. At the same time, the head is also the only part that does not change significantly its relative position in relation to the torso. Therefore, we found it reasonable to define constraints for the size and shape parameters of superquadrics, in order to reject as many unsuitable parts as possible. The values for the size and shape parameters were defined on the basis of thirty reconstructions.

Matching the Torso. Superquadrics reconstructed on the torso region differ the most from the torso's model superquadric. On this region several possibly overlapping superquadrics can be recovered, which can partially cover even regions belonging to extremities. Different volume interval was used for the cases of a single superquadric and the case of overlapping superquadrics

If there is also a head in the interpretation, one can more precisely compute the position and (although not complete) rotation parameters for the whole model. This information can be used to further constrain the search for interpretations.

The analysis of the reconstruction showed, that the centers of superquadrics from the torso region were fairly close to the real vertical axis of the torso, whereas the distance to the real center of the torso was not. This can be corrected by taking into account the distance between the centers of the head and the torso. Orientation parameters can be resolved by computing the rotation, that transforms the z-axis into the vector $c_0 - c_1$. The parameter ψ_1 defines the rotation around the z-axis of the object itself.

Let c_0 be the center of the head's superquadric, and c_1' approximate center of the torso, as mentioned earlier. Then a better approximate center of the torso c_1 is

$$c_1 = c_0 - |s_0|s \tag{1}$$

$$\phi_1 = \arctan \frac{-s_x}{-s_y} \tag{2}$$

$$\theta_1 = -\arctan \frac{\sqrt{s_x^2 + s_y^2}}{s_z}, \tag{3}$$

where

$$s_0 = r_h - p_h$$
$$s = \frac{c_0 - c_1'}{|c_0 - c_1'|}. \tag{4}$$

c_1', c_1 and c_0 are 3D vectors of the form $[x, y, z]^T$, s is a 3D vector $[s_x, s_y, s_z]^T$, whereas ϕ_1, θ_1 and ψ_1 are the angles that define the torso's orientation and thus the object's orientation.

Matching the Extremities. For consistency check of matches for upper arms and legs, two cases were distinct. In the first, there is a real match for the head as well as for the torso included in the interpretation. Therefore the parameters for position and (partially) orientation for the whole model is known and can be used for checking the consistency of a given match. When a superquadric from a scene is in the same volume range as the upper arm or leg part model, parameter ψ_1 can be computed. Parameter ψ_1 rotates the object part model so that the joint position on the object model approximately overlaps with the possible upper arm or leg

$$\psi_1 = \arctan \frac{s_y}{s_x} \qquad (5)$$

$$s = R^{-1}(c_x - c_1),$$

where c_x is the superquadric center, whose match with the upper arm or leg part model is being checked, and R is a ZYZ rotation matrix with parameters $(\phi_1, \theta_1, 0)$.

If the part in question is really an extremity, its center has to be approximately as distant from the joint as is the case with the object model. If the conditions are met, the match is consistent and rotation parameters for the part can be computed by

$$\phi_x = \arctan \frac{-s_x}{-s_y}, \qquad (6)$$

$$\theta_x = -\arctan \frac{\sqrt{s_x^2 + s_y^2}}{s_z}, \qquad (7)$$

$$\psi_x = 0. \qquad (8)$$

In the second case, there is a wildcard match for head or torso, so there is no information about the position and orientation of the model (or at least about orientation, if there is a real match for the torso in the interpretation).

The lower extremities were not included in the interpretation tree search because the interpretation can be verified quite well without them. Inclusion of lower extremities in the tree search would only increase the search space.

Interpretation Verification. The checking of the global consistency of an interpretation means that the system should answer the question: *"Does the given set of parts really represent object X?"*

We decided that the system should reject all interpretations that include less than four real matches. The system may, therefore, reject some correct interpretations (false negatives), but it will reject many more wrong ones (false positives), since there is a low probability that some parts will "randomly" form a structure similar to the structure of the human body.

To verify a hypothesis obtained by the interpretation tree we fitted individual superquadrics of the stored model to corresponding regions of the range image, which were defined by the superquadrics included into the tested interpretation. To fit individual superquadric models to such part regions the standard fitting method was used [13]. The fitting function was minimized only for the position and orientation parameters, the size and shape parameters were fixed to the values of the tested model part superquadric.

As in consistency check, there are two cases of interpretation verification that need to be considered. In the first case, there are real matches for both the head and the torso, and in the second case, at least one of those matches is a wild card. In the first case the position and orientation parameters for the model can easily be computed (Eqs. 1, 5). It turns out that those parameters fairly accurately describe the position of the model. In the second case, the position and orientation of the model could be computed based on the position of joints, that connect the extremities to the torso. But, since the position of the joints cannot be accurately defined based on the reconstructions, the position of the joints cannot be accurately computed as well. In fact, the computation did not deliver any reasonable results and thus we decided that the interpretation is only good if it includes real matches for both, the torso and the head.

Let us return to the interpretation verification by fitting the model parts to the corresponding regions of the range image. The model part parameters were used as initial parameters. The fitting is only performed for superquadrics that model the extremities, since they are elongated, and the fitted model defines the position of the joints well. When the superquadric is fitted, the distance between the joint position on the fitted superquadric and the initial one can be computed as

$$d = |(\mathbf{R}_i r_i + c_i) - (\mathbf{R}_f r_i + c_f)|, \tag{9}$$

where $\mathbf{R}_i$ is a rotation matrix, that rotates part i into initial position, r_i is the relative position of the joint, c_i is the center of initial superquadric, $\mathbf{R}_f$ is the rotation matrix, that rotates part i into the fitted position, and c_f is the center of the fitted superquadric. The computed distance d is compared with a threshold p, which was set to value $p = 11.0$, based on the analysis of reconstructions. If the distance d is greater than the threshold with at least one extremity, the interpretation is rejected.

Interpretation tree search does not include matches for lower extremities, that is lower arms and shanks. The presence of those parts is searched for only if the part attached to it (upper arm, thigh) is present, otherwise the results cannot be verified well. This parts are matched based on volume and joint distance.

3 Experimental Results

Range images which were obtained by a structured light range scanner and processed with the *Segmentor* system [5]. The resulting set of superquadric models were processed as described in the previous sections.

First, we tried to systematically test the system's performace for isolated figurines. A single figure was configured in seven different poses and eight range images were taken from different viewpoints. Figure 2(a-c) shows one of the results.

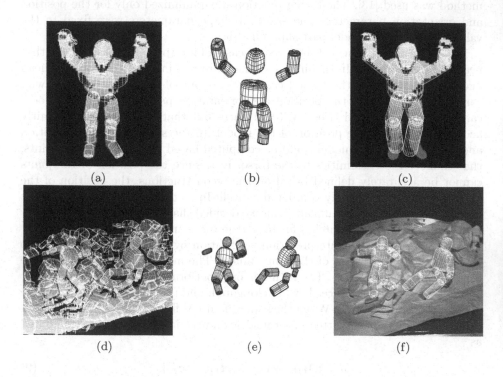

(a)   (b)   (c)

(d)   (e)   (f)

Fig. 2. Interpretation of a single-figurine scene (a,b,c) and a complex scene (d,e,f): (a,d) input range image with superimposed reconstructed superquadrics, (b,e) superquadrics from two hypothesis interpretations, (c,f) verified interpretations.

The object was detected in 39 out of 56 cases. In 24 of those 39 cases, the model computed from the best interpretation fitted the object very well. Interpretation included on the average 7.2 real matches. The object was not detected in 17 cases. In 9 of those cases, the reason for that was an occluded head or torso, so that the reconstructed superquadrics on those regions were not even close to the ones from the model. In 8 other cases the best interpretation found included less than four real matches, and was therefore rejected.

The system's performance was also tested on 20 different complex scenes. Complex scenes included several appearances of the toy figure, as well as many unknown objects (Fig. 2d-f). Nevertheless, there were no false positive recognitions of the human form, although there were many at least partially misleading configurations. It is much harder to test a complex scene in a systematic fashion because of so many possible variables. One can observe that the reconstructions of the supporting surfaces in complex scenes were not appropriate, because such surfaces cannot be modeled well by superquadrics.

4   Conclusions

In this paper, we have investigated if superquadric based shape decomposition can be used for object recognition. The system is based on interpretation trees. We have shown, that despite very rough and somewhat unstable part description, superquadrics can be used in an object recognition scheme. The system can handle flexible articulated objects that cannot be perfectly modeled by superquadrics which is demonstrated by the recognition of the human figure.

References

1. Barr, A.H.: Superquadrics and Angle-Preserving Transformations. IEEE Computer Graphics and Applications, 1 (January 1981) 11–23
2. Bolles R.C., Horaud. P.: 3DPO: A Three-Dimensional Part Orientation System. The International Journal of Robotic Research, 5 (1986) 3–26
3. Dickinson, S.J., Pentland, A.P., Rosenfeld, A.: From volumes to views: An approach to 3-D object recognition. CVGIP: Image Understanding, 55 (1992) 130–154
4. Grimson W.E.L.: Object Recognition by Computer. MIT Press, Cambridge, MA (1990)
5. Jaklič A., Leonardis A., Solina F.: Segmentation and Recovery of Superquadrics. Kluwer Academic Publishers, Dordrecht (2000)
6. Jojić N., Huang T.S.: Computer vision and graphics techniques for modeling dressed humans. In: Leonardis A., Solina F., Bajcsy R. (eds.): The confluence of computer vision and computer graphics. Kluwer, Dordrecht (2000) 179–200
7. Chella A., Frixione M., Gaglio S.: Understanding dynamic scenes. Artificial Intelligence, 123 (2000) 89–132
8. Leonardis A., Gupta A., Bajcsy R.: Segmentation of range images as the search for geometric parametric models. International Journal of Computer Vision, 14 (1995) 253–277
9. Leonardis A., Jaklič A., Solina F.: Superquadrics for segmentation and modeling range data. IEEE Transactions on Pattern Recognition and Machine Intelligence, 19 (1997) 1289–1295
10. Nevatia R., Binford T.: Description and Recognition of Curved Objects. Artificial Intelligence, 38 (1977) 77–98
11. Pentland A.P.: Perceptual organization and the representation of natural form. Artificial Intelligence, 28 (1986) 293–331
12. Raja N.S., Jain A.K.: Recognizing geons from superquadrics fitted to range data. Image and Vision Computing, 10 (1992) 179–190
13. Solina F., Bajcsy R.: Recovery of parametric models from range images: The case for superquadrics with global deformations. IEEE Trans. Pattern Anal. Machine Intell., 12 (1990) 131–147
14. Trucco E., Verri A.: Introductory techniques for 3-D computer vision. Prentice Hall, NJ (1998)

Weighted Graph-Matching Using Modal Clusters

Marco Carcassoni and Edwin R. Hancock

Department of Computer Science, University of York, York Y01 5DD, UK
{marco,erh}@cs.york.ac.uk

Abstract. This paper describes a new eigendecomposition method
for weighted graph-matching. Although elegant by means of its matrix
representation, the eigendecomposition method is proved notoriously
susceptible to differences in the size of the graphs under consideration. In this paper we demonstrate how the method can be rendered
robust to structural differences by adopting a hierarchical approach.
We place the weighted graph matching problem in a probabilistic
setting in which the correspondences between pairwise clusters can
be used to constrain the individual correspondences. By assigning
nodes to pairwise relational clusters, we compute within-cluster and
between-cluster adjacency matrices. The modal co-efficients for these
adjacency matrices are used to compute cluster correspondence and
cluster-conditional correspondence probabilities. A sensitivity study on
synthetic point-sets reveals that the method is considerably more robust than the conventional method to clutter or point-set contamination.

Keywords: pattern recognition, eigendecomposition method, weighted
graph matching

1 Introduction

Matrix factorisation and eigendecomposition methods [4,11,8,13] have proved to
be alluring yet elusive tools for weighted graph matching. Stated simply, the
aim is to find the pattern of correspondence matches between two sets of objects
using the eigenvectors of a weighted adjacency matrix. The problem has much
in common with spectral graph theory [1] and has been extensively studied for
both the general problem of graph-matching [13,10], and for the more specific
problem of point pattern matching [11,8]. Both problems commence from an
adjacency matrix whose elements represent the strength or weight of the pairwise
relations between nodes. In the case of point-sets, these pairwise relations are
based on interpoint distance, and the weights may be generated using a Gaussian
function or a similarly shaped function. The modal structure of the adjacency
matrix is used to locate correspondence matches between nodes. The method
may be implemented in a number of ways. The simplest of these is to follow
Shapiro and Brady by minimizing the distance between the modal co-efficients
[11]. Umeyama has suggested a more sophisticated approach which uses a matrix

W. Skarbek (Ed.): CAIP 2001, LNCS 2124, pp. 142–151, 2001.

factorization method to find the node permutation matrix which minimizes the distance between the weighted adjacency matrices for the graphs being matched [13]. Unfortunately, both methods fail when there are differences in the sizes of the graphs being matched. The reason for this is that the pattern of eigenvectors is unstable when structural differences are present.

In a recent paper [2] we have revisited the method of Shapiro and Brady. Our aim was to use the correspondence information delivered by the method to develop an EM algorithm for point-set alignment. For structurally intact point-sets subject to positional jitter, we showed that the performance of the Shapiro and Brady method could be improved using ideas from robust statistics to compute the weighted adjacency matrix and to compare the modal co-efficients. To overcome the difficulties encountered with point-sets of different size, an explicit alignment process was required.

The aim in this paper is to return to weighted graph-matching problem and to focus on how the method can be rendered robust to size differences. We adopt a hierarchical approach. The method is based on the observation that the modes of the adjacency matrix can be viewed as pairwise clusters. Moreover, the modal co-efficients represent the affinity of nodes to clusters. Similar ideas have been exploited by several authors to develop image segmentation [12] and grouping methods [9,7,5].

Our approach is as follows. Each mode of the adjacency matrix represents a pairwise cluster of nodes. While the pattern of modal co-efficients of the adjacency matrix may be disturbed by structural differences in the graphs, the pairwise clusters may be more stable. Hence, we can use the cluster adjacency matrix to improve the correspondence process. Here we use an evidence combining method which is posed in a hierarchical framework. We compute the probability that pairs of nodes are in correspondence by developing a mixture model over the set of possible correspondences between the pairwise clusters. In this way, the cluster correspondences weight the node-correspondence probabilities.

To provide a concrete example of the method we focus on the problem of finding point correspondences. We discuss various alternative ways in which the correspondence process may be modelled using the modal co-efficients of the point and cluster adjacency matrices. We compare these alternatives with both the Shapiro and Brady method and our previously reported method.

2 Background

We are interested in the problem of weighted graph-matching. To be more formal let $G_D = (\mathcal{D}, H_D)$ and $G_M = (\mathcal{M}, H_M)$ represent weighted data and model graphs to be matched. Here $\mathcal{D}$ and $\mathcal{M}$ represent the node-sets of the two graphs. The connectivity structure of the two graphs are represented using the $|\mathcal{D}| \times |\mathcal{D}|$ weighted data graph adjacency matrix H_D and the $|\mathcal{M}| \times |\mathcal{M}|$ weighted model graph adjacency matrix H_M. To provide a concrete illustration we will consider the problem of point-pattern matching. Here the weighted adjacency matrix is computed using the Euclidean distance between pairs of points. In particular, we are interested in finding the correspondences between two point-sets, a model point-set $\mathbf{z}$ and a data point-set $\mathbf{w}$. Each point in the image data set is represented by an position vector co-ordinates $\underline{w}_i = (x_i, y_i)^T$ where i is the point index. In the interests of brevity we will denote the entire set of image points by $\mathbf{w} = \{\underline{w}_i, \forall i \in \mathcal{D}\}$ where $\mathcal{D}$ is the point set. The corresponding fiducial points constituting the model are similarly represented by $\mathbf{z} = \{\underline{z}_j, \forall j \in \mathcal{M}\}$ where $\mathcal{M}$ denotes the index-set for the model feature-points $\underline{z}_j$. Although the standard weighting function used to compute the elements of the adjacency matrices H_D and H_M is the Gaussian, in our previous work we have found that better results can be obtained using weighting functions suggested by robust statistics. In particular, we have found that a log-cosh function gives very good results. If i and i' are two nodes from the data graph, then the corresponding element of the weighted adjacency matrix is given by

$$H_D(i, i') = \frac{2}{\pi ||\underline{w}_i - \underline{w}_{i'}||} \log \cosh \left[\frac{\pi}{s} ||\underline{w}_i - w_{i'}|| \right] \tag{1}$$

[11] and aim to perform weighted graph-matching using the modal structure of the adjacency matrices. The modal structure of the adjacency matrix is found by solving the eigenvalue equation $det[H - \lambda I] = 0$ together with the associated eigenvector equation $H\phi_l = \lambda_l \phi_l$, where λ_l is the lth eigenvalue of the matrix H and ϕ_l is the corresponding eigenvector. The vectors are ordered according to the magnitude of the associated eigenvalues. The ordered column-vectors are used to construct a modal matrix $\Phi = (\phi_1|\phi_2|\phi_3|.....)$. The column index of this matrix refers to the magnitude order of the eigenvalues while the row-index is the index of the original point-set. This modal decomposition is repeated for both the data and transformed model point-sets to give a data-graph modal matrix $\Phi_D = (\phi_1^D|\phi_2^D|\phi_3^D|...|\phi_{|\mathcal{D}|}^D)$ and a model-graph modal matrix $\Phi_M = (\phi_1^M|\phi_2^M|\phi_3^M|...|\phi_{|\mathcal{M}|}^M)$. Since the two point-sets are potentially of different size, the modes are truncated of the larger graph. This corresponds to removing the last $||\mathcal{D}| - |\mathcal{M}||$ rows and columns of the larger modal matrix. The resulting matrix has $o = \min[|\mathcal{D}|, |\mathcal{M}|]$ rows and columns.

Shapiro and Brady [11] find correspondences by seeking the most similar rows of the model matrices for the data and the model graph adjacency matrices using a Euclidean distance norm. However, we have found that a better strategy is to adopt a probabilistic method to compare the elements of the modal matrices. When there is a significant difference between one or more of the components of the eigenvectors, then these errors dominate the Euclidean distance measure used by Shapiro and Brady. One way to make the computation of correspondences robust to outlier measurement error is to accumulate probability on a component by component basis over the eigenvectors. To do this we assume that the individual elements of the modal matrix are subject to Gaussian measurement errors. Under this assumption the probability that the data-graph node $i \in \mathcal{D}$ is in correspondence with th model-graph node $j \in \mathcal{M}$ is

$$\zeta_{i,j} = \frac{\sum_{l=1}^{o} \exp\left[-k||\Phi_D(i,l) - \Phi_M(j,l)||^2\right]}{\sum_{j' \in \mathcal{M}} \sum_{l=1}^{o} \exp\left[-k||\Phi_D(i,l) - \Phi_M(j',l)||^2\right]} \qquad (2)$$

where k is a constant. In this way large measurement errors contribute insignificantly through the individual exponentials appearing under the summation over the different eigenmodes. These probabilities can be used to locate significantly better correspondences than those obtained using the method of Shapiro and Brady when the elements of the adjacency matrix are noisy. However, the method also fails as soon as there are structural differences in the graphs.

3   Modal Clusters

The idea underpinning this paper is that the coefficients of the modal matrix Φ can be viewed as providing information concerning pairwise clusters of nodes under the pattern of adjacency weights. Each mode, i.e. each column of the modal matrix Φ_D, is represented by an orthogonal vector in a $|\mathcal{D}|$ dimensional space. The columns associated with the eigenvalues of largest magnitude represent the most significant pairwise arrangements of nodes, while those associated with the eigenvalues of smallest magnitude represent insignificant structure. For a given node i the different modal co-efficients $\Phi_D(i,l)$, $l = 1, ..., |\mathcal{D}|$ represent the affinity of the node to the different clusters. The larger the magnitude of the co-efficient, the greater the cluster affinity of the node. In other words, the entries in the columns of the modal matrix represent the membership affinities for the different clusters. The row-entries, on the other hand represent the way in which the individual nodes are distributed among the different clusters. Here we aim to exploit this property of the modal matrix to develop a fast and robust matching method.

Our idea is based on the simple observation, that while the modal coefficients, i.e. the entries in the columns of the modal matrix, may not be stable under the addition of extra nodes, the associated pairwise cluster of nodes will be relatively robust to the addition of outliers.

3.1   Within-Cluster Modal Matrices

We commence by defining the pairwise clusters of the adjacency matrix. To do this we select the S largest eigenvalues, i.e. the first S columns of Φ_D. There are a number of ways of choosing S. Here we set the value of S so that the co-efficients of the subsequent columns of Φ_D are insignificant. If T is a threshold, then the condition is that $|\Phi_D(i,l)| < T$ for $i = 1, ..., |\mathcal{D}|$ and $l > S$.

We are interested in using the modal co-efficients to compute the probability $P(i \in \omega_d)$ that the node i belongs to the cluster associated with mode ω_d of the original node-set. We use the co-efficients of the first S columns of the modal matrix Φ_D to compute the cluster membership probability. Here we assume that cluster membership probability is proportional to the magnitude of the entry in the row indexed i and column indexed ω_d of the modal matrix Φ_D and write

$$P(i \in \omega_d) = \Phi_D^*(i, \omega_d) = \frac{|\Phi_D(i, \omega_d)|}{\sum_{l=1}^{S} |\Phi_D(i, \omega_d)|} \tag{3}$$

construct a within-cluster weighted node adjacency matrix using the weighting function given in Equation (1). To do this we first identify the nodes which belong to each modal cluster. This is done of the basis of the cluster-membership probabilities $P(i \in \omega_d)$. The set of nodes assigned to the cluster ω_D is $\mathcal{C}_{\omega_d}^D = \{i | P(i \in \omega_d) > T_c\}$ where T_c is a membership probability threshold. To construct this matrix we will need to relabel the nodes using a cluster node index which runs from 1 to $|\mathcal{C}_{\omega_d}|$. Accordingly we let δ_{i,ω_d}^D denote the node-index assigned to the node i in the cluster ω_d. The weighted adjacency matrix for the nodes belonging to this cluster is denoted by F_{ω_D} and the corresponding modal matrix is $\Theta_{\omega_d}^D$. The modal matrix for the cluster indexed ω_m in the model node-set is denoted by $\Theta_{\omega_m}^M$.

3.2   Between Cluster Modal Matrix

We also construct an adajcency matrix for the cluster-centres. In our experiments later on we will consider point-sets. For the mode with eigenvalue λ_l, the position-vector for the cluster centre is

$$\underline{c}_l^D {}^{(n)} = \frac{\sum_{i=1}^{|\mathcal{D}|} |\Phi_D(i, l)| \underline{w}_i}{\sum_{i=1}^{|\mathcal{D}|} \Phi_D(i, l)|} \tag{4}$$

The positions of the cluster-centres for the S most significant modes of the data-graph adjacency matrix H_D are used to compute a $S \times S$ cluster-centre adjacency matrix C_D using the log-cosh weighting function given in Equation (1). Our idea is to use the modes of the cluster-centre adjacency matrices C_D for the data-graph and C_M for the model-graph to provide constraints on the pattern of correspondences for the purposes of matching. Accordingly, we solve the equation $det(C_D - \Lambda^D I) = 0$ to locate the eigenvalues of the modal cluster adjacency matrix. The eigenvectors ψ_L, $L = 1, .., S$ of the cluster adjacency matrix are found by solving the equation $C_D \psi_l^D = \Lambda_l^D \psi_l^D$ As before, these eigenvectors can be used to construct a modal-matrix for the pairwise clusters.

The matrix has the eigenvectors of G as columns, i.e. $\Psi_D = \left(\psi_1^D | \psi_2^D | ..... | \psi_S^D \right)$.

This procedure is repeated to construct a second $S \times S$ cluster modal matrix Ψ_M for the set of model nodes. Since the principal modal-clusters are selected on the magnitude-order of the associated eigenvalues, there is no need to re-order them.

4   Matching

The aim in this paper is to explore whether the additional information provided by the modal clusters can be used to improve the robustness of the matching process to node addition and dropout. We would like to compute the probability $P(i \leftrightarrow j)$, that the data-graph node $i \in \mathcal{D}$ is in correspondence with the model-graph node $j \in \mathcal{M}$. To do this we construct a mixture model over the set of possible correspondences between the set of S modal clusters extracted from the data-graph and their counterparts for the model graph. Suppose that ω_d and ω_m respectively represent node labels assigned to the modal clusters of the data and model graphs. Applying the Bayes formula, we can write

$$P(i \leftrightarrow j) = \sum_{\omega_d=1}^{S} \sum_{\omega_m=1}^{S} P(i \leftrightarrow j | \omega_d \leftrightarrow \omega_m) P(\omega_d \leftrightarrow \omega_m) \tag{5}$$

where $P(i \leftrightarrow j | \omega_d \leftrightarrow \omega_m)$ represents the cluster-conditional probability that the node i belonging to the data-graph cluster ω_d is in correspondence with the node j that belongs to the model-graph cluster ω_m. The quantity $P(\omega_d \leftrightarrow \omega_m)$ denotes the probability that the data graph cluster indexed ω_d is in correspondence with the model graph cluster indexed ω_m.

4.1 Cluster Conditional Correspondence Probabilities

To compute the cluster-conditional correspondence probabilities we use the modal structure of the within-cluster adjacency matrices. These correspondence probabilities are computed using the method outlined in Equation (2). As a result, we write

$$P(i \leftrightarrow j | \omega_d \leftrightarrow \omega_m) = \frac{\sum_{l=1}^{O_{\omega_d,\omega_m}} \exp\left[-k_w \|\Theta_{\omega_d}^D(\delta_{i,\omega_d}^D, l) - \Theta_{\omega_m}^M(\delta_{j,\omega_m}^M, l)\|^2\right]}{\sum_{j' \in \mathcal{M}} \sum_{l=1}^{O_{\omega_d,\omega_m}} \exp\left[-k_w \|\Theta_{\omega_d}^D(\delta_{l,\omega_d}^D, l) - \Theta_{\omega_m}^M(\delta_{j',\omega_m}^M, l)\|^2\right]} \tag{6}$$

where $O_{\omega_d,\omega_m} = min[|\mathcal{C}_{\omega_m}|, |\mathcal{C}_{\omega_d}|]$ is the size of the smaller cluster.

4.2 Cluster Correspondence Probabilities

We have investigated two methods for computing the cluster correspondence probabilities $P(\omega_d \leftrightarrow \omega_m)$:

– **Modal eigenvalues:** The first method used to compute the cluster correspondence probabilities relies on the similarity of the normalized eigenvalues of the cluster modal matrix. The probabilities are computed in the following manner

$$P(\omega_d \leftrightarrow \omega_m) = \frac{\exp\left[-k_e\left\{\frac{|\Lambda_{\omega_d}^D|}{\sum_{\omega_d=1}^{S} |\Lambda_{\omega_d}^D|} - \frac{|\Lambda_{\omega_m}^M|}{\sum_{\omega_m=1}^{S} |\Lambda_{\omega_m}^M|}\right\}^2\right]}{\sum_{\omega_m=1}^{S} \exp\left[-k_e\left\{\frac{|\Lambda_{\omega_d}^D|}{\sum_{\omega_d=1}^{S} |\Lambda_{\omega_d}^D|} - \frac{|\Lambda_{\omega_m}^M|}{\sum_{\omega_m=1}^{S} |\Lambda_{\omega_m}^M|}\right\}^2\right]} \tag{7}$$

– **Modal co-efficients:** The mode correspondence probabilities have also been computed by performing a robust comparison of the co-efficients of the modal matrices of the cluster adjacency matrix. This is simply an application of the method outlined in Equation (2) to the modal co-efficients of the between-cluster adjacency matrix. We therefore set

$$P(\omega_d \leftrightarrow \omega_m) = \frac{\sum_{L=1}^{S} \exp\left[-k_b ||\Psi_D(\omega_d, L)| - |\Psi_M(\omega_m, L)||^2\right]}{\sum_{\omega_m=1}^{S} \sum_{L=1}^{S} \exp\left[-k_b ||\Psi_D(\omega_d, L)| - |\Psi_M(\omega_m, L)||^2\right]} \quad (8)$$

Note that we no-longer have to truncate the number of modes of the larger graph since we have chosen only the S principal clusters from both the model and data.

5   Experiments

In this section we describe our experimental evaluation of the new modal correspondence method. This is based on the matching of point-sets. The experimentation is divided into two parts. We commence with a sensitivity study on synthetic data. This is aimed at measuring the effectiveness of the method when the point sets under study are subject to clutter and positional jitter. The second part of the study focuses on real world data. Here we investigate the method when applied to finding point correspondences between curvature features in gesture sequences.

5.1   Sensitivity Study

In our sensitivity study, we have compared the new correspondence method with those of Shapiro and Brady [11] and our previous work [2]. The Shapiro and Brady method is based purely on modal correspondence analysis, while our previous method uses modal correspondence probabilities to weight the estimation of affine alignment parameters in a dual-step EM algorithm.

Our sensitivity study uses randomly generated point-sets. We ensure that the point-sets have a clump structure by sampling the point positions from six partially overlapping Gaussian distributions with controlled variance. We have then added both new points at random positions, and, random point-jitter to the synthetic data. The randomly inserted points have been sampled from a uniform distribution. The positional jitter has been generated by displacing the points from their original positions by Gaussian measurement errors. The displacements have been randomly sampled from a circularly symmetric Gaussian distribution of zero mean and controlled standard deviation.

In Figure 1a we show the effect of increasing the number of randomly added points. In this experiment, we commence with a point-set of size 100. The plot shows the fraction of points correctly matched as a function of the number of randomly added points. The long-dashed curve, i.e. the one with gives the

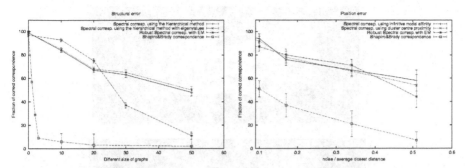

Fig. 1. Experimental results: (a) structural error, (b) position error.

consistently poorest results, is the result of applying the Shapiro and Brady algorithm. Here the fraction of correct correspondences falls below 25% once the fraction of added clutter exceeds 2%. The dual-step EM algorithm used in our previous work which finds correspondences by explicitly aligning the points, is shown as a dot-dashed curve and performs best of all when the level of clutter is less than 20%. The remaining two curves show the results obtained with the two variants of our hierarchical correspondence algorithm detailed in Section 4.3. In the case of the dotted curve the cluster correspondences are computed using only the modal co-efficients of the between-cluster adjacency matrix. The solid curve shows the results obtained if the eigenvalues are used instead. There is little to distinguish the two methods. Both perform rather more poorly than the dual-step EM algorithm when the level of clutter is less than 20%. However, for larger clutter levels, they provide significantly better performance. The additional use of the eigenvalues results in a slight improvement in performance.

Figure 1b investigates the effect of positional jitter. Here we plot the fraction of correct correspondence matches as a function of the standard deviation of the Gaussian position error added to the point-positions. We report the level of jitter using the ratio of the standard deviation of the Gaussian error distribution to the average closest inter-point distance. Here there is nothing to distinguish the behaviour of our hierarchical correspondence method from the dual-step alignment method. In each case the fraction of correct correspondences degrades slowly with increasing point-position jitter. However, even when the standard deviation of the position errors is 50% of the average minimum interpoint-distance then the fraction of correct correspondences is still greater than 50%. By contrast, the accuracy of the Shapiro and Brady method falls below 50% once the standard deviation of the positional error exceeds 10% of the minimum interpoint distance.

5.2   Real World Data

In this section we provide some experiments with real world data. We have taken images from a gesture sequence of a hand. The images used in this study are shown in Figure 6. We commence by running the Canny edge detector over the

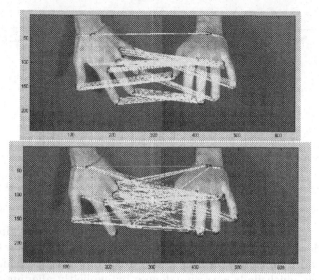

Fig. 2. Experimental results: a) real data experimentation, b) with Shapiro and Brady's method

images to locate the boundary of the hand. From this edge data, point features have been detected using the corner detector of Mokhtarian and Suomela [6]. The raw points returned by this method are distributed relatively uniformly along the outer edge of the hand and are hence not suitable for cluster analysis. We have therefore pruned the feature points using a curvature criterion. We have removed all points for which the curvature of the outline is smaller than a heuristically set threshold. Initially there are some 800 feature points, but after pruning this number is reduced to 271. The pruned feature-points are shown in blue in the figure. They are clustered around the finger-tips and the points at which the fingers join the hand. After applying the clustering method, the set of centres shown in red is obtained. There are ten clusters in both images. The yellow lines between the two images show the detected correspondences. The fraction of correct correspondences is 81.2%. Moreover, from the pattern of correspondence lines it is clear that the clusters are consistently matched. In Figure 2 we show the result ontained with the Shapiro and Brady algorithm. By comparison, the pattern of correspondences is considerably less consistent. There are many matches between incorrect pairs of clusters.

6   Conclusions

In this paper we have shown how constraints provided by the arrangement of modal groups of points can be used to improve the matching of weighted graphs via eigendecomposition. The idea has been to use the modal co-efficients of the adjacency matrix to establish significant pairwise clusters of nodes. We exploit these pairwise clusters to develop a hierarchical correspondence method. The method has been illustrated on the practical problem of point pattern matching. Here, we have shown that while the Shapiro and Brady method fails once more tha a few percent of clutter is added, the new method degrades more gracefully. There are a number of ways in which the method described in this paper could be extended. One of the most important of these is to extend the method to line-pattern matching.

References

1. Chung, F.R.K.: Spectral Graph Theory. CBMS series **92**. AMS Ed. (1997)
2. Carcassoni, M., Hancock, E.R.: Point Pattern Matching with Robust Spectral Correspondence. CVPR (2000)
3. Dempster, A.P., Laird, N.M., Rubin, D.B.: Maximum-likelihood from incomplete data via the EM algorithm. J. Royal Statistical Soc. Ser. B (methodological) **39** (1977) 1–38
4. Horaud, R., Sossa, H.: Polyhedral Object Recognition by Indexing. Pattern Recognition **28** (1995) 1855–1870
5. Inoue, K., Urahama, K.: Sequential fuzzy cluster extraction by a graph spectral method. Pattern Recognition Letters, **20** (1999) 699–705
6. Mokhtarian, F., Suomela, R.: Robust Image Corner Detection Through Curvature Scale Space. IEEE PAMI, **20** (December 1998) 1376–1381
7. Perona, P., Freeman, W.: A Factorisation Approach to Grouping. ECCV 98, Vol 1 (1998) 655–670
8. Scott, G.L., Longuet-Higgins, H.C.: An algorithm for associating the features of 2 images. In: Proceedings of the Royal Society of London Series B (Biological) **244** (1991) 21–26
9. Sengupta, K., Boyer, K.L.: Modelbase partitioning using property matrix spectra. Computer Vision and Image Understanding, **70** (1008) 177 196
10. Shokoufandeh, A., Dickinson, S.J., Siddiqi, K., Zucker, S.W.: Indexing using a spectral encoding of topological structure. In: Proc. of the IEEE Conf. on Computer Vision and Pattern Recognition (1999) 491–497
11. Shapiro, L.S., Brady, J.M.: Feature-based correspondence - an eigenvector approach. Image and Vision Computing, **10** (1992) 283–288
12. Shi, J., Malik, J.: Normalized cuts and image segmentation. In: Proc. of the IEEE Conf. on Computer Vision and Pattern Recognition (1997)
13. Umeyama, S.: An eigen decomposition approach to weighted graph matching problems. IEEE PAMI **10** (1988) 695–703

Discovering Shape Categories by Clustering Shock Trees

B. Luo, A. Robles-Kelly, A. Torsello, R.C. Wilson, and E.R. Hancock

Department of Computer Science
University of York, York, Y01 5DD, UK

Abstract. This paper investigates whether meaningful shape categories can be identified in an unsupervised way by clustering shock-trees. We commence by computing weighted and unweighted edit distances between shock-trees extracted from the Hamilton-Jacobi skeleton of 2D binary shapes. Next we use an EM-like algorithm to locate pairwise clusters in the pattern of edit-distances. We show that when the tree edit distance is weighted using the geometry of the skeleton, then the clustering method returns meaningful shape categories.

Keywords: clustering shock trees, EM algorithm, Hamilton-Jacobi skeleton

1 Introduction

There has recently been considerable interest in the use of the reaction-diffusion equation as a means of representing and analysing both 2D and 3D shapes [1, 2,3]. In a nutshell, the idea is to extract a skeletal representation by evolving the shape-boundary inwards until singularities appear. Through the analysis of the differential properies of the singularities, a structural abstraction of the skeleton known as the shock-graph may be extracted. Although this abstraction has been widely used for shape-matching and recognition [2], its use as a means of learning shape categories has attracted less attention. The aim in this paper is to investigate whether graph-clustering can be used as a means of partitioning shock-trees into shape classes via unsupervised learning.

Graph clustering is an important yet relatively under-researched topic in machine learning [4,5]. The importance of the topic stems from the fact that it is an important tool for learning the class-structure of data abstracted in terms of relational graphs. Problems of this sort are posed by a multitude of unsupervised learning tasks in knowledge engineering, pattern recognition and computer vision. The process can be used to structure large data-bases of relational models [6] or to learn equivalence classes. One of the reasons for limited progress in the area has been the lack of algorithms suitable for clustering relational structures. In particular, the problem has proved elusive to conventional central clustering techniques. The reason for this is that it has proved difficult to define what is meant by the mean or representative graph for each cluster. However, Munger,

W. Skarbek (Ed.): CAIP 2001, LNCS 2124, pp. 152–160, 2001.
© Springer-Verlag Berlin Heidelberg 2001

Bunke and Jiang [7] have recently taken some important steps in this direction by developing a genetic algorithm for searching for median graphs. A more fruitful avenue of investigation may be to pose the problem as pairwise clustering. This requires only that a set of pairwise distances between graphs be supplied. The clusters are located by identifying sets of graphs that have strong mutual pairwise affinities. There is therefore no need to explicitly identify an representative (mean, mode or median) graph for each cluster. Unfortunately, the literature on pairwise clustering is much less developed than that on central clustering.

When posed in a pairwise setting, the graph-clustering problem requires two computational ingredients. The first of these is a distance measure between relational structures. The second is a means of performing pairwise clustering on the distance measure. There are several distance measures available in the literature. For instance, in the classical pattern recognition literature, Haralick and Shapiro [8] have described a relational distance measure between structural descriptions, while Eshera and Fu [9] have extended the concept of edit distance from strings to graphs. More recently, Christmas, Kittler and Petrou [10], Wilson and Hancock [11] and Huet and Hancock [12] have developed probabilistic measures of graph-similarity. Turning our attention to pairwise clustering, there are several possible routes available. The simplest is to transform the problem into a central clustering problem. For instance, it is possible to embed the set of pairwise distances in a Euclidean space using a technique such as multi-dimensional scaling and to apply central clustering to the resulting embedding. The second approach is to use a graph-based method [13] to induce a classification tree on the data. Finally, there are mean-field methods which can be used to iteratively compute cluster-membership weights [14]. These methods require that the number of pairwise clusters be known a priori.

Our appraoch is as follows. Commencing from a data-base of silhouettes, we extract the Hamilton-Jacobi skeleton and locate the shocks which correspond to singularities in the evolution of the object boundary under the eikonal equation. We compute the similarity of the shapes using weighted and un-weighted tree edit distance. With the set of pairwise edit-distances between the shock-graphs to hand, we use a maximum-likelihood method for pairwise clustering. Our experiments show that when used in conjunction with the weighted tree edit distance, the pairwise clustering process locates meaningful shape categories.

2   Shock Tree Edit Distance

The practical problem tackled in this paper is the clustering of 2D binary shapes based on the similarity of their shock-trees. The idea of characterizing boundary shape using the differential singularities of the reaction equation was first introduced into the computer vision literature by Kimia, Tannenbaum and Zucker [3]. The idea is to evolve the boundary of an object to a canonical skeletal form using the reaction-diffusion equation. The skeleton represents the singularities in the curve evolution, where inward moving boundaries collide. The reaction component of the boundary motion corresponds to morphological erosion of the

	＼	／	＼	／	⤜	＼	⤙	＼	／	⬅	⬅	🐎	🐎	✋	✋	✋
＼	0.981	0.844	0.434	0.604	0.562	0.600	0.497	0.554	0.422	0.511	0.557	0.484	0.427	0.428	0.415	0.402
／	0.844	0.981	0.548	0.685	0.683	0.720	0.559	0.509	0.381	0.621	0.629	0.534	0.447	0.474	0.517	0.434
＼	0.446	0.543	1.000	0.569	0.554	0.643	0.637	0.475	0.651	0.436	0.559	0.428	0.395	0.406	0.414	0.373
／	0.604	0.685	0.809	1.000	0.663	0.731	0.713	0.592	0.496	0.507	0.624	0.404	0.418	0.443	0.449	0.404
⤜	0.562	0.684	0.531	0.666	1.000	0.857	0.714	0.459	0.385	0.588	0.622	0.456	0.497	0.524	0.541	0.521
＼	0.571	0.650	0.643	0.750	0.857	1.000	0.793	0.584	0.505	0.580	0.670	0.456	0.515	0.514	0.480	0.503
⤙	0.502	0.559	0.648	0.727	0.703	0.796	1.000	0.533	0.503	0.438	0.562	0.433	0.457	0.467	0.506	0.482
＼	0.554	0.606	0.475	0.592	0.459	0.531	0.554	1.000	0.415	0.434	0.405	0.443	0.459	0.419	0.400	0.408
／	0.441	0.356	0.676	0.506	0.388	0.472	0.530	0.394	0.981	0.384	0.438	0.396	0.386	0.310	0.353	0.310
⬅	0.516	0.626	0.479	0.507	0.593	0.586	0.487	0.434	0.379	1.000	0.825	0.556	0.520	0.634	0.630	0.555
⬅	0.556	0.636	0.559	0.626	0.622	0.670	0.578	0.405	0.428	0.820	1.000	0.449	0.590	0.542	0.532	0.518
🐎	0.496	0.449	0.417	0.388	0.462	0.429	0.475	0.397	0.403	0.492	0.449	1.000	0.627	0.496	0.519	0.560
🐎	0.395	0.447	0.395	0.441	0.495	0.488	0.445	0.497	0.371	0.602	0.573	0.693	0.992	0.711	0.687	0.699
✋	0.443	0.520	0.334	0.436	0.570	0.536	0.465	0.422	0.335	0.632	0.554	0.494	0.696	0.976	0.895	0.840
✋	0.431	0.450	0.409	0.425	0.520	0.520	0.480	0.445	0.378	0.629	0.543	0.572	0.719	0.847	0.982	0.771
✋	0.441	0.447	0.380	0.447	0.560	0.520	0.493	0.434	0.346	0.563	0.558	0.541	0.765	0.864	0.817	1.000

Fig. 1. Pairwise edit distances computed using un-weighted trees.

boundary, while the diffusion component introduces curvature dependent boundary smoothing. Once the skeleton is to hand, the next step is to devise ways of using it to characterize the shape of the original boundary. Here we follow Zucker, Siddiqi, and others, by labeling points on the skeleton using so-called shock-classes [2]. We abstract the skeletons as trees in which the level in the tree is determined by their time of formation [15,2]. The later the time of formation, and hence their proximity to the center of the shape, the higher the shock in the hierarchy. While this temporal notion of relevance can work well with isolated shocks (maxima and minima of the radius function), it fails on monotonically increasing or decreasing shock groups. To give an example, a protrusion that ends on a vertex will always have the earliest time of creation, regardless of its relative relevance to the shape.

To overcome this drawback, we augment the structural information given by the skeleton topology and the relative time of shock formation, with a measure of feature importance. We opt to use a shape-measure based on the rate of change of boundary length with distance along the skeleton. To compute the measure we construct the osculating circle to the two nearest boundary points at each location on the skeleton.

This measurement has previously been used in the literature to express *relevance* of a branch when extracting or pruning the skeleton, but is has recently been shown that its geometric and differential properties make it a good measure of shape similarity [16].

	⟨s1⟩	⟨s2⟩	⟨s3⟩	⟨s4⟩	⟨s5⟩	⟨s6⟩	⟨s7⟩	⟨s8⟩	⟨s9⟩	⟨s10⟩	⟨s11⟩	⟨s12⟩	⟨s13⟩	⟨s14⟩	⟨s15⟩	⟨s16⟩
⟨r1⟩	1.000	1.000	0.774	1.000	0.889	0.889	0.706	0.643	0.850	0.818	0.889	0.635	0.640	0.706	0.694	0.684
⟨r2⟩	1.000	1.000	0.774	1.000	0.889	0.889	0.706	0.643	0.850	0.818	0.889	0.635	0.640	0.706	0.694	0.684
⟨r3⟩	0.774	0.774	1.000	0.774	0.833	0.833	0.676	0.714	0.800	0.773	0.694	0.615	0.620	0.676	0.667	0.658
⟨r4⟩	1.000	1.000	0.774	1.000	0.889	0.889	0.706	0.643	0.850	0.818	0.889	0.635	0.640	0.706	0.694	0.684
⟨r5⟩	0.889	0.889	0.694	0.889	1.000	1.000	0.765	0.730	0.950	0.808	0.778	0.673	0.680	0.765	0.750	0.737
⟨r6⟩	0.889	0.889	0.833	0.889	1.000	1.000	0.765	0.730	0.950	0.808	0.778	0.673	0.680	0.765	0.750	0.737
⟨r7⟩	0.706	0.706	0.676	0.706	0.765	0.765	1.000	0.782	0.794	0.674	0.680	0.730	0.692	0.765	0.801	0.669
⟨r8⟩	0.643	0.643	0.714	0.643	0.730	0.730	0.847	1.000	0.771	0.649	0.639	0.769	0.724	0.651	0.698	0.682
⟨r9⟩	0.850	0.850	0.800	0.850	0.950	0.950	0.794	0.857	1.000	0.764	0.739	0.692	0.700	0.794	0.778	0.763
⟨r10⟩	0.818	0.818	0.773	0.818	0.808	0.808	0.599	0.731	0.764	1.000	0.909	0.647	0.720	0.749	0.806	0.789
⟨r11⟩	0.889	0.889	0.694	0.889	0.778	0.778	0.080	0.548	0.739	0.909	0.778	0.598	0.680	0.595	0.667	0.737
⟨r12⟩	0.635	0.635	0.615	0.635	0.673	0.673	0.778	0.769	0.692	0.647	0.598	1.000	0.785	0.730	0.752	0.729
⟨r13⟩	0.640	0.640	0.620	0.640	0.680	0.680	0.741	0.724	0.700	0.655	0.680	0.824	1.000	0.840	0.764	0.834
⟨r14⟩	0.706	0.706	0.676	0.706	0.765	0.765	0.706	0.651	0.715	0.824	0.765	0.730	0.840	1.000	0.972	0.947
⟨r15⟩	0.694	0.694	0.667	0.694	0.750	0.750	0.801	0.698	0.778	0.659	0.583	0.752	0.812	0.915	1.000	0.920
⟨r16⟩	0.684	0.684	0.658	0.684	0.737	0.737	0.669	0.682	0.763	0.718	0.573	0.729	0.741	0.947	0.920	0.947

Fig. 2. Pairwise edit distances computed using weighted trees.

Given this representation we can cast the problem of computing distances between different shapes as that of finding the tree edit distance between the weighted graphs for their skeletons.

With our measure assigned to each edge of the tree, we define the cost of matching two edges as the difference of the total length ratio measure along the branches. The cost of eliminating an edge is equivalent to the cost of matching it to an edge with zero weight, i.e. one along which the total length ratio is zero.

Using the edit distance of the shock trees we generate two similarity measures for a pair of shapes.

- The first measure is obtained weighting the nodes with the border length ratio normalized by the total length of the border of the shape. That is the length of the fraction of the border spanned by the shock group divided by the total length of the border. In this way the sum of the weights in a tree is 1 and the measure is scale invariant. The similarity of the shapes is computed by adding the minimum weight for each matched node, that is $d_{w,w'} = \sum_i min(w_i, w'_i)$, where w_i and w'_i are the weight of the nodes that are matched together by our tree edit distance algorithm.
- The second measure of shape similarity is computed from the unweighted structure: We assign a uniform edit cost of 1 to each node and we compute the average ratio of matched nodes: $d_{w,w'} = \frac{1}{2}\left(\frac{\#\hat{T}}{\#T_1} + \frac{\#\hat{T}}{\#T_2}\right)$, where T_1 and

T_2 are the two trees to be matched, $\hat{T}$ is the median of the two trees obtained through cut operations only, and $\#$ indicates the number of nodes in the tree.

3   Graph-Clustering

We pose the problem of learning the set of shape-classes as that of finding pairwise clusters in the distribution of tree-edit distance. The process of pairwise clustering is somewhat different to the more familiar one of central clustering. Whereas central clustering aims to characterise cluster-membership using the cluster mean and variance, in pairwise clustering it is the relational similarity of pairs of objects which are used to establish cluster membership. Although less well studied than central clustering, there has recently been renewed interest in pairwise clustering aimed at placing the method on a more principled footing using techniques such as mean-field annealing [14].

To commence, we require some formalism. We are interested in grouping a set of graphs $\mathcal{G} = \{G_1, ..., G_{|M|}\}$ whose index set is M. The set of graphs is characterised using a matrix of pairwise similarity weights. The elements of this weight matrix are computed using tree-edit distance $d_{i,j}$ between the graphs indexed i and j. Here we use the exponential similarity function $W_{i,j}^{(0)} = \{\exp[-kd_{i,j}]$ if $i \neq j$, 0 otherwise$\}$ to generate the elements of the weight-matrix, where k is a constant which is heuristically set. The aim in graph-clustering is to locate the updated set of similarity weights which partition the set of graphs into disjoint subsets. To be more formal, suppose that Ω is the set of graph-clusters and let S_ω represent the set of the graphs belonging to the cluster indexed ω. Further, let $s_{i\omega}^{(n)}$ represent the probability that the graph indexed i belongs to the cluster indexed ω at iteration n of the algorithm. We are interested in posing the clustering problem in a maximum likelihood setting. Under the assumption that the cluster memberships of the graphs follow a Bernoulli distribution with the link-weights as parameters, the likelihood-function for the weight matrix W is given by

$$P(W) = \prod_{\omega \in \Omega} \prod_{(i,j) \in M \times M} W_{i,j}^{s_{i\omega} s_{j\omega}} (1 - W_{i,j})^{1 - s_{i\omega} s_{j\omega}} \qquad (1)$$

The corresponding log-likelihood function is

$$\mathcal{L} = \sum_{\omega \in \Omega} \sum_{(i,j) \in M \times M} \left\{ s_{i\omega} s_{j\omega} \ln W_{ij} + (1 - s_{i\omega} s_{j\omega}) \ln(1 - W_{i,j}) \right\} \qquad (2)$$

We have recently, shown how this log-likeihood function can be iteratively optmised using an EM-like process. In the E (expectation) step, the cluster membership probabilities are updated according to the formula

$$s_{i\omega}^{(n+1)} = \frac{\prod_{j\in M}\left\{\frac{W_{i,j}^{(n)}}{1-W_{ij}^{(n)}}\right\}^{s_{j\omega}^{(n)}}}{\sum_{i\in M}\prod_{j\in M}\left\{\frac{W_{ij}^{(n)}}{1-W_{ij}^{(n)}}\right\}^{s_{j\omega}^{(n)}}} \tag{3}$$

Once the revised cluster membership variables are to hand then we apply the M (maximisation) step of the algorithm to update the similarity-weight matrix. The updated similarity-weights are given by $W_{ij}^{(n+1)} = \sum_{\omega\in\Omega} s_{i\omega}^{(n)} s_{j\omega}^{(n)}$. These two steps are interleaved and iterated to convergence.

To set the number of clusters we perform a modal analysis on the initial similarity matrix $W^{(0)}$. Here we use a result due to Sarkar and Boyer [17] who have shown that the positive eigenvectors of the matrix of similarity-weights can be used to assign nodes to clusters. Using the Rayleigh-Ritz theorem, they observe that the scalar quantity $x^t W^{(0)} x$ is maximised when x is the leading eigenvector of $W^{(0)}$. Moreover, each of the subdominant eigenvectors corresponds to a disjoint perceptual cluster. They confine their attention to the same-sign positive eigenvectors (i.e. those whose corresponding eigenvalues are real and positive, and whose components are either all positive or are all negative in sign). If a component of a positive eigenvector is non-zero, then the corresponding node belongs to the perceptual cluster associated with the associated eigenmodes of the weighted adjacency matrix. The eigenvalues $\lambda_1, \lambda_2....$ of $W^{(0)}$ are the solutions of the equation $|W^{(0)} - \lambda I| = 0$ where I is the $|V| \times |V|$ identity matrix. The corresponding eigenvectors $x_{\lambda_1}, x_{\lambda_2},...$ are found by solving the equation $W^{(0)} x_{\lambda_i} = \lambda_i x_{\lambda_i}$. Let the set of positive same-sign eigenvectors be represented by $\Omega = \{\omega | \lambda_\omega > 0 \wedge [(x_\omega^*(i) > 0 \forall i) \vee x_\omega^*(i) < 0 \forall i])\}$. Since the positive eigenvectors are orthogonal, this means that there is only one value of ω for which $x_\omega^*(i) \neq 0$. In other words, each node i is associated with a unique

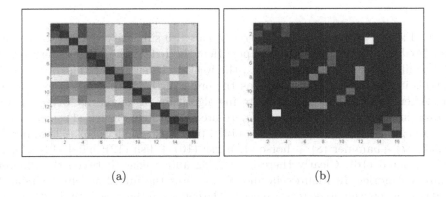

(a)   (b)

Fig. 3. (a) Initial similarity matrix for the unweighted tree edit distances; (b)Final similarity matrix for the unweighted tree edit distances.

cluster. We denote the set of nodes assigned to the cluster with modal index ω as $V_\omega = \{i | \boldsymbol{x}_\omega^*(i) \neq 0\}$. Hence each positive same-sign eigenvector is associated with a distinct mixing component. We use the eigenvectors of the initial affinity matrix to initialise the cluster membership variables. This is done using the magnitudes of the modal co-efficients and we set $s_{iw}^{(0)} = \frac{|\boldsymbol{x}_\omega^*(i)|}{\sum_{i \in V_\omega} |\boldsymbol{x}_\omega^*(i)|}$.

4   Experiments

The 16 shapes used in our study are shown in Figures 1 and 2. In Figure 1 we show the pattern of unweighted edit distances between the shock-trees for the shapes, while Figure 2 shows the corresponding weighted tree edit distances.

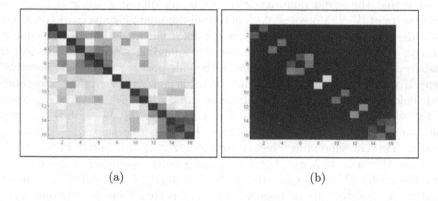

(a)   (b)

Fig. 4. (b) Initial similarity matrix for the weighted tree edit distances; (b) Final similarity matrix for the weighted tree edit distances.

In Figures 3a we show the matrix of pairwise similarity weights for the unweighted trees for the different shapes. Here the redder the entries, the stronger the similarity; the bluer the entries, the weaker the similarity. The order of the entries in the matrix is the same as the order of the shapes in Figures 1a and 1b. After six iterations of the clustering algorithm the similarity weight matrix shown in Figure 3b is obtained. There are six clusters (brush (1) + brush (2) + wrench (4); spanner (3) + horse (13) ; pliers (5) + pliers (6) + hammer (9) ;pliers (7) +hammer (8) + horse (12); fish (10) + fish (12); hand (14) + hand (15) + hand (16). Clearly there is merging and leakage between the different shape categories. In Figures 4a and 4b we show the initial and final similarity matrices when weighted trees are used. The entries in the initial similarity matrix are better grouped than those obtained when the unweighted tree edit distance is used. There are now seven clusters. brush (1) + brush (2) ; spanner (3) + spanner (4); pliers (5) + pliars (6) + pliers (7); hammer (8) + hammer (9); fish

(10) + fish (11); horse (12) + horse (13); hand (14) + hand (15) + hand (16)). These correspond exactly to the shape categories in the data-base.

5 Conclusions

This paper has presented a study of the problem of clustering shock-trees. We gauge the similarity of the trees using weighted and unweighted edit distance. To idetify distinct groups of trees, we develop a maximum likelihood algorithm for pairwise clustering. This takes as its input a matrix of pairwise similarities between shock-trees computed from the edit distances. The algorithm is reminiscent of the EM algorithm and has interleaved iterative steps for computing cluster-memberships and for updating the pairwise similarity matrix. The number of clusters is controlled by the number of same-sign eigenvectors of the current similarity matrix. Experimental evaluation of the method shows that it is capable of extracting clusters of trees which correspond closely to the shape-categories present.

References

1. Siddiqi, K., Bouix, S., Tannenbaum, A., Zucker, S.W.: The hamilton-jacobi skeleton. ICCV (1999) 828–834
2. Siddiqi, K., Shokoufandeh, A., Dickinson, S.J., Zucker, S.W.: Shock graphs and shape matching. International Journal of Computer Vision, **35** (1999) 13–32
3. Kimia, B.B., Tannenbaum, A.R., Zucker, S.W.: Shapes, shocks, and deforamtions i. International Journal of Computer Vision, **15** (1995) 189–224
4. Rizzi, S.: Genetic operators for hierarchical graph clustering. Pattern Recognition Letters, **19** (1998) 1293–1300
5. Segen, J.: Learning graph models of shape. In: Laird, J. (ed.): Proceedings of the Fifth International Conference on Machine Learning (1988) 29–25
6. Sengupta, K., Boyer, K.L.: Organizing large structural modelbases. IEEE Transactions on Pattern Analysis and Machine Intelligence, **17** (1995)
7. Munger, A., Bunke, H., Jiang, X.: Combinatorial search vs. genetic algorithms: A case study based on the generalized median graph problem. Pattern Recognition Letters, **20** (1999) 1271–1279
8. Shapiro, L.G., Haralick, R.M.: Relational models for scene analysis. IEEE Transactions on Pattern Analysis and Machine Intelligence, **4** (1982) 595–602
9. Eshera, M.A., Fu, K.S.: A graph distance measure for image analysis. IEEE Transactions on Systems, Man and Cybernetics, **14** (1984) 398–407
10. Kittler, J., Christmas, W.J., Petrou, M.: Structural matching in computer vision using probabilistic relaxation. IEEE PAMI, **17** (1995) 749–764
11. Wilson, R., Hancock, E.R.: Structural matching by discrete relaxation. IEEE Transactions on Pattern Analysis and Machine Intelligence, **19** (1997) 634–648
12. Huet, B., Hancock, E.: Relational histograms for shape indexing. In: IEEE International Conference of Computer Vision (1998)
13. Sengupta, K., Boyer, K.L.: Modelbase partitioning using property matris spectra. Computer Vision and Image Understanding, **70** (1998)

14. Hofmann, T., Buhmann, M.: Pairwise data clustering by deterministic annealing. IEEE Tansactions on Pattern Analysis and Machine Intelligence, **19** (1997)
15. Shokoufandeh, A., et al.: Indexing using a spectral encoding of topological structure. In: Conference on Computer Vision and Pattern Recognition (June 1999)
16. Torsello, A., Hancock, E.R.: A skeletal measure of 2d shape similarity. In: Visual Form 2001. LNCS 2059 (2001)
17. Sarkar, S., Boyer, K.L.: Quantitative measures of change based on feature organization: Eigenvalues and eigenvectors. Computer Vision and Image Understanding, **71** (1998) 110–136

Feature Selection for Classification Using Genetic Algorithms with a Novel Encoding

Franz Pernkopf and Paul O'Leary

University of Leoben, Peter-Tunner-Strasse 27, A-8700 Leoben, Austria
Franz.Pernkopf@unileoben.ac.at

Abstract. Genetic algorithms with a novel encoding scheme for feature selection are introduced. The proposed genetic algorithm is restricted to a particular predetermined feature subset size where the local optimal set of features is searched for. The encoding scheme limits the length of the individual to the specified subset size, whereby each gene has a value in the range from 1 to the total number of available features.

This article also gives a comparative study of suboptimal feature selection methods using real-world data. The validation of the optimized results shows that the true feature subset size is significantly smaller than the global optimum found by the optimization algorithms.

Keywords: pattern recognition, feature selection, genetic algorithm

1 Introduction

In real-world classification problems the relevant features are often unknown a priori. Thus, many features are derived and the features which do not contribute or even worsen the classification performance have to be discarded. Therefore, many algorithms exist which typically consist of four basic steps [3]:

1. a generation procedure to generate the next subset of features X.
2. an evaluation criterion J to evaluate the quality of X.
3. a stopping criterion for concluding the search. It can either be based on the generation procedure or on the evaluation function.
4. a validation procedure for verifying the validity of the selected subset.

The task of feature selection is to reduce the number of extracted features to a set of a few significant features which optimize the classification performance. The best subset

$$X = \{x_i | i = 1, \dots, d; x_i \in Y\} \tag{1}$$

is selected from the set

$$Y = \{y_i | i = 1, \dots, D\}, \tag{2}$$

where D is the number of extracted features and $d \leq D$ denotes the size of the feature subset [4]. A feature selection criterion function $J(X)$ evaluates a chosen

W. Skarbek (Ed.): CAIP 2001, LNCS 2124, pp. 161–168, 2001.

subset X, whereby a higher value of J indicates a better subset. The problem of feature selection [7] is to find a subset $X \subseteq Y$ such that the number of chosen features $|X|$ is d and J reaches the maximum

$$J\left(X^{opt}\right) = \max_{X \subseteq Y, |X|=d} J\left(X\right). \tag{3}$$

The evaluation criterion J is proposed to be the performance of a statistical classifier (classification rate) used as decision rule. Other evaluation measures are available, but Dash [3] showed that the accuracy of selecting the best subset by applying the later used classifier is the best. Unfortunately, this is computationally very demanding.

Exhaustive search for feature selection is too time consuming even for a small number of features. Hence, suboptimal algorithms are treated in this article.

2   Genetic Algorithms

Genetic algorithms [10][1] are a stochastic optimization procedure which have been successfully applied in many feature selection tasks [14][6]. Siedlecki and Sklansky [14] introduced the genetic algorithms for feature selection to find the smallest subset for which the optimization criterion (e.g. performance of a classifier) does not deteriorate below a certain level. So they deal with a constrained optimization problem, where a *penalty function* with additional parameters is used. To avoid this, a new encoding scheme is proposed to constrain the algorithm to a particular subset size. The notation for the description of the algorithm is taken from Bäck [1].

2.1   Encoding

All genetic algorithms employed for feature selection tasks use a binary encoding where the length of the individual is determined by the number of available features D. In our encoding the length of the individuals in the population is given by a predetermined subset size d. Thus, an optimum for a particular size of subset is computed.

Each gene, possessed by an individual a_i, has a value between 1 and the maximum number of features D. A single individual is given as

$$a_i = (a_{i,1}, \dots, a_{i,d}), \quad i \in \{1, \dots, \mu\}; a_{i,1}, \dots, a_{i,d} \in \{1, \dots, D\}, \tag{4}$$

where μ denotes the population size.

2.2   Selection

The selection directs the search toward promising regions in the search space. Stochastic universal sampling [2] is a widely used selection method due to the advantages of zero bias and minimum spread. To each individual a_i a reproduction probability $\eta\left(a_i\right)$ according to

$$\eta\left(a_i\right) = \frac{\phi\left(a_i\right)}{\sum_{j=1}^{\mu} \phi\left(a_j\right)} \tag{5}$$

is assigned, where $\phi\left(a_i\right)$ is the fitness value which is computed as $C(a_i) - C_w$. $C(a_i)$ denotes the classification rate of the individual a_i and C_w is the worst classification result within the population. A model spinning wheel is constructed with a slot size corresponding to the individuals reproduction probability $\eta\left(a_i\right)$ and μ equally spaced pointers select the individuals for the subsequent population. For implementation details refer to Baker [2] and Bäck [1].

2.3   Recombination

Recombination performs an exchange of alleles between individuals. The idea is that useful segments of different parents are combined in order to yield a new individual that benefits from both parents. The crossover operator $r'_{\{p_c\}}$ selects with probability p_c two parent individuals randomly from the population and recombines them to form two new individuals. A lot of different crossover operators are suggested in the literature [1].

Extensive experiments have shown that single-point crossover yields comparable results to other recombination mechanisms utilized for this feature selection task. The crossover position within the individual is determined at random.

2.4   Mutation

The mutation operator $m'_{\{p_m\}}$ chooses a new value ($a'_{i,j} \in \{1, \dots, D\}$) for each gene ($i \in \{1, \dots, \mu\}$; $j \in \{1, \dots, d\}$) in the population that undergoes mutation. Mutation reintroduces alleles into the population, i.e. features which have vanished throughout the complete population and therefore can never be regained by other operators.

2.5   Framework of the Genetic Algorithm

During the initialization of the population $P(0)$, to each individual d different features out of the pool of D features are assigned. The following conceptual algorithm has been slightly modified by a step enforced mutation which avoids a repeated selection of the same feature in one individual. The enforced mutation always ensures the correction of individuals. Each individual, which has selected the same features more often, is enforced to a further mutation of the identical alleles after the crossover and mutation operation. In the following, the framework of the genetic algorithm is shown.

> **procedure** GeneticAlgorithm
> **begin**
> $t = 0$
>   Initialize $P(t) = \{a_1(t), \dots, a_\mu(t)\} \in I^\mu$

where $I \in \{1, \ldots, D\}^d$
Evaluate $P(t) : \{\phi(a_1(t)), \ldots, \phi(a_\mu(t))\}$
while $(\iota \neq t)$
    Select: $P(t+1) = s(P(t))$
    $t = t + 1$
    Recombine: $a'_k(t) = r'_{\{p_c\}}(P(t))$ $\forall k \in \{1, \ldots, \mu\}$
    Mutate: $a_k(t) = m'_{\{p_m\}}(a'_k(t))$ $\forall k \in \{1, \ldots, \mu\}$
    Enforced Mutation:
        $a_k(t) = m'_{\{p_m=1 \; \forall \text{identical alleles in } (a_k(t))\}}(a_k(t))$
        $\forall k \in \{1, \ldots, \mu\}$
    Evaluate: $P(t) : \{\phi(a_1(t)), \ldots, \phi(a_\mu(t))\}$
  end
end

Symbol s denotes the stochastic universal sampling and ι is the number of generations.

3 Sequential Feature Selection Algorithms

Sequential feature selection algorithms search in a sequential deterministic manner for the suboptimal best feature subset. Basically, forward and backward algorithms are available. The forward methods start with an empty set and add features until a stopping criterion concludes the search. The backward algorithms are the counterpart. They begin with all features selected and remove features iteratively. The well-known suboptimal sequential algorithms are listed and a comparative study with real-world data is given in Section 4.

- Sequential forward selection (SFS): With each iteration one feature among the remaining features is added to the subset, so that the subset maximizes the evaluation criterion J.
- Sequential backward selection (SBS): In each step one feature is rejected so that the remaining subset gives the best result.
- Plus l-take away r selection (PTA(l, r)): This iterative method enlarges the subset by adding l features with the help of the SFS algorithm in the first step. Afterwards r features are removed with the SBS algorithm.
- Generalized sequential forward selection (GSFS(r)): At each stage r features are added simultaneously instead of adding just one feature to the subset at a time like the SFS method.
- Generalized sequential backward selection (GSBS(r)) is the counterpart of the GSFS method.
- Generalized plus l-take away r selection (GPTA(l, r)): The difference between the PTA and the GPTA method is that the former approach employs the SFS and the SBS procedures instead of the GSFS and GSBS algorithms.
- Sequential forward floating selection (SFFS): The SFFS includes new features with the help of the SFS procedure. Afterwards conditional exclusions of the worst features in the previously updated subset take place. Therefore, the parameters l and r are superfluous.

- Sequential backward floating selection (SBFS).
- Adaptive sequential forward floating selection (ASFFS(r, b, d)): This algorithm is similar to the SFFS procedure where the SFS and the SBS methods are replaced by their generalized versions GSFS(r) and GSBS(r). The level of generalization r is determined dynamically. For a detailed description refer to [15].
- Adaptive sequential backward floating selection (ASBFS(r, b, d)).

The SFFS and SBFS algorithms are described in Pudil et. al [12]. The recently published adaptive floating algorithms (ASFFS, ASBFS) are represented in [15], and all other sequential methods are explained in [4][8].

4   Experimental Results

The feature selection algorithms were applied to a data set of 540 flaw images gained from a surface inspection task. Five different classes of flaws have been established, and a total of 54 features were computed from each image. The algorithms for computing the different features are summarized in [11].

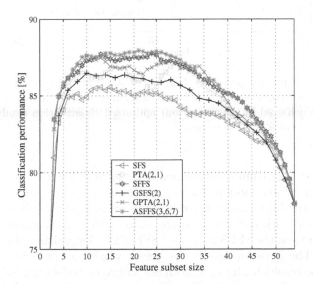

Fig. 1. Results of feature selection obtained by different sequential optimization methods for a subset size up to $d = 54$ using the 3NN classification result as evaluation criterion. The performance is achieved by averaging four optimized classification results gained over a rotation scheme of four data set parts.

First, the sequential feature selection algorithms were compared in terms of the achieved classification result. The performance of the 3-nearest neighbor decision rule[1] [5] serves as optimization criterion. The evaluation of the sequential forward algorithms is shown in Figure 1. The optimized values are averaged over a rotation scheme of four data set parts similar to cross-validation.

Generally, the floating algorithms perform better than their non-floating counterparts. The adaptive floating method (ASFFS(3,6,7)) performs best in the given subset around 7^2. Whereby, there is only a marginal difference to the result of the classical floating algorithm (SFFS). For a small subset size the generalized PTA method achieves roughly the same performance as the ASFFS algorithm.

Figure 2 compares the result of the proposed genetic algorithm (see Section 2) with the results of the ASFFS(3,6,7), the ASFFS(3,3,d)[3], and the SFFS algorithm.

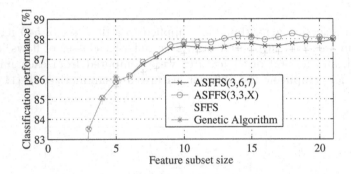

Fig. 2. Results of feature selection obtained by ASFFS(3,6,7), ASFFS(3,3,d), SFFS, and genetic algorithms using the 3NN classification result as evaluation criterion. The performance is achieved by averaging four optimized classification results gained over a rotation scheme of four data set parts.

The genetic algorithm optimization is performed for a subset size of 5, 10, 15, and 20. Therefore, 3 independent optimization runs for each subset size were executed and the best classification result is averaged over 4 best classification results obtained with different training data. The genetic algorithm performs slightly better than the SFFS and the ASFFS(3,6,7) algorithms for all optimized subset sizes. The performance of the genetic algorithm and the ASFFS(3,3,d) achieve similar results both in performance and computational efficiency. In comparison to the sequential algorithms, the genetic algorithm supplies many possi-

---

[1] The optimal number of neighbors k were found by estimating the classification error for different values of k [13].

[2] With the given parameter setting of the ASFFS algorithm the classification performance of a subset size of 7 within a neighborhood of 6 is optimized more thoroughly.

[3] For the ASFFS(3,3,d) method different optimization runs for each subset size d were performed and the best achieved results are summarized.

ble solutions with approximately the same performance. This can facilitate the choice of the most useful features.

The chosen parameter setting for the genetic algorithm is summarized in Table 1.

Table 1. Parameter setting for genetic algorithm.

Subset size	μ	p_m	p_c	ι
5	100	0.03	0.95	100
10	100	0.01	0.95	150
15	100	0.01	0.95	200
20	100	0.01	0.95	200

5   Verification of the Feature Subsets

Due to overfitting of the training set the selected features for an optimal classification of the training data produce inferior accuracy on independent test data. In practice the result of the feature selection optimization must be validated with a test set of objects. Therefore, several methods are suggested in the literature [9] [13].

The four performances obtained by optimizing with four data set parts (rotation scheme) and the classification estimate of the best feature subsets found for the remaining test data are averaged and shown in Figure 3.

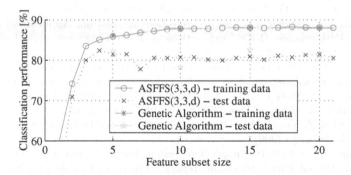

Fig. 3. Verification of the results of the ASFFS(3,3,d) and the genetic algorithm obtained by using the 3NN decision rule as evaluation criterion.

The effect that a small subset of features yields a better classification performance on test data is apparent. The best classification performance of the test data set is achieved for a feature subset size around 5, whereby the best classification performance of the training data is obtained with a feature subset size of 18.

6   Summary

Genetic algorithms with a novel encoding scheme for feature selection are presented. Conventional genetic algorithms employed for feature selection tasks use binary encoding. Due to the fact that a big number of features overfit the training data and consequently decrease the predictive accuracy on independent test data, the algorithm has to be constrained to find an optimal subset for a small number of features. To avoid a penalty function with additional parameters, the proposed encoding restricts the search to a particular subset size.

Additionally, a comparative study of suboptimal feature selection methods using real-world data is treated.

References

1. Bäck, T.: Evolutionary Algorithms in Theory and Practice. Oxford University Press (1996)
2. Baker, L.E.: Reducing bias and inefficiency in the selection algorithm. Proceedings of the 2nd International Conference on Genetic Algorithms and Their Applications. Lawrence Erlbaum Associates (1987) 14–21
3. Dash, M., Liu, H.: Feature Selection for Classification. Intelligent Data Analysis, 1 (1997) 131–156
4. Devijver, P.A., Kittler, J.: Pattern Recognition: A statistical approach. Prentice Hall International (1982)
5. Duda, R.O., Hart, P.E.: Pattern Classification and Scene Analysis. Wiley-Interscience Publications (1973)
6. Ferri, F.J., Pudil, P., Hatef, M., Kittler, J.: Comparative study of techniques for large-scale feature selection. In: Gelsema, E.S., Kamal, L.N. (eds.): Pattern Recognition in Practice IV. Elsevier Science B.V. (1994) 403–413
7. Jain, A., and Zongker, D.: Feature Selection: Evaluation, Application, and Small Sample Performance. IEEE Transactions on Pattern Analysis and Machine Intelligence, 19 (1997) 153–158
8. Kittler, J.: Feature set search algorithms. In: Chen, C.H. (ed.): Pattern Recognition and Signal Processing. Sijthoff and Noordhoff, Alphen aan den Rijn (1978) 41–60
9. Kohavi, R.: A Study of Cross-Validation and Bootstrap for Accuracy Estimation and Model Selection. In: Proc. IJCAI-95 (1995) 1146–1151
10. Michalewicz, Z., Fogel, D.B.: How to solve it: Modern Heuristics. Springer-Verlag, Berlin Heidelberg (2000)
11. Pernkopf, F.: Automatic Inspection System for Detection and Classification of Flaws on turned Parts. Technical Report. University of Leoben (2000)
12. Pudil, P., Novovičová, J., Kittler, J.: Floating search methods in feature selection. Pattern Recognition Letters, 15 (1994) 1119–1125
13. Raudys, S.J., Jain, A.K.: Small sample size effects in statistical pattern recognition: Recommendations for practitioners. IEEE Transactions on Pattern Analysis and Machine Intelligence, 13 (1991) 252–264
14. Siedlecki, W., and Sklansky, J.: A note on genetic algorithms for large-scale feature selection. Pattern Recognition Letters, 10 (1989) 335–347
15. Somol, P., Pudil, P., Novovičová, J., Paclík, P.: Adaptive floating search methods in feature selection. Pattern Recognition Letters, 20 (1999) 1157–1163

A Contribution to the Schlesinger's Algorithm Separating Mixtures of Gaussians

Vojtěch Franc and Václav Hlaváč

Czech Technical University, Faculty of Electrical Engineering,
Center for Machine Perception
121 35 Praha 2, Karlovo náměstí 13, Czech Republic
http://cmp.felk.cvut.cz, {xfrancv,hlavac}@cmp.felk.cvut.cz

Abstract. This paper contributes to the statistical pattern recognition problem in which two classes of objects are considered and either of them is described by a mixture of Gaussian distributions. The components of either mixture are known, and unknown are only their weights. The class (state) of the object k is to be found at the mentioned incomplete a priori knowledge of the statistical model and the known observation x. The task can be expressed as a statistical decision making with non-random interventions. The task was formulated and solved first by Anderson and Bahadur [1] for a simpler case where each of two classes is described by a single Gaussian. The more general formulation with more Gaussians describing each of two classes was suggested by M.I. Schlesinger under the name generalized Anderson's task (abbreviated GAT in the sequel). The linear solution to GAT was proposed in [5] and described recently in a more general context in a monograph [4].

This contribution provides (i) a formulation of GAT, (ii) a taxonomy of various solutions to GAT including their brief description, (iii) the novel improvement to one of its solutions by proposing better direction vector for next iteration, (iv) points to our implementation of GAT in a more general Statistical Pattern Recognition Toolbox (in MATLAB, public domain) and (v) shows experimentally the performance of the improvement (iii).

Keywords: pattern recognition, Gaussian separation

1 Definition of the Generalized Anderson's Task

Let X be a multidimensional linear space. The result of object observation is a point x in the n-dimensional feature space X. Let k be an unobservable state which can have only two possible value $k \in \{1, 2\}$. It is assumed that conditional probabilities $p_{X|K}(x|k), x \in X, k \in K$ are multidimensional Gaussian distributions. Mathematical expectations μ_k and covariance matrices $\sigma_k, k = 1, 2$, of these distributions are not known. The only knowledge available is that parameters (μ_1, σ_1) belong to a certain known set of parameters $\{(\mu^j, \sigma^j)|j \in J_1\}$ and similarly for (μ_2, σ_2) it is set $\{(\mu^j, \sigma^j)|j \in J_2\}$ (J_1, J_2 denote set of indexes). Parameters μ_1 and σ_1 denote real but unknown statistical parameters of an object

W. Skarbek (Ed.): CAIP 2001, LNCS 2124, pp. 169–176, 2001.
© Springer-Verlag Berlin Heidelberg 2001

in the state $k = 1$. Parameters $\{\mu^j, \sigma^j\}$ for a certain upper index j represents one pair from possible pairs of values.

The goal is to find a decision strategy $q: X \to \{1, 2\}$ mapping feature space X to space of the classes K that minimizes the value

$$\max_{j \in J_1 \cup J_2} \varepsilon(q, \mu^j, \sigma^j), \tag{1}$$

where $\varepsilon(q, \mu^j, \sigma^j)$ is a probability that the Gaussian random vector x with mathematical expectation μ^j and covariance matrix σ^j fulfills either constraint $q(x) = 1$ for $j \in J_2$ or $q(x) = 2$ for $j \in J_1$. In other words, it is the probability that the random vector x will be classified to the different class then it actually belongs to.

We are interested in the solution of the formulated task under an additional constraint on the decision strategy q. The requirements is that the discriminant function should be linear, i.e. a hyperplane $\langle \alpha, x \rangle = \theta$ and

$$q(x, \alpha, \theta) = \begin{cases} 1, \text{ if } \langle \alpha, x \rangle > \theta, \\ 2, \text{ if } \langle \alpha, x \rangle < \theta, \end{cases} \tag{2}$$

for certain vector $\alpha \in X$ and the scalar θ. The expression in angle brackets $\langle \alpha, x \rangle$ denote scalar product of vectors α, x.

The task (1) satisfying condition (2) minimizes probability of classification error and can be rewritten as

$$\{\alpha, \theta\} = \operatorname*{argmin}_{\alpha, \theta} \max_{j \in J_1 \cup J_2} \varepsilon(q(x, \alpha, \theta), \mu^j, \sigma^j). \tag{3}$$

This is a generalization of the known Anderson's and Bahadur's task [1] that was formulated and solved for a simpler case, where each class is described by only one distribution, i.e. $|J_1| = |J_2| = 1$. A toy example is shown in Figure 1.

2   Solution to the Generalized Anderson's Task

There are several approaches how to solve GAT. They are thoroughly analyzed in [4]. First, we will list them and (in next section) we will focus on one of them which we have improved.

- **General Algorithm Framework.** The general method based on proofs which leads to the optimal solution defined by the criterion (3). We shell devote our attention to this approach in Section 3.
- **Solution by the help of optimization using general gradient theorem.** The criterion (3) defining GAT is unimodal but it is neither convex nor differentiable [4]. Thus standard hill climbing methods cannot be used but so called *generalized gradient optimization theorem* can be used instead.
- **ε-optimal solution.** The ε-solution method finds such a decision hyperplane (α, θ) that the probability of wrong classification is smaller than a given limit ε_0, i.e.

$$\max_{j \in J_1 \cup J_2} \varepsilon((\alpha, \theta), \mu^j, \sigma^j) < \varepsilon_0.$$

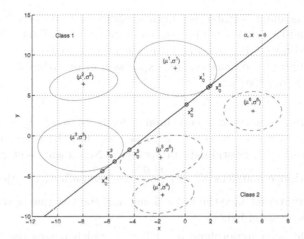

Fig. 1. An example of GAT. The first class is described by Gaussians with parameters $\{(\mu^1, \sigma^1), (\mu^2, \sigma^2), (\mu^3, \sigma^3)\}$ and the second class by $\{(\mu^4, \sigma^4), (\mu^5, \sigma^5), (\mu^6, \sigma^6)\}$. Mean values μ^j are denoted by crosses and covariance matrices σ^j by ellipsoids. The line represents found linear decision rule which maximizes Mahalanobis distance from the nearest distribution $\{(\mu^1, \sigma^1), (\mu^3, \sigma^3), (\mu^4, \sigma^4)\}$. The points x_0^j laying and the decision hyperplane have the nearest Mahalanobis distance from given distribution.

The optimal solution (3) does not need to be found so that the problem is thus easier. The task is reduced to splitting two sets of ellipsoids their radius is determined by the ε_0. This task can be solved by Kozinec's algorithm [4] which is similar to Perceptron learning rule.

3   Algorithm Framework

In this section we will introduce the general algorithm framework which solves GAT. Our contribution to this algorithm will be given in Section 4. The algorithm framework as well as concepts we will use are thoroughly analyzed and proved in [4]. We will introduce them without proofs.

Our goal is, in accordance with the definition of GAT (see Section 1), to optimize following criterion

$$\{\alpha, \theta\} = \operatorname*{argmin}_{\alpha, \theta} \max_{j \in J_1 \cup J_2} \varepsilon(q(x, \alpha, \theta), \mu^j, \sigma^j). \tag{4}$$

Where α, θ are parameters of a decision hyperplane $\langle \alpha, x \rangle = \theta$ we are searching for. Vectors $\mu^j, j \in J_1 \cup J_2$ and matrices $\sigma^j, j \in J_1 \cup J_2$ are parameters of Gaussians describing the first class J_1 and the second class J_2. This optimization task can be transformed to equivalent optimization task

$$\alpha = \operatorname*{argmax}_{\alpha} \min_{j \in J} r(\alpha, \mu^j, \sigma^j). \tag{5}$$

The task (5) is more suitable for both analysis and computation. The transformation consists of (i) introducing homogeneous coordinates by adding one constant coordinate, (ii) merging both the classes together by swapping one class along origin of coordinates, (iii) expressing of probability $\varepsilon(q(x,\alpha),\mu^j,\sigma^j)$ using number $r(\alpha,\mu^j,\sigma^j)$.

(i)   Introducing homogenous coordinates leads to formally simpler problem since only vector $\alpha' = [\alpha, -\theta]$ is looked for and the threshold θ is hidden in the $(n+1)$-th cooridinate of vector α. New mean vectors $\mu'^j = [\mu^j, 1]$ and covariance matrices $\sigma'^j = \begin{bmatrix} \sigma^j & 0 \\ 0 & 0 \end{bmatrix}$ are used after it. Notice that new covariance matrices have the last column and the last row zero since constant coordinate was added.

(ii) Having decision hyperplane $\langle \alpha', x' \rangle = 0$, which passes the origin of coordinates, it holds that $\varepsilon(q(x',\alpha'),\mu'^j,\sigma'^j) = \varepsilon(q(x',\alpha'),-\mu'^j,\sigma'^j), j \in J_2$. It allows us to merge the input parameter sets into one set of parameters $\{(\mu''^j,\sigma''^j)|j \in J\} = \{(\mu'^j,\sigma'^j)|j \in J_1\} \cup \{(-\mu'^j,\sigma'^j)|j \in J_2\}$. To make notation simpler we will use further on notation x,α,μ^j,σ^j instead of $x'', \alpha'', \mu''^j, \sigma''^j$.

(iii) The number $r(\alpha,\mu^j,\sigma^j)$ is the Mahalanobis distance between the normal distribution $N(\mu^j,\sigma^j)$ and a point x_0^j laying on the hyperplane $\langle \alpha, x \rangle = 0$ which has the smallest distance. It has been proven [4] that $\varepsilon(q(x,\alpha,\theta),\mu^j,\sigma^j)$ monotonically decreases when $r(\alpha,\mu^j,\sigma^j)$ incereases which allows us exchange minmax criterion (4) to maxmin (5). The point with the smallest distance is

$$x_0^j = \underset{x|\langle\alpha,x\rangle=0}{\text{argmin}} \; \langle(\mu^j - x),(\sigma^j)^{-1} \cdot (\mu^j - x)\rangle = \mu^j - \frac{\langle\alpha,\mu^j\rangle}{\langle\alpha,\sigma^j \cdot \alpha\rangle}\sigma^j \cdot \alpha.$$

The number $r(\alpha,\mu^j,\sigma^j)$ can be computed as

$$r(\alpha,\mu^j,\sigma^j) = \langle(\mu^j - x_0^j),(\sigma^j)^{-1} \cdot (\mu^j - x_0^j)\rangle = \frac{\langle\alpha,\mu^j\rangle}{\sqrt{\langle\alpha,\sigma^j \cdot \alpha\rangle}}.$$

The objective function in criterion (5) is unimodal and monotonically decreasing function. The algorithm which solves the criterion is similar to hill climbing methods but the direction in which the criterion improves cannot be computed as a derivative since it is not differentiable. The main part of algorithm consists of (i) finding of an improving direction $\Delta\alpha$ in which criterion descends and (ii) determining how much to move in the direction $\Delta\alpha$. The convergence of the algorithm crucially depends on the method of improving direction $\Delta\alpha$.

First, we will introduce the algorithm framework, then the original method finding $\Delta\alpha$ will be given and finally we will introduce our improvement which concerns finding $\Delta\alpha$.

3.1   General Algorithm Framework for GAT

Algorithm:

1. **Transformations.** First, as we mentioned above, we have to perform transformations of $(\mu^j, \sigma^j), j \in J_1 \cup J_2$. Then we obtain one set $(\mu^j, \sigma^j), j \in J$. The algorithm processes the transformed parameters and its result is vector α also in the transformed space. When the algorithm exits we can easily transform the α back into the original space.
2. **Initialization.** Such a vector is found that all scalar products $\langle \alpha_1, \mu^j \rangle, j \in J$ are positive. If such α_1 does not exist then the algorithm exits and indicates that there is not a solution with probability of wrong classification less than 50%. Lower index t of the vector α_t denotes iteration number.
3. **Improving direction.** The improving direction $\Delta\alpha$ is found which satisfies

$$\min_{j \in J} r(\alpha_t + k \cdot \Delta\alpha, \mu^j, \sigma^j) > \min_{j \in J} r(\alpha_t, \mu^j, \sigma^j), \tag{6}$$

   where k is a positive real number. If no vector $\Delta\alpha$ satisfying (6) is found then the current vector α_t solves the task and algorithm exits. In the opposite case the algorithm proceeds to the following step.
4. **Movement in the improving direction.** A positive real number k is looked for which satisfies

$$k = \operatorname*{argmax}_{k>0} \min_{j \in J} r(\alpha_t + k \cdot \Delta\alpha, \mu^j, \sigma^j).$$

   A new vector α_{t+1} is calculated as $\alpha_{t+1} = \alpha_t + k \cdot \Delta\alpha$.
5. **Additional stop condition.** If a change in criterial function value during t_{hist} iterations is less than giving limit Δ_r, i.e.

$$\left| \min_{j \in J} r(\alpha_t, \mu^j, \alpha^j) - \min_{j \in J} r(\alpha_{(t-t_{hist})}, \mu^j, \alpha^j) \right| \leq \Delta_r,$$

   then the algorithm exits else continues in iterations by jumping to step 3.

   The algorithm can exit in two cases. The first possibility is in the step 3 when the improve is not found then the vector α_t corresponds to the optimal solution (proof in [4]).

   The second possibility can occur when a change in criterial function value after t_{hist} iterations is smaller than prescribed threshold Δ_r. This phenomenon is checked in step 5. Ideally, this case should not occur but due to numerical solution during optimization it is possible. The occurance of this case means that the algorithm got stuck in some improving direction $\Delta\alpha$ and the current solution α_t does not need to be optimal. This case is undesirable and thus we intended to find suitable method that finds improving direction in the step 3 and avoids this case.

   The main part of the algorithm is step 3 and step 4. The improving direction is searched for in the step 3. Having found the improving direction we should decide how much to move in this direction, it is solved in the step 4. Following subsections deal with these two subtasks.

3.2 Numerical Optimization of the Criterion Depending on One Real Variable

Having finished the step 3 the current solution α and the improving direction $\Delta\alpha$ are available. The aim is to find the vector $\alpha_{t+1} = \alpha_t + k \cdot \Delta\alpha$ which determines the next value of the solution. This vector has to maximize $\min_{j \in J} r(\alpha + k \cdot \Delta\alpha, \mu^j, \sigma^j)$, so we have new optimization problem

$$k = \operatorname*{argmax}_{k>0} \min_{j \in J} r(\alpha + k \cdot \Delta\alpha, \mu^j, \sigma^j),$$

where k is a real positive number. To solve this optimization task we have to find a maximum of a real function of one real variable. This task we solve numerically (details are given in [4]).

3.3 Search for an Improving Direction $\Delta\alpha$

Here we will describe step 3 of the algorithm, that finds a direction in which the error decreases. Overall effectivity of the algorithm crucially depends upon this direction as the performed experiments have shown. Such vector $\Delta\alpha$ must ensure that the classification error decreases in this direction, i.e.

$$\min_{j \in J} r(\alpha_t + k \cdot \Delta\alpha, \mu^j, \sigma^j) > \min_{j \in J} r(\alpha_t, \mu^j, \sigma^j), \tag{7}$$

where k is any positive real number. It is proved in [4] that the vector $\Delta\alpha$ satisfying the condition (7) must fulfill

$$\langle \Delta\alpha, x_0^j \rangle > 0, j \in J^0. \tag{8}$$

The set J^0 contains the distributions with highest error or lowest Mahalanobis distance, i.e. $\{j | j \in J^0\} = \operatorname{argmin}_{j \in J} r(\alpha, \mu^j, \sigma^j)$.

The original approach, proposed in [4], determines improving direction as

$$\Delta\alpha = \operatorname*{argmax}_{\Delta\alpha} \min_{j \in J^0} \frac{\langle \Delta\alpha, y^j \rangle}{|\Delta\alpha|}, \tag{9}$$

where $y^j = \frac{x_0^j}{\sqrt{\langle \alpha, \sigma^j \cdot \alpha \rangle}}$, then $\Delta\alpha$ is a direction in which the classification error for the worst distributions $j \in J^0$ decreases the quickest. The task (9) is equivalent to the separation of finite point set with maximal margin. We used linear Support Vector Machines (SVM) algorithm [2].

Following section describes the new method, which approximates Gaussian distribution with an identity covariance matrix, tries to improve the algorithm.

4 Local Approximation of the Gaussian Distribution by the Identity Covariance Matrix

The main contribution of the paper is described in this section. We have proposed the new approach how to find the improving direction in which the error of the optimized criterion decreases (see Section 3.3).

Each distribution $N(\mu^j, \sigma^j)$ is approximated in the point x_0^j by the Gaussian distribution $N(\mu^j, E)$, where E denotes the identity matrix. In the case when all the covariance matrixes are identity, GAT is equivalent to the optimal separation of finite point sets. So we determine the improving vector $\Delta\alpha$ as the optimal solution for the approximated distributions.

The points x_0^j for all the distributions are found first as

$$r^* = \min_{j \in J} r(\mu^j, \sigma^j, \alpha) , \quad x_0^j = \mu^j - \frac{r^*}{\sqrt{\langle \alpha, \sigma^j \alpha \rangle}} \cdot \sigma^j \cdot \alpha ,$$

then the improving direction $\Delta\alpha$ is computed as

$$\Delta\alpha = \underset{\Delta\alpha}{\mathrm{argmax}} \min_{j \in J} \frac{\langle \Delta\alpha, x_0^j \rangle}{|\Delta\alpha|} . \tag{10}$$

The optimization problem (10) is solved by linear SVM algorithm.

5   Experiments

The aim of the experiments was to compare several algorithms solving GAT. We have tested algorithms on synthetic data. Experiments on real data are foreseen.

The experiments have to determine (i) the ability of algorithms to find an accurate solution (close to the optimal one) and (ii) their robustness with regard to various input data. We created 180 data sets corresponding to task with known solutions.

We tested three algorithms. The first two algorithm GAT-ORIG and GAT-NEW fulfill the general algorithm framework (see Section 3) and the third one GAT-GGRAD uses the generalized gradient theorem (see Section 2). The algorithm GAT-ORIG uses the original method (see Section 3.3) and the algorithm GAT-NEW uses our improvement (see Section 4). All the algorithms mentioned in this paper are implemented in the Statistical Pattern Recognition Toolbox for Matlab [3].

For each algorithm we had to prescribe a stopping condition. The first stop condition is given by a maximal number of algorithm steps which was set to 10000. The second one is a minimal improvement in the optimized criterion which was set to $1e - 8$.

Using of synthetically generated data allows us to compare the solution found by an algorithm to the known optimal solution. Moreover, we could control complexity of the problem: (1) the number of distributions describing classes varying from 4 to 340; (2) dimension of data varying from 2 to 75; (3) the number of additional distributions which do not affect the optimal solution but which make work of the tested algorithm harder. Total number of randomly generated testing instances was 180 used.

Having results from the algortihms tested on the synthetic data we computed following statistics from: (i) mean deviation between the optimal and the found solution $E(|\varepsilon_{found} - \varepsilon_{optimal}|)$ in [%], (ii) maximal deviation between the optimal and the found solution $\max(\varepsilon_{found} - \varepsilon_{optimal})$ in [%], (iii) number of wrong solutions, i.e. their probability of wrong classification is worse by 1% compared to the optimal solution. When this limit is exceeded we consider the solution as wrong.

The following table summarizes the results. We conclude that the algorithm GAT-NEW appeared as the best. This algorithm found the optimal solution in all tests. The algorithm GAT-GRAD failed in 7% in our tests. The algorithm GAT-ORIG failed in 91.5% in our tests.

	GAT-ORIG	GAT-NEW	GAT-GGRAD
Mean deviation in [%]	8.96	0	0.29
Maximal deviation in [%]	35.14	0	10.56
Wrong solutions in [%]	91.5	0	7

6   Conclusions

We have proposed an improvement of the Schlesinger's algorithm separating the statistical model given by the mixture of Gaussians (Generalized Anderson's task, GAT) [4]. We composed and extensively tested three algorithms solving GAT. One of them was our improvement. The tests were performed on 180 test cases given by synthetic data as needed the ground truth. Our improvement outperformed the other algorithms. All the tested methods are implemented in the Statistical Pattern Recognition Toolbox for Matlab [3] that is free for use.

Acknowledgements. V. Franc acknowledges the support from the Czech Ministry of Education under Research Programme J04/98:212300013. V. Hlaváč was supported by the Czech Ministry of Education under Project LN00B096.

References

1. Anderson, T.W., Bahadur, R.R.: Classification into two multivariate normal distributions with differrentia covariance matrices. Annals Math. Stat. **33** (Jun 1962) 420–431
2. Burges C.J.: A tutorial on support vector machines for pattern recognition. Data Mining and Knowledge Discovery, **2** (1998) 121–167
3. Franc, V.: Statistical pattern recognition toolbox for Matlab, Master thesis, Czech Technical University in Prague (2000). http://cmp.felk.cvut.cz
4. Schlesinger, M.I., Hlaváč, V.: Deset přednášek z teorie statistického a strukturního rozpoznávání, in Czech (Ten lectures on statistical and structural pattern recognition). Czech Technical University Publishing House, Praha, Czech Republic (1999). (English version is supposed to be published by Kluwer Academic Publishers (2001))
5. Schlesinger, M.I., Kalmykov, V.G., Suchorukov, A.A.: Sravnitelnyj analiz algoritmov sinteza linejnogo reshajushchego pravila dlja proverki slozhnych gipotez, in Russian (Comparative analysis of algorithms synthesising linear decision rule for analysis of complex hypotheses). Automatika, **1** (1981) 3–9

Diophantine Approximations of Algebraic Irrationalities and Stability Theorems for Polynomial Decision Rules

Vladimir M. Chernov

Image processing Systems Institute of RAS
151 Molodogvardejskaya St., IPSI RAS, 443001, Samara, Russia
vche@smr.ru

Abstract. The theoretical aspects of the decision rules stability problem are considered in the article. The new metric theorems of the stability of the polynomial decision rules are proven. These theorems are sequent from the well-known results of approximating irrationalities by rational numbers obtained by Liouville, Roth and Khinchin. The problem of optimal correlation between deterministic and stochastic methods and quality criterion in pattern recognition problems is also discussed.

Keywords: Diophantine approximations, decision rules

1   Introduction

We announce the results of the research on the following problem.

Main problem. Let X and Y be two finite disjoint subsets of the space $\mathbf{V} = \mathbf{R}^n$ (classes of objects), and let a function $F(v)$ separate the classes, i.e.,

$$F(v) > 0 \quad \text{at} \quad \nu \in X \quad \text{and} \quad F(v) < 0 \quad \text{at} \quad \nu \in Y. \tag{1}$$

Do there exist subsets $X^*, Y^* \in \mathbf{R}^n$ such that $X^* \supset X, Y^* \supset Y$, and $F(v)$ separates X^* and Y^*?

We consider the metric aspect of this problem; i.e., our goal is to determine *how much* we can distort the sets X and Y (i.e., to find $X^* \supset X$, and $Y^* \supset Y$ and the *metric* relations between these sets) so that the polynomial functions $F(v)$ from a given finite set keep separating X^* and Y^*. In other words, we want to *evaluate* the stability of the decision rules associated with the separation functions from a given class.

Stability of decision rules plays a fundamental role in pattern recognition. It is stability that determines eventually the reliability of solutions of applied problems by pattern recognition methods. However, nearly all results in this field are of distinct *statistical* character. Namely, the feature space is endowed with a metric (such as the Mahalanobis metric) associated to some probability measure, and all algorithmic constructions are performed relative to this measure. The results are usually stated in probabilistic terms; certainly, they can be interpreted

W. Skarbek (Ed.): CAIP 2001, LNCS 2124, pp. 177–182, 2001.

as metric results, but with respect to a special non-Euclidean metric. We are unaware of "Euclidean" metric studies of stability of decision rules. At the same time, the mathematical theory (namely, the theory of Diophantine approxima-tions) - the framework, in which such a study can be performed - not only exists but has a long history. For this reason, the main goal of this report is to obtain corollaries to well-known theorems of Diophantine approximation theory in the form sufficient for analyzing the "determinate" stability of decision rules.

2   Basic Definition and Notation

We use the notation $\mathbf{Q}$ for the field of rational numbers, $\mathbf{R}$ for the field of real numbers, and $\mathbf{K}$ for an arbitrary field of algebraic numbers; $E_{\mathbf{Q}}$, $E_{\mathbf{R}}$, $E_{\mathbf{K}}$ are the open n-dimensional unit cubes in $\mathbf{Q}, \mathbf{R}$, and $\mathbf{K}$, respectively;

$$E_{\mathbf{Q}}(q) = \left\{ \mathbf{r} \in E_{\mathbf{Q}} : \quad \mathbf{r} = \left(\frac{a_1}{q}, ..., \frac{a_n}{q} \right) \right\};$$

and $S(\mathbf{x}, \rho)$ is the full sphere centered at $\mathbf{x}$ of radius ρ.

Definition 1 *Let $\nu = (\nu_1, ..., \nu_n) \in \mathbf{R}^n$. A polynomial $F(\nu)$ of degree $d > 1$ in n variables with rational coefficients is called a separating polynomial for sets $X, Y \subset E_{\mathbf{Q}}(q)$ (where $X \cap Y = \emptyset$) if it satisfies relation 1.*

Definition 2 *Let*

$$S = \left(\bigcup_{\mathbf{x} \in X} S(\mathbf{x}, \rho) \right) \cup \left(\bigcup_{\mathbf{y} \in Y} S(\mathbf{y}, \rho) \right) = S_X^{(\rho)} \cup S_Y^{(\rho)} \subset E_{\mathbf{R}}.$$

A separating polynomial $F(v)$ for sets $X, Y \subset E_{\mathbf{Q}}(q)$ is called locally stable on the set $E_{\mathbf{Q}}$ if, for any positive integer q there exists a real $\rho = \rho(q) > 0$ such that $F(v)$ separates the sets $S_X^{(\rho)}$ and $S_Y^{(\rho)}$.

Definition 3 *The least upper bound $\rho^*(q)$ of the numbers ρ such that a polynomial $F(v)$ separates $S_X^{(\rho)}$ and $S_Y^{(\rho)}$ for a fixed q is called the radius of local stability of the polynomial $F(v)$. If $\rho^*(q) = \beta q^{-t}$ (where $\beta > 0$ is an absolute constant) then the number t is called the index of local stability.*

The larger the index of local stability, the "less stable" the polynomial deci-sion rule.

Definition 4 *A separating polynomial $F(v)$ is weakly locally stable if the sets $S_X^{(\rho)}$ and $S_Y^{(\rho)}$ exist for all sufficiently large $q > q_0$.*

The index and radius of weak local stability are defined similarly to the index and radius of local stability, respectively.

If the sets X and Y are finite, then the (weak) local stability of a polynomial $F(v)$ is a trivial corollary to textbook theorems of calculus. In general for the continuous functions, these theorems state only the existence of the sets $S_X^{(\rho)}$ and $S_Y^{(\rho)}$. The goal of this work is to obtain estimates of the radius and index of local stability valid for the entire class of discriminant polynomials. The ground for optimism is the unique arithmetic nature of polynomial functions with integer coefficients, in particular, the fact that the roots of polynomials in one variable are algebraic numbers.

Let us mention two conjectures on the index of (weak) local stability, which are easily substantiated from "general considerations".

Proposition 1 *The index of (weak) local stability is no smaller than one.*

Indeed, if $t < 1$, then the cube $E_{\mathbf{Q}}$ is covered by stability full spheres at some q, and all recognition problems are reduced to learning for finite sets $X, Y \subset E_{\mathbf{Q}}(q)$. Fortunately for pattern recognition theory (and unfortunately for applications), this is impossible.

Proposition 2 *At a fixed q, the index of (weak) local stability is a non-decreasing function of the degree d of the separating polynomial.*

Indeed, the larger value of d, the larger number of "degrees of freedom" (coefficients) of the polynomial $F(\nu)$ and, therefore, the larger the number of separating polynomials specific to given finite sets X and Y.

Certainly, these considerations are not formal proofs of the properties stated. Rigorous proofs and precise quantitative formulations can be obtained based on the known facts in the Diophantine approximation theory.

3   Diophantine Theorems on Stability for Polynomial Decision Rules

In this section, we assume that, first, $X, Y \subset E_{\mathbf{Q}}(q)$ and, secondly, the algebraic manyfold

$$\Im = \{\nu \in \mathbf{Q}^n : \quad F(\nu) = 0\}$$

is irreducible [1]. The former constraint is a natural "user" assumption. The hard-to-verify later constraint can be replaced in practice by a prior reduction of the dimension of the feature space at the expense of eliminating inessential features.

Let $F(\nu)$ be a separating polynomial for $X, Y \subset E_{\mathbf{Q}}(q)$. Suppose that, for some

$$\mathbf{r} = (r_1, ..., r_n) \in E_{\mathbf{Q}}(q), \tag{2}$$

the polynomials

$$F_j(x) = F(r_1, ..., r_{j-1}, x, r_{j+1}, ..., r_n) \tag{3}$$

are defined. Then, the roots of these polynomials are algebraic numbers; their "distance" from the rationales is an object of study of one of the directions in Diophantine approximation theory. The formulations and proofs of the theorems of Liouville, Roth and Khinchin, and LeVeque that are used in this paper and belong to the "classics" of theory of approximations of algebraic irrationalities by rational numbers are given in, e.g., monographs [2], [3], [4] and original papers [5], [6]. The first theorem states that algebraic irrationalities cannot be approximated by rationales "too well", and the remaining theorems are its quantitative refinements.

Theorem 1 *(The corollary to the Liouville theorem). If a separating polynomial does not vanish at points of the set $E_\mathbf{Q}$, then the index of local stability does not exceed d.*

The following assertion seems paradoxical: the degree of the polynomial does not appear (!) in the statement of the Theorem 2.

Theorem 2 *(The corollary to the Roth theorem). If a separating polynomial does not vanish at points of the set $E_\mathbf{Q}$, then the index of weak local stability does not exceed $2 + \delta$ for any $\delta > 0$.* $\Diamond$

The next theorem asserts the existence of lattices in $E_\mathbf{Q}$ so "fine" that the index of weak local stability of a polynomial separating sets X and Y of these lattices can be arbitrary close to 1.

Theorem 3 *(The corollary to the Khinchin theorem). Let $\varphi(q)$ be an arbitrary positive function of integer argument q such that $\varphi(q) \to 0$ as $q \to \infty$. then, there exists a sequence $\{m_q\}$ (where $q \le m_q$) such that, for a separating polynomial not vanishing at rational points of the set $E_\mathbf{Q}$, the inequality $\rho^*(m_q) \ge \varphi(q)(m_q)^{-1}$ holds.* $\Diamond$

The informal interpretation of the last theorem in terms of recognition theory is as follows: Does it make practical sense to increase the learning sample size? Does learning drastically increase the "determinate" stability? The answer is yes. However, the learning process may be so complex that it cannot be implemented in practice.

The main idea of the proofs of the theorems stated above is illustrated in Fig. 1 for $n = 2$. If the function $F(\nu)$ is polynomial, then the line $F(\nu) = 0$ cannot be "too curved". If, in addition, h_1 and h_2 are "sufficiently large" (i.e., the roots of the polynomials $F(r_1, x)$ and $F(x, r_2)$ are "poorly" approximated by the rationales r_1 and r_2), which guaranteed by the corresponding Diophantine theorems, then the line $F(\nu) = 0$ cannot be "too close" to nodes of the lattice $E_\mathbf{Q}(q)$. Therefore, for a suitable value of ρ, which is determined by the parameters r_1, r_2, h_1 and h_2, there exists a full sphere of radius ρ centered at (r_1, r_2) such that, at all its interior points, the function $F(\nu)$ takes values of the same sign as at the point (r_1, r_2).

As an example let us consider the proof of the Theorem 1 for the case $n = 2$.

Proof. Let $F(\nu)$ be a separating polynomial for sets X and Y. Let $X, Y \subset E_\mathbf{Q}(q)$, $\mathbf{r} = (r_1, r_2) \in E_\mathbf{Q}(q)$, and $F(\mathbf{r}) > 0$. Let

$$F(r_1, x_2), F(x_1, r_2) > 0$$
$$\text{for} \quad r_1 \le x_1 < h_1 + r_1, \quad r_2 \le x_2 < h_2 + r_2,$$
$$\text{and} \quad F(r_1, r_2 + h_2), F(r_1 + h_1, r_2) = 0.$$

Let us show that for any $\nu \in S(\mathbf{r}, \rho) \subset E_\mathbf{R}$ the inequality $F(\nu) > 0$ is true, where $\rho \ge q^{-d}$.

Indeed, let us consider the line $\nu_2 = r_2 + t_2$ for rational t_2 ($0 \le t_2 < q^{-d}$). The inequality $F(\nu) > 0$ holds in the point $\nu = (r_1, r_2 + t_2)$. Let $(r_1 + \xi_1, r_2 + t_2)$ be the intersection point of the line $\nu_2 = r_2 + t_2$ and $F(\nu) = 0$ function graph. Since $r_1 + \xi_1$ is the algebraic number, according to the Liouville theorem the following relation is true:

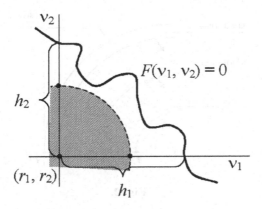

Fig. 1.

$$|(r_1 + \xi_1) - \xi_1| = |\xi_1| > q^{-d}.$$

In the same manner it can be shown that $|\xi_1| > q^{-d}$. Thus for all rational ν_1 $(r_1 < \nu_1 < h_1 + q^{-d})$ the distance from $F(\nu) = 0$ function graph to the line $\nu_2 = r_2$ is greater than q^{-d}. Since the rational numbers are dense in $\mathbf{R}$ the graph of the continuous curve $F(\nu) = 0$ can not be "far" from the line $\nu_2 = r_2$ for rational ν_1 and "near" to this line for irrational ν_1.◇

The assumption that the coordinates of all points in the sets X and Y are natural if the sets X and Y obtained from observations. In practice, these sets are often formed as results of some computational procedures and may be irrational. For instance, if the features are the components of discrete Fourier spectra of multidimensional signals (images), then the coordinates of points in X and Y are algebraic irrationalities. A computer representation of algebraic numbers uses their rational approximations; for this reason, it is natural to look for metric results on a sufficient accuracy of such approximations when the "scattering" of the points in the learning sample is known *a priori*. This makes it necessary to use Diophantine results on approximations of algebraic numbers by algebraic numbers. Among the diversity of known theorem of this kind, we mention LeVeque's generalization ([3], Theorem 8A) of the Roth theorem. It has the following corollary (an analog of Theorem 2).

Theorem 4 *Suppose that $\mathbf{K}$ is a real algebraic field; $X, Y \subset E_{\mathbf{K}}$; and H is the maximum absolute height of minimal polynomials of the coordinates of points in X and Y (see [3], Chapter VIII). Then, for any $\delta > 0$, the radius of weak local stability of a separating polynomials does not exceed $\rho^* = \beta H^{-(2+\delta)}$.* ◇

4 Conclusion

The theorems stated above for a certain impression of the expediency of applying determinate and/or statistical methods to recognition problems and of their optimal relation.

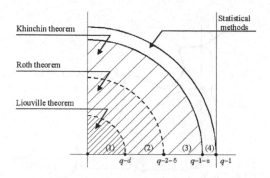

Fig. 2.

Indeed (see Fig. 2), for zone (1) of local stability, the certainty of recognition with the use of polynomial discriminant function is of determinate character; it is ensured by the Liouville theorem and does not require employment of probabilistic methods.

Zone (4) is an exclusive sphere of statistical methods. In zones (2) and (3) of weak local stability, statistical methods and quality criteria for recognition algorithms are a palliative making it possible to reduce learning at the expense of passing to an "indeterminate", probabilistic concept of certainty of the results of recognition algorithms.

Let us also mention that in the zones of weak local stability, combined methods that use probabilistic information on the distribution of "poorly approxiable" irrationalities [8] can be applied.

References

1. Lang, S.: Algebra. Reading, Mass. (1965)
2. Cassels, J.: An Introduction to Diophantine Approximation. Cambridge (1957)
3. Schmidt, W.: Diophantine Approximation. Springer (1980)
4. LeVeque, W.J.: Topics in Number Theory. Addison-Wesley, Reading, Mass. (1955)
5. Roth, K.F.: Rational Approximations to Algebraic Numbers. Mathematika, **2** (1955) 1–20
6. Khintchine, A.J.: Uber eine Klasse linearer Diophantische Approximationen. Rendiconti Circ. Mat. Palermo, **50** (1926) 170–195
7. LeVeque, W.J.: On Maler's U-numbers. J.Lond. Math. Soc., **28** 220–229
8. Kargaev, P.P., Zhigljavsky, A.: Approximation of Real Numbers by Rationals: Some Metric Theorems. J. Number Theory, **65** (1996) 130-149

Features Invariant Simultaneously to Convolution and Affine Transformation

Tomáš Suk and Jan Flusser

Institute of Information Theory and Automation
Academy of Sciences of the Czech Republic
Pod vodárenskou věží 4
182 08 Praha 8
Czech Republic
{suk, flusser}@utia.cas.cz

Abstract. The contribution is devoted to the recognition of objects and patterns deformed by imaging geometry as well as by unknown blurring. We introduce a new class of features invariant simultaneously to blurring with a centrosymmetric PSF and to affine transformation. As we prove in the contribution, they can be constructed by combining affine moment invariants and blur invariants derived earlier. Combined invariants allow to recognize objects in the degraded scene without any restoration.

Keywords: pattern recognition, image invariants

1 Introduction

Recognition of objects and patterns that are deformed in various ways has been a goal of much recent research. The degradations (geometric as well as radiometric) are introduced during the image acquisition process by such factors as imaging geometry, illumination changes, wrong focus, lens aberration, systematic and random sensor errors, object occlusion, etc. Finding a set of invariant descriptors is a key step to recognizing degraded objects regardless of the particular deformations.

Many papers have been devoted to the invariants to transform of spatial coordinates, like rigid-body affine and projective transform (see [1], [2] for a survey and other references). Moment invariants [3], [4], Fourier-domain invariants [5], differential invariants [6] and point sets invariants [7], [8] belong to the most popular classes of geometric invariants.

Much less attention has been paid to invariants with respect to changes of the image intensity function (we call them radiometric invariants) and to combined radiometric-geometric invariants. Van Gool et al. introduced so-called affine-photometric invariants of graylevel [9] and color [10] images. These features are invariant to the affine transform and to the change of contrast and brightness of the image simultaneously. Some other authors used various local features (mostly derivatives of the image function) to find invariants to rigid-body transform and contrast brightness changes [11]. This technique has become popular in image

W. Skarbek (Ed.): CAIP 2001, LNCS 2124, pp. 183–190, 2001.
© Springer-Verlag Berlin Heidelberg 2001

retrieval, but it is far beyond the scope of this contribution. Numerous references can be found in [11].

An important class of radiometric degradations we are faced with often in practice is image blurring. Blurring can be caused by camera defocus, atmospheric turbulence, vibrations, and by sensor or/and scene motion, to name a few. If the scene is flat, blurring can be usually described by a convolution $g(x, y) = (f * h)(x, y)$, where f is an original (ideal) image, g is an acquired image and h is a point spread function (PSF) of the imaging system. Since in most practical tasks the PSF is unknown, having the invariants to convolution is of prime importance when recognizing objects in a blurred scene. A pioneer work on this field was done by Flusser and Suk [12] who derived invariants to convolution with an arbitrary centrosymmetric PSF. From the geometric point of view, their descriptors were invariant to translation only. Despite of this, the invariants have found successful applications in face recognition on defocused photographs [13], in normalizing blurred images into the canonical forms [14], in template-to-scene matching of satellite images [12], in blurred digit recognition [15], and in focus/defocus quantitative measurement [16]. Other sets of blur invariants (but still only shift- invariant) were derived for some particular kinds of PSF – axisymmetric blur invariants [17] and motion blur invariants [18]. A significant improvement motivated by a problem of registration of blurred images was made by Flusser and Zitová. They introduced so-called combined blur-rotation invariants [19] and reported their successful usage in satellite image registration [20].

However, in a real world, the imaging geometry is projective rather than rigid-body. If the scene is flat and the camera far from the object in comparison to its size, the projective transform can be well approximated by an affine transform. Thus, having combined affine-blur invariants is in great demand.

In this contribution, we introduce combined invariants to affine transform and to convolution with an arbitrary centrosymmetric PSF, which are based on image moments. Since the affine transform does not preserve the shape of the objects, this cannot be done via straightforward generalization of previous work. In Section 2, we briefly recall the basic terms and earlier results of the theory of invariants that will be used later in the contribution. Section 3 performs the core of this work – we present there the central theorem on combined invariants. Section 4 demonstrates the numerical properties and recognition power of the invariants.

2   Recalling the Theory of the Invariants

In this Section we introduce some basic terms and briefly recall the theorems on the invariants which will be used later in this contribution.

2.1   Basic Terms

Definition 1: By *image function* (or *image*) we understand any real function $f(x, y) \in L_1$ having a bounded support and a nonzero integral.

Definition 2: *Central moment* $\mu_{pq}^{(f)}$ of order $(p+q)$ of the image $f(x,y)$ is defined as

$$\mu_{pq}^{(f)} = \int\limits_{-\infty}^{\infty} \int\limits_{-\infty}^{\infty} (x-x_c)^p (y-y_c)^q f(x,y)dxdy, \tag{1}$$

where the coordinates (x_c, y_c) denote the centroid of $f(x,y)$.

Definition 3: *Affine transform* is a transform of spatial coordinates (x,y) into (u,v) defined by equations

$$\begin{aligned} u &= a_0 + a_1 x + a_2 y, \\ v &= b_0 + b_1 x + b_2 y. \end{aligned} \tag{2}$$

Proposition 1: Every PSF mentioned in this contribution is assumed to be *centrosymmetric* and *energy-preserving*, i.e.

$$h(x,y) = h(-x,-y),$$

$$\int\limits_{-\infty}^{\infty} \int\limits_{-\infty}^{\infty} h(x,y)dxdy = 1.$$

The assumption of centrosymmetry is not a significant limitation of practical utilization of the method. Most real sensors and imaging systems, both optical and non-optical ones, have the PSF with certain degree of symmetry. In many cases they have even higher symmetry than central one, such as axial or radial symmetry. Thus, the central symmetry is general enough to handle almost all practical situations.

Note that because of centrosymmetry all moments of odd orders of the PSF equal zero.

2.2 Blur Invariants

In our earlier paper [12], the following theorem of blur invariants was proven.

Theorem 1: Let $f(x,y)$ be an image function. Let us define the following function $C^{(f)} : \mathcal{N}_0 \times \mathcal{N}_0 \to \mathcal{R}$.

If $(p+q)$ is even then

$$C(p,q)^{(f)} = 0.$$

If $(p+q)$ is odd then

$$C(p,q)^{(f)} = \mu_{pq}^{(f)} - \frac{1}{\mu_{00}^{(f)}} \sum_{\substack{n=0 \\ 0<n+m<p+q}}^{p} \sum_{m=0}^{q} \binom{p}{n}\binom{q}{m} C(p-n,q-m)^{(f)} \cdot \mu_{nm}^{(f)}. \tag{3}$$

Then $C(p,q)$ is invariant to convolution with any centrosymmetric function $h(x,y)$, i.e.

$$C(p,q)^{(f)} = C(p,q)^{(f*h)}$$

for any p and q.

2.3   Affine Moment Invariants

Affine moment invariants (AMI's) were introduced independently by Reiss [21] and Flusser and Suk [22]. They originate from the classical theory of algebraic invariants [23], [24]. The fundamental theorem, which was used for the derivation of the explicit forms of the invariants, can be formulated as follows.

Theorem 2: If the binary form of order p has an algebraic invariant of weight w and order k

$$I(a'_{p,0}, \cdots, a'_{0,p}) = \triangle^w I(a_{p,0}, \cdots, a_{0,p})$$

($\triangle$ denotes the determinant of the respective affine transform) then the moments of order p have the same invariant but with the additional factor $|J|^k$:

$$I(\mu'_{p0}, \cdots, \mu'_{0p}) = \triangle^w |J|^k I(\mu_{p0}, \cdots, \mu_{0p}),$$

where $|J|$ is the absolute value of the Jacobian.

We refer to [22], [21] for the proof of Theorem 2 and for a deeper explanation of the theory of the AMI's.

3   Combined Blur and Affine Invariants

By combined blur-affine invariants (CBAI's) we understand any functional defined on the set of image functions whose value does not change if the image function is convolved with a centrosymmetric PSF and transformed by an affine transform. Note, that the convolution and the affine transform are commutative here. The following theorem not only guarantees the existence of the CBAI's but also provides an explicit algorithm how to construct them.

Theorem 3: Let $I(\mu_{00}, \cdots, \mu_{PQ})$ be an affine moment invariant. Then $I(C(0,0), \cdots, C(P,Q))$, where $C(0,0) = \mu_{00}$ and all other blur invariants $C(p,q)$ are defined by Theorem 1, is a combined blur-affine invariant.

The proof can be carried out in the following way. Since $I(C(0,0), \cdots, C(P,Q))$ is a function of blur invariants $C(p,q)$ only, it is also a blur invariant. To prove its invariance to affine transform, it is sufficient to prove that partial derivatives with respect to each parameter $a \in \{a_0, a_1, a_2, b_0, b_1, b_2\}$ of the affine transform equal zero. Since it holds for $I(\mu_{00}, \cdots, \mu_{PQ})$, it is sufficient to prove that this property is preserved when substituting $C(p,q)$ for μ_{pq}. To do that, the affine transform (2) can be decomposed into six one-parametric transformations: horizontal translation, vertical translation, uniform scaling, stretching, horizontal skewing and vertical skewing. It is sufficient to prove the invariance to each of these transformations separately, which is easier than to prove it directly to (2).

To express the simplest CBAI's in explicit forms, we take five affine moment invariants of the 3rd and 5th orders:

$$I_1 = (\mu_{30}^2 \mu_{03}^2 - 6\mu_{30}\mu_{21}\mu_{12}\mu_{03} + 4\mu_{30}\mu_{12}^3 + 4\mu_{21}^3\mu_{03} - 3\mu_{21}^2\mu_{12}^2)/\mu_{00}^{10},$$

$$I_2 = (\mu_{50}^2\mu_{05}^2 - 10\mu_{50}\mu_{41}\mu_{14}\mu_{05} + 4\mu_{50}\mu_{32}\mu_{23}\mu_{05} + 16\mu_{50}\mu_{32}\mu_{14}^2 - 12\mu_{50}\mu_{23}^2\mu_{14}$$
$$+16\mu_{41}^2\mu_{23}\mu_{05} + 9\mu_{41}^2\mu_{14}^2 - 12\mu_{41}\mu_{32}^2\mu_{05} - 76\mu_{41}\mu_{32}\mu_{23}\mu_{14} + 48\mu_{41}\mu_{23}^3$$
$$+48\mu_{32}^3\mu_{14} - 32\mu_{32}^2\mu_{23}^2)/\mu_{00}^{14},$$

$$I_3 = (\mu_{30}^2\mu_{12}\mu_{05} - \mu_{30}^2\mu_{03}\mu_{14} - \mu_{30}\mu_{21}^2\mu_{05} - 2\mu_{30}\mu_{21}\mu_{12}\mu_{14} + 4\mu_{30}\mu_{21}\mu_{03}\mu_{23}$$
$$+2\mu_{30}\mu_{12}^2\mu_{23} - 4\mu_{30}\mu_{12}\mu_{03}\mu_{32} + \mu_{30}\mu_{03}^2\mu_{41} + 3\mu_{21}^3\mu_{14} - 6\mu_{21}^2\mu_{12}\mu_{23}$$
$$-2\mu_{21}^2\mu_{03}\mu_{32} + 6\mu_{21}\mu_{12}^2\mu_{32} + 2\mu_{21}\mu_{12}\mu_{03}\mu_{41} - \mu_{21}\mu_{03}^2\mu_{50} - 3\mu_{12}^3\mu_{41}$$
$$+\mu_{12}^2\mu_{03}\mu_{50})/\mu_{00}^{11},$$

$$I_4 = (2\mu_{30}\mu_{12}\mu_{41}\mu_{05} - 8\mu_{30}\mu_{12}\mu_{32}\mu_{14} + 6\mu_{30}\mu_{12}\mu_{23}^2 - \mu_{30}\mu_{03}\mu_{50}\mu_{05}$$
$$+3\mu_{30}\mu_{03}\mu_{41}\mu_{14} - 2\mu_{30}\mu_{03}\mu_{32}\mu_{23} - 2\mu_{21}^2\mu_{41}\mu_{05} + 8\mu_{21}^2\mu_{32}\mu_{14} - 6\mu_{21}^2\mu_{23}^2$$
$$+\mu_{21}\mu_{12}\mu_{50}\mu_{05} - 3\mu_{21}\mu_{12}\mu_{41}\mu_{14} + 2\mu_{21}\mu_{12}\mu_{32}\mu_{23} + 2\mu_{21}\mu_{03}\mu_{50}\mu_{14}$$
$$-8\mu_{21}\mu_{03}\mu_{41}\mu_{23} + 6\mu_{21}\mu_{03}\mu_{32}^2 - 2\mu_{12}^2\mu_{50}\mu_{14} + 8\mu_{12}^2\mu_{41}\mu_{23}$$
$$-6\mu_{12}^2\mu_{32}^2)/\mu_{00}^{12},$$

$$I_5 = (\mu_{30}\mu_{41}\mu_{23}\mu_{05} - \mu_{30}\mu_{41}\mu_{14}^2 - \mu_{30}\mu_{32}^2\mu_{05} + 2\mu_{30}\mu_{32}\mu_{23}\mu_{14} - \mu_{30}\mu_{23}^3$$
$$-\mu_{21}\mu_{50}\mu_{23}\mu_{05} + \mu_{21}\mu_{50}\mu_{14}^2 + \mu_{21}\mu_{41}\mu_{32}\mu_{05} - \mu_{21}\mu_{41}\mu_{23}\mu_{14} - \mu_{21}\mu_{32}^2\mu_{14}$$
$$+\mu_{21}\mu_{32}\mu_{23}^2 + \mu_{12}\mu_{50}\mu_{32}\mu_{05} - \mu_{12}\mu_{50}\mu_{23}\mu_{14} - \mu_{12}\mu_{41}^2\mu_{05} + \mu_{12}\mu_{41}\mu_{32}\mu_{14}$$
$$+\mu_{12}\mu_{41}\mu_{23}^2 - \mu_{12}\mu_{32}^2\mu_{23} - \mu_{03}\mu_{50}\mu_{32}\mu_{14} + \mu_{03}\mu_{50}\mu_{23}^2 + \mu_{03}\mu_{41}^2\mu_{14}$$
$$-2\mu_{03}\mu_{41}\mu_{32}\mu_{23} + \mu_{03}\mu_{32}^3)/\mu_{00}^{13}.$$

and substitute $C(p,q)$ for μ_{pq}. Since the even-order $C(p,q)$'s equal zero by definition, only those AMI's composed from odd-order moments are meaningful for this purpose. The explicit forms of $C(p,q)$'s can be found in [12] or derived directly from Theorem 1. Those needed for substitution into the above AMI's are listed below.

$$C(3,0) = \mu_{30}, \qquad C(2,1) = \mu_{21}, \qquad C(1,2) = \mu_{12}, \qquad C(0,3) = \mu_{03},$$

$$C(5,0) = \mu_{50} - \frac{10\mu_{30}\mu_{20}}{\mu_{00}},$$

$$C(4,1) = \mu_{41} - \frac{2}{\mu_{00}}(3\mu_{21}\mu_{20} + 2\mu_{30}\mu_{11}),$$

$$C(3,2) = \mu_{32} - \frac{1}{\mu_{00}}(3\mu_{12}\mu_{20} + \mu_{30}\mu_{02} + 6\mu_{21}\mu_{11}),$$

$$C(2,3) = \mu_{23} - \frac{1}{\mu_{00}}(3\mu_{21}\mu_{02} + \mu_{03}\mu_{20} + 6\mu_{12}\mu_{11}),$$

$$C(1,4) = \mu_{14} - \frac{2}{\mu_{00}}(3\mu_{12}\mu_{02} + 2\mu_{03}\mu_{11}),$$

$$C(0,5) = \mu_{05} - \frac{10\mu_{03}\mu_{02}}{\mu_{00}}.$$

4   Numerical Experiment

Two sequences of images (the comb and the picture of the Earth) of size 512×256 pixels were created to test the numerical properties of the new invariants. They were captured and affinely deformed by weak and strong affine transformations. Each image was then blurred by two square masks, 5×5 and 11×11. The values of five combined invariants from the previous section were computed for all combinations of blur and affine transformation (see Fig. 1). The values of the invariants are shown in Table 1.

We can see that the values of the invariants within one class are very close to each other. Slight deviations are caused by quantization and discretization errors. On the other hand, the difference between the invariant values of the comb and Earth pictures is much bigger than the intra-class deviations, what illustrates a good discrimination power of the invariants.

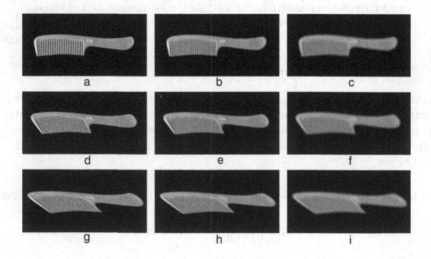

Fig. 1. The object "Comb", (a) original, (b) blurred by square mask 5×5, (c) blurred by square mask 11×11, (d) weakly affinely deformed, without blurring, (e) weakly affinely deformed and blurred by square mask 5×5, (f) weakly affinely deformed and blurred by square mask 11×11, (g) strongly affinely deformed, without blurring, (h) strongly affinely deformed and blurred by square mask 5×5 and (i) strongly affinely deformed and blurred by square mask 11×11.

5   Conclusion

The paper was devoted to the image features which are invariant simultaneously to blurring by a filter with centrally symmetric PSF and to the affine transformation. The major theoretical result of this work is Theorem 3, showing that

Table 1. The values of the combined affine and blur invariants

Object	Transform	Blur	$I_1[10^{-10}]$	$I_2[10^{-11}]$	$I_3[10^{-11}]$	$I_4[10^{-11}]$	$I_5[10^{-12}]$
Comb	-	-	-15810	-17820	2520	56040	-1641
Comb	weak	-	-15730	-17760	2514	55780	-1647
Comb	strong	-	-15730	-17770	2514	55780	-1642
Comb	-	5×5	-16150	-18450	2584	57650	-1692
Comb	weak	5×5	-16080	-18400	2578	57390	-1697
Comb	strong	5×5	-16080	-18410	2582	57400	-1695
Comb	-	11×11	-16190	-18510	2591	57810	-1699
Comb	weak	11×11	-16110	-18460	2582	57560	-1702
Comb	strong	11×11	-16110	-18470	2588	57550	-1.700
Earth	-	-	-2.095	6.952	3.758	5.844	3.067
Earth	weak	-	-2.074	6.852	3.593	5.791	3.082
Earth	strong	-	-2.045	6.387	3.712	5.903	3.224
Earth	-	5×5	-2.164	7.420	3.935	6.014	3.228
Earth	weak	5×5	-2.143	7.326	3.755	5.951	3.249
Earth	strong	5×5	-2.115	6.831	3.886	6.079	3.398
Earth	-	11×11	-2.171	7.462	3.945	6.016	3.242
Earth	weak	11×11	-2.143	7.383	3.767	5.920	3.258
Earth	strong	11×11	-2.122	6.891	3.903	6.069	3.415

the combined invariants can be constructed by substituting blur invariants into the affine moment invariants. The numerical experiment verified the theoretical results and also illustrated the discrimination power of the invariants.

Practical applications of the new invariants may be found in object recognition in blurred and noisy environment, in template matching, image registration, etc.

Acknowledgement. This work has been supported by the grants No. 102/00/1711 and No. 102/98/P069 of the Grant Agency of the Czech Republic.

References

1. Mundy, J.L., Zisserman, A.: Geometric Invariance in Computer Vision. MIT Press (1992)
2. Reiss, T.H.: Recognizing Planar Objects using Invariant Image Features. Lecture Notes in Computer Science, Vol. 676. Springer, Berlin (1993)
3. Hu, M.K.: Visual pattern recognition by moment invariants. IRE Trans. Information Theory, **8** (1962) 179–187
4. Belkasim, S.O., Shridhar, M., Ahmadi, M.: Pattern recognition with moment invariants: a comparative study and new results. Pattern Recognition, **24** (1991) 1117–1138
5. Arbter, K., Snyder, W.E., Burkhardt, H., Hirzinger, G.: Application of affine-invariant Fourier descriptors to recognition of 3-D objects. IEEE Trans. Pattern Analysis and Machine Intelligence, **12** (1990) 640–647

6. Weiss, I.: Projective invariants of shapes. In: Proc. Image Understanding Workshop. Cambridge, Mass. (1988) 1125–1134
7. Suk, T., Flusser, J.: Vertex-based features for recognition of projectively deformed polygons. Pattern Recognition, **29** (1996) 361–367
8. Lenz, R., Meer, P.: Point configuration invariants under simultaneous projective and permutation transformations. Pattern Recognition, **27** (1994) 1523–1532
9. van Gool, L., Moons, T., Ungureanu, D.: Affine/photometric invariants for planar intensity patterns. In: Proc. 4th ECCV'96, LNCS Vol. 1064. Springer (1996) 642–651
10. Mindru, F., Moons, T., van Gool, L.: Recognizing color patterns irrespective of viewpoint and illumination. In: Proc. IEEE Conf. Computer Vision Pattern Recognition CVPR'99, Vol. 1 (1999) 368–373
11. Schmid, C., Mohr, R.: Local grayvalue invariants for image retrieval. IEEE Trans. Pattern Analysis and Machine Intelligence, **19** (1997) 530–535
12. Flusser, J., Suk, T.: Degraded image analysis: An invariant approach. IEEE Trans. Pattern Analysis and Machine Intelligence, **20** (1998) 590–603
13. Flusser, J., Suk, T., Saic, S.: Recognition of blurred images by the method of moments. IEEE Transactions on Image Processing, **5** (1996) 533–538
14. Zhang, Y., Wen, C., Zhang, Y.: Estimation of motion parameters from blurred images. Pattern Recognition Letters, **21** (2000) 425–433
15. Lu, J., Yoshida, Y.: Blurred image recognition based on phase invariants. IEICE Trans. Fundamentals of El. Comm. and Comp. Sci., **E82A** (1999) 1450–1455
16. Zhang, Y., Zhang, Y., Wen, C.: A new focus measure method using moments. Image and Vision Computing, **18** (2000) 959–965
17. Flusser, J., Suk, T., Saic, S.: Image features invariant with respect to blur. Pattern Recognition, **28** (1995) 1723–1732
18. Flusser, J., Suk, T., Saic, S.: Recognition of images degraded by linear motion blur without restoration. Computing Suppl., **11** (1996) 37–51
19. Flusser, J., Zitová, B.: Combined invariants to linear filtering and rotation. Intl. J. Pattern Recognition Art. Intell., **13** (1999) 1123–1136
20. Flusser, J., Zitová, B., Suk, T.: Invariant-based registration of rotated and blurred images. In: Tammy, I.S. (ed.): IEEE 1999 International Geoscience and Remote Sensing Symposium. Proceedings. IEEE Computer Society, Los Alamitos (June 1999) 1262–1264
21. Reiss, T.H.: The revised fundamental theorem of moment invariants. IEEE Trans. Pattern Analysis and Machine Intelligence, vol. 13 (1991) 830–834
22. Flusser, J., Suk, T.: Pattern recognition by affine moment invariants. Pattern Recognition, **26** (1993) 167–174
23. Schur, I.: Vorlesungen über Invariantentheorie. Springer, Berlin (1968)
24. Gurevich, G.B.: Foundations of the Theory of Algebraic Invariants. Nordhoff, Groningen, The Netherlands (1964)

A Technique for Segmentation of Gurmukhi Text

G.S. Lehal[1] and Chandan Singh[2]

[1]Department of Computer Science and Engineering, Thapar Institute of Engineering & Technology, Patiala, India. gslehal@mailcity.com
[2]Department of Computer Science and Engineering, Punjabi University, Patiala, India.

Abstract. This paper describes a technique for text segmentation of machine printed Gurmukhi script documents. Research in the field of segmentation of Gurmukhi script faces major problems mainly related to the unique characteristics of the script like connectivity of characters on the headline, two or more characters in a word having intersecting minimum bounding rectangles, multi-component characters, touching characters which are present even in clean documents. The segmentation problems unique to the Gurmukhi script such as horizontally overlapping text segments and touching characters in various zonal positions in a word have been discussed in detail and a solution has been proposed.

Keywords: text segmentation, Gurmukhi script

1 Introduction

Text segmentation is a process in which the text image is segregated into units of patterns that seem to form characters. All recognition algorithms depend on the segmentation algorithm to break up the image into individual characters. Many papers concerning the segmentation of strings consisting of English letters and numerals have been published. The recent surveys on this topic can be found in references [1-2]. Some papers dealing with segmentation of different Indian language scripts such as Bangla, Devnagri and Gurmukhi have also appeared in literature [3-7], but none of the paper has dealt in detail with the practical problems peculiar to the Indian language scripts such as horizontally overlapping text segments and touching characters in various zonal positions in a word.

2 Characteristics of Gurmukhi Script

Gurmukhi script is used primarily for the Punjabi language, which is the world's 14th most widely spoken language. Some of the major properties of the Gurmukhi script are:
➢ Gurmukhi script alphabet consists of 41 consonants, 12 vowels and 3 half characters which lie at the feet of consonants (Fig 1).

W. Skarbek (Ed.): CAIP 2001, LNCS 2124, pp. 191–200, 2001.
© Springer-Verlag Berlin Heidelberg 2001

> A majority of the characters have a horizontal line at the upper part (Fig 1). The characters of words are connected mostly by this line called head line and so there is no vertical inter-character gap in the letters of a word and formation of merged characters is a norm rather than an aberration in Gurmukhi script The words are, however, separated with blank spaces.

> A word in Gurmukhi script can be partitioned into three horizontal zones (Fig 2). The upper zone denotes the region above the head line, where vowels reside, while the middle zone represents the area below the head line where the consonants and some sub-parts of vowels are present. The lower zone represents the area below middle zone where some vowels and certain half characters lie in the feet of consonants. A statistical analysis of Punjabi corpus has shown the zone-wise percentage distribution of Gurmukhi symbols in printed text as: upper zone (25.39%), middle zone (70.41%) and lower zone (4.20%).

> The bounding boxes of 2 or more characters in a word may intersect or overlap vertically.

> The half characters in the lower zone frequently touch the above lying consonants in the middle zone. Similarly closely lying upper zone vowels frequently touch each other.

> There are many multi-component characters in Gurmukhi script. A multi-component character is a character, which can decompose into isolated parts (e.g. S, K, z, Z, F, <).

ੳ	ਅ	ੲ	ਸ	ਹ	ਕ	ਖ	ਗ	ਘ	ਙ	
ਚ	ਛ	ਜ	ਝ	ਞ	ਟ	ਠ	ਡ	ਢ	ਣ	
ਤ	ਥ	ਦ	ਧ	ਨ	ਪ	ਫ	ਬ	ਭ	ਮ	
ਯ	ਰ	ਲ	ਵ	ੜ	ਸ਼	ਜ਼	ਖ਼	ਗ਼	ਫ਼	ਲ਼
ਾ	ਿ	ੀ	ੁ	ੂ	ੇ	ੈ	ੋ	ੌ	–	=
ੰ	ੱ	੍								

Fig 1: Gurmukhi script character set

Upper Zone
Middle Zone
Lower Zone

Fig 2: Three zones of a Gurmukhi word

3 Proposed Technique

After digitization of the text, the text image is subjected to pre-processing routines such as noise removal, thinning and skew correction. The thinned and cleaned text image is then sent to the text segmenter, which segments each uniform text zone into text lines and text lines into words. Words are further segmented into characters and sub-characters. To simplify character segmentation, since it is difficult to separate a cursive word directly into characters, a smaller unit than a character is preferred. In our current work, we have taken an 8-connected component as the basic image representation throughout the recognition process and thus instead of character segmentation we have performed *connected component segmentation*. A combination of statistical analysis of text height, horizontal projection and vertical projection and connected component analysis is performed to segment a text image into connected components.

3.1 Line Segmentation

Horizontal projection, which is most commonly employed to extract the lines from the document [3-5], fails in many cases when applied to Gurmukhi text and results in over segmentation or under segmentation. Over segmentation occurs when the white space breaks a text line into 2 or more horizontal text strips (Fig 3a). Under segmentation occurs when one or more vowel symbols in upper zone of a text line overlap with modifiers present in lower zone of previous line. As a result, white space no longer separates 2 consecutive text lines and two or more text lines may be fused together (Fig 3b). So special care has to be taken for these cases.

The text image is broken into horizontal text strips using horizontal projection in each row. The gaps on the horizontal projection profile are taken as separators between the text strips. Each text strip could represent a) Core zone of one text line consisting of upper, middle zone and optionally lower zone (core strip). b) Upper zone of a text line (upper strip). c) Lower zone of a text line (lower strip). d) Core zone of more than one text line (multi strip). The next task is to identify the type of each strip.

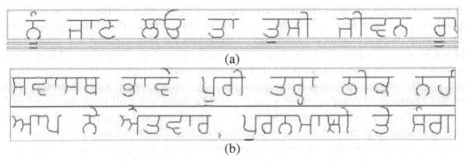

(a)

(b)

Fig. 3. Example of a) Over Segmentation b)Under Segmentation. Red line indicates the overlap region.

For this we first calculate the estimated average height of the core strip. We found that we cannot directly take the arithmetic mean of all the strips as the average height of the core strip, since the height of the upper strips, lower strips and multi strips present in the document image can greatly influence the overall figure. From experiments (table 1), it was observed that more than 60% of the text strips were core strips and thus 75 percentile of height of all the strips closely represented the average height of a core strip. We call this height as *AV*. Once the average height of core strip is found, then the class of the strips is identified. If the height of a strip is lesser than 33% of *AV*, then the strip is a lower strip or an upper strip. If the height is greater then 150% of *AV*, then the strip is a multi strip. Otherwise the strip is core strip. To distinguish between upper and lower strips, we look at the immediate next core strip. We determine the spatial position of headline in the immediate next core strip, where the headline is found by locating the row with maximum number of black pixels. If the headline is present in upper 10% region of the core strip, then the previous strip is an upper strip else it is a lower strip. Next determine the accurate average height of a core strip (ACSH) by calculating the arithmetic mean of all core strip. This information will be used to dissect the multi strip into constituent text lines. The average consonant height (ACH) is also estimated by calculating the average height of the text lying below the headline in each core strip. This information is needed in the other segmentation phases. Some statistics of horizontal strips generated from 40 document images from books, laser print outs and newspaper are tabulated in table 1.

Table 1. Statistics of Horizontal Strips

Strip Type	Min height (Pixels)	Max height (Pixels)	Average Height (Pixels)	Percentage of occurrence
Lower Strip	1	13	2.6	35.7%
Upper Strip	6	8	6.5	0.2%
Core Strip	36	63	46.7	60.5%
Multi Strip	73	101	87.2	3.6%

3.2 Sub Division of Strips into Smaller Units

In the next stage, all the text strips are processed from top to bottom in the order in which they occur in text and divided into smaller components. If a strip is an upper or lower strip, then it is entirely made up of disconnected characters or sub characters. These characters can easily be isolated by scanning from left to right till a black pixel is found and then using a search algorithm finding all the black pixels connected to it. Each such connected component represents one character or sub character. The smallest connected component of a core strip is a word since all the consonants and majority of upper zone vowels in the word are glued with the headline, so there is no inter character gap and white space separates words. The word may not contain the complete character images, as some of the characters or their parts may be present in the neighbouring strips. For segmentation of the strip into words vertical projection is employed by counting the number of black pixels in each vertical line, and a gap of 2 or more pixels in the histogram is taken to be the word delimiter. The word is then broken into sub-characters. First the position of the headline in the word is found by looking for the most dominant row in the upper half of the word. The connected component segmentation process proceeds in 3 stages. In the first stage the connected component touching the headline and present in the middle and upper zone are isolated. In the second stage the character sub-parts not touching the headline and lying in upper zone are segmented while the characters in lower zones are isolated in third stage. The black pixels lying on the headline are not considered while calculating the connected components, otherwise all the characters glued along the headline will be treated as a single connected component. Each connected component now represents a) a single character or b) a part of character lying in one of upper, middle or lower zone. The zonal position of the connected components, coordinate of the left most pixel in the bitmap and the amount of overlapping with other components in other zones is later used to cluster the connected components into characters.

A multiple core strip is made of multiple overlapping text lines, which cannot be separated by horizontal projection. To segregate the text lines, the statistics generated in first pass will be used. The zonal height is divided by average core strip height (ACSH) to get an idea about the number of text lines present in the strip. To extract the first row, an imaginary cut is made at 0.75*ACSH. We have deliberately made a cut at 0.75*ACHS instead of ACSH, so that by chance the line does not cross any character lying in next text row. This cut will be slicing most of the words into two parts, but that does not create any problem since we are looking for connected components only. The portions of the words which have been sliced and are lying in next sub-strip will also be added to the connected component of current word, since physically they are still connected. The sub-strip is then split into words by vertical projection analysis using the same method as used in segmentation of core strip. For the next sub-strip a cut is made at ACSH height and the words are extracted in that sub-strip. This process continues till all the sub-strips have been segmented into words. Next the connected components present in the upper zone of the word are identified. The upper zone can also contain lower zone vowels of words lying in previous line. So a distinction is made using the distance of minimum bounding rectangle with the headline. We do not search for connected components in the lower zone, since if accidentally a connected component of upper zone of a word present in next line is encountered, then the search will lead to words present in next line and they will all be identified as lower zone symbols of current word.

4 Touching Characters

Segmentation of touching characters has been the most difficult problem in character segmentation. Various papers have appeared concerning the segmentation of touching characters [1, 8-9]. All these papers have dealt with Roman script text only. As already mentioned segmentation process for Gurmukhi script proceeds in both x and y directions since two or more characters of a word may be sharing the same x coordinate. So for segmentation of touching characters in Gurmukhi script, the merging points of the touching characters have to be determined both along the x and y axes. During our experiments we found a large percentage of touching characters in even clean machine printed Gurmukhi text. These touching characters can be categorized as follows as a)Touching characters in upper zone b)Touching characters in middle zone c)Lower zone characters touching with middle zone characters and d)Lower zone characters touching with each other (Fig. 4)

4.1 Touching Characters/Connected Components in Upper Zone
Closely lying upper zone vowels frequently touch each other in even clean documents. Another common problem encountered is the merging of the dot symbol with other vowels or headline. In our experiments we found that about **6.9%** of the upper zone vowels were touching other vowels or merging with the headline.

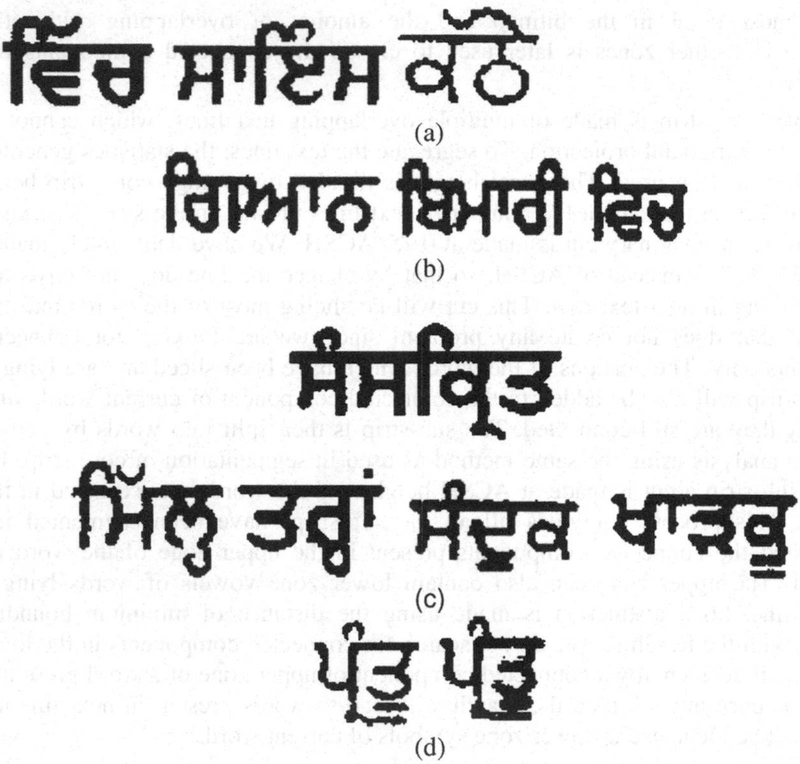

ਵਿੱਚ ਸਾਇੰਸ ਕੰਠ

(a)

ਗਿਆਨ ਬਿਮਾਰੀ ਵਿੱਚ

(b)

ਸੰਸਕ੍ਰਿਤ

ਮਿਲ਼ੁ ਤਰ੍ਹਾਂ ਸੰਦੂਕ ਖਾੜ੍ਹੁ

(c)

ਪੂੜ੍ਹ ਪੂੜ੍ਹ

(d)

Fig 4. Examples of touching characters a) touching characters in upper zone b)touching characters in middle zone c) Lower zone characters touching with middle zone characters d) Lower zone characters touching with each other

A connected component (CC) in upper zone is a candidate for further splitting if it satisfies one of the following two conditions:
a) Width of the CC is more than 75% of ACH: In normal cases the width of a vowel in upper zone is less than 75% of average height of a consonant (ACH). So if the width of the CC exceeds 75% of ACH then it is assumed that we have multiple fused vowels. If the width of the CC is x, then the potential cutting point is found in the region x/4 to 3x/4. For determining the cutting point, the algorithm suggested by Kahan at al [8] is used.
b) Presence of an eastward oriented stroke in the second half of the CC along the x-axis: A careful analysis of the shapes of upper zone vowels (Fig 1) reveals that there is no vowel in upper zone which has a junction in the second half along the x-axis and a stroke originating from that junction which is oriented in eastern direction and not touching the headline. Thus if there exists one such stroke, then it is not part of the vowel. We use this property to search for a joint in the second half of the CC. If it is present then check for existence of a stroke emerging from this joint which is oriented in eastern direction and not touching the headline. If such a stroke is present then it is disconnected from the main connected component. Using this technique we were able to successfully separate the merged dot symbol from the CC of an upper vowel (Fig. 5).

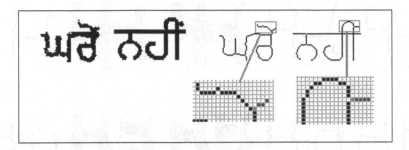

Fig 5. Separation of touching connected components in upper zone. On the left side are the scanned images of text and their thinned versions are on the right. Blue segments represent the detected touching connected components

4.2 Touching Characters/Connected Components in Middle Zone

On clean machine printed text, the frequency of occurrence of touching characters in the middle zone is found to be quite low **(0.12%)**. This is also the region, where the touching characters, if present, are most difficult to detect and split. Since the chances of touching characters in middle zone are very low, so we normally do not test for occurrence of touching characters in this region. The testing is done only if :
i) The width of a CC is more than 175% of ACH ii) The classifier fails to recognize the CC. To detect the cutting point, the method suggested by Kahan et al[8] is used. But a new problem was faced, as we are dealing with thinned images, it was found that sometimes a) The cutting point was not correctly found. This is because of thinned image the number of pixels is reduced and there is a shift in the direction of pixels because of thinning. This is illustrated in fig 6, where the image of touching character pair mA is incorrectly segmented . b) The cutting point may be correctly found but after separation into two CCs, the shape of the CC is so badly deformed that the classifier fails to recognize it. This can also be seen in fig 6, where the CC of the thinned image of touching character pair bm sans headline, is correctly segmented but the shape of the CC corresponding to m is disfigured and it cannot be correctly recognized.
These twin problems were overcome by considering the unthinned version of the CCs. The original image is retained along with its thinned version and as there are sufficient number of pixels in original scanned images, so the touching point is more easily identified. The image is then split at the touching point and the separated images are then thinned and sent to the classifier for identification (Fig 6).

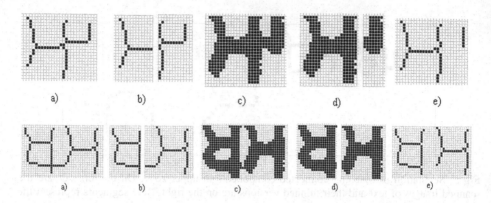

Fig 6. a) Touching Connected components b) Components separated by the algorithm c) Unthinned versions of connected components of fig 6a. d) Images of fig 6c split by Kahan method e) Thinned images of segmented images of fig 6d.

4.3 Connected Components in Lower Zone Touching the Above Lying Connected Components

It was observed that the half characters and the vowels lying in lower zone frequently touch the above lying consonants. The frequency of lower zone characters touching the middle zone characters is found to be **19.1%**. For segmentation of these half characters, the average height of a consonant was used. It can be observed that for same font and size, the height of all consonants is almost same. So if the height of a consonant is found to be more than 120% of ACH then the consonant is topologically disconnected into two parts at the row near ACH with minimum number of black pixels. By default the cutting point is taken as the immediate row below the ACH number of pixels in the y-axis and we look up and below 10% of ACH number of rows for any row with fewer number of pixels. In case there exists such row, that row is taken as the cutting row. This method works well for most of the cases but fails in case where the touching lower zone character (such as vertical line like character such as the character ⁻ ▪)does not much increase the height of the middle zone consonant (Fig 7).

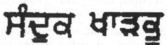

Fig 7. Scanned touching images of words ਸੰਦੂਕ ਖਾੜਕੂ

4.4 Connected Components in Lower Zone Touching Each Other

In some rare instances the lower zone characters were found touching each other. The frequency of such occurrence though is very low **(0.031%)**. For segmentation of such character pairs, the simple technique of splitting the connected component at the middle of the horizontal axis served the purpose. For identification of such merged character pairs, the same criteria as used for upper zone vowels is used, that is if the width of a connected component in lower zone exceeds 75% of ACH then it is assumed that we have multiple fused vowels/half characters.

5 Conclusion

We have presented a scheme for decomposing a text image in Gurmukhi script into sub-characters or connected components. These connected components are then recognized by the classifier and merged to form characters. The various complexities such as absence of vertical inter-character gap in a word, horizontally overlapping text lines, multi-component characters, overlapping and intersecting circumscribing rectangles of characters, touching characters in various zonal positions in a word, presence of a lower character of a word in upper zone of a word in next line etc. have been taken care of by using this scheme. This is the first time that the problem of touching characters in Gurmukhi text has been studied in detail and solutions suggested for tackling the touching characters in various zonal positions of a word. Table 2 shows the accuracy rate of detecting and correctly segmenting the touching characters. These results are obtained by implementing and testing the proposed technique on about 40 machine printed Gurmukhi documents.

Table 2. Accuracy rate of detection and segmentation of touching characters

Type of touching characters/connected components	% of correct detection and segmentation
Touching/merging upper zone vowels	92.5%
Touching middle zone consonants	72.3%
Touching middle zone and lower zone characters	89.3%
Touching lower zone characters	95.2%

References

[1] Lu, Y.: Machine Printed Character Segmentation – an Overview. Pattern Recognition, **28** (1995) 67-80
[2] Casy, R.G., Lecolinet, E.: A survey of methods and strategies in character segmentation. IEEE Transactions on Pattern Analysis and Machine Intelligence, **18** (1996) 690-706
[3] Chaudhuri, B.B., Pal, U.: A complete printed Bangla OCR system. Pattern Recognition, **31** (1998) 531-549
[4] Pal, U., Chaudhuri, B.B.: Printed Devnagri Script OCR System. Vivek, **10** (1997) 12-24

[5] Bansal, V.: Integrating knowledge sources in Devanagri text recognition. Ph.D. thesis, IIT Kanpur, INDIA (1999)
[6] Goyal, A.K., Lehal, G.S., Deol, S.S.: Segmentation of Machine Printed Gurmukhi Script. In: Proceedings 9th International Graphonomics Society Conference, Singapore (1999) 293-297
[7] Lehal, G.S., Singh, S.: Text segmentation of Machine Printed Gurmukhi Script. Document Recognition and Retrieval VIII, Kantor, P.B., Lopresti, D.P., Jiangying Zhou, (eds.): Proceedings SPIE, USA, Vol. 4307 (2001) 223-231
[8] Kahan, S., Pavlidis, T., Baird, H.S.: On the recognition of printed characters of any font and size. IEEE Transactions on Pattern Analysis and Machine Intelligence, 9 (1987) 274-287
[9] Liang, S., Shirdhar, M., Ahmed, M.: Segmentation of touching characters in printed document recognition. Pattern Recognition, 27 (1994) 825-840

Efficient Computation of Body Moments

Alexander V. Tuzikov*, Stanislav A. Sheynin, and Pavel V. Vasiliev**

Institute of Engineering Cybernetics
Academy of Sciences of Republic Belarus
Surganova 6, 220012 Minsk, Belarus
{tuzikov,sheynin,vasiliev}@mpen.bas-net.by

Abstract. We describe an efficient algorithm for a calculation of 3D body volume and surface moments. The algorithm is based on implicit formulae for moment calculation and takes advantages of a polygonal representation. It uses only coordinates of the body vertices and facets orientation.

Keywords: moment, polygonal representation, implicit formulae

1 Introduction

It is well known that such important geometric characteristics of 3D bodies as volume and orientation can be defined in terms of moments [1]. Ellipsoid of inertia corresponding to the body has, for example, the same zero, first and second moments as the body itself. The orientation of this ellipsoid in space usually is taken for the orientation of the body.

Ellipsoid of inertia provides also a very useful information for defining symmetry planes and axes of rotation symmetry for almost symmetrical bodies [10].

Note also that moments and moment invariants are widely used in object recognition (see, for example, [6,7]).

A moment computation depends greatly on a body representation. In a discrete case when a body is given by all its voxels there is sufficient to perform one scan of body elements to compute the moments. However, the problem becomes much more complicated if the body is given by its surface. Even in the discrete case one needs a procedure to get coordinates of all body voxels.

One of the popular representations in computer graphics is a polygonal representation. In this case an object is represented by a mesh of polygonal facets. In the simplest case a polygon mesh is a structure that consists of polygons represented by a list of linked (x, y, z) coordinates that are the polygon vertices [11].

The aim of this report is to describe an efficient algorithm for a 3D body moment calculation. The algorithm takes the advantages of a polygonal representation. It uses only coordinates of the body vertices and facets orientation.

---

* A. Tuzikov and S. Sheynin were supported by INTAS 00-397 grant.
** P. Vasiliev was supported by INTAS YSF 00-72 grant.

W. Skarbek (Ed.): CAIP 2001, LNCS 2124, pp. 201–208, 2001.

A number of papers were devoted to developing fast algorithms of moment computation [2,3]. Explicit formulae for 2D polygonal shape moment calculation based on boundary vertex coordinates were derived in [9]. Related results are obtained also for 3D polyhedral shapes [4,5]. However, the formulae or recurrent procedures developed there for polyhedral shape moment calculation are not convenient for applications.

2   Tetrahedron Moments

Let us suppose that body P is a tetrahedron $T = T(\mathbf{a}, \mathbf{b}, \mathbf{c})$ defined by the coordinates origin and three points $\mathbf{a} = (a_1, a_2, a_3)$, $\mathbf{b} = (b_1, b_2, b_3)$, $\mathbf{c} = (c_1, c_2, c_3)$ (see Figure 1). Denote also by $S_0(\mathbf{a}, \mathbf{b}, \mathbf{c})$ a tetrahedron facet that is opposite to the coordinates origin. We assume that a triple of vectors $\boldsymbol{a} = (a_1, a_2, a_3)$, $\boldsymbol{b} = (b_1, b_2, b_3)$, $\boldsymbol{c} = (c_1, c_2, c_3)$ has a right orientation. It means that points $\mathbf{a}, \mathbf{b}, \mathbf{c}$ have a counter-clockwise order with respect to the outward normal of the facet $S_0(\mathbf{a}, \mathbf{b}, \mathbf{c})$.

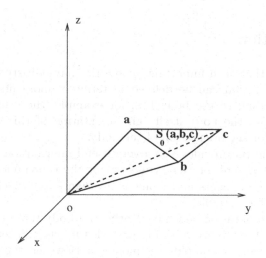

Fig. 1. Tetrahedron $T(\mathbf{a}, \mathbf{b}, \mathbf{c})$.

In this section we are interested in computation of tetrahedron volume and surface moments denoted by $m_{k_1 k_2 k_3} V(T)$ and $m_{k_1 k_2 k_3} S_0(T)$, respectively

$$m_{k_1 k_2 k_3} V(T) = \int_T x^{k_1} y^{k_2} z^{k_3} \, dx \, dy \, dz \tag{1}$$

$$m_{k_1 k_2 k_3} S_0(T) = \int_{S_0(T)} x^{k_1} y^{k_2} z^{k_3} \, dS \tag{2}$$

Here the first integral is taken on the volume of T and the second one – on the facet $S_0(T)$ of T.

We have derived in [8] the explicit formulae for computation of arbitrary order volume moments for a tetrahedron. Let us introduce some notations used in these formulae. Denote by $A = (a_{ij})$ a matrix consisting of $\mathbf{a}, \mathbf{b}, \mathbf{c}$ coordinates, i.e. $a_{i1} = a_i$, $a_{i2} = b_i$, $a_{i3} = c_i$, $i = 1, 2, 3$.

Given integers k_1, k_2, k_3 denote by $\mathcal{K}$ a set of such 3×3 matrices (k_{ij}) with integer values $0 \leq k_{ij} \leq k_i$ that $\sum_{j=1}^{3} k_{ij} = k_i$, $i = 1, 2, 3$.

Now we present an implicit formula for a volume moment [8] (here it is considered a 3D case instead of the general n-dimensional one discussed there).

Theorem 1. *Given a tetrahedron $T = T(\mathbf{a}, \mathbf{b}, \mathbf{c})$ the following formula is true for a volume moment $m_{k_1 k_2 k_3} V(T)$ of order $k = k_1 + k_2 + k_3$*

$$m_{k_1 k_2 k_3} V(T) = \frac{|A| k_1! k_2! k_3!}{(k+3)!} \sum_{(k_{ij}) \in \mathcal{K}} \frac{\prod_{j=1}^{3} \left(\left(\sum_{i=1}^{3} k_{ij} \right)! \right)}{\prod_{i,j=1}^{3} (k_{ij}!)} \prod_{i,j=1}^{3} a_{ij}^{k_{ij}}, \qquad (3)$$

where $|A|$ denotes the determinant of A.

A similar result is valid for the surface moment $m_{k_1 k_2 k_3} S_0(T)$.

Theorem 2. *Given a tetrahedron $T = T(\mathbf{a}, \mathbf{b}, \mathbf{c})$ the following formula is true for a surface moment $m_{k_1 k_2 k_3} S_0(T)$ of order $k = k_1 + k_2 + k_3$*

$$m_{k_1 k_2 k_3} S_0(T) = \frac{2 Ar(S_0) k_1! k_2! k_3!}{(k+2)!} \sum_{(k_{ij}) \in \mathcal{K}} \frac{\prod_{j=1}^{3} \left(\left(\sum_{i=1}^{3} k_{ij} \right)! \right)}{\prod_{i,j=1}^{3} (k_{ij}!)} \prod_{i,j=1}^{3} a_{ij}^{k_{ij}}, \qquad (4)$$

where $Ar(S_0)$ denotes the area of the facet $S_0(T)$.

For moments of order k for one coordinate only the formulae (3) and (4) have a simpler form (below they are given for the first coordinate):

$$m_{k00} V(T) = \frac{|A| k!}{(k+3)!} \sum_{(k_1, k_2, k_3)} a_1^{k_1} b_1^{k_2} c_1^{k_3},$$

$$m_{k00} S_0(T) = \frac{2 Ar(S_0)}{(k+2)!} \sum_{(k_1, k_2, k_3)} a_1^{k_1} b_1^{k_2} c_1^{k_3},$$

where the summation is done for such triples (k_1, k_2, k_3) of integer numbers that $k_1 + k_2 + k_3 = k$, $k_j \geq 0$, $j = 1, 2, 3$.

Let us now derive as an illustration the formula for the volume moment $m_{101} V(T(\mathbf{a}, \mathbf{b}, \mathbf{c}))$. The summation in (3) should be done for the following matrices (k_{ij}):

$$\begin{pmatrix} 1 & 0 & 0 \\ 0 & 0 & 0 \\ 1 & 0 & 0 \end{pmatrix} \quad \begin{pmatrix} 1 & 0 & 0 \\ 0 & 0 & 0 \\ 0 & 1 & 0 \end{pmatrix} \quad \begin{pmatrix} 1 & 0 & 0 \\ 0 & 0 & 0 \\ 0 & 0 & 1 \end{pmatrix} \quad \begin{pmatrix} 0 & 1 & 0 \\ 0 & 0 & 0 \\ 1 & 0 & 0 \end{pmatrix} \quad \begin{pmatrix} 0 & 1 & 0 \\ 0 & 0 & 0 \\ 0 & 1 & 0 \end{pmatrix}$$

$$\begin{pmatrix} 0 & 1 & 0 \\ 0 & 0 & 0 \\ 0 & 0 & 1 \end{pmatrix} \quad \begin{pmatrix} 0 & 0 & 1 \\ 0 & 0 & 0 \\ 1 & 0 & 0 \end{pmatrix} \quad \begin{pmatrix} 0 & 0 & 1 \\ 0 & 0 & 0 \\ 0 & 1 & 0 \end{pmatrix} \quad \begin{pmatrix} 0 & 0 & 1 \\ 0 & 0 & 0 \\ 0 & 0 & 1 \end{pmatrix}$$

One gets the following expression for the first matrix

$$\frac{\prod_{j=1}^{3}(\sum_{i=1}^{3} k_{ij})!}{\prod_{i,j=1}^{3} k_{ij}!} \prod_{i,j=1}^{3} a_{ij}^{k_{ij}} = 2a_{11}a_{31}$$

Adding results for other matrices gives the moment formula

$$m_{101}V(T(\mathbf{a},\mathbf{b},\mathbf{c})) = \frac{|A|}{120}(2a_{11}a_{31} + a_{11}a_{32} +$$

$$a_{11}a_{33} + a_{12}a_{31} + 2a_{12}a_{32} + a_{12}a_{33} + a_{13}a_{31} + a_{13}a_{32} + 2a_{13}a_{33}) =$$

$$\frac{|A|}{120}(2a_1a_3 + a_1b_3 + a_1c_3 + a_3b_1 + 2b_1b_3 + b_1c_3 + c_1a_3 + c_1b_3 + 2c_1c_3)$$

Similarly we can get the formulae for moments $m_{000}V(T)$ (tetrahedron volume), $m_{100}V(T)$, $m_{200}V(T)$. The formulae for other moments till order 2 follow straightforward from them by the symmetry property.

$$m_{000}V(T) = \frac{1}{6}|A|, \qquad m_{100}V(T) = \frac{1}{24}|A|(a_1 + b_1 + c_1),$$

$$m_{200}V(T) = \frac{1}{60}|A|(a_1^2 + b_1^2 + c_1^2 + a_1b_1 + a_1c_1 + b_1c_1)$$

3   Body Moment Computation

In this section we consider a moment computation for arbitrary body P bounded by polygonal facets. Let us partition every face of P into oriented triangles. Suppose that vertices of every face are ordered counter-clockwise while looking from the outward normal to the face. When choosing arbitrarily a starting point $\mathbf{v}_0$ on a face with n vertices one gets an ordered sequence of vertices $\mathbf{v}_1, \mathbf{v}_2, \dots, \mathbf{v}_n = \mathbf{v}_0$. Then the i-th triangle is defined by vertices $\mathbf{v}_0, \mathbf{v}_i, \mathbf{v}_{i+1}$, $i = 1, \dots, n-2$. Denote by $T(\mathbf{v}_0, \mathbf{v}_i, \mathbf{v}_{i+1})$ a tetrahedron defined by i-th triangle and the coordinate origin (see Fig. 2).

Then the following formula is true for a volume moments of arbitrary orders

$$m_{k_1 k_2 k_3}V(P) = \int_P x^{k_1} y^{k_2} z^{k_3} dx dy dz =$$

$$\sum_i \int_{T_i} x^{k_1} y^{k_2} z^{k_3} dx dy dz = \sum_i m_{k_1 k_2 k_3}V(T_i), \quad (5)$$

where the sum is taken for all tetrahedra T_i of all faces and $m_{k_1 k_2 k_3}V(T_i)$ is the volume moment of tethahedron T_i. Note that we consider oriented moments

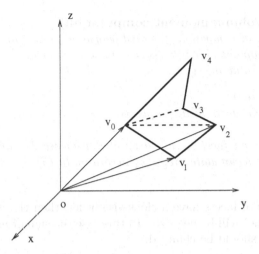

Fig. 2. Partition of a body facet $v_0 v_1 v_2 v_3 v_4$ into triangles.

(i.e. the sign of this moment for tetrahedron T_i depends on the orientation of the vertices of facet $S_0(T_i)$). Thus a body volume moment equals to the sum of oriented moments of the corresponding tetrahedra. It is clear that moments for some tetrahedra will be added and for others will be subtracted. Since the sum is computed for all triangles of all facets, then the formula is independent on the position of the coordinate origin.

One can check that a similar formula is true for the computation of surface moments of body P, i.e.

$$m_{k_1 k_2 k_3} S(P) = \int_{S(P)} x^{k_1} y^{k_2} z^{k_3} \, dS = \sum_i m_{k_1 k_2 k_3} S_0(T_i). \tag{6}$$

Here $S(P)$ and $S_0(T_i)$ denote the surface of P and the facet S_0 of tetrahedron T_i, respectively.

4   Algorithm Implementation

Due to the nature of formulae (5) and (6), the computational algorithm may benefit from the following features:

- Polygonal facets may be processed independently one from another. Therefore the set of facets may be considered as a sequence, or a "stream".
- A body moment can be computed as a sum of corresponding moments calculated in several computational threads, running in parallel, each of them processing independently a subsequence of facets.
- Formulae (5) and (6) for volume and surface moments differ only in a constant factor and volume/area multipliers. However, the computation of volume moments requires that all facets have a compatible orientation. This requirement is not necessary for surface moments.

Algorithm 1 (Volume moment computation)
Input: a sequence of compatibly oriented polygonal facets (for example, counter-clockwise) and moment order (k_1, k_2, k_3) to be computed.
Output: moment value m_{k_1,k_2,k_3}.

1. *Set m_{k_1,k_2,k_3} to 0.*
2. *Iterate over all facets in the sequence.*
   {

   > *For a chosen facet iterate over all tetrahedra T_i corresponding to this facet and accumulate m_{k_1,k_2,k_3} according to (5).*

   }

If the vertices of facets have a clockwise order, then the calculated moment $m_{000}V(P)$ (volume) will be negative. In this case the sign of m_{k_1,k_2,k_3} calculated by the algorithm should be changed.

Algorithm 2 (Surface moment computation)
Input: a sequence of polygonal facets and moment order (k_1, k_2, k_3) to be computed.
Intermediate: facet moment value m and facet area Ar.
Output: moment value m_{k_1,k_2,k_3}.

1. *Set m_{k_1,k_2,k_3} to 0.*
2. *Iterate over all facets in the sequence.*
   {
   - *For a chosen facet set m to zero and iterate over all triangles corresponding to this facet.*
     {
       - *For the first triangle defined by vertices $\mathbf{v}_0, \mathbf{v}_1, \mathbf{v}_2$ compute the facet normal vector as $\mathbf{n} = [(\mathbf{v}_1 - \mathbf{v}_0) \times (\mathbf{v}_2 - \mathbf{v}_0)]$ and the area $Ar_1 = |\mathbf{n}|/2$. Here $[\mathbf{a} \times \mathbf{b}]$ denotes the outer product of vectors $\mathbf{a}$ and $\mathbf{b}$. Set $Ar = Ar_1$.*
       - *For the i-th triangle defined by vertices $\mathbf{v}_0, \mathbf{v}_i, \mathbf{v}_{i+1}$, $i \geq 2$, calculate vector $\mathbf{n}_i = [(\mathbf{v}_i - \mathbf{v}_0) \times (\mathbf{v}_{i+1} - \mathbf{v}_0)]$ and oriented area $Ar_i = (|\mathbf{n}_i|/2) \cdot sign(< \mathbf{n}_i, \mathbf{n}_1 >)$. Here $< \mathbf{a}, \mathbf{b} >$ denotes the inner product of vectors $\mathbf{a}$ and $\mathbf{b}$. Set $Ar = Ar + Ar_i$ and accumulate m according to (6).*
     }
   - *If Ar is negative then change the sign of the computed m.*
   - *Add m to m_{k_1,k_2,k_3}.*
   }

Let us point out that while computing a surface moment the requirement of compatible orientation for all facets is not important. It means that some facets can have a counter-clockwise order of their vertices and the others – a clockwise order of vertices. The different orders result only in a positive or a negative value of the facet area. Therefore, checking the sign of facet areas allows to compute a correct value of a surface moment.

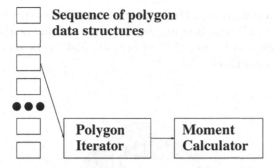

Fig. 3. A single-threaded (non-parallel) moment computation engine. "Moment Calculator" uses "Polygon Iterator" to scan a sequence of polygons (facets), one polygon at a time.

Our implementation of the algorithms is written in C++. We use the following approach to design a computational engine (see Figures 3 and 4). Body moments are calculated by the "Moment Calculator". This is a C++ object, that computes and accumulates moments for facets, one at a time, that are supplied in a predefined format by another object "Polygon Iterator". In our implementation the "Moment Calculator" is represented by a template class that accepts "Polygon Iterator" classes as a template parameter. This scheme allows to use a single implementation of "Moment Calculator" for different polygon representations, provided that a specific "Polygon Iterator" class is implemented for the data format under consideration.

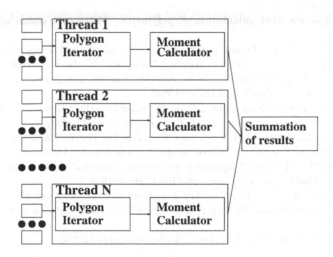

Fig. 4. A multithreaded (parallel) moment computation engine. There are several threads, each processing a subsequence of polygons. A separate "Moment Calculator" and "Polygon Iterator" are used in every thread.

The algorithm computes volume and surface moments very fast. For example, the computation of all zero, first and second order volume (surface) moments for the object presented in Figure 5 (320 vertices, 320 facets and 640 edges) takes 0.36ms (0.64ms) on PII-400 PC.

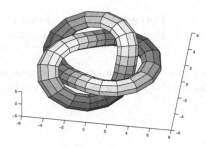

Fig. 5. Polyhedron "trefoil" (320 vertices, 320 facets, 640 edges).

References

1. Horn, B.K.P.: Robot Vision. MIT Press, Cambridge (1986)
2. Jiang, X.Y., Bunke, H.: Simple and fast computation of moments. Pattern Recognition, **24** (1991) 801–806
3. Leu, J.-G.: Computing a shapes's moments from its boundary. Pattern Recognition, **24** (1991) 949–957
4. Li, B.: The moment calculation of polyhedra. Pattern Recognition, **26** (1993) 1229–1233
5. Liggett, J.A.: Exact formulae for areas, volumes and moments of polygons and polyhedra. Communications in Applied Numerical Methods **4** (1988) 815–820
6. Mukundan, R., Ramakrishnan, K.R.: Moment Functions in Image Analysis: Theory and Applications. World Scientific (1998)
7. Reiss, T.H.: Recognizing Planar Objects Using Invariant Image Features. Lecture Notes in Computer Science, Vol. 676. Springer-Verlag (1993)
8. Sheynin, S.A., Tuzikov, A.V.: Exact formulae for polyhedra moments. Preprint 11. Institute of Engineering Cybernetics NANB, Minsk (1999)
9. Singer, M.H.: A general approach to moment calculation for polygons and line segments. Pattern Recognition, **26** (1993) 1019–1028
10. Sun, C., Sherrah, J.: 3-D symmetry detection using the extended Gaussian image. IEEE Transactions on Pattern Analysis and Machine Intelligence, **19** (1997) 164–168
11. Watt, A.: 3D Computer Graphics. Addison-Wesley (1993)

Genetic Programming with Local Improvement for Visual Learning from Examples

Krzysztof Krawiec

Institute of Computing Science, Poznań University of Technology,
Piotrowo 3A, 60965 Poznań, Poland
krawiec@cs.put.poznan.pl

Abstract. This paper investigates the use of evolutionary programming for the search of hypothesis space in visual learning tasks. The general goal of the project is to elaborate human-competitive procedures for pattern discrimination by means of learning based on the training data (set of images). In particular, the topic addressed here is the comparison between the 'standard' genetic programming (as defined by Koza [13]) and the genetic programming extended by local optimization of solutions, so-called *genetic local search*. The hypothesis formulated in the paper is that genetic local search provides better solutions (i.e. classifiers with higher predictive accuracy) than the genetic search without that extension. This supposition was positively verified in an extensive comparative experiment of visual learning concerning the recognition of handwritten characters.

Keywords: visual learning, genetic local search, learning from examples

1 Introduction

The search for an appropriate processing and representation of the visual information data is the most complex part of the design of computer vision systems. This task is weekly formalized so far, therefore in most cases the human designer is made responsible for it (see [8], p.657). Requiring significant body of knowledge, experience and even intuition, this work is usually tedious and expensive. The solutions (representations, algorithms, etc.) chosen by the human expert, although usually useful, are subjective and sub-optimal in terms of system performance. For instance, if a machine learning classifier [18, 21] is used for the decision making based on image representation (what is a common approach), the hypothesis space searched during its training is limited by that representation. This phenomenon may disable the training process from discovering beneficial solutions for the considered vision task (e.g. recognition, analysis, or interpretation).

To overcome these difficulties, in this study we aim at synthesizing the complete image analysis and interpretation program without splitting it explicitly into stages of feature extraction and classification. As a result, the learning process is no more limited by the representation predefined by the human expert, but encompasses also the

W. Skarbek (Ed.): CAIP 2001, LNCS 2124, pp. 209–216, 2001.

image processing and analysis. Thus, we follow here the paradigm of *direct* approach to pattern recognition, occupied so far mostly by the artificial neural networks. However, instead of subsymbolic reasoning and gradient-based learning, we employ the paradigm of evolutionary computation for the search of the space of image analysis programs.

2   Evolutionary Computation and Genetic Programming

Evolutionary computation (EC) [4,9] has been used in machine learning (ML) and pattern recognition community for quite a long time [20, 21]. Now it is widely recognized as a useful metaheuristics or even as one of ML paradigms [18, 21]. It is highly appreciated due to its ability to perform global parallel search of the solution space with low probability of getting stuck in local optima of evaluation function. Its most renowned ML-related applications include feature selection [25] and concept induction [5,6]. In this paper, we focus on the latter of the aforementioned tasks, with individual solutions implementing hypotheses considered by the system being trained (from now on, the terms 'solution' and 'hypothesis' will be used interchangeably). For the sake of brevity we skip the presentation of EC fundamentals, assuming reader's familiarity with them.

Genetic programming (GP) is a specific paradigm of evolutionary computation proposed by Koza [13] employing more sophisticated representation of solutions than 'plain' genetic algorithms, which use strings over binary alphabet. Thus, solutions in GP, usually LISP expressions, implement the problem's specificity in a more direct way. GP is reported to be very effective in solving a broad scope of learning and optimization problems, including the impressive achievement of creating human-competitive designs for the controller synthesis problems, some of which have been even patented [14].

3   Genetic Programming for Visual Learning

Evolutionary computation found some applications in image processing and analysis (e.g. [2]). However, there are relatively few research projects, which aim at combining image processing and analysis with learning, i.e. the *visual learning*, meant as the search for pattern recognition programs [12, 23, 22, 15, 17]. Only a small fraction of research focuses on inducing the complete image analysis programs based on training examples (images), a paradigm which is very promising and universal in our opinion and which is subject of our research on *genetic programming-based visual learning* [15, 16, 17]. From the machine learning viewpoint [18, 21], we focus on the paradigm of supervised learning from examples, employing the genetic programming for the hypothesis representation and for the search of hypothesis space. The candidate programs performing image analysis and recognition are evaluated on a set of training cases (images). In particular, the solution's fitness reflects the accuracy of classification it provides on the training set.

The solutions (pattern recognition procedures) are expressed in GPVIS [16], an image analysis-oriented language encompassing a set of operators implementing simple feature extraction, region-of-interest selection, and arithmetic and logical operations. GPVIS allows formulating the complete pattern recognition program unnecessarily of an external machine learning classifier. This is the opposite of conventional approach, where the processing is divided into the stages of feature extraction and the reasoning.

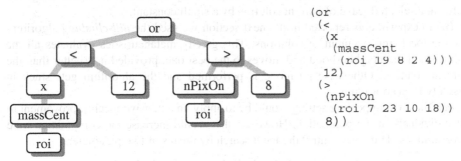

Fig. 1. Tree-like and textual representations of an exemplary solution formulated in GPVIS language (numerical values omitted in the former representation).

Figure 1 shows a simple example of image analysis program formulated in GPVIS, which has the following interpretation: *if* the x-coordinate of the visual mass center of the rectangular region of interest (*roi*) with coordinates of upper left corner (19,8) and lower right corner (2,4) is less than 12, *or* there are more than 8 pixels in the 'on' state in another region of interest, *then* the return value is *true*. The returned value implies decision in the context of particular task (e.g. classification).

4   Extending Genetic Programming by Local Search

The evolutionary computation carries out random parallel search and therefore does not require any neighborhood definition in the solution space. As a result, solution having high value of evaluation function may be overlooked even being very similar to the solution visited in the search. Starting from the late 80's, several attempts have been made to prevent that negative phenomenon. One of the considered remedies was to combine the genetic search with other, more systematic, metaheuristics. In particular, hybridizing GA with various types of local search was reported to improve significantly the effectiveness of the search in combinatorial optimization [1]. Such approaches, known in the literature as Genetic Local Search (GLS), Memetic Algorithms or Hybrid Genetic Algorithms are now intensively studied in single- and multi-objective metaheuristic combinatorial optimization (for review, see [10]).

This paper follows these directions in the context of GP by intertwining the actions of genetic operators with iterative improvement of solutions. Our expectation is that such a local improvement can make the search more biased toward good (optimal or suboptimal) solutions. The particular implementation of this idea may be different as far as the following features are concerned: (a) neighborhood definition, (b) local search algorithm and (c) the extent of the local optimization.

(a) A natural way is to base the neighborhood definition on the representation of solutions in the GPVIS language. As the neighbor solutions should be similar, we generate neighbors of a given solution by introducing minor changes in leaves of its expression tree, which, according to GPVIS syntax, contain exclusively numerical values. In particular, we obtain a single neighbor by increasing (or decreasing) the value at a selected leave of current solution by a small constant.

(b) In experiments reported in the next section we used the *hill-climbing* algorithm for the local improvement of solutions. This greedy metaheuristics evaluates all the solutions in the neighborhood and moves to the best one, provided it is better than the current solution. Otherwise no move is performed and the algorithm gets stuck in (usually local) optimum.

(c) The most appealing setting would be to apply an extensive local improvement to all individuals in the population. However, that would increase the computation time several times. Thus, we control the local search by means of two parameters:

- the percentage of best (fittest) individuals that undergo the local improvement,
- the limit of hill-climbing steps.

In particular, in this study we chose the minimal option, improving locally only *one best solution* in each generation and allowing only *one step* of the hill-climbing algorithm. Seemingly, such a setting limits seriously the local search extent, as it stops prematurely the hill-climbing for solutions, which could continue with improvement. In fact, many solutions that *could* be further optimized actually *will* continue with local search in next generations of evolution, as only a part of the population is being modified in the recombination process.

5   The Computational Experiment

5.1   Offline Handwritten Character Data

As a test bed for the method, we chose the problem of off-line handwritten character recognition, often referred to in the literature due to the wide scope of real-world applications (see [19] for review of methods). The data source was the MNIST database of handwritten digits provided by LeCun et al. [19]. MNIST consists of two subsets, training and test, containing together images of 70,000 digits written by approx. 250 persons, each represented by a 28 28 halftone image (Fig. 2). Characters are centered and scaled with respect to their horizontal and vertical dimensions, however, not 'deskewed'.

5.2   Experiment Design and Presentation of Results

The primary goal of the computational experiment was to compare the search effectiveness of the 'plain' genetic programming (GP) and genetic programming extended by the local search (GPLS). To ensure comparability of results, the runs of GP and GPLS algorithms were paired, i.e. they used the same initial population, training and test sets as well as the values of parameters: population size: 50; probability of mutation: .05; tournament selection scheme [7] with tournament size equal to 5. In each generation, half of the population was retained unchanged, whereas the other fifty percent underwent recombination.

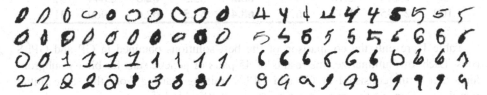

Fig. 2. Selected difficult examples from the MNIST database.

An evolutionary experiment started from a randomly generated initial population. Each solution was evaluated by computing ist accuracy of classification on the training (fitness) set, containing 50 images from each of considered decision (digit) classes. In the recombination process, the offspring solutions were created by means of the crossover operator, which randomly selects subexpressions (corresponding to subtrees in Fig. 1) in the two parent solutions and exchanges them. Then, for a small (.05) fraction of the population, the mutation operator randomly selected an individual's subexpression and replaced it by expression generated at random. The operators obeyed the so-called *strong typing* principle [13], i.e. they yielded correct expressions w.r.t. to GPVIS syntax.

As the evolution proceeds, the GP solutions tend to grow, because large expressions are more resistant to the performance deterioration, which often results from recombination. This phenomenon is conducive to inconvenient overfitting of hypotheses to the training data. To prevent that tendency, the fitness function was extended by an additional penalty term implementing the so-called *parsimony pressure*. Solutions exceeding 100 terms (nodes of expression tree) were linearly penalized with the evaluation decreasing to 0 when the threshold of 200 terms was reached.

To simplify the task, the problem was decomposed into 10 9/2=45 binary problems, each for a unique pair of decision classes. The evolution was carried out for each binary subproblem separately, based on the training set limited to the examples representing the two appropriate decision classes. The triangular matrix of 45 independently induced classifiers form the so-called *metaclassifier*, in particular the n^2 (or *pairwise coupling*) type of it [11]. The metaclassifier was then verified on an independent test set, containing 2000 cases, i.e. 200 images for each of 10 decision classes, acquired from a different group of people [19].

Table 1. Comparison of the pattern recognition programs evolved in GP and GPLS runs (accuracy of classification expressed in percents, superior results in bold).

Max # of generations	GPLS better	Average inc. Of accuracy	Metaclassifier accuracy		Metaclassifier accuracy		Classifier size (# of terms)	
			GP	GPLS	GP	GPLS	GP	GPLS
20	34/45	3.64	57.4	**64.3**	52.0	**57.4**	2353	2203
40	33/45	3.58	60.6	**66.8**	55.1	**61.8**	2669	2719
60	31/45	3.02	62.4	**66.9**	58.3	**62.5**	2627	2927
80	25/45	2.16	64.7	**68.3**	62.5	**62.8**	2844	3139
100	24/45	1.47	67.7	**68.4**	**64.5**	62.6	2969	3131

Table 1 presents the comparison of the best solutions obtained in GP and GPLS runs. Each row summarizes the results of 45 pairs of genetic runs for binary subproblems. Each run consisted of GP and GPLS training starting from the same initial population. Consecutive rows describe experiments with different maximal number of generations. To provide better insight, apart from the complete 10-class classification problem, the 2nd and 3rd columns of the table contain also results for binary subproblems. In particular, the table includes:

- the maximal number of generations (iterations) for the search process (run length), as set for a single binary learning task ('*Max # of generations*'),
- the number of pairs of GP and GPLS runs (per total of 45) for which the best solution evolved in GPLS yielded strictly better accuracy of classification on the training set than the best one obtained from 'plain' GP ('*GPLS better*'),
- the average increase (GPLS minus GP) of accuracy of classification for binary subproblems, obtained on the training set ('*Average inc. Of acc.*'),
- the accuracy of classification of the compound n^2 metaclassifier on the training set and test set for GP and GPLS ('*Metaclassifier accuracy*'),
- the size of the metaclassifier, measured as the number of terms of GPVIS expression ('*Classifier size*').

6   Conclusions and Future Research Directions

The main qualitative result obtained in the experiment is that evolutionary search involving local improvement of solutions (GPLS) outperforms on average the 'plain' genetic programming (GP).

As far as the binary classification problems are concerned (2nd and 3rd columns of Table 1), GPLS provided positive gain in terms of accuracy of classification on the training set, for all settings of maximal number of iterations. Each increase shown in 3rd column of the table is statistically significant at 0.05 level with respect to the Willcoxon's matched pairs signed rank test, computed for the results obtained by particular binary classifiers. These improvements seem to be attractive, bearing in mind the complexity of the visual learning task in the direct approach (see Section 3) and the fact, that the accuracy provided by both the methods for binary problems was at the end of runs close to 100%.

The results for the binary classification tasks propagate to the metaclassifiers (columns 4-7 of Table 1). For all run lengths set in the experiment, GPLS metaclassifiers are superior on the training set. On the test set, that superiority is also observable, but it significantly decreases with time and when the maximal number of iteration reaches 100, the metaclassifier evolved by 'plain' GP becomes better. This observation indicates that some of the GPLS solutions probably get stuck in the local optima of the fitness function due to the local optimization.

It should be stressed that these improvements have been obtained by means of very limited extent of local optimization, which was applied only to the best solution and lasted only for one step in each GP generation. Note also that the obtained GPLS solutions have similar size to those reached by GP (last two columns of Table 1).

Further work on this topic may concern different aspects of the approach. In particular, it seems to be interesting to consider other local search metaheuristics as tools for local improvement of solutions or to look for the compromise between the extent of the local optimization and the computation time. We will also try to improve the performance of the search (the accuracy of classification of induced classifiers), for instance by preventing the premature convergence to local optima.

Acknowledgements. The author would like to thank Yann LeCun for making the MNIST database available to the public and to KBN for support (grant # 8T11F 006 19).

References

1. Ackley, D.H.: A connectionist machine for genetic hillclimbing. Kluwer Academic Press, Boston (1987)
2. Bala, J.W., De Jong, K.A., Pachowicz, P.W.: Multistrategy learning from engineering data by integrating inductive generalization and genetic algorithms. In: Michalski, R.S., Tecuci, G.: Machine learning. A multistrategy approach. Vol. IV. Morgan Kaufmann, San Francisco (1994) 471-487
3. Chan, P.K., Stolfo, S.J.: Experiments on multistrategy learning by meta-learning. Proceedings of the Second International Conference on Information and Knowledge Management (1993)
4. De Jong, K.A.: An analysis of the behavior of a class of genetic adaptive systems. Doctoral dissertation, University of Michigan, Ann Arbor (1975)
5. De Jong, K.A., Spears, W.M., Gordon, D.F.: Using genetic algorithms for concept learning. Machine Learning, **13** (1993) 161-188
6. Goldberg, D.: Genetic algorithms in search, optimization and machine learning. Addison-Wesley, Reading (1989)

7. Goldberg, D.E., Deb, K., Korb, B.: Do not worry, be messy. In: Proceedings of the Fourth International Conference on Genetic Algorithms. Morgan Kaufmann, San Mateo (1991) 24-30
8. Gonzalez, R.C., Woods, R.E.: Digital image processing. Addison-Wesley, Reading (1992)
9. Holland, J.H.: Adaptation in natural and artificial systems. University of Michigan Press, Ann Arbor (1975)
10. Jaszkiewicz, A.: On the performance of multiple objective genetic local search on the 0/1 knapsack problem. A comparative experiment. Research report, Institute of Computing Science, Poznań University of Technology, RA-002 (2000)
11. Jelonek, J., Stefanowski, J.: Experiments on solving multiclass learning problems by n^2-classifier. In: C. Nedellec, C. Rouveirol (eds.) Lecture Notes in Artificial Intelligence 1398, Springer Verlag, Berlin (1998) 172-177
12. Johnson, M.P.: Evolving visual routines. Master's Thesis, Massachusetts Institute of Technology (1995)
13. Koza, J.R.: Genetic programming - 2. MIT Press, Cambridge (1994)
14. Koza, J.R., Keane, M., Yu, J., Forrest, H.B., Mydlowiec, W.: Automatic Creation of Human-Competetive Programs and Controllers by Means of Genetic Programming. Genetic Programming and Evolvable Machines, 1 (2000) 121-164
15. Krawiec, K.: Constructive induction in picture-based decision support. Doctoral dissertation, Institute of Computing Science, Poznań University of Technology, Poznań (2000)
16. Krawiec, K.: Constructive induction in learning of image representation. Research Report RA-006, Institute of Computing Science, Poznań University of Technology (2000)
17. Krawiec, K.: Pairwise Comparison of Hypotheses in Evolutionary Learning. In: Proceedings of The Eighteenth International Conference on Machine Learning (accepted) (2001)
18. Langley, P.: Elements of machine learning. Morgan Kaufmann, San Francisco (1996)
19. LeCun, Y., Jackel, L. D., Bottou, L., Brunot, A., et al.: Comparison of learning algorithms for handwritten digit recognition. In: International Conference on Artificial Neural Networks (1995) 53-60
20. Mitchell, T.M.: An introduction to genetic algorithms. MIT Press, Cambridge, MA (1996)
21. Mitchell, T.M.: Machine learning. McGraw-Hill, New York (1997)
22. Poli, R.: Genetic programming for image analysis. Technical Report CSRP-96-1. The University of Birmingham (1996)
23. Teller, A., Veloso, M.: A controlled experiment: evolution for learning difficult image classification. Lecture Notes in Computer Science, Vol. 990, Springer (1995) 165-185
24. Vafaie, H., Imam, I.F.: Feature selection methods: genetic algorithms vs. greedy-like search. In: Proceedings of International Conference on Fuzzy and Intelligent Control Systems (1994)
25. Yang, J., Honavar, V.: Feature subset selection using a genetic algorithm. In: Motoda, H., Liu H. (Eds.): Feature extraction, construction, and subset selection: A data mining perspective. Kluwer Academic, New York (1998)

Improved Recognition of Spectrally Mixed Land Cover Classes Using Spatial Textures and Voting Classifications

Jonathan C.-W. Chan[1], Ruth S. DeFries[2], and John R.G. Townshend[1]

[1] Laboratory for Global Remote Sensing Studies, Geography Department, University of Maryland, College Park MD 20742, USA
[2] Earth System Interdisciplinary Science Center and Geography Department, University of Maryland, College Park MD 20742, USA

Abstract. Regenerating forest is important to account for carbon sink. Mapping regenerating forest from satellite data is difficult because it is spectrally mixed with natural forest. This paper investigated the combined use of texture features and voting classifications to enhance recognition of these two classes. Bagging and boosting were applied on Learning Vector Quantization (LVQ) and decision tree. Our results show that spatial textures improved separability. After applying voting classifications, class accuracy of decision tree increased by 5-7% and that of LVQ by approximately 3%. Substantial reduction (between 23% to 40%) of confusions between regenerating forest and natural forest were recorded. Comparatively, bagging is more consistent than boosting. An interesting observation is that even LVQ, a stable learner, was able to benefit from both voting classification algorithms.

Keywords: pattern recognition, spatial textures, rating classifier

1 Introduction

Land cover information is readily extractable from remotely sensed data. Since reflectance of different land covers may behave very differently at different wavelengths, classification of land cover types is possible using multi-spectral bands as inputs. For example, typical land and water surface are almost instantly distinguishable by visible bands. However, land covers can be spectrally mixed. In such cases, even in high dimensional spectral space, they are still unseparable. For examples, natural forest and logged forest, or a semi-complete construction site and built-up land, can be spectrally identical. This is a classic problem of mixed classes that class boundary is not able to find. One way to improve classification of mixed classes is to increase input features in the hope that class boundary can be more distinctive in the new input space. Apart from spectral features, spatial texture is another information that can be obtained from satellite images.

In this paper, we investigated the use of texture measures with two voting classification algorithms, bagging and boosting, to improve recognition of two

W. Skarbek (Ed.): CAIP 2001, LNCS 2124, pp. 217–227, 2001.

spectrally mixed classes: natural forest and regenerating forest. First we generated an arbitrary number of texture features based on Haralick's Gray Level Co-occurrence Matrix (GLCM) [1]. Then, we ran feature subset selection to select the texture measures that are most useful to enhance separability of these two classes. The chosen textures would then be combined with spectral information for classification. We compared two learning algorithms, decision tree and Learning Vector Quantization, and the behavior of two voting classification algorithms, bagging and boosting, being applied to them [2], [3].

2   Background

Increasingly, forest re-growth has gained more attentions because it is important for explaining the global carbon cycle. Recently generating forests during the first 30 years accumulate biomass faster than mature forests and some believe that regenerating forests pay a key role in offsetting the carbon emitted from prior clearings [4], [5], [6]. Hence, regenerating forest is a process of carbon sequestering. For instance, one ton of carbon is sequestered to produce 2.2 tons of wood. To quantify the carbon sink related to the regeneration of tropical forests, efforts were made to map the spatial extent of regenerating forests and forest cover changes using AVHRR 1km data [7], [8], [9]. A study in Brazil found that accuracies of mapping different stages of forest regeneration were low and it was suggested that finer resolution data such as Landsat Thematic Mapper might help to solve the problem [10]. For this study, a scene used in the NASA Landsat Pathfinder Humid Tropical Deforestation Project was made available.

The Pathfinder project focused on the tropical deforestation in the Amazon Basin, Central Africa, and Southeast Asia [11]. Around three-quarters of the tropical rain forests worldwide were mapped using multitemporal Landsat images and it is believed that the majority of deforestation activities in closed tropical forests have been accounted for. Since *regenerating forest* was included in the classification scheme of the Pathfinder project, it naturally shed lights on measuring the magnitude of forest re-growth. The Landsat scene was acquired on October 16, 1996 (path/row: 006/066) near Pucallpa, Peru. The dimensions of the scene are 8192 by 7200 pixels. With expert knowledge and visual interpretation, six classes were identified: 1) forest, 2) water, 3) cloud, 4) cloud shadow, 5) regenerating forest and 6) non-forest vegetation. Regenerating forest is defined here as secondary forest that used to be abandoned farmland or timber cuts and has been since grown back to forest but has not fully recovered. Initial experiments to map these land cover classes using spectral features have shown serious confusions between forest and regenerating forest. Advanced techniques are needed to better define these two classes.

Texture is the visual impression of roughness, smoothness or regularity created by objects. To identify certain land cover types from satellite images, texture analysis has been reportedly useful. Among different statistical texture measures, GLCM have been widely examined with promising results. In [12], GLCM textures were applied to synthetic aperture radar (SAR) and Landsat

Thematic Mapper for discriminating between lithologies. It was reported that adding GLCM textures to spectral data enhanced accuracy by almost 10% for certain geological types. Using a Landsat TM scene for land cover inventory of a complex rural region, it was found in [13] that GLCM textures are superior to ancillary data when they are combined with spectral features. GLCM was also found to be useful in updating urban land cover map [14]. To investigate the problem of classifying regenerating forest, we decided to use GLCM texture measures because of its popularity, though using other texture measures, such as Texture Spectrum [15], are possible.

3   Voting Classification and Learning Algorithms

Previous studies showed that combining predictions of multiple classifiers into one classifier is generally more accurate than any of the individual classifier making up the ensemble [16], [17]. Bagging and Boosting are two popular approaches to generate different classifiers by re-sampling from the original training data. Bagging involves the procedure to form new training sets that are re-sampled from the original training set with replacement. For each new training set, a new classifier is formed. Boosting attaches a weight to each training instance after every training cycle to reflect the importance of the instance. Subsequent trainings are tuned to the more important/*difficult* instances and produce different classifiers. Majority voting is a common way to combine multiple classifiers to one classifier. For bagging, each classifier carries equal weight in voting; boosting, however, attaches a weight to the vote of each classifier according to their accuracy attainments.

3.1   Bagging

For unstable learning algorithms, a small change in the training data set can produce a very different classifier. Bagging, or bootstrap **aggregating**, is a procedure to take advantage of this instability [2]. For a data set $D = \{(y_n, \boldsymbol{x}_n), n = 1, \ldots, N\}$ where y is the label and $\boldsymbol{x}$ is the input vector, a classifier $C(\boldsymbol{x}, \boldsymbol{D})$ is produced from a base learning algorithm. For each trial $t = 1, \ldots, T$, a new training set of size $\boldsymbol{N}$, is sampled from the original set with replacement. As such, some of the instance in the original set will not be present, while some instances will be repeatedly presented in the new training set. To determine a class $k \in \{1, \ldots, K\}$ for any unknown case a, every classifier created after each trial has its own record of $\boldsymbol{C}_t(a) = k$ and a final classifier $\boldsymbol{C}_A$ is formed by aggregating $\boldsymbol{C}_t$ using majority vote. Bagging can improve accuracy only if instability exists in a learning algorithm meaning that a very different classifier is generated by a small change in $\boldsymbol{D}$. For stable algorithms, bagging could degrade performance.

3.2 Boosting

Boosting produces multiple classifiers by readjusting the weight of instances in the original training set. Instances incorrectly classified will have their weights increased so that they are chosen more often than the correctly predicting cases in a subsequent trial. New classifiers are made to focus on these *difficult* cases and become more capable of correctly predicting the cases that caused the poor performance of the present classifier. Different boosting methods such as Arc-x4 and Adaboost have been proposed and we have investigated the latter in this study [18], [19] .

The idea of boosting is to create new classifiers that are specialized in classifying difficult cases. It is achieved by adjusting the probability of sampling from the original data set. Initially, all instances in a training set of size N have the same probability of being chosen $(1/N)$. After each trial, the probability will be recalculated. Suppose α_t is the sum of the misclassified case probabilities of a current classifier C_t at trial t. The probability of misclassified cases will be changed by the factor $\beta_t = (1 - \alpha_t)/\alpha_t$ and the total sum of the probability is normalized to 1. The parameter β_t is used to update the weight vector with the motivation of reducing the probability assigned to cases on which the previous trial makes a good prediction and increasing the probability assigned to cases on which the prediction was poor. If α_t is greater than 0.5, the trials will terminate and trial T becomes $t - 1$. If $\alpha = 0$, then trial T becomes t. For Adaboost, the classifiers $C_1, \ldots, C_t$ are combined with weighted voting by $\log(\beta_t)$. Provided that α_t is always less than 0.5 (better than a random guess) it is proved that training errors drop exponentially fast to zero [19]. In practice, zero training error does not always occur in boosting because of extreme difficult cases [20]. Furthermore, since boosting is applied to the training data set, there is no guarantee for improvement in classification of unseen cases. Though empirical studies have shown that boosting is effective in improving classification rates in different data domains, overfitting and unpredicted behaviors have been reported [16], [21].

3.3 Decision Trees and Learning Vector Quantization

Decision tree classifiers partition the feature space into homogenous subsets recursively according to certain predefined rules. Advantages with decision tree classifier are that it does not assume a distribution on the data and it is robust in training. Learning Vector Quantization (LVQ) is a supervised version of Self-Organizing Maps. Codebook representatives are tuned to represent different classes through training. Classification of unknowns is decided by the nearest-neighbor rule. While its training is not as speedy as decision trees (number of codebooks needed to be determined by experiments), it can be considered as a robust learner [22]. Apart from the popularity of these two algorithms, the reason of choosing them is also that each of them represents a major category of learning algorithms: decision trees belong to the category that builds a classification rule using function estimation; LVQ belongs to an alternative paradigm based on density estimation [23]. Furthermore, decision trees are considered unstable while LVQ stable algorithms. As such, the degree to which they benefit from bagging and boosting would vary. In theory, unstable and weak learners will benefit more from bagging and boosting. For this study we have employed C5.0 [24] for decision tree experiments and the LVQ_PAKv3.1 from the Helsinki University of Technology for LVQ experiments.

4 Methodology

One subscene, 750 by 750 pixels, was extracted from the image and 5,506 training cases and 55,993 testing cases were randomly selected which represent 1% and 10% of the total instances in each scene, respectively. Texture features used in this study include eight features generated from the GLCM: *Homogeneity, Contrast, Dissimilarity, Mean, Variance, Entropy, Angular Second Moment and Correlation*, and four features generated from Gray Level Difference Vector (GLDV): *Angular Second Moment, Entropy, Mean and Contrast*. Given a displacement vector $\delta = (\Delta_x, \Delta_y)$, if N is the number of gray levels and p(i, j) is the (i, j)th element of the given matrix $(N \times N)$ divided by sum of all the matrix elements, then the texture measures are defined as follows:

1) *Homogeneity:* $\sum [p(i,j)/1 + (i-j)^2]$
2) *Contrast:* $\sum (i-j)^2 p(i,j)$
3) *Dissimilarity:* $\sum [p(i,j) * |i-j|]$
4) *Mean: Mean of i* $= \sum (i * p(i,j))$
5) *Variance: Variance of i* $= \sum [p(i,j) * (i - \text{Mean of } i)^2]$
6) *Angular Second Moment:* $\sum p(i,j)^2$
7) *Entropy:* $-\sum p(i,j) \log p(i,j)$
8) *Correlation:* $\sum [ijp(i,j) - \mu_x \mu_y]/\sigma_x \sigma_y$

where μ_x and σ_x are the mean and standard deviation of the row sum of GLCM, and μ_y and σ_y are the same for the column sum of GLCM. GLDV counts the occurrence of the absolute difference between the reference pixel and its neighbor. Let $f_\delta(x,y) = |f(x,y) - f(x + \Delta x, y + \Delta y)|$ and p_δ the probability density of $f_\delta(x,y)$. The four GLDV features are defined as follows:

9) GLDV *Angular Second Moment:* $\sum p_\delta(i)^2$
10) GLDV *Entropy:* $-\sum p_\delta(i) \log p_\delta(i)$
11) GLDV *Mean:* $\sum i p_\delta(i)$
12) GLDV *Contrast:* $\sum i^2 p_\delta(i)$

We generated the texture measures with a 31 by 31 window from the infrared band (Channel 4) since infrared band is reportedly more effective for forest applications [25]. The displacement vector for texture generation is $\delta = (0, 1)$ at all four directions (0, 45, 90, 135 degrees). To avoid negative effects of irrelevant features, feature subset selection algorithm was used to get the best features. Since we want to get the features that can best separate the mixed classes, a training set with only *forest* and *regenerating forest* was prepared for feature selection.

The selection procedure was done using the FSS program in the Machine Learning C++ library [26]. Best-first search with forward expansion selection was adopted. Finally, six features were chosen and they included GLCM *Homogeneity*, GLCM *Mean*, GLCM *Variance*, GLCM *Entropy*, GLDV *Mean* and GLDV *Contrast*.

Since the accuracy boosted by bagging and boosting reportedly peaks in the first 10 to 30 trials, we have applied 10 ensemble classifiers for both algorithms [21]. As for the number of codebooks for LVQ, we tried from 20 to 280 per class. The accuracy peaked at 180. We have used 180 × 4 classes = 720 codebooks for our experiments (The training data randomly selected included only 4 classes out of six classes).

5   Results and Discussions

When only spectral features are used, overall accuracies of decision tree and LVQ are 83% and 80.36%, respectively. Confusion matrices in Table 1 show that both algorithms have similar performance on individual classes except that decision tree is about 2% better for *non-forest vegetation*. Serious confusions were observed between *regenerating forest* and *forest*. This can be explained by their similar brightness values for most of the spectral bands. Figure 1a is a biplot of spectral bands 4 ($0.76 - 0.90\mu m$) and 5 ($1.55 - 1.75\mu m$) for these two classes. Almost identical clusters were generated. Class separability measured in Bhattacharrya Distance was 0.66 indicating a very poor separability. After texture features was added, Bhattacharrya Distance improved to 1.004 but this is still considered as poor. Figure 1b is a bi-plot of band 5 and texture measure GLCM *Homogeneity*. It does show slight improvement as visually the two classes have more distinct clusters. Table 2 lists the recognition rates of *forest* and *regenerating forest* after the inclusion of selected textures and the application of bagging and boosting. Both decision trees and LVQ were able to tap the additional texture information as the accuracies of both classes were raised by around 3%. The only exception is the classification of *forest* with LVQ. Applying voting classifications brought further improvements. Bagging and boosting the decision trees enhanced accuracies by 5 to 7%. Though LVQ did not have the same level of improvements as decision trees, bagging LVQ still brought a 3% improvement to both classes.

Table 3 presented the results for confusions between *forest* and *regenerating forest*. Substantial improvements were observed from adding texture features. Voting classifications improved the results further. Boosting decision tree brought enormous improvement to the misclassification of 'Regenerating Forest to Forest', which was down by 40.67%, and the confusion of 'Forest to Regenerating Forest' was down by 26.27%. Bagging decision tree reduced both categories by more than 25%. Apply bagging and boosting to LVQ has similar effects. Misclassification of 'Regenerating Forest to Forest' was reduced by around 30%, and 'Forest to Regenerating Forest' by around 23%.

Table 1. Confusion matrices (in no. of cases) and accuracy distributions (%) of decision tree and Learning Vector Quantization using only spectral features

Decision Tree

	Forest	Water	Regenerating Forest	Non-forest Vegetation	Class Accuracy (%)
Forest	12863	27	3711	292	76.11
Water	27	3428	32	186	93.20
Regenerating Forest	3123	6	18291	1286	80.56
Non-forest Vegetation	388	82	1806	10395	82.01

Learning Vector Quantization

	Forest	Water	Regenerating Forest	Non-forest Vegetation	Class Accuracy (%)
Forest	13031	23	3647	200	77.10
Water	54	3474	39	111	94.45
Regenerating Forest	3215	8	18337	1146	80.75
Non-forest Vegetation	423	102	2024	10126	79.89

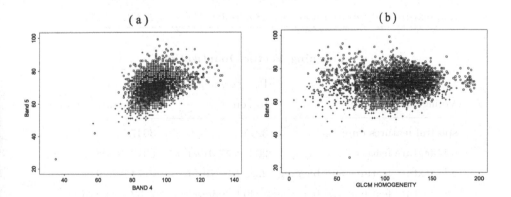

Fig. 1. Bi-plots of Forest (+) and Regenerating Forest (○) with (a) spectral bands 4 and 5, and (b) spectral band 5 and GLCM Homogeneity

Table 2. Accuracy enhancements using texture features and voting classifications

Input Features and Voting Classifications	Decision Tree		LVQ	
	Forest	Regenerating Forest	Forest	Regenerating Forest
spectral features only	76.11%	80.56%	77.10%	80.76%
add texture features	+3.48%	+2.70%	no change	+3.71%
add texture features and bagging	+5.84%	+5.34	+3.70%	+3.29%
add texture features and boosting	+5.28%	+7.17%	+3.10%	+1.32%

Table 3. Class confusions (no. of cases) between Regenerating Forest and Forest after incorporating texture features and voting classifications. The figures in the brackets represent per cent changes as compared to using spectral features alone

Decision Tree

Input Features and Voting Classifications	Regenerating Forest to Forest	Forest to Regenerating Forest
spectral features only	3123	3711
add texture features	2641 (-15.43%)	3047 (-17.89%)
add texture features and bagging	2322 (-25.65%)	2650 (-28.59%)
add texture features and boosting	1853 (-40.67%)	2736 (-26.27%)

Learning Vector Quantization

Input Features and Voting Classifications	Regenerating Forest to Forest	Forest to Regenerating Forest
spectral features only	3215	3647
add texture features	2025 (-37.01%)	3316 (-9.08%)
add texture features and bagging	2225 (-30.79%)	2774 (-23.94%)
add texture features and boosting	2540 (-30.00%)	2752 (-24.54%)

Figure 2 shows the trend of improvement of overall accuracy as related to the number of ensemble classifiers. Bagging (in solid lines) increased the accuracy of both learning algorithms but the rate of increase apparently reached a plateau at the 10th trials. Boosting (in dotted lines) continuously improved accuracy with decision tree but degraded that of LVQ after the 5th trial. This echoed with previous reports that boosting could have negative results because of overfitting [21]. It has also showed that LVQ, a stable algorithm, can be benefited from voting classifications, especially from bagging ($>$ 3% improvement). As for decision tree, boosting is more effective than bagging.

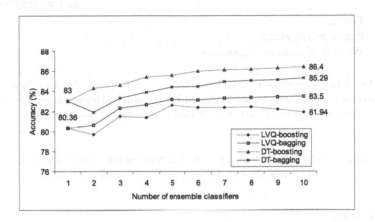

Fig. 2. Behaviour of bagging and boosting as related to the number of ensemble classifiers

Final maps were produced in Figure 3 for comparison. The map with spectral features alone was generated by a single decision tree. The map using combined features was generated by a 10-round boosting of the decision trees. No postclassification filtering was applied to the final maps (otherwise, the resultant maps should have less speckles). When compared to the map using only spectral features, the map produced using combined features and 10-round boosting shows apparent improvements with less speckles and fragmentation, and confusions between *forest* and *regenerating forest* are significantly reduced.

This study examined the incorporation of texture features and voting classifications to separate two spectrally mixed classes. The techniques were tested using two learning algorithms, decision tree and LVQ. While texture features only increase class accuracy modestly, it has significant effect on reducing confusions between mixed classes. The effect was further propagated when bagging and boosting were applied. Comparatively, bagging behavior is consistent with both algorithms since boosting has degraded the performance of LVQ. The combination of voting classifications and spatial textures has helped separate two mixed classes to the extent where either using texture analysis or voting classi-

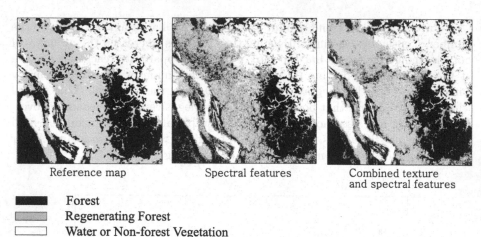

| Reference map | Spectral features | Combined texture and spectral features |

■ Forest
▨ Regenerating Forest
☐ Water or Non-forest Vegetation

Fig. 3. Final maps produced by different inputs and voting classifications. The map using spectral features alone was produced by one single decision tree. The map using combined features was produced from a 10-round boosting decision tree

fication alone could not have achieved. An interesting observation is that even a stable learner (LVQ) was able to benefit from the voting classifications.

References

1. Haralick, R.M.: Statistical and Structural Approaches to Texture. Proc. IEEE, **67** (1979) 786–803
2. Breiman, L.: Bagging Predictors. Machine Learning, **24** (1996) 23–140
3. Kohonen, T.: Self-Organizing Maps. Springer-Verlag, Berlin (1995)
4. Intergovernmental Panel on Climate Change: Land Use, Land-Use Change, and Forestry. A Special Report of IPCC, Summary for Policymakers (2000)
5. Houghton, R.A., Boone, R.D., Fruci, J.R., Hobbie, J.E., Melillo, J.M., Palm, C.A., Peterson, B.J., Shaver, G.R., Woodwell, G.M., Moore, B., Skole, D.L., Meyers, N.: The Flux of Carbon from Terrestrial Ecosystems to the Atmosphere in 1980 due to Changes in Land Use: Geographic Distribution of the Global Flux. Tellus, **39B** (1987) 122–139
6. Lugo, A.E., Brown, S.: Tropical Forests as Sinks of Atmospheric Carbon. Forest Ecology and Management, **54** (1992) 239–255
7. Lucas, R.M., Honzak, M., Curran, P.J., Foody, G.M., Nguele, D.T.: Characterizing Tropical Forest Regeneration in Cameroon Using NOAA AVHRR Data. Int. J. Remote Sensing, **21** (2000) 2831–2854
8. Richards, T., Gallego, J., Achard, F.: Sampling for Forest Cover Change Assessment at the Pan-Tropical Scale. Int. J. Remote Sensing, **21** (2000) 1473–1490
9. Laporte, N., Justice, C., Kendall, J.: Mapping the Dense Humid Forest of Cameroon and Zaire Using AVHRR Satellite Data. Int. J. Remote Sensing, **16** (1995) 1127–1145

10. Lucas, R.M., Honzak, M., Curran, P.J., Foody, G.M., Milne, R., Brown, T., Amaral, S.: Mapping the Regional Extent of Tropical Forest Regeneration Stages in the Brazilian Legal Amazon Using NOAA AVHRR Data. Int. J. Remote Sensing, **21** (2000) 2855–2881

11. Townshend, J.R.G., Justice, C.O.: Spatial Variability of Images and the Monitoring of Changes in the Normalized Difference Vegetation Index. Int. J. Remote Sensing, **16** (1995) 2187–2196

12. Mather, P.M., Tso, B.C.K., Koch, M.: An Evaluation of Landsat TM Spectral Data and SAR - Derived Textural Information for Lithological Discrimination in the Red Sea Hills, Sudan. Int. J. Remote Sensing, **19** (1998) 587–604

13. Bruzzone, L., Conese, C., Maselli, F., Roli, F.: Multi-Source Classification of Complex Rural Areas by Statistical and Neural-Network Approaches. Photogram. Eng. & Remote Sensing, **63** (1997) 523–533

14. Smits, P.C., Annoni, A.: Updating Land-cover Maps by Using Texture Information From Very High-resolution Space-borne Imagery. IEEE Trans. Geosci. Remote Sensing, **37** (1999) 1244–1254

15. He, D.C., Wang, L.: Texture Unit, Texture Spectrum, and Texture Analysis. IEEE Trans. Geosci. Remote Sensing, **28** (1990) 509–512

16. Maclin, R., Opitz, D.: An Empirical Evaluation of Bagging and Boosting. In: Proc. of the 14th Nat. Conf. on Artificial Intelligence, 27-31 July, Providence, Rhode Island, AAAI Press (1997) 546–551

17. Quinlan, R.J.: Bagging, Boosting, and C4.5. In: Proc. of the Thirteenth Nat. Conference on Artificial Intelligence. Cambridge, MA. AAAI Press/MIT Press (1996) 725–730

18. Breiman, L.: Arcing classifiers. Technical Report, Statistics Department, University of California at Berkeley (1996)

19. Freund, Y., Schapire, R.E.: Experiments with a New Boosting Algorithm. Machine Learning: Proc. of the 13th International Conference, 3-6 July, Bari (1996) 148–156

20. Bauer, E., Kohavi, R.: An Empirical Comparison of Voting Classification Algorithms: Bagging, Boosting, and Variants. Machine Learning, **36** (1998) 105–139

21. Chan, J.C.W., Huang, C., DeFries, R.S.: Enhanced Algorithm Performance for Land Cover Classification from Remotely Sensed Data Using Bagging and Boosting. IEEE Trans. Geosc. Remote Sensing, **39** (2001) 693–695

22. Chan, J.C.W., Chan, K.P., Yeh, A.G.O.: Detecting the Nature of Change in an Urban Environment - A Comparison of Machine Learning Algorithms. Photo. Eng. & Remote Sensing, **67** (2001) 213–225

23. Friedman, J.H.: On Bias, Variance, 0/1 - loss, and the Curse-of-Dimensionality. Data Mining and Knowledge Discovery, **1** (1997) 55–77

24. Quinlan, R.J.: C4.5: Programs for Machine Learning. Morgan Kaufmann Publishers, Inc., San Mateo, California (1993)

25. Riou, R., Seyler, F.: Texture Analysis of Tropical Rain Forest Infrared Satellite Images. Photogramm. Engineering and Remote Sensing, **63** (1997) 515–521

26. Kohavi, R., Sommerfield, D., Dougherty, J.: Data mining using MLC++: a machine learning library in C++, Tools with Artificial Intelligence, IEEE Computer Society Press (1996)

Texture Feature Extraction and Classification

B. Verma and S. Kulkarni

School of Information Technology
Griffith University, Gold Coast Campus, Qld 9726, Australia
{B.Verma, S.Kulkarni}@gu.edu.au, http://intsun.int.gu.edu.au

Abstract. This paper describes a novel technique for texture feature extraction and classification. The proposed feature extraction technique uses an Auto-Associative Neural Network (AANN) and the classification technique uses a Multi-Layer Perceptron (MLP) with a single hidden layer. The two approaches such as AANN-MLP and statistical-MLP were investigated. The performance of the proposed techniques was evaluated on large benchmark database of texture patterns. The results are very promising compared to other techniques. Some of the experimental results are presented in this paper.

Keywords: pattern recognition, feature extraction, neural networks

1 Introduction

Texture is an important characteristic for the analysis of various types of the images including natural scenes, remotely sensed data and biomedical modalities. It is believed that the texture plays an important role in the visual systems for recognition and interpretation of data. Texture analysis is an important research field in computer vision, image processing and pattern recognition. A number of techniques have been developed for texture feature extraction, segmentation, classification and synthesis [1]. In recent years, big efforts have been devoted to the attempt to improve the performance of image retrieval systems and research has explored many different directions trying to use with profit results achieved in other areas [2].

There are a number of statistical techniques such as spectral method [3], wavelet theory [4], 2-D Wold [5] and MRF technique exist in the literature which have been recently developed for texture analysis, segmentation, extraction, etc. There are also a number of papers published describing neural based texture feature extraction and classification. However, the most of research using neural networks was mainly based on Kohonen neural network [7]. In this paper we focus on feature extraction process using MLP type neural networks. Research shown [10] that the neural networks can be good feature extractors and classifiers. We propose and investigate an auto-associator and classifier to extract features and classify texture images.

The rest of the paper is organised as follows: Section 2 describes in detail the proposed feature extraction process using an AANN and classification using MLP. The statistical techniques, the preparation of training and testing data sets from Brodatz texture patterns and texture classes in detail are presented in Section 3. The experimental results are presented in Section 4. The results, analysis and comparisons are discussed in Section 5. The conclusion is presented in Section 6.

W. Skarbek (Ed.): CAIP 2001, LNCS 2124, pp. 228–235, 2001.
© Springer-Verlag Berlin Heidelberg 2001

2 Proposed Technique

This section describes in detail the proposed technique for feature extraction and classification. The overall block diagram of the proposed technique is presented in Figure 1 as follows.

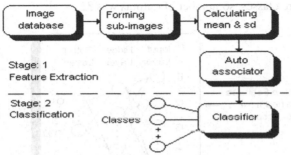

Fig. 1. Block diagram of the proposed technique

The proposed technique is divided into two stages. Stage 1 deals with feature extraction from texture sub-images. An auto-associator was designed to extract features. Stage 2, deals with classification of features into texture classes. A Multi-Layer Perceptron (MLP) was designed to classify texture classes. The auto-associator feature extractor and MLP texture feature classifier are described below in detail.

2.1 Auto-Associator Feature Extractor

The main idea of auto-associator feature extractor is based on input:hidden:output mapping, where input and outputs are the same patterns. AAFE learns the same patterns and provides a characteristic through its hidden layer as a feature vector. As shown in Figure 2, we designed an auto-associator feature extractor using a single hidden layer feed-forward neural network. It has n inputs, n outputs and p hidden units. The input and output of the AAFE are the same texture patterns and the network is trained using a supervised learning. After training is finished, the values of the hidden units extracted and taken as a feature vector. The feature vector is fed to the MLP feature classifier, which is described in the next section.

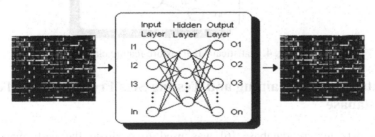

Fig. 2. Auto-associator as a feature extractor

2.2 MLP Texture Feature Classifier

An MLP texture feature classifier is shown in Figure 3. It has *n* inputs same as the number of hidden units in auto-associator feature extractor. The output of the hidden layer that was obtained from auto-associator was used as input to classifier. There were 32 texture classes, so number of outputs was 32.

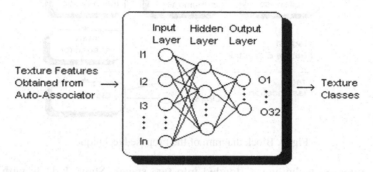

Fig. 3. MLP texture feature classifier

2.3 Statistical-MLP Texture Feature Classifier

As shown in Figure 4, statistical techniques were used to extract the texture features from the images. The mean and standard deviation form the properties of texture, were calculated for the each row of the image. These were treated as the features of the texture patterns and applied to the texture feature classifier, as input.

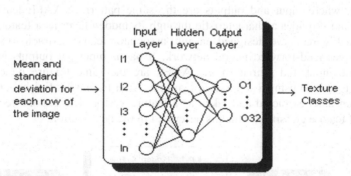

Fig. 4. Statistical-MLP texture feature classifier

3 Extraction of Training and Testing Sets From Brodatz Texture Database

The Brodatz texture database [6] was used to evaluate the performance of the proposed techniques detailed in previous section for texture features extraction and classification. The database contains 96, 512 $\times$ 512 texture images. In order to create a number of small images which belong to the same class, we partition each of the 512

512 images into 128   128 sub-images, thus forming 16 sub-images from each image. To reduce the size of input vector to neural network, the mean and standard deviation was calculated for each row (128 pixels) as follows:

$$\mu = \frac{1}{n}\sum_{i=1}^{n} x_i$$

(1)

$$\sigma = \sqrt{\frac{1}{n}\sum_{i=1}^{n}(x_i - \mu)^2}$$

(2)

where μ and σ are the mean and standard deviations, n is the number of pixels and in our case it was 128. First 12 sub-images were used for the training of auto-associator and the last 4 images were used as testing data set. These images were normalized in the range of 0 and 1.

There were total 96 texture patterns in the database which were grouped into 32 similar classes, each of them containing 1-5 texture classes. All the texture sub-images belonging to the same similarity class are visually similar. This classification was done manually [7] and the following table shows these various similarity classes and the corresponding textures.

Table 1. Texture clusters used for classifier

Cluster	Texture Class	Cluster	Texture Class
1	D1, D6, D14, D20, D49	17	D69, D71, D72, D93
2	D8, D56, D64, D65	18	D4, D29, D57, D92
3	D34, D52	19	D39, D40, D41, D42
4	D18, D46, D47	20	D3, D10, D22, D35, D36, D87
5	D11, D16, D17	21	D48, D90, D91
6	D21, D55, D84	22	D43, D44, D45
7	D53, D77, D78, D79	23	D19, D82, D83, D85
8	D5, D33	24	D66, D67, D74, D75
9	D23, D27, D28, D30, D54	25	D2
10	D7, D58, D60	26	D86
11	D59, D61, D63	27	D37, D38
12	D62, D88, D89	28	D9
13	D24, D80, D81	29	D12, D13
14	D50, D51, D68, D70, D76	30	D15
15	D25, D26, D96	31	D31
16	D94, D95	32	D32

4 Experimental Results

The experimental results were conducted separately for an auto-associator-classifier technique and statistical-neural technique. The following Sections 4.1 and 4.2 explain the results obtained on Brodatz texture patterns.

4.1 Auto-Associator and Classifier

The experiments were conducted in two stages, firstly the training of the auto-associator and then the training of the classifier. The auto-associator was trained for the different number of hidden units and iterations to improve the feature extraction. Table 2 shows some of the results obtained after the training of the auto-associator. The number of inputs and outputs were 256. The values of momentum and learning rate used for the experiments were 0.7 and 0.8 respectively. The number of pairs for training and testing were 1152 and 384 respectively. The experiments were conducted by varying the number of hidden units and iterations to achieve better results.

Table 2. Auto-associator feature extractor

Hidden Units	Iterations	RMS Error
10	1000	0.005698
10	10000	0.003166
15	10000	0.002160
18	10000	0.001778
22	10000	0.001471
30	100000	0.000832

The classifier was trained after obtaining the output from the hidden layer from auto-associator. Experiments were conducted for different number of hidden units and iterations and the results are shown in Table 3.

Table 3. MLP texture feature classifier

Hidden Units	Iterations	RMS Error	Classification Rate	Classification Rate [%]
32	125000	0.008312	339/384	88.28
32	100000	0.008604	336/384	87.50
32	50000	0.008807	331/384	86.19
30	50000	0.009102	286/384	74.47
28	50000	0.009331	267/384	69.53

The following graphs show the difference in the amplitude of the features extracted from different classes.

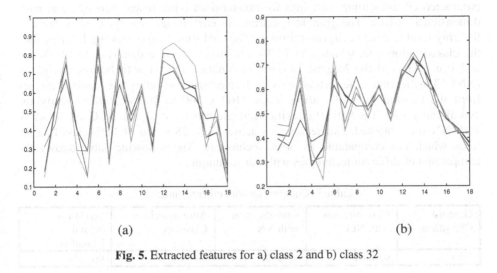

(a)   (b)

Fig. 5. Extracted features for a) class 2 and b) class 32

4.2 Statistical-MLP Texture Feature Classifier

The 16 non-overlapping sub-images were formed from the image of size 512 x 512. The size of the sub-images were 128 x 128. The mean and standard deviation were calculated for each row of the image, i.e. for 128 pixels, thus forming 256 as the number of inputs. The number of classes was 32, same as it was used in auto-associator and classifier experiments. The results obtained from statistical-texture feature classifier are presented in Table 4. The values of momentum and learning rate used for the experiments were 0.7 and 0.8 respectively. The number of training pairs was 1152 for training and 384 for testing.

Table 4. Statistical-MLP texture feature classifier

Hidden Units	Iterations	RMS Error	Classification Rate	Classification Rate [%]
18	100000	0.00517	369	96.06
18	50000	0.00520	367	95.57
18	40000	0.00532	366	95.31
18	30000	0.00556	364	94.79
18	20000	0.00592	355	92.44
18	10000	0.00693	338	88.02
15	10000	0.00810	327	85.15

5 Analysis and Comparison of the Results

The classification results obtained by our techniques were compared with other techniques. In [8] Jones and Jackway used granold technique for texture classification. A granold is a texture representation technique that uses two parameterised monotonic mappings to transform an input image into two and half-dimensional surface. The granold spectrum for each image was then calculated and the gray level and size marginals formed. The confusion matrix was used to calculate the classification rate, which was 76.9% on Brodatz texture database. In [9], Wang and Liu compared the Nearest Linear Combination (NLC) with Nearest Neighbor (NN) Classification. The testing set was formed from selecting the random images from the training set with various sizes (100 x 100 to 200 x 200) to verify the classification rate. The highest classification rate was 95.48% with NLC and 93.47 % with NN was obtained. Our techniques achieved 88.28% and 96.06% classification rates which are comparable to other techniques. The following table shows the comparison of different techniques with our technique.

Table 5. Comparison with other techniques

Granold Classification	Classification with NLC	Classification with NN	Auto-associator-Classifier	Statistical-Neural Classifier
76.9 %	95.48%	93.47%	88.28%	96.06%

6 Conclusions

In this paper, we have presented a novel auto-associator texture feature extractor and also we have described two techniques for classification of the texture patterns such as an auto-associator-MLP texture feature classifier and statistical-MLP texture feature classifier. The auto-associator seems to be a promising feature extractor. We have presented some feature graphs in Figure 5, which show that the auto-associator is capable of separating two classes very well and without any feedback from user. The feature extraction and classification techniques were tested on large database of texture patterns such as Brodatz texture database. The results obtained by our techniques were analysed and compared with other intelligent and conventional techniques. The highest classification rate was obtained using our statistical-neural technique. Our neural technique outperformed other statistical techniques such as MRF, NN, etc and neural network based techniques such as Kohonen neural network and learning vector quantization. Our auto-associator feature extractor improves the results in every new experiment with different number of iterations and hidden units. Currently, the investigations are underway to find out the optimal parameters for our auto-associator feature extractor.

References

1. Manjunath, B., Ma, W.: Texture Features for Browsing and Retrieval of Image Data. IEEE Transaction on Pattern Analysis and Machine Intelligence, **8** (1996) 837-842
2. Niblack, W.: The QBIC Project: Querying Images by Content using Color, Texture and Shape. SPIE Proceedings of Storage and Retrieval for Color and Image Video Databases, (1993) 173-187

3. Jain, A., Farrokhnia, F.: Unsupervised Texture Segmentation Using Gabor Filters. Journal of Pattern Recognition, **24** (1991) 1167-1186
4. Rubner, Y., Tomasi, C.: Texture Matrices. In: Proceedings of IEEE International Conference on Systems, Man and Cybernetics, San-Diego, USA (1998) 4601-4607
5. Lui, F., Picard, R.: Periodicity, directionality and Randomness: Wold Features for Image Modelling and Retrieval. IEEE Transactions on Pattern Analysis and Machine Intelligence, **18** (1996) 722-733
6. Brodatz, P.: Textures: A Photographic Album for Artists and Designers. Dover Publications, New York (1996)
7. Ma, W., Manjunath, B.: Texture Features and Learning Similarity. In: Proceedings of IEEE International Conference on Computer Vision and Pattern Recognition, San Francisco, USA (1996)
8. Jones, D., Jackway, P.: Using Granold for Texture Classification. In: Fifth International Conference on Digital Image Computing, Techniques and Applications Perth, Australia (1999) 270-274
9. Wang, L., Liu, J.: Texture Classification using Multiresolution Markov Random Field Models. Journal of Pattern Recognition Letters, **20** (1999) 171-182
10. Lerner, B., Guterman, H., Aladjem, M., Dinstein, H.: A Comparative Study of Neural Network based Feature Extraction Paradigms. Journal of Pattern Recognition Letters, **20** (1999) 7-14

Today's and Tomorrow's Medical Imaging

Artur Przelaskowski

Institute of Radioelectronics
Warsaw University of Technology
Nowowiejska 15/19, 00-665 Warszawa
arturp@ire.pw.edu.pl

Abstract. Biomedical Engineering in present form started its developing since the late 1960's and includes engineering applications in physiology and medicine, such as Biomechanics, Biomedical Instrumentation, Bioelectrical processes, Biocontrol systems, Biomedical signal and
image processing, Medical informatics and others. In last decades
Medical Imaging (MI) started to play important role in innovatory solutions and applications of biomedical engineering. In our presentation
current trends of medical imaging development are considered. We
mean an interesting projects, the ideas currently developed in labs and
many research centers. Underlying our research leaded in many areas of
medical imaging, nuclear and medical engineering, in collaborations
with several medical and biomedical centers and institutes of physics
and nuclear science, we intended to present a quick review of the most
hopeful research directions. -What is important, and worth of work
with? -Is the medical imaging dynamically developing science of the
useful applications, truly important in an information society development, able to cumulate the resources and interests of youth? Subjectively, we tried to find the answers considering the following topics:

-   functional imaging of organs and tissues: PET (brain), SPECT (circula-
    tory system, organs), MRI (brain, circulatory system, organs),
    CT(circulatory system, organs); dynamic imaging of heart and blood
    vessels, blood supply of liver and kidneys, etc., 2-D and even 3-D perfu-
    sion maps, statistical flow models and objective computable parameters
    required to be standardized (EBCT, dynamic MRI, even US Power
    Doopler);

-   image detectors (PET, radiography, CT), detection systems (SPECT),
    detectors (scintillators), sensors with amorphous silicon and selenium in
    digital radiography, x-ray tubes with laser beam irradiation;

-   virtual endoscopy (bronchoscopy, gastroscophy);

-   telemedicine, means protocols, network switches, satellite connectors,
    and PACS, DICOM servers, indexed data basis, Hospital Information
    Systems, remote health care, interactive consultations, patient and fam

W. Skarbek (Ed.): CAIP 2001, LNCS 2124, pp. 236–237, 2001.

- ily education, structure of safety access, hierarchical exam evaluation, teleconferences, inspection and quality control, etc.;

- medical image compression, JPEG2000 and other encoding lossy and lossless techniques necessary for efficient data storing and progressive transmission,

- computer-aided diagnosis: the examples of improvements in digital mammography, ultrasound systems; image-guided surgery and therapy, multimodal systems (PET, MRI, CT), 3-D imaging (acquisition, reconstruction) for various medical systems;

- physics of medical imaging, image and signal processing, physiology and function from multidimensional images, visualization and display procedures, image perception, observer performance, image quality evaluation tests and technology assessment;

and others.

Because of such wide range of these image engineering applications it is very difficult to select the most important perspective research. Presented ones were chosen to show the important from our point of view proofs of MI support necessity in modern diagnosis and therapy. Therefore, more elements of MI should be included in medical education at the Universities and propagated by Society organizations. An important conclusions derived from our study depicts predicted sources of increasing industrial development of MI and a role of MI which is expected to play in a future hospital clinical service.

Keywords: medical imaging systems, image processing and analysis, telemedicine

A New Approach for Model-Based Adaptive Region Growing in Medical Image Analysis

Regina Pohle and Klaus D. Toennies

Otto-von-Guericke University Magdeburg, Department of Simulation and Graphics,
Postfach 4120, D-39016 Magdeburg, Germany
{regina,klaus}isg.cs.uni-magdeburg.de

Abstract. Interaction increases flexibility of segmentation but it leads to unde-sired behaviour of an algorithm if knowledge being requested is inappropriate. In region growing, this is the case for defining the homogeneity criterion as its specification depends also on image formation properties that are not known to the user. We developed a region growing algorithm that learns its homogeneity criterion automatically from characteristics of the region to be segmented. It produces results that are only little sensitive to the seed point location and it allows a segmentation of individual structures. The method was successfully tested on artificial images and on CT images.

Keywords: medical imaging, image segmentation, region growing

1 Introduction

The analysis of medical images often requires segmentation prior to visualization or quantification. For segmentation of structures in CT images many different approaches exist [1] among which region growing is popular having the advantage of letting the user specify just one region that he/she is interested in [2]. Location and homogeneity criterion for region growing have to be supplied by the user. The former poses no problems as it can be expected that the user possesses sufficient anatomical knowledge to pinpoint a structure that he/she wants to segment. Specifying homogeneity, however, is difficult because the user's concept of homogeneity is often vague and fuzzy and it is not translated easily into a computable criterion. Thus research in regionbased segmentation has focused on the design of the growing criteria as well as on algorithm efficiency [3]. Methods can be categorized into:

- Criterion selection based on gray-level properties of the current points [4,5,6].
- Comparison of segmentations with different homogeneity criterions [7,8,9]. Methods are often slow because of the large number of segmentations and they require distinguishing the true result from segmentations with slightly different homogeneity criteria. These methods are dependent on seed point location and search order.
- Criterion selection for a complete segmentation of the scene with potentially varying criterion for different regions [10,11]. The complete image has to be segmented being based on a notion of overall optimality.

W. Skarbek (Ed.): CAIP 2001, LNCS 2124, pp. 238–246, 2001.

We designed a new process of region growing for segmenting single structures that overcomes the limitations listed above. The process estimates the homogeneity criterion from the image itself, produces results that are far less sensitive to the seed point selection, and allows a segmentation of individual structures. The performance of our method is compared with the adaptive moving mean value region growing method [5] because we found it is the most similar method.

2 The Homogeneity Model

Homogeneity at a pixel x of a region R defined by $a_1, ..., a_n$ can be described as a function h of that pixel and pixels x_i in a neighbourhood of x and within R:

$$h(x; x_i; a_i, ..., a_n) = \begin{cases} 1, & \text{if } x \text{ given pixels } x_i \text{ is part of R, defined by } a_i, .., a_n \\ 0, & \text{otherwise} \end{cases}$$

Getting a reliable estimate of $a_1, ..., a_n$ should be possible provided that a sufficient number of pixels can be found that are part of the region. For CT images - which were the primary application of our segmentation method - we used a simple homogeneity model. Values represent average x-ray absorption distorted by noise and artefacts. The absorption is assumed to be constant for a given anatomical structure. The noise is assumed to be zero mean gaussian noise with an unknown standard deviation. The main artefact in the image is assumed to be the partial volume effect (PVE). Homogeneity in this case may be defined as likelihood of belonging to a gaussian distribution of grey values with a given mean and standard deviation. This criterion was also used by other authors, e.g. [4,5,6]. The effects of the PVE can be captured in an approximative fashion by assuming different "standard deviations" for grey values that are higher or lower than the mean. The homogeneity function h is then

$$h(x; \mu, ud, ld) = \begin{cases} 1, & \text{if } \mu - w \cdot ld < x < \mu + ud \\ 0, & \text{otherwise} \end{cases} \tag{1}$$

with ld and ud being the standard deviations of two different gaussian distributions. The appropriateness of this model was tested by investigating grey level distributions of CTs of the abdomen from manually segmented regions (see Fig. 1). The asymmetry of the distribution due to the PVE can be seen clearly in the image. Its degree depends on the grey values of the surrounding tissue of a region.

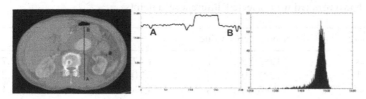

Fig. 1. CT image of the abdomen (left), grey levels along line A-B (centre) and grey level histogram for the passable lumen of the aortic aneurysm

3 A New Adaptive Region Growing Algorithm

Finding a connected region by region growing requires the homogeneity criterion to remain constant during search. It changes, however, if region growing is also applied to learn the criterion. Thus, two runs of region growing are necessary. Homogeneity parameters are estimated in the first run and they are applied in a second run for extracting the region. Learning our homogeneity criterion requires to estimate mean and two different standard deviations for grey values from a number of pixels of the region. As the combination of two halves of two different gaussian functions is not a gaussian, the mean is estimated from the median instead of the average of the pixel values. The two standard deviations are then estimated separately. During learning process, the range of grey value from the current estimate for homogeneity is extended by a relaxation function $c(n)$ in order to avoid premature termination of the process (see Table 1). $c(n) = {}^{50}/_{\sqrt{n}}$ decreases with number of pixels n that are used for the estimate. Given m and deviations ld and ud that were computed at a previous step (initially from 3x3 neighbourhoods around the seed point), lower and upper thresholds are computed for determining region membership:

$$T_{lower} = \mu(n) - [ud(n) \cdot w + c(n)] \quad \text{and} \quad T_{upper} = \mu(n) + [ud(n) \cdot w + c(n)] \quad (2)$$

The weight w was set to 1.5 to include approximately 86% of all region members if the distribution were truly gaussian. It was set that low in order to avoid leaking out of the region. Estimating region characteristics of a yet undetermined region requires a reliable estimate of the homogeneity parameters before the first nonregion pixel is encountered. Three assumptions were made for the search process:
- The user sets the seed point not close to the region boundary.

- The region consists of a sufficient number of pixels.

- The region is compact.

Under these assumptions the standard region growing technique is inappropriate for the first run and we chose the order of visiting neighbours each time at random. It produces a search order which closely resembles a random walk (Fig. 2). Parameters of the homogeneity are recomputed each time the number of pixels in the training region has doubled. After termination, the two deviations ld and ud are adjusted in order to account

Table 1. Mean, ld and ud computed for subset of a manually segmented liver in CT. Standard deviation shown resulted from selected different seed points.

region size	average grey value	ld	ud
9 pixels	1288.72 ± 5.11	4.47 ± 2.94	3.87 ± 2.75
20 pixels	1287.80 ± 7.46	7.70 ± 4.29	6.12 ± 4.57
100 pixels	1285.36 ± 8.33	5.73 ± 1.56	6.40 ± 3.14
500 pixels	1286.48 ± 3.40	7.60 ± 3.58	7.82 ± 2.48
complete region	1287.00	10.14	7.94

Fig. 2. Randomised region growing after seed point setting and visiting 9, 40, 65, 100, 150, 200, 400 and 475 pixels

for the constant underestimation of the standard deviation by using only pixels that deviate by less than $1.5 \cdot w$ from the current mean. Adjustment factors can be computed from tabulated integrals of the gaussian curve (see Table 2). For the second run, the region growing process is repeated using the same seed point, the estimated mean and corrected standard deviations. We set $w = 2.58$ in order to include 99% of the pixels if the grey level variation were truly gaussian.

Table 2. Adjustment factor after each update when the region size has doubled. The true standard deviation is more and more underestimated, because new pixels are included only if they deviate by less than 1.5 times the standard deviation

number of updates	1	2	3	4	5	6	7	8
Correction factor for std dev	1.0	1.14	1.29	1.43	1.58	1.72	1.90	2.05

4 Performance Evaluation of the Algorithm

Empirical discrepancy methods - comparing results with gold standard - were used for evaluation of our model-based adaptive region growing (MBA). We selected discrepancy methods because they allow an objective and quantitative assessment of the segmentation algorithm with a close relationship to a concrete application. We subdivided our tests into two categories. First, we tested how much deviation from our homogeneity model the algorithm would tolerate. For this purpose we used artificial test images that let us control the model parameters. Then, we investigated whether the outcome on test images helped predict the performance on CT images if compared to a manual segmentation. The tests were carried out using five to six different seed point locations. We have compared the results of our method with the results of the adaptive moving mean value region growing (AMM) [5]. The following metrics we used that allow a characterisation of the error level that is independently of the region characteristics and the segmentation method:
1. Average deviation from the contour in the gold standard [12]
2. Hausdorff distance, calculated as in [13]
3. Number of over-segmented pixels
4. Number of under-segmented pixels
The average deviation value describes the general quality of the result. The Hausdorff distance tells whether the region has leaked. Finally, analysis of under- and oversegmented pixels serves to assess whether du and ld were too high or too low.

4.1   Evaluation on Artificial Test Images

Variation of the signal-noise ratio (SNR), modification of the edge gradient for the simulation of the partial volume effect and modification of the shading were analysed for segmentation on artificial test images. The first feature captures variation within our region homogeneity model. The modification of the edge gradient is partly included in the model using separate upper and lower standard deviations. Shading is not part of the model but may well occur in reality.

Fig. 3. Test image generation, from left to right: source image, image of the noise portion, test image with an SNR of 1.5:1, segmentation result of the first run, final segmentation result.

- **Evaluation of the influence of the signal-noise-ratio (SNR)**
  For a close simulation of the real conditions, we have extracted noise from the homogeneous areas in CT images, rather than using artificial noise. This noise was overlaid on test images with different contrast in order to achieve different SNRs (Fig. 3). The image size was 13.786 pixels and the object size was 1.666 pixels. The results in Fig. 6 show that the average deviation for our MBA method from the true boundary was below one pixel at an SNR of 1:1 or higher. The maximum deviation was two pixels on average. Oversegmentation was below 0.5% while undersegmen-tation was below 1%. For the AMM method all error metrics have had higher values with larger variance. Segmentation with AMM was not possible for a SNR of 1:1 because of region leaking.

- **Evaluation of the influence of the edge gradient**
  Influence of variation of the edge gradient was tested on images that were blurred with boxcar filters of size 3x3 to 9x9 pixels prior to adding noise with SNR of 1.5:1 (Fig. 4). The error as well as its variance increased with the amount of edge blurring (see Fig. 7). The latter indicates an increasing dependence of the segmentation result on the initial positioning of the seed point. Segmentation was not possible using AMM because the region leaked into the background.

- **Evaluation of the influence of shading effects**
  The influence of shading effects was investigated for testing the appropriateness of MBA for segmentation on images from other modalities such as MRI. We added a shading of 0.14, 0.28, 0.42 and 0.56 grey values per pixel at a SNR of 1.5:1. Variance of the average and maximum deviation increased significantly with a gradient of 0.42 or more (Fig. 8). Using AMM, three out of six segmentations leaked into the background at a gradient value of 0.56.

Fig. 4. Test image (left) and result image with MBA method (right) for the evaluation of the influence of the edge gradient. Source image was blurred with a boxcar filter of size 9x9 pixels.

4.2 Evaluation of the Segmentation in CT Images

In the CT images, we segmented the kidney cortex, the liver and the passable lumen in an aortic aneurysm. SNR estimates were 1:1.75 for the liver and 1:5 for the kidneys and the passable lumen. The liver exhibited only little PVE, while the PVE for the other structures was quite significant. Segmentation was repeated five times with different seed points. The following homogeneity parameters were estimated for the liver region: mean grey value 1292.00±0.0, lower deviation 8.69±0.55 and upper deviation 7.80±0.08. The variation due to different seed points was very low. For the region of the passable lumen and the left kidney cortex we obtained similar results. They were compared with a manual segmentation that was carried out by a practicing surgeon (see Table 3 and Figure 5). Segmentation with AMM depended on the seed point location. For the liver, an overspill happened each time using five different seed points. In order to compare automatic computation of the homogeneity criterion with influences from erroneous manual definition of it, we repeated the process using intentionally deviated thresholds from its estimated value. At ±1% of the total grey level range (i. e., ±15 grey values) it led to a doubling or tripling of all error metrics.

Table 3. Measured values for a comparison of the segmentation results with the regions manually segmented.

Metrics for error evaluation	liver	passable lumen	left kidney cortex	right kidney cortex
average error	1.1 ± 0.1	1.4 ± 0.2	1.3 ± 0.1	0.6 ± 0.2
Hausdorff distance	3.6 ± 0.3	4.6 ± 0.6	6.0 ± 0.0	2.1 ± 0.1
oversegmented pixels %	1.4 ± 0.3	0.1 ± 0.0	0.1 ± 0.0	0.2 ± 0.1
not segmented pixels %	5.6 ± 0.7	10.2 ± 1.2	23.1 ± 2.0	14.1 ± 3.1

5   Discussion

Results from test images indicated that the homogeneity criterion can be computed very accurately using our method provided that the model of homogeneity was appropriate for the segmentation task. Even some deviations due to shading and PVE were tolerated. Results on CT images were similar to those on the test images. The error estimates for the liver segmentation were in the expected range except for the number of non-segmented pixels which was higher than expected. However, this may also be due to errors in manual segmentation. There was no leakage of the region if the seed point was positioned at least three pixels off the segment boundary. For the passable lumen of the aortic aneurysm and for the kidney cortex, the Hausdorff distance and the number of oversegmented pixels were in the expected range. The average deviation and the number of non-segmented pixels were higher than in the test runs. In summary, the tests have shown the practicability of our method for segmentation of organs in CT images. The advantages over other region growing methods are that it does not require the error-prone specification of parameters of a homogeneity criterion, it does not require full segmentation of the scene, and it is less sensitive to the seed point selection than other methods that adapt their parameters.

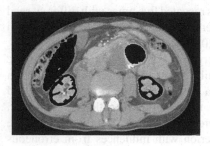

Fig. 5. Segmentation results of the adaptive region growing method for the liver, passable lumen of the aortic aneurysm and for the kidney cortex as a dark overlay

6   Conclusions

We developed a new process of automatically finding the homogeneity criterion in region growing based on a simple model of region homogeneity. The method was shown to deliver a good approximation of the homogeneity parameters provided that the region to be segmented followed the model. Furthermore, the method is rather robust towards deviations from this model. In future work we will extend this model in order to accommodate more complex notions of homogeneity by including spatial dependency into the model.

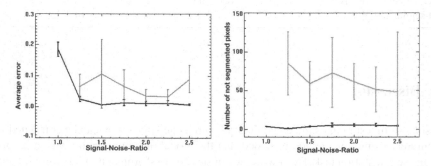

Fig. 6. Results of the discrepancy measurement for the average error and for the number of undersegmented pixels (black: MBA method, gray: AMM method), average of each time six runs per measuring point for the segmentation of test images with variation of the SNR.

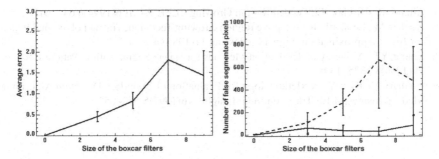

Fig. 7. Results of the discrepancy measurement for average error and for number of incorrectly segmented pixels (solid: under-segmentation, dotted: over-segmentation), average of each time six runs per measuring point for the segmentation of test images with variation of the edge gradient.

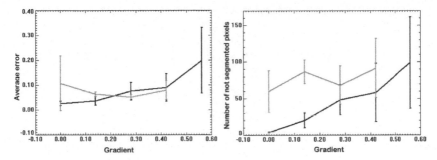

Fig. 8. Results of the discrepancy measurement for the average error and number of undersegmented pixels (black: MBA method, gray: AMM method), average of each time six runs per measuring point for the segmentation of test images with variation of the shading.

References

1. Sonka M., Fitzpatrick J.M.: Handbook of Medical Imaging. Vol. 2, SPIE Press (2000)
2. Gonzales R.C., Woods R.E.: Digital Image Processing. Addison Wesley, Reading (1993)
3. Wan S.Y., Higgins W.E.: Symmetric Region Growing. ICIP 2000, Vancouver, (2000)
4. Chang Y.L., Li X.: Adaptive Image Region-Growing. IEEE Trans. on Image Processing, Vol. 3 (1994) 6 868–872
5. Levine M., Shaheen M.: A Modular Computer Vision System for Image segmentations. IEEE PAMI-3(1981)5 540–554
6. Haralick, R.M., Shapiro, L.G.: Image Segmentation Techniques. CVGIP 29(1985) 1100–132
7. Law, T.Y., Heng, P.A.: Automated extraction of bronchus from 3D CT images of lung based on genetic algorithm and 3D region growing. Proc. SPIE Vol. 3979 (2000) 906–916
8. Siebert, A.: Dynamic Region Growing. Vision Interface, Kelowna (1997)
9. Hojjattolesami, S.A., Kittler, J.: Region Growing - A New Approach. VSSP-TR-6/95 (1995)

10. Adams, R., Bischof, L.: Seeded Region Growing. IEEE-PAMI 6(1994)6 641–647
11. Kocher, M., Leonardi, R.: Adaptive Region Growing Techniques using Polynomial Functions for Image Approximation. Signal Processing 11(1986) 47–60
12. Zhang, Y.J.: A Survey on Evaluation Methods for Image Segmentation. Pattern Recognition 29(1996)8 1335–1346
13. Chalana, V., Kim, Y.: A Methodology for Evaluation of Boundary Detection Algorithms on Medical Images. IEEE Trans. on Med. Imaging 16(1997)5 642–652

Attempts to Bronchial Tumor Motion Tracking in Portal Images during Conformal Radiotherapy Treatment

Maciej Orkisz[1], Anne Frery[1], Olivier Chapet[2],
Françoise Mornex[2], and Isabelle E. Magnin[1]

[1] CREATIS, CNRS Research Unit (UMR 5515) affiliated to INSERM, Lyon, France,
INSA de Lyon, Bat. Blaise Pascal, 7 rue J. Capelle, F-69621 Villeurbanne Cedex,
maciej.orkisz@creatis.insa-lyon.fr, http://www.creatis.insa-lyon.fr
[2] Dept. of Radiotherapy Oncology, Lyon-South Hospital, 69310 Pierre Bénite, France,
mornex@radiotherapie.univ-lyon1.fr

Abstract. This is a feasibility study of tumor motion tracking in images generated by radiotherapy treatment beam. The objective is to control the beam during free breathing so as to reduce the irradiated zone and thus preserve healthy tissues. Two algorithms were tested on portal images sequences. Optical flow estimation (standard Horn and Schunck's algorithm), applied to images from a patient, gave poor results because of low contrast and absence of texture in these images. Target tracking algorithm (block-matching), tested on images of a phantom with a radio-opaque marker, gave satisfactory results : mean absolute error was less than 1 mm. Hence, tumor tracking in portal images is possible, provided that a marker can be implanted in tumor's vicinity. For images without markers, further work is necessary to assess if the small amount of motion information contained in such images can be reliably exploited.

Keywords: medical imaging, motion tracking, portal image

1 Introduction

This work relates to the treatment of bronchial cancers by conformal radiotherapy. To sterilize these cancers effectively, it is necessary to deliver a high amount of irradiation, in a reduced volume. However, to ensure the survival and the best quality of life of the patient after the treatment, it is necessary to avoid irradiating the neighboring healthy parenchyma. The conformal radiotherapy makes it possible to destroy cancerous cells and to protect neighboring healthy tissues [9]. In order to concentrate the radiation energy in the tumoral volume determined by an expert from scanner images [2], the treatment beam orientation varies during the radiotherapy session (Fig. 1a). Moreover, accelerator's collimator is provided with mobile blades (leaves) which mask the healthy zones (Fig. 1b). Their position is recomputed for each angle of the beam. To make coincide the planned irradiation zone with the actual tumor, at the beginning of each radiotherapy session the patient is repositioned using an immobilizing matrix and

W. Skarbek (Ed.): CAIP 2001, LNCS 2124, pp. 247–255, 2001.
© Springer-Verlag Berlin Heidelberg 2001

external marks. Accuracy of the repositioning can be verified using an electronic portal imaging device (EPID) [8]. The portal images are generated by the treatment beam itself, on a sensor placed behind the patient's body. However, few sensors accept the very high energy level of the ionizing beam (typically between 6 and 15 MeV) and provide a sufficiently contrasted image. Only some natural landmarks, such as bones, are approximately perceptible. To facilitate the locating of soft organs, *e.g.* prostate, which can move slightly from one day to another compared to the bones, an artificial marker can sometimes be implanted in the tumor or in its vicinity.

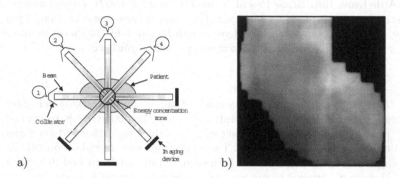

Fig. 1. a) Schematic representation of a radiotherapy treatment setup: an electronic portal imaging device (EPID) can be placed in front of the collimator and moved around the patient, following the beam orientation changes. b) Bronchial zone portal image obtained by IRIS (Bio-Scan s.a., Geneva, Switzerland). Teeth-like black forms visible in the image correspond to the regions protected by collimator leaves.

The above described technique suits the treatment of the tumors located on organs which do not move during the radiotherapy session. For moving organs such as lungs, either the irradiated zone is enlarged to cover all the amplitude of displacement [7], thus inevitably exposing some healthy tissues, or a monitoring device is used to detect the patient's state (apnea or breathing) and to automatically start and stop the irradiation in consequence [6],[7]. However, only a minority of patients with pulmonary lesions is able to correctly perform the breath-holds necessary in the latter case. The objective of our study is to develop a system capable of tracking the tumor motion in real time, so as to automatically control the position of the beam or of the patient. We took as a starting point the existing techniques of image processing, in particular a software developed within our laboratory, called Echotrack [11],[12]. Echotrack uses ultrasound images to track in real time the motion (induced by breathing and by patient's movements) of renal stones and automatically adjusts the position of lithotriptor's shock waves generator, according to the displacements of the stone. However, ultrasound is not suitable for pulmonary imaging, due to air contained in lungs. Instead, the use of portal images seems to be a good solution, since it avoids additional equipment and additional amounts of ionizing radiation that would occur when using fluoroscopy, for example. Although conventional EPIDs provide very low-contrast images and their imaging rate is not sufficient, compared with the breathing rate, our project is based on the prospect for acquisition of a new-generation system called IRIS (Interactive Radiotherapy Imaging System, Bio-Scan s.a., Geneva, Switzerland). This imaging device performs frame grabbing of better quality and at a much higher acquisition rate (up to 12 images a second) than the conventional EPIDs.

In the sequel of this paper, firstly two approaches based on optical flow estimation and on target tracking will be justified, then the tests carried out will be described, lastly the results and perspectives will be discussed.

2 Methods

Our project is at the stage of a feasibility study. Let us note that portal images have photometrical characteristics (contrasts, dynamics, noise...) which strongly differ from the video images (Figs. 1b, 3b). Given that motion estimation and tracking algorithms reported in literature generally were developed in the context of video image sequences, the first step is to test the applicability of the "gold standard" algorithms from the video field, to the portal image sequences. In our research, we are exploring two directions. The first one is based on a realistic assumption according to which the tumor is not visible in the portal images because of an insufficient contrast between soft tissues. Provided that the motion of the surrounding organs can be determined, it should be possible to determine, by interpolation, the displacement of the tumor. This assumption leads to explore the algorithms of dense motion field (optical flow) estimation, i.e. algorithms which provide an estimate of the motion for each point of the image. The second direction is based on the assumption according to which the tumour can be highlighted in the images, *e.g.* by implanting gold markers, like in the case of the prostate. Although currently this process is not clinically available for the lungs, such an attempt was recently reported [13]. The use of target tracking algorithms based on block-matching should then be possible.

2.1 Optical Flow Estimation

In the field of the optical flow estimation, the reference method is the one proposed by Horn and Schunck [4]. Generally, each new algorithm is compared to the Horn and Schunck's one, so as to show the improvements achieved [3]. Most of these improvements attempt to cope with motion field discontinuities which are frequent in video images [10]. However, in a first approximation, motion of different organs and tissues within the human body can be considered as continuous. Indeed, "static" organs (liver, kidneys, ...) undergo movements induced by the neighboring "moving" organs (heart, lungs) and the amplitude of this motion "continuously" decreases, as the distance from the heart or lungs increases and the distance from the bones decreases. Moreover, given the low contrasts in the portal images of the thorax, motion discontinuities (if any) would not appear as clean breaks. That is why we chose to test the Horn and Schunck's method, before seeking its improvements.

Let us remind the foundations of this method. It is based on two constraints: 1) brightness invariance of the moving points and 2) smoothness of the motion field. The first-order Taylor series development of the brightness invariance equation gives rise to the so-called motion constraint equation:

$$I_x u + I_y v + I_t = 0 \ , \tag{1}$$

where I_x, I_y and I_t are the components of the spatio-temporal gradient of the brightness. This single equation is not sufficient to find two unknowns, u and v, which are the components of the velocity vector. That is why a second constraint is necessary. In the Horn and Schunck's method, the smoothness constraint is expressed using the spatial derivatives of u and v:

$$\varepsilon_c^2 = u_x^2 + u_y^2 + v_x^2 + v_y^2 \ . \tag{2}$$

The motion estimation becomes an energy minimization problem with two energy terms weighted by a coefficient α, the smoothing (regularizing) term ε_c^2 and a data attachment term :

$$\varepsilon_d^2 = (I_x u + I_y v + I_t)^2 \ . \tag{3}$$

The motion field is estimated by iterative minimization of the following energy functional on the image support S:

$$E = \sum_{s \in S} \varepsilon_d^2(s) + \alpha \varepsilon_c^2(s) \ . \tag{4}$$

2.2   Target Tracking

Usually, the target is an icon, *i.e.* a small fragment of an image. An icon is first selected (automatically or interactively) in an initial image of the considered sequence. Then a similarity criterion is used to find, in each image of the sequence, the fragment which matches the target [1]. High-level criteria of similarity, such as the criteria based on lengths and orientations of line segments, usually are more reliable than low-level ones, provided that structural information characterizing the icon can be extracted from the image data. In our case no structural information is available. Hence we are forced to choose low-level methods. These methods are generally referenced as "block matching", where a block is simply the brightness pattern of the icon. Their implementations differ from each other by the search strategy. There are also different possible choices of the similarity measure. We implemented the cross-correlation centered on the mean value and normalized by the variance (CNV), defined as follows:

$$C(x,y) = \frac{\sum_{i=1}^{M} \sum_{j=1}^{N} [R(i,j) - \mu_r] \times [P(i,j) - \mu_p]}{\sqrt{\sum_{i=1}^{M} \sum_{j=1}^{N} [R(i,j) - \mu_r]^2} \times \sqrt{\sum_{i=1}^{M} \sum_{j=1}^{N} [P(i,j) - \mu_p]^2}} \ . \tag{5}$$

In this formula (x, y) represent the location of the currently tested block, $P(i, j)$ is the brightness of the pixel (i, j) in this block, $R(i, j)$ is the brightness of the corresponding pixel in the reference block, μ_p, μ_r are the mean values of these blocks and M, N represent the block size. This measure was also implemented in our Echotrack software. Its numerical implementation is computationally faster when using the following transformation of (5):

$$C(x, y) = \frac{\sum_{i=1}^{M} \sum_{j=1}^{N} R(i, j) P(i, j) - MN\mu_r\mu_p}{\sqrt{\sum_{i=1}^{M} \sum_{j=1}^{N} R^2(i, j) - MN\mu_r^2} \times \sqrt{\sum_{i=1}^{M} \sum_{j=1}^{N} P^2(i, j) - MN\mu_p^2}}.$$

(6)

Instead of searching for a maximum of similarity, one can seek a minimum of disparity. We implemented two well-known disparity measures: sum of absolute differences (SAD: $k = 1$) and sum of squared differences (SSD: $k = 2$):

$$D(x, y) = \sum_{i=1}^{M} \sum_{j=1}^{N} |R(i, j) - P(i, j)|^k.$$

(7)

As for the search strategy, we implemented the exhaustive search, which scans all the possible locations within a search window, and thus guarantees that the global optimum is found. We also tested a much faster, but potentially sub-optimal method, used in the Echotrack: a multi-grid strategy with progressive mesh-size reduction, called log-D step [5]. The search is carried out in a small number L of iterations. At each iteration $l + 1$, the similarity criterion is calculated for eight locations around the best match location (x_l, y_l) from the iteration l: $(x_l + \Delta_{l+1}, y_l + \Delta_{l+1}), \Delta_{l+1} \in \{-d_{l+1}, 0, d_{l+1}\}$. Distance between these locations (mesh-size) is first equal to $d_1 = 2^{L-1}$, then $d_{l+1} = d_l/2$, $l + 1 \in \{1, .., L\}$. At the first iteration, the similarity criterion is also calculated for the location (x_0, y_0) corresponding to the target position in the previous image or to its predicted position, in the case of predictable movements. Three iterations are sufficient to estimate the displacements for normal breathing and usual imaging conditions. Indeed, the in-plane movement of lungs can be approximately considered as sinusoidal with a typical frequency of 12 breaths per minute. The maximum peak-to-peak displacement (observed for the lower part of the lungs) reaches 40 mms. For an acquisition rate of 5 images per second available with IRIS, it can be easily calculated that the maximum displacement between two consecutive images is equal to 5 mms. Given the typical size of the field of view and the resolution of the image acquisition matrix, we have 1 mm = 1 pixel. Hence, we need to estimate displacements up to 5 pixels. With $L = 3$ the maximum displacement which can be estimated, is equal to 7 pixels ($d_1 + d_2 + d_3 = 4 + 2 + 1$). The number of scanned locations is then only 25, while 225 locations would be scanned for the exhaustive search within a window of equivalent size.

To facilitate the tests, the above described algorithms were implemented within a user-friendly interface designed to easily accept new algorithms in future. It enables the user to open image sequences and to display them, to select a sub-sequence and a region of interest (or an icon), to choose an algorithm, to tune its parameters and to run it, to display the results and to save them.

3   Tests

The optical flow estimation algorithm was applied to images from a patient with bronchial cancer, acquired during a short trial period when IRIS was available in Lyon South Hospital. True displacements were unknown. Hence two strategies were used to assess the consistence of the results. Firstly, motion fields obtained for each pair of consecutive images were visually inspected and compared with the perceptible movements. Secondly, the algorithm was tested on sequences with synthetic motion generated by taking one image from the original sequence and by applying to it known translations.

Obviously, it was not possible to test the target tracking algorithm on images from patients, since implanting markers in the bronchial region is not yet clinically available. However, it was necessary to test it on image sequences where the location of he target is perfectly known. Moreover, the images were to show the same characteristics (contrast, noise, resolution) as real images and the movement was to resemble breathing (same amplitude, speed and acceleration). The best means of acquiring images under conditions close to the clinical reality, while controlling the displacements which one wishes to estimate, is the use of a phantom. We used a phantom (Fig. 2) having the same average density as the human body, and density variations simulating the internal organs (namely lungs). A metal implant approximately 3 mms in diameter was placed inside. Breathing was simulated by table displacements controlled by electric drives. We simulated a sinusoidal movement along the head-feet line, with 40 mms of amplitude and a frequency of 12 breaths per minute. A series of static images was acquired using a conventional EPID (Varian, Palo Alto, USA), in table positions corresponding to the acquisition rate of 5 images per second. Then the user manually selected, in the first image, rectangles of various sizes, framing the marker. These rectangles, each one in his turn, were considered as a target and tracked using the implemented algorithms.

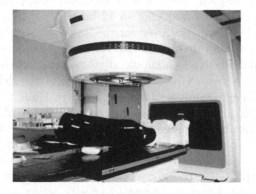

Fig. 2. Experimental setup for image acquisition of human-body phantom. Top down : the collimator of the accelerator, the phantom, the mobile table with position controlled by electric drives. The imaging device is placed below the table.

4   Results and Discussion

The optical flow estimation method, applied to images from a patient, gave poor results for both real and synthetic motion. Significant velocity vectors were concentrated in a few regions having enough contrast to carry the motion information (Fig. 3a). For the majority of pixels, the spatio-temporal gradients were close to zero, which explains the null amplitudes of the velocity vectors. The smoothness constraint involves a kind of averaging: although it propagates the motion information towards the homogeneous zones, it also reduces the amplitude of estimated displacements where the motion information is present. The conventional approach is thus unsuited. However, one can consider a similar method where the points used for calculations would be selected according to their spatio-temporal gradients. Instead of the smoothness constraint applied to dispersed points, one could use a model of movement common to all the points, *e.g.*: translation combined with homothety, to take into account the expansion/contraction of the lungs.

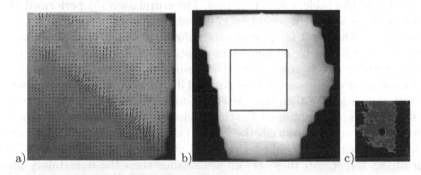

Fig. 3. a) Fragment of motion field obtained for a pair of IRIS images from a patient. The only region with enough contrast to carry the motion information is the boundary of the heart (near the diagonal line of this sub-image). b) Portal image of the phantom, obtained by a conventional EPID. The implanted marker (center of the rectangle) is hardly perceptible in the image. c) The marker highlighted by histogram equalization of the image fragment bounded by the rectangle.

Unlike the Horn and Schunck's algorithm, the target tracking algorithms gave satisfactory results, although the marker was hardly perceptible in the conventional EPID images (Fig. 3b). The most important result is the fact that each algorithm followed the target all along its displacements, although the estimated position was not always perfectly centered on the theoretical position of the marker. This means that portal images can be used for the purpose of pulmonary tumor tracking, provided that stable (non-migrating) markers can be implanted in clinical practice. The tumor-tracking system presented in [13] was based on the same assumption, but used an additional imaging system with fluoroscopic cameras instead of an EPID.

Table 1. Estimation errors in target tracking. First three columns: exhaustive search with different similarity / disparity measures and constant block-size; SSD = Sum of Squared Differences, SAD= Sum of Absolute Differences (7), $k = 2$ and 1 respectively, CNV = Correlation Normalized by Variance (5). Next three columns: exhaustive search with CNV and different block-sizes. The last column: log-D step search with CNV.

error [mms]	SSD	SAD	CNV	10×10	16×16	40×40	log-D step
maximum	3.16	2.24	2.24	1.0	1.0	2.24	1.0
mean	0.91	0.84	0.65	0.52	0.56	0.64	0.56
cumulated	1.0	2.0	2.24	1.0	0.0	4.12	0.0

The precision of the tracking for different combinations of search strategy and similarity/disparity measure is reported in (Tab. 1). With an image resolution of 1 mm = 1 pixel, the mean value of absolute errors was comprised between 0.52 mm and 0.91 mm. These values are to be compared with the uncertainty of positioning of the table, equal to 1 mm. The correlation (5) performed slightly better than the disparity measures (7), hence it can be recommended for this application. Although it is approximately three times slower than the disparity measures, it still can be computed in real-time when combined with the log-D step strategy which gave as good as results as the exhaustive search. The cumulated error corresponds to the residual distance between the estimated and theoretical positions of the target after a complete breathing cycle. As expected, the best results were obtained for the blocks big enough to contain the entire target, but not too big. Indeed, the largest cumulated error (4.12 mm) occurred for a great block which could be regarded as model of the background rather than of the target. All these errors are smaller than the uncertainty margins (5-15 mm) added by the expert to perceptible tumoral volume, when defining the irradiation zone in static cases.

5   Conclusion

In our study, we applied conventional motion estimation and tracking algorithms in a new context of portal image sequences, to assess the possibility of tracking bronchial tumors during radiotherapy treatment. Unlike the existing methods designed to take into account these tumors' motion, our approach does not require any additional equipment compared to the standard radiotherapy setup. Despite very low contrasts in the portal images, our results show that tracking of bronchial tumors' motion is feasible, using block-matching algorithms, provided that a radio-opaque marker is implanted in the tumor's vicinity. Such markers were already used by a competing Japanese team, but in a more complex setup with an additional fluoroscopic imaging system instead of the portal imaging device. Let us note however, that marker implanting is not yet available in the clinical context. Motion estimation without markers is much more difficult. The conventional optical flow estimation approach using the smoothness constraint is not appropriate. However, our results suggest that a specially tailored algorithm, using a model of the movement and a selection of points based on the spatio-temporal gradient values, could do better.

Acknowledgements. This work is in the scope of the scientific topics of the GDR-PRC ISIS research group of the French National Center for Scientific Research (CNRS). The authors are grateful to Arnaud Ruillère for his help in acquisition of images of phantom.

References

1. Aggarwal, J.K., Davis, L.S., Martin, W.N.: Correspondance processes in dynamic scene analysis. Proc. IEEE **69**(5) (1981) 562–572
2. Armstrong, J.G.: Target volume definition for three-dimensional conformal radiotherapy of lung cancer. British Journal of Radiology **71** (1998) 587–594
3. Barron, J.L., Fleet, D.J., Beauchemin, S.S.: Systems and experiment : performance of optical flow techniques. Int. J. Computer Vision **12**(1) (1994) 43–77
4. Horn, B.K.P., Schunck, B.G.: Determining Optical Flow. Artif. Intell. **17** (1981) 185–203
5. Jain, J.R., Jain, A.K.: Displacement measurement and its application in interframe image coding. IEEE Trans Comm. **COM.29** (1981) 1799–1808
6. Kubo, H.D., Len, P.M., Wang, L., Leigh, B.R., Mostafavi, H.: Clinical experience of breathing synchronized radiotherapy procedure at the university of California Davis Cancer Center. Proceedings of the 41st Annual ASTRO Meeting , Int. J. Radiation Oncology **45**(3) supplement (1999) 204
7. Mah, D., Hanley, J., Rosenzweig, K.E., Yorke, E., Ling, C.C., Leibel, S.A., Fuks, Z., Mageras, G.S.: The deep inspiration breath-hold technique for the treatment of lung cancer: the first 200 treatments. Proceedings of the 41st Annual ASTRO Meeting, Int. J. Radiation Oncology **45**(3) supplement (1999) 204
8. Mornex, F.: L'imagerie portale, pour quoi faire ?. Bull. Cancer / Radiother. **83** (1996) 391–396 (in French)
9. Mornex, F., Giraud, P., Van Houtte, P. Mirimanoff, R., Chapet, O., Loubeyre, P.: Conformal radiotherapy of non-small-cell lung cancer. Cancer Radiother. **2**(5) (1999) 425–436 (in French)
10. Orkisz, M., Clarysse P.: Estimation du flot optique en présence de discontinuités : une revue. Traitement du Signal **13**(5) (1996) 489–513 (in French)
11. Orkisz, M., Farchtchian, T., Saighi, D., Bourlion, M., Thiounn, N., Gimenez, G., Debré, B., Flam, T.A.: Image-based renal stone tracking to improve efficacy in extra-corporeal lithotripsy. J. Urol. **160** (1998) 1237–1240
12. Orkisz, M., Bourlion, M., Gimenez, G., Flam, T.A.: Real-time target tracking applied to improve fragmentation of renal stones in extra-corporeal lithotripsy. Machine Vision and Applications **11** (1999) 138–144
13. Shirato, H., Shimizu, S., Shimizu, T., Akita, H., Kurauchi, N., Shinohara, N., Ogura, N.S., Harabayashi, T., Aoyama, H., Miyasaka, K.: Fluoroscopic real-time tumor tracking radiotherapy. Proceedings of the 41st Annual ASTRO Meeting, Int. J. Radiation Oncology **45**(3) supplement (1999) 205

Color Thinning with Applications to Biomedical Images

A. Nedzved[1], Y. Ilyich[1], S. Ablameyko[2], and S. Kamata[3]

[1] Minsk Medical State Institute, Laboratory of the Information Computer Technologies,
ul. Leningradskaya 6, Minsk 220050 Belarus
bigbear@itlab.unibel.by
[2] Institute of Engineering Cybernetics, Academy of Sciences of Belarus,
ul. Surganova 6, Minsk, 220012 Belarus
Tel: +375 (17) 284-21-44 Fax: +375 (17) 231-84-03
abl@newman.bas-net.by
[3] Department of Intelligent Systems, Kyushu University,
6-10-1 Hakozaki, Higashi-ku, Fukuoka 812-8581,
Japan Tel: +81-92-642-4070 Fax: +81-92-632-5204
kamata@is.kyushu-u.ac.jp

Abstract. A scheme for cell extraction in color histological images based edge detection and thinning is considered. An algorithm for thinning of color images is proposed that is based on thinning of pseudo gray-scale image. To extract accurately gray-scale levels, we propose a new coordinate system for color representation: system PHS, where P is a vector of color distance, H is a hue (chromaticity), S is a relative saturation. This coordinate system allows one to take into account specifics of histological images. Comparison of image thinning in other coordinate color systems is given that shows the image thinning in PHS system produces a rather high-quality skeleton of the objects in a color image. The proposed algorithm was tested on the histological images.

Keywords: image thinning, color spaces, biomedical images

1 Introduction

Color image processing is rapidly developing area of research and applications due to widely used color image acquisition devices now. However, many algorithms that have been successfully used in grey-scale and binary image processing are not immediately applicable to color images.

In difference with gray-scale images where each pixel is represented by three coordinates (x, y and brightness), color image is exemplified by five coordinates: three characteristics of the color (hue, luminance, saturation), and two coordinates of the spatial location of the object (x, y). Mutual influence of these coordinates is very complex and its analysis is usually time-consuming. Based on color coordinates, we may consider that in color image we deal with three

W. Skarbek (Ed.): CAIP 2001, LNCS 2124, pp. 256–263, 2001.

gray-scale images. Correspondingly it is possibly to build color image processing algorithms proceeding from corresponding gray-scale algorithms. However, we should answer on the question: what does correspond to a grey level in a color image ?

One of the most significant operations in image processing is thinning that transforms original "thick" object in lines of one-pixel thickness. Thinning is very well developed for binary images [8]. During last decades, there appeared quite many papers with algorithms for thinning of grey-scale images [1,3] but there are no practically algorithms for thinning of color images. However, color always gives specifics in image processing and object recognition that leads to necessity to develop special algorithms for color image processing.

Here, we consider processing of color histological images that usually contain cells, kernels, etc. Cells can be classified by using their shape and structure. The cell shape can be treated as almost round-shaped (circular-based primitive) and not elongated. The cell structure is characterized by its kernels and kernels' position. Another important cell feature is its position in a tissue.

It is practically impossible to select or develop automatic segmentation methods that can automatically extract required cells and compute their characteristics. That is why most of the papers consider particular features of cell images and methods of their segmentation. Among the main cell segmentation approaches are those based on edge detection, thresholding, snakes, Hough transform, morphological operations, neural networks [10,4,6].

There are several papers analyzing color cell images [11,2]. In paper [5], for cell image segmentation, RGB space and Lab space are combined. By this method, both the nucleus and cytoplasm of cancer cells can be separated from background. Then, the candidate cancer cells are selected using some morphological features of nuclei, the purpose of this step is to pick out most of the non cancer cells and leave a few doubtful cells for further verification, therefore improving the efficiency of the whole recognition process. As the last step, for all the candidate cells, some statistic parameters in different color space are calculated, which are used as features for recognition. Paper [7] proposes to segment color cell images by Constraint Satisfaction Neural Network (CSNN) algorithm. This is accomplished by incorporating in the CSNN algorithm multiresolution scheme using pyramid structure.

The results of segmentation often depend on a cell image quality and when difference between cell or kernel and background is small, most of the methods do not work properly. We analyzed a wide range of cell images and consider that cell shapes can be correctly extracted through edge detection and thinning approach.

In this paper, we propose the algorithm for thinning of color images that is based on thinning of pseudo gray-scale images. To extract accurately gray-scale levels, we propose a new coordinate system for color representation and propose algorithm to thin separate gray-scale images. Examples of thinning for color histological images are given.

2   Coordinate System for Thinning Histological Color Images

Pseudo-colors usually require introduction of special coordinate system where they will be reflected. From other side, new coordinate system should reflect specifics of images and algorithms. Such as we orient to morphological algorithms and histological images that have special features, we will build the corresponding coordinate system.

One of the most important characteristics of an object in a histological color image is its chromaticity. If image fragments have an abrupt jump in chromaticity, they are either the object or its part. However, chromaticity almost does not influence the object topology. From other side, if we process one coordinate instead of two, we get a twofold gain in the speed. In most operations of the thinning, the processing of the image consists of many iterations, hence this gain is essential.

Thus, processing of vector of color distance between the origin of coordinates and the desired point can be an advantageous. It equals the sum of the vectors of the luminance and saturation or the sum of the basis RGB vectors which directly specify the color. Therefore, it contains information about both the luminance and saturation and is a gray-scale value, which is most appropriate for the thinning on color images. That is why we propose to introduce the coordinate system where one of the axes is the vector of the color distance, and another, the quantity which features chromaticity.

Let us consider a cylindrical coordinate system whose central axis is a result of the vector summation of axes of RGB system and has a meaning of the monochromatic component. First of all, it should be noted that chromaticity, which is characterized by the angle of rotation, is an independent feature of the object, whereas saturation and luminance depend upon the external conditions. Thus, the most efficient for analysis will be employment of the vector of the color distance, which is equal to the sum of vectors of luminance and saturation or the sum of the basis RGB vectors, which directly exemplify the color. It remains to introduce a quantity, which will characterize a correlation of the luminance and saturation. Such a quantity is the angle between the vector of the color distance and central axis of the frame of reference. This quantity has a meaning of the relative saturation and should not vary during image processing. The purpose of this quantity is to reconstruct the relation of the luminance and saturation in inverse transformations. Thus, in the following, we shall call this spherical system of coordinates as PHS, where P is the vector of the color distance, H is hue (chromaticity), S is relative saturation.

In order to obtain a desired system, it is necessary to carry out two rotations of the coordinate axes about 45°. We have carried out a rotation in the plane RB, and then in the plane RG. This resulted in the following transformations of the coordinates:

$$\begin{cases} Z = \frac{1}{2}(R + \sqrt{2}G + B) \\ Y = -\frac{\sqrt{2}}{2}(R - B) \\ X = -\frac{1}{2}(R - \sqrt{2}G + B). \end{cases}$$

Passing from Cartesian coordinates ZYX to the spherical ones, we obtain PHS:

$$\begin{cases} P = \sqrt{R^2 + G^2 + B^2} \\ H = \arctan\left(\frac{\sqrt{2}(B-R)}{\sqrt{2}G-R-B}\right) \\ S = \arccos\left(\frac{R+\sqrt{2}G+B}{2P}\right). \end{cases}$$

For operations of the gray-scale morphology in this system, P is used. For separation, H is employed, and S is always invariable because this value corresponds to the relation of the luminance and saturation.

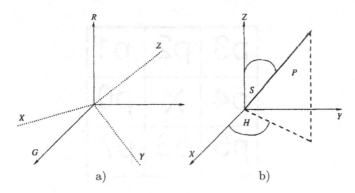

Fig. 1. Coordinate systems: a) RGB and ZYX systems, b) ZYX and PHS.

Thus, we have obtained a system of color coordinates, which can be profitably employed for thinning of color images.

3   Thinning of Color Images

We propose to use pseudoimage, each pixel of which corresponds to the vector of the color distance of initial image. Pseudoimage corresponds to a gray-scale image, thus we can apply our algorithm of gray-scale thinning [9] for thinning of color image.

In the pseudoimage, which consists of the vector of color, all gray-scale levels are analyzed simultaneously and radius-vector of the distance of the color of a pixel which satisfies the above-mentioned conditions is diminished by 1. In order that diagonal pixels would not be processed twice during one run, they should be marked; for this purpose, an additional array is generated with dimension that corresponds to the dimension of the image. After the cycle is passed, the array is zeroed. This guarantees that pixels are processed only once in a single cycle. The number of iterations of tail removal is equal to the maximum length of the branches, which should be removed.

When each cycle is executed, the branches that arise due to roughness of the object are removed. The following condition is checked for every pixel:

If ($p0 >= x$ OR $p1 >= x$ OR $p2 >= x$ OR $p3 >= x$ OR $p4 >= x$ OR $p5 >= x$ AND $p6 < x$)

AND ($p2 >= x$ OR $p3 >= x$ OR $p4 >= x$ OR $p5 >= x$ OR $p6 >= x$ OR $p7 >= x$ AND $p0 < x$)

AND ($p4 >= x$ OR $p5 >= x$ OR $p6 >= x$ OR $p7 >= x$ OR $p0 >= x$ OR $p1 >= x$ AND $p2 < x$)

AND ($p6 >= x$ OR $p7 >= x$ OR $p0 >= x$ OR $p1 >= x$ OR $p2 >= x$ OR $p3 >= x$ AND $p4 < x$) then the central pixel is diminished by 1.

Each pixel in the image has eight neighbors, which are numbered according to the following scheme (Fig. 2).

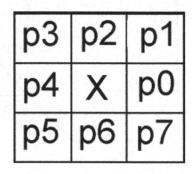

Fig. 2. 8-neighborhood of a central pixel x

The result is a skeleton of the image, which has fragments with different hues and smoothly varying luminance and saturation.

4   Binarization of Color Skeleton

The binarization is carried out at a single run of the image. If the pixel to be processed has at least one of the four neighbors ($p0$, $p2$, $p4$, $p6$) with a radius-vector of the color distance less than its own, or all diagonal octo-neighbors have a radius-vector with the color distance less than its own, its value is set to 1, otherwise to 0. That is:

$$X = \begin{cases} 1, & \text{if } p0 < X \text{ OR } p2 < X \text{ OR } p4 < X \text{ OR } p6 < X \text{ OR} \\ & (p1 < X \text{ AND } p3 < X \text{ AND } p5 < X \text{ AND } p7 < X) \\ 0, & \text{otherwise.} \end{cases}$$

5   Experimental Results

The developed algorithm has been applied for processing of histological images of a human liver, where the cell nucleus had to be extracted (Fig. 3). The algorithm of cell extraction includes cell edge detection for which Sobel gradient has been applied first (Fig. 3b). Then, color thinning allowed us to obtain one-pixel thickness contours of the cell edges (Fig. 3c). Subsequent hue binarization allowed us to extract precise contours of the cell nucleus (Fig. 3d).

Practical verification has demonstrated that the algorithm allows one to obtain a good-quality skeleton of the objects in the color image. However, employment of the harmonic functions decelerates the process of the preparation of the image for processing, therefore, subsequent refinement of the algorithm through the optimization of the coordinate transformation is possible.

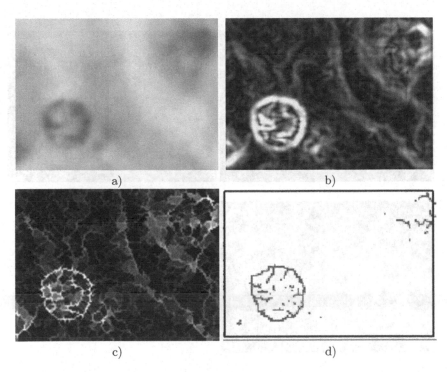

Fig. 3. The process of separation of the nucleus of the cell from a histological image: (a) initial image; (b) Sobel gradient transformation with respect to the radius-vector of the color distance; (c) thinning with respect to the radius-vector of the color distance; (d) threshold segmentation with respect to the hue and binarization.

6   Comparisons

Let us compare thinning in the proposed coordinate system with that in other systems. The result of the processing of pseudoimages, which consist of the vectors L and S of HLS system, is presented in Fig. 4a. There are many superfluous ridges in this system, although processing here may be accepted as satisfactory.

Systems MKO with adapted to human vision monochromatic luminance Y (YUV, YXZ, YIQ, etc.) are also well accepted. Thinning is not very effective in these systems because they do not take into account correlation between the luminance and saturation. The result of the processing of pseudoimages, which consist of vectors Y of YIQ system is presented in Fig. 4c. In this coordinate system, there is a distortion of colors. Similar operations in the RGB system lead to occurrence of false branches, ridge bifurcation, distortion of the geometric colorimetric properties (Fig. 4b) and do not lead to a result of practical significance.

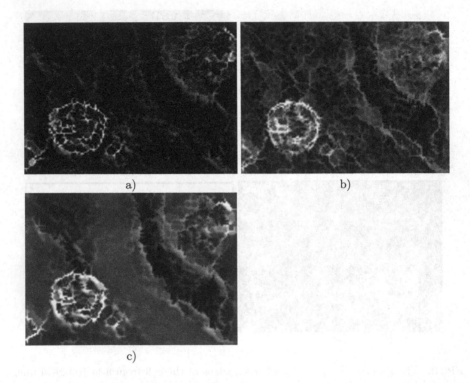

a)   b)

c)

Fig. 4. The results of image thinning for extraction of a cell nucleus in hystology image: (a) in HLS system; (b) in RGB system; (c) in YIQ system.

7   Conclusion

Algorithm for thinning of color images has been proposed. Special color coordinate system PHS has been proposed that allows to take into account specifics of images and algorithm. The proposed algorithm was tested on the biomedical images. Comparison with other coordinate color system was given that shows the image thinning in PHS system produces a rather high-quality skeleton of objects in a color image.

References

1. Abe, K., Mizutani, E., Wang, C.: Thinning of gray-scale images with combined sequential and parallel conditions for pixel removal. IEEE Transactions on Systems Man and Cybernetics **24** (1994) 294–299
2. Ablameyko, S., Lagunovsky, D.: Image processing: technologies, methods and applications. Minsk, IEC (1999)
3. Arcelli, C., Ramella, G.: Finding gray-skeletons by iterated pixel removal. Image and Vision Computing **13** (1995) 159–167
4. Garrido, A., Perez de la Blanca, N.: Applying deformable templates for cell image segmentation. Pattern Recognition Letters **22** (2000) 821–832
5. Jianfeng, L., Shijin, L., Jingyu, Y., Leijian, L., Anguang, H., Keli, L.: Lung cancer cell recognition based on multiple color space. Proceedings of SPIE **7** (1998) 378–386
6. Kim, J., Gillies, D.: Automatic morphometric analysis of neural cells. International Journal of Machine graphics and vision **7** (1998) 693–710
7. Kurugollu, F., Sankur, B.: Color cell image segmentation using pyramidal constraint satisfaction neural network. Proceedings of IAPR Workshop on Machine Vision Applications (1998) 85–88
8. Lam, L., Lee, S.W., Suen, C.Y.: Thinning methodologies-a comprehensive survey. IEEE Trans. on Pattern Analysis and Machine Intelligence **14** (1992) 869–885
9. Nedzved, A.M., Ablameyko, S.V.: Thinning of gray scale images in medical image processing. Pattern Recognition and Image Analysis **8** (1998) 436–438
10. Pardo, X.M., Cabello, D.: Biomedical active segmentation guided by edge saliency. Pattern Recognition Letters **21** (2000) 559–572
11. Plataniotis, K.N., Lacroix, A., Venetsanopoulos, A.: Color image processing and applications. Springer-Verlag (2000)

Dynamic Active Contour Model for Size Independent Blood Vessel Lumen Segmentation and Quantification in High-Resolution Magnetic Resonance Images

Catherine Desbleds-Mansard[1], Alfred Anwander[1], Linda Chaabane[2],
Maciej Orkisz[1], Bruno Neyran[1], Philippe C. Douek[1], Isabelle E. Magnin[1]

[1] CREATIS, CNRS Research Unit (UMR 5515) affiliated to INSERM, Lyon, France
[2] Laboratory RMN, CNRS Research Unit (UMR 5012) , Lyon, France
Correspondence address: CREATIS, INSA de Lyon, bat. Blaise Pascal,
7 rue J. Capelle, 69621 Villeurbanne cedex, France
catherine.mansard@creatis.insa-lyon.fr

Abstract. We are presenting a software tool developed for the purpose of atherosclerotic plaque study in high resolution Magnetic Resonance Images. A new implementation of balloon-type active contour model used for segmentation and quantification of blood vessel lumen is described. Its originality resides in a dynamic scaling process which makes the influence of the balloon force independent of the current size of the contour. The contour can therefore be initialized by single point. Moreover, system matrix inversion is performed only once. Hence computational cost is strongly reduced. This model was validated in ex vivo vascular images from Watanabe heritable hyperlipidaemic rabbits. Automatic quantification results were compared to measurements performed by experts. Mean quantification error was smaller than average intra-observer variability.

Keywords: medical imaging, active contour model, magnetic resonance images

1 Introduction

Atherosclerosis is a disease of the vascular wall, which can cause hemodynamic troubles and acute accidents due to rupture of the atherosclerotic plaque. Its two well known manifestations are stroke (acute cerebrovascular disease) and myocardial infarction (acute manifestation of cardiac ischaemia). There is a growing field of investigation dealing with the atherosclerotic plaque identification and characterization before plaque rupture and severe lumen obstruction. Animal models of pathologies met in humans are used to evaluate both the therapeutic methods and exploration techniques, Magnetic Resonance Imaging (MRI) in our case. Watanabe heritable hyperlipidaemic (WHHL) rabbit was described as a well suited model for the atherosclerosis. Characterization of the

W. Skarbek (Ed.): CAIP 2001, LNCS 2124, pp. 264–273, 2001.

plaque structure and composition has been performed by means of *in vitro* high resolution MRI. These studies have shown good correlation with histology [1] - [5] (fig.1). Among other tasks, such studies require quantification of the vascular lumen and of the plaque in numerous images. This tedious task needs to be computer-assisted. First-level assistance is an interactive region drawing and automatic calculation of the corresponding areas, diameters and perimeters. The second level is a user-initialized automatic segmentation, with eventual interactive corrections in difficult cases where the user disagrees with the automatically found outlines. The objective of the hereafter presented work was to realize an assistance tool with a maximum degree of automation, so as to save the user's time and to improve the reproducibility of the results.

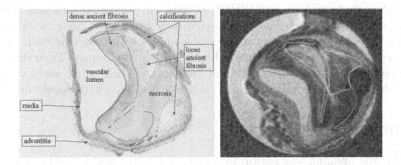

Fig. 1. Histological section (left) and in vitro MR image (right) of a human artery with a large atheroscerotic plaque. Expert hand-drawn regions representing different plaque components in the histological section were simply superimposed on the MR image. Slight differences are mostly due to tearing of the artery during manipulations

One of the main functionalities of this tool is an automatic extraction of the vessel lumen contour. In medical images, smooth closed boundaries are expected in most cases, in particular in the case of the vascular lumen, while simple edge detection often leads to discontinuous lines with gaps and protrusions, because of contrast variations and noise. One of the methods suggested to obtain smooth continuous contours is based on the minimum-cost path search formalism [6]. However, the technique most used today in the medical image processing, is based on deformable models (see [7] for a survey) or on deformable templates [8]. This technique consists in using an initial form (curve, surface . . .), provided with certain properties of flexibility and elasticity, which is gradually deformed to coincide with the boundaries present in the image. The deformations occur under influence of external forces which attract the form towards image points the likely to belong to a boundary (typically maximum of the gradient), and of internal forces which attract it towards an expected reference form. Since the very first implementation of planar deformable models called active contours or snakes [9] many improvements have been proposed in order to make the result initialization-independent. We proposed an original implementation called

Dynamic Active Contour (DAC) which only needs a single initialization point inside the object [10] and which can propagate on a whole sequence of images representing contiguous cross-sections. The algorithm was implemented within a user-friendly graphical interface nicknamed ATHER.

This paper describes a validation of this tool applied to the vessel lumen quantification in *ex vivo* MR images from WHHL rabbits. In the sequel, firstly the animals and the image acquisition protocol will be described. Then the DAC model, the features of the software and the validation method will be detailed. Lastly, the obtained results will be presented and discussed.

2   Animals

The entire study includes six WHHL rabbits, three homozygous and three heterozygous, two of which had a lipid-rich diet. First, two kinds of *in vivo* MR images were acquired: one set of images focused on the vascular wall, the other one used a contrast agent in order to highlight the vessel lumen. Then the animals were sacrificed at age between 13 and 18 months. Their entire aorta with heart and kidneys were removed after fixation under perfusion with paraformaldehyde.

3   Images

In vitro MR imaging was performed with a 2 T horizontal Oxford magnet and a SMIS console. The specimens were placed on a half-birdcage RF coil of 25 mm diameter working at 85.13 MHz. High resolution axial images of the thoracic and abdominal aorta were taken using a multislice 2D spin-echo sequence with a 256×128 matrix. Slice thickness was 0.8 to 1 mm and the pixel size was 58 to 78 μm. T1 and T2 weighted images were obtained with TR/TE = 600/21 ms and 1800/50 ms respectively. Six territories have been explored from the aortic arch down to the iliac bifurcation. The acquired images were organized into series, one per territory (fig.2).

4   Dynamic Active Contours

Let us first remind some fundamentals on active contours. A deformable contour model is defined by a parametric curve $\nu(s,t) = (x(s,t), y(s,t))^T$, which evolves in time t and space, and by its energy, $E(\nu) = E_{int}(\nu) + E_{ext}(\nu)$. The energy $E_{int}(\nu)$, corresponding to internal forces, imposes constraints on the first and second derivatives of the curve:

$$E(\nu) = E_{elast}(\nu) + E_{flex}(\nu) = \alpha \int_0^1 \left| \frac{\partial \nu(s,t)}{\partial s} \right|^2 ds + \beta \int_0^1 \left| \frac{\partial^2 \nu(s,t)}{\partial s^2} \right|^2 ds \ , \ (1)$$

where α controls its elasticity while β controls its flexibility and $s \in [0,1]$ is the arc length. There are two kinds of external forces represented by a potential

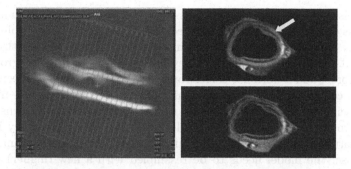

Fig. 2. Examples of high resolution cross-sectional MR images of a rabbit aorta. Longitudinal section with lines showing the positions of cross sections from the corresponding series (left). Cross-sectional images weighted T1 and T2 (right). Vessel lumen appears in black, surrounded by a thick vessel wall layer (media), which appears in grey (better delineated on top). Atherosclerotic plaque is perceptible as an additional layer between the media and the vessel lumen in the right part of each image (arrow). In fact, the plaque is a pathological thickening of a thin internal vessel wall layer (intima).

energy $E_{ext}(\nu) = \int_{s=0}^{1} P(\nu(s,t))ds$: image forces and a balloon force. The link between potentials $P_i(\nu)$ and forces is given by : $F_i(\nu) = -\nabla P_i(\nu)$. The image forces are usually designed to attract the snake toward strong intensity gradients. One can also use prior knowledge about the vessel outline as function of the intensity, so that the potential energy function has a minimum at the vessel boundary. With an initialization far from any edge, the active contour would collapse. One solution consists in adding an external balloon force F_b [11], which acts like an inflating pressure, and makes the active contour model move without any image gradient. It is expressed as $F_b = b\mathbf{n}(s,t)$, where $\mathbf{n}(s,t)$ is the normal unitary vector oriented outward at the point $\nu(s,t)$, and b is a coefficient which controls the inflation. The total external force $F_{ext}(\nu)$ is a weighted sum of these forces.

The parametric curve $\nu(s)$ is fitted to the image by minimization of the energy $E(\nu)$. The minimization is a temporal and spatial discrete process, using a finite number N of snake points $\nu(i,k)$ that approximate the snake as a polygon. The partial derivatives in (1) are approximated by finite differences $\partial\nu(s,t)/\partial s \approx (\nu((i+1),k)-\nu((i-1),k)/2h$, with h the discretization step of the snake points on the contour. The snake is iteratively deformed, explicitly using the external forces $F_{ext}(\nu)$ at each point of the snake, while implicitly minimizing the internal energy. The new position $\nu(i,k)$ at time step k is computed by solving the associated Euler equations. These equations can be solved by matrix inversion :

$$\nu(i,k) = [\gamma\mathbf{I} + \mathbf{A}]^{-1} \cdot [\gamma \cdot \nu(i,k-1)] + F_{ext}(\nu(i,k-1))] . \qquad (2)$$

This equation is referred to as the evolution equation of the snake. The matrix $\mathbf{A}$ represents a discretized formulation of the internal energy. The damping parameter γ controls the snake deformation magnitude at each iteration.

The discretization step h depends on the number of snake points N and, initially, is equal to the distance between the snake points. The evolution equation (2) is only valid if the snake points are equally spaced and the step h is unchanged. After each iteration however, the length of a snake grows due to external force, and h does not correspond to the real distance between the snake points. In conventional implementations of the active contours, the internal energy, especially the term $E_{elast}(\nu)$ associated with the contour's tension, grows with the snake length and may stop the snake before the boundary is reached. This is particularly annoying if the initialization is far from the final position. In this case, the model needs to be resampled with a new (higher) number of snake points N, or/and a new step h. The N by N matrix $[\gamma \mathbf{I} + \mathbf{A}]$, as well as its inverse have to be recomputed. This is a time-consuming task which limits the classical snake model application in many cases.

In order to make the snake scale-independent and allow an initialization by a single pixel in the image, we proposed [10] a new numerical scheme for the snake energy minimization. It was designed to preserve the validity of the equation (2) in spite of the evolution of the snake's size. It only needs a single computation of the matrices $[\gamma \mathbf{I} + \mathbf{A}]$ and $[\gamma \mathbf{I} + \mathbf{A}]^{-1}$, and uses a fixed number of points for all the iterations. The snake size is scaled at each iteration, in order to normalize the internal energy of the model with a fixed discretization step $h' = 1$. The scaled snake $\nu'(i, k - 1)$ has a normalized length equal to N. It is obtained from the actual snake and the average distance $\overline{h}_{k-1}$ between its points:

$$\nu'(i, (k - 1)) = \nu(i, k - 1) / \overline{h}_{k-1} \ . \tag{3}$$

The scaled snake is deformed according to the evolution equation (2), using the external forces F_{ext} from the non-scaled snake. Hence, the new (scaled) positions for the snake points $\nu'(i, k)$ are:

$$\nu'(i, k) = [\gamma \mathbf{I} + \mathbf{A}]^{-1} \cdot [\gamma \cdot \nu'(i, k - 1)] + F_{ext} (\nu(i, k - 1))] \ . \tag{4}$$

Let $\triangle \nu'(i, k)$ be the snake deformation vectors for each snake point:

$$\triangle \nu'(i, k) = \nu'(i, k) - \nu'(i, k - 1) \ . \tag{5}$$

The deformation vectors are computed with a normalized internal energy, and the deformation is directly applied onto the non-scaled snake $\triangle \nu(i, k) = \triangle \nu'(i, k)$. Therefore, the new position of the snake, after the normalized deformation is:

$$\nu(i, k) = \nu(i, k - 1) + \triangle \nu(i, k) \ . \tag{6}$$

In this new position, the snake points are redistributed along the polygonal outline, to ensure equal distances h_k for the next iteration.

5   Graphical Interface: ATHER

ATHER stands for Atherosclerosis MR image visualization and quantification software. It displays one entire image series at a time. Each slice can be zoomed, its histogram can be interactively stretched and signal to noise ratio as well as the standard deviation of noise can be measured in user-selected regions. Vessel lumen contours can both be manually drawn and automatically extracted, with a possibility of interactive refinement if necessary. The automatic contour extraction needs no parameter tuning, since the parameters of the DAC model have been experimentally optimized for the available vascular images. Firstly, we set the parameters of the general shape of the boundary, *i.e.* weighting coefficients of the flexibility and elasticity terms. Secondly, we set the coefficients of the image force (gradient) term and of the stop criterion. We thus obtained a class-of-image dependent model. The same set of parameters is applicable for example to MRI of rabbits (*in vitro* and *in vivo*) and to human MR angiography of aorta and carotid arteries. Hence, the only necessary user interaction is a double-click in the vessel lumen in the selected slice. For further speed-up, the contour extraction can be performed in the entire series. The extracted contour's gravity center becomes the starting point for the contour extraction in the next slice and so on. This method is applicable because each series represents a short, approximately straight, portion of the aorta (fig.2). Quantitative results (area, perimeter and diameter) are displayed and stored in a computer file. ATHER also features such functionalities as automatic extraction of the external contour of the vessel wall, as well as semi-automatic plaque quantification (fig.3). They will not be detailed in this paper, because their validation is still ongoing.

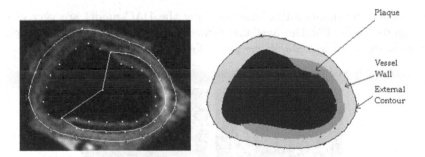

Fig. 3. Left: cross-section of an artery with internal (dots) and external contours (line) of the vessel wall superimposed and plaque delineated. Right: schematic representation explaining the image structure.

6 Measurement Protocol

Validation of the DAC model was performed on a subset of three from six rabbits. The measurements were done on 121 slices subdivided into series corresponding to different territories (contiguous slices).

Quantitative validation was based on area measurement. The areas of the automatically extracted contours were compared with the areas of the contours previously drawn by two experts using Osiris software (University Hospital Geneva Swizterland). Let us note that Osiris also offers the possibility of an automatic contour extraction. However, it needs to be done in each slice separately, after an interactive selection of a region of interest and an adjustment of a tolerance parameter (percentage of image dynamics). In practice, the experts used this automatic extraction and stored the result only when, in their opinion, the contour perfectly fit the boundary. Otherwise they drew it manually. In order to quantify intra-observer variability of the measurements, one expert re-measured the areas in 45 slices, 6 months later. Inter-observer variability was calculated for the same set of 45 slices. ATHER's intra-variability was calculated after applying the automatic contour extraction twice to all the data set. The user was authorized to change the initialization point in single images within a series but not to modify the resulting contours.

For a qualitative comparison between ATHER and the experts, contours' forms were visually inspected. However, distances between contours could not be measured except for a couple of slices for which the contours hand-drawn by the experts were stored as curves. All the other results of the expert-performed measurements were only stored as area values.

7 Results

The contours automatically extracted using the DAC model are very close to the experts' ones. For the available (stored as curves) experts' contours, the maximum distance was about one pixel (fig. 4).

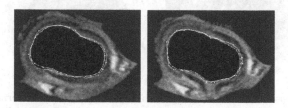

Fig. 4. Qualitative comparison between the contours drawn by ATHER and by the Expert. Maximum distance is about one pixel. The curves' thickness, in this picture, is 1/3 of the pixel size.

This qualitative result (see also fig.5) agrees with the quantitative comparison. Average absolute difference between the experts' measurements and ATHER measurements, calculated over the slices i and over the experts j:

$$\mu = \frac{|Area_{ATHER}(i) - Area_{expert_j}(i)|}{Area_{\overline{expert}}(i)} \times 100 \ , \tag{7}$$

was equal to 4% for the entire set of data. It is to be compared with the variability of the expert-performed measurements calculated in a similar way. The intra-observer variability was as large as 6,4% and the inter-variability between the two experts was equal to 4%. The high intra-observer variability, compared to the inter-observer variability, can be explained by the very long delay between the two measures and by the fact the the second series of expert's measurements was carried out on a different screen.

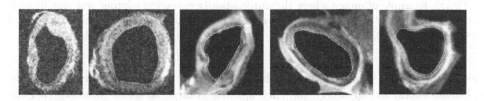

Fig. 5. Examples of results obtained with the DAC model in different territories (we note the variable quality (signal to noise ratio) of images).

However, the maximum difference between ATHER and the experts was as large as 19%. Large errors like this occurred in a few images of particularly poor quality (high level of noise) and in case of images with residual quantities of formol in the vessel lumen (fig.6). In the latter case, even an experienced human eye is sometimes unable to distinguish the boundary between these residues and the vessel wall. As confirmed by the second expert, human diagnosis based on these images cannot be reliable. We also calculated the intra-variability of the results obtained with ATHER. This variability was not more than 1,2%. When the images are of good quality, the automatic contour gives very good results and is very insensitive to the position of the initialization point. The error increases moderately for reasonable noise levels. In one of the rabbits, the measurement error increased from 6.7% to 9.5%, as the noise standard deviation increased from 0 to 6% of the image dynamics.

Let us underline time savings obtained with ATHER. The experts used approximately 1 hour to obtain the measurements for 45 slices, with their usual operating mode. It is to be compared with execution time by ATHER, *e.g.*: for a series of 17 slices, all the operations from image file opening to 3D surface rendering of the automatically extracted contours, took 30 s on the same PC.

8   Discussion and Conclusion

The main characteristics of our active contour implementation are fast computation and results stability with different starting points within the vessel lumen. The latter characteristic reduces to minimum the user interaction. In particular, when propagating the boundary extraction from one slice to another, the previous contour's center can be used as starting point. This is very useful when slice

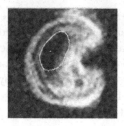

Fig. 6. Example of image where the automatically extracted contour was significantly different from the expert's one, due to poor image quality and a residue of formol in vessel lumen.

spacing is large and the contour forms may vary significantly between neighboring slices. In our study the slice spacing is large so as to coincide with the thickness of histological slices. However, with smaller spacing, the current contour can be initialized by the previous contour, slightly reduced. In this case, the contour extraction is even faster and more robust. The extracted planar contours are not only used for the purpose of quantification. They are also exploited for the sake of 3D visualization. From the detected boundary points, ATHER generates a shaded surface display (fig.7) which can be interactively animated. This is of great interest for visual assessment of the overall intraluminal morphology of the plaque. The herein presented work is a part of a wider project. ATHER will be validated on in vivo images from the same rabbits, also for the external boundaries of the vessels. In parallel, the same model was implemented in another graphical interface, named MARACAS, specially designed for the study of MR angiography images, i.e. images of the vascular lumen obtained after a contrast agent injection. It was already validated on images from phantoms [10]. Now, it is applied to images from our rabbits and from patients.

Acknowledgements. This work is in the scope of the scientific topics of the GDR-PRC ISIS research group of the French National Center for Scientific Research (CNRS). The authors are grateful to Emmanuelle Canet for her help in image acquisition and assessment.

References

1. Asdente, M., et al.: Evaluation of atherosclerotic lesions using NMR microimaging. Atherosclerosis **80**(3) (1990) 243–53
2. Yuan, C., et al.: Techniques for high-resolution MR imaging of atherosclerotic plaque. J Magn Reson Imaging **4**(1) (1994) 43–49
3. Toussaint, J.F., et al.: Magnetic resonance images lipid, fibrous, calcified, hemorrhagic, and thrombotic components of human atherosclerosis in vivo. Circulation **94**(5) (1996) 332–338

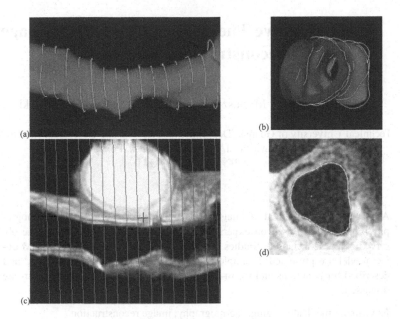

Fig. 7. 3D rendering of the internal surface of a vessel wall (a) and (b), generated from the extracted boundary points (d). Corresponding longitudinal (c) and cross-sectional 2D images (d).

4. Serfaty, J.M., Chaabane, L., Tabib, A., Chevallier, J.M., Briguet, A. and Douek, P.: The value of T2-Weighted High Spatial Resolution Magnetic Resonance Imaging in Classifying and Characterizing Atherosclerotic Plaques: An In Vitro Study. Radlology (2001) (in press)
5. Chaabane, L., Canet, E., Serfaty, J.M., Contard, F., Guerrier, D., Douek, P., Briguet, A.: Microimaging of atherosclerotic plaque in animal models. Magnetic Resonance Materials in Physics, Biology and Medicine 11 (2000) 58–60
6. Wink, O., et al.: Semi-automated quantification and segmentation of abdominal aorta aneurysins from CTA volumes. CARS Paris, France (1999) 208–212
7. Mc Inerney, T. and Terzopoulos, D.: Deformable models in medical image analysis: a survey. Med. Image Analysis 1(2) (1996) 91–108
8. Rueckert, D., et al., Automatic tracking of the aorta in cardiovascular MR images using deformable models. IEEE Trans. Med. Imaging 16(5) (1997) 581–590
9. Kass, M., Witkin, A., Terzopoulos, D.: Active contour models. Int. J Computer Vision 1 (1988) 321–331
10. Hernández-Hoyos, M., Anwander, A., Orkisz, M., Roux, J.-P., Douek, P.C., Magnin, I.E.: A Deformable Vessel Model with Single Point Initialization for Segmentation, Quantification and Visualization of Blood Vessels in 3D MRA. MICCAI, Pittsburgh, Oct. 9-13 (2000) 735–745
11. Cohen, L.D.: On Active Contour Models and Balloons. Computer Vision, Graphics and Image Processing: Image Understanding, 53(2) (1991) 211–218

Medical Active Thermography – A New Image Reconstruction Method

Jacek Rumiński, Mariusz Kaczmarek, and Antoni Nowakowski

Technical University of Gdańsk, Department of Medical and Ecological Electronics,
Narutowicza 11/12, 80-952 Gdańsk, Poland
jwr@eti.pg.gda.pl

Abstract. A new method of image reconstruction for active, pulse thermography is presented. Based on experimental results the thermal model of the observed object is proposed. Studies on thermal transients basing on the FEM object model are presented. Examples of reconstructed images are presented and described for phantoms and for in-vivo measurements. Possible applications are discussed.

Keywords: medical imaging, thermography, image reconstruction

1 Introduction

Non-destructive and non-invasive methods of object inspection are under studies since years. Imaging methods are especially under development because of easy interpretation of an image by a human. In medicine such techniques as Computed X-ray Tomography or Magnetic Resonance Imaging are currently in a common practice. Thermography has been introduced as a new medical diagnostic method by the late fifties. The first studies were dedicated to the breast cancer detection based on static and unprocessed thermal images [5]. During next decades some methods have been proposed to enhanced quality of thermal images or to create quantitative detection rules. Some of them improved contrast by application of microwave irradiation applied to tumor detection [12][13]. Other introduced statistical description and differential analysis of thermal image regions [1][8]. However, because of overuse resulting in high number of false negative diagnosis produced, lack of standard procedures and sensitivity to local conditions, medical applications of thermography have been criticized and not fully accepted in clinical practice. In the passing decade thermal imaging was under continuous development resulting in the new technology of uncooled focal-plane-array thermal detectors [7]. New thermal cameras were introduced and applied, especially for non-destructive testing (NDT) of materials. New measurement techniques proposed for NDT has been found also interesting for medical applications [9]. In this paper a new image reconstruction technique for pulsed thermography [2] is proposed.

W. Skarbek (Ed.): CAIP 2001, LNCS 2124, pp. 274–281, 2001.
© Springer-Verlag Berlin Heidelberg 2001

2 Method

In the pulsed thermography a target object is excited during a given time period t_1

$$U_p(t) = A*(\Phi(t) - \Phi(t - t_1)) , \qquad (1)$$

where: A – amplitude of the excitation, $\Phi(t)$ - Heaviside, step function.

Different methods of excitations can be applied including optical heating, used in the reported study. As a result of the heating the object temperature varies in time according to thermal properties of the object. Taking into account a single point of the object its temperature can be represented as a set of samples

$$S_k^p = S_{n+c}^p = S_n^p \quad S_c^p = \left\{ s_1^p, s_2^p, s_2^p, ..., s_n^p \right\} \left\{ s_{n+1}^p, s_{n+2}^p, s_{n+3}^p, ..., s_{n+c}^p \right\} \qquad (2)$$

where: n – total number of samples measured during heating, c – total number of samples measured during cooling, p – index of a measurement point.

In thermography a set of object points are measured giving a rectangular grid of values – image pixels. The geometrical size of the measurement area depends on the spatial resolution of the thermal camera and the measurement setup (distance, lens, etc.). In our study the Agema Thermovision 900 camera was situated 55 cm far from the front of the object. The camera spatial resolution is 204x128, which gives 26112 measurement points for a one image. Sequence of images was measured to calculate a set of samples for each point in time (Fig. 1.).

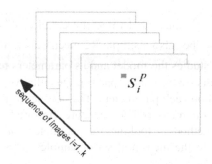

Fig. 1. Sequence of measured images for pulsed thermography with indicated single measurement point (pixel).

Images can be recorded by thermal camera with frequency adjusted to the object properties and heating conditions. Usually it can be set up to 50 frames per second. In the reported study we heat the object during 30 s and record images every 1 s during 3 minutes. Some measurements were done for a fast acquisition (30 fps) however it did not change the final results. As a result of the measurement during cooling phase a set of 180 samples for each measurement point is created. Samples can be presented at a graph to analyze their distribution, e.g., Fig.2. The heating phase was omitted since high influence of heating source on the measured signal.

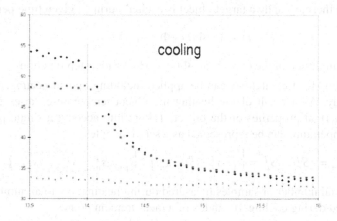

Fig. 2. Distribution of measured samples for different materials (a quantitative comparision).

The character of distribution is very similar to exponential distribution and can be described by the formula

$$\hat{s}_i^p = B_0^p + B^p * \exp\left(-\frac{t_{i-1}}{\tau^p}\right) ; i = 1..c , \tag{3}$$

where: B – amplitudes, $\hat{s}_i^p$ - analytical value of a sample, τ^p - time constant.

The equation (3) describes also a simple RC electrical circuit with the time constant τ^p =R*C. For thermal studies the model and its parameters represent thermal properties of the object (conductivity, heat capacity). Quantitative description for the thermography can be based on such parameters. The new set of parametric images should be constructed basing on measured temperature (infrared radiation) samples. The image reconstruction method is looking for parameters of the equation (3) which minimize fitting errors to the measured set of samples. This can be achieved by the application of the χ^2 test

$$\chi^2 = \sum_{i=1}^{c} \left(\frac{s_i^p - \hat{s}_i^p}{\delta_i}\right)^2 , \tag{4}$$

and the fitting algorithm. We used Marquardt method with the modification given by Bevington [3].

For each pixel and the corresponding set of temperature samples (180) model parameters are reconstructed. As a result, for a one sequence of images a set of two parametric images is reconstructed. The time constant parametric image is especially important since it does not depend on such critical parameters as object emissivity, variations in excitation amplitude and environment conditions. Moreover, it is possible to expand the proposed model to a multilayer one, which produce a rich set of parameters. We explore one and two layers model (two time constants), however, if required for a particular object, it can be expanded to the general form:

$$\hat{s}_i^p = B_0^p + \sum_{j=1}^{N} B_j^p * \exp\left(-\frac{t_{i-1}}{\tau_j^p}\right) ; i = 1..c \; , \tag{5}$$

where: N – total number of layers in the equivalent RC model.

Such a description is directly related to the model proposed by Buettner (in Shitzer [11]) considering changes in skin temperatures that occur subsequently to removing a heat source.

The reconstruction algorithm was implemented in Java programming language, so for input sequence of thermograms, a set of parametric images is automatically calculated. With the Pentium II 400MHz (128MB RAM) processor it takes about 50 seconds.

Verification of the proposed method was performed on phantoms and in-vivo, on pigs. Phantoms were constructed using materials which thermal properties are similar to thermal properties of biological tissues. We created a gelatine object with some embedded materials as it is presented in Fig. 3.

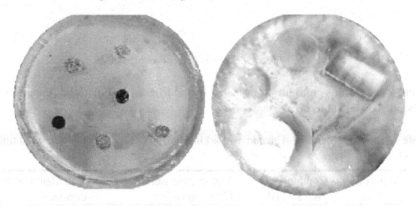

Fig. 3. Phantoms used in the study: a gelatine phantom with metal embeddings below the surface, a gelatine   phantom with different materials embedded (rectangle – plexiglass, circle below the rectangle – silicon, white circle – acryl).

Extended description of phantoms is presented in [10]. In-vivo measurements were performed on anesthetized pigs, which were locally burned with controlled temperatures, time and location.

3   Results

The verification of the method requires to test if reconstructed parameters varies for different materials. This was achieved with the FEM model studies of heat transfer in the object. In the environment of I-DEAS [6] simulation software the FEM model was constructed. The model was build as a homogenous object with an embedded disk (Fig. 4.).

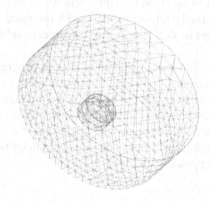

Fig. 4. The FEM model of the object (about 5000 elements).

There were various configurations tested by modification of material properties of the embedded disk (Table 1).

Table 1. Thermal properties of materials used in FEM model studies, and calculated time constants [4].

Material	Conductivity [W/(K*m)]	Volumetric heat capacity [J/(K*m3)]	Calculated time constant
Gelatine	0.490	2510000	49.33
Plexiglass	0.151	1556400	70.37
Silicon	0.334	1630000	59.19
Acryl	0.600	927100	30.23

Table 1 presents also calculated time constants for various materials, as results of the FEM model studies. Time constants were calculated as results of temperature distribution in the model after excitation by heat flux (0.2 W/cm2). In Fig. 5. simulation results after 120 seconds since removing the heat source are presented.

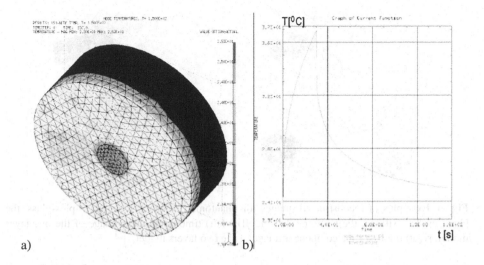

a)   b)

Fig. 5. Simulation results a) temperature distribution in the model b) temperature change graph for a chosen FEM element.

Calculated time constants are different for chosen materials so it should be possible to distinguish those elements in reconstructed, parametric images. More than ten separate measurements were performed on phantoms. Acquired sequences of thermograms were processed to reconstruct parametric images. All results are repeatable and are summarized in Table 2 (reconstructed mean time constants taken over set of all one material pixels in the image). Some reconstructed images are presented in Fig. 6.

Table 2. Calculated and reconstructed mean time constants for phantoms.

Material	Reconstructed mean time constant	Calculated time constant
Gelatine	48.19 (6,7%)	49.33
Plexiglass	58.82 (6,4%)	70.37
Silicon	59.17 (5,8%)	59.19
Acryl	30.76 (6,2%)	30.23

High correlation between reconstructed and simulated (calculated) values of time constants for chosen materials is observed. Big divergence between only plexiglass results can be justified because we did not know exact thermal parameters of this material, so we used those found in tables. The qualitative examination of original thermograms is very difficult since for steady state conditions the "temperature fog" is present in the image. However time constant images are clear giving evident localization of embedded materials in the uniform phantom.

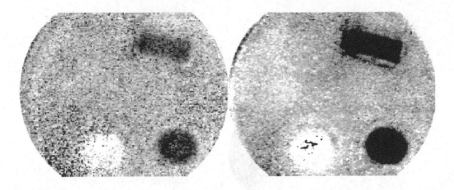

Fig. 6. Examples of reconstructed images for phantoms (the black rectangle – plexiglass, the black circle – silicon, the white circle – acryl): left) time component image of the one layer model, right) the first time component image of the two layers model.

Reconstructed results for in-vivo experiments are also very promising. In Fig. 7. some parametric (time constant) images are presented. Images were reconstructed for two different pigs during two days. There are clear differences between healthy skin and burns created by applying 2.5 cm by 2.5 cm by 9 cm aluminum bar (100 C, 10 sec.). Observed results are very similar for two animals, so using the same experiment conditions parametric images are repeatable.

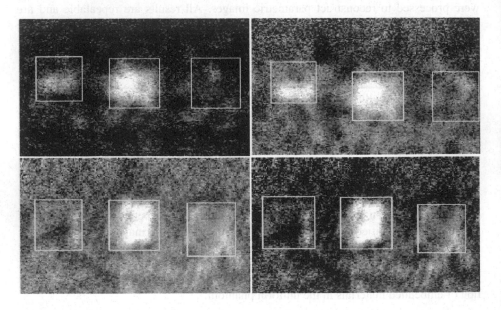

Fig. 7. Examples of reconstructed images for skin burns (two upper images: pig No. 3, two lower images: pig No. 4) with indicated aluminum bar location.

4 Conclusion

The new technique of parametric image reconstruction in active pulse thermography since to be vary effective. Both, the FEM model studies as well as phantoms studies gave similar and very promising results. Using new parametric images it is possible to distinguish different objects according to their thermal properties. Different applications of the parametric images are possible including medical diagnostics and non-destructive testing of materials. With further research on pigs it is expected to establish standard procedure for diagnostics of skin burns.

References

1. Anbar, M., D'Arcy, S.: Localized regulatory frequencies of human skin temperature derived from the analysis of series of infrared images. Proc. IV Annual IEEE Symposium Computer-Based Medical Systems (1991) 184-190
2. Balageas, D.L.: Characterization of Living Tissues from the Measurement of Thermal Effusivity. Innov. Tech. Biol. Med., vol. 12, No. 1 (1991) 145-153
3. Bevington, P.R., Robinson, D.K.: Data Reduction and Error Analysis for The Physical Sciences. McGraw-Hill Higher Education (1991)
4. Edwards, A.L.: A Compilation of Thermal Property Data for Computer Heat-Conduction Calculations. UCRL-50589, University of California Lawrence Radiation Laboratory (1969)
5. Gautherie, M., Gros, C.: Contribution of infrared thermography to early diagnosis, pre-therapeutic prognosis, and post-irradiation follow-up of breast carcinomas. Medicamundi 21(35) (1976)
6. Lin, J., Shiue, T.: Mastering I-DEAS. Scholars International Publishing Corp. (2000)
7. Lipari, C.A., Head, J.F.: Advanced Infrared Image Processing for Breast Cancer Risk Assesment. Proc. of the 19th International Conference IEEE/EMBS, Chicago (1997) 673-676
8. Mabuchi, K., Chinzei T., Fujimasa I., Haeno S., Abe Y., Yonezawa T.: An image-processing program for the evaluation of asymmetrical thermal distributions. Proc. 19th International Conference IEEE/EMBS, Chicago, USA (1997) 725-728
9. Rumiński, J., Kaczmarek, M., Nowakowski, A., Hryciuk, M., Werra, W.: Differential Analysis of Medical Images in Dynamic Thermography. Proc. of the V Conference on Application of Mathematics in Biology and Medicine, Ustrzyki, Poland (1999) 126-131
10. Rumiński, J., Kaczmarek, M., Nowakowski, A.: Data Visualization in dynamic thermography. Journal of Medical Informatics and Technologies 5 (2000) 29-36
11. Shitzer, A., Eberhart, R.C.: Heat transfer in medicine and biology. Plenum Press, New York, USA (1985)
12. Steenhaut, O., van Denhaute, E., Cornelis, J.: Contrast enhancement in IR-thermography by application of microwave irradiation applied to tumor detection. Proc. IV Mediterranean Conference on Medical and Biological Engineering - MECOMBE '86, Sevilla-Spain (1986) 485-488
13. Thomson, J.E., Simpson, T.L., Caulfield, J.B.: Thermographic tumor detection enhancement using microwave heating. IEEE Trans. on Microwave theory and techniques 26 (1978) 573-580

3-D Modeling and Parametrisation of Pelvis and Hip Joint

Czesław Jędrzejek[1], Andrzej Lempicki[2], Rafał Renk[1], and Jakub Radziulis[1]

[1] Institute of Telecommunications, University of Technology and Agriculture,
Al. Prof. S. Kaliskiego 7, 85-796 Bydgoszcz, Poland;
[2] Department of Orthopaedic Surgery, University of Medical Sciences,
ul. 29 czerwca 1956 no. 135, 60-545, Poznań, Poland

Abstract. In this work we model human hip joint. We determine certain parameters of the hip joint and make relation to existing parametrization based on two-dimensional radiological pictures. Initial proposals as to usefulness of these parameters are made. We develop a Java 3-D tool that facilitates obtaining 3D based parameters.

Keywords: medical imaging, 3D modeling, human hip point

1 Introduction

Since the discovery of x-rays, flat images have characterized the medical practice. Film images viewed on a light box or held up to an outside window have provided the comfort of a "hands-on" experience. Recently digital 2-dimensional (2D) radiography together with advanced imaging techniques such as CAT scan (Computerized Axial Tomography), magnetic resonance imaging (MRI), and ultrasound are leading diagnosticians into the future. Large progress has already taken place in developing methods and systems for 3D visualization and image processing for medicotechnical applications. Ultimately one will like to build computer systems which can create interactive models from medical images for visualization and modelling of clinical information.

In certain areas 3D images are already indispensable. MR and CT scanners acquire and reconstruct blood vessels in three dimensions, demonstrating pinched stenoses and ballooning aneurysms. A 3-D presentation of a right renal artery can show a collapsed segment not visible in the projected plane of an x-ray arteriogram. Surgeons can then plan their approach to the pathology. Many companies are putting products on the market that use technology of 3-D visualization to surgical specialities, principally in head, neck and spine surgery, general and cardiac surgery, gynecology and urology.

However, in radiology, many prospective users of the technology remain unimpressed. Some radiologists claim they not need 3-D reconstruction, because their mind's eye has been trained to visualize two-dimensional images rather than in 3-D.

W. Skarbek (Ed.): CAIP 2001, LNCS 2124, pp. 282–289, 2001.

Research aimed at finding clinical value in performing on-screen 3-D reformating has uncovered only a few cases in which 3-D is actually useful in diagnostic interpretation. These involved complex cases, such as:

1. analysing traumas of the skull using MRI or CAT based reconstructions,
2. subtle signs of fetal malformations using ultrasonography based 3D reconstructions.

Also being considered are computer-assisted diagnostic aids that might highlight certain findings indicative of disease. Such advanced interpretation methods, however, are barely more than concepts at present. Nevertheless, there is general consensus that advanced methods are needed to handle and benefit from the information overload. In certain areas of radiology apparatus has been developed for quantitative 2-D determination of parameters useful in diagnosis. These parameters are not valid for every image modality because they see somewhat different features.

This work investigates 3-D modeling of pelvis and hip joint. We use 3-D model of human hip joint obtained form the Visible Human data. In 1989, the US National Library of Medicine (NLM - a division of the U.S. government's National Institute of Health) embarked on a prototype project of building digital image library of volumetric data, representing a complete, normal adult male and female. This library was to include imaging from computerized tomography (CT scans) and magnetic resonance imaging (MRI scans) of cadavers. This image data acquisition project was awarded to a team at the University of Colorado at Denver [1]. The complete male data set, 15GB in size, was made available in November 1994. The Female data set was released in November 1995, at a size of 40GB. Subsequently, research and industrial teams segmented the Visible Human data and obtained three dimensional reconstruction of anatomical structures. Among these were a company Medical Gold [2] and several groups in the USA and Europe [3] [4]. Subsequently, tools were developed that used as input the Visible Human data and were able to perform 3D image processing [5], notably 3DVIEWNIX, coming from Udupa, Department od Radiology, University of Pennsylvania Group. This is a most popular software system for the visualization, manipulation and analysis of multidimensional, multi-parametric, multi-modality images. An alternate solution is 3D reconstruction from two radiologies in orthogonal views [6] [7] [8]. Segmenting the Visible Human data is quite a task and in addition at the beginning, some groups may have used misaligned data.

We find relevant parameters of the 3-D that can be compared with the 2-D data. Since it is difficult to mark a point on a 3-D object we developed a tool for allowing certain operations. This Java 3D based tool facilitates variety of operations: such as marking a point on a 3-D object, cut away, calculating angles, drawing planes, etc. In Fig.1 we show surface of a hip joint. In Fig. 2 an angle between femur and axis of hip is displayed.

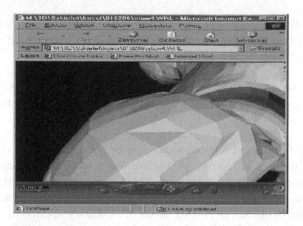

Fig. 1. Triangularised hip-joint as seen in a browser.

2   Anatomy of the Normal Hip Joint

The hip joint is located where the thigh bone (femur) meets the pelvic bone. It is a "ball and socket" joint. The upper end of the femur is formed into an incomplete ball (the "head" of the femur). A cavity in the pelvic bone forms the socket (acetabulum). The ball is normally held in the socket by very powerful ligaments that form a complete sleeve around the joint (the joint capsule). The capsule has a delicate lining (the synovium), that produces lubricant - synovial fluid. The head of the femur is covered with a layer of smooth cartilage, which is a fairly soft, but strong and elastic white substance about 2 mm thick. The socket is also lined with cartilage (also about 3-4 mm thick). The cartilage cushions the joint, and allows the bones to move on each other with very little friction. An x-ray of the hip joint shows a "space" between the ball and the socket because due to its radiolucence the cartilage is not visible on x-rays. Normally surfaces of the head and acetabulum are congruent, that means that forces are distributed equally on the whole area of contacting surfaces.

3   Computerized Modeling of Pelvis and Hip Joint

Computerized modeling of anatomical structures is becoming useful for visualizing complex three-dimensional forms. However, quantitative results are scarce and hardly any comparison with to existing parametrisation based on two-dimensional radiological pictures. We use 3-D model of human pelvis and hip joint to investigate usefulness of different methods of descriptions. Initially we modelled these bones using mathematical surface descriptions. Due to highly irregular shapes we quickly discovered limited applicability of methods widely used in simpler settings (CAD etc.). Moreover having found 80 parameters of quadrics it is extremely difficult to tie them to medically meaningful parameters. Therefore we describe these bones by geometrical distances that are pretty

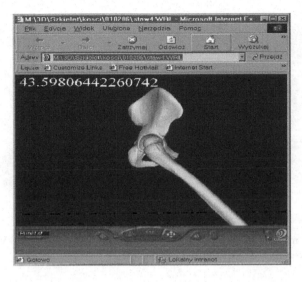

Fig. 2. Demonstration of functionality of our tool.

general (i.e. also seen in radiological 3-D pictures) and identifiable (e.g. the outermost point of a certain region - not necessarily the whole structure).

Throughout the paper we do not use absolute values because these vary depending on age, race, genre. We also do not employ coefficients because there does exist a standard of a "healthy" pelvis and hip joint in a quantitative sense. However, we find that it is advantageous to use some characteristic distances that were previously used in literature [9] [10]

The parameters are:

I Distance between the upper parts of the innominate bone
II Tranverse diameter of the pelvis outlet
III Tranverse diameter of the pelvis Intel
IV Distance between the lower parts of the innominate bone
V Anteroposterior diameter of the pelvis Intel

These parameters correspond to distances of some characteristic points in Fig. 3(a) of [7]. We claim that many points used in that work, such as 13, 20 etc, are not good anatomical landmarks. For example, the Kohler drop is purely a 2D feature associated with X-ray radiology.

This work is preliminary, because uses limited set of data coming only from two groups. One is easily accessible data from NPAC laboratory [11]. The other data comes from reputable Vesalius group [3]. The following Figures represent shapes of pelvis and hip joint as segmented by various groups: Fig. 3(a) pertains to NPAC work, and Fig. 3(b) to Columbia University Vesalius group.

Because the algorithm used to generate the images preserves the surface texture and color of the segmented object, the surface of the pelvic bones in Fig. 3(b) was disfigured by portions of muscle tissue that were also segmented

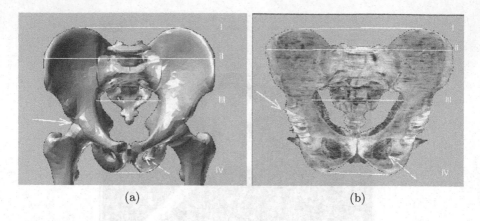

(a)   (b)

Fig. 3. Pelvis model: (a) from NPAC Laboratory [11]; (b) from Vesalius project, Columbia University [3] - disfigured.

because of their proximity to the bony surface. Further processing needs to be done to obtain bony pelvis.

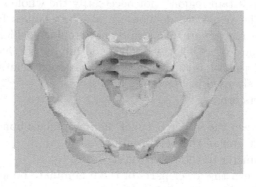

Fig. 4. Bony pelvis model from Vesalius project, Columbia University [3].

Based on these models we employ ratios of previously listed characteristic distances (parameters) for the same human. Table 1. represents these ratios from several sources

Figs. 5 and 6 show NPAC data at different views (view in Fig. 5 tilted a few degrees with respect to Fig. 3(a)). This is to investigate sensitivity of results on alignment.

Except for one entry all other entries in Table 1 pertain to male pelvises. Zsebok data [9] represent average from 1500 measured (X-ray photos) pelvises of young humans at age 18. Since one cannot be sure that the front view is ideally 90^0 view we tilt the model but this changes the NPAC results slightly.

Table 1. Comparison of characteristic ratios of pelvis sizes from radiological and Visible Human data.

Surce of data	Zsebok (male) [9]	Zsebok (female)	NPAC (90°) Fig. 1	NPAC tilted Fig. 4	Columbia-disfigured	Columbia-bony pelvis
Ratio II:III	1.87	1.93	2.60	2.60	2.06	2.00
Ratio II:IV	2,83	2.94	5.24	4..90	2.87	3.06
Ratio III:IV	1.51	1.52	2.01	1.70	1.39	1.53

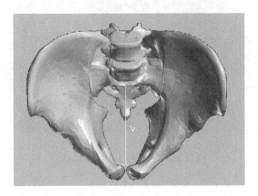

Fig. 5. Anteroposterior diameter of the pelvis Intel (front view).

Surprisingly, NPAC segmentation data cannot be reconciled with classical 2D radiological data of Zsebok (despite the difference in the age of humans and time when the humans in question lived), because the discrepancy is too large.

Fig. 6. Anteroposterior diameter of the pelvis Intel (side view).

Although the Columbia data can be disfigured (arrows point problem areas) both disfigured and bony pelvis data are consistent with 2D radiological data. Finally, using our developed Java 3D tools we find birth canal (shaded area in Fig. 7).

This differs from the previous work [12] [13] that pelvis data do not come from 3D MRI and CT reconstruction but from the Visible Human data.

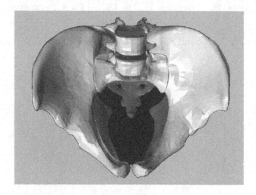

Fig. 7. The birth canal.

4   Conclusions

In this work we model human hip joint by determining certain characteristic parameters of the hip joint and making relation to existing parametrisation based on two-dimensional radiological pictures. The unexpected result is quite a large difference of certain ratios from Visible Human and classic radiological data in case of NPAC segmentation. This stresses need of verifying numerical results with medical knowledge. Analysis of 2D radiological data led surgeons to correlate many parameters of hip joint such as Wiberg angle, Cramer-Haike angle, Sharp angle, neck-shaft angle, neck-acetabulum angle, epiphysis-triradiate cartilage angle with hip and joint status (dysplastic hip) [10]. This knowledge is difficult to transfer to 3 dimensions.

Altogether there is need for second generation more sophisticated segmentation methods. More accurate models, i.e. due to hybrid method combining region based and contour-based segmentations are beginning to appear [14].

Acknowledgments. The authors thank Prof. Celina Imielinska and Małgorzata Wierusz-Kozłowska for enlighteming discussions and providing us with data. The authors acknowledge the support of the KBN project 8 T11E 035 10.

References

1. Ackerman, M.J.: Fact Sheet: The Visible Human Project. National Library of Medicine (1995)
2. http://www.medicalgold.com/

3. Venuti, J., Soliz, E., Imielinska, C., Molholt, P., Wacholder, N.: Development of a Pelvic Anatomy Lesson: Innovation in Electronic Curriculum for Medical Students. Proceedings of the Second User Conference of the National Library of Medicine's Visible Human Project. Bethesda, MD, Oct 1 (1998)
4. Schiemann, T., Nuthmann, J., Tiede, U., Hohne, K.H.: Segmentation of the Visible Human for High Quality Volume based Visualization in Visualization in Biomedical Computing. In: Höhne, K.H., Kikinis, R. (eds.): Lecture Notes in Computer Science **1131**. Springer-Verlag (1996) 13-22
5. http://mipgsun.mipg.upenn.edu/Vnews/info/features.html
6. Safont, L., Martínez, E.: 3D reconstruction of third proximal femur (31-) with active contours. Proceedings of the International Conference on Image Analysis and Processing (ICIAP '99). Venice, Italy (1999)
7. Delorme, S., Petit, Y., de Guise Jaubin, C., Labelle, H., Landry, C., Dansereau, J.: Three-Dimensional Modelling and Rendering of the Human Skeletal Trunk from 2D Radiographic Images. Second International Conference on 3-D Imaging and Modeling. Ottawa, Canada, IEEE (Oct 1999) 497-505
8. http://www.cineca.it/HPSystems/Vis.I.T/Researches/hipop.html
9. Zsebok, Z.: in Handbuch der medizinischen radiologie. Springer (1968) 702-768
10. Wierusz–Kozłowska, M.: Rozwój i przebudowa przebiegu wzrostu dysplastycznego stawu biodrowego leczonego zachowawczo z powodu wrodzonego zwichnięcia i prognozowanie tej przebudowy. Praca habilitacyjna. Poznań (1995)
11. http://www.npac.syr.edu/projects/vishuman/paper/index.htm
12. Liu, Y., Scudder M., Gimovsky, M.L.: CAD Modeling of the Birth Process. II Proceedings Medicine Meets Virtual Reality. San Diego (Jan 1996)
13. Geiger, B.: Three-Dimensional Modelling of Human Organs and its Application to Diagnosis and Surgical planning. Ph.D. thesis. INRIA, Sophia Antipolis, France (Apr 1993)
14. Imielinska, C., Metaxas, D., Udupa, J., Jin, Y., Cheng, T.: Hybrid Sementation of Visible Human Data (to be published)

Cardiac Rhythm Analysis Using Spatial ECG Parameters and SPART Method

Henryk A. Kowalski[1], Andrzej Skorupski[1], Zbigniew Szymański[1],
Wojciech Ziembla[1], Dariusz Wojciechowski[2], Piotr Sionek[2], and Piotr Jędrasik[3]

[1] Warsaw Technical University, Institute of Computer Science, Nowowiejska 15/19,
00-665 Warsaw, Poland
kowalski@ii.pw.edu.pl
[2] Department of Clinical Bioengineering IBBE PAS, Wolski Hospital, Kasprzaka 17
01-211 Warsaw, Poland
sionek@ibib.waw.pl
[3] Department of Paediatric Cardiology, Warsaw University of Medicine, Marszałkowska 24
00-576 Warsaw, Poland

Abstract. Investigation of heart rate variability is of considerable interest in physiology, clinical medicine and drug development. HRV analysis requires accurate rhythm classification. The well-known ARGUS system defines the useful method of rhythm classification. The original set of features used in this method contains 4 parameters: QRS duration, QRS height, QRS offset and QRS area. Zhou at al. showed, that the spatial features: T wave amplitude in lead V2, QRS and T axes angles in frontal plane, and QRS-T spatial angle are of utmost value for diagnostic classification of ventricular conduction defects. We studied usefulness of spatial features instead of original ones in the ARGUS system classification method. The spatial features were computed using SPART method developed by the authors. Classification results for spatial and original features are similar and close to those obtained by the original ARGUS system. The study results confirm usefulness of spatial features for automatic rhythm analysis.

Keywords: cardiac rhythm analysis, ECG parameters

1 Introduction

The pattern recognition is the science that concerns the description or classification (recognition) of measurements. It may be characterized as an information reduction, mapping or labeling process. The structure of typical recognition system consists of sensor, feature extraction mechanism (algorithm) and classification (description) algorithm [16]. A pattern may be a set of measurements represented in vector notation. Features are any extractable measurement used. Feature selection is the process of choosing elements of feature vector used in the pattern recognition system. It is important that the extracted features are relevant to the application. Features are arranged

W. Skarbek (Ed.): CAIP 2001, LNCS 2124, pp. 290–297, 2001.

in a multidimensional feature vector, which yields a multidimensional measurement space or feature space [6].

Accurate identification (classification) of ectopic beats is important problem during frequency HRV analysis in presence of ECG signal artifacts related to body position changes and respiration [3]. Electrical activity of the heart has a spatial character. The orthogonal electrocardiographic (ECG) lead system is used for recording the surface spatial X, Y and Z signals, defined spatial vector [17]. Conventionally, vectorcardiogram is presented as three plane projections of spatial vector rotation: frontal, horizontal and sagittal. The surface ECG signal is affected by the position of the heart in the thorax. This signal may vary between patients and within patients due to changes in posture and body position [3], [10], [13]. Respiratory action produces a rotation of cardiac vector [1], [3]. These effects result in changes in the morphology of the surface electrocardiogram and may introduce artifacts during ECG features measurement [3]. The spatial loops of normal subjects form almost a perfect plane, named optimal plane [5], [13]. The cardiac activity can be adequately represented by planar vector loop in the optimal plane [2], [13].

The ARGUS system was developed by Nolle for one lead ECG signal monitoring [7]. The rhythm classification algorithm was based on four features: duration, height, area and offset of QRS complex of a single lead [7].

ARGUS rhythm classification method with original set of features (feature vector) was implemented in the ICAR system [8]. We proposed new feature vector. It consists of one original and four new spatial features. These new features were primary used for classification of ventricular conduction defects [4]. The spatial features were obtained using SPART method [2], [3], [9], [18]. We study the usefulness of the spatial features for rhythm classification. The objective of the research was to compare the ectopic beat classification accuracy of ARGUS classification method with standard and modified feature vector. We performed research using CSE and ICS-WH electrocardiographic databases.

2 Methods

We studied selected signals from CSE and ICS-WH database. First we use the ARGUS system classification method with original feature vector and next with modified feature vector.

2.1 ICAR System

ICAR System for ECG analysis was developed in Institute of Computer Science at Warsaw University of Technology. Modularized construction of the system allows QRS complex detection, beat segmentation, spatial analysis, HRV analysis and spatial loop visualization [8]. New modules can be easily added. The method of QRS detection and ECG waveform boundary estimation is insensitive for high noise level and baseline drift due to respiration [9]. The SPART method is used for spatial analysis of orthogonal XYZ leads ECG signal [2]. In the first stage SPART method define optimal plane of QRS loop [14]. In the next stage SPART method provides feature extraction and classification of QRS loop morphology using Fourier descriptors [2]. One of the HRV analyses is done by Berger method [11], [12]. Different methods for spatial visualization can be used in ICAR System: standard planar presentation, spatial (3D) and stereoscopic [14].

2.2 Patients

The standard CSE database consists of short (ca 10 sec) ECG signals with the XYZ and standard leads [15]. In the multilead library 25% of 125 ECG signals were normal, the remaining were abnormal [9], [14].

ICS-WH database was developed in the Institute of Computer Science and Wolski Hospital in Warsaw (Poland) [3]. It contains short-term (ca 5-30 min) recordings of multilead orthogonal XYZ and V2 leads and other simultaneously recorded signals.

The low noise, high-resolution signal is digitized at 1000 samples/sec with 12-bit resolution. Signal was recorded during spontaneous and controlled respiration at rest and after +60 passive head-up tilt [2]. The ICS-WH database contains recordings for normal subject, patients after corrective surgery of tetralogy of Fallot (TOF), patients with syncope, AV block, chronotropic incompetence and sick sinus [11], [12]

The XYZ recordings from multilead ICS-WH database of patients from 5 to 15 years (9.7   2.3 years) after corrective surgery of tetralogy of Fallot (TOF) were analyzed [11], [12]. All patients were in good hemodynamic status assessed by clinical and non-invasive evaluation. We studied 20 patients and 7 normal subjects during supine controlled respiration (6/min) and after +60 passive head-up tilt (spontaneous respiration). There were about 12500 normal and abnormal beats annotated by cardiologist. The pathological signals included multiple ventricular and supraventricular beats.

The XYZ recordings from multilead CSE database of 23 patients were analyzed [15]. There were about 250 normal and abnormal beats annotated by cardiologist.

2.3 Extended ARGUS Method

Extracted features vector used for rhythm classification should be relevant to the application [16]. The ARGUS system rhythm classification method with original feature vector: QRS duration, QRS height, QRS offset and QRS area was implemented in the ICAR system [7], [8]. In the original ARGUS system the feature values have been scaled in reference to calibration pulse.

The five features used by Zhou at al. for diagnostic classification of ventricular conduction defects were: QRS duration, T wave amplitude in lead V2, QRS axes angle in frontal plane, T axes angle in frontal plane and QRS-T spatial angle [4]. This features were carefully selected from seventeen-feature initial set using disjoint cluster analysis method (FASTCLUS procedure in SAS) [4]. Selection of features shows that spatial information is very important.

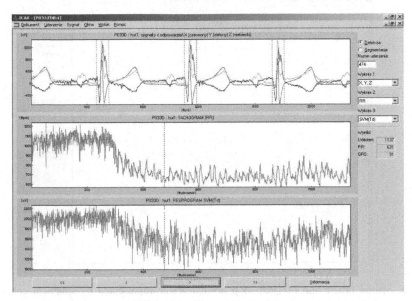

Fig. 1. Segment of X, Y, Z signal (upper panel), RR sequence (middle panel) and respiration sequence (lower panel) of normal subject from ICS-WH database with spontaneous respiration in supine position and after +60 passive head-up tilt (at about #290 beat).

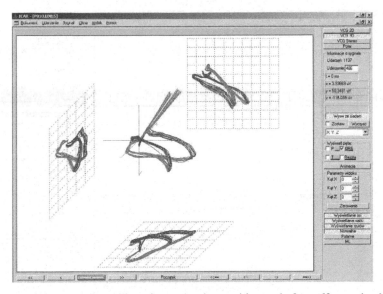

Fig. 2. The influence of body position change (supine position and after +60 passive head-up tilt) and deep breathing on the QRS loops spatial orientation and morphology. Drawn for selected beats from Fig. 1. The straight lines represent the QRS loops normal (polar) vector.

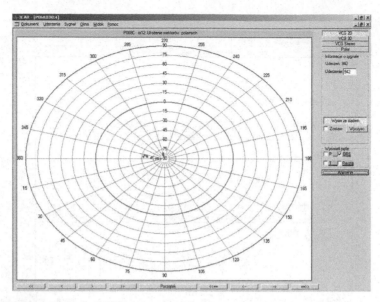

Fig. 3. Spatial orientation of QRS loop normal vector represented by spherical coordinates: azimuth and elevation (see text). TOF subject from ICS-WH database in supine position during spontaneous respiration.

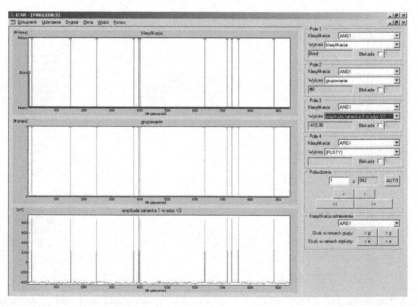

Fig. 4. ARGUS method with modified angular features. Result of rhythm classification (upper panel), grouping (middle panel) for all beats of signal as in Fig.3. Lower panel shows amplitude sequence of T wave in V2 lead.

We studied usefulness of these spatial features for rhythm classification. Modified feature vector was created which consists of one original feature (QRS duration) and four new spatial features listed above. The two loops can be compared by distance measure as the Euclidean metric in the feature space [2], [16]. We used weighted distance measure of loops similarity.

The dimensions of the feature space must be homogeneous. The feature normalization in such a way as to have zero mean and unity variance is not optimal representation of data, because the values are not normally distributed [10]. We use normal subjects from ICE-WH database to obtain the upper and lower limits of feature ranges based on 90 per cent of each distribution [13]. Next we calculate the scaling factor dividing the obtained range into eight segments [7]. The scaling factors are: 10 ms for QRS duration, 45 μV for T wave amplitude in lead V2 and 45° for all angle features.

We applied the new method for frequency HRV analysis. Recordings selected from ICS-WH database contain artifacts related to the body position changes and respiration.

Fig. 1 shows an example of XYZ signals segment (upper panel) as a function of time (ca 3 sec.) for one normal subject selected from ICS-WH database. The subject lays in supine position and passed the +60 passive head-up tilt during spontaneous respiration. The middle and lower panel of Fig. 1 shows tachogram and respirogram (respiration signal sample sequence) for the same signal as function of beat number. We can observe big increase in the heart rate and respiration depth after head-up tilt.

Fig. 2 shows the influence of changes in body position and deep breathing on QRS loops spatial orientation and morphology in XYZ space and plane standard projections (XY, XZ, YZ). The normal (polar) vectors of optimal QRS loops plane, obtained by SPART method, are also shown. Some beats are selected before (#240-250) and after (#460-490) passive head up tilt. The spatial loops and their plane projections were rotated, as indicated by the change in normal vector spatial orientation. The size changes of QRS loops are mainly along Z-axis direction. This changes lead to morphology distortion in the planar projections which can not be eliminated. We can observe increase in orientation dispersion after head-up tilt.

Fig. 3 shows azimuth and elevation of QRS loop normal vector for all beats of signal for one TOF subject in supine position during spontaneous respiration. The periphery located number represent azimuth: 90° - front, 180° - left, 270° – back and 360° (0°) – right direction. The center point represents the head of patient with elevation angle of -90°. The opposite direction is represented by outer circle of diagram with elevation angle of +90°. The upper, right group of points represents normal vector direction of sinus (normal) beats and left group represents 11 ventricular (abnormal) beats. We can observe big dispersion of QRS loop normal vector spatial orientation for all ventricular beats with similar QRS loops morphology. It indicates influence of respiration.

Fig.4 shows result example of rhythm classification and grouping by ARGUS method with modified feature vector for TOF patient signal from Fig. 3. All ventricular beats are grouped correct (group #1) and classified as abnormal. Rhythm classification results of analysis with standard feature vector for the same patient are identical as in Fig. 4.

3 Results

Rhythm classification results by original ARGUS system developed by Nolle [7] are shown in Tab.1. Rhythm classification results obtained in ICAR system with standard and modified feature vectors for ICS-WH and CSE databases are shown in next rows of Tab.1. We used statistical indices recommended by CSE standard [15]. The prevalence of abnormal beats was low: 0.82% for ICS-WH and 16.08% for CSE database. The results estimated for standard and modified feature vectors were similar. The negative predictive value shows good result. It indicates the accurate identification of ectopic beats. Accuracy, sensitivity and specificity for ICS-WH database were better than result obtained by original ARGUS system. Sensitivity for CSE database was better than result obtained by original ARGUS system, but accuracy shows lower value.

Table 1. Results of rhythm analysis obtained using ARGUS method with standard and modified feature vector for ICS-WH and CSE databases (ICAR system). Also, results of the original ARGUS system by Nolle [7]. A - accuracy, P - prevalence, SE - sensitivity, SP – specificity, PPV – positive predictive value, NPV – negative predictive value.

Analysis	A [%]	P [%]	SE [%]	SP [%]	PPV [%]	NPV [%]
ARGUS (Nolle)	98.24	8.13	86.45	99.29	91.47	98.81
ARGUS (ICS-WH)	99.96	0.82	98.06	99.98	97.12	99.98
ARGUS (CSE)	97.99	16.08	87.50	100.00	100.00	97.66
Spatial (ICS-WH)	99.97	0.82	98.06	99.98	98.06	99.98
Spatial (CSE)	95.48	16.08	93.75	95.81	81.08	98.77

There are different numbers of beats in presented studies: Nolle 46.000, ICS-WH 12.500, CSE 250. The small number of abnormal QRS waves was limitation of the study.

4 Conclusions

We studied usefulness of spatial features for accurate identification of ectopic beats in frequency HRV analysis. We used spatial features, instead of original ones. The spatial features were computed using SPART method developed by the authors. Application of the new spatial features for ARGUS classification method is original contribution of the authors.

Classification results for spatial and original features were similar and close to those obtained by the original ARGUS system. The study results confirm usefulness of spatial features for automatic rhythm analysis. The results also indicate the possibility to use of this new method for frequency HRV analysis in presence of ECG signal artifacts related to the body position changes and respiration. Further investigations for optimization of classification algorithm are needed.

References

1. Pallas-Areny, R., Colominas-Balague, J., Rosell, F.J.: The Effect of Respiration-Induced Heart Movements on The ECG. IEEE Trans. Biomed. Eng. **36** (1989) 585-590
2. Kowalski, H.A., Skorupski, A.: SPART – Real Time Spatial Analysis of ECG Signal. In: Computers in Cardiology 1995. Vol. 22. IEEE Computer Society Press, Piscataway (1995) 357-360.
3. Kowalski, H.A., Skorupski, A., Wojciechowski, D., Szymański, Z., Jędrasik, P., Kacprzak, P., Zych, R.: Use of SPART Method for Automatic Rhythm Analysis in Presence of Body Position Changes and Respiration Artifacts. In: Chicago 2000 World Congress on Medical Physics and Biomedical Engineering. Medical Physics 27 (2000) 1394
4. Zhou, S.H., Raoutaharaju, P.M., Calhoun, H.P.: Selection of Reduced Set of Parameters for Classification of Ventricular Defects by Clasters Analysis. Vol. 20. IEEE Computer Society Press, Piscataway (1993) 879-882
5. Rubel, P.: Past And Future Of Quantitative 3-D Electrocardiography And Vectorcardiography. In: IEEE Engineering in Medicine and Biology 14th Annual Conference. Paris (1992) 2768D-2768F
6. Ciacio, E.J., Dunn, S.M., Akay, M., Biosignal Pattern Recognition and Interpretation Systems, Part. 2: Methods for Feature Extraction and Selection. IEEE Eng. in Med. & Biol. **12** (1993) 106-113
7. Nolle, F.M.: ARGUS, a Clinical Computer System for Monitoring Electrocardiographic Rhythms. D.Sc.dissertation, Washington Univ., St. Louis Mo. (1972)
8. Kowalski, H.A., Skorupski, A.: Prototyping of ECG Analysis Methods for Real-Time Applications. In: VII Mediterranean Conference on Medical and Biological Engineering Medicon'95. Jerusalem (1995) 54
9. Kowalski, H.A., Skorupski, A., Szymański, Z., Jędrasik, P., Wojciechowski, D.: Computer Analysis of Electrocardiograms: Validation with the CSE Database. In: 2nd International Congress Polish Cardiac Society. Polish Heart Journal **49** (1998) Sup.I:156.
10. Shapiro, W., Berson, A.S., Pipberger, H.V: Differences Between Supine and Sitting Frank-Lead Electrocardiograms. J. Electrocardiol. **9** (1976) 303-308
11. Jędrasik, P., Kowalski, H.A., Wojciechowski, D., Skorupski, A., Gniłka, A., Mikołajewski, A., Sionek, P., Ozimek, W., Wróblewska-Kałużewska, M.: Heart Rate Variability in Children After Total Repair of Tetralogy of Fallot: An Autonomic Nervous System Dysfunction?. In: The Second World Congress of Pediatric Cardiology and Cardiac Surgery Honolulu (1997) 921-923
12. Jędrasik, P., Wojciechowski, D., Lewandowski, Z., Kowalski, H.A., Wróblewska-Kałużewska, M.: Does Time After Repair of Fallot Tetralogy Increase the Risk of Occurrence of Factors Having Influence on Electrical Function of the Heart?. In: 11th World Symposium on Cardiac Pacing and Electrophysiology. Pace **22** (1999) A35
13. Pipberger, H.V., Carter, T.N.: Analysis of the Normal and Abnormal Vectorcardiogam in its own Reference Frame. Circulation **25** (1962) 827-840
14. Kowalski, H.A., Skorupski, A., Jędrasik, P., Wojciechowski, D., Kacprzak, P., Zych, R.: Automatic Rhythm Analysis Using SPART Method. In: Proceedings Computers in Cardiology 1999, Vol. 26. IEEE Computer Society Press, Piscataway (1999) 515-518
15. Commission of the European Communities, Medical and Public Health Research (Willems J.L. CSE Project Leader).: Common Standards for Quantitative Electrocardiography, CSE Multilead Atlas. CSE Ref. 88-04.15 Acco Publ., Leuven Belgia, (1988)
16. Schalkoff, R.J.: Pattern Recognition: Statistical, Structural and Neural Approaches. John Wiley&Sons, New York (1992)

Deformable Contour Based Algorithm for Segmentation of the Hippocampus from MRI

Jan Klemenčič[1], Vojko Valenčič[1], and Nuška Pečarič[2]

[1] Faculty of Electrical Engineering, Laboratory for Biomedical Visualisation
and Muscle Biomechanics, Tržaška 25, 1000 Ljubljana, Slovenia
{jan.klemencic, vojko}@fe.uni-lj.si
[2] Clinical Institute of Radiology, University Clinical Centre,
Zaloška 7, 1000 Ljubljana, Slovenia
nuska.pecaric@kclj.si

Abstract. Automatic segmentation of MR images is a complex task, particularly for structures which are barely visible on MR. Hippocampus is one of such structures. We present an active contour based segmentation algorithm, suited to badly defined structures, and test it on 8 hippocampi. The basic algorithm principle could also be applied for object tracking on movie sequences. Algorithm initialisation consists of manual segmentation of some key images. We discuss and solve numerous problems: partially blurred or discontinuous object boundaries; low image contrasts and S/N ratios; multiple distracting edges, surrounding the correct object boundaries. The active contours' inherent limitations were overcome by encoding *a priori* geometric information into the deformation algorithm. We present a geometry encoding algorithm, followed by specializations needed for hippocampus segmentation. We validate the algorithm by segmenting normal and atrophic hippocampi. We achieve volumetric errors in the same range as those of manual segmentation ($\pm 5\%$). We also evaluate the results by false positive/negative errors and relative amounts of volume agreements.

Keywords: image segmentation, active contour method, MRI imaging

1 Introduction

Semi automatic or automatic segmentation of medical images is a relatively hard task, due to low signal-to-noise (S/N) ratios, low contrasts, and multiple or discontinuous object edges. Therefore, standard edge detecting algorithms rarely give satisfactory results. A successful medical image segmentation algorithm treats medical structures as unit objects, to produce closed contours regardless of input data. This is achieved by deformable models approach. To make the segmentation effective, it is also of advantage to implement *a priori* knowledge of the structures' shapes in the algorithm. We present an algorithm for segmentation of badly defined structures on medical images. To evaluate the algorithm, we perform the segmentation of the hippocampus. With some

W. Skarbek (Ed.): CAIP 2001, LNCS 2124, pp. 298–308, 2001.

modifications, the algorithm could also be applied for object tracking on movie sequences.

Hippocampus is a small temporal lobe brain structure. Its atrophy is related to some brain diseases, such as temporal lobe epilepsy (eg. [1], [2]). Atrophy detection is crucial for correct diagnosis and often for making surgical decisions. Traditionally, atrophy is subjectively evaluated from coronal T1 weighted MR cross-sections. Alternatively, segmentation allows accurate, objective and reproducible volume and shape measurements.

Hippocampus segmentation is a difficult task for a number of reasons: (1) its edges have relatively low contrast; (2) there are other, often more pronounced edges in its close vicinity; (3) the resolution of MR images is not adequate, as hippocampus is a relatively small structure; (4) the edges are often partially blurred and/or discontinuous, making them only subjectively perceptible. Therefore, we argue that fully automatic segmentation is not feasible, and choose the semi automatic approach. This requires certain amount of operator interaction for algorithm initialization. We designed the algorithm in such a way to make this interaction familiar and intuitive to the operator - neuroradiologist.

To solve the problem of discontinuous edges, we used the active contours as a basic principle of our algorithm. However, active contours usually fail to give satisfactory results when faced with low contrast edges. Also, they are very sensitive to false edges and noise. We present a specialized version of the algorithm, solving the above problems. In particular, we use *a priori* knowledge about expected object shape and encode it into the deformation algorithm.

2   Background

Active contours were first introduced by Kass et al. [3]. They have already been used for hippocampus segmentation by Ghanei et al. [4]. Ghanei proposed some specific improvements, which assist the operator by the segmentation of each volume slice. However, the operator must still deal with each slice separately. Our method allows skipping the majority of the slices, but doesn't deal with single slice segmentation automatisation. Therefore, these two methods can be considered as complementary.

Hippocampal atrophy is also dealt with by Webb et al. [5]. Here, a method for automatic hippocampus atrophy detection is presented. However, it is primarily based on image intensity measurements in the average hippocampal region, which is achieved by registering the volume of interest to representative manually segmented brain. Therefore, no actual segmentation takes place.

A whole brain segmentation was introduced by Christensen et al. [6] by mapping a deformable anatomical atlas to patient specific images. This algorithm was already used for hippocampus segmentation with the help of manually positioned landmarks [7]. It is highly computationally intensive and thus time consuming.

There are many other adaptations of deformable contours principle for segmentation of brain MRI, but in most part these are not sensitive and selective enough for weakly defined structures such as hippocampus. The main contributions of this paper are: (1) presenting a simple and effective way of encoding *a priori* geometry knowledge into active contours; (2) specializing the algorithm for best MR segmentation results; (3) illustrating the power of the algorithm on hippocampus segmentation examples with clinical validation.

3   The Segmentation Algorithm

3.1   Initialization

Current medical practice involves visual analysis of coronal, T1 weighted MR images of the brain. Hippocampus size and shape is evaluated by the neuroradiologist, slice by slice. In order to make the required interaction (algorithm initialization) familiar and intuitive to the operator - neuroradiologist, we utilise his experience by letting him draw the hippocampus boundaries on some of the key slices. We call these slices *reference slices*, and the rest are *in-between slices*. Of course, this poses a task on the operator to decide which slices to choose as reference ones. In our experience, the decision is relatively quick and simple. The operator slides through all the slices, inspecting them visually, and marks the slices where the object starts to grow, shrink or the ones with abrupt shape changes. These, together with the first and last slice containing the object, form the reference slice set.

3.2   Tracking Object Edges

We implemented the well-known object tracking principle: a *reference contour* (manually drawn on each reference slice) is copied to the neighbouring in-between slice, deformed to fit to the edges, and the result is then passed to the next in-between slice as a new initial guess. To make the results tracking-direction independent, we initiate the tracking in both directions from each reference slice - see Fig. 1(a). In this way, the algorithm attempts to find object boundaries on in-between slices, based on the geometry information from manually segmented reference slices. However, faced with difficult conditions such as MR images, active contours usually fail to find the correct edges. Therefore, we must make better use of the *a priori* geometry information contained in reference contours.

Using *a priori* Geometry Information. The algorithm must utilise the geometry information, contained in reference slices, as thoroughly as possible. We extend each reference contour into 3D surface by drawing straight lines between the corresponding vertices of each reference contour - see Fig. 1(b). Thus, we create a full surface approximation of the object, which we call the *linearly approximated object model (LAOM)*. Then we cut the *LAOM* at positions, corresponding to in-between slices, obtaining approximate shapes of object cross-sections for each in-between slice. We call these shapes *transition models*, and use them as built-in geometric models for deforming contour on each in-between slice. The deforming contour then tends to converge to the built-in model shape, which in turn lies close to the correct object boundaries. Such model-assisted deformation gives superior segmentation results.

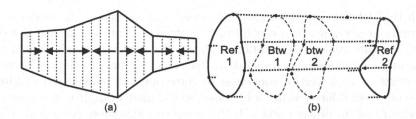

Fig. 1. (a) Tracking the object on slice sequence (side view). Solid vertical lines represent reference slices; dashed vertical lines represent in-between slices; arrows show the contour travel direction. (b) Creating the linearly approximated object model (*LAOM*) from reference contours (*Ref1,2*). The in-between contours (dashed lines, *btw1,2*) or *transition models* are computed from reference contours by connecting the corresponding reference contours' vertices through 3D space (horizontal dotted lines) and cutting them at appropriate positions

3.3   Encoding *a priori* Knowledge into the Deformation Algorithm

The active contour deformation is defined as energy minimization process. The contour energy is, in its basic form, a sum of two independent contributions - internal and external. Internal energy is responsible for contour's continuity and smoothness. Its strength is adjusted with two elasticity parameters. External energy pulls the contour towards chosen features of the image, which are contained in the image's potential distribution (for details see [3]).

Standard active contours have some commonly known weaknesses. These include contour shrinking due to internal energy, clustering of contour vertices in areas with low energy potential, limited range of external forces which drive the contour towards image edges, contour's inability to align to sharp corners and to reach concave portions of object edges. Also, standard active contours offer no means of using *a priori* shape information about the target object. Various authors have addressed these issues, some of them allowing the contours to exploit some means of *a priori* shape or position information (eg. [8], [9], [10]). We base our work on active contours described in [8]. Here, it is shown that active contour, defined with vector v (or v_k for the k-th iteration step) doesn't shrink on a uniform image (that is, in the absence of external forces), if the deformation iterations are defined as

$$v_k = (A + \gamma I)^{-1}(\gamma v_{k-1} + A v_{k-1} - f(v_{k-1})) \ . \tag{1}$$

Matrix A is elasticity matrix, computed from elasticity parameters; I is identity matrix; γ is reciprocal of iteration step size, k is the iteration number and $f(v)$ is external force (as a result of external energy). For details, see [3].

We modify the equation (1) by introducing the geometric model v_r into the deformation algorithm. This is achieved by the following equation:

$$v_k = (A + \gamma I)^{-1}(\gamma v_{k-1} + A v_r - f(v_{k-1})) \ . \tag{2}$$

With this definition of deformation process, each vertice of the model v_r corresponds to a vertice of the deforming contour v. This simple fact solves all active contour weaknesses mentioned above. The contour doesn't shrink, as its minimum energy state corresponds to its model shape v_r. There's no vertice clustering, as each vertice converges to corresponding model vertice. Limited range of external forces is not a big issue, as the model can drive the contour to the vicinity of the object's edges. If the model contains sharp corners or concave shapes, the contour can easily follow them. Most significantly, vector v_r is used to encode *a priori* geometry information in the deformation algorithm. An example of how the contour deforms into the built-in model shape on the uniform image is shown on Fig. 2(a).

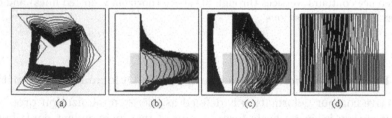

Fig. 2. Deformations of contours with built-in geometric models. (a) Deformation steps of the contour towards its built-in shape on an uniform image (the larger outside contour is the initial position, while the inner M-shaped result is the built-in shape). (b-d) Deformation of a contour on potential distribution image, using low (b), medium (c) and high (d) elasticity parameters. The built-in model is the same as contour's initial position - a rectangle (best visible on (b)). The grey ribbon on the lower right part of each image represents image features, which drive the contour to the right. On Fig. (b), only the part of the contour that lies on the grey ribbon moves right. On Fig. (c), external and model forces are balanced. On Fig. (d), the initial rectangle doesn't deform; it only moves right. We observe that by raising the elasticity, we raise the strength of the built-in model

A similar algorithm is described in [10], where v_0 (initial contour) was used in place of v_r, effectively making the contour's initial shape also its built-in model. However, this only solves the shrinking and vertice clustering problems.

For hippocampus segmentation, we use transition models to define vector v_r, for each slice in turn. The contours then tend to converge towards expected object shape on each corresponding slice.

Behaviour of Active Contours with Geometric Models. The deformation of a contour with built-in geometric model is quite different from a regular active contour. Traditionally, the elasticity parameters control the smoothness of the contour and make its shrinking tendency stronger. In our version, they control the tendency of the contour to acquire the shape of the model. We can use them as a means of regulating the model's strength - see Fig. 2(b-d). If elasticity parameters are low, the contour only roughly follows the model shape, which allows for greater fluctuations in shape due to external forces - Fig. 2(b). If elasticity parameters are high, the contour holds tight to the model shape, and external forces are only able to change this shape by a small extent - Fig. 2(d). With geometric models, both elasticity parameters have very similar effect, so we can only use one and set the other to zero, thus simplifying the calibration. The ability to easily adjust the extent of geometric model influence on the deformation process is an important strength of the method.

The deforming contour will follow the model's angle and size, but not its position. Depending on the application, this can be either beneficial or distracting.

3.4 Segmenting the Hippocampus

To create effective algorithm based on transition models idea, we implemented some further additions to the basic definition of active contours with geometric models. The additions include contour's position and scale adjustments, as well as the appropriate definition of external energy, which drives the contour towards the image features.

The Object Bounding Volume. We defined three discrete scale and position displacement zones (the displacements are defined in terms of the mismatch between the deforming contour and the built-in model). In the first zone (small mismatch), the deforming contour is left as it is. In the second zone (medium mismatch), a new deformation force is introduced, which attempts to match the contour's scale and position to the model's. The force is linearly dependant on the mismatch. If the mismatch nevertheless reaches the third zone (high mismatch), the position and/or size of the contour are corrected in a single iteration step to force the mismatch level back into the second zone.

By implementing the scale and position correcting forces, we in fact create an object bounding volume with 'soft' boundaries, based on the reference contours. We can adjust the size and 'softness' of this volume by adjusting the range of the mismatch zones. Optimal size (defined by the third zone margin) depends on the maximum expected range of the object shapes. Optimal softness (defined by the second zone margin) depends on the probability of the object shape variation between a pair of reference slices.

External Energy. Gradient operator is most commonly used with active contours for energy potential computation. However, there are other edge-detecting operators, which are better suited to MR images [4]. We used the Step Expansion Filter (SEF), introduced by Rao et al. [11].

Usually, external forces are linearly dependent on the local gradient magnitude of potential distribution. Therefore, low contrast edges can result in weak external forces. We used an exponential weighting curve, achieving a compromise between noise sensitivity and inappropriate linear weighting. In this way, we balanced the power of strong and weak edges. This is important, as correct object boundaries can be surrounded by other, more pronounced edges. This is often the case with MR images, especially by hippocampus segmentation.

3.5   Results

The algorithm was evaluated on 8 hippocampi, 3 of which were atrophic. From Fig. 3, the dependence of the contour's behaviour on the elasticity parameter is evident. Fig. 3(a) shows the shape of the built-in model used for segmentation of the exemplar slice. This slice (T1 weighted MR cross-section) is shown on Figs. 3(b-d), with the white line showing segmentation results. On Fig. 3(b), high elasticity was used, and the contour's shape stays strongly tied to the model's shape. Image influence is too weak to deform the contour. On Fig. 3(c), low elasticity was applied, and the contour deforms mainly according to the underlying image, disregarding the built-in model. Therefore, it partially loses hippocampus' edges. On Fig. 3(d), balanced elasticity was chosen (elasticity parameters $\alpha = 0.003$, $\beta = 0$). The model and the image influence are of similar strength, and we get the best segmentation result.

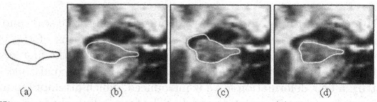

(a)   (b)   (c)   (d)

Fig. 3. Hippocampus segmentation and contour elasticity. (a) Shape of the geometric model. (b) High elasticity (strong model): the contour stays tied to the model shape. (c) Low elasticity (weak model): the contour loses the correct hippocampus edges. (d) Balanced elasticity gives best result. The contour stays close to the model, but still aligns to the local edges

When segmenting the 8 hippocampi, we used identical values of all parameters in each case, to prove that their adjustment is only necessary once. Numerical results of these segmentations are presented in Table 1. The first 5 rows correspond to normal hippocampi, and the last 3 rows to atrophic ones. The column legends are as follows: N - full number of slices, over which each hippocampus spans; N_{ref} - number of reference slices used for algorithmic segmentation; V_{man} - volume obtained by manual segmentation; V_{alg} - volume obtained by the algorithm; V_{LAOM} - volume of linearly approximated object model ($LAOM$).

Table 2 lists voxel-based error estimates for each hippocampus. False Positive Error is the percent of voxels marked by algorithm, but not by manual segmentation. False Negative Error is the percent of voxels marked by manual segmentation, but not by algorithm. Relative amounts of agreement and disagreement between the voxels of V_{man} and V_{alg} are computed by:

$$A_{agr} = 2\,\frac{V_{man} \cap V_{alg}}{V_{man} + V_{alg}} \quad , \quad A_{dis} = 2\,\frac{V_{man} \cup V_{alg} - V_{man} \cap V_{alg}}{V_{man} + V_{alg}} \; . \quad (3)$$

Fig. 4 shows some examples of the final algorithmic segmentation results (full white line) in comparison to manual segmentation (dotted line). Figs. 4(a-c) show relatively good match between the two, while there is some considerable mismatch on Fig. 4(d). However, the area surrounded by both contours is still approximately the same. Such good match of the covered area in cases, when the contour loses exact hippocampus boundaries, is a direct consequence of geometric model and scale-adjusting forces. This leads to good overall volume matching.

Table 1. Segmentation results for 8 hippocampi. N and N_{ref} give the number of all hippocampus slices and the number of reference slices used. Column 6 shows the mismatch between manually segmented volume (V_{man}) and the algorithmically computed volume (V_{alg}). Column 7 shows the mismatch between manually segmented volume and $LAOM$ volume (V_{LAOM}). For details see main text

$Hip.$	N	N_{ref}	$V_{man}\,[cm^3]$	$V_{alg}\,[cm^3]$	$\frac{V_{alg}}{V_{man}} - 1\,[\%]$	$\frac{V_{LAOM}}{V_{man}} - 1\,[\%]$
1	42	10	2.577	2.623	1.8	-5.9
2	36	9	2.549	2.597	1.9	-1.9
3	40	10	2.240	2.235	-0.2	-6.6
4	38	10	2.442	2.348	3.8	-9.1
5	42	10	2.464	2.588	5.0	-1.5
6	34	10	0.953	0.925	-2.9	-10.4
7	31	9	1.027	1.010	-1.7	-6.1
8	39	10	1.293	1.334	3.2	-3.2

3.6 Validation

The sixth column of Table 1 shows a good match between manually and algorithmically segmented volumes, especially considering the fact that intra-operator hand segmentation error also reaches about 5%, sometimes even more. Evaluating the inter-operator error is not really possible for hippocampus. Neuroradiologists generally don't agree about the correct hippocampus boundaries. With years of experience, each neuroradiologist develops his own sense of which small structures surrounding the hippocampus' core actually belong to the hippocampus, and which don't. As a result, the inter-operator hippocampal volume errors easily exceed 200% [1]. Here lies another strength of our approach: due to the appropriate initialisation step, each operator can include or exclude various small structures on reference slices, and the algorithm will follow given guidelines when segmenting the in-between slices. Consequently, the results are independent of the neuroradiologist's individual perception of the hippocampus' structure.

Table 2. Voxel-based evaluation of the algorithm results in comparison to manual segmentation, for each of the 8 hippocampi. FPE/FNE - False Positive/Negative Error; A_{agr} and A_{dis} - relative amounts of volume agreement and disagreement. For details see main text

$Hip.$	$FPE\,[\%]$	$FNE\,[\%]$	A_{agr}	A_{dis}
1	9.1	6.8	0.92	0.18
2	8.4	5.9	0.93	0.17
3	7.5	7.7	0.92	0.18
4	3.7	8.8	0.94	0.16
5	10.4	3.7	0.93	0.16
6	8.7	12.6	0.89	0.25
7	10.7	11.8	0.89	0.25
8	12.7	8.7	0.89	0.24

To appreciate the effect of the contour deformation, we need to compare the results in column 6 to column 7 (Table 1). Column 7 shows the match between manually segmented volume and $LAOM$ volume. We can see that our algorithm offers remarkable improvement over the simple $LAOM$ computation. The $LAOM$ volume mismatch is actually more or less random, depending on the chosen reference slices.

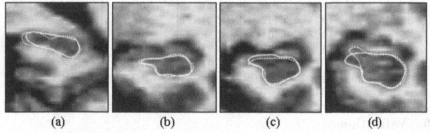

(a)   (b)   (c)   (d)

Fig. 4. Examples of hippocampus segmentation: algorithmic results (full white line) and manual segmentations (dotted white line)

Table 2 lists some common voxel based (as opposed to volume based) segmentation quality estimates. The errors obtained may seem large, but they are in most part a consequence of the fact that intra-operator manual segmentations also yield similar errors and disagreements. The main reason for this is the low resolution of MR scans. Hippocampus is a small structure, and removing a single layer of voxels from its surface can change its volume by up to 30% [2]. Thus, we get high FPE, FNE, A_{dis} and low A_{agr} at each slice, even in cases when the contour lies very close to the optimal position. However, due to error averaging through slice sequence, the full volume computations remain accurate.

A trained neuroradiologist inspected the results of our segmentation algorithm visually. She found all of the 8 segmentations acceptable. Nevertheless, if a badly segmented slice (or region) is found in the resulting volume, our approach offers simple remedy: it is only necessary to manually segment the slice in question (effectively creating another reference slice) and to re-run the algorithm.

3.7 Discussion and Conclusion

We described an effective algorithm for tracking objects on image sequences. We demonstrated the algorithm strength on MR images, by successfully segmenting the hippocampus, which is a badly defined structure, featuring low contrast, discontinuous edges and many spurious edges in close vicinity. With proper modifications, the same basic principle could also be used on video image sequences.

To initialize the hippocampus segmentation, we manually delineated about a quarter of all hippocampus slices, which is evident from the second and third column of Table 1. This effectively reduces the workload on the operator by 75%. Depending on the accuracy requirements, the number of manually delineated cross-sections can be increased or decreased.

The deformation computation took around two minutes to complete on 180 MHz MIPS R10000 processor, the code being written in Matlab (which is slow compared to C/C++). The algorithm features high parallelism, and could run extremely quickly on a multi-processor computer. Still, even two minutes is little in comparison to about half an hour, needed for full manual segmentation.

Acknowledgements. This work is supported by Ministry of Science and Technology, Trg OF 13, 1000 Ljubljana, Slovenia, and Hermes SoftLab, Litijska 51, 1000 Ljubljana, Slovenia.

References

1. Jack, C.R.: Mesial Temporal Sclerosis: MR based Hippocampal Volume Measurements. In: Cascino, G.D., Jack, C.R. (eds.): Neuroimaging in Epilepsy. Butterworth-Heinemann (1996) 111-119
2. Pfluger, T., Weil, S., Weis, S., Vollmar, C., Heiss, D., Egger, J., Scheck, R., Hahn, K.: Normative Volumetric Data of the Developing Hippocampus in Children Based on Magnetic Resonance Imaging. Epilepsia 40(4) (1999) 414-423
3. Kass, M., Witkin, A., Terzopoulus, D.: Snakes: Active Contour Models. Proc. of 1st Int. Conf. Comp. Vision, London (1987) 259-269
4. Ghanei, A., Soltanian-Zadeh, H., Windham, J.P.: Segmentation of the Hippocampus from Brain MRI Using Deformable Contours. Comp. Med. Img. Graph., vol. 22 (1998) 203-216
5. Webb, J., Guimond, A., Eldridge, P., Chadwick, D., Meunier, J., Thirion, J.P., Roberts, N.: Automatic Detection of Hippocampal Atrophy on Magnetic Resonance Images. Mag. Reson. Imaging, vol. 17, no. 8 (1999) 1149-1161
6. Christensen, G.E., Rabbitt, R.D., Miller, M.I.: Deformable Templates Using Large Deformation Kinematics. Tr. Img. Proc., vol. 5, no. 10 (1996) 1435-1447

7. Haller, J.W., Christensen, G.E., Joshi, S.C., Newcomer, J.W., Miller, M.I., Csernansky, J.G., Vannier, M.W.: Hippocampal MR imaging morphometry by means of general pattern matching. Radiology, 199 (3) (1996) 787-791
8. Klemenčič, A., Kovačič, S., Pernuš, F.: Automated Segmentation of Muscle Fiber Images Using Active Contour Models. Cytometry 32 (1998) 317-326
9. Olstad, V., Torp, A.H.: Encoding of a priori Information in Active Contour Models. Tr. Pat. Anal. Mach. Intellig., vol. 18, no. 9 (1996) 836-872
10. Radeva, P., Serrat, J., Marti, E.: A Snake for Model-Based Segmentation. Proc. of 5th Int. Conf. Comp. Vision, Massachusetts (1995) 816-821
11. Rao, K.R., Ben-Arie, J.: Optimal Edge Detection Using Expansion Matching and Restoration. Tr. Pat. Anal. Mach. Intellig., vol. 16 (1994) 1169-1182

Edge-Based Robust Image Registration for Incomplete and Partly Erroneous Data

Piotr Gut[1], Leszek Chmielewski[1], Paweł Kukołowicz[2], and Andrzej Dąbrowski[2]

[1] Institute of Fundamental Technological Research, PAS,
Świętokrzyska 21, PL 00-049 Warsaw
{pgut,lchmiel}@ippt.gov.pl
[2] Holycross Cancer Centre,
Artwińskiego 3, PL 25-734 Kielce
{PawelKu,AndrzejDa}@onkol.kielce.pl

Abstract. In image registration it is vital to perform matching of those points in a pair of images which actually match each other, and to postpone those which do not match. It is not always known in advance, however, which points have their counterparts, and where are they located. To overcome this, we propose to use the Hausdorff distance function modified by using a voting scheme as a fitting quality function. This known function performs very well in guiding the matching process and supports stable matches even for low quality data. It also makes it possible to speed up the algorithms in various ways. An application to accuracy assessment of oncological radiotherapy is presented. Low contrast of images used to perform this task makes this application a challenging test.

Keywords: image restoration, Hausdorff distance

1 Introduction

The availability of numerous imaging modalities makes it possible to visualise in a single image the phenomena available in different modalities, if the images coming from them are precisely overlaid on one another, or *registered*. The choice of landmarks in the images to be used as nodes for registration seems to be the most fundamental task. Contents of the two registered images can differ in such a way that it is difficult to state which particular pixel in one image should be matched with which one in the second image.

Quality assessment of the teleradiotherapy of cancer is an important application which has all the features of a very challenging image registration task. In an ideal treatment, the positions of the irradiation field and the shields with respect to the anatomical structures of the patient conform to the planned ones and are constant in all the radiotherapy sessions. A change of the applied dose by one per cent can lead to a change in time to local recurrence of the disease by as much as several percents [1]. Imprecise realisation of the geometry can imply the loss of chance for permanent remission of the disease or to severe post-irradiation effects.

W. Skarbek (Ed.): CAIP 2001, LNCS 2124, pp. 309–316, 2001.

To assess the accuracy of the treatment of cancer it is necessary to compare the planned geometry with the actual location of the irradiation field and anatomical structures. The actual geometry in a specified therapeutical session can be recorded in the *portal image*. The planned geometry is recorded in the *simulation image*, routinely made during therapy planning. The *simulation image* should be registered with each of the *portal images* made. The simulation image is an X-ray of high quality (see Fig. 1a for an example). The portal image is produced by the therapeutical beam of the ionising radiation, and is inherently of low contrast due to that different tissues, like bones and muscles, attenuate the radiation very similarly (Fig. 1b).

The literature on image registration in the context of radiotherapy refers to the portal images made with beams generated in accelerators rather than with the cobalt apparata. In Poland more than a half of patients are treated with cobalt, which produces the portal images of even worse quality. In any case, the features to be compared are difficult to find in the portal images, and the problem of missing and spurious features arises. In consequence, the comparison of geometries in these images is difficult and time-consuming.

Although full automation of the method has been attempted, as for example in [9,10,11,12], the majority of algorithms lack generality. In [9] the rectilinearity of edges of the irradiation field is used. In [12] the mutual rotation of images is not taken into account. Satisfactory results were received only in the case of pelvis [10], where edges of thick bones are clearly visible. In other localisations the registration of the simulation and portal images is generally a difficult task, even for humans, due to lack of clear natural landmarks. Algorithms in which landmarks to be matched should be shown manually are also in use [5,6]. Apart from the case of the portal films, the image registration literature is extremely broad. The surveys can be found in [4,13,19].

In the literature, the main effort seems to go to the proper choice of landmarks. The approach proposed here resolves this by applying the partial Hausdorff distance [14,18] as the registration accuracy measure. This measure removes the necessity of finding the corresponding landmarks. All that is needed is to specify the sets of pixels to be matched in both images. Moreover, it is allowed for these sets to contain incomplete and partly erroneous data. Such an approach makes it easier to build an image registration system with more adaptivity than systems using strict correspondence. In the method proposed here, it is not necessary for the software to exhibit a deep understanding of the contents of the images, like the ability to recognise any specific anatomical details. Hence, the complexity of the system is not excessively large.

In the present clinical practice the assessment of the accuracy of radiotherapy is done manually by an experienced observer. As a rule, this tedious procedure is not performed routinely. There is a need to support and objectify this process. The methodology presented below is general and requires little user intervention. However, full control of the physician over the process will be maintained, according to the requirement of a human decision in the therapeutical process.

At the time of preparation of the paper the studies were at the stage of stabilisation of the algorithms and the first series of experiments. The methodology will be thoroughly tested, first in laboratory tests, with phantoms used in radiological practice, then in clinical tests in the Holycross Cancer Centre in Kielce.

2   The Method

The following description of the proposed method will be structured according to the classification used in [4]. The practical information, like the hints on the values of parameters, will be given in Chapt. 4.

2.1   Features to Be Matched

Edges can be considered as the most natural and easily detectable features to be used as landmarks for matching. Edges, classically understood as loci of largest brightness gradient, can be very efficiently found with the *zero-second-derivative* filter. The location is the pixel where the second derivative along the gradient direction changes its sign, and the intensity is the gradient modulus. This typical filter was modified by scaling up its masks and convolving them with a circular-symmetrical Gaussian, with the deviation chosen such that the 3σ circles are tangent to each other. The scale should be chosen suitably to the level of noise in the images and the scale of the edges sought.

The resulting edges are one-pixel wide. The thresholding was complemented with an edge-following algorithm which lets the thresholded edges prolong to their neighbouring nonzero pixels from the non-thresholded image, with the ability to jump over gaps up to two pixels wide, along the direction close to that of the preceding edge fragment. The threshold is selected manually. There is a possibility of manually selecting the most relevant parts of edges of the chosen structures – anatomical structures or irradiation field. In the portal image, the edges of the anatomical structures are more easily found if the image is enhanced with a typical histogram equalisation procedure (see Fig. 1b, d).

2.2   Transformation Space

Affine transformation is used. This resolves to five parameters. According to the day-to-day radiological experience, this mathematically simple transform is enough for the application. An iterative algorithm described below is used to find the optimum transformation. Throughout the iteration process the current transformation is always calculated starting from the original image, which reduces the geometrical distortions of the images to a negligible level.

2.3   Measure of the Registration Accuracy

For the sake of robustness of the whole registration process, the partial Hausdorff distance measure was used. This measure was proposed in [14,18], and then used in various applications, including those closely related to image registration (*e.g.* [7,16]). This unsymmetrical measure of similarity of two sets of points has an intrinsic property of neglecting the *outliers* without biasing the result. Let B be the base set and O the overlaid set, and let $d(o, b)$ be the Euclidean distance between points $o \in O$ and $b \in B$. The partial Hausdorff distance is

$$H^r(O, B) = Q^r_{o \in O} \min_{b \in B} d(o, b) , \qquad (1)$$

where $Q^r_{x \in X} g(x)$ is the *quantile* of rank r of $g(x)$ over the set X, $r \in (0, 1)$. In the considered application, B and O are the discrete sets of edge pixels in the simulation and the portal image. Let us explain the notion of the quantile for this discrete case in th following way. Note that for each pixel of the overlaid set $o \in O$ we have one measurement of distance; let us call it $m(o) = \min_{b \in B} d(o, b)$. We can sort these measurements for all the set O in an ascending order. Now, the quantile of rank r can be defined as the s-th smallest element $m(o), o \in O$, where $s = r/|O|$, and $|O|$ is the power of the set O – the number of measurements.

For example, setting r to 0.60 in Eq. (1) can be understood as follows: "see what is the maximum distance for which 60% of the points in the overlaid set vote". Hence, the partial Hausdorff distance can be considered as a kind of a *voting scheme*, in which the *outliers* are rejected and the *inliers* are taken into account in a natural way.

A formal definition of the quantile for the discrete case is as follows. Let v be a discrete random variable with the probability density $P(v)$. Then,

$$Q^r v = q : P(v \leq q) \geq r \land P(v \geq q) \geq 1 - r . \qquad (2)$$

Here, $v = m(o)$. In the calculations, the density $P(v)$ is replaced by the frequency $F(v) = F[m(o)], o \in O$, which is represented by a histogram. Finding the histogram is more effective than sorting. Moreover, the minimum distance of a given pixel $o \in O$ can be directly read out from the pre-calculated distance transform of the edge image B ([2,8,17], see *e.g.* [7,15] for similar applications).

2.4   Strategy of Search in the Space of Transformations

At present we apply a maximum gradient optimisation, starting from a reasonable initial state, provided manually by the user by "dropping" the overlaid image onto the base one. In spite of the obvious drawback of possibly falling into local minima, this simple method performs sufficiently well. The algorithm starts with the quantile rank $r = 1.0$. Together with the application of the distance transform, this resembles the chamfer matching technique [3], applied in [7], but the similarity ends here. After a minimum is reached, the rank is reduced by a small value (say, 0.01), and the algorithm goes on, until the rank reaches zero or the distance becomes zero. This procedure exhibits some analogy to simulated annealing, with the quantile rank related to temperature.

As the quantile rank r is reduced in the optimisation algorithm, the registering transformation stabilises and the result tends to an optimum. However, when the rank approaches zero, the number of edge pixels considered in the matching is going down, which can degrade the result. The approach we chose to find the optimum is as follows. During iterations, a series of registration versions is received, each of them easy to store in a form of the transformation matrix. The set of results for the last iterations of the optimisation process is presented to the user, and this is the user who chooses the best one. This eliminates the unreasonable results and also has a very important *virtue*: the final choice is made by the user. This is in line with the requirement of letting the most important decision be made by the clinician, not the software.

3   Example

Let us now present a practical example of calculations performed for a pair of images of the pelvis, shown in Fig. 1, both 500 × 500 pixels large.

The simulation image of Fig. 1a is a relevant fragment of the therapy simulation image, and the portal image b shows the irradiation field and some of the anatomical details inside it. Figs. c and d show the respective edge images, both received with the scale 4. This corresponds to $3\sigma = 4$ pixels, and the filter mask of 9×9 pixels. The edge images were thresholded. For registration, the edges of the anatomical structures (bones) considered as relevant were selected by the user. In Fig. f there is the result of registration which started from the initial state of Fig. e, with the reference image taken as the base one, and the portal image as the overlaid one (see Eq. (1)). Note that both images contain the edge fragments having no counterpart in the other image, which did not influence the quality of the result.

The result of the 174th iteration of 177 was chosen as the best one, with the distance measure 2.00 and the rank $r = 0.54$. The final 177th iteration gave zero distance at $r = 0.20$ after about 50 s, on an 800 MHz Pentium III processor.

4   Practical Hints

The following practical hints can be formulated on the basis of the experience with the described method gathered until now.

Roles of images. In general, not always the more accurate simulation image should be used as the base one, in the meaning of Eq. (1). As the base image the one having a larger number of edge pixels *presumably* having the counterparts, in reference to all its edge pixels, should be taken.

Scale and thresholds of the edge detector. The scale should be chosen suitably to the scale of the edges sought. For the structures in the simulation image and the shields in the portal image the scales can be small (2-3). For the edges of bones in the portal image, scale should be larger (*cf.* [9]). Obviously, larger scale means less noise and fewer details.

The thresholds should be chosen to receive a maximum subjective ratio of visibility of important edges with respect to irrelevant ones, to reduce further

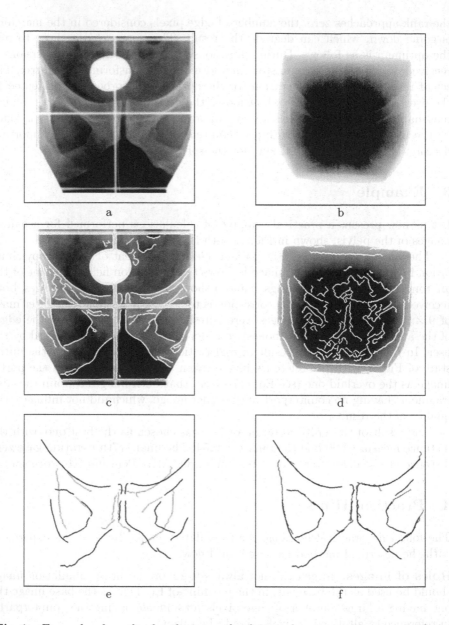

Fig. 1. Example of results for the case of pelvis. a, b: source images – simulation and portal, respectively (portal image enhanced by histogram equalisation). c, d: edges after thresholding, overlaid on source images (portal image Gaussian filtered). e, f: input and result of registration of the relevant fragments of edges of anatomical structures manually selected from those of c and d. Base, portal image: blue; overlaid, simulation image: green – matched pixels (*inliers*), red – pixels rejected from matching (*outliers*). Contrast enhanced and edges thickened for better visualisation.

manual editing of the edge images. High thresholds are used to find the edges of the irradiation field, which is very clear in the portal image. In the simulation image the edges of the irradiation field is represented by wires, also clearly seen (see Fig. 1a). For anatomical structures lower thresholds are used.

For a given anatomical localisation and constant imaging parameters, the scales and thresholds found will be valid for large series of images.

5   Further Developments

Some of the above hints and the hitherto experiments with the method lead to the concepts which will be considered for use in the further work.

Directional information on edges. This will be utilised in the distance function to improve the reliability of matches. It seems that this enhancement will be necessary in the case of more complex images like those for the head, neck or breast, but only if the portal images are of sufficiently good quality.

Pre-calculated distance maps for virtual elementary transformations. Assuming the centre of rotation and scaling is fixed with respect to the base image, from such maps the maximum distances for pixels of transformed images can be read, without performing the transformations. This is profitable if a number of steps made is enough to compensate for the difference in time of calculating a transformed image, which is quick, and pre-calculating a distance map, which is slower. The described speed-up was already implemented and tested, but the above mentioned assumption is not always true. Further investigation of the applicability of this concept is necessary.

Hierarchical algorithm. This speed-up and the pre-calculation of the distance maps for virtual transformations, proposed above, will be used only if calculation time with a processor available when the final version of the software is delivered is excessively long.

6   Conclusion

The image registration algorithm using edges as landmarks and the modified Hausdorff distance as the measure of the registration accuracy performs reasonably well in the application to the quality assessment of the teleradiotherapy of cancer. Required user intervention is reduced to a necessary minimum. The present experience and the concepts worked out in the hitherto experiments make it possible to develop a software tool to support and objectify the otherwise difficult and tedious process of precisely comparing the location of anatomical structures and irradiation field during the radiotherapy sessions with the planned geometry. The software tool emerging as the method matures will be tested first with images received from human-like phantoms, and later in clinical practice. Detailed information on the accuracy of the method will be available then.

Acknowledgement. The work is supported by the Committee for Scientific Research, Poland, under the grant no. KBN 4 P05B 064 18.

References

1. Aaltonen-Brahme, P., Brahme, A. et al.: Specification of dose delivery in radiation therapy. Acta Oncologica, **36**(Supplementum 10) (1977)
2. Borgefors, G.: Distance transformations in digital images. Comp. Vision, Graph., and Image Proc. **34**(3) (1986) 344–371
3. Borgefors, G.: Hierarchical chamfer matching: A parametric edge matching algorithm. IEEE Trans. PAMI **10**(6) (1988) 849–865
4. Gottesfeld Brown, L.: A survey of image registration techniques. ACM Computing Surveys **24**(4) (1992) 325–376
5. Cai, J., Chu, J.C.H., Saxena, A., Lanzl, L.H.: A simple algorithm for planar image registration in radiation therapy. Med. Phys. **25**(6) (1998) 824–829
6. Cai, J., Zhou, S-Q., Hopkinson, J., Saxena, A.V., Chu, J.: Alignment of multi-segmented anatomical features from radiation therapy images by using least squares fitting. Med. Phys. **23**(13) (1996) 2069–2075
7. Chetverikov, D., Khenokh, Y.: Matching for shape defect detection. Proc. Conf. Computer Analysis of Images and Patterns (CAIP'99, Ljubljana, Slovenia). Lecture Notes on Computer Science, vol. 1689. Springer Verlag (Sept 1999) 367–374
8. Danielsson, P.E.: Euclidean distance mapping. Comp. Graph. and Image Proc. **14** (1980) 227–248
9. Eilertsen, K., Skretting, A., Tennvassas, T.L.: Methods for fully automated verification of patient set-up in external beam radiotherapy with polygon shaped fields. Phys. Med. Biol. **39** (1994) 993–1012
10. Gilhuijs, K.G.A., El-Gayed, A.A.H., van Herk, M., Vijlbrief, R.E.: An algorithm for automatic analysis of portal images: clinical evaluation for prostate treatments. Radiotherapy and Oncology **29** (1993) 261–268
11. Gilhuijs, K.G.A., van Herk, M.: Automatic on-line inspection of patient setup in radiation therapy using digital portal images. Med. Phys. **20**(3) (1993) 667–677
12. Giraud, L.M., Pouliot, J., Maldague, X., Zaccarin, A.: Automatic setup deviation measurements with electronic portal images for pelvic fields. Med. Phys. **25**(7) (1998) 1180–1185
13. Arrige, S.R., Lester, H.: A survey of hierarchical non-linear medical image registration. Pattern Recognition **32** (1999) 129–149
14. Huttenlocher, D.P., Rucklidge, W.J.: A multi-resolution technique for comparing images using the Hausdorff distance. Proc. IEEE Conf. on Computer Vision and Pattern Recognition. New York (Jun 1993) 705–706
15. Kozińska, D., Tretiak, O.J., Nissanov, J., Oztruk, C.: Multidimensional alignment using the Euclidean distance transform. Graphical Models and Image Proc., **59**(6) (1997) 373–387
16. Mount, D.M., Netanyahu, N.S., Le Moigne, J.: Efficient algorithms for robust feature matching. Pattern Recognition **32** (1999) 17–38
17. Rosenfeld, A., Pfaltz, J.: Distance functions on digital pictures. Pattern Recognition **1** (1968) 33–61
18. Rucklidge, W.J.: Efficiently locating objects using the Hausdorff distance. Int. J. Comput. Vision **24**(3) (1997) 251–270
19. van Elsen, P.A., Pol, E.J.D., Viergever, M.A.: Medical image matching – a review with classification. ACM Computing Surveys **24**(4) (1992) 325–376

Real Time Segmentation of Lip Pixels for Lip Tracker Initialization

Mohammad Sadeghi, Josef Kittler, and Kieron Messer

Centre for Vision, Speech and Signal Processing
School of Electronics, Computing and Mathematics
University of Surrey, Guildford GU2 7XH, UK
{M.Sadeghi,J.Kittler,K.Messer}@surrey.ac.uk
http://www.ee.surrey.ac.uk/CVSSP/

Abstract. We propose a novel segmentation method for real time lip tracker initialisation which is based on a Gaussian mixture model of the pixel data. The model is built using the **Predictive Validation** technique advocated in [4]. In order to construct an accurate model in real time, we adopt a quasi-random image sampling technique based on **Sobol** sequences. We test the proposed method on a database of 145 images and demonstrate that its accuracy, even with a few number of samples, is satisfactory and significantly better than the segmentation obtained by k-means clustering. Moreover, the proposed method does not require the number of segments to be specified a priori.

Keywords: gaussian mixture modelling, lip tracking

1 Introduction

Real time lip tracking is an attractive research area in the computer vision community. The overwhelming interest in this topic stems from the numerous applications in which the visual information extracted from the mouth region of the human face could improve the performance of the vision system [5], especially in noisy environments where the acoustic source of information becomes unreliable [2]. These applications include audio-visual speech recognition, audio-visual person identification, lip synchronisation, speech-driven talking heads and speech-based image coding. In most applications, visual measurements need to be reliably estimated in the real time.

Lip tracking is a complex problem which involves many stages of processing. In the first instance the face of a subject has to be detected and its main facial features, including the mouth region, localised. Once the mouth region is localised the lip tracker is initialised by segmenting out the lip region pixels and detecting the boundary of this region. The actual tracker then attempts to follow the changes in the boundary without necessarily performing segmentation.

A main obstacle to full automation of the lip tracking systems is the lip tracker initialisation which usually involves operator supervision. By initialisation we mean the process of producing the first model of the lip region. Clustering

W. Skarbek (Ed.): CAIP 2001, LNCS 2124, pp. 317–324, 2001.

of the mouth area is a popular solution to this problem. Lip image clustering is usually performed under the assumption that the number of clusters (e.g. "skin" and "lip" clusters) is given. However, factors such as facial hair and the visibility of the teeth in the mouth opening, demand that the number of clusters has to be selected adaptively. This is the main reason why the current approaches fail to operate fully automatically. In this paper we depart from the conventional clustering methods and propose an approach based on Gaussian mixture modelling of the pixels around the mouth area. In our approach the number of functions is selected adaptively using objective criteria inherent to the Predictive Validation principle on which the mixture architecture selection is based.

When we are dealing with a large data set, it is desirable to select an efficient data sub set to reduce the computational complexity without affecting the accuracy of the estimated model. In our application to image data modelling, Sobol sampling is presented as a good solution here.

The lip region segmentation is then based on the inferred model. Unfortunately there is no guarantee that the lip pixels correspond to a single component of the mixture. For this reason, a grouping of the components is performed first. The resulting groups are then the basis of a Bayesian decision making system which labels the pixels in the mouth area as *lip* or *non lip*. The lip region segmentation scheme based on the proposed method has been tested on mouth region images extracted from the xm2vts database [6]. It compares favourably with those yielded by the k-means clustering procedure.

The rest of the paper is organised as follows. Section 2 gives a brief description of the predictive validation technique. The proposed method of image segmentation based on the Gaussian mixture model is described in Section 3. The experiments carried out on images of the xm2vts database to examine the segmentation performance of the method are discussed in Section 4 together with the results obtained. Finally in Section 5 we draw some conclusions.

2   Model Selection by Predictive Validation

The key step in our approach to image segmentation is to build a probability density function model of the image data. We adopt a semi-parametric modelling method. Accordingly we approximate the image distribution of pixel values using a mixture of Gaussian functions. Consider a finite set of data points $\mathcal{X} = \mathbf{x}_1, \ldots \mathbf{x}_N$, where $\mathbf{x}_i \in \Re^d$ and $1 \leq i \leq N$, that are identically distributed samples of the random variable $\mathbf{x}$. Our density model is then defined as

$$p(\mathbf{x}) = \sum_{j=1}^{M} P_j \frac{1}{\sqrt{(2\pi)^2 |\boldsymbol{\Sigma}_j|}} \exp\left\{-\frac{1}{2}(\mathbf{x} - \boldsymbol{\mu}_j)^T \boldsymbol{\Sigma}_j^{-1}(\mathbf{x} - \boldsymbol{\mu}_j)\right\} \qquad (1)$$

where the coefficient, P_j, is the weight associated with the jth d-dimensional component, $\boldsymbol{\mu}_j$ and $\boldsymbol{\Sigma}_j$ are the mean vector and covariance matrix of the component. For a given number of components, P_j, $\boldsymbol{\mu}_j$, and $\boldsymbol{\Sigma}_j$ are estimated via a standard maximum likelihood procedure using the EM algorithm [1]. There are several methods for selecting the architecture, i.e. the number of components, M, but here we advocate the use of the predictive validation method. This technique has the advantage over the Minimum Description Length [8] and the Bayesian Information Criterion [9] method in that it prevents both over fitting and under fitting. The latter can happen easily, especially when the data is correlated. The goal is to find the least complex model that gives a satisfactory fit to the data.

The basis of the validation method is that a good model can predict the data. A validation test is performed by placing hyper-cubic random size windows in the observation space and comparing the empirical and predicted probabilities. The former is defined as $p_{emp}(\mathbf{x}) = \frac{k(W)}{N}$, where $k(W)$ is the number of training points falling within window W, and the latter $p_{pred}(\mathbf{x}) = \int_W p(\mathbf{x})dx$.

For a mixture of diagonal covariance matrix functions the predicted probability can be determined easily using the standard error function [7]. In full covariance matrix modelling, we use an adaptive and recursive Monte Carlo method to find $p_{pred}(\mathbf{x})$. The agreement between the empirical and predicted frequency is checked by a weighted linear least square fit of p_{emp} against p_{pred}. The model selection algorithm is a bottom up procedure which starts from the simplest model, a one component model and keeps adding components until the model is validated. For details of the method the reader is referred to [4].

3   Image Segmentation

The *pdf* modelling technique discussed in the previous section has been applied to the mouth region image data. First, the original RGB colour space has been transformed into the chromaticity space, $(r, g, b) = (\frac{R}{I}, \frac{G}{I}, \frac{B}{I})$ where $I = \frac{1}{3}(R + G + B)$. In the chromaticity space, only two of the three illumination invariant [10] values are used to obtain the *pdf* models.

The execution time of the density estimation method is highly dependent on the number of samples used, especially in the validation step. The reason why we need a large number of samples is that the random sampling of the validation set to build the model and place validation window can result in an uneven coverage of the definition domain of the estimated density. In order to avoid this problem we propose to adopt a "quasi-random" sampling method based on Sobol sequences. Quasi-random or "sub-random" sequences are n-tuples sequences that fill n-space more uniformly than uncorrelated random points. A Sobol sequence is a quasi-random sequence of numerically generated numbers maximally spread out over a given hyper-cube [7] [3]. A sampling technique based on these sequences is commonly used in Monte Carlo integration to increase the convergence rate.

The constructed model is then used to aid an un-supervised segmentation of the images to extract lip pixels. As we cannot automatically consider each Gaussian component as a separate class, we propose a grouping algorithm which merges components into a pre-determined number of groups. It should be emphasised that having to specify a priori the number of groups in our approach is completely different from having to specify the number of clusters in conventional clustering. In the former case the number of groups is clear. We wish to segment the image into two categories: *lip* and *non lip* pixels. In the clustering case the number of clusters is subject dependent as some people may have a beard, the teeth may show, etc. Depending on these situations different numbers of clusters may be required. Thus, in our approach we capture the specificity of the data by building a separate model for each image. If the *non lip* pixels include several clusters, this will be reflected in our model. We then need to identify which components of the model belong to each of the two categories.

The grouping process is based on a global model for each group. The global models can be constructed using prior knowledge about the physical or other characteristics of each group or by statistical modelling using a training data set for each group. The mixture components can then be grouped according to their distance from the global models. We adopted single component Gaussian functions as the global models, $G(\mathbf{x}|\hat{\boldsymbol{\mu}}_i, \hat{\boldsymbol{\Sigma}}_i)$ $i = 1, \ldots, g$ where g is the number of groups. The distance between each component and the global models is then calculated the Bhattacharyya criterion function

$$B_{ji} = \frac{1}{8}(\boldsymbol{\mu}_j - \hat{\boldsymbol{\mu}}_i)^T \left(\frac{\boldsymbol{\Sigma}_j + \hat{\boldsymbol{\Sigma}}_i}{2}\right)^{-1} (\boldsymbol{\mu}_j - \hat{\boldsymbol{\mu}}_i) + \frac{1}{2}\ln\frac{|\frac{(\boldsymbol{\Sigma}_j + \hat{\boldsymbol{\Sigma}}_i)}{2}|}{|\boldsymbol{\Sigma}_j|^{\frac{1}{2}}|\hat{\boldsymbol{\Sigma}}_i|^{\frac{1}{2}}} \qquad (2)$$

where B_{ji} represents the distance between j^{th} component of the image model and i^{th} global model. If $B_{lk} < B_{li}$ where $i = 1, \ldots, g$ and $i \neq k$ then l^{th} component is assigned to the kth group.

4   Experiments

The experiments summarised in this article were performed on 145 colour images taken from the xm2vts database [6]. The images were picked at random from the second half of the video sequence recording the utterance of a specified text by each speaker. Only one image per speaker was included in the database. A rectangular block was selected around the mouth region of each subject. Figure 1 shows three examples of the selected images and the associated ground truth. The segmentation results of the proposed method using different sampling rates are compared with the results of the ISODATA clustering routine. Our investigation of the scatter plots of the pixels chromaticity values using the *rg*, *gb*, and *br* spaces showed that the lips and skin data are better separated in the *gb* space. Therefore, the *gb* chromaticity space is considered as the feature space.

A Gaussian mixture model was obtained using the predictive validation method using all image data samples. For Figure 1(a) and 1(b) 4-component models were selected, whereas for image 1(c) a 5-component mixture model was acquired. The first row of figure 2 shows the segmentations produced considering each mixture component as a separate class in a Bayes classifier. As we expected these results were not satisfactory. Next the grouping algorithm was used to merge components into two groups (lip and non-lip). We selected at random 10% of the database subjects to construct two Gaussian functions as the lip and non-lip global models. The segmentations resulting from this grouping are shown in the second row of figure 2. The third row of the figure contains the segmentation results output by the ISODATA clustering method for $k = 3$.

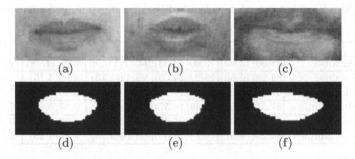

Fig. 1. (Top:)Three examples of the rectangular blocks taken from the xm2vts database images. (Below:) The associated Ground Truth.

Table 1 shows the average segmentation results for all the 145 subjects along with the corresponding results obtained by the ISODATA clustering approach with $k = 3$. The average error has been reduced from 11.84% to 7.12% using our method when all samples of each subject were used to build the model. [1]

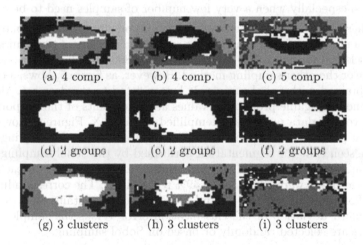

Fig. 2. Segmentation results of figure 1 (**Top row**: Before grouping, **Middle**: After grouping and **Below**: k-means clustering).

In order to obtain the required *pdf* model in a real time, we then investigated the effects of data sampling. Figure 3 shows the average computational time required to build a validated image model versus the number of samples drawn. As the number of samples increases the computational time increases exponentially. The required time using all samples is about 67 CPU units whilst using 100 and 50 samples it decreases to 0.15 and 0.08 CPU units respectively.

---

[1] True-positive: lip pixels are classified to the lip cluster. False-negative: lip pixels are classified to the non-lip cluster. True-negative: non-lip pixels are classified to the non-lip cluster. False-positive: non-lip pixels are classified to the lip cluster.

Table 1. Mean (and standard deviation) of the classification performance of different classifiers.

	True Pos.%	False Neg.%	True Neg.%	False Pos.%	Error%
k-means3	68.00(13.03)	32.00(13.03)	94.12(11.78)	5.87(11.78)	11.84(10.29)
Diagonal	84.98(14.85)	15.02(14.85)	95.42(4.15)	4.58(4.15)	7.12(4.17)
100 random samples	87.53(16.92)	12.47(16.92)	93.80(5.11)	6.20(5.11)	7.83(5.66)
100 Sobol samples	87.51(14.57)	12.49(14.57)	94.36(5.56)	3.37(5.56)	7.34(5.54)
50 random samples	84.12(21.96)	15.88(21.96)	93.02(7.13)	6.98(7.13)	9.49(7.04)
50 Sobol samples	84.29(19.07)	15.71(19.07)	95.14(4.28)	4.86(4.28)	7.61(5.54)

Next, we wanted to investigate the effect of the sampling methods. n random samples were taken from figure 1(a). These samples were then used to train and validate the mixture model. This experiment was then repeated 120 times using different starting seeds for the random generator. Figure 4(a) is the plot of the average number of components accepted, with its associated standard deviation, versus the sample size when the points are selected randomly. Figure 4(b) shows the results of the same experiment based on Sobol sampling. These results indicate that more stable models are obtained by sampling via Sobol sequences, especially when a very few number of samples need to be chosen.

Furthermore, for the same number of components accepted, Sobol samples produce a more satisfactory model. For example a mixture of two Gaussian functions is selected for figure 1(c) when 100 of the 1800 available samples are drawn using one or the other sampling method. However, as figure 5 shows, a calibrated model obtained with Sobol samples is better fitted to the data set. Model 5(a) predicts nonnegligable probability values at some points of the support domain which is void of data points, as exemplified by point X. Figure 6 shows the segmentation produced based on the models obtained via each sampling method. It can be seen that the segmentation performed by the Sobol sampling model is more desirable. Using 100 and 50 uncorrelated random samples for each subject, the average errors were 7.83% and 9.49% respectively. The corresponding results using Sobol sampling were 7.34% and 7.61%. Figure 7 contains plots of the mean and standard deviation of the classification error versus the sample size when the points are selected randomly or based on Sobol sampling.

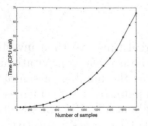 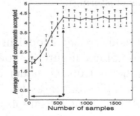 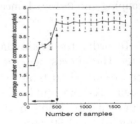

Fig. 3. The average computational time vs the number of samples

Fig. 4. The average number of components accepted for various sample size (left: (a)random sampling, right: (b)Sobol sampling)

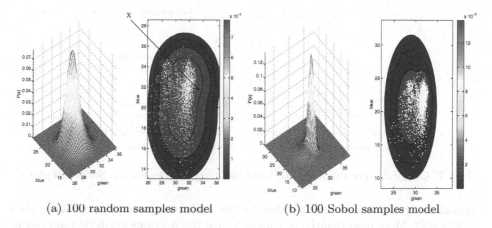

(a) 100 random samples model   (b) 100 Sobol samples model

Fig. 5. Validated model and data distribution of figure 1(c). Distribution of the data has been plotted on the model contours.

These results demonstrate that Sobol sampling is more efficient than random sampling, especially when a few number of samples need to be selected to build the model which is an essential requirement in real time applications. Also, the results prove the superiority of the method over the k-means clustering.

5   Conclusion

The problem of image segmentation for real time initialisation of an automatic lip tracker has been considered. We have proposed an approach based on Gaussian mixture modelling of mouth area images. The lip region segmentation is based on the estimated model. First a grouping of the model components is performed using a novel approach. The resulting groups are then the basis of a Bayesian decision making system which labels the pixels as *lip* or *non-lip*.

The experimental results show that the segmentation errors obtained with the proposed scheme even when a few number of samples are used, is

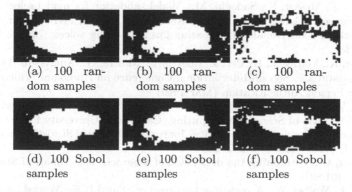

(a) 100 ran-   (b) 100 ran-   (c) 100 random samples   dom samples   dom samples

(d) 100 Sobol   (e) 100 Sobol   (f) 100 Sobol
samples   samples   samples

Fig. 6. Segmentation results of figure 1 (**Top**: Random and **Below**: Sobol sampling).

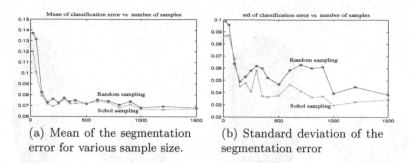

(a) Mean of the segmentation error for various sample size.

(b) Standard deviation of the segmentation error

Fig. 7. Comparison of random and Sobol sampling (*-: random, x- Sobol sampling)¿

significantly less than those yielded by the k-means clustering procedure used previously. Most importantly, in contrast with the k-means clustering approach, the number of segments in the proposed scheme was determined completely automatically. Thus, the proposed lip pixel segmentation method enables a fully automatic lip tracking system. Moreover, we demonstrated that image modelling can be improved by Sobol sampling. Using 50 Sobol samples of 1800 available samples the classification time decreased from 67 to 0.08 CPU unit whilst the average error increased from 7.12% to 7.61%.

Acknowledgements. The financial support from the EU project Banca and from the Ministry of Science, Research and Technology of Iran is gratefully acknowledged.

References

1. Dempster, A., Laird, N., Rubin, D.: Maximum likelyhood from incomplete data via the em algorithm. Journal of the Royal Statistical Society **39**(1) (1977) 1–38
2. Dupont, S., Luettin, J.: Audio-visual speech modeling for continuous speech recognition. IEEE Transactions on Multimedia **2**(3) (Sep 2000) 141–151
3. Galambos, C., Kittler, J., Matas, J.: Sampling techniques for the progressive probabilistic Hough transform. (2001) submitted to Pattern Recognition Letters
4. Kittler, J., Messer, K., Sadeghi, M.: Model validation for model selection. Second Conference on Advances in Pattern Recognition. Brazil, 11-14 March (2001)
5. McGurk, H., MacDonald, J.: Hearing lips and seeing voices. Nature **264** (1976) 746–748
6. Messer, K., Matas, J., Kittler, J., Luettin, J., Maitre, G.: Xm2vtsdb: The extended m2vts database. Second International Conference on Audio and Video-based Biometric Person Authentication (Mar 1999)
7. Press, W.H., Flanney, B.P., Teukolsky, S.A., Vetterling, W.T.: Numerical Recipes in C : The Art of Scientific Computing. Cambridge University Press (1992)
8. Rissanen, J.: Stochastic complexity. Journal of The Royal Statistical Society, Series B **49**(3) (1987) 223–239, 252–265
9. Schwarz, G.: Estimating the dimension of a model. The Annals of Statistics **6**(2) (1978) 461–464
10. Yang, J., Waibel, A.: A real-time face tracker. Third IEEE Workshop on Applications of Computer Vision. Sarasota, Florida, USA (1996) 142–147

Particle Image Velocimetry by Feature Tracking

Dmitry Chetverikov

Computer and Automation Research Institute
1111 Budapest, Kende u.13-17, Hungary
csetverikov@sztaki.hu

Abstract. Particle Image Velocimetry (PIV) is a popular approach to
flow visualisation in hydro- and aerodynamic studies and applications.
The fluid is seeded with particles that follow the flow and make it visible.
Traditionally, correlation techniques have been used to estimate the displacements of the particles in a digital PIV sequence. In this paper, two
efficient feature tracking algorithms are customised and applied to PIV.
The algorithmic solutions of the application are described. Techniques for
coherence filtering and interpolation of a velocity field are developed. Experimental results are given, demonstrating that the tracking algorithms
offer Particle Image Velocimetry a good alternative to the existing techniques. [1]

Keywords: motion analysis, particle image velocimetry, feature tracking

1 Introduction

In this paper we apply feature based tracking algorithms to flow measurement
with Particle Image Velocimetry [7]. Flow visualisation and measurement appear in many scientific and industrial tasks, including the studies of combustion
processes, hydrodynamic and aeronautical phenomena, flame propagation and
heat exchange problems. PIV refers to a particular method of flow visualisation, when the flow is seeded with particles that reflect light and make the
motion visible. Digital PIV, or DPIV, is the recently emerged technique of using
high-performance CCD cameras and frame-grabbers to store and process the
digitised PIV sequences by computer. Cross-correlation methods implemented
via the Fast Fourier Transform have been conventionally used to estimate the
flow velocity [14].

Particle Image Velocimetry is one of the techniques used for flow visualisation
and measurement. It is applicable in laboratory conditions, when the flow can
be seeded with particles. Otherwise, alternative methods are used, such as those
based on a dedicated fluid motion models developed in fluid mechanics [6]. These
models and computational methods are beyond the scope of this paper.

Motion estimation and tracking have a long history in machine vision, where
numerous efficient optical flow [2,8] and feature based [13,5] algorithms have
been developed. However, only recently attempts have been made to adapt

---

[1] This work was supported by the grants OTKA T026592 and M28078.

W. Skarbek (Ed.): CAIP 2001, LNCS 2124, pp. 325–332, 2001.

machine vision techniques to PIV. In particular, G. Quénot, J. Pakleza, and T. Kowalewski [9] presented a dynamic programming based optical flow algorithm customised to PIV. The optimal matching is searched that minimises a distance between two images. This is achieved using the Orthogonal Dynamic Programming (ODP) technique that slices each image into two orthogonal sets of parallel overlapping strips. The strips are then matched as one-dimensional signals.

Experimental studies show that the ODP techniques compare favourably to the classical correlation methods in all aspects except the computational load, which is very high. This limits the potential application area, because online processing, flow monitoring and analysis of time-varying, non-stationary flows are not feasible.

Our search for alternatives to both conventional and ODP approaches aims at developing a fast algorithm that would give reasonable accuracy when applied to complex, possibly time-varying flows. Recently, we have successfully applied to DPIV two feature based tracking techniques: the KLT Tracker [13] and our algorithm called IPAN Tracker [5]. (IPAN stands for Image and Pattern ANalysis group.) Some experimental results demonstrating the PIV-efficiency of the trackers were presented in the conference paper [4], in which the algorithmic solutions were not described. This is done in the current paper, which is structured as follows. First, we discuss the feature tracking approach to PIV. Then, techniques are proposed for coherence filtering and interpolation of a velocity field. Finally, quantitative results of flow estimation are presented. The tests compare three groups of approaches: those based on correlation, ODP, and feature tracking.

2   Feature Tracking Algorithms Applied to PIV

Feature tracking techniques extract local regions of interest (features) and identify the corresponding features in each image of a sequence. In this section, we outline two particular algorithms, the KLT Tracker [13] and the IPAN Tracker [5], which we apply to Particle Image Velocimetry. Particle flow is usually a coherent motion: spatially close particles tend to have similar velocity vectors. A velocity vector field is typically quite smooth, allowing for detection and correction of a wrong measurement based on a sound and coherent neighbourhood. Resampling non-uniform measurements to a regular grid is also possible. Such resampling is normally needed for better visualisation of a flow and for comparison of velocity fields obtained in different ways.

2.1   The KLT Tracker

The KLT Tracker [13] selects features which are optimal for tracking, and keeps track of these features. A good feature is a textured patch with high intensity variation in both x and y directions, such as a corner. The algorithm defines a measure of dissimilarity that quantifies the change in appearance of a feature between the first and the current frame, allowing for affine distortions. At the same time, a pure translational model of motion is used to estimate the position in the next frame. To cope with relatively large motion, the algorithm is implemented in a multiresolution way.

Each feature being tracked is monitored to determine if its current appearance is still similar, under affine distortion, to the initial appearance observed in the first frame. When the dissimilarity exceeds a predefined threshold, the feature is considered to have been lost and will no longer be tracked. Since the KLT algorithm incorporates an analytical solution to motion estimation, it is much faster than the methods that use explicit region matching, such as the conventional cross-correlation and the ODP techniques. The source code of the tracker can be downloaded from the web site [3].

2.2 The IPAN Tracker

The IPAN Tracker is a non-iterative, competitive feature point linking algorithm. Its original, application-independent version [5] tracks a moving point set, tolerating point entry, exit and false and missing measurements. Position is the only data assigned to a point. The algorithm is based on a repetitive hypothesis testing procedure that switches between three consecutive image frames and maximises the smoothness of the evolving trajectories. When applied to PIV, the algorithm is modified as follows: 1. a PIV-specific feature selection mechanism is added; 2. the cost function is modified to include feature appearance.

A PIV image $g(x, y)$ is smoothed by a 3×3 mean filter and the (real-valued) maxima of the smoothed image $s(x, y)$ are selected as the features. Each bright particle is represented by a maximum of $s(x, y)$. For more precise motion estimation, the position of each maximum is corrected by parabolic interpolation in x and y directions, separately. This results in the corrected position (x_f, y_f). A feature $P(x_f, y_f)$ is then assigned a dominance value $f(x_f, y_f)$. The features (particles) are ranked according to their dominance.

The IPAN Tracker processes sequentially each of three consecutive frames F_{n-1}, F_n and F_{n+1}, where n is the current frame. Consider a feature in each of the 3 frames, $A \in F_{n-1}$, $B \in F_n$ and $C \in F_{n+1}$. When applied to PIV, the tracker minimises the cost function $\delta(A, B, C)$ which accounts for changes in velocity and appearance of a feature:

$$\delta(A, B, C) = w_1 \Theta(A, B, C) + w_2 \Lambda(A, B, C) + w_3 \Phi(A, B, C), \qquad (1)$$

where

$$\Theta(A, B, C) = 1 - \frac{\overline{AB} \cdot \overline{BC}}{\|\overline{AB}\| \cdot \|\overline{BC}\|}, \qquad \Lambda(A, B, C) = 1 - \frac{2\left[\|\overline{AB}\| \cdot \|\overline{BC}\|\right]^{\frac{1}{2}}}{\|\overline{AB}\| + \|\overline{BC}\|}$$

$\overline{AB}$ and $\overline{BC}$ are the vectors pointing from A to B and from B to C, respectively, $\overline{AB} \cdot \overline{BC}$ is their scalar product. Θ and Λ, the trajectory smoothness terms, penalise changes in the direction and the magnitude of the velocity vector.

$\Phi(A, B, C)$, the appearance term of $\delta(A, B, C)$, accounts for changes in feature appearance in a 3×3 neighbourhood. It is defined as

$$\Phi(A, B, C) = \frac{1}{18g_{max}} \sum_{k=1}^{9} [|s(A_k) - s(B_k)| + |s(B_k) - s(C_k)|],$$

where $s(P_k)$ are the gray values of the feature point P and its 8 neighbours. The weights are set as $w_1 = 0.1$, $w_2 = w_3 = 0.45$. The cost function $\delta(A, B, C)$ is minimised using the repetitive hypothesis testing procedure presented in [5].

3   Post-processing of Velocity Vector Field

3.1   Coherence Filtering

Feature trackers may occasionally yield completely wrong velocity vectors. To enhance the result of measurement, coherence based post-processing is applied to the 'raw' velocity field obtained by the trackers. The coherence filter modifies a velocity vector if it is inconsistent with the dominant surrounding vectors. The solution we use is a modified version of the vector median filter [1]. The procedure operates as follows.

Given a feature point P_c with the velocity vector $\mathbf{v}_c$, consider all features P_i, $i = 1, 2, \ldots, p$, lying within a distance S from P_c, including P_c itself. Let their velocities be $\mathbf{v}_i$. Due to coherent motion, these vectors are assumed to form a cluster in the velocity space. Introduce the mean cumulative difference between a vector $\mathbf{v}_i$ and all other vectors $\mathbf{v}_j$, $j \neq i$:

$$\Delta_i = \frac{\sum_{j \neq i} \|\mathbf{v}_i - \mathbf{v}_j\|}{p - 1} \tag{2}$$

The median vector is the vector that minimises the cumulative difference. Its index is $med = \arg\min_i \Delta_i$. Δ_{med}, the mean cumulative difference of the median velocity, characterises the spread of the velocity cluster. The standard median filter [1] substitutes $\mathbf{v}_c$ by the median $\mathbf{v}_{med}$. In our implementation, $\mathbf{v}_c$ is substituted by $\mathbf{v}_{med}$ only if the difference between $\mathbf{v}_c$ and $\mathbf{v}_{med}$ is significant: $\|\mathbf{v}_c - \mathbf{v}_{med}\| > \Delta_{med}$. The standard median filter tends to modify most of the measurements and introduce an additional error. The conditional median filter only modifies the vectors that are likely to be imprecise or erroneous measurements. Our tests show that, as far as the overall accuracy is concerned, the conditional median is superior to the standard version.

The above coherence filter is applied iteratively until any of the stopping conditions is fulfilled. Denote by V_k the number of vectors modified at the kth iteration. The conditions for stopping after the kth iteration are as follows: 1. $V_k = 0$ OR 2. $V_k < V_{min}$ and $V_k > V_{k-1}$ OR 3. $k = k_{max}$. We use $V_{min} = 20$ and $k_{max} = 30$. Figure 1 shows an example of coherence filtering.

3.2   Resampling

Uniform sampling of the measured velocity field is normally required. A number of techniques [8] are available for resampling the results of a measurement. However, most of them perform resampling from one regular grid to another. We use the following procedure.

Given a point, G, on the required regular grid, consider all feature points P_i, $i = 1, 2, \ldots, p$, lying within a certain distance S from G. (S is selected adaptively in the way already discussed.) Let their velocity vectors be $\mathbf{v}_i$. Denote by d_i^2 the squared distance from P_i to G and by $\widehat{d^2}$ the mean of d_i^2 over all i. Introduce

$$\alpha_i = \exp\left(-\frac{d_i^2}{\widehat{d^2}}\right), \qquad \beta_i = \frac{\alpha_i}{\sum_{i=1}^k \alpha_i},$$

where $0 < \alpha_i, \beta_i \leq 1$ and $\sum_{i=1}^k \beta_i = 1$. The interpolated velocity vector in G is calculated as $\mathbf{v}_G = \sum_{i=1}^k \beta_i \mathbf{v}_i$. Figure 1 shows an example of resampling.

Fig. 1. Example of vector field post-processing. Left: the original field. Center: coherence filtering. Right: resampling.

4   Tests

The KLT and IPAN Trackers were run on a number of standard test sequences available on the Internet. The post-processing algorithms described in section 3 were applied to the 'raw' velocity fields obtained by the KLT and IPAN Trackers. In this section, the two algorithms complemented by the post-processing are referred to as KLT-PIV and IPAN-PIV, respectively. They are compared to the correlation (CORR-PIV [14]) and the Orthogonal Dynamic Programming (ODP-PIV [9]) algorithms.

The experimental results for CORR-PIV and ODP-PIV given in this section are cited from the papers [9,11]. The results refer to two sets of synthetic data with the ground truth available, which were downloaded from the Japanese PIV-STD Project web site [12] and the web site [10]. In the tables below, the error is the mean deviation from the ground truth; the variance of the absolute deviation is also given.

4.1   The PIV-STD Data

The Visualisation Society of Japan has launched a project named PIV-STD [12] whose goal is to generate and distribute standard synthetic test sequences for performance evaluation of PIV algorithms. The web site [12] offers a collection of pregenerated sequences as well as a program allowing to produce sequences with desired parameters. The character of flow is the same in all PIV-STD sequences. Relatively simple, 'wavy' flows without vortices are generated. The images are 256 × 256 pixel size. The standard average speed is 7.39 pixel/frame. Figure 2 shows a sample image with the standard density of 4000 particles. A typical flow is also visualised as a velocity vector field.

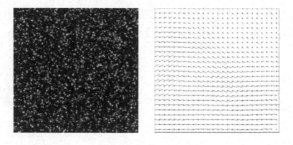

Fig. 2. A sample PIV-STD image and a typical flow.

Table 1 compares displacement errors of the three methods, IPAN-PIV, KLT-PIV and ODP-PIV2, for six STD-PIV sequences. The results of ODP-PIV are cited from [11]. Results of CORR-PIV were not available for this dataset. ODP-PIV2 is the most precise of the three variants presented in [9]. For the STD-PIV dataset, the performance of KLT-PIV is very close to that of ODP-PIV. The accuracy of IPAN-PIV is lower; in particular, a poor result was obtained for the frequently disappearing particles. The high density and the indistinguishable particles also pose some problem to IPAN-PIV. Note that the ODP-PIV algorithm is much slower than the trackers, as discussed below.

Table 1. Relative displacement errors (%) for PIV-STD

	s01	s03	s04	s05	s06	s08
IPAN-PIV	4.72 ± 4.3	10.3 ± 4.5	4.86 ± 6.2	5.41 ± 4.9	4.59 ± 5.7	23.8 ± 30
KLT-PIV	3.37 ± 2.3	9.82 ± 4.0	2.02 ± 2.1	3.51 ± 2.7	2.02 ± 1.8	4.18 ± 3.9
ODP-PIV2	3.52 ± 2.9	9.82 ± 5.1	1.97 ± 2.7	2.68 ± 3.1	1.50 ± 1.6	5.05 ± 4.6

4.2   The CYLINDER Data

Table 2 compares IPAN-PIV, KLT-PIV, ODP-PIV and the correlation method CORR-PIV for a set the synthetic flow sequences called CYLINDER [10]. ODP-PIV1 is the fastest and least precise of the three variants presented in [9]. CORR-PIV is a 32×32 window size correlator. N5, N10 and N20 are noisy versions of the original noise-free sequence N0 visualised in figure 3. The numbers indicate the degrees of noise varying from 5% to 20%. CYLINDER is a complex flow with the mean displacement 7.6 pixel/frame. For the CYLINDER dataset, the feature trackers outperform the correlation algorithm in both accuracy and execution time. For noisy (and more realistic) data, they compete with the fastest variant of ODP-PIV in accuracy and are definitely superior in processing speed.

Table 2. Displacement errors for CYLINDER, in pixels

	IPAN-PIV	KLT-PIV	CORR-PIV	ODP-PIV1	ODP-PIV2
N0	0.42 ± 0.5	0.35 ± 0.6	0.55 ± 1.0	0.13 ± 0.1	0.07 ± 0.1
N5	0.50 ± 0.7	0.35 ± 0.6	0.61 ± 1.2	0.21 ± 0.5	0.08 ± 0.1
N10	0.55 ± 0.7	0.33 ± 0.5	0.77 ± 1.6	0.53 ± 1.4	0.11 ± 0.1
N20	0.90 ± 1.2	0.44 ± 0.6	3.11 ± 4.1	0.88 ± 1.6	0.20 ± 0.1
Time	20 sec	15 sec	10 min	20 min	200 min

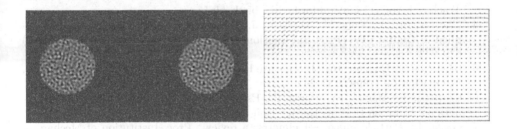

Fig. 3. First frame and flow visualisation of CYLINDER N0.

5   Conclusion

We have presented a novel approach to Particle Image Velocimetry based on feature tracking. Two efficient algorithms have been customised to PIV by adding the procedures for coherence filtering and resampling. The coherence filtering improves the accuracy of velocity estimation. The resampling provides a uniform sampling at the expense of a moderate decrease of the accuracy. The results of the tests demonstrate that the proposed approach offers a good alternative to both correlation and ODP techniques. KLT-PIV and IPAN-PIV provide higher flow estimation accuracy and are faster than the conventional correlation techniques. For noisy images, the feature tracking PIV provides accuracy comparable with that of the precise ODP algorithms, while requiring much less computation. The processing speed of the trackers can potentially make them suitable for fast flow visualisation, qualitative estimation, and analysis of time-varying flows.

The most interesting, open question is that of flow complexity. Simple, smooth flows like PIV-STD can be in most cases reliably measured by all methods considered in this paper. The real differences in performance become clear as the complexity grows. Future research should focus on flow complexity and involve creation of realistic but complex test sequences with the ground truth provided.

Online demonstrations of IPAN-PIV and KLT-PIV are available on the Internet at the web site: http://visual.ipan.sztaki.hu/demo/demo.html. A remote user can select an algorithm, set the parameters and run the algorithm on a short PIV sequence. Alternatively, the user can upload and process his/her own data.

References

1. Astola, J., Haavisto, P., Neuvo, Y.: Vector Median Filters. Proceedings of the IEEE **78** (1990) 678–689
2. Barron, J.L., Fleet, D.J., Beauchemin, S.S.: Performance of optical flow techniques. International Journal of Computer Vision **12**(1) (1994) 43–77
3. Birchfield, S.: KLT: An Implementation of the Kanade-Lucas-Tomasi Feature Tracker. http://vision.stanford.edu/~birch/klt/
4. Chetverikov, D., Nagy, M., Verestóy, J.: Comparison of Tracking Techniques Applied to Digital PIV. Proc. International Conf. on Pattern Recognition **4** (2000) 619–622
5. Chetverikov, D., Verestóy, J.: Feature Point Tracking for Incomplete Trajectories. Computing **62** (1999) 233–242
6. Corpetti, T., Mémin, É., Pérez, P.: Estimating Fluid Optical Flow. Proc. International Conf. on Pattern Recognition **3** (2000) 1045–1048
7. Grant, I.: Particle image velocimetry: a review. Proc. Institution of Mechanical Engineers, **211** Part C (1997) 55–76
8. Jähne, B.: Digital Image Processing. Springer (1997)
9. Quénot, G., Pakleza, J., Kowalewski, T.: Particle image velocimetry with optical flow. Experiments in Fluids **25** (1998) 177–189
10. Quénot, G.: Data and procedures for development and testing of PIV applications. ftp://ftp.limsi.fr/pub/quenot/opflow/
11. Quénot, G.: Performance evaluation of an optical flow technique for particle image velocimetry. Proc. Euromech 406 Colloquim. Warsaw (1999) 177–180
12. Standard images for particle imaging velocimetry. http://www.vsj.or.jp/piv/
13. Shi, J., Tomasi, C.: Good features to track. Proc. IEEE Conf. on Computer Vision and Pattern Recognition (CVPR94). Seattle (Jun 1994)
14. Tokumaru, P.T., Dimotakis, P.E.: Image correlation velocimetry. Experiments in Fluids **19** (1995) 1–15

Estimation of Motion through Inverse Finite Element Methods with Triangular Meshes

J.V. Condell, B.W. Scotney, and P.J. Morrow

School of Information and Software Engineering, University of Ulster at Coleraine,
Cromore Road, Coleraine, BT52 1SA, Northern Ireland
Tel. +44 (0) 28 7032 4698 Fax. +44 (0) 28 7032 4916
{jv.graham, bw.scotney, pj.morrow}@ulst.ac.uk

Abstract. This paper presents algorithms to implement the estimation of motion, focusing on the finite element method as a framework for the development of techniques. The finite element approach has the advantages of a rigorous mathematical formulation, speed of reconstruction, conceptual simplicity and ease of implementation via well-established finite element procedures in comparison to finite volume or finite difference techniques. The finite element techniques are implemented as a triangular discretisation, and preliminary results are presented. An important advantage is the capacity to tackle problems in which non-uniform sampling of the image sequence is appropriate, which will be addressed in future work.

Keywords: motion estimation, inverse finite element methods, triangular meshes

1 Introduction

The estimation of grey-level derivatives is of great importance to the analysis of image sequences. Intensity-based differential techniques are frequently used to compute velocities from spatio-temporal derivatives of image intensity or filtered versions of the image, [5]. By assuming that image intensity is conserved, a gradient constraint equation may be derived; approximation of the image intensity derivatives in this equation, together with smoothness constraints, forms the basis of algorithms for the computation of the optical flow field describing the transformation of one grey-level image in a sequence to the next.

For convenience, it is straightforward to begin with a problem where the apparent velocity of intensity patterns can be directly identified with the movement of surfaces in the scene. *Horn* and *Schunck* [5] present an iterative algorithm to find the optical flow pattern, assuming the apparent velocity of the brightness pattern varies smoothly almost everywhere in the image, in which the derivatives of image intensity are approximated using a finite volume approach. The approach described here approximates these derivatives of image intensity using a finite element technique. The finite element algorithm is described in Section 2. Experimental results with synthetic and real grey-scale images are shown in Section 3. Conclusions are discussed in Section 4.

W. Skarbek (Ed.): CAIP 2001, LNCS 2124, pp. 333–340, 2001.

2   Finite Element Formulation

The finite volume approach used to develop the *Horn* and *Schunck* algorithm requires a regular sampling of image intensity values stored in simple rectangular arrays. Computational efficiency can be greatly increased by focusing only on those regions in which motion has been detected to be occurring. This requires the ability to work with images in which the intensity values are sampled more frequently in areas of interest, such as localised regions of high intensity gradient or rapid variation of intensity with time, than in areas likely to correspond to scene background. For example, in [6] image variation is used to facilitate the adaptive partitioning of the image for the efficient estimation of optical flow.

Implementations based on finite difference or finite volume methods become complicated to generalise when the image intensity values are not uniformly sampled because they are based on point differences along regular co-ordinate directions. In contrast, finite element methods are ideally suited for use with variable and adaptive grids, and hence provide a framework for developing algorithms to work with non-uniformly sampled images. Such images may arise from the application of image compression techniques, [2]. Appropriate image partitioning may lead to the use of a variety of element sizes and orientations. It is thus necessary to be able to implement neighbourhood operators systematically in situations involving an irregular discretisation of the image.

2.1   A Finite Element Formulation in Two Dimensions

It has previously been shown, [3], that the algorithm of *Horn* and *Schunck* for optical flow estimation may be considered as one of a family of methods that may be derived using an inverse finite element approach. In [3], [4], the two-dimensional Galerkin bilinear finite element technique (FETBI) has been developed and extensively compared with the *Horn* and *Schunck* algorithm (HS), as well as the best of the existing motion estimation methods reported in the literature, and is shown to perform well.

The image intensity at the point (x, y) in the image plane at time t is denoted by $u(x, y, t)$ which is considered to be a member of the Hilbert image space H^1 at any time $t > 0$. The optical flow is denoted by $\underline{b} \equiv (b_1(x, y, t), b_2(x, y, t))$ where b_1 and b_2 denote the x and y components of the flow $\underline{b}$ at time $t > 0$. In the *inverse* problem with which we are concerned, the image intensity values are known and it is the *velocity function* $\underline{b}$, or *optical flow*, that is unknown and is to be approximated. The optical flow constraint equation is

$$u_x b_1 + u_y b_2 + u_t = 0, \tag{1}$$

where u_x, u_y, u_t are the partial derivatives of the image intensity with respect to x, y and t respectively. This equation cannot fully determine the flow but can give the component of the flow in the direction of the intensity gradient. An additional constraint must be imposed, to ensure a smooth variation in the flow across the image, which is formed by minimising the (square of) the magnitude of the gradient of the optical flow velocity components. It provides a smoothness measure, and may be implemented by setting to zero the Laplacian of b_1 and b_2. The computation of optical flow may then be treated as a minimisation problem for the sum of the errors in the equation for the rate of change of image intensity and the measure of the departure from smoothness in the velocity field. This results in a pair of equations, which along with an approximation of laplacians of the velocity components, allow the optical flow to be computed.

The infinite dimensional function space H^1 cannot be used directly to provide a tractable computational technique, however a function in the image space H^1 may be approximately represented by a function from a finite dimensional subspace $S^h \subset H^1$. From S^h a finite basis $\{\phi_j\}_{j=1}^N$ is selected, where the support of ϕ_i is restricted to a neighbourhood Ω_i (the domain in which its value is non-zero). The image u may thus be approximately represented at time t_n by a function $U^n \in S^h$, where $U^n = \sum_{j=1}^N U_j^n \phi_j$ and in which the parameters $\{U_j^n\}_{j=1}^N$ are associated with the image intensity values at time $t = t_n$. A basis for S^h may be formed by associating with each node i, a trial function $\phi_i(x,y)$ such that $\phi_i(x,y) = 1$ at node i, $\phi_i(x,y) = 0$ at node j, $j \neq i$, and $\phi_i(x,y)$ is polynomial on each element. The basis function $\phi_i(x,y)$ is thus a *"tent shaped"* function with support limited to elements that have node i as a vertex.

A Euler forward finite difference $\frac{U^{n+1}-U^n}{\Delta t}$ approximates the partial derivative $\frac{\partial U_j}{\partial t}$, where U_j^n and U_j^{n+1} are image intensity values in the nth and $(n+1)^{th}$ images respectively. If the components of the optical flow are considered to be piecewise constant on each element, then the finite element formulation

$$\sum_{j=1}^N [\frac{U_j^{n+1} - U_j^n}{\Delta t}] \int_\Omega \phi_j \phi_i \, d\Omega + \theta \sum_{j=1}^N U_j^n \int_\Omega [b_1 . \nabla \phi_j \phi_i + b_2 . \nabla \phi_j \phi_i] \, d\Omega$$

$$+ (1-\theta) \sum_{j=1}^N U_j^{n+1} \int_\Omega [b_1 . \nabla \phi_j \phi_i + b_2 . \nabla \phi_j \phi_i] \, d\Omega = 0 \qquad (2)$$

provides an approximation to the flow constraint equation in which each equation contains unknown velocity component values. These equations, along with the smoothing constraint, may be rearranged to provide an iteration scheme similar to the *Horn-Schunck* equations to approximate the optical flow for each element.

2.2 Triangular Grid Implementation of Finite Element Technique

This paper focuses on the development of techniques (FETTRI) based on a triangular finite element mesh. This is to facilitate the development of algorithms that are sufficiently general to be capable of addressing optical flow estimation using image sequences that are non-uniformly sampled.

It is straightforward to begin with a discretisation of right-angled isosceles triangles. Consider a finite element discretisation of the image domain Ω based on this triangular array of pixels. Nodes are placed at the pixel centres, and this array of nodes is used by a 2D Delaunay algorithm, applied to the row and column co-ordinates of the nodes, to produce a triangular mesh. Given a set of data points the Delaunay triangulation provides a set of lines connecting each point to its natural neighbours. The resulting mesh allows the partitioning of the image into elements, as illustrated in Fig. 1, with nodes located at the three vertices of the triangular elements. Fig. 2 shows how the element numbering provided by the Delaunay algorithm is locally random, with a section of a typical triangular grid of elements and nodes. When considering the flow on a particular element e_k, we must take account of the elements which share a node with e_k (i.e. the "surrounding elements"). For example, in Fig. 2, only the velocity calculations from elements $e_6, e_3, e_5, e_1, e_{11}, e_7, e_9, e_{10}, e_8, e_{12}, e_{16}, e_{17}$ are included in the velocity equation for solution over element e_k.

The motion estimation equations may be formed by associating with each element e, a trial function $\Phi_e(x,y) = \phi_1(x,y) + \phi_2(x,y) + \phi_3(x,y)$ such that $\Phi_e(x,y) = 1$ over element e, $\Phi_e(x,y) = 0$ over all elements other than those surrounding element e, and $\Phi_e(x,y)$ falls linearly from 1 to 0 over the elements that surround element e. The basis function $\Phi_e(x,y)$ is thus a *"plateau"* function and has support limited to those triangular elements which share one of these three vertices of element e (locally labelled nodes 1, 2 and 3). With the triangular mesh computed, it is possible to compute the 2D velocity over the image by considering the components of the optical flow to be piecewise constant over each element. The motion estimation equations are as shown in Eqn. 2 with ϕ_i substituted by the new trial function Φ_e.

It is relatively simple to calculate and assemble the relevant element mass and stiffness matrix for this problem as the equations are still based on typical finite element matrices. The element contributions are assembled into a global matrix to be used in the solution of the optical flow equations. Isoparametric mappings were used to evaluate the integrals to map from the arbitrary element isosceles triangle to the standard element triangle on the grid.

Using the motion equation in a *Horn* and *Schunck* style iteration, that includes a smoothing term based on laplacians of the velocity components, results in a pair of equations for the optical flow for each triangular element in the mesh. For plotting, a grid which is twice as fine as the pixel grid is used, and velocities for each element are plotted at the grid point closest to the centroid of the element, see Fig. 3, (v_k is the optical flow for element k, that is e_k).

3   Experimental Results

Results for image sequences are obtained, primarily for the proposed triangular implementation of the finite element technique (FETTRI), and comparisons made to the FETBI and HS algorithms, [4]. **Synthetic** image sequences are used for which the correct 2D motion fields are known, with the *angular* and *absolute* errors measures used for quantitative comparisons. **Real** image sequences are also used for analysis, [1], [4], [7]. Velocity is written as displacement per unit time; pixels/frame (*pix/fr*).

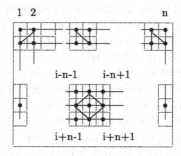

Fig. 1. Regular triangular array of pixels and nodes for finite element discretisation

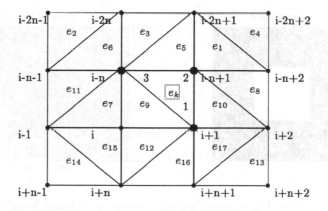

Fig. 2. A section of a triangular grid of elements and nodes

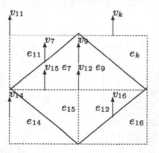

Fig. 3. Plotting of velocity values

3.1 Synthetic Image Sequence

Translating Square sequence. This simple test case is shown in Fig. 4 along with the corresponding actual flow. The sequence involves a translating square with a width of 40 pixels on a dark background with uniform motion of $v_1 = (1, 1)$. The obtained optical flow fields for a tolerance of 0.00001 and $\alpha = 600$ are shown in Fig. 5 for the FETBI and FETTRI algorithms respectively. Table 1 shows the algorithms' performance in terms of the absolute and angular error values, where it can be seen that the FETTRI outperforms the other two algorithms significantly. The FETTRI flow (Fig. 5) may seem to be rather 'spread out'. This is due to the plotting technique of the FETTRI algorithm as described in Section 2.2. Values are not located on a regular rectangular grid therefore spaces are left where velocity is not computed, though velocity values are always calculated for every element in the triangular discretisation. This could be post-processed to fill in the flow field, though it is preferable at this stage to just show the 'pure' flow for all experiments.

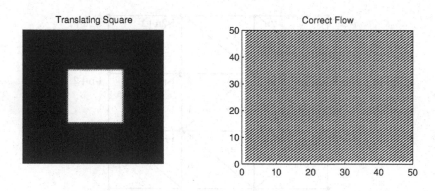

Fig. 4. *Translating Square* Image and Correct Flow

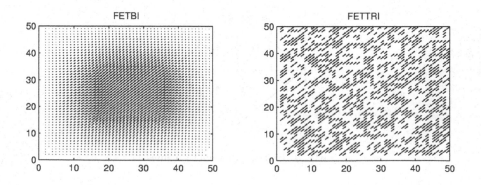

Fig. 5. *Translating Square* FETBI and FETTRI results

Table 1. Performance of FET (TRIangular, BIlinear) and HS algorithms with Absolute and Angular Error Values for *Translating Square* Image Sequence

Algorithm	Absolute Error Mean	SD	Angular Error Mean	SD	Percentage of flow vectors with Angular Error less than: < 1°	< 3°	< 5°	< 10°
Tolerance = 0.00001; α = 600								
FETTRI*	0.04	0.04	0.13	0.27	98.36%	99.96%	100.00%	100.00%
FETBI	0.98	0.38	6.38	4.25	10.60%	27.89%	41.23%	76.91%
HS	1.00	0.37	6.31	4.15	10.65%	27.84%	41.12%	77.88%

* For FETTRI the error values shown are worst case bounds, since they include optical flow values of zero at grid points where optical flow is not computed.

Rotating Rubic

FETTRI

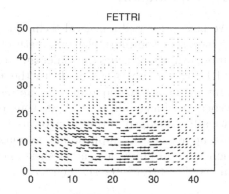

Fig. 6. *Rotating Rubic* Image and FETTRI results

3.2 Real Image Sequence

Rotating Rubic cube. A frame from this sequence is shown in Fig. 6. The Rubic cube is rotating counter-clockwise on a turntable involving velocities less than *2 pix/fr*. The obtained flow field for a tolerance of 0.0001 and α = 800 for the FETTRI algorithm is shown in Fig. 6, which is satisfactory for this sequence.

Hamburg Taxi scene. This sequence is a street scene involving four moving objects - a car on the left driving left to right, *3 pix/fr*, a taxi near the centre turning the corner at *1.0 pix/fr*, a pedestrian in the upper left moving down, *0.3 pix/fr*, and a van on the right moving to the left, *3 pix/fr*. A frame from the sequence and a result from the FETTRI algorithm (tolerance 0.001 and α = 200) are shown in Fig. 7. This result is promising for the *Hamburg Taxi* sequence.

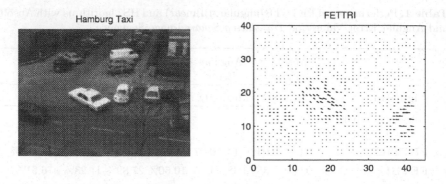

Fig. 7. *Hamburg Taxi* Image and FETTRI results

4   Conclusion

We have presented a general technique for the computation of optical flow that is based on the finite element method. The approach has been illustrated by implementations using the Galerkin triangular finite element formulation (FET-TRI) with comparisons made to the FETBI and HS algorithms. Our inverse finite element technique may be applied in general using a variety of element types and image discretisations, thus providing a family of algorithms systematically developed within a rigorous framework. An important contribution of our approach is its applicability to images that are sampled non-uniformly using non-uniform finite element meshes, allowing computational activity to focus on areas of greatest interest, such as those where motion is detected.

References

1. Barron, J.L., Fleet, D.J., Beauchemin, S.S.: Performance of Optical Flow Techniques. Int. J. Comput. Vis. **12** 1 (1994) 43–77
2. Garcia, M.A., Vintimilla, B.X., Sappa, A.D.: Approximation of Intensity Images with Adaptive Triangular Meshes: Towards a Processable Compressed Representation. Proc. Third Irish Machine Vision and Image Process. Confer. (1999) 241–249
3. Graham, J.V., Scotney, B.W., Morrow, P.J.: Inverse Finite Element Formulations for the Computation of Optical Flow. Proc. Fourth Irish Machine Vision and Image Process. Confer. (2000) 59-66
4. Graham, J.V., Scotney, B.W., Morrow, P.J.: Evaluation of Inverse Finite Element Techniques for Gradient Based Motion Estimation. Proc. Third IMA Confer. on Imaging and Digital Image Process. (2001)
5. Horn, B.K.P., Schunck, B.G.: Determining Optical Flow. Artificial Intelligence **17** (1981) 185–203
6. Kirchner, H., Niemann, H.: Finite Element Method for Determination of Optical Flow. Pattern Recognition Letters **13** (1992) 131–141
7. Ong, E.P., Spann, M.: Robust Optical Flow Computation Based on Least-Median-of-Squares Regression. Int. J. Comput. Vis. **31** 1 (1999) 51–82

Face Tracking Using the Dynamic Grey World Algorithm*

José M. Buenaposada, David Sopeña, and Luis Baumela

Departamento de Inteligencia Artificial, Universidad Politécnica de Madrid
Campus de Montegancedo s/n, 28660, Madrid, SPAIN
{jmbuena,dsgalindo}@dia.fi.upm.es, lbaumela@fi.upm.es

Abstract. In this paper we present a colour constancy algorithm for real-time face tracking. It is based on a modification of the well known Grey World algorithm in order to use the redundant information available in an image sequence. In the experiments conducted it is clearly more robust to sudden illuminant colour changes than popular the rg-normalised algorithm.

Keywords: face tracking, Grey World algorithm, colour constancy

1 Introduction

In this paper we will study the problem of face tracking using colour. Trackers based on this feature, which is the most frequently used feature for face tracking [14], are used as an initial estimate or follow-up verification of face location in the image plane.

The primary problem in automatic skin detection is colour constancy. The [RGB] colour of an image pixel depends not only on the imaged object colour, but also on the lighting geometry, illuminant colour and camera response. For example [6], if the scene light intensity is scaled by a factor s, each perceived pixel colour becomes $[sR, sG, sB]$. The rg-normalisation algorithm provides a colour constancy solution which is independent of the illuminant intensity by doing: $[sR, sG, sB] \mapsto [sR/s(R + G + B), sG/s(R + G + B)]$. On the other hand, a change in illuminant colour can be modelled as a scaling α, β and γ in the R, G and B image colour channels. In this case the previous normalisation fails. The Grey World (GW) algorithm [6] provides a constancy solution independent of the illuminant colour by dividing each colour channel by its average value.

In this paper we introduce a colour constancy algorithm, based on GW, that can be used for real-time colour-based image segmentation which is more robust to big sudden illuminant colour changes than the popular rg-normalised algorithm.

---

* Work funded by CICyT under project number TIC1999-1021

W. Skarbek (Ed.): CAIP 2001, LNCS 2124, pp. 341–348, 2001.

2 Grey World-Based Colour Constancy

Colour constancy is the perceptual ability to assign the same colour to objects under different lighting conditions. The goal of any colour constancy algorithm is to transform the original $[RGB]$ values of the image into constant colour descriptors. In the case of Lambertian surfaces, the colour of an image pixel (ij) can be modelled by a lighting geometry component s_{ij}, which scales the $[rgb]$ surface reflectances of every pixel independently, and three colour illuminant components (α, β, γ), which scale respectively the red, green and blue colour channels of the image as a whole [6]. The lighting geometry component accounts for surface geometry and illuminant intensity variations, while the colour illuminant components account for variations in the illuminant colour. According to this model, two pixels $I(ij)$ and $I(kl)$ of an image would have the following $[RGB]$ values: $[s_{ij}\alpha r_{ij}, s_{ij}\beta g_{ij}, s_{ij}\gamma b_{ij}]$, $[s_{kl}\alpha r_{kl}, s_{kl}\beta g_{kl}, s_{kl}\gamma b_{kl}]$, where $[r_{ij}, g_{ij}, b_{ij}]$ and $[r_{kl}, g_{kl}, b_{kl}]$ represent surface reflectances; i.e. real object colour, independent of the illuminant.

The GW algorithm proposed by Buchsbaum [2] assumed that the average surface reflectances in an image with enough different surfaces is grey. So, the average reflected intensity corresponds to the illuminant colour, which can be used to compute the colour descriptors. This algorithm was refined in [8] by actually obtaining an average model of surface reflectances and proposing a procedure to compute the average image reflectance.

Let us define the *image average geometrical reflectance*, $\bar{\mu}$, as

$$\bar{\mu} = \frac{1}{n} \sum_{ij \in I} [s_{ij} r_{ij}, s_{ij} g_{ij}, s_{ij} b_{ij}],$$

where n is the number of image pixels. It represents the average $[RGB]$ image values, once we have eliminated the colour illuminant component.

If we assume that the average geometrical reflectance is constant over the image sequence, then the image average $[RGB]$ variation between two images is proportional to the illuminant colour variation. On the basis of this, a colour normalisation invariant to illuminant colour changes can be devised:

$$[\frac{I_r(ij)}{\frac{1}{n}\sum_{ij \in I} I_r(ij)}, \frac{I_g(ij)}{\frac{1}{n}\sum_{ij \in I} I_g(ij)}, \frac{I_b(ij)}{\frac{1}{n}\sum_{ij \in I} I_b(ij)}] = [\frac{s_{ij}r_{ij}}{\bar{\mu}_r}, \frac{s_{ij}g_{ij}}{\bar{\mu}_g}, \frac{s_{ij}b_{ij}}{\bar{\mu}_b}], \quad (1)$$

where, if x represents the colour channel ($x \in [r, g, b]$), $I_x(ij)$ is the value of the channel x for pixel $I(ij)$, and $\bar{\mu}_x$ is the image channel x average geometrical reflectance.

The previous normalisation is what we call basic GW algorithm. It is robust to illuminant colour variations, but it only works for sequences with constant image average geometrical reflectance. In consequence, basic GW fails when a new object appears in the image or when the illuminant geometry changes. In the next section we propose an extension to the basic GW algorithm that solves these problems using redundant information available in an image sequence.

3   Face Tracking Using Dynamic Grey World

In this section we present a colour-based face tracking algorithm. First we will briefly describe how to track a coloured patch using simple statistics, afterwards the Dynamic GW (DGW) algorithm is presented.

3.1   Face Segmentation and Tracking Using a Skin Colour Model

Given a sequence of colour images, building a face tracker is straightforward if we have a reliable model of the image colour distributions. Let I_{rgb} be the $[RGB]$ channels of image I, and let $p(I_{rgb}|skin)$ and $p(I_{rgb}|back)$ be the conditional colour probability density functions (pfds) of the skin and background respectively (we assume that background is anything that is not skin). Using the Bayes formula, the probability that a pixel with colour I_{rgb} be *skin*, $P(skin|I_{rgb})$, can be computed as follows:

$$P(skin|I_{rgb}) = \frac{p(I_{rgb}|skin)P_s}{p(I_{rgb}|skin)P_s + p(I_{rgb}|back)P_b},$$

where P_s and P_b are the a priori probabilities of *skin* and *background*. The transformation $\mathcal{T}(I_{rgb}) - 255 \times P(skin|I_{rgb})$ returns an image whose grey values represent the probability of being skin (see first column in Fig. 3). Face tracking on this image can be performed with a mode seeking algorithm, like [4], by computing the position and orientation of the face colour cluster in each frame [3].

The problem now is how to make the previous statistical model invariant to variations in the scene illumination. In most real-time systems this invariance is achieved by working in a rg normalised chromaticity space. As we previously mentioned, this method fails when there is a sudden change of the illuminant colour. In our segmentation algorithm we propose using the GW colour space, $[\hat{r}\hat{g}\hat{b}]$, defined in section 2:

$$I_{\hat{r}\hat{g}\hat{b}}(ij) = n \times \left[\frac{I_r(ij)}{\sum_I I_r} \frac{I_g(ij)}{\sum_I I_g} \frac{I_b(ij)}{\sum_I I_b} \right].$$

We model the *skin* GW colour distribution with a continuous Gaussian model. As can be seen in Fig. 1, $p(I_{\hat{r}\hat{g}\hat{b}}|skin)$ is approximately Gaussian. On the left are shown the Chi-square and Gaussian plots of the $I_{\hat{r}}$, $I_{\hat{g}}$ and $I_{\hat{b}}$ marginals and the $I_{\hat{r}\hat{g}\hat{b}}$ multivariate distribution. From the analysis of these plots we can verify that the assumption $p(I_{\hat{r}\hat{g}\hat{b}}|skin) \sim N(\bar{m}_s, \Sigma_s, I_{\hat{r}\hat{g}\hat{b}})$ can not be rejected. On the other hand, it is not possible to find an analytic model for the *background* pdf, so we will model it with a uniform distribution, $h_b(I_{\hat{r}\hat{g}\hat{b}})$. Other authors have indicated different preferences for modelling the colour distributions. In [11] Gaussian mixture models, whereas in [5] and [12] pure histogram-based representations are chosen. In our experiments we found that using a a continuous model yields better results because of the high space dimensionality (3D).

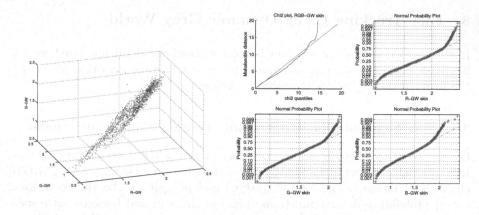

Fig. 1. Skin colour pdf in GW space. On the left is shown the skin colour cluster in GW colour space. On the right are shown the Chi-square plot for the multivariate distribution and the Normal plots for the marginals.

If we approximate the priors $P_s \approx n_s/n$ and $P_b \approx n_b/n$, where n_s and n_b are respectively the number of the *skin* and *background* pixels, then

$$P(skin|I_{rgb}) = \frac{n_s \, N(\bar{m}, \, \Sigma, I_{rgb})}{n_s \, N(\bar{m}, \, \Sigma, I_{rgb}) + n_b \, h_b(I_{rgb})}.$$

3.2 The Dynamic Grey World Algorithm

The main problem of the basic GW algorithm is that it was conceived for static images; i.e. it fails when there is a big change in the image average geometrical reflectance. In this section we propose a dynamic extension to GW (DGW) which will detect this situation and update the GW model (see Fig. 2).

In the following we assume that there exists a partition of the image sequence into a set of image subsequences such that the image average geometrical reflectance is constant over each subsequence; i.e. the basic GW algorithm can be used as a colour constancy criterion over each subsequence. We will use the first image of each subsequence as a *reference image*. The other images of the subsequence will be segmented using the colour descriptors of the reference image.

Let I_{rgb}^r, I_{rgb}^t and I_{rgb}^{t-1} be respectively the reference image, the present and the previous image, $F_{\hat{r}\hat{g}\hat{b}}^r$ be the face pixels in GW space, $\bar{\mu}_{rgb}^{I^t}$ be the average value for each colour channel in I_{rgb}^t, $\bar{\mu}_{\hat{r}\hat{g}\hat{b}}^{F^r}$ and $\bar{\mu}_{\hat{r}\hat{g}\hat{b}}^{F^t}$ be the average GW descriptors for the face pixels in the reference and present image, and $\bar{m}_s$, Σ_s, h_b, P_s, P_b be the GW colour descriptors statistical distribution for the reference image.

The problem now is how to segment each reference image and how to detect a change of subsequence. Reference images can be segmented with the average [RGB] values of the previous image ($\bar{\mu}_{rgb}^{I^{t-1}}$), provided that the change in average

```
Initialisation
  /*Initialise the reference image model using motion
  segmentation and a precalculated colour model*/
  [m̄ₛ,Σₛ,Pₛ,P_b,μ̄_{r̂ĝb̂}^{Fʳ}] = InitTracking();
While (true)   /* tracker main loop */
  μ̄_{rgb}^{Iᵗ} = Mean(I_{rgb}^t); /* image mean rgb values */
  I_{r̂ĝb̂}^t = \frac{I_{rgb}^t}{μ̄_{rgb}^{Iᵗ}}   /* GW normalisation */
  F_{r̂ĝb̂}^t = ProbabilisticSegment(I_{r̂ĝb̂}^t,m̄ₛ,Σₛ,Pₛ,P_b); /* segment img */
  μ̄_{r̂ĝb̂}^{Fᵗ} = ComputeAvgFaceGW(F_{r̂ĝb̂}^t); /* face avg GW descriptors */
  If ‖μ̄_{r̂ĝb̂}^{Fʳ} - μ̄_{r̂ĝb̂}^{Fᵗ}‖ > Δ then /* change of subsequence */
    I_{r̂ĝb̂}^t = \frac{I_{rgb}^t}{μ_{rgb}^{Iᵗ⁻¹}}   /* GW normalise with previous mean */
    F_{r̂ĝb̂}^{'t} = ProbabilisticSegment(I_{r̂ĝb̂}^t,m̄ₛ,Σₛ,Pₛ,P_b); /* segment image */
    I_{r̂ĝb̂}^r = I_{r̂ĝb̂}^t /* update reference image */
    μ̄_{r̂ĝb̂}^{Fʳ} = ComputeAvgFaceGW(F_{r̂ĝb̂}^t); /* face avg GW descriptors */
    [m̄ₛ,Σₛ,Pₛ,P_b] = ColourDistrib(F_{r̂ĝb̂}^t); /* ref. colour distrib */
  end /* if */
end /* while */
```

Fig. 2. Dymanic Grey World Algorithm

geometrical reflectance is caused mainly by the appearance of new objects in the scene.

A change of subsequence is detected just by detecting a change in the average geometrical reflectance. This can not be accomplished on the basis of analysing $\bar{\mu}_{rgb}^I$, as $\bar{\mu}_{rgb}^I$ also changes with the illuminant colour. We solve this problem by monitoring the average GW descriptors of the face pixels. As they are invariant to illuminant colour changes, a change in these descriptors is necessarily caused by a change in average geometrical reflectance.

4 Experiments

In our experiments we used a VL500 Sony colour digital camera at 320×240 resolution, iris open, no gain, no gamma correction. Images were taken with regular roof fluorescent lights and variations in illumination colour were obtained using a controled tungsten light, a green color filter, and turning on and off roof fluorescent lights.

In the first experiment we validate the DGW algorithm hypothesis: variations in the average geometrical reflectance can be detected, and the reference image of each subsequence can be segmented. We acquired a sequence of 200 images with a green object appearing at one point and illuminant geometrical variations taking place at different moments. The result of this experiment is shown, from left to right, in Fig. 3: the first image of the sequence (image 1), a change in the illuminant (roof lights turned off) (image 26), and the appearance and disappearance of an object (images 88 and 139). In this experiment the system detects three subsequences (1 to 87, 88 to 138, and 139 to 200). This is clearly visible in the plot at the bottom of Fig. 3. In image 26 the roof fluorescent lights are turned off. This geometrical illumination variation can be perceived again in the face GW descriptors plot. In this case the segmentation is good. This is an example of "worst case" test. In similar situations with stronger variations in the illuminant geometry, the system may not be able to segment the image and eventually may loose the target.

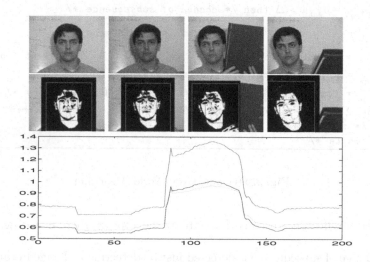

Fig. 3. Hypothesis validation experiment. On the first row four images of a sequence are shown. Their segmentation with the DGW algorithm is presented on the second row. The average r,g and b face GW descriptors (in red, green and blue color respectively) are shown in the third row.

The goal of the next experiment is to check that the dynamic extension to GW is necessary; i.e. to see what would happen if we segment the previous sequence with the basic GW algorithm. In Fig. 4 are shown the same images as in Fig. 3. We can clearly perceive that without the dynamic extension, the initial colour model is invalid when a change in the image average geometrical reflectance (caused by the appearance of an object) takes place. The initial model gradually becomes valid again as the object disappears (see last column).

In the following experiment we compare the performance of the DGW algorithm with the rg-normalised algorithm. We have created a sequence with a set of images with "difficult" background (i.e. brownish door and shelves to distract the segmentation). In Fig. 5 four frames of the sequence are shown in each colum representing: initial image, red object appears, tungsten frontal light turns on, green filter is introduced. DGW segmentation results are shown in the second

Fig. 4. DGW algorithm versus basic GW. DGW algorithm segmentation results are shown in first row and basic GW in the second one.

row and rg-normalised results in the third one. Visual inspection of these results show that both algorithms have similar results in the least favourable cases for the DGW algorithm (second and third columns) and a clear success of the DGW compared to the rg-normalisation when the illuminant colour abruptly changes (fourth column).

Fig. 5. Comparison of DGW and RG-normalisation colour constancy for face tracking.

5   Conclusions

We have introduced the Dynamic Grey World algorithm (DGW) a colour constacy algorithm based on an extension of the well known GW algorithm. It was designed to work in real-time with sequences of images with varying environmental conditions. In the experiments conducted it performed better than the rg-normalised algorithm when sudden changes in the illuminant colour take place. The least favorable case for our algorithm occurs when changes in the illuminant geometry take place. In this paper we have analysed some of the weak points of the rg-normalised algorithm. The DGW algorithm is not perfect either, as its performance can be seriously affected by strong and fast changes in the illuminant geometry. In spite of these limitations, colour-based trackers are good as a fast initial estimate or follow-up verification of face location.

References

1. Berwick, D., Lee., S.W.: A chromaticity space for specularity-, illumination color- and illumination pose invariant 3-d object recognition. Proc. of the Int. Conf. on Computer Vision. Bombay, India (1998)
2. Buchsbaum, G.: A spatial processor model for object colour perception. Journal of the Fanklin Institute **310** (1980) 1–26
3. Bradski, G.: Computer Vision face tracking for use in a perceptual user interface. Proc. of Workshop on applications of Computer Vision, WACV'98 (1998) 214–219
4. Cheng, Y.: Mean shift, mode seeking and clustering. IEEE Trans. on Pattern Analysis and Machine Intelligence **17** (1995) 790–799
5. Crowley, J.L., Schwerdt, J.: Robust tracking and compression for video communication. Proc. of the Int. Workshop on Recognition, Analysis and Tracking of Faces and Gestures in Real-Time (RATFG'99). Corfu, Greece (1999) 2–9
6. Finlayson, G.D., Shiele, B., Crowley, J.L.: Comprehensive colour normalization. Proc. European Conf. on Computer Vison (ECCV). Vol. I. Freiburg, Germany (1998) 475–490
7. Finlayson, G.D. Shaefer, G.: Constrained dichromatic colour constancy. Proc. ECCV. Vol. II. Dublin, Ireland (2000) 342–358
8. Gershon, R., Jepson, A.D., Tsotsos, J.K.: From [R,G,B] to surface reflectance: Computing color constant descriptors in images. Proc. Int. Joint Conf. on Artificial Intelligence (1987) 755–758
9. Klinker, G.J., Shafer, S.A., Kanade, T.: A physical approach to color image understanding. International Journal of Computer Vision **4** (1990) 7–38
10. Lee, H.: Method for computing the scene illuminant chromaticity from specular highlights. Journal of the Optical Society of America A **3** (1986) 1694–1699
11. Raja, Y., McKenna, S.J., Gong, S.: Colour model selection and adaptation in dynamic scenes. Proc. ECCV. Vol. I (1998) 460–474
12. Soriano, M., Martinkauppi, B., Huovinen, S., Laaksonen, M.: Skin detection in video under changing illumination conditions. Proceedings of the Int. Conference on Automatic Face and Gesture Recognition (FG'00). Grenoble, France (2000) 839–842
13. Störring, M., Andersen, H.J. Granum, E.: Estimation of the illuminant colour from human skin colour. Proceedings of the Int. Conference on Automatic Face and Gesture Recognition (FG'00). Grenoble, France (2000) 64–69
14. Toyama, K.: Prolegomena for robust face tracking. MSR-TR-98-65. Microsoft Research (Nov 1998)
15. Wu, Y., Liu, Q., Huang, T.S.: Robust real-time hand localization by self-organizing color segmentation. Proceedings RATFG'99 (1999) 161–166
16. Yang, J., Lu, W., Waibel, A.: Skin-color modeling and adaptation. Proceedings Third Asian Conference on Computer Vision, Vol. II (1998) 142–147
17. D'Zmura, M., Lennie, P.: Mechanisms of colour constancy. Journal of the Optical Society of America A **3** (1986) 1662–1672

Fast Local Estimation of Optical Flow Using Variational and Wavelet Methods

Kai Neckels*

Institut für Informatik und Praktische Mathematik
Christian-Albrecht-Universität Kiel
Preußerstraße 1–9
24105 Kiel, Germany
kn@ks.informatik.uni-kiel.de

Abstract. We present a framework for fast (linear time) local estimation of optical flow in image sequences. Starting from the commonly used brightness constancy assumption, a simple differential technique is derived in a first step. Afterwards, this approach will be extended by the application of a nonlinear diffusion process to the flow field in order to reduce smoothing at motion boundaries. Due to the ill-posedness of the determination of optical flow from the related differential equations, a Wavelet-GALERKIN projection method is applied to regularize and linearize the problem.

Keywords: optical flow estimation, wavelet methods, variational methods

1 Introduction

The main purpose of this article is the presentation of a fast algorithm for the determination of optical flow fields from given image sequences, which is a very important task in image processing. The question of a *reliable estimation* has been adressed by several authors and just as many different approaches were made (see, e.g. [9], [20], [7], [23], [1] and many others mentioned therein). Our approach falls into the large group of differential methods and the flow is computed locally, because here we are mostly interested in fast computations — nonlocal approaches lead to very large equation systems, that can only be solved iteratively, which is usually more expensive. On the other hand, nonlocal methods give commonly better results and are better suited to handle large displacements. Howerver, the presented methods are not limited to the local case and might also be applied to global flow calculations.

The paper is organized as follows: in Section 2, we will briefly recall the differential flow model and the most important invariant, which occurs as *brightness constancy assumption*. Additionally, we will show, how a Wavelet-GALERKIN

---

* The author is supported by the DEUTSCHE FORSCHUNGSGEMEINSCHAFT (DFG) within the Graduiertenkolleg 357, "Effiziente Algorithmen und Mehrskalenmethoden".

W. Skarbek (Ed.): CAIP 2001, LNCS 2124, pp. 349–356, 2001.

procedure helps to linearize and regularize the associated flow equations in order to get stable estimations ([11]) and we argue, how certain properties of MRA-Wavelets may be utilized in this context. Since this simple model has some drawbacks, in particular blurring at motion edges and numerical instabilites at image points with very slow motion, we derive a certain extension in the following section; this extension is closely related to nonlinear diffusion processes, which are also briefly presented there. Finally, we will show some experimental results of the implementations in Section 4 and we discuss the advantages and disadvantages of this method in comparison to existing approaches of optical flow calculation.

2   A Simple Differential Model

As mentioned above, the starting point to calculate the optical flow will be the assumption, that the image brightness between distinct frames of the considered sequence is pointwise time-invariant under motion. This idealization has shown to be reliable for most real world image sequences unless no rapid illumination changes occur. A simple TAYLOR expansion of first order of the statement $I(x(t), y(t)) = \text{const}$ leads to the famous differential flow equation

$$I_x \cdot u + I_y \cdot v = -I_t, \tag{1}$$

where the vector field (u, v) contains the optical flow information. Unfortunately, there are several problems concerned with this formulation, starting from the problem to determine two unknowns from one equation or from the difficulty to find good approximations of the partial derivatives of the image sequence in order to achieve credible solutions of (1). Many of these and other aspects of model problems are intensively discussed in [2] and [12].

We propose the following method to regularize the ill-posed optical flow equation: by projecting (1) into several subspaces V_i of the signal space $L_2(\mathbb{R})$, we obtain a number of equations, that contain the whole flow information and we may hope to get a solution to (1) by an optimization procedure of the received equation system. In more detail, we use a wavelet basis (or frame) to obtain a multiresolution analysis (MRA) of $L_2(\mathbb{R})$ (many details may be found in [6]). Now, the projection can be easily done by taking the inner products of (1) with the building functions of the V_i (sometimes called *test functions*), which are the scaling functions and wavelets the MRA stems from. Moreover, we may also embed the image representation into this multiscale framework, by approximating $I(x, y)$ as well as the temporal derivation $I_t(x, y)$ by series of the type

$$I(x, y) \approx \sum_i \sum_{k_x, k_y} c_{k_x, k_y} \cdot \varphi_i(x - k_x, y - k_y) \qquad \text{and}$$

$$I_t(x, y) \approx \sum_i \sum_{k_x, k_y} t_{k_x, k_y} \cdot \varphi_i(x - k_x, y - k_y).$$

To complete our modelling, the optical flow is assumed to be locally polynomial, i.e. we assume

$$u(x,y)|_{B(l_x,l_y)} = u_{00} + u_{10} \cdot (x - l_x) + u_{01} \cdot (y - l_y) + \dots$$
$$v(x,y)|_{B(l_x,l_y)} = v_{00} + v_{10} \cdot (x - l_x) + v_{01} \cdot (y - l_y) + \dots$$

for a capable surrounding $B(l_x, l_y)$ of a considered image point $I(l_x, l_y)$. After this procedure, we arrive at the system

$$\sum_{i,j,k,l,k_x,k_y,l_x,l_y} c_{k_x,k_y} \cdot \left(u_{k,l} \cdot \Gamma_{i,j,x}^{k,l}(k_x - l_x - s_x, k_y - l_y - s_y) + \right.$$

$$\left. v_{k,l} \cdot \Gamma_{i,j,y}^{k,l}(k_x - l_x - s_x, k_y - l_y - s_y) \right)$$

$$= - \sum_{i,j,k_x,k_y,l_x,l_y} t_{k_x,k_y} \cdot \Gamma_{i,j,1}^{0,0}(k_x - l_x - s_x, k_y - l_y - s_y)$$

for $(l_x + s_x, l_y + s_y) \in B(l_x, l_y)$. Hereby, the $\Gamma_{(\cdot,\cdot,\cdot)}^{(\cdot,\cdot)}(\cdot, \cdot)$ denote the *generalized wavelet connection coefficients*, which are given by

$$\Gamma_{i,j,\varsigma}^{k,l}(m_x, m_y) = \int_{\mathbb{R}^2} x^k \cdot y^l \cdot \varphi_i(x - m_x, y - m_y) \cdot \frac{\partial}{\partial\varsigma} \varphi_j(x - m_x, y - m_y).$$

As was shown in [10] for 1D and in [15] for the general multidimensional case, these connection coefficients can be evaluated (which might be quite expensive) by a finite-dimensional linear system, if the filter mask of the scaling function is finite and the scaling matrix for the associated MRA satisfies a certain mild criterion. However, one obviously sees, that the connection coefficients are independent of the data to be processed and can thus be computed offline and stored in look-up-tables afterwards. Therefore, the linear equation system determining the optical flow can be built up by simple discrete convolutions of the image data with the connection coefficient matrices. Moreover, since we use compactly supported wavelets (see e.g. [14] for design principles of compactly supported multidimensional wavelets), only a rather small number of nontrivial linear equations has to be solved for each image point; the whole processing can be done in linear time and is in addition massively parallelisable.

One might see this approach as a kind of advanced finite difference method, which might be enlightened before the background, that connection coefficients are closely related to discrete differentiation schemes [18]. This is very much in the spirit of [20] and [3], where somewhat similar approaches were made. However, in [20] a scale-space-embedding using GAUSSians and GAUSSian derivatives on several scales were used to obtain a linear equation system from (1) and in [3], partial integration and analytical wavelets were employed instead of the connection coefficient framework.

3   Application of Nonlinear Diffusion

The algorithm derived in the previous section has some nice features, in particular, it is very simple to implement and it is also very fast. But on the other hand there are also some deficiencies. The linear filtering, which is implicitly performed by using the MRA framework leads to blurring that especially detoriorates the flow estimation at motion boundaries and moreover, the equations become numerically instable for small displacement areas; this is a general problem in optimization of overdetermined linear systems with entries of small magnitude. Here, the consequence are estimation outliers.

To overcome these problems, several authors ([13], [23]) proposed the usage of some additional nonlinear flow-based functional, that shall be minimized in order to reduce smoothing at motion boundaries and to stabilize the numerical robustness of the estimations [4]. One very well-known approach is the minimization of the functional

$$\int_\Omega \lambda \cdot W_\sigma(|\nabla u|^2 + |\nabla v|^2) + (I_x \cdot u + I_y \cdot v + I_t)^2 \, dx \, dy, \tag{2}$$

where $W_\sigma(|\nabla u|^2 + |\nabla v|^2)$ is some potential function, that shall guarantee piecewise smooth flow fields as solutions. Here, λ is just a weighting factor and σ is a steering parameter, that thresholds motion boundaries. A necessary condition for (u, v) to be a minimizing solution to (2) is the satisfaction of the related EULER differential equations

$$\lambda \cdot \mathrm{div}\left(W_\sigma'(|\nabla u|^2 + |\nabla v|^2) \cdot \nabla u\right) = I_x \cdot (I_x \cdot u + I_y \cdot v + I_t),$$
$$\lambda \cdot \mathrm{div}\left(W_\sigma'(|\nabla u|^2 + |\nabla v|^2) \cdot \nabla v\right) = I_y \cdot (I_x \cdot u + I_y \cdot v + I_t), \tag{3}$$

which are closely related to nonlinear diffusion processes ([16], [21]). In this context, the solutions are obtained by an evolutionary iteration of the PDEs using e.g. GAUSS-SEIDEL iterations or some advanced semi-implicit numerical schemes like additive operator splitting [22]. Here, we want to go a different way, by solving the EULER differential equations not by temporal evolution, but directly with the connection coefficient framework presented in Section 2. Since we want to use non-polynomial potentials W_σ (e.g. like in [17]), the PDEs (3) cannot by linearized directly by the usage of connection coefficients — aiming to receive a system of linear equations, we have to do some modification first. The technique, we use here is called *half-quadratic regularization* [5], its starting point is the fact, that W_σ may be rewritten as

$$W_\sigma(x^2) = \inf_\gamma(\gamma \cdot x^2 + \rho(\gamma))$$

with some well-chosen convex function ρ depending on W_σ. Rewriting (2) leads to the minimization of

$$\int_\Omega \lambda \cdot (\gamma(x,y) \cdot (|\nabla u|^2 + |\nabla v|^2) + \rho(\gamma(x,y))) + (I_x \cdot u + I_y \cdot v + I_t)^2 \, dx \, dy,$$

which is done in a two-step-way. First, the functional is minimized with respect to γ, while (u, v) are fixed. Under certain conditions, which are fulfilled by the used potential, the solution is given by $\gamma = W'_\sigma$. In the second step, we keep this γ fixed and minimize with respect to (u, v) which leads to the linear EULER differential equations

$$
\begin{aligned}
\lambda \cdot \operatorname{div}\left(\gamma \cdot \nabla u\right) &= I_x \cdot \left(I_x \cdot u + I_y \cdot v + I_t\right), \\
\lambda \cdot \operatorname{div}\left(\gamma \cdot \nabla v\right) &= I_y \cdot \left(I_x \cdot u + I_y \cdot v + I_t\right).
\end{aligned}
\tag{4}
$$

These linear PDEs are again solved by the Wavelet-GALERKIN projection method under application of the connection coefficient framework as described before. The processing is still feasible in linear time, but requires about six times more operations than the *simple* method. We close this section with one final remark about the half-quadratic regularization utilized here: obviously, in our case this method is equivalent to a BANACH-type iteration of the system (3) with $W'_\sigma(|\nabla u_{k-1}|^2 + |\nabla v_{k-1}|^2)$ as linearizer from the previous step. Thus, a good preprocessing for the initial guess of (u, v) is very desirable and in our implementation realized by the simple differential approach described before. With this initializing step, very few iterations of the system (4) are sufficient to achieve a stable solution.

4   Implementation and Experimental Results

First, we will give a brief overview of our implementation. We used symmetric, nonorthogonal, twodimensional, nonseparable wavelets with at least two vanishing moments as generators of an MRA. The same wavelets (scaling functions) and their connection coefficients were used to build the linearized flow equations, which are solved by QR-algorithm. In order to handle large displacements, a multiscale coarse-to-fine strategy with subsequent image warping is utilized, motion outliers are identified via thresholding the temporal derivatives.

Table 1. Various results for the *Yosemite fly-through.*

Method	Av. ang. err.	Density	Type
FLEET & JEPSON	4.36°	34.1%	phase-based
WEBER & MALIK	4.31°	64.2%	local / differential
SINGH	12.90°	97.8%	region-based
URAS	10.44°	100%	local / differential
ALVAREZ ET AL.	5.53°	100%	global / differential
NAGEL	11.71°	100%	global / differential
Section 2	6.94°	96.1%	local / differential
Section 3	6.88°	100%	local / differential

We applied our methods to the famous *Yosemite* sequence (synthetic), see Figure 1. Additionally, in Table 1, we present the results that were achieved by the described methods regarding the average angular error from the true flow field.

One sees, that both described methods perform competitively well in comparison with other methods, especially to those with high estimation densities.

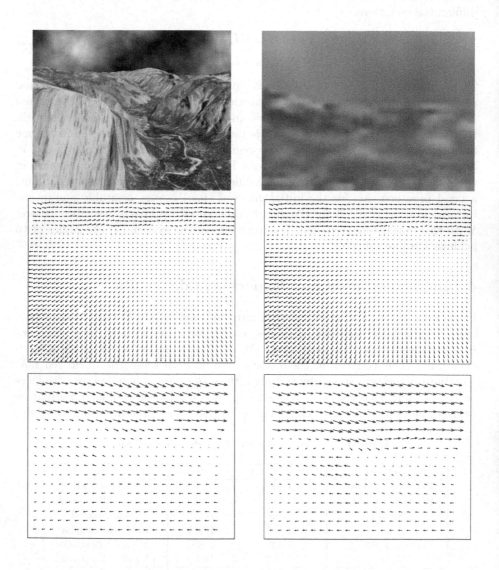

Fig. 1. Top: The data. **Left:** One frame of the *Yosemite* sequence. **Right:** A little section of the ridge in focus. **Middle left:** Calculated flow field for the method described in Section 2. **Middle Right:** Results belonging to the extended nonlinear method. **Bottom:** Improvement in the recovery of motion boundaries and supression of motion outliers — a little section of the ridge in the upper left from the *Yosemite* sequence is focussed. **Left:** Calculated flow by the simple differential method. **Right:** Results of nonlinear diffusion method.

5 Discussion

We presented a new very fast (i.e. linear time) algorithm for the estimation of optical flow, that mainly bases on the usage of wavelet connection coefficients. By adding a nonlinear diffusion term, flow blurring at motion boundaries could be reduced. However, as the experiments show, there is still some smoothing at such boundaries. This is mainly due to the fact, that the discrete filters, that are used to approximate the partial derivatives, cause some overlapping of the spatial information and thus, different motion information are mixed near singular curves (motion boudaries). This is a general problem with differential techniques and cannot be completely overcome. Therefore, if one is interested in exact detection of motion boundaries, treatments that try to minimize some kind of displacement energy functional are superior. On the other hand such approaches are nonlocal and require more computational amount. Which method one should finally use, depends surely on the application, that stands behind. For tracking or video compression purposes, a fast method is certainly preferrable, while for object detection or segmentation, nonlocal but more exact algorithms might be the better choice. However, as already mentioned, the wavelet connection coefficient framework could also be applied to nonlocal approaches, but the price of a higher computational cost has to be paid also. Nevertheless, this is one of the authors purposes for the next time in order to improve his flow calculations further.

Acknowledgements. The author would like to thank G. SOMMER and S. VUKOVAC.

References

[1] Alvarez, L., Weickert, J., Sánchez, J.: Reliable Estimation of Dense Optical Flow Fields with Large Displacements. International Journal of Computer Vision **1** (2000) 205–206

[2] Barron, J.L., Fleet, D.J., Beauchemin, S.S.: Performance of Optical Flow Techniques. Int. Journal of Computer Vision **12** (1994) 43–77

[3] Bernard, C.: Fast Optic Flow with Wavelets. Proceedings Wavelets and Applications Workshop (WAW '98) (1998)

[4] Black, M.J., Anandan, P.: The Robust Estimation of Multiple Motions: Parametric and Piecewise-Smooth Flow Fields. Computer Vision and Image Understanding **63**(1) (1996) 75–104

[5] Charbonnier, P., Blanc-Féraud, L., Aubert, G., Barlaud, M.: Deterministic Edge-Preserving Regularization in Computing Imaging. IEEE Transactions on Image Processing **6**(2) (1997) 298–311

[6] Daubechies, I.: Ten Lectures on Wavelets. CBMS-NSF Regional Conference Series in Applied Mathematics **61**. SIAM Publishing, Philadelphia (1992)

[7] Fleet, D.J., Langley, K.: Recursive Filters for Optical Flow. IEEE Trans. Pattern Analysis and Machine Intelligence **17**(1) (1995) 61–67

[8] Haußecker, H., Jähne, B.: A Tensor Approach for Local Structure Analysis in Multi-Dimensional Images. In: Girod, B., Niemann, H., Seidel, H.-P. (eds.): Proceedings 3D Image Analysis and Synthesis '96. Proceedings in Artifical Intelligence **3** (1996) 171–178

[9] Horn, B., Schunck, B.: Determining Optical Flow. Artifical Intelligence **17** (1981) 185–203

[10] Latto, A., Resnikoff, H.L., Tenenbaum, E.: The Evaluation of Connection Coefficients of Compactly Supported Wavelets. In: Maday, Y. (ed.): Proceedings of the French-USA Workshop on Wavelets and Turbulence. Springer Verlag, New York (1996)

[11] Maaß, P.: Wavelet-projection Methods for Inverse Problems. Beiträge zur angewandten Analysis und Informatik. Shaker Verlag, Aachen (1994) 213–224

[12] Mitchie, A., Bouthemy, P.: Computation and Analysis of Image Motion: A Synopsis of Current Problems and Methods. International Journal of Computer Vision **19**(1) (1996) 29–55

[13] Nagel, H.-H., Enkelmann, W.: An Investigation of Smoothness Constraints for the Estimation of Displacement Vector Fields from Image Sequences. IEEE Transactions on Pattern Analysis and Machine Intelligence **8** (1986) 565–593

[14] Neckels, K.: Wavelet Filter Design via Linear Independent Basic Filters. In: Sommer, G., Zeevi, Y.Y. (eds.): Algebraic Frames for the Perception Action Cycle 2000. Springer Verlag, Berlin (2000) 251–258

[15] Neckels, K.: Wavelet Connection Coefficients in Higher Dimensions. Paper draft, Cognitive Systems, CAU Kiel (2000)

[16] Nordström, K.N.: Biased Anisotropic Diffusion — A Unified Regularization and Diffusion Approach to Edge Detection. Technical Report. University of California, Berkeley (1989) 89–514

[17] Perona, P., Malik, J.: Scale Space and Edge Detection Using Anisotropic Diffusion. IEEE Trans. Pattern Analysis and Machine Intelligence **12** (1990) 629–639

[18] Resnikoff, H.L., Wells, R.O., Jr.: Wavelet Analysis — The Scalable Structure of Information. Springer Verlag, Berlin — Heidelberg — New York (1998)

[19] Schar, H., Körkel, S., Jähne, B.: Numerische Isotropieoptimierung von FIR-Filtern mittels Querglättung. In: Wahl, F., Paulus, E.(eds.): Proceedings Mustererkennung 1997. Springer Verlag, Berlin (1997)

[20] Weber, J., Malik, J.: Robust Computation of Optical Flow in a Multi-Scale Differential Framework. International Journal of Computer Vision **2** (1994) 5–19

[21] Weickert, J.: Anisotropic Diffusion in Image Processing. B.G. Teubner, Stuttgart (1998)

[22] Weickert, J., ter Haar Romeny, B.M., Viergever, M.A.: Efficient and Reliable Schemes for Nonlinear Diffusion Filtering. IEEE Transactions on Image Processing **7**(3) (1998) 398–410

[23] Weickert, J., Schnörr, C.: Räumlich-zeitliche Berechnung des optischen Flusses mit nichtlinearen flußabhängigen Glattheitstermen. In: Förstner, W., Buhmann, J.M., Faber, A., Faber, P. (eds.): Mustererkennung 1999. Springer Verlag, Berlin (1999) 317–324

A Method to Analyse the Motion of Solid Particle in Oscillatory Stream of a Viscous Liquid

Witold Suchecki[1] and Krzysztof Urbaniec[2]

Warsaw University of Technology, Plock Branch, Department of Process Equipment,
09-402 Plock, Jachowicza 2/4, Poland,
tel.:+48 24 262 2610; fax: +48 24 262 6542
[1] suchecki@pl.onet.pl; [2] gstku@coi.pw.edu.pl

Abstract. A mathematical model of the motion of a solid particle in oscillatory stream of a viscous liquid was set up and analytically solved for Reynolds number in the relative motion Re < 2.0 (Stokes flow) and Reynolds number characterising the liquid stream Re* < 2100 (laminar flow). A computer aided method based on video image processing was applied as an experimental tool to analyse the particle motion and verify the mathematical model.

Keywords: motion analysis, viscous liquid

1 Introduction

The motion of solid particles in a flowing liquid is of key importance to such processes as hydraulic transport, crystallisation, fluidisation etc. To control the velocity of solid phase relative to liquid phase, an oscillatory component is sometimes added to the two-phase flow [2].

The present work is devoted to studying the motion of a solid particle in oscillatory stream of a viscous liquid in the Stokes range, that is, relative motion characterised by Reynolds number Re < 2.0. It is also assumed that the fluid flows in a vertical channel under laminar conditions corresponding to Re* < 2100. In this paper, the case just described is called two-phase laminar flow.

The work is aimed at setting up, solving and experimentally verifying the mathematical model of the motion of a solid particle. An experimental method making it possible to avoid any measurement-related disturbances in the flow field is applied. Following a successful experimental verification, the possibility is studied of using the model to determine advantageous oscillatory-flow characteristics that could increase the particle velocity relative to the liquid.

2 Theoretical Model

The system under consideration is illustrated in Fig. 1. A viscous liquid flows in vertical test tube in upward direction. A solid particle is placed in the center

W. Skarbek (Ed.): CAIP 2001, LNCS 2124, pp. 357–365, 2001.

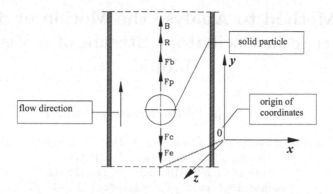

Fig. 1. Basic assumptions and forces acting on a solid particle in flowing liquid

of the section of the test tube where the laminar flow is fully developed. The flow is subject to periodic changes resulting from sinusoidal oscillations and consequently, the particle velocity also changes periodically. As a consequence of the laminar flow velocity profile, if the particle position does not coincide with the tube center, then the gradient of flow velocity causes the particle to rotate, and the Magnus effect brings the particle back to the center. The time-averaged flow velocity in the tube center is equal to the settling velocity of the solid particle.

The momentaneous flow velocity in the tube center is twice as large as the flow velocity averaged over the tube cross-section. It can be considered as a sum of constant component (balancing the settling velocity of the particle) and sinusoidally changing component (resulting from flow oscillation), and can be expressed as:

$$u(t) = u_0 + 4a_0\pi f \sin 2\pi ft \left(1 + \frac{a_0}{l} \cos 2\pi f\right) \tag{1}$$

where: u_0 – settling velocity of the particle, a_0 – piston oscillation amplitude, f – piston oscillation frequency, t – time, l – connecting-rod length.

The following assumptions were adopted for the theoretical model: the fluid flow and the particle motion can be considered in one dimension along the y axis, the influence of particle rotation on the drag coefficient can be neglected [1, 7], flow velocity is always considered outside the boundary layer surrounding the particle, the origin of the coordinate system ($y = 0$) coincides with the average position of the solid particle.

The following forces are taken into account: inertial force on the particle Fc, drag force R, force resulting from the pressure gradient in flowing liquid Fp, inertial force on the liquid layer surrounding the particle Fb, potential force equal to the difference between gravity and buoyancy forces Fe, force resulting from non-steady motion of the particle in the viscous fluid B (Basset force). In Fig. 1, forces acting on the particle are shown under the assumption that the fluid is accelerated upwards and the particle velocity relative to the fluid is directed downwards.

The mathematical model is set up on the basis of d'Alembert principle and includes the balance of forces acting on the solid particle [5, 6]. The resulting equation, which can be regarded as a special case of the Navier–Stokes equation, is

$$\frac{\pi d^3}{6} \rho_p \frac{du_p}{dt} = 3\pi\mu V d - \frac{\pi d^3}{6} \rho \frac{du}{dt} + \frac{\pi d^3}{12} \frac{dV}{dt} \rho + \frac{\pi d^3}{6}(\rho_p - \rho)g$$

$$+ \frac{3d^2}{2}\sqrt{\pi\rho\mu} \int_{t_0}^{t} \frac{\frac{dV}{d\tau}}{\sqrt{t-\tau}} \, d\tau. \tag{2}$$

The initial conditions are:

$$u_p(0) = 0 \quad \text{and} \quad u(0) = u_0, \quad V(0) = u_p(0) - u(0) = -u_0 \tag{3}$$

where: t – time, g – acceleration of gravity, d – particle diameter, ρ_p – particle density, u_p – particle velocity, u – fluid flow velocity, u_0 – settling velocity of the particle, ρ – fluid density, μ – fluid viscosity, $V = u_p - u$ – velocity of particle relative to fluid.

From the mathematical viewpoint, this is the differential-integral equation of the Volterra type with weakly singular kernel. It can be analytically solved [5] to yield a function which reflects the relationship between the relative velocity of the particle and the amplitude and frequency of fluid flow oscillation.

$$V(t) = c_1 M_1(t) + c_2 M_2(t) +$$
$$+ A_1 \cos 2\pi ft + A_2 \sin 2\pi ft + B_1 \cos 4\pi ft + B_2 \sin 4\pi ft + A_3 +$$
$$+ \cos(\text{Im}\,\lambda_1 t)\left(-\frac{c_0}{\text{Im}\,\lambda_1}\right) e^{\text{Re}\,\lambda_1 t}\left\{2\sqrt{t}\sin(\text{Im}\,\lambda_1 t) + \right.$$
$$\left. - 2\int_0^t \sqrt{\tau}\, e^{\text{Re}\,\lambda_1(t-\tau)}[-\text{Re}\lambda_1 \sin(\text{Im}\,\lambda_1\tau) + \text{Im}\lambda_1 \cos(\text{Im}\,\lambda_1\tau)]\,d\tau\right\} +$$
$$+ \sin(\text{Im}\,\lambda_1 t)\frac{c_0}{\text{Im}\,\lambda_1}\left\{2\sqrt{t}\cos(\text{Im}\,\lambda_1 t) + \right.$$
$$\left. + 2\int_0^t \sqrt{\tau}\, e^{\text{Re}\,\lambda_1(t-\tau)}[\text{Re}\lambda_1 \cos(\text{Im}\,\lambda_1\tau) + \text{Im}\lambda_1 \sin(\text{Im}\,\lambda_1\tau)]\,d\tau\right\} +$$
$$+ k_1 k_9 \int_0^t \sqrt{y}\, e^{\text{Re}\,\lambda_1(t-y)}\,\text{Re}\lambda_1 \sin[\text{Im}\,\lambda_1(t-y)]\,dy +$$
$$- k_2 k_9 \int_0^t \sqrt{y}\, e^{\text{Re}\,\lambda_1(t-y)}\cos[\text{Im}\,\lambda_1(t-y)]\,dy +$$
$$- k_3 k_9 \int_0^t \sqrt{y}\,\text{Re}\,\lambda_1 \sin[2\pi f(t-y)]\,dy + k_4 k_9 \int_0^t \sqrt{y}\cos[2\pi f(t-y)]\,dy +$$
$$+ k_5 k_{10} \int_0^t \sqrt{y}\, e^{\text{Re}\,\lambda_1(t-y)}\,\text{Re}\lambda_1 \sin[\text{Im}\,\lambda_1(t-y)]\,dy +$$
$$- k_6 k_{10} \int_0^t \sqrt{y}\, e^{\text{Re}\,\lambda_1(t-y)}\cos[\text{Im}\,\lambda_1(t-y)]\,dy +$$
$$- k_7 k_{10} \int_0^t \sqrt{y}\,\text{Re}\,\lambda_1 \sin[4\pi f(t-y)]\,dy + k_8 k_{10} \int_0^t \sqrt{y}\cos[4\pi f(t-y)]\,dy \tag{4}$$

The procedure of solution finding consists of the following steps: 1) Obtaining the non-homogeneous, linear differential equation of the second order with constant coeficients, 2) Solving the homogeneous equation, 3) Solving the non-homogeneous equation (by of revaluation of constants), 4) Simplifying the double integral to single integral (by changing the order of integration and taking advantage of special properties of the integrand).

The equation (4) is the general solution of the mathematical model of the motion of a solid particle in oscillatory laminar flow of a liquid. (The constants appearing in the above expression, $k_1 \div k_{10}$, c_1, c_2, c_0, A_1, A_2, A_3, B_1, B_2, $M_1(t)$, $M_2(t)$, λ_1, are explained elsewhere [5].) The values of single integrals in equation (4) can be numerically determined using general purpose mathematical software like MathCAD, and integration constants c_1 and c_2 can be found from the initial conditions (3).

Using the MathCAD package, the described mathematical model was implemented [5] to determine particle velocity in the oscillatory flow of a liquid – both absolute and relative to the liquid. It can be proved that the mathematical model is correctly formulated as it has a unique solution which is stable, that is, continuously dependent on boundary conditions and initial conditions.

3   Experimental Investigations

3.1   Experimental Apparatus

In Fig. 2, the experimental rig is schematically shown. It includes test tube 1 in which oscillatory flow of a liquid can be obtained by imposing oscillatory component on steady flow. The liquid is pumped by centrifugal pump 7 from vessel 9 through supply pipe to toroidal channel, from which it flows via distribution holes to the test tube. The supply pipe is connected with the toroidal channel via two symmetrical nozzles placed below flow straightener 12 in the test tube. From the test tube, the liquid flows through overflow and outlet pipe to vessel 9. The centrifugal pump is driven by electric AC motor whose speed is controlled by frequency converter to balance the flow velocity of the liquid with the settling velocity of solid particle 2. Flow oscillations are obtained through reciprocating movement of a piston and piston rod that are driven by a connecting rod. The oscillation device is set in motion by rotation of a disc that is driven via reduction gear by electric AC motor 8 whose speed is controlled by frequency converter and measured by tachometer 11. The slot along disc radius makes it possible to set the oscillation amplitude, while the frequency can be set by adjusting motor speed.

For a convenient visualisation of piston movement, the piston and the piston rod are connected with a slider that can move parallel to the piston. Indicator 10 on the slider can be used for checking the momentaneous position of the piston and its stroke. In order to precisely determine the reference position of the piston, the rig includes infra-red transducer and a position indicator placed on the piston rod. When the reference position is detected, a light-emitting diode is turned on.

The motion of the solid particle is recorded using CCD camera 3 so positioned that it also can record the status of diode and the movements of upper

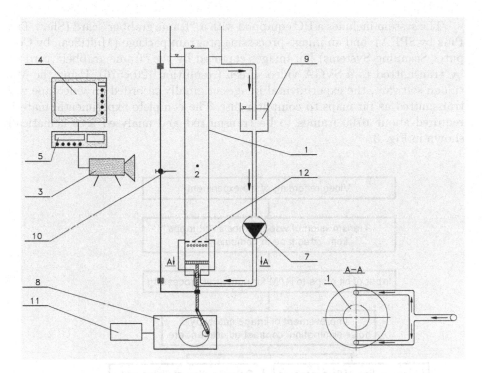

Fig. 2. Scheme of the experimental rig

indicator of piston position 10. A screen placed behind the test tube and properly selected light sources make it possible to maintain optimum conditions for high-quality video recording of the experiments using video recorder 5. The camera is also connected with video screen 4 for convenient continuous monitoring of experiments.

The rig makes it possible to determine the liquid flow by volumetric method, as the liquid can be directed to a measuring vessel.

The solid particle used in the experiments was an agalit sphere 2.43 mm in diameter, and the fluid used was a mixture of water and glycerol. The parameters of flow oscillations were varied by setting piston stroke to 6, 10 or 15 mm and varying the frequency between 0.5 s^{-1} and 3.7 s^{-1}. 48 series of experiments were carried out.

3.2 Method of Measurement

The experimental method is based on video recording of piston indicator and particle images. When operating the experimental rig, it is only required that stable flow conditions are maintained and video recording on tape is not disturbed. Instead of making direct measurements on the rig, the video tape is subsequently searched for images to be processed using an image processing system.

The system includes a PC equipped with a "frame grabber" card (Show Time Plus by SPEA), and an image-processing program package (MultiScan by Computer Scanning Systems) [3]. Images captured by the "frame grabber" card can by transmitted to a SVGA video screen (resolution 720×576). Using the MultiScan software, the experimental images originally recorded on videotape were transmitted as bit maps to computer disc. The complete experimental material required about 6700 frames to be transmitted and analysed as schematically shown in Fig. 3.

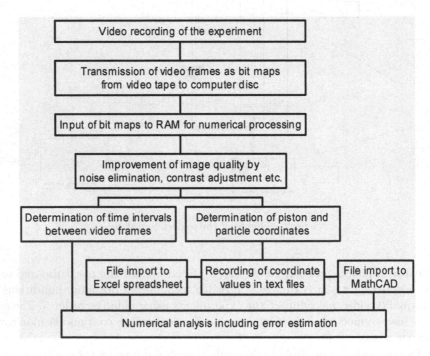

Fig. 3. Flow diagram of computer-aided processing of experimental data

To determine particle and piston coordinates, a macrocommand was programmed for the MultiScan software. By numerical processing of coordinate values taken from a sequence of video frames, momentaneous values of particle and liquid-flow velocities, as well as those of particle velocity relative to the liquid were determined.

In Fig. 4, theoretical and experimental values of velocities are shown for the experiment in which piston amplitude was 10 mm and frequency was 0.714 s^{-1}. The symbols denote: U – experimental values of flow velocity; U_t – theoretical values of flow velocity; Up – experimental values of particle velocity; Up_t – theoretical values of particle velocity; V – particle velocity relative to the liquid based on experiment; V_t – particle velocity relative to the liquid based on theory.

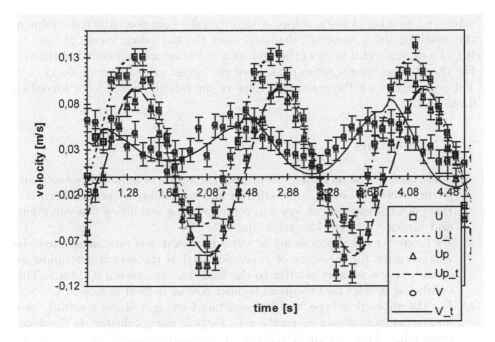

Fig. 4. Comparison of theoretical and experimental values of particle velocity, flow velocity and particle velocity relative to the liquid; experimental values are shown together with estimated error margins. Piston amplitude 10 mm, frequency 0.714 s^{-1}

4   Evaluation of the Practicability of Theoretical Model

Assuming that the experimental flow conditions are compatible with the assumptions of the theoretical model, one can compare the experimental results with theoretical predictions and evaluate the suitability of the model for determining momentaneous values of particle velocity, fluid flow velocity and relative particle velocity. All the experimental series were carried out at Reynolds number values characterizing fluid flow in the test tube well within the laminar range (Re* ≤ 787). However, Reynolds number characterizing the motion of the solid particle relative to the fluid was found to be contained in the Stokes range (Re ≤ 2) only in 13 out of a total of 48 experimental series. These 13 series were taken as a basis for comparison between theory and experiments.

The error of verification of a theoretical model is always greater than the error of experimental determination, that is, the accuracy of the experimental method sets a limit for the accuracy of model verification [4]. As a measure of model inaccuracy, the average absolute error of a series of experiments was adopted

$$\frac{1}{n} \sum_{i=1}^{n} |V(t_i) - W(t_i)| \tag{5}$$

where: n – number of momentaneous velocity values considered, $V(t_i)$ – value of the relative particle velocity calculated using the theoretical model at time t_i, $W(t_i)$ – experimental value of the relative particle velocity measured at time t_i. For the 13 experimental series considered the largest error value was 0.011 m/s. This compares with the maximum value of the relative velocity not exceeding 0.085 m/s.

5   Conclusions

A. The experimental method to record instant values of fluid flow velocity and particle velocity proved to be efficient and reasonably accurate. When carrying out the experiment, one can concentrate on stabilizing flow conditions and ensuring correct video recording.
B. By computer aided processing of video images it was established that the average error (in the sense of expression (5)) of theoretical determination of the particle velocity relative to the fluid does not exceed 0.011m/s. This conclusion is valid for two-phase laminar flow as defined in section 1.
C. For the above flow type, oscillation-induced changes of the resulting force and its components acting on the solid particle were evaluated. In the maximum value of the resulting force, the largest contribution (about 90%) comes from combined drag and potential forces, while the inertial forces, pressure-gradient induced force and Basset force add up to less than 10% .
D. The maximum velocity of the particle relative to the fluid increases with increasing amplitude and frequency of oscillation.

Acknowledgement. The research reported here was partly funded by the State Committee for Scientific Research, Warsaw, Poland.

References

1. Clift, R., Grace, J.R., Weber, M.E.: Bubbles, Drops, and Particles. Academic Press, New York, San Francisco, London (1978)
2. Levesley, J.A., Bellhouse, B.J.: The Retention and Suspension of Particles in a Fluid Using Oscillatory Flow. Trans IChemE **75** Part A (1997) 288–297
3. Handbook of MultiScan (in Polish). Computer Scanning Systems. Warszawa (1995)
4. Squires, G.L.: Practical Physics. Cambridge University Press, Cambridge (1986)
5. Suchecki, W.: Analysis of the motion of a solid particle in oscillatory flow of a viscous liquid (in Polish). PhD Dissertation. Plock (1997)
6. Tchen, C.M.: Mean value and correlation problems connected with the motion of small particles suspended in a turbulent fluid. Dissertation, Hague (1947)
7. White, B.R., Schulz J.C.: Magnus effect in saltation. Journal of Fluid Mechanics **81** (1977) 497–512

An Optimization Approach for Translational Motion Estimation in Log-Polar Domain*

V. Javier Traver and Filiberto Pla

Dept. Llenguatges i Sistemes Informàtics · Universitat Jaume I
Edifici TI · Campus Riu Sec · E12071 Castelló (Spain)
{vtraver|pla}@uji.es

Abstract. Log-polar imaging is an important topic in space-variant active vision, and facilitates some visual tasks. Translation estimation, though essential for active tracking, is more difficult in (log-)polar coordinates. We propose here a novel, conceptually simple, effective, and efficient method for translational motion estimation. It is based on a gradient-based minimization procedure. Experimental results with log-polar images using a software-based log-polar remmapper are presented.

Keywords: motion estimation, log-polar domain

1   Introduction

Recently, an increasing attention has being paid to the concept of *active* vision, and to the closely related topic of space-variant imaging. Space-variant images have an area of high-visual acuity at its center, and a decreasing resolution towards the periphery. This results in a trade-off between a wide field of view, a high resolution, and a small fast-to-process output. The log-polar geometry is by far the most often used space-variant model [4] because of its interesting properties in fields such as pattern recognition and active vision. In the latter case, it has been proven its usefulness for time-to-impact computation [9], for active docking in mobile robots [2], and for vergence control [3,5] to name but a few.

Despite its obvious advantages in some problems, log-polar space also complicates some other visual tasks. Estimating image-plane translation, for instance, becomes more difficult in log-polar domain than in cartesian coordinates. Many researchers use stereo configurations (e.g., [3], [5]), and a few of them address the problem of motion estimation using a single camera for tracking in the log-polar domain [8,1]. These approaches are computationally expensive or conceptually difficult (Okajima *et al.* use complex wavelets for motion estimation [8]), or have some limitations (Arhns and Neumann control only the pan angle of a pan-tilt

---

* This work has been partially supported by projects GV97-TI-05-27 from the *Conselleria d'Educació, Cultura i Ciència, Generalitat Valenciana*, and CICYT TIC98-0677-C02-01 from the Spanish *Ministerio de Educación y Cultura*.

W. Skarbek (Ed.): CAIP 2001, LNCS 2124, pp. 365–373, 2001.

head [1]). We propose a simple and effective method for translation estimation, which is based on a gradient-based minimization procedure. In the following sections we show the method and the experimental results obtained. The technique is intended to be used for active object pursuit with monocular log-polar images.

2   Optimization-Based Motion Estimation

The log-polar transform. The log-polar transform we use here [7] defines the log-polar coordinates as $(\xi, \eta) = \left(\log_a \left(\frac{\rho + \rho_0}{\rho_0}\right), \theta\right)$, where (ρ, θ) are the polar coordinates, defined as usual, and a and ρ_0 being parameters, which are found from the selected log-polar image resolution $(R \times S)$, i.e., the number of rings (R) and sectors (S). The log-polar transformation of a cartesian image I will be denoted by $\mathcal{L}(I)$ or $\mathcal{I}$. An example of log-polar images can be seen in figure 1.

The role of the correlation measure. The correlation measure between two stereo log-polar images for varying values of the vergence angle has a profile with interesting properties, which are not present when correlation is computed between cartesian images [5,3]. On the one hand, a deep global minimum occurs at the correct vergence configuration. On the other hand —and more importantly— the correlation profile over the vergence angle range is unaffected by false local minima in the case of log-polar mapping while it behaves poorer in uniformly sampled cartesian images.

In the context of monocular fixation, Ahrns and Neumann present a similar idea [1] using log-polar images too. In this case, a gradient-descent control is proposed for controlling the pan degree of freedom of a camera mount. However, they do not show why a gradient-based search is appropriate. Moreover, at some point in their algorithm, they use the original cartesian images rather than the log-polar remapped ones. This has, at least, two inconveniences. On the one hand, it is not a biologically plausible option[1]. On the other hand, doing this may involve an additional computational cost which is opposed to the interesting data reduction advantage attributed to discrete log-polar images.

In contrast, our contributions are as follows:

1. We first show that the idea used for vergence control can also be exploited in monocular tracking in log-polar space. The choice of a steepest-descent control is therefore justified by the shape of the resulting correlation surfaces.
2. We stick to the log-polar images as if they were directly grabbed from a hardware log-polar sensor. Therefore, after the log-polar mapping, no use at all is made of the cartesian images.
3. We propose an algorithm for translational motion estimation which lends itself for controlling both the pan and tilt angles of a camera mount.

---

[1] One may simulate by software the output of a retina-like sensor, but it is probably less acceptable to make use of the original cartesian images for purposes other than the log-polar mapping.

(a) A (b) B (c) $I_t = A_B(x_0, x_0)$ (d) $\mathcal{I}_t$ (e) $\mathcal{I}_{t+1}$

Fig. 1. A foveated target has undergone a retinal shift $(x_0, y_0) = (5, 5)$

In [5,3] the correlation index C for the vergence angle ψ in a certain range, gives rise to a 1-D function, $C(\psi)$. Our approach extends this idea to 2-D function (the correlation surface), which is dependent on the translational components in x and y direction, x_0 and y_0, respectively, $C(x_0, y_0)$. We can compute the value of a correlation measure C between one image and versions of this one shifted by (x_0, y_0). Figure 2 shows an example of the resulting surface $C(x_0, y_0)$ computed in cartesian (figure 2(a)) and log-polar (figure 2(b)) domains for the same images. As can be seen in the figure, the surfaces shape is quite smooth. Both surfaces also exhibit a distinguishable minimum. In the case of log-polar domain, the location of such a minimum is very close to the actual translational motion undergone in image plane ($x_0 = 5, y_0 = 5$). In the case of cartesian images, however, this minimum is near $(0,0)$, which are clearly wrong motion parameters. These surfaces have been obtained from the images shown in figure 1, where a small image patch (figure 1(b)) is pasted on the image in figure 1(a) at its center (figures 1(c) and 1(d)) and in a shifted position (figure 1(e)). This simulates the projection onto the image plane of a target motion.

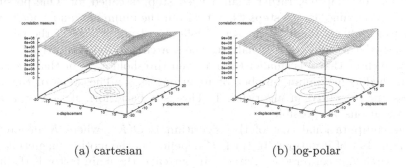

(a) cartesian   (b) log-polar

Fig. 2. Example of correlation surface in cartesian and log-polar spaces

As we have shown, the minimum of the correlation surface computed on log-polar images occurs at the correct displacement of the target. Additionally, in log-polar domain, when compared to cartesian case, the surface is somewhat smoother and free of local minima and inflection points (which are present in the cartesian case, see figure 2(a)). This shows one of the most interesting properties of the log-polar geometry, which makes it advantageous over cartesian images for tracking purposes. The logarithmically sampled radial coordinate offers a built-in *focus-of-attention* mechanism, by means of which pixels at the center of the image are emphasized over outer pixels. As a result, information far from the fovea are much less distracting in log-polar case than in uniformly sampled images. This is so because, in uniformly resolved images, every pixel contributes the same to the correlation function, regardless of their position. In log-polar coordinates, however, Bernardino and Santos-Victor showed [3] that each pixel contribution is weighted by an amount inversely proportional to the squared distance of the pixel to the image centre. Thus, in the example shown in figure 1, the static background influences little over the shifted, but foveated, object.

The algorithm. To estimate the translation parameters (x_0, y_0), our approach consists of finding the location of the minimum of the correlation measure $C(x'_0, y'_0)$. The shape of the correlation surfaces seen in section 2 suggests that a gradient-based search could be an adequate search algorithm. The estimation at iteration k, (x_0^k, y_0^k), is updated from the estimation at the previous iteration, (x_0^{k-1}, y_0^{k-1}), by using the gradient, ∇C, of the correlation measure, as the most promising direction to move, i.e.:

$$(x_0^k, y_0^k) = (x_0^{k-1}, y_0^{k-1}) - g(\nabla C) \tag{1}$$

A common definition for $g(\cdot)$ is $g(\nabla C) = \delta \cdot \frac{\nabla C}{|\nabla C|}$, i.e., consider the unit vector in the direction of the gradient and move a certain amount δ in that direction. The question of choosing the value for δ usually involves a trade-off. Then, an adaptive, rather than a fixed step, is called for. One possibility consists of moving "large" steps when far from the minimum, and "small" steps closer to it. This is the idea we use: initially $\delta = 1$ and whenever the minimum is surpassed the value of δ is halved. The search may be stopped using some criteria such as that δ is smaller than a given threshold δ_{min}, or that the search has reached a maximum number of iterations k_{max}. The steps of the whole process are presented in algorithm 1. This is a basic formulation over which some variations are possible.

The computational cost of the algorithm is $\mathcal{O}(K)$, where K denotes the actual number of iterations. In turn, K depends on the motion magnitude, and on the particular convergence criteria. In our case, the main factor is the image displacement. Therefore, as we expect only small translations of the target onto the image plane, the efficiency of the algorithm is guaranteed. Regarding the initial guess, (x_0^0, y_0^0), in the absence of any other information, it is sensible to use $(0,0)$. However, during an active tracking, it may be expected the target to move following some dynamics. Therefore, if the target motion can be predicted, we can use the expected future target position as the initial guess for the algorithm. This would help speed up the motion estimation process.

Algorithmus 1 Gradient-descent-based translation estimation of log-polar images

---

Input: Two log-polar images $\mathcal{I}$ and $\mathcal{I}'$
Output: The estimated translation vector (x_0, y_0)
1: $(x_0^0, y_0^0) \leftarrow (0, 0)$ { *initial guess* }
2: $k \leftarrow 0$ { *iteration number* }
3: $\delta \leftarrow 1$
4: **while** $(\delta > \delta_{\min}) \wedge (k < k_{\max})$ **do**
5:    $k \leftarrow k + 1$
6:    $(x_0^k, y_0^k) \leftarrow (x_0^{k-1}, y_0^{k-1}) - g(\nabla \mathcal{C})$ { *estimation update rule* }
7:    **if** minimum surpassed **then**
8:       $\delta \leftarrow \delta/2$
9:    **end if**
10: **end while**
11: $(x_0, y_0) \leftarrow (x_0^k, y_0^k)$

Computing the correlation gradient. To evaluate the equation 1 we need a way to compute the gradient. In the case of the well-known SSD (sum of squared differences) correlation measure [6], the gradient $\nabla \mathcal{C} = (\mathcal{C}_{x_0}, \mathcal{C}_{y_0})$ becomes:

$$
\begin{cases}
\mathcal{C}_{x_0} = \dfrac{\partial \mathcal{C}}{\partial x_0} = 2 \displaystyle\sum_{(\xi, \eta) \in \mathcal{D}} \left\{ (I'(\xi', \eta') - I(\xi, \eta)) \cdot I'_{x_0}(\xi', \eta') \right\} \\[4mm]
\mathcal{C}_{y_0} = \dfrac{\partial \mathcal{C}}{\partial y_0} = 2 \displaystyle\sum_{(\xi, \eta) \in \mathcal{D}} \left\{ (I'(\xi', \eta') - I(\xi, \eta)) \cdot I'_{y_0}(\xi', \eta') \right\}
\end{cases}
\tag{2}
$$

with $\mathcal{D}$ being a certain set of image pixels (usually, the entire image), and where $I'_{x_0} = I'_{x_0}(\zeta', \eta')$ and $I'_{y_0} = I'_{y_0}(\xi', \eta')$ are:

$$
\begin{cases}
I'_{x_0} = I'_{\xi'} \cdot \xi'_{x_0} + I'_{\eta'} \cdot \eta'_{x_0} \\
I'_{y_0} = I'_{\xi'} \cdot \xi'_{y_0} + I'_{\eta'} \cdot \eta'_{y_0}
\end{cases}
\tag{3}
$$

with $\xi'_{x_0} = \xi_x$, $\xi'_{y_0} = \xi_y$, $\eta'_{x_0} = \eta_x$, and $\eta'_{y_0} = \eta_y$. The common notation for partial derivatives, $f_z = \partial f/\partial z$, is used. On the other hand, $(\xi', \eta') = (\xi + \xi_0, \eta + \eta_0)$, where the increments in log-polar coordinates, ξ_0 and η_0, due to a translational cartesian displacement (x_0, y_0) depend on the position in the log-polar space, and can be approximated as:

$$
\begin{cases}
\xi_0 \approx \xi_x \cdot x_0 + \xi_y \cdot y_0 \\
\eta_0 \approx \eta_x \cdot x_0 + \eta_y \cdot y_0
\end{cases}
\tag{4}
$$

By taking the partial derivatives of ξ and η as defined above, with respect to x and y, we get ξ_x, ξ_y, η_x, and η_y, as follows:

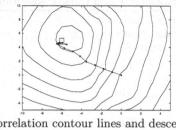

iso-correlation contour lines and descent path

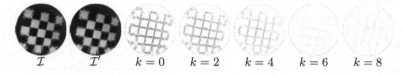

$\mathcal{I}$ $\qquad$ $\mathcal{I}'$ $\qquad$ $k=0$ $\qquad$ $k=2$ $\qquad$ $k=4$ $\qquad$ $k=6$ $\qquad$ $k=8$

Fig. 3. The input images, $\mathcal{I}$ and $\mathcal{I}'$, to the algorithm $(I'(x,y) = I(x-x_0, y-y_0))$; the iso-correlation contour lines, and the descent followed by the algorithm (the square represents the true motion parameters, $(x_0 = -6, y_0 = 5)$); and the absolute difference $|\mathcal{I}_k - \mathcal{I}'|$ at selected iterations k.

$$
\begin{cases}
\xi_x = \dfrac{\partial \xi}{\partial x} = \dfrac{\cos\theta}{(\rho+\rho_0)\ln a} \\[3mm]
\xi_y = \dfrac{\partial \xi}{\partial y} = \dfrac{\sin\theta}{(\rho+\rho_0)\ln a}
\end{cases}
\qquad
\begin{cases}
\eta_x = \dfrac{\partial \eta}{\partial x} = -\dfrac{\sin\theta}{\rho} \\[3mm]
\eta_y = \dfrac{\partial \eta}{\partial y} = \dfrac{\cos\theta}{\rho}
\end{cases}
\tag{5}
$$

3 Experimental Results

Qualitative analysis. Figure 3 illustrates graphically the workings of the technique. Let $\mathcal{I}_k = \mathcal{L}(I_k)$ be the image at iteration k of the process (see algorithm 1), i.e., $I_k(x,y) = I(x-x_0^k, y-y_0^k)$. As the descent proceeds, $\mathcal{I}_k$ is more and more similar to $\mathcal{I}'$. This can be appreciated by observing that the difference image $|\mathcal{I}_k - \mathcal{I}'|$ becomes whiter and whiter (the whiter a pixel is, the less the gray level difference between the two images). Notice that $\mathcal{I}_k$ is computed here for illustrating purposes, but it is not computed during the actual process.

Quantitative analysis. For each image I_i in our image test set, a set of displacements $\{(x_0, y_0)\}$ were applied. Each of these translations correspond to a translation magnitude m_j (cartesian pixels) in a set M_t and to a translation orientation o_k (direction) in a set O_t. Let n_m and n_o be the size of sets M_t and O_t, respectively. The elements $m_j \in M_t$ are selected to range from small to large motions. Then, the elements $o_k \in O_t$, $k \in \{0, 1, \dots, n_o - 1\}$ are equally

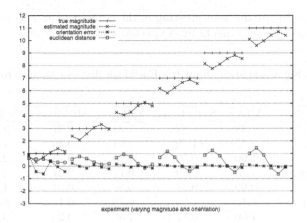

Fig. 4. True and estimated motion parameters, and error measures between them

spaced and selected to cover the span of possible orientations from 0 to 2π radians, i.e., $o_k = 2\pi k/n_o$ rad. For each combination of magnitude and direction of motion, the translated version of I_i, $I_i^{j,k}$, is computed, and their log-polar transformations, $\mathcal{I}_i$ and $\mathcal{I}_i^{j,k}$, are input to the algorithm. The output of the algorithm (the estimated translation) is compared to the known motion parameters.

Figure 4 plots the estimation errors for image *Barche* (one of the test images). The tested translations are computed from the particular set $M_t = \{1, 3, 5, 7, 9, 11\}$ and $n_o = 6$. The horizontal axis of the plot represents each experiment with a given combination of motion magnitude and direction. In the vertical axis, different (ground-truth and estimated) parameters and error measures are represented. From this graphic, several observations can be made. First of all, the estimated motion magnitude differs little from the true motion magnitude. The difference is less than a pixel for the smaller displacements, and a little bigger for larger ones. Secondly, the orientation error becomes smaller with increasing magnitude. Big angular errors (in radians) occur only when translations are small. Lastly, the Euclidean distance between the true and estimated (x_0, y_0), reveals how small the overall motion estimation error is. The error is less than a pixel for small-medium displacements (up to 5 pixels), and about one pixel and a half for big displacements (more than 5 pixels).

Table 1 gives some statistical results of the Euclidean distance as an error measure. The average error ranges from less than one pixel to ≈ 2.5 pixels in an extreme case, depending on the image tested. It is important to notice that the mean errors in table 1 are somewhat biased by the big estimation errors resulting in the case of large motion displacements. Thus, the median is a better global measure. Maximum errors can be large; but these errors occur when translations are also large. In these cases, the global minimum in the correlation surfaces is far from the initial guess (x_0^0, y_0^0). The shape of the correlation surface far from the global minimum may make difficult for a gradient-descent search to succeed.

The overall results are quite interesting in the context of an active tracking task, where only small displacements are expected within the fovea.

Table 1. Some statistics about the Euclidean distance as an error measure

IMAGE	MEAN	STD. DEV.	MEDIAN	MINIMUM	MAXIMUM
Barche	1.33	0.62	1.24	0.27	2.70
Lena	2.46	1.62	2.07	0.30	7.11
Grid	1.00	0.52	0.95	0.25	2.49
Oranges	1.13	0.99	0.77	0.21	4.54
Lab	0.90	0.38	0.96	0.04	1.84
Rubic	1.10	1.35	0.79	0.14	6.50
Face	1.60	0.87	1.56	0.23	3.76
Corridor	1.07	0.42	1.00	0.43	2.33
Scream	0.98	0.69	0.85	0.13	3.32

4   Conclusions

We have shown that the correlation surface computed between two log-polar images (in which a foveated target has moved) over a range of displacements along the x and y axes, presents a minimum at the actual displacement undergone by that target. This interesting property does not hold in case of cartesian images, which demonstrates the superiority of log-polar domain over cartesian geometry for monocular object tracking.

We have developed a gradient-descent-based algorithm for searching the global minimum of the correlation function, as a means to estimate a translational motion. Experimental results of this approach demonstrate its feasibility. An analysis of the estimation errors reveals that the method works best with small-medium displacements, which is the case in foveated active tracking.

References

1. Ahrns, I., Neumann, H.:   Real-time monocular fixation control using the log-polar transformation and a confidence-based similarity measure. In: Jain, A.K., Venkatesh, S., Lovell, B.C. (eds.): Intl. Conf. on Pattern Recognition (ICPR). Brisbane, Australia (Aug. 1998) 310–315
2. Barnes, N., Sandini, G.: Direction control for an active docking behavior based on the rotational component of log-polar optic flow. In: Tsai, W.-H., Lee, H.-J. (eds): European Conf. on Computer Vision, vol. 2. Dublin, Ireland (Jun 2000) 167–181
3. Bernardino, A., Santos-Victor, J.: Visual behaviors for binocular tracking. Robotics and Autonomous Systems **25** (1998) 137–146
4. Bolduc, M., Levine, M.D.: A review of biologically motivated space-variant data reduction models for robotic vision. Computer Vision and Image Understanding (CVIU) **69**(2) (Feb 1998) 170–184

5. Capurro, C., Panerai, F., Sandini, G.: Dynamic vergence using log-polar images. Intl. Journal of Computer Vision **24**(1) (1997) 79–94
6. Brown, L.G.: A survey of image registration techniques. ACM Computing Surveys **24**(4) (Dec 1992) 325–376
7. Jurie, F.: A new log-polar mapping for space variant imaging. Application to face detection and tracking. Pattern Recognition **32** (1999) 865–875
8. Okajima, N., Nitta, H., Mitsuhashi, W.: Motion estimation and target tracking in the log-polar geometry. 17th Sensor Symposium. Kawasaki, Japan (May 2000) `chihara3.aist-nara.ac.jp/gakkai/sensor17`
9. Tistarelli, M., Sandini, G.: On the advantages of polar and log-polar mapping for direct estimation of time-to-impact from optical flow. IEEE Trans. on Pattern Analysis and Machine Intelligence (PAMI) **15** (1993) 401–410

Tracking People in Sport: Making Use of Partially Controlled Environment*

Janez Perš and Stanislav Kovačič

Faculty of Electrical Engineering, University of Ljubljana
Tržaška 25, SI-1000 Ljubljana, Slovenija
{janez.pers,stanislav.kovacic}@fe.uni-lj.si,
http://vision.fe.uni-lj.si

Abstract. Many different methods for tracking humans were proposed in the past several years, yet surprisingly only a few authors examined the accuracy of the proposed systems. As the accuracy analysis is impossible without the well-defined ground truth, some kind of at least partially controlled environment is needed. Analysis of an athlete motion in sport match is well suited for that purpose, and it coincides with the need of the sport research community for accurate and reliable results of motion acquisition. This paper presents a development of a two-camera people tracker, incorporating two complementary tracking algorithms. The developed system is suited for simultaneously tracking several people on a large area of a handball court, using a sequence of 384-by-288 pixel images from fixed cameras. We also examine the level of accuracy that this kind of computer vision system setup is capable of.

Keywords: motion tracking, sports events, partially controlled environment

1 Introduction

People tracking is a rapidly developing field of computer vision. The most interesting situations to deploy computer vision based people trackers usually represent highly uncontrollable environments (railway stations, crowded halls, etc.), which poses significant difficulty in evaluating the performance of developed systems. On the other hand, many sport researchers struggle to obtain reliable data about movement of athletes, especially when sport activity covers a large area, for example in team sports. Sport matches, especially indoor ones, represent *partially controlled environment*, and are as such highly suitable as a testing ground for development and testing of new people tracking methods.

The research in the fields of people tracking and analysis of sports-related video has flourished in the past several years [1,2,3,4,5,6,7]. However, the emphasis is still on development of tracking methods and improvement of reliability

---

* This work was supported by the Ministry of Science and Technology of the Republic of Slovenia (Research program 1538-517)

W. Skarbek (Ed.): CAIP 2001, LNCS 2124, pp. 374–382, 2001.

of the tracking itself. Only a few authors (for example [4]) examined the accuracy of their tracking systems or suggested both the method for evaluating the accuracy and obtaining the ground truth (for example [5]). On the other hand, use of computers in gathering and analyzing the sport data is an established practice in sport science [8,9]. One of important aspects of football, handball or basketball match analysis is the information about player movement [10], but due to limitations in available technology, the results obtained were often coarse and only approximate.

In this article, we present the method for tracking known number of people in a partially controlled environment - the handball court inside the sports hall. First, problems associated with image acquisition are discussed. Next, two algorithms for tracking athletes during the match are presented, and their combination which yields best results in terms of reliability and accuracy is presented. Next, the required post-processing of trajectories is briefly discussed. The collaboration with sports scientists enabled thorough evaluation of the accuracy of the developed system, which is described in a separate section. Finally, some conclusions about tracker performance are drawn.

2   Image Acquisition

Proper image acquisition significantly influences the performance of the tracking algorithms. In case of player motion acquisition and analysis, where certain measurements are performed and degree of uncertainty has to be specified, careful planning of image acquisition proves to be crucial for the success of the whole system [3]. As camera movement requires continuous calibration, two stationary cameras with wide-angle lenses were chosen in our case. Their placement on the ceiling of the sports hall and the resulting combined image is shown in Figure 1.

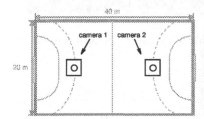 

Fig. 1. Handball playing court and camera placement (left). Example of combined image from two cameras, taken at the same instant of time (right).

The whole handball match that lasted for about an hour was recorded using two PAL cameras and two S-VHS videorecorders. A transfer to digital domain was carried out using the Motion-JPEG video acquisition hardware, at 25 frames per second and image resolution of 384x288 pixel.

3 Camera Calibration

To perform position *measurements* based on the acquired images, the relations between pixel coordinates in each of the images and world (court) coordinates have to be known. These relations are obtained by the camera calibration. The procedure is simplified due to rigid sport regulations, which precisely specify the locations and dimensions of various landmarks on the playing court. Unfortunately, due to the large radial distortion otherwise widely used calibration technique [11] fails to produce satisfactory results. We decided to build the model of radial image distortion, and couple it with simple linear camera model.

Fig. 2 illustrates the problem of radial distortion. For illustrative purposes only, let us imagine an ideal pinhole camera, mounted on a pan-tilt device. Point 0 is the point of intersection of optical axis of the camera with the court plane, when the pan-tilt device is in its vertical position. Point C denotes the location of the camera, and X is the observed point on the court plane, at distance R from the point 0. H is the distance from the camera to the court plane. Angle α is the angle of the pan-tilt device when observing the point X. The differential dR of radius R is projected to the differential dr, which is parallel to the camera image plane. The image of dr appears on the image plane. Relations between dR, dr and α are given within the triangle on the enlarged part of Fig. 2 (left).

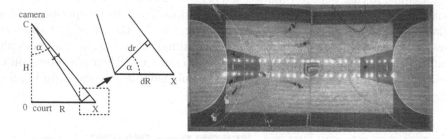

Fig. 2. A model of radial distortion (left). A combined image from both cameras after the radial distortion correction (right).

Thus, we can write the following relations:

$$dr = \cos(\alpha) \cdot dR, \quad \alpha = \operatorname{arctg}(\frac{R}{H}), \quad dr = \cos(\operatorname{arctg}(\frac{R}{H}))dR. \tag{1}$$

Let us substitute the pan-tilt camera with a fixed camera, equipped with wide-angle lens. The whole area, which is covered by changing the angle α of the pan-tilt camera, is captured simultaneously to the single image of the stationary camera. Additionally, let us assume that the scaling factor between the dr and the image of dr on the image plane equals 1. Therefore, we can obtain the length of the image of radius R on the image plane by integration:

$$\int\limits_{0}^{r1} dr = \int\limits_{0}^{R1} \cos(\arctan(\frac{R}{H}))dR, \tag{2}$$

R_1 being the distance from the observed point X to the point 0 and r_1 being the distance from the image of point X to the image of point 0 on the image plane. This yields the solution of the inverse problem $r_1 = r_1(R_1)$. By solving this equation for R_1 we obtain the formula $R_1 = R_1(r_1)$, which describes the radial distortion.

$$r_1 = H \cdot \ln \left(\frac{R_1}{H} + \sqrt{1 + \frac{R_1^2}{H^2}} \right), \quad R_1 = \frac{H}{2} \frac{(e^{-\frac{2r_1}{H}}) - 1}{e^{-\frac{r_1}{H}}}. \tag{3}$$

Parameters were obtained with the help of various marks, which are part of the standard court marking for handball matches (boundary lines, 6 and 9 m lines, etc.), and non-linear optimization. For illustrative purposes, a result of radial distortion correction is shown in Fig. 2 (right). Nevertheless, we decided to perform tracking on uncorrected images, and to correct the obtained player positions thereafter.

4   Player Tracking

Many general-purpose tracking algorithms could be used for the player tracking. However, in our setup, players are small objects, typically only 10-15 pixels in diameter. They cast shadows, causing trouble for simple background subtraction techniques. Any placement of markers is forbidden during the European Handball Federation (EHF) matches. However, players of different teams wear differently colored dresses.

4.1   Color-Based Tracking

Color identification and localization, based on color histograms, was reported by Swain and Ballard [12]. However, given a small number of pixels that comprise each of the players, this technique is not appropriate. In most cases, there are only a few (3-6) pixels that closely resemble the reference color of the player's dress. The situation is illustrated by Fig. 4b. Therefore, different approach was needed.

The algorithm searches for the pixel most similar to the recorded color of the player. The search is performed in a limited area (9-by-9 pixels) around the previous player position. The three-dimensional RGB color representation was chosen instead of HSI, as some players wear dark dresses, which would result in undefined values of hue. The similarity measure is defined as euclidean distance $S_{color}(x,y)$ between the image $I(x,y)$ and the predefined color of the player C in the RGB space.

The advantage of described algorithm is high reliability. The algorithm tracks players successfully even when the apparent player color is changed due to signal distortion during tape recording or lossy compression. The main problem is caused by diverse background with colored areas, which closely correspond to the color of the player's dress. The disadvantage of this method is also a high amount of jitter in the resulting player trajectories, which makes it inappropriate for a stand-alone use.

4.2   Template Tracking

Visual differences between the players and the background are exploited to further improve the tracking process.

Due to low resolution and rapidly changing appearance of the players it is extremely difficult to build an accurate model of a handball player. Instead, we have used a subset of modified Walsh functions and their complements, i.e. "templates", shown in Fig. 3, which extract the very basic appearances of the players.

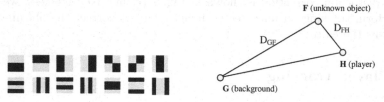

Fig. 3. Basic templates of the player - modified Walsh functions. Black areas represent zeros, while gray areas denote ones (left). Classification of an unknown object (right).

First, the region of interest (ROI) which surrounds the position of the player is defined. Considering the size of the players in captured image, the region size was set to 16x16 pixels, with player position in the center of the region. Each channel of the RGB color image is processed separately and the vector **F**, consisting of 14 features for each channel, is obtained using the following formula:

$$F_{i+14j} = \sum_{x=1}^{16} \sum_{y=1}^{16} K_i(x,y) \cdot I_j(x,y), \tag{4}$$

where K_i is one of the 14 template functions   ($i = 0 \ldots 13$), and I_j is one of the three RGB channels   ($j = 0, 1, 2$), obtained with respect to ROI from the current image. Each channel yields 14 features, which results in 42-dimensional feature vector **F**.

Let vector **H** represent the estimated appearance of the player, and vector **G** represent the appearance of the background (empty playing court) at the same coordinates. Our goal is to classify the unknown object in the region of interest I, represented as vector **F**, either as a "player" or a "background". The simplified, two-dimensional case is shown in Fig. 3, right.

Vector of features **G** is calculated from the image of the empty playing court at the same coordinates as **F**. The reference vector **H** is obtained by averaging the last n vectors of features for a successfully located player, which allows certain adaptivity, as the player appearance changes over time. The value of n depends on the velocity at which the players move, but best results were obtained with the value of $n = 50$ which corresponds to two seconds of video sequence.

Similarity measure S is obtained using the following formula:

$$S_{\text{template}} = \frac{D_{\text{FH}}}{D_{\text{GF}} + D_{\text{FH}}}, \qquad S \in [0, 1], \tag{5}$$

where D_{GF} and D_{FH} are Euclidean distances. Low value of S_{template} means high similarity between the observed object $\mathbf{F}$ from the region of interest I and stored appearance of the player in the vector $\mathbf{H}$. Fig. 4 (c) shows the result on a single player and demonstrates the ability of this technique to locate an object.

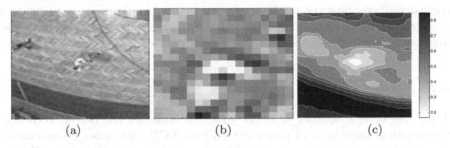

(a)   (b)   (c)

Fig. 4. Locating the player wearing a yellow dress. (a) The player is shown in the center of the image. (b) Distance S_{color} to the yellow color for a close-up view of a player from the image (a). (c) Similarity measure S_{template} of the whole image (a), as defined in (5). Feature vector $\mathbf{H}$ (player reference) was obtained from the subimage of the same player, captured at different instants of time. In both cases white pixels denote high similarity.

4.3   Tracking

At the very beginning of the tracking, human operator initializes player positions. Initial estimates for player positions for each subsequent frame are derived from the preceding image from image sequence. The color tracking method is used to roughly estimate player position. Next, a 3-by-3 pixel neighborhood of estimated position is examined for the minimum of similarity measure (5). The position of the minimum is used as the next estimate and the process is iteratively repeated up to 10 times. Maximum number of iterations is defined by the maximum expected player movement. Initial estimate, provided by the color tracking algorithm is saved and used as the starting position for the next frame. This ensures both high reliability provided by the color tracking mechanism, and low amount of trajectory jitter, due to use of template tracking to correct initial estimates.

However, some amount of smoothing has to be applied to trajectories before path length and velocity calculation. We post-process x and y components of the trajectory separately, treating them as one-dimensional, time-dependent signals, using a Gaussian filter.

5   Evaluation and Results

5.1   Reliability and Efficiency

A sequence of 750 images (30 seconds of the match), was used to test tracker reliability. Tracking was performed simultaneously for the all 14 players, present at the playing court. Human supervisor had to intervene 14 times during the tracking process, which is considerably less than 10500 (750*14) annotations, which would be needed for full manual tracking of the sequence. Intervention consisted of stopping the tracking process, manual marking of the player which was tracked incorrectly and restarting the tracking process. Processing speed averaged 4.5 frames per second.

5.2   Accuracy

There are several sources of errors that can influence the overall accuracy of the tracking: movement of player extremities, VCR tape noise, image compression artifacts, imperfect camera calibration and quantization error. However, instead of analysis of the error propagation, a set of experiments was designed to determine tracker accuracy. All experiments included several handball players, differently dressed, which performed various activities, according to the purpose of each experiment.

Ground truth was obtained simply by drawing a pattern of lines near the middle of the one half of the handball court. The pattern, shown in Fig. 5 was measured using a measuring tape.

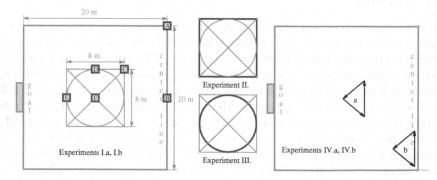

Fig. 5. Setup for experiments I.-IV. on the one half of the handball court. Left: player positions during the experiment I. are marked with black boxes. Middle: Reference player trajectories for experiments II. and III, shown with thick lines. Right: Approximate player trajectories for the experiment IV.

In each experiment, we observed path length error and RMS (Root Mean Square) error in player position (and velocity, where applicable). The following experiments were performed:

Experiment I. Players were instructed to stay still at the predefined places. In the second part of experiment, they were instructed to perform various activities (passing ball, jumping on the spot, etc.) but they were not allowed to move across the court plane. Reference position was obtained from the drawn pattern. Reference velocity and path length were exactly zero, since the players never left their positions. Effect of trajectory smoothing was also evaluated.

Experiment II. Players were instructed to run and follow the square trajectory. Influence of trajectory filtering was observed. The results confirmed that heavy smoothing hides rapid changes in player trajectory and is therefore inappropriate from this viewpoint.

Experiment III. Players were instructed to run and follow the circular trajectory with constant velocity. Reference velocity was simply calculated from the length of the circular path and the time each player needed for one round.

Experiment IV. We compared our system to widely used manual, video-based kinematic analysis tool - APAS (Ariel Performance Analysis System, [9]), which was used as a ground truth this time. Results were consistent with previous experiments, except in the level of detail that APAS captured. Velocity graph obtained using APAS has clearly shown accelerations and deccelerations of the player, associated with each of his steps. This is the feature that our system was designed to avoid, since it is not needed in match analysis.

Overall accuracy. of the designed system can be summarized as follows. Position: 0.2 - 0.5 m RMS for stationary player and 0.3 - 0.6 m RMS for active player, depending on player distance from optical axis. Velocity (uniform motion): 0.2 - 0.4 m/s (7 - 12%) RMS, depending on amount of trajectory smoothing. Path length was always overestimated, the error was +0.6 to +0.9 m/player/minute for still players and ten times more for active players.

6   Conclusion

Use of the controlled environment for our people tracker has enabled us to develop a tracking system which suits its purpose and, most importantly, it has enabled us to test its accuracy. We are confident that the obtained accuracy does not hit the limits of our tracking system, but rather the limits of possible definition of human body position, velocity and path length itself, when observing people on a such a large scale. Our system observes the movement of the people across a plane at the level of detail it was designed for. In the development of our system we deliberately sacrificed some of accuracy to enable it to observe a large area.

References

1. Aggarval, J.K., Cai, Q.: Human motion analysis: A review. IEEE Nonrigid and Articulated Motion Workshop. (1997) 90–102
2. Haritaoglu, I., Harwood, D., Davis, L.S.: An appearance-based body model for multiple people tracking. ICPR 2000, vol. 4 (2000) 184–187
3. Intille, S.S., Bobick, A.F.: Visual tracking using closed-worlds. ICCV '95 (1995) 672–678
4. Pingali, G., Opalach, A., Jean. Y.: Ball tracking and virtual replays for innovative tennis broadcasts. ICPR 2000, vol. 4 (2000) 152–156
5. Pujol, A., Lumbreras, F., Varona, X., Villanueva, J.: Locating people in indoor scenes for real applications. ICPR 2000, vol. 4 (2000) 632–635
6. Boghossian, B.A., Velastin, S.A.: Motion-based machine vision techniques for the management of large crowds. ICECS 99, Cyprus (1999)
7. Murakami, S., Wada, A.: An automatic extraction and display method of walking person's trajectories. ICPR 2000, vol. 4 (2000) 611–614
8. Ali, A., Farrally, M.: A computer-video aided time motion analysis technique for match analysis. The Journal of Sports Medicine and Physical Fitness 31(1) (March 1991) 82–88
9. Ariel dynamics worldwide. http://www.arielnet.com
10. Erdmann, W.S.: Gathering of kinematic data of sport event by televising the whole pitch and track. Proceedings of 10th ISBS symposium. Rome (1992) 159–162
11. Tsai, Y.R.: A versatile camera calibration technique for high-accuracy 3d machine vision metrology using off-the-shelf TV cameras and lenses. IEEE Journal of Robotics and Automation, RA-3(4) (1987) 323–344
12. Swain, M.J., Ballard, D.H.: Color indexing. International Journal of Computer Vision 7(1) (Nov 1991) 11–32

Linear Augmented Reality Registration

Adnan Ansar and Kostas Daniilidis

GRASP Lab, University of Pennsylvania, Philadelphia, PA 19104
{ansar,kostas}@grasp.cis.upenn.edu

Abstract. Augmented reality requires the geometric registration of virtual or remote worlds with the visual stimulus of the user. This can be achieved by tracking the head pose of the user with respect to the reference coordinate system of virtual objects. If tracking is achieved with head-mounted cameras, registration is known in computer vision as pose estimation. Augmented reality is by definition a real-time problem, so we are interested only in bounded and short computational time. We propose a new linear algorithm for pose estimation. Our algorithm shows better performance than the linear algorithm of Quan and Lan [14] and is comparable to the non-predicted time iterative approach of Kumar and Hanson [8].

Keywords: augmented reality, linear algorithm, pose estimation

1 Introduction

Augmented Reality assumes the observation of the real world either by use of a see-through head mounted display (HMD) or by cameras using a video-based HMD. The goal is to superimpose synthetic objects onto the image stream at desired positions and times. A critical issue in augmented reality is the registration or tracking problem, which refers to the proper alignment of the real and virtual worlds. This requires that the camera's pose, its location and orientation in the world reference frame, be known at all times.

Imagine, for example, that a user wants to test visually and haptically how a new laptop design feels and how it can be inserted into a docking or multiport station. We have developed a working system [1] to facilitate this by using an HMD display with a mounted camera to track marked reference points on a rectangular laptop mock-up and a WAM robot arm to supply force feedback (see Fig. 1). The transformations from the moving target frame to the head-mounted camera frame and from the camera frame to the world frame lead to a double pose estimation problem requiring both target and world reference points. Pose estimates for the target-camera mapping enable augmentation with a synthetic laptop image while pose estimates for the world-camera mapping enable the projection of a stationary docking station. The novel aspect of our approach is that pose estimates are used not only for graphics but also for haptic interaction: collision feedback requires the relative position of the virtual laptop with respect to the virtual docking station.

W. Skarbek (Ed.): CAIP 2001, LNCS 2124, pp. 383–390, 2001.
© Springer-Verlag Berlin Heidelberg 2001

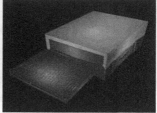

Fig. 1. On the left a user holds a rectangular plate on which the WAM robot "displays" haptic feedback. Based on pose estimates the image of a laptop is projected into the HMD, and its interaction with the docking station is sensed both visually and haptically via collision feedback.

Most virtual and augmented reality systems use magnetic or infrared trackers [16]. We employ purely visual sensors which offer a direct feedback: the image data used for visualization and merging are also used for solving the registration problem. However, the haptic aspect prevents us from using projective [15] or affine [9] representations and forces us to use Euclidean external orientation.

Pose estimation appears in multiple instances in computer vision, from visual servoing over 3D input devices to head pose computation. Like augmented reality, most of these systems must operate in real-time. For this reason we are interested in non-iterative algorithms with guaranteed convergence in near video frame performance. Since we use reference points on targets of limited extent and cannot anticipate the user's motion, we require robust performance with few fiducials and no initialization.

In this paper, we propose a novel algorithm for pose estimation. The solution is based on the kernel estimation of a system whose unknowns are an over-dimensioned mapping of the original depth scalings. The solution is unique for four or more points, provided they do not lie in a critical configuration [13] and even if they are coplanar. In Sect. 4 we study the behavior of our algorithm in synthetic experiments and compare with three other algorithms. We also show result of real world experiments.

2   Related Work

A similar approach to ours is taken by Quan and Lan [14]. Our algorithm, like theirs, is based on point depth recovery, but unlike theirs, avoids a polynomial degree increase, couples all n points in a single system of equations and solves for all n simultaneously. These factors contribute to more robust performance in the presence of noise as seen in Sect. 4. Recently Fiore [5] has produced another linear algorithm requiring little computational overhead, which is not a consideration for the limited number of points needed in our applications. He compares his algorithm to that of Quan and Lan and shows performance similar to theirs in noisy conditions.

There are many closed form solutions for 3 and 4 point or line correspondences, possibly with finite multiplicities, such as [4,6]. These closed form methods can be extended to more points by taking subsets and solving algebraically for common solutions to several polynomial systems, but the results are susceptible to noise and the solutions ignore much of the redundancy in the data.

Many iterative solutions exist. These are based on minimizing the error in some nonlinear geometric constraint, either in the image or target, by some variation on gradient descent or Gauss-Newton methods. Typical of these is the work of Lowe [11]. There are also approaches which more carefully incorporate the geometry of the problem directly into the update step, such as that of Lu et al. [12] or of Kumar and Hanson [8], which we compare to our algorithm in Sect. 4. Dementhon and Davis [3] initialize their iterative scheme by relaxing the camera model to scaled orthographic, and Bruckstein deals with the issue of optimal placement of fiducials in this context [2]. These iterative approaches typically suffer from slow convergence for bad initialization, convergence to local minima and the requirement for a large number of points for stability. Our algorithm requires no initialization, can be used for a small number of points, and guarantees a unique solution when one exists.

Another approach is to recover the world to image plane projection matrix and extract pose information. This is the basis for the calibration technique of Lenz and Tsai [10]. We compare the projection matrix approach to our algorithm in Sect. 4. It is inherently less stable for pose estimation because of the simultaneous solution for the calibration parameters, and it requires a large data set for accuracy.

3   Pose Estimation Algorithm

Throughout this paper, we assume a calibrated camera and a perspective projection model. If a point has coordinates $(x, y, z)^T$ in the coordinate frame of the camera, its projection onto the image plane is $(x/z, y/z, 1)^T$. We assume that the coordinates of n points are known in some global frame, and that for every reference point in this frame, we have a correspondence to a point on the image plane. Our approach is to recover the depths of points by using the geometric rigidity of the target in the form of the $\frac{n(n-1)}{2}$ distances between n points. Let $\{\mathbf{w}_i\}$ be a set of n points with projections $\{\mathbf{p}_i\}$ and let $\{t_i\}$ be positive real numbers such that $\mathbf{w}_i = t_i\mathbf{p}_i$. We indicate by d_{ij} the Euclidean distance between $\mathbf{w}_i$ and $\mathbf{w}_j$. Then $d_{ij} = |t_i\mathbf{p}_i - t_j\mathbf{p}_j|$. This is the fundamental constraint we use throughout (see Fig. 2).

Let $c_{ij} = d_{ij}^2$. Then we have

$$c_{ij} = (t_i\mathbf{p}_i - t_j\mathbf{p}_j)^T(t_i\mathbf{p}_i - t_j\mathbf{p}_j) = t_i^2 p_{ii} + t_j^2 p_{jj} - 2t_it_j p_{ij} \tag{1}$$

where we use the notation p_{ij} to indicate $\mathbf{p}_i^T\mathbf{p}_j$, the inner product of the two vectors. Note that the right hand side of (1) contains only quadratic terms in the unknown depth scalings. Let $t_{ij} = t_it_j$ and let $\rho = 1$. We rewrite (1) as

$$t_{ii}p_{ii} + t_{jj}p_{jj} - 2t_{ij}p_{ij} - \rho c_{ij} = 0 \tag{2}$$

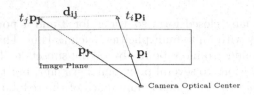

Fig. 2. The fundamental geometric constraint used in our algorithm relates the distance between points in the world d_{ij} and the scale factors t_i and t_j associated with the projections $\mathbf{p}_i$ and $\mathbf{p}_j$.

This is a homogeneous linear equation in 4 unknowns. Since $t_{ij} = t_{ji}$, observe that for n points there are $N_r = \frac{n(n-1)}{2}$ equations of the form (2) in $N_c = \frac{n(n+1)}{2} + 1$ variables. Define $\bar{t}$ as

$$\bar{t} = (t_{12} \ ... \ t_{1n} \ \ t_{23} \ ... \ t_{2n} \ ... \ t_{(n-1)n} \ \ t_{11} \ \ ... \ t_{nn} \ \ \rho)^T \tag{3}$$

Then we have $\mathbf{M}\bar{t} = (\mathbf{M}_d | \mathbf{M}')\bar{t} = 0$. Where $\mathbf{M}_d$ is an $N_r \times N_r$ diagonal matrix with $\{-2p_{11}, \dots, -2p_{N_r N_r}\}$ along the diagonal and $\mathbf{M}'$ is $N_r \times (n+1)$ dimensional. It follows immediately that the $\mathbf{M}$ has rank $N_r = \frac{n(n-1)}{2}$. Hence, it has kernel $Ker(\mathbf{M})$ of dimension $n + 1 = N_c - N_r$ with $\bar{t} \in Ker(\mathbf{M})$. We compute $Ker(\mathbf{M})$ by singular value decomposition. If $\mathbf{M} = \mathbf{U}\Sigma\mathbf{V}^{\mathbf{T}}$, $Ker(\mathbf{M})$ is spanned by the columns of $\mathbf{V}$ corresponding to the zero or (in the case of image noise) $n + 1$ smallest singular values.

We now use the quadratic constraints imposed by linearizing the original system to solve for $\bar{t}$. Let $\{\lambda_1, ..., \lambda_{n+1}\}$ be the specific values for which

$$\bar{t} = \sum_{i=1}^{n+1} \lambda_i \mathbf{v}_i \tag{4}$$

Where $\{\mathbf{v}_i\}$ are such that $Ker(\mathbf{M}) = span(\{\mathbf{v}_i\})$. Recall the form of $\bar{t}$ in (3). For any integers $\{i, j, k, l\}$ and any permutation $\{i', j', k', l'\}$, we have $t_{ij}t_{kl} = t_{i'j'}t_{k'l'}$. Substituting individual rows from the right hand side of (4) into expressions of this sort leads, after some algebraic manipulation, to constraints on the λ_i of the form

$$\sum_{a=1}^{n+1} \lambda_{aa}(v_a^{ij}v_a^{kl} - v_a^{i'j'}v_a^{k'l'}) + \sum_{a=1}^{n+1}\sum_{b=a+1}^{n+1} 2\lambda_{ab}(v_a^{ij}v_b^{kl} - v_a^{i'j'}v_b^{k'l'}) = 0 \tag{5}$$

where we use the notation $\lambda_{ab} = \lambda_a\lambda_b$ for integers a and b, and v_a^{ij} refers to the row of $\mathbf{v}_a$ corresponding to the variable t_{ij} in $\bar{t}$. Again, we have the obvious relation $\lambda_{ab} = \lambda_{ba}$. It follows that equations of the form (5) are linear and homogeneous in the $\frac{(n+1)(n+2)}{2}$ variables λ_{ab}. The quadratic constraints determine the t_i, hence λ_a, up to a uniform scale. Observe that there are $O(n^3)$ such constraints, from which we select the $\frac{n^2(n-1)}{2}$ relations of the form

$t_{ii}t_{jk} = t_{ij}t_{ik}$. The linear system can be expressed in matrix form as $\mathbf{K}\bar{\lambda} = 0$. Where $\bar{\lambda} = (\lambda_{11} \quad \lambda_{12} \quad ... \quad \lambda_{(n+1)(n+1)})^T$ and $\mathbf{K}$ is an $\frac{n^2(n-1)}{2} \times \frac{(n+1)(n+2)}{2}$ matrix, each row of which corresponds to an equation of the form (5). Since we can recover the λ_{ab} up to scale, we expect that $\mathbf{K}$ will have at most one zero singular value. Let $\mathbf{K} = \mathbf{U}'\mathbf{\Sigma}'\mathbf{V}'^T$ be the singular value decomposition of $\mathbf{K}$. It follows that $Ker(\mathbf{K})$ is spanned by the column of $\mathbf{V}'$ corresponding to the smallest singular value in $\mathbf{\Sigma}'$. However, $\bar{\lambda} \in Ker(\mathbf{K})$ implies that we have recovered λ_{ab} up to scale. Now we find a unique solution by using the constraint implied by the last row of (4), specifically $\lambda_1 v_1^L + \lambda_2 v_2^L + ... + \lambda_{n+1}v_{n+1}^L = \rho = 1$, where v_i^L refers to the last row of vector $\mathbf{v}_i$. In practice we pick an arbitrary scale, compute $\bar{t}$ from (4) and the SVD as above, then normalize by the last row of $\bar{t}$. Depth recovery in the camera frame now amounts to computing $t_i = \sqrt{t_{ii}}$ for $i = 1...n$.

Having recovered the 3D camera coordinates of the points, we find the transformation between the world and camera frames. The optimal rotation is recovered by SVD of the cross-covariance matrix of the two sets of points translated to their respective centroids [7]. Once the rotation is computed, translation between the centroids, hence the two frames, is immediately recovered.

Let $HQ(\mathbf{R}^n)$ and $HL(\mathbf{R}^n)$ be the set of homogeneous quadratic and linear equations on $\mathbf{R}^n$, respectively. Our approach was to linearize the quadratic system, say QS, in (1) to the linear one, say LS, in (2) by applying the map $f : HQ(\mathbf{R}^n) \to HL(\mathbf{R}^{\frac{n(n+1)}{2}+1})$ defined by $f(t_i t_j) = t_{ij}$, $f(1) = \rho$ with $f(QS) = LS$, by abuse of notation. As we have seen, this increases the dimension of the solution space to $n+1$ by artificially disambiguating related quadratic terms. Let $V_0 = Var(LS) = Ker(LS)$ be the $n+1$ dimensional affine variety in $\mathbf{R}^{\frac{n(n+1)}{2}+1}$ corresponding to the kernel of LS. To recover the original solution, we impose additional constraints of the form $t_{ij}t_{kl} = t_{i'j'}t_{k'l'}$ for $\{i', j', k', l'\}$ a permutation of $\{i, j, k, l\}$. Let e_1 be one such equation. Then $V_1 = Var(\{LS, e_1\}) = Var(LS) \cap Var(e_1)$ is a proper subvariety of V_0, since $Var(e_1)$ is not in any linear subspace of $R^{\frac{n(n+1)}{2}+1}$. Given a sequence of such constraints $\{e_i\}$ with e_i independent of $\{e_j | j < 1\}$, we obtain a nested sequence of varieties $V_0 \supset V_1 \supset V_2 ...$ of decreasing dimension. Since we have more quadratic constraints than the dimension of V_0, we eventually arrive at the desired solution. Observe that this procedure is entirely generic and does not depend on the coefficients of the original system QS. It follows that an abstract description of the subspace $S = Var(\{e_i\})$, which we do not yet have, would allow us to eliminate the second, more computationally intensive SVD from our algorithm.

4   Results

We conduct a number of experiments, both simulated and real, to test our algorithm (hereafter referred to as **NPL**, for n-point linear). We compare to three other algorithms in simulation. The first, indicated by **PM**, is direct recovery and decomposition of the full projection matrix from 6 or more points by standard SVD methods. The second, indicated by **KH**, is the so-called "R_and_T" algorithm of Kumar and Hanson [8] applied to the $\frac{n(n-1)}{2}$ line segments defined by n points. The third, indicated by **QL**, is the n-point linear algorithm due to Quan and Lan [14].

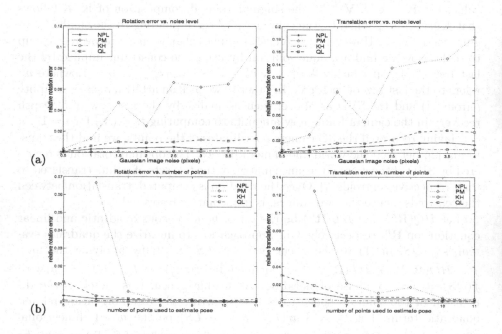

Fig. 3. Rotation and Translation errors vs. (a) noise level for 6 points and (b) number of points with fixed 1.5×1.5 pixel Gaussian noise. We plot results for the four algorithms, NPL, PM, KH and QL. Note that NPL outperforms all but the iterative KH with ground truth initialization. The performance difference is greatest for a smaller number of points.

Simulations: All simulations are performed using MATLAB. We report errors in terms of relative rotation and relative translation error. Each pose is written as $(\bar{q}, T) \in SE(3)$ where $\bar{q}$ is a unit quaternion representing a rotation and T is a translation. For recovered values $(\bar{q}_r, T_r)$, the relative translation error is computed as $2\frac{|T-T_r|}{|T|+|T_r|}$ and the relative rotation error as the absolute error $|\bar{q}-\bar{q}_r|$ in the unit quaternion. Noise level in image measurements is given in terms of the standard deviation σ of a zero mean Gaussian. We add Gaussian noise with equal σ independently to both the x and y coordinates but only admit noise between -3σ and 3σ.

We generate 31 sets of 6 points with distance between 0 and 200 from the camera and 31 random rotations and translation ($|T| \leq 100$). For each of these 961 pose-point combinations, we run the four algorithms and add Gaussian noise varying from 0.5 to 4 pixels in each direction. We see in Fig. 3(a) that NPL outperforms both PM and QL.

All 4 algorithms perform better as the number of points used for pose estimation increases. Points and poses are generated as before, but the number of points is varied from 4 to 11 with Gaussian noise fixed at $\sigma = 1.5$ pixels in each direction. For each run, we use 20 sets of points for each of 20 poses for a total of 400. Note in Fig. 3(b) that NPL outerpforms the other linear algorithms in each case. The performance difference is largest for a smaller number of points, which is our primary concern. Note that we do not compute a result for 5 points in PM because the projection matrix cannot be recovered with only 5 points.

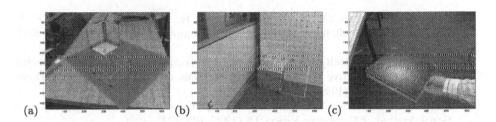

(a)   (b)   (c)

Fig. 4. Reprojection of virtual objects into a real scene. Camera pose is estimated using NPL from 8 points in (a) and (b) and 5 coplanar points in (c). Observe that all virtual objects have the correct scale and are properly oriented.

Real Experiments: All images were captured with a Matrox Meteor II frame grabber and Sony XC-999 camera calibrated with Tsai's algorithm [10]. Processing was done offline using MATLAB. With our current approach, the second SVD for the system cannot be computed in real-time for large values of n. However, our primary concern is with a small number of targets. On a 600 MHz PIII using the implementation of SVD from Numerical Recipes in C, we obtained rates for SVD computation for matrices of the size of K in Sect. 3 ranging from < 0.5 msec. for 4 points to < 35 msec. for 8 points, which is adequate for our purposes.

We demonstrate that virtual objects are correctly registered into real scenes using our algorithm. In Fig. 4(a) we determine camera pose from 8 points (marked in the image) using NPL and transform the coordinates of virtual objects from the world frame to the camera frame. The reprojected virtual box and edges are correctly aligned with real world objects. In Fig. 4(b) we repeat the experiment but on a scale approximately 3 times larger. The virtual boxes are correctly stacked on and next to the real one. In Fig. 4(c) we estimate the pose of a rectangular plate (the laptop mock-up in our description in Sect. 1) and reproject a virtual laptop onto the mock-up. In this case, we use 5 markers (at the vertices and the center of the rectangle) to estimate pose. Observe that the algorithm performs well even for these coplanar points.

5   Conclusion

Our goal was to develop a fast, accurate pose estimation algorithm for a limited number of points in close range, consistent with the type of augmented reality applications in which we are interested. We have produced an algorithm which runs in real-time under these conditions and which guarantees a solution when one exists. The only other algorithms which satisfy our requirements are those of Fiore [5]and of Quan and Lan [14], which have performances comparable to each other. We have seen in the results section that our algorithm outperforms that of Quan and Lan.

References

1. Ansar, A., Rodrigues, D., Desai, J.P., Daniilidis, K., Kumar, V., Campos, M.F.M.: Visual and haptic collaborative tele-presence. Computers & Graphics 15(5) (2001) (to appear)
2. Bruckstein, A.M., Holt, R.J., Huang, T.S., Netravali, A.N.: Optimum fiducials under weak perspective projection. Proc. Int. Conf. on Computer Vision. Kerkyra, Greece, Sep. 20-23 (1999) 67–72
3. Dementhon, D., Davis, L.S.: Model-based object pose in 25 lines of code. International Journal of Computer Vision 15 (1995) 123–144
4. Dhome, M., Richetin, M., Lapreste, J.T., Rives, G.: Determination of the attitude of 3-D objects from a single perspective view. IEEE Trans. Pattern Analysis and Machine Intelligence 11 (1989) 1265–1278
5. Fiore, P.D.: Efficient linear solution of exterior orientation. IEEE Trans. Pattern Analysis and Machine Intelligence 23 (2001) 140–148
6. Horaud, R., Conio, B., Leboulleux, O., Lacolle, B.: An analytic solution for the perspective 4-point problem. Computer Vision, Graphics, and Image Processing 47 (1989) 33–44
7. Horn, B.K.P., Hilden, H.M., Negahdaripour, S.: Closed-form solution of absolute orientation using orthonormal matrices. Journal Opt. Soc. Am. A, A5 (1988) 1127–1135
8. Kumar, R. Hanson, A.R.: Robust methods for estimaging pose and a sensitivity analysis. Computer Vision and Image Understanding 60 (1994) 313–342
9. Kutulakos, K., Vallino, J.: Calibration-free augmented reality. IEEE Trans. on Visualization and Computer Graphics 4(1) (1998) 1–20
10. Lenz, R., Tsai, R.Y.: Techniques for calibration of the scale factor and image center for high accuracy 3-d machine vision metrology. IEEE Trans. Pattern Analysis and Machine Intelligence 10 (1988) 713–720
11. Lowe, D.G.: Fitting parameterized three-dimensional models to images. IEEE Trans. Pattern Analysis and Machine Intelligence 13 (1991) 441–450
12. Lu, C.-P., Hager, G., Mjolsness, E.: Fast and globally convergent pose estimation from video images. IEEE Trans. Pattern Analysis and Machine Intelligence 22 (2000) 610–622
13. Maybank, S.: Theory of Reconstruction from Image Motion. Springer-Verlag, Berlin et al. (1993)
14. Quan, L., Lan, Z.: Linear n-point camera pose determination. IEEE Trans. Pattern Analysis and Machine Intelligence 21 (1999) 774–780
15. Seo, Y., Hong, K.-S.: Weakly calibrated video based augmented reality. Proc. Int. Symposium on Augmented Reality. Munich, Germany, Oct. 5-6 (2000) 129–136
16. Welch, G., Bishop, G.: Scaat: Incremental tracking with incomplete information. ACM SIGGRAPH (1997)

Shape and Position Determination Based on Combination of Photogrammetry with Phase Analysis of Fringe Patterns

Michał Pawłowski[1] and Małgorzata Kujawińska[2]

Warsaw University of Technology, Institute of Micromechanics and Photonics
8 Chodkiewicza St., 02-525 Warsaw, Poland,
{mpzto, m.kujawinska}@mech.pw.edu.pl

Abstract. The basic methodologies used in animation are presented and their most significant problems connected with combining real and virtual worlds are recognised. It includes the creation of virtual object and description of its realistic movement. To perform these tasks an optical method based on fringe projection and marker tracking is applied. The spatial analysis of a projected fringe pattern delivers an information about shape of an object and its out-of-plane displacement. Additionally the analysis of positions of fiducial points at the object observed by two CCDs provides an on-line information about (x,y,z) co-ordinates of these points. Combining the information about object shape and 3D co-ordinates of fiducial points during the measurement enables to generate a virtual model of the object together with the description of its movement. The concept described above is experimentally tested and the exemplary results are presented. The further works to implement this technique are discussed.

Keywords: augmented reality, photogrammetry, fringe patterns

1 Introduction

Combining real and virtual worlds requires capturing and transferring the shape of 3D object and its movement/deformation parameters into 3D space. This task may be performed by computer animation techniques. The animation is understood as a process of automatic generation of a sequence of images (scenes), in which next image contains an alternation in relation to the previous one. To carry out such a process, the information about the global and local shape of a 3D object, its surface properties (texture, reflectance, shines, etc.) and mechanical properties (degrees of freedom of an object or its parts, allowed velocities and range of movements, possible directions of movement, etc.) are required.

Considerable work has been invested to create a description of 3D object shapes. A good example is the viewpoint catalogue [1] in which the most popular objects can be found and their geometric descriptions are given. From an experimental point of view, the most useful methods, which enable measurement of shape of various classes of objects [2-5] are optical techniques. Often full-field structured light techniques are

W. Skarbek (Ed.): CAIP 2001, LNCS 2124, pp. 391–399, 2001.

applied. In this case the shape information is coded into fringe pattern with a spatial carrier frequency [6,7].

If the knowledge of the geometry of an object and its possible movement/alternation is available, the animation may be calculated by one of six cardinal algorithms:
- actual parameters values,
- basic parameter values,
- procedural animation,
- animation language,
- simulation,
- behaviour simulation.

If the experimental data has to support an animation process, the most useful are the first two algorithms, which refer to a cinematic description of an object and its movement. The data for these algorithms may be delivered by monitoring of an object with projected fringe pattern.

In such case, the intensity observed on the object surface is given by:

$$I(x, y) = a(x, y) + b(x, y)\cos[2\pi f_{0x} x + \phi_0(x, y)] \tag{1}$$

where: $a(x,y)$ and $b(x,y)$ are background and local modulation functions, f_{0x} – is the spatial carrier frequency of the projected grid (number of lines/mm), $\phi_0(x,y)$ – represents a phase with shape information $h(x,y)$ coded i.e. $h(x,y) = f[\phi_0(x,y)]$.

In applications which require combined shape and measurement of movement parameters (shifts in 3D space), the time of data processing is important. In order to decrease the measurement time the use of spatio-temporal approach is suggested [10]. While using spatio-temporal approach, the measured objects can be classified to one of the following classes:
- objects which spatial bandwidth which fulfils the condition

$$B_0 < f_{0x}/2 \tag{2}$$

where: B_0 is the object initial spatial frequencies bandwidth of the object.

In such a case the shape changes may be monitored by analysis of sequential fringe patterns by means of spatial carrier phase shifting method or spatial Fourier transform method [8,9,10],
- object which spatial bandwidth does not fulfils the condition (2). Such complex and/or non-continuous shapes may be analysed using multiply grids projection techniques supported by more sophisticated phase unwrapping algorithms including temporal unwrapping [11] or hierarchical unwrapping [12] which require capturing of several images.

The variations in time of the initial shape obtained by one of the above mentioned methods have to be determined on the base of a sequence of frames captured during object changes.

Additionally, together with shape variations the object may move globally in the measurement volume. This movement description may be referred to the regions which are not varying their shape in time or to the global object representation e.g. in the form of localisation of its centre of gravity. In general case the rigid body movement is difficult to detect by fringe projection method only, therefore it has to be supported strongly by photogrammetric methods, namely by monitoring of the position of markers attached to the object [13].

Also, if the true 3D object analysis has to be performed, the system should enable the simultaneous monitoring and analysis from multiple directions and combining this information to form full representation of an object. This will require a multiple projection–detection system preferably working on the base of autocalibration principle [14], or in reference to the precalibrated measurement volume [15]. In the paper below we will present principles of combined spatio-temporal [14] and photogrametric approach to dynamic object analysis. At the moment, we assume that the object is subjected to shape variations (z direction) and rigid body motion. The concept presented will not solve all the problems connected with true 3D object monitoring, as it refers to a certain restricted class of object and their deformations/movements. However, the methodology shows the possibility of the efficient use of the spatio-temporal bandwidth of the measurement system in reference to a variable object bandwidth $[B_0 + \Delta B(t)]$.

2 Basic Concept

To obtain full information about varying 3D object we recommend the use of the spatio-temporal approach in which a fringe pattern is given by the equation:

$$I(x,y;t) = a(x,y;t) + b(x,y;t) \cos[2\pi f_{0x}x + \phi_0(x,y) + \phi(x,y;t)] \tag{3}$$

where: $a(x,y;t)$ and $b(x,y;t)$ are background and modulation functions which may vary in time respectively; f_{0x} – spatial carrier frequency; $\phi_0(x,y)$ – initial phase of the object; $\phi(x,y;t)$ – phase variable in space and time.

The varying shape of the object is coded in deformation of a grid projected on the object during measurement in a set of sequential frames. From these images the geometry of the object can be reconstructed correctly by use of Spatial Carrier Phase Shifting (SCPS) method [8] if the sum of an object initial bandwidth and the bandwidth changes gained by the object shape variations fulfils the condition

$$B_0 + \Delta B(t) < f_{0x}/2 \tag{4}$$

The condition (4) can be easily fulfilled if an active fringe projection is used. Active fringe projector enables also the dynamic change of measurement volume depending on the object size and assumed movement range.

The shape reconstructed by SCPS method is given in the form of height values $h(x,y)$ and does not include the information about the position of the object within the measurement volume. The position of an object is calculated based on the marker tracing technique known from photogrammetry. Two CCD cameras observe fiducial points attached to the object surface before measurement. The information about opto-mechanical parameters of the measurement system together with information about parallax of the marker on two CCDs enables calculation of 3D co-ordinates of fiducial points. Combination of shape information with 3D co-ordinates of markers determines the position of object points in 3D space. Fig.1 presents the general scheme of measurement.

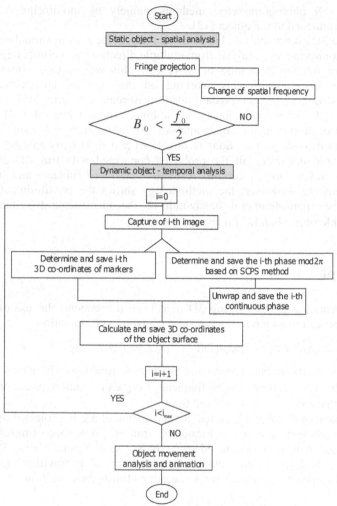

Fig. 1. The scheme of spatio-temporal analysis of dynamic 3D objects; B_0,–spatial frequency bandwidths of an object, f_{0x}-spatial carrier frequency of the projected grid

3   Measurement System and Test Object

The experiments were conducted using a fringe projection system, shown in Fig.2. A Digital Light Projector (DLP) (3) projects the fringes onto the surface of an object (4). CCD1 and CCD2 observe deformation of the linear fringe pattern. The optical axes of CCDs are parallel to the optical axis of the fringe projector. The DLP unit is placed non-centrally in order to provide large measurement base for fringe projection system. It results in high sensitivity of the shape determination (Fig.2). Measurement volume is created within the common part of the field of views of DLP projector and two cameras.

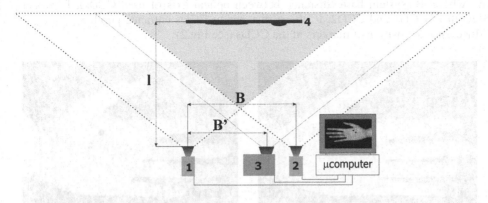

Fig. 2. The scheme of the measurement system: 1), 2) CCD camera, 3) DLP projector, 4) object under test

During the object movement a pre-set fringe pattern is continuously projected onto the measured object surface. The distorted fringe pattern is observed by two CCDs and saved in computer memory with a frequency of 50 frames per second. Standard cameras are used and each half-frame image is saved with a resolution of 256x256 pixels. To evade problems connected with significantly smaller speed of saving data on a hard disk all images are saved in RAM. This extracts a use of approximately 3MB of RAM for one second of movement recording. By use of computer with 512MB it is possible to record approximately two minutes of recording (depending on the installed operating system and software). After the measurement the analysis procedure is performed.

3.1 Determination of Rigid Body Motion

Before the measurement the markers are attached to the object surface. Markers are made from black non-reflective circular pieces of velour. As this material very poorly reflects light markers are visible on the object surface as black dots (Fig. 3). Presented method requires at least two fiducial points. Since during object movement/shape deformation some regions may by hidden in the shade or invisible the use of more markers is required.

At the beginning of the marker analysis procedure, all saved images are searched for fiducial points. Regions that fulfil the fiducial point conditions are flagged. The feature vector of an marker is defined by: region intensity, region size, intensity difference between closest neighbourhood and marked area, aspect ratio height to width and fill factor (height*width/number of points). Each fiducial point recognised from the first saved frame is flagged with a unique identification number. The flagged regions are tracked in sequential frames. Knowing the opto-mechanical parameters of the system the centre of gravity of a marker can be calculated from

$$Z = \frac{B*f*l}{(1-f)*(x_{CCD\,1} - x_{CCD\,2})} \tag{5}$$

where: B- system base (distance between optical axis of used CCDs), f – focal length of CCD1 and CCD2, l- object distance to CCDs entrance pupil, x_{CCD1}-x_{CCD2} – difference in position of markers at the CCDs (see Fig.2)

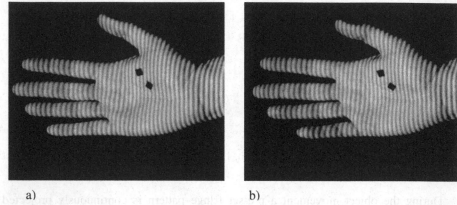

a)   b)

Fig. 3. Object illuminated by fringe pattern at the beginning a) and the end b) of the measurement

3.2 Determination of Temporary Shape

After the marker processing procedure is finalised the phase analysis of the fringe pattern images saved by CCD1 is performed by means of 5 points spatial carrier phase shifting (SCPS) algorithm [16]:

$$\Phi(x,y) = \tan^{-1}\left(\frac{\sqrt{4[I(x-1,y)-I(x+1,y)]^2 - [I(x-2,y)-I(x+2,y)]^2}}{2I(x,y)-I(x-2,y)-I(x+2,y)}\right) \tag{6}$$

where: I(x,y) –intensity at co-ordinates (x,y).

Due to the periodic properties of cosine function, the phase is reconstructed from (6) in the form of mod2π (Fig.4a). The map mod2π is later unwrapped by modified quality-guided path following algorithm[17]. The linear phase term $2\pi f_{ox}$ is removed from the unwrapped continuous phase distribution as shown in Fig.4b. After that the scaling of phase into physical quantity (mm) based on (7) is performed.

$$h(x,y) = \frac{ld'\phi(x,y)}{2\pi B' - d'\phi(x,y)} \tag{7}$$

where: h – height in mm, l – object distance to entrance pupil of CCD1, d' – projected fringe period in the object plane, φ(x,y) – phase value, B' – fringe projection system base (distance between the optical axis of DLP and CCD).

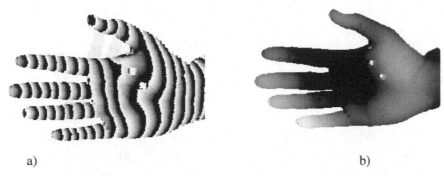

a)   b)

Fig. 4. The result of fringe pattern analysis a) wrapped and b) unwrapped

3.3 Calculation of Absolute Shape and Position

Calculated shape maps (one per image) include information about the height relation between the object points only. In other words h(x,y) is fixed with an object surface. To determine the position of the object points in the measurement volume the transformation from object surface co-ordinates into measurement volume co-ordinates is required. This process is conducted based on z values calculated according to equation (5). In the middle of two closest fiducial points the z co-ordinate of the surface is estimated by calculation of mean value of centres of gravity of the fiducial points. The difference between the height of the object and average z value produces a transformation vector **T** of the surface. The 3D co-ordinates of the object within the measurement volume are calculated by applying the shift **T** to the h(x,y). The exemplary 3D visualisation of the object 3D co-ordinates in two selected stages of movement are shown in Fig.5.

The errors in height reconstruction is at the level of d'/10. In the case reported the grid period equals d'=3 mm. The positioning of the markers within the CCD matrix is 1 pixel [10]. The joined error of measurement depends strongly from the measurement volume (here 240x240x200mm^3) and is estimated to be 10^{-3} of the max dimension of the measurement volume. So finally the combined error for shape and position determination can be estimated for 0.4mm.

4 Conclusions

The presented methodology gives the possibility to obtain full-field information about actual 3D co-ordinates of the object during the measurement. The shape obtained for the sequential stages of movement, together with the information about the position within the measurement volume can be delivered in the form compatible with the formats required by computer graphics and animation systems. The accuracy reached by the presented system is fully acceptable in the multimedia applications, where realistic behaviour of an object is important. The future works will be focused on making an interface between the presented system and 3D Max animation software.

a)   b)

Fig. 5. 3D visualisation of the measured object within the measurement volume at the beginning a) of the measurement and at the end b) In the figure to enhance the visualisation the z-dimension was rescaled

Acknowledgements. This work has been financially supported by the State Committee for Scientific Research, Poland KBN T 10C 018 20 and 8T10C 015 19

References

1. Viewpoint Catalog, Spring '97 Edition, Viewpoint Datalabs International, 1997.
2. Tiziani, H.-J.: Optical techniques for shape measurement. Proc. of the 2nd International Workshop on Automatic Processing of Fringe Patterns, Fringe'93 pp165-174, 1993.
3. Demers, M.H., Hurley, J.D., Wulpern, R.C.: Three Dimensional Surface Capture for Body Measurement using Projected Sinusoidal Patterns. Proc. SPIE, 3023, 13-25, 1997.
4. El-Hakim, S.F., Boulanger, P., Blais, F., Beraldin, J.-A.: A system for Indoor 3-D Mapping and Virtual Enviroments. Proc. SPIE, 3174, 21-35, 1997.
5. Kowarschik, R., Kuehmstedt, P., Schreiber, W.: 3-coordinate measurement with structured light. Proc. 2nd int. Workshop Fringe'93, eds. W. Jüptner, W. Osten, Akademie Verlag, 204-208, 1993.
6. Takeda, M.: Fourier transform speckle profilometry; three dimensional shape measurements of diffuse objects with large height steps and/or isolated surfaces. Appl. Opt. 33,7829-7837, 1994.
7. Pirga, M., Kujawińska, M.: High sensitivity fringe projection microshape measurement system. Proc. SPIE 2248, 162-166, 1994.
8. Kujawińska, M.: Spatial phase measurement methods in Interferogram Analysis and digital processing techniques for fringe pattern analysis. D.Robinson, G.Reid (eds.), Institute of Physics Publishing, London, 141-193, 1993J.
9. Pawłowski, M., Kujawińska, M., Węgiel M.: Monitoring and measurement of movement of objects by fringe projection method. Proc. SPIE, v 3958, 116-125, 2000.
10. Kujawińska, M., Pawłowski. M.: Spatio-temporal approach to shape and motion measurement of 3D objects. Proc. SPIE, v.4101, 21-28, 2000.
11. Huntley, M., Saldner, H.: Temporal phase-unwrapping algorithm for automated interferogram analysis. Appl. Opt., 32, 3047-3052, 1993.
12. Osten W., Nadeborn, W., Andra, P.: General hierarchical approach in absolute phase measurement. Proc. SPIE, 2860, 2-13, 1996.
13. Kujawińska, M. Pawłowski, M.: Application of shape measuring methods in animation. Proc. SPIE, 3407, 522-527, 1998.

14. Schreiber, W., Notni, G.: Theory and arrangements of self-calibrating whole-body 3D-measure,ment systems using fringe projection technique. Opt. Eng., 39, 159-169, 2000.
15. Kujawińska, M., Sitnik, R.: Three-dimensional objects opto-numerical acquisition and processing method for CAD/CAM and multimedia applications. Proc. SPIE, 3958, 2000.
16. Pirga M.: Doctoral dissertation: Phase measurement methods in adaptive 3D shape determination by fringe projection system. Warsaw University of Technology, 1998.
17. Ghiglia, D.C., Pritt, M.D.: Two-dimensional phase unwrapping. Wiley Interscience, 19998.

Automated Acquisition of Lifelike 3D Human Models from Multiple Posture Data

Jochen Wingbermühle[1], Claus-Eberhard Liedtke[1], and Juri Solodenko[1]

Institut für Theoretische Nachrichtentechnik und Informationverarbeitung,
Universität Hannover, Appelstr. 9a,
D-30167 Hannover, Germany
Phone: +49 511 762-5323, Fax: +49 511 762-5333
{wingber, liedtke}@tnt.uni-hannover.de

Abstract. In this paper we propose a method for the automated acquisition of 3D human models for real-time animation. The individual to be modelled is placed in a monochrome environment and captured simultaneously by a set of 16 calibrated cameras distributed on a metal support above and around the head. A number of image sets is taken from various postures. A binary volume model will then be reconstructed from each image set via a shape-from-silhouette approach. Based on the surface shape and a reliable 3D skeletonisation of the volume model, a parametric human body template is fitted to each captured posture independently. Finally, from the parameter sets obtained initially, one unique set of posture-invariant parameters and the corresponding multiple sets of posture-dependent parameters are estimated using iterative optimisation. The resulting model consists of a fully textured triangular surface mesh over a bone structure, ready to be used in real-time applications such as 3D video-conferencing or off-the-shelf multi-player games.

Keywords: human models, shape-from-silhouette, 3D videoconferencing

1 Introduction

Today's rapidly growing number of virtual-reality applications, e.g. 3D video-conferencing, multi-player games, simulations etc., leads to an increasing demand in virtual representations of the human individual, i.e. in *3D human models* or *avatars* that look and behave as human beings.

These requirements are currently fulfilled using a textured polygonal surface mesh of which vertices are assigned to an internal articulated bone structure. Animation occurs as the surface mesh is deformed by shifting the vertices relatively to specified movements of the inside bone structure. The dependency of vertex shifts on bone movements is described by *deformation-* or *skin-models*. Various deformation-models with different levels of complexity have been developed over the last years and are readily available in commercially distributed animation software packages [1,2].

W. Skarbek (Ed.): CAIP 2001, LNCS 2124, pp. 400–409, 2001.

Adequate devices for the representation of the human being in virtual settings exist already; however, the automated acquisition of this type of model is still the subject of ongoing research activities. The human individual is not a rigid object: the acquisition system, therefore, has to cope with the problem of a shape varying over time. For one single time sequence, i.e. to capture a 'frozen' model, so to speak, the problem has been solved using fast capturing devices. Two categories of devices can be distinguished: on one side, laser or structured light based, active 3D sensors [3,4] which provide accurate surface data but, being rather expensive, are a deterrent to a broader usage; on the other side, passive, camera based systems which are much cheaper, yet provide less accurate data. To improve the unsatisfactory accuracy of passive system data capture, various approaches for a model-based analysis of captured data sets have been proposed recently [5,6].

Working towards the automated acquisition of *moving* human models implies that the various positions of the model's internal bone structure must be determined in such a way that, in conjunction with the respective deformation-models, realistic motion-dependent deformations during animation be achieved. Recent approaches to estimate suitable positions define a single standard posture with as few self-occlusions as possible for capturing a human being [7,8] and calculate joint positions relatively to some feature points of that particular posture, e.g. armpits etc.. These approaches allow the reliable and fast segmentation of the scanned dataset and the obtained joint positions are suitable for smaller movements around the respective standard postures. However, due to the limited accuracy of the joint positions that are determined, moving the model into completely different postures often lacks the desired lifelike impression. Better results can be expected when analysing multiple data sets captured from varying postures. In [9], a comparable approach employing image data from a sequence of *pre-defined motions* is presented. In [10], starting from an interactive initialisation, an approach is being presented to track and refine parameters of an articulated skeleton from image sequences showing multiple views of arbitrary motions. We aim at achieving an automated, easy-to-use, process, to acquire a unique lifelike human model from the discrete observations of a human being in different *a priori unknown postures* without any interaction.

The setup used for data acquisition is described in the following section. In section 3 the parametric human body template is introduced and in section 4 the multiple posture adaptation method that we are proposing is outlined. First results are given in section 5, followed by concluding remarks and possible directions for future work in section 6.

2   Data Acquisition

The human being to be modelled is placed in a monochrome environment and captured simultaneously by a set of 16 calibrated standard digital cameras distributed above and around the head (see Fig. 1). For details on this acquisition setup see [8]. Several image sets are taken for different postures. From each ac-

Fig. 1. Body-scanner Setup: The person to be modelled is captured by 16 calibrated digital cameras in a monochrome environment.

quired image set, a binary volume model is reconstructed through a shape-from-silhouette approach [11] as depicted in Fig. 2. Firstly, the person's silhouette is extracted from each image using a chroma-keying technique. The calibrated setup means that the position, orientation and inner parameters of each camera are known. Thus, in a second step, the obtained silhouettes can be projected back into space, creating conic volumes that contain the individual. The spatial sections of these volumes approximate the individual's shape in the respective posture. Eventually, applying this technique to all acquired image sets gives us a set of binary volume models describing the varying shape of a given individual in different postures.

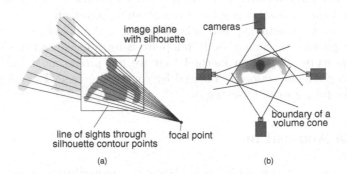

Fig. 2. Shape-from-Silhouette: a) For each image, the person's silhouette is determined using chroma-keying. b) The 3D shape is reconstructed by intersecting all cones obtained as the silhouettes are being projected back into space.

3   Parametric Body Template

The following section outlines the parametric model that we use to describe the shape of the human body, its appearance and movability. The latter element is controlled by an internal bone structure while shape and appearance are approximated with a textured triangular mesh.

3.1   Skeleton

The model used consists of a hierarchy of skeleton parts as depicted in the left column of Fig. 3. The root of the hierarchy is situated in the lower torso. Each skeleton part has exactly one parent and one or more children (e.g. the upper torso is connected to three children, the neck and the left and right upper arm). Each skeleton part is connected to its parent by a joint, where its own local coordinate system is situated [12]. A joint can handle up to three degrees of freedom. Every motion occurs within the local coordinate systems allowing a hierarchical motion transmission. Moving the shoulder, for instance, leaves everything else in the hierarchy (upper/lower arm, hand etc.) untouched.

3.2   Deformation Model

Together with the skeleton, a surface mesh is created. For each part of the skeleton, the vertices of this initial triangle mesh are shaped into rigid rings which will be lined up along the bone (Fig. 3a). Each vertex-ring has a relative position and orientation within the local coordinate system of its corresponding part. Thus, the shape of the rigid rings may be controlled independently for each part of the body, giving a rough approximation of the shape to be utilised in the following shape adaptation. If the skeleton is bent or twisted at a given joint, the transformation is decreasingly propagated to the vertex-rings on both sides of the joint, i.e. the rings tilt on the bone as shown in Fig. 3b. At present, we use a simple linear propagation function. The propagation range varies depending on the joints and may even vary from parent-side to child-side. Thus, the appropriate deformation model can be chosen for different joints, e.g. the turning of the head would be interpolated solely within the neck. Depending on the number of children, skeleton parts are either standard limb-like parts or special elements. The mesh in Fig. 3, which consists of simple rigid vertex rings en-casing the internal bone structure, is only usable for limb-like parts of the body, i.e. parts connected to their parent and one single child. Special parts are connected to several children, and require therefore a more complicated mesh: the mesh will be created first, and its vertices affixed to a rigid ring structure (necessary for the deformation of body parts) later on (see [6] for details).

4   Multiple Posture Adaptation

4.1   Overview

This section outlines the evaluation strategy that we are suggesting for multiple posture analysis. The human body template that we introduced in the previous section presents a default posture, i.e. joint angles, a default size, i.e. bone

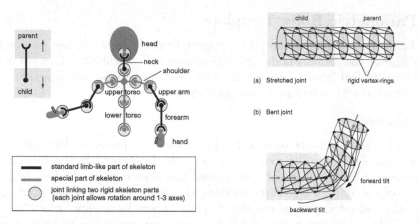

Fig. 3. A simplified hierarchical skeleton is used to control the surface deformation (left). Each deformation is based on the linear interpolation of the relative motion over rigid vertex rings (right).

lengths, and a default shape, all of which will have to be adapted to the measured data. Additionally, the photo-texture has to be extracted from the input images and assigned correctly to the model-surface. Whereas the model's size and surface texture are assumed to be constant, posture and shape will vary. Consequently, multiple posture analysis aims at determining one unique set of *posture-invariant* parameters and the corresponding multiple sets of *posture-dependent* parameters in such a way that all postures be represented as realistically as possible. Since the captured postures are unknown a priori, a reliable scale and posture estimation will be required initially for each input posture. Details on our approach for this initialisation, which is based on the 3D skeletonisation of the volume models, appear in the following subsection. Once initial parameters have been determined independently for all input data sets, an initial unique set of posture invariant parameters is calculated and the corresponding first posture-dependant parameters are determined, giving us the initial parameters that we need for subsequent optimisation. During the iterative optimisation described in the third subsection, the posture-invariant parameters are varied in a hierarchical order, we try and find the best corresponding posture-dependant parameters for each input data set. Both posture-invariant and posture-dependant parameters are evaluated using criteria that analyse the remaining differences between the model shape and the measured volume model. Some aspects of texture mapping are briefly discussed at the end of this section.

4.2   Initial Scaling and Posture Adjustment

In order to determine the initial parameters that describe the respective postures, i.e. joint positions, we use a four-step approach. First step: the 3D thinning algorithm proposed in [13] is applied to each volume model. Due to the limited number of cameras, the shape of the binary volume models is locally erroneous, causing a noisy skeleton structure with many erroneous branches. Second step: the relevant branches of the obtained skeleton are extracted by matching the skeleton against a target structure (see Fig. 4). This is done for arms, legs and torso by searching the combination of 4 branches that maximises the matching-criterion given in (1) *and* at the same time forms a *double-Y-structure* like the one obtained when removing the head-branch of the target structure in Fig. 4.

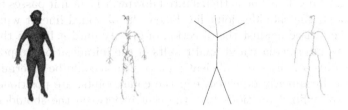

Fig. 4. From left to right: Initial volume model, raw skeleton obtained by thinning, target structure (double-Y and head) and extracted relevant skeleton parts.

$$q_{match} = \frac{min(l_1, l_2)}{max(l_1, l_2)} + \frac{min(L_1, L_2)}{max(L_1, L_2)} + 2 \cdot \frac{S_h}{S} \tag{1}$$

With l_1 and l_2 being the respective lengths of the current arm candidates, L_1 and L_2 the respective lengths of the leg candidates, q_{match} gives us a high rating for pairs of equal-size leg and arm branches. The third term expresses the ratio between the size S_h of the double-Y-structure already obtained and the overall size of the raw skeleton S, i.e. it privileges solutions that cover bigger parts of the raw skeleton. After extracting arm-, leg- and torso-branches, the head-branch has to be determined. To that end, we will look for the longest remaining branch of the raw skeleton around the arm junctions.

Third step: using the junction-points of arms and legs for an initial segmentation of the skeleton, joint positions are determined with regards to the skeleton of our human body template. Whereas hips and ankles can easily be detected by locating significant angles in the skeleton, shoulder-joints, knees and elbows require advanced techniques: to determine the shoulder-joint positions, the torso of the corresponding volume model is approximated by a cone and the sections between this cone and the arm branches of the skeleton are used as joint positions (see middle row of Fig. 5). To determine elbow and knee positions, the respective skeleton branches, i.e. 80 percent of the distance between shoulder joint and fingertip for the arm, 80 percent of the distance between hip and ankle for the leg, are divided into two parts of equal length, and the corresponding parts of the volume model are approximated by cones. The dividing point will then be shifted inwards and outwards along the skeleton within a pre-defined range; the position in which the obtained cones approximate the volume model best is finally used as joint position.

The last step aims at compensating the effect of clothes that, particularly in the upper arm region, tend to make the skeleton 'collapse', affecting the joint positions. As we use a calibrated setup, we know the direction of gravity. We can thus define a plane spanned by the difference vector between the determined shoulder and elbow joint positions and the direction of gravity, i.e. a section plane cutting through the upper arm that includes the direction of gravity. The upper contour of the arm within this section plane provides the most reliable information on the true upper arm direction and will be therefore approximated by a straight line. This line is then shifted downwards until it passes through the elbow joint. If the shoulder joint lies below the obtained line, it will be shifted upwards, i.e. raised against the direction of gravity until it lies on the line.

Fig. 5 shows intermediate and final results of the initialisation proposed for different postures. In all cases sufficient joint positions could be determined. Joints that cannot be directly detected, e.g. the collar joints, are positioned last, using fixed pre-defined relations like 'the centre between the shoulder joint and the arm junction'. Once we know the joint positions for every posture, we will calculate the mean bone lengths and respective joint angles.

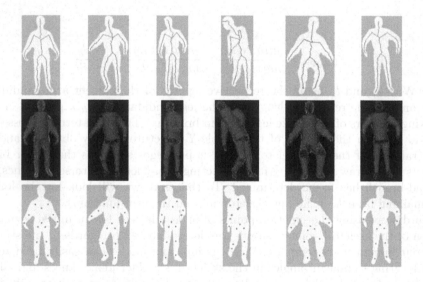

Fig. 5. Skeletons in varying postures. Top row: after the clothes weighing down effect has been corrected. Middle row: conic surface approximation. Bottom row: estimated joint positions.

4.3   Iterative Shape and Skeleton Optimisation

Shape Update. At the beginning of each iteration cycle, the local surface shape of the model will be updated. For each vertex, in every posture, we will determine the value δ of the shifting along the local surface normal required to move the vertex to the surface of the volume model (see left of Fig.6). From this analysis, we get a set of shifting values for each vertex. During the shape update, the model is moved back into its default posture and the respective mean shifting values are applied to each vertex.

Skeleton Update. Once the shape of the model has been updated, the bone lengths of our internal skeleton are optimised. This is done in a hierarchical order, starting from the root joint. Hence, for the upper body: the length of the lower torso will be optimised first, then the distances between the back joint and neck as well as between the left and right collar joints, and so on. After varying the length of a bone, new optimal angles are determined for all the adjacent joints in each posture before re-evaluating the bone length. Several approaches to estimate the joint angles in an articulated object have recently been proposed [15]. However, since we aim at the optimal combination between an internal skeleton and a surface shape with respect to a given deformation-model, we use a specific optimisation that relies on an evaluation based on the remaining shape error. The shape error is measured integrating the local vertex shifting value δ necessary to fit the model surface to the volume model surface in a given posture. Once the length of a bone has been changed, the angles of all adjacent joints will be updated (see right of Fig. 6). As our deformation-model changes, any joint angle will influence the model shape on both sides of the joint. Thus, to evaluate any given angle, the corresponding parent and child have to be considered. In the case of limb-like children, the grandchild will also be considered to determine the child's twist correctly. The mean square sum of the vertex shifting values (δ) of all vertices of the parent, the child and in some cases, the grandchild, is therefore used as our error criterion to optimise the angles of a joint for a particular posture. Once the angles of the adjacent joints of a bone have been updated in all available postures, the length of the given bone will be evaluated by calculating the variance of the shifting values for each vertex in all postures. The mean variance of the given bone, its parent and child, tells us how accurately the surface deformation in this area is described by the deformation-model, and we shall therefore use it as our shape error criterion.
Shape and skeleton update will then be iteratively repeated until no significant changes occur.

4.4   Texture Mapping

Once suitable size and shape parameters for the model have been determined, the texture information collected from the original images has to be assigned to the model surface. Mapping texture from multiple calibrated views to a rigid object model using *uv-mapping* has been described in [14], and further applied to deformable objects in [6]. Although we basically follow the same approach, some additional issues should perhaps be mentioned: when texturing an object from multiple camera images, changing illumination, shape or calibration errors may cause texture seams to appear along the borders between surface parts

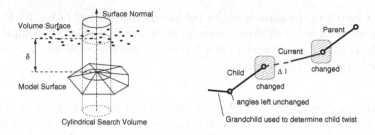

Fig. 6. Determination of local shape error δ (left) and evaluation scheme for bone length update (right).

textured from different images. Blending techniques may be employed to reduce this problem; however, we recommend that as few different camera images as possible be used for the texturing process, and in particular in the case of images of different postures, as the shape variation between the different postures is not completely described by the simple deformation model that we are using. Therefore, we choose a posture with few self-occlusions, move the model into that posture and texture all visible surface parts from images of that posture. Only the remaining occluded surface parts should be textured using images from different postures, once the model has been moved into the respective posture.

5   Results

Fig. 7 shows eight frames of an animation obtained by creating a model from data sets of six different postures (see Fig. 5) and applying a captured motion data set of a long jump. The resulting deformation of the model looks quite realistic and demonstrates potential performances of the method that we are proposing. However, these results are first results and have to be verified against other input data sets.

Fig. 7. Several frames of an independently captured long jump motion data set applied to a human model created with the method proposed here.

6   Conclusions

We have described a method to fit a parametric human body template to the human body using several sets of multiple camera images that show an individual in different postures. 3D skeletonisation of volume models obtained through shape-from-silhouette has been used to estimate initial postures. Subsequently, postures, size and shape of the model have been optimised iteratively before mapping photo-texture extracted from the input images to the model surface. The method that we are proposing shows promising results, yet remains computationally very time-consuming. Future work, therefore, should concentrate on reducing process-time through improved optimisation, and on conducting detailed comparison with results obtained through single posture adaptation. Further, we shall be working on improving the rough shape obtained from shape-from-silhouette by adding a predefined facial mask to our body template.

References

1. Maestri, G.: Digital Character Animation 2: Essential Techniques Vol. 1. New Riders Publishing, Rome (1999)
2. Thalmann, D., Shen, J., Chauvineau, E.: Fast Realistic Human Body Deformation for Animation, VR Applications. Proc. Computer Graphics International 1996, Pohang, Korea. IEEE Computer Society Press (1996) 166–174
3. Wilson, Wicks, GB, Company-URL: http://www.wwl.co.uk/triform.htm
4. Cyberware, US, Company-URL: http://www.cyberware.com
5. Kakariadis, I.A., Metaxas, D.: Three-Dimensional Human Body Model Acquisition from Multiple Views. Int. Journal of Computer Vision 30(3) (1998) 191–218
6. Wingbermühle, J., Weik, S.: Towards Automatic Creation of Realistic Anthropomorphic Models for Realtime 3D Telecommunication. Journal of VLSI Signal Processing Systems, Vol. 20 (1998) 81–96
7. Hilton, A., Gentils, T.: Popup People: Capturing human models to populate virtual worlds. Proc. SIGGRAPH (1998)
8. Weik, S.: A Passive Full Body Scanner Using Shape From Silhouettes. Proc. ICPR 2000. Barcelona, Spain (2000) 99–105
9. Kakariadis, I.A., Metaxas, D.: Model-Based Estimation of 3D Human Motion. IEEE Trans. PAMI 22(12) (2000) 1453–1459
10. Bregler, C., Malik, J.: Tracking People with Twists, Exponential Maps. Proc. CVPR 98. Santa Barbara, California (1998)
11. Niem, W.: Robust and Fast Modelling of 3D Natural Objects from Multiple Views. SPIE Proc., Image and Video Processing II, Vol. 2182 (1994) 388–397
12. Badler, N., Philips, C.B., Webber, B.L.: Simulating Humans. Oxford University Press, Oxford/UK (1993)
13. Min Ma, C., Sonka, M.: A Fully Parallel 3D Thinning Algorithm and Its Applications. Comp. Vis. And Image Understanding, Vol. 64, No. 3, (1996) 420–433
14. Niem, W., Broszio, H.: Mapping Texture from Multiple Camera Views onto 3D-Object Models for Computer Animation. Proc. International Workshop on Stereoscopic and Three Dimensional Imaging. Santorini, Greece (1995) 99–105
15. Gavrila, D.M.: The Visual Analysis of Human Movement: A Survey. Computer Vision and Image Understanding, Vol. 73, No. 1 (1999) 82-98

Augmented Reality and Semi-automated Landmarking of Cephalometric Radiographs

B. Romaniuk[1], M. Desvignes[1], J. Robiaille[1], M. Revenu[1], and M.J. Deshayes[2]

[1] GREYC-CNRS 6072
Bd Maréchal Juin
14050 Caen Cedex, France
{bromaniu, michel, jrobiail, mrevenu}@greyc.ismra.fr
http://www.greyc.ismra.fr/~bromaniu/Pres.html
[2] Société Télécrâne Innovation
http://www.cranexplo.net

Abstract. In this paper, we propose computer assisted visualization for manual landmarking of specific points on cephalometric radiographs. The signal to noise ratio of radiographs is very low, because of superimposing of anatomical structures, dissymetries or artefacts. On radiographs of children, the localization of cephalometric points presents a great inter-subject, and inter- and intra-expert varibility, which is considerably reduced by considering an adaptative coordinates space. This coordinates space allows us to obtain statistical landmarking of cephalometric points used to define regions of interest. Each region takes advantage of a specific image processing, to enhance local and particular features (bone or suture). An augmented reality image is presented to the human expert, to focus on main sutures and bones in a small region of interest. This method is applied to the nettlesome problem of the interpretation of cephalometric radiographs, and provides satisfying results according to a cephalometric expert.

Keywords: statistical shape model, augmented reality, cephalometry

1 Introduction

The goal of cephalometry is the study of the skull growth of young children in order to improve orthodontic therapy. It is based on the landmarking of cephalometric points on radiographs, two-dimensional X-ray images of the sagittal skull projection [2]. These points are used for the computation of features, such as the length or angles between lines. The interpretation of these features is used to diagnose the deviation of the patient form from the ideal one [5]. It is also used to evaluate the results of different orthodontic treatments [3] [9]. However, the anatomical definition of cephalometric points is difficult to apply directly on radiographs. The landmarking is hard and an important inter- and intra-expert variability is to be noticed. It is necessary to reduce this variability.

W. Skarbek (Ed.): CAIP 2001, LNCS 2124, pp. 410–418, 2001.
© Springer-Verlag Berlin Heidelberg 2001

Cephalometric points are usually precisely defined in relation to skull and bones which are three-dimensional objects. Unfortunately it is sometimes impossible to apply directly these definitions on radiographs, which are bidimensional projections of the three-dimensional object. Most cephalograms are defined as an intersection between bones and sutures; this appears as a curvilinear white (bone) and black (suture) structure on the radiograph. They are often difficult to locate because of the superimposition of many other bones on the 2D projection of the radiograph [6], and the definition of cephalometric points is specific for each point: we must decompose the problem into many sub-problems, specific to each landmark.

The goal of this work is a computer assisted landmarking using augmented reality. On a cephalogram, a small region of interest is automatically defined for each cephalometric point. The contrast of this region is modified to enhance feature used to locate cephalometric points. These regions are the support for manual landmarking of cephalometric points on radiographs by experts.

The first step of the work consists of finding a coordinates space that will evolve with the shape, the size and the orientation of each skull. The inside outline of the skull is the main component of this coordinates space because of its adaptivity to the morphology of each patient [8]. The second step is the statistical landmarking of cephalometric points. Training has been done on 58 samples which provide statistical coefficients used to determine the statistical position of cephalometric points. On new radiographs, the statistical landmarking of cephalometric points gives an approximation of the final position of the points. Using the standard deviation of the position, we defined small regions around each cephalometric point. Some of the points, with the help of the expert, are gathered, allowing the definition of regions of interest. The regions of interest contain the cephalometric points as well as anatomic structures, that the expert needs for landmarking. For each region of interest, we use augmented reality to enhance the contrast of the main structures used to locate the cephalometric landmarks.

In the first part of this paper we will present the automatic localization of regions of interest that is composed of the definition of the coordinates space, the statistical landmarking of cephalograms, and the definition of regions of interest. The second part deals with the augmented reality process in the standard case for the first and sixth regions. Lastly the results and conclusion of our study will be presented.

2   Automatic Localization of Regions of Interest

2.1   An Adaptative Coordinates Space

The idea is to extract a simple form that could characterize the dimensions of the skull. Then we decide to consider the internal surface of the skull (endocranial contour) that will bring us a simple numeric model for the position and the form of the skull.

The main problem of the contours detection is the difference of the intensity of contours which depends on the regions in which they lie. The bone parts of the superior part of the skull are more pronounced than those located in the vicinity of the sphenoid, where the low contrast makes the detection difficult.

Deriche's gradient is used to detect edges [4], and the endocranial contour is detected by active contours, based on the Cohen's model of balloons [1]. The result is visible on Figure 1. The curve (endocranial contour) is sampled by only 16 points. The first point is the one with the higher curvature, that is near the intersection of the curve and the nasal bone. This point is stable, *i.e.* it is always present in the same area of any image. The other 15 are regularly distributed on the endocranian curve, starting with the point of maximal curvature. 5 points are also considered: the isobarycenter of all the points of the endocranial curve, and the 4 corners of the smallest rectangle containing the curve. Figure 2 presents those 21 points.

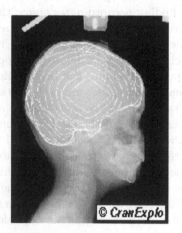 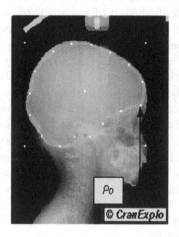

Fig. 1. Endocranial contour detection   **Fig. 2.** Sampled endocranial contour

Let ζ be the set of vectors :

$$\zeta = \{(O_i, \vec{v_i}) \mid \forall \, (j,k) \in \{1...n\}, \, j < k, \, (\exists \, i) \, O_i = P_j \text{ and } \vec{v_i} = \overrightarrow{P_i P_k}\} \quad (1)$$

The points P_i are the characteristic points extracted from the endocranial contour curve. We then obtain the set of $p = n(n-1)/2$ couples, made of one point and a vector. ζ is the set of all vectors that can be made between the characteristic points. For the given point M, we define the set of parameters $\{\alpha_1, ..., \alpha_p\}$ in the following equation :

$$\alpha = \langle \overrightarrow{OM} \mid \vec{v} \rangle \quad (2)$$

The change of the cartesian coordinates to the set of $\{\alpha_1, ..., \alpha_p\}$ can be written as a linear system. We define the vector X that represents the point M using homogeneous coordinates. Then we obtain:

$$X = \begin{pmatrix} M_x \\ M_y \\ 1 \end{pmatrix} \tag{3}$$

The computation of the coefficient α_i, associated to the couple $(O_i, \vec{v_i}) \in \zeta$ can be written as the equation :

$$\alpha_i = \begin{pmatrix} v_{ix} & v_{iy} & -(O_{ix}v_{ix} + O_{iy}v_{iy})) \end{pmatrix} \begin{pmatrix} M_x \\ M_y \\ 1 \end{pmatrix} \tag{4}$$

Then, when we have the set ζ, we can associate to every point M its parameters α_i by the matrix multiplication $\alpha = AX$.

The coordinates α_i are the projection of the point M on each vector built using the endocranial contour. This system of coordinates allows every point of the cephalometric radiograph to evolve with the shape and the morphology of the patient. This is our coordinate space.

2.2   Statistical Landmarking of Cephalograms

The method of global landmarking of cephalometric points is based on the estimation of the mean value of α, $\hat{E}[\alpha]$, on a training set of evaluated radiographs, where we know the exact position of the cephalometric point. Then, when we analyse a new radiograph, we assess α by $\hat{E}[\alpha]$.

This estimation is made on a training set of 58 evaluated radiographs. We know how to determine the set ζ by means of the endocranial contour. This allows us to find the matrix A. To locate an unknown landmark Y on a new radiograph, we need to inverse the system $\hat{E}[\alpha] = AY$.

The resolution of this system is obtained by using a modified version of the least squares method. The idea is to consider the variance $\sigma_i = \hat{E}[(\alpha_i - \hat{E}[\alpha_i])^2]$ as a balancing coefficient. Indeed, if the variance $\hat{\sigma}_i$ of the coefficient α_i remains low during the calculation of $\hat{E}[\alpha]$, it means that the information contained by this coefficient is more precise than the one contained by another coefficient, with a higher variance. We introduce the ponderation matrix:

$$P = \begin{pmatrix} \frac{1}{\hat{\sigma}_0} & 0 & \cdots & 0 \\ 0 & \frac{1}{\hat{\sigma}_1} & \cdots & 0 \\ \vdots & \vdots & \ddots & \vdots \\ 0 & 0 & \cdots & \frac{1}{\hat{\sigma}_p} \end{pmatrix} \tag{5}$$

The cost function in the least squares inversion becomes: $J = \| P\hat{E}[\alpha] - APX \|^2$, because not every coefficient has the same importance in the resolution.

Finally, we will only retain the highest coefficients of P, and the others will take the value 0. Only 20 coefficients are kept, because the cost function is minimum for this number.

The position of the unknown landmark X is given by the equation:

$$\hat{X}_{MC} = (A^t P^t P A)^{-1} A^t P^t \hat{E}[\alpha] \tag{6}$$

When we want to compute the position of a point on a new cephalometric radiograph, we compute the adaptative coordinates space. In this way we obtain the vectors $\vec{v_i}$ that define the internal surface of the skull. Then we may compute the matix A, and finally solve the equation (6) that gives us the coordinates of the landmark.

2.3 Regions of Interest

The next step in this study is the definition of regions of interest. This definition is based on a statistical analysis of landmarking of cephalometric points computed in the previous phase.

The windows of interest are radiography regions. They gather some cephalometric points and structures (bone or suture), that are necessary for a good localization of a cephalometric point by a human expert. Six regions of interest have been defined by an orthodontist and an image processing expert. For example, to find the real position of the point M, we need the direction of the ascending branch of the maxillary. Then the window number one contains three cephalometric points and the ascending branch of maxillary.

The statistical position of cephalometric points allows us to compute the errors between the computed position and the real position obtained by the expert. The computation is done on the training sample of radiographs. The standard deviation is computed between the statistical coordinates and the expert coordinates according to the X-axis and the Y-axis. The limits of the windows use the standard deviations previously computed on the training sample. Assuming gaussian hypotheses for the distribution of the position, a cephalometric point is located in a three standard deviation radius (according to the X-axis, and the Y-axis) from the statistical location of this point with an error of 0.5%. This area delimits the probability zone for every cephalometric point.

The limits of the regions of interest require the definition of the probability zone of every cephalometric point they contain. The frontiers of those probability zones are then used to delimit the final interest windows. Figure 3 presents an example of the first window. The three white frames represent the probability zones of the three cephalometric points included in the first region of interest. On this image, the extreme frontiers 1, 2, 3 and 4, of the three probability windows are used to define the frontiers of the final window. These frontiers are adjusted by some expert constraints. For example, in the first window, the orthodontist needs the direction of the ascending branch of the maxillary to find the right position of cephalometric points: the height of this window is then increased. The 14 points require six windows that contain between one to four points to locate. Figure 4 represents the result of an automatic localization of these windows on the radiograph.

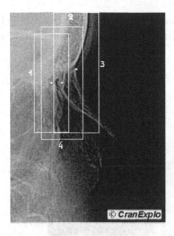 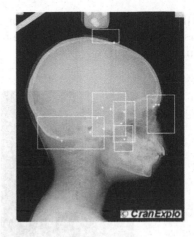

Fig. 3. Probability zones used to define the final interest window   **Fig. 4.** Automatic localization of interest windows on a radiograph

3   Augmented Reality

3.1   Standard Processing

The next step is to provide a better visualization to help the orthodontist in the labelling of cephalometric points. It is implemented by the enhancement of the main structures that are needed to locate each cephalometric point [7].

Each of the regions of interest present different particularities, consequently every window will take profit of a specific processing. The windows contain much visual information, but only a few of them are important during the localization. Thus, we try to have these information more visible. The window one and six take profit of very specific treatments. The others present the same characteristics and a very simple treatment was judged satisfactory by the human expert. This treatment is a histogram equalization.

3.2   First Region of Interest

The ascending branch of maxillary, which is a black structure on the radiograph, has to be enhanced. The solution of this problem requires the detection of thin lines, and the enhancement of the contrast in the regions which contain black small structures [6].

For the detection of fine curves, we compute the Deriche's gradient image. This image contains the fine structures that we want to detect, but also noise. This noise is the result of soft tissue and useless structures. To eliminate this additional information, we apply a hysteresis thresholding on the gradient image. The lower threshold is equal to two standard deviations computed on the first interest window in the gradient image. The higher threshold equals three standard deviations.

All contour chains that are no longer than three pixels are eliminated. Those chains can not be considered as the ascending branch of the maxillary, which is a thin and long curve.

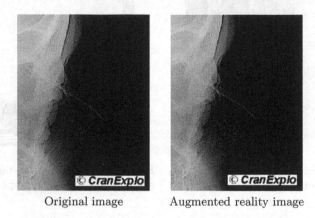

Original image   Augmented reality image

Fig. 5. Results obtained for the first region of interest

Finally, the treated gradient image is superimposed on the original one. Pixels of the detected contour (i.e. the maxillary) are set to 0, as well as their neighbours, other values are unchanged. We then obtain the augmented reality image, with the ascending branch of maxillary. Figure 5 presents the results obtained.

3.3   Sixth Region of Interest

The fronto-parietal suture, which is usually undetectable, has to be enhanced. It is located in a zone of the image, in which the cephalostat adds some information. The second problem is that this suture disappears for older children. The suture is a curvilinear structure, which appears darker than the neighbourhood. It is located on the exocranian contour, and lies towards the center of the skull. The angle between the suture and the exocranial contour is always around 60°. The detection of this suture is made using these features. The detection scheme is an iterative one: we start with a curve on the exocranial contour. Several positions of the suture are obtained by fitting an intensity model of the suture along the previously defined curve. These positions must be linked to the previously detected points, except for the 3 first iterations. The curve is then displaced toward the barycenter of the skull, and the process is iterated.

The structures detected by this method are thin, dark structures, connected to the exocranial contour. The three longest are kept by discarding all the little noisy structures. We conserve the three longest, because the cephalostat in this window often appears as two "artificial" sutures. Figure 6 presents the results obtained for the sixth window.

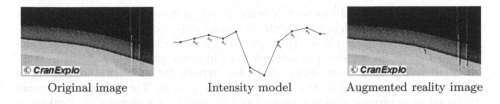

| Original image | Intensity model | Augmented reality image |

Fig. 6. Results obtained for the sixth interest window

4   Results

Landmarking

An evaluation of the error between the results of the statistical landmarking and the expert one was completed. In the region near the fronto-nasal region, in which we can found the first region, the mean error is about 4.5 mm, with the maximum error about 12.5 mm. The other regions, in which most of cephalometric points where found, the mean error is about 3.5 mm, and the maximum is reduced to 6.5 mm. The error values must be compared to intra and extra expert variabilities, which has been reported to be higher than these values.

Landmarking

The efficiency of the proposed processing for the interest windows one and six was evaluated as well. Specific image processing of windows one and six are considered as unsatisfying result, if they enhance false structures, or if they does not enhance the true structure. 90% of images of augmented reality obtained by the process of the first window give satisfying results. The proposed process for the sixth window is less efficient: only 85% of result images are interesting for the practitionner. However, the reason of these failures is often the abcence of anatomical structures in the original image. Another reason of failure for the sixth region of interest can be the superimposition of the anatomical structures with the artefact (cephalostat) in the window.

5   Conclusion

This paper deals with the difficult problem of landmarking X-Ray cephalograms. We have proposed an augmented reality scheme, for computer assisted landmarking. The first step has been to locate automatically and to design regions of interest. In these regions, a specific image processing enhances local features, and helps the practitioner to locate precisely the cephalometric points.

Our methodology allows us to obtain an adaptative coordinates space, that evolves with the growth of each skull. In this specific coordinates space, cephalometric points are statistically landmarked. Then the location leads to the determination of the windows of interest, on which the practitioner focuses. Augmented reality images are proposed to the expert, for manual landmarking of cephalometric points. The results are quite promising. Those treatments were included in a softwear developped by the TCI society (Cranexplo©). Regions of interest are also presented in a new pedagogical publication [7]. Future investigations will deal with a large database of cephalometric radiographs and with building deformable models of suture and bone to fit into the regions of interest.

References

1. Cohen, L.D., Cohen I.: Finite-Element Methods for Active Contour Models and Balloons for 2-D and 3-D Images. IEEE Transactions on PAMI, Vol 15 (1993) 1131–1147
2. Cretot, M.: L'image téléradiographique en céphalométrie. Éditions cpd (1989)
3. Davis D.N., Taylor, C.J.: A Blackboard Architecture for Automatic Cephalogram Analysis. Medical Informatics, Vol 16 (1991) 137–149
4. Deriche, R.: Fast algorithms for low-level vision. IEEE Transactions on PAMI 1(12) (1990) 78–88
5. Deshayes, M.J.: Nouvelle approche de céphalometrie: le projet télcrâne international. L'orthodontie française (1997)
6. Deshayes, M.J.: Cranofacial Morphogenesis. CD-ROM (1998)
7. Deshayes, M.J.: Repérages crâniens Cranial Landmarks. Editions CRANEXPLO (2000)
8. Desvignes, M., Romaniuk, B., Robialle, J., Revenu, M.,Deshayes, M.J.: Computer Assisted Landmarking of Cephalometric Radiographs. IEEE Southwest Symposium on Image Analysis and Interpretation. Austin, Texas (April 2000)
9. Ratter, A., Baujard, O., Taylor, C.J., Cootes, T.F.: A Distributed Approach to Image Interpretation Using Model Based Spacial Reasoning. BVMA (1993) 476–481

On Restoration of Degraded Cinematic Sequences by Means of Digital Image Processing

Slawomir Skoneczny and Marcin Iwanowski

Institute of Control and Industrial Electronics (ISEP)
Warsaw University of Technology
00-662 Warszawa ul. Koszykowa 75 POLAND
slaweks@isep.pw.edu.pl

Abstract. There are thousands of old black and white movies that are the cultural heritage of nations. These films are quite often seriously degraded. This is a problem of significant importance especially in Poland, where most of cinematic heritage was damaged during and after World War II. There is a wide spectrum of defects of different kinds and various complexity, which is a serious challenge for image processing scientists. In this paper a systematic methodology for solving these difficult problems is proposed. It contains an analysis of most common defects and introduces their taxonomy. The most important part of the work is devoted to the detection and removal of degradations. For this purpose different tools of image processing are applied, especially based on mathematical morphology. Considering the diversity and complexity of the defects one can easily observe that there is no uniform methodology that could be successfully applied to all degradation types. Unfortunately it does not seem to be possible to detect and remove all of them completely automatically. Therefore the whole system for semi-automatic treatment (with limited human interaction) is proposed.

Keywords: image restoration, degraded image sequence

1 Introduction

For at least one decade a need for automatic restoration of old degraded motion pictures has been growing. Classical Optical Film Processing (OFP) methods perform well in very simple cases of small degradation artifacts and to some extend if there is only a little chemical damage of the film material. On the other hand digital image processing techniques offer much more flexibility especially if they are slightly supported by interactively acting human computer operator. Recently some papers on restoration of degraded sequences have been appeared, concentrating attention on particular defects [1,2,3]. In our paper we propose a methodology for treatment of destroyed image sequences in order to obtain significant quality improvement. It is achieved in two stages: defects detection phase and defects removal step (spatiotemporal filtering plus interpolation).

W. Skarbek (Ed.): CAIP 2001, LNCS 2124, pp. 419–426, 2001.

2   Taxonomy of Defects

Many different defects in old motion pictures are encountered but we have concentrated on such degradations like big blotches, spots, crosses, wide white or dark scratches.

- Big white or black crosses that are present in the same place in one frame (but not in the first frame of the scene) only -*nonstationary blotches*. The main reason of such defects was caused by human activity for film editing purposes. It usually marks the end of the scene, but it also decreases the visual quality of the movie.
- Big white or black blotches that are present in the same place or in neighboring places in a few consecutive frames -*stationary blotches*. The assumption here is that we operate within the same visual scene and that the first two frames of the scene are defect–free. These types of markers were done on purpose as an indication for film projector operator working in cinema to change the film reel (actually to use another film projector with another reel).
- White (black) vertical lines (nonstationary ones). The main reason for this type of defect is a mechanical damage caused by film projector to a film medium. The celluloid tape is destroyed mainly in the direction parallel to the direction of its transportation.
- "stochastic" noise – caused by time and improper film storage condition, too high or too low humidity or temperature.

3   Detection

3.1   Big, Single Black, or White Crosses

1. **Motion independent gradient calculation.** The defect is present in a single frame. It can be found easily using the image gradient calculated along the time axis. Traditional gradient in time-axis has one important disadvantage. It doesn't distinguish between image defects and motion. Considering the defects which appear in one frame, one can see that they have the same gradient image as moving object. But the gradient peaks resulting from movement permanently change their position. Gradient peaks connected with one-frame image defect are stable. Both features can be joined and new gradient can be introduced. We call it *motion independent gradient*. It has a form of the following equation:

$$g^*(I_t) = \inf(|I_{t-1} - I_t|, |I_{t+1} - I_t|)$$

In such a way gradient peaks connected with movement are removed. Alternatively, this type of temporal gradient may be used in the defects detection process.

2. **Cleaning the gradient.** The gradient image obtained in the previous step contains not only the required areas, but also some other, smaller ones being the result of some specific motion of people, objects and camera. These motion areas were impossible to remove. In order to get rid of these areas of gradient some filtering is necessary. The criterion of this filtering is based on assumption that the blotch to be found is bigger than other areas visible on the gradient image. The morphological filters [4] are especially well suited to this purpose because one can easily define the size of objects to be removed. In our case the areas brighter than the background have to be removed, what implies that the correct morphological filter is an *opening* defined as follows:

$$\gamma^{(n)}(f) = \delta^{(n)}(\varepsilon^{(n)}(f)) \tag{1}$$

where $\delta^{(n)}$ and $\varepsilon^{(n)}$ are the operations of dilation and erosion respectively. However, this type of filter introduces also a distortion of contours of objects. Fortunately mathematical morphology offers more complex filter - *the opening by reconstruction*. This operation belongs to group called geodesic morphological transformations [5] and is based on the geodesic dilation used in the operation of reconstruction. Geodesic dilation of size 1 of the marker image f with respect to the mask image g where $f \le g$ is defined as infimum image of mask image and marker image after dilation of size one - with elementary structuring element, and is denoted as: $\delta_g^{(1)}(f)$ and defined by :

$$\delta_g^{(1)}(f) = \delta^{(1)}(f) \wedge g$$

Geodesic dilation of size n consists of n successive geodesic dilations of size 1:

$$\delta_g^{(n)}(f) = \underbrace{\delta_g^{(1)}(\delta_g^{(1)} \ldots \delta_g^{(1)}(f) \ldots))}_{n-times}$$

Reconstruction by dilation of a mask image f from a marker image g where $f \le g$ is defined as successive geodesic dilation of f with respect to g performed until idempotence and is denoted by $R_g(f)$:

$$R_g(f) = \delta_g^{(i)}(f)$$

where i is such that (idempotence):

$$\delta_g^{(i)}(f) = \delta_g^{(i+1)}(f)$$

The opening by reconstruction of a gray-level image g is defined as follows:

$$\gamma_R^{(n)}(g) = R_g(\varepsilon^{(n)}(g))$$

Unlike traditional opening and closing, opening by reconstruction and closing by reconstruction preserve shapes in the image. It is very important while using these operations as a first step of segmentation. It makes possible the removal of the local peaks of gray-intensity without changing the borders of regions (which happens in traditional opening and closing).

3. **Binarisation.** The result of detection should contain only these areas which have been found. So it should be a binary image of value 1 for the pixels of detected object, and value 0 for others. The binarisation of the filtered image is performed by using the thresholding with a given threshold.

4. **Cleaning the binary image.** It could happen that the binary image after binarisation contains some unwanted regions, which does not belong to the defect. This is a similar situation to that after the gradient calculation. The solution is also similar in this case. It consists in the calculation of the opening by reconstruction, which removes the areas smaller than a given size, without deforming the detected objects.

5. **Region enlargement.** The binary image should contain the detected damaged areas. In order to obtain later better reconstruction, the binary shapes of the objects have to be slightly enlarged. The enlargement is obtained by the dilation of small size (1 or 2).

3.2   Big Multiple Black or White Blotches

In the current case the situation is different from the one presented in the previous section. The problem is in the length of the part of a sequence containing the object to be detected. In the former case we had both neighboring frames, in this one - both neighbors of the defected frame are also degraded. So the motion-independent gradient cannot be calculated. In this case the classical gradient must be generated. Fortunately the defects are in this case of different kind than in the former one. They have the form of a blob - relatively large area of a regular shape close to the oval. The detection is in this case relatively simpler. The complete detection algorithm is the following:

1. **Gradient calculation.**
   As it was written before, the classical temporal gradient is used instead of the motion-independent one. This type of gradient is defined as follows:

$$grad = abs(I_n - I_{n-1}) \tag{2}$$

2. **Gradient cleaning.**
   This step is similar to step 2 in the previous method. The same type of filtering is applied - the opening by reconstruction. The size of the operation could be in this case bigger than in the previous one because the defect to be detected is bigger.

3. **Binarisation.**
   The binarisation step is based on the thresholding of the filtered image, as in the previously described algorithm. Due to the fact that the size of opening by reconstruction is bigger than in the previous case, it removes bigger unwanted areas. Due to that the additional opening by reconstruction of the binary image is not necessary in this case.

4. **Region enlargement.**
   The binary image should contain the detected damaged areas. In order to

obtain later better reconstruction, the binary shapes of the objects have to be slightly enlarged. This enlargement is obtained by the dilation of small size (1 or 2).

3.3 Vertical Lines

Another often encountered damage is the case of vertical lines –scratches. Classical morphological operators consider the closest neighborhood of a pixel by using the elementary structuring element. It does not allow to analyse the image features in a given direction - all the directions are of the same importance. The directional operators use the structuring element consisting of the neighbors in one, particular direction. Due to this fact directional filtering allows to remove the areas of a particular orientation in the image. This feature is successfully used in the current case. The algorithm is as follows:

1. **Directional top-hat operation.** The top hat operation is defined in the following way:

$$WTH(f) = f - \gamma(f) \tag{3}$$

This operation allows to detect the areas on the initial image f, which are brighter than the background. The opening removes these areas and the difference with the initial image makes them visible. The scratches are brighter than the background so this operation is reasonable. But, when performed with an elementary structuring element, it detects all the areas darker than the background. In order to find these scratches only an directional opening in the direction perpendicular to the scratches may be applied. It finds successfully the appropriate areas.

2. **Cleaning the detected areas.**
   The images containing detected areas, being the result of the previous step, must be cleaned. Unfortunately apart from the appropriate regions the preceding operation found also some other areas which are not the scratches. These areas are removed taking into consideration the fact that the scratches themselves are relatively long, while other areas are much shorter. In order to filter the image in such a way that only the longer lines are left, one applies the directional opening but in this case in the direction parallel to the scratch of a relatively large size.

3.4 Noise

Concerning noise it is possible to use in such case a variety of morphological filters i.e. alternating sequence filters, morphological filters with opening by reconstruction or closing by reconstruction. In many cases spatio–temporal median type filters are sufficient.

4 Reconstruction

There exists many methods for detected defects reconstruction (removal). We assume that a defect is marked and its location is known having a proper reconstructing mask. Examples of such approaches are given below:

- Copying to detected areas of current frame defects (without motion compensation) from the previous frame area of the same location —this approach is acceptable for long blotch sequences in unchanged areas.
- Copying with motion compensation (bidirectional). We calculate motion vectors between I_{n-2} and I_{n-1} so that $[\Delta X, \Delta Y] = blockmatch(I_{n-2}, I_{n-1})$. We apply this vector to I_{n-1} in order to obtain warped image $I_{n-1}^{wrp}(x, y) = I_{n-1}(x + \delta x, y + \delta y)$. Then we copy pixels to marked area of I from area of the same location of I_{n-1}^{wrp}. The same technique is possible by taking into account frames I_{n+1}, I_{n+2}—this approach is acceptable for moderate long blotch sequences.

Fig. 1. An image with a blotch defect before and after reconstruction

5 An Example

In this case it is necessary to remove a big white blotch in upper right part of the frame. The parameters for operations applied is this example (Fig.1) are: opening by reconstruction of radius = 5, threshold = 30, dilation radius = 1.

6 Restoration System

The restoration system can be presented using the theoretical model presented in Fig. 2. The process of old movie restoration is divided into two main phases: automatic phase and manual phase. Although the fully-automatic system is the most interesting, the existence of manual phase seems to be difficult to avoid.

It is due to the fact that because of the variety and complexity of the defects in old films the fully-automatic restoration of degraded cinematic sequence is not possible in all of the cases. The automatic restoration methods can repair the damages using the information included in the not-damaged parts of the sequence. Unfortunately in some of the cases the amount of information in the not-damaged parts is not sufficient to restore the damaged parts. In such a case the human assistance in the reconstruction process is necessary.

The first, automatic phase consists of two parts. The first one - a detection stage applies various detection algorithms in order to find the destroyed areas of the damaged sequence. Since different defects could require different detection methods, the proposed model allows to use paralelly more detection algorithms. The areas detected by these algorithms are not only marked, but also labelled. The label of each defected area includes the information regarding the kind of the particular defect, which in some case can be obtained from the type of algorithm used to detect the particular defect. Such information can be helpful in the next, automatic part - reconstruction. All of the defects - marked and labelled - are stored in a separated sequence. It contains the defected areas and the appropriate labels. This sequence and of course the initial, damaged one make the input for the automatic reconstruction phase. In this phase the areas where the damages are present are treated by the reconstruction algorithms according to the label appropriate to a particular defect. As a result of this phase, the whole restored sequence is produced.

At this moment the automatic phase stops its main activity, and the manual phase starts. Three sequences build the input: the original sequence, the sequence with labelled defects and the automatically restored sequence. In this phase two operators are involved. The first one evaluates the quality of automatic reconstruction. The second one who is a computer artist repairs manually, using the painting computer programs, the parts of the sequence which cannot be repaired automatically.

The first operator works on three sequences mentioned above. He evaluates the precision of detection of the destroyed areas. If the detection was not correct he can manually mark the appropriate damaged region and decides in such a case to reapply the automatic reconstruction algorithm. He also evaluates the quality of the automatic reconstruction. If the quality is not satisfactory, he can decide to apply another algorithm of automatic reconstruction. In case when none of the automatic reconstruction algorithms gives a satisfactory results, he can classify the damage to be intended for a manual reconstruction, performed by the second operator.

The second operator works only on these parts of the sequence, which were classified by the first one, as unqualified for being reconstructed automatically. He is a computer artist, who paints manually the content of the these part of the sequence. He can also make all the necessary retouches of the finally restored sequence. After that the part reconstructed automatically is joined with the part restored and retouched manually in such a way the finally restored sequence is produced.

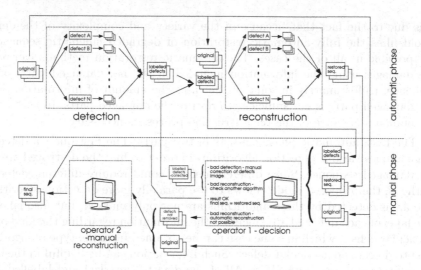

Fig. 2. The restoration system

7   Conclusions

The proposed methodology allows to apply the image sequence processing techniques based on mathematical morphology (used for frame defects detection) and taking advantage of motion vectors estimation to semiautomatic (with limited assistance of human operator) system for removal of certain typical classes of old movies degradations.

References

1. Kokoram, A., Morris, R., Fitzgerald, W., Rayner, P.: Detection of Missing Data in Image Sequences. IEEE Trans. Image Processing **11** (1997) 1496–1508
2. Kokoram, A., Morris, R., Fitzgerald, W., Rayner, P.: Interpolation of Missing Data in Image Sequences. IEEE Trans. Image Processing **11** (1997) 1509–1519
3. Fernandiere, E.D., Marshall, S., Serra, J.: Application of the Morphological Geodesic Reconstruction to Image Sequence Analysis. IEEE Proceedings: Vision, Image and Signal Proc. **6** (1997) 339–344
4. Serra, J.: Image Analysis and Mathematical Morphology, vol.1. Academic Press (1982)
5. Soille, P: Morphological Image Analysis. Springer–Verlag (1999)

Vision Based Measurement System to Quantify Straightness Defect in Steel Sheets

Rafael C. González[1], Raul Valdés[1], and Jose A. Cancelas[1]

University of Oviedo, ISA
33204 Gijón, Spain
{corsino,rvaldes,cancelas}@isa.uniovi.es
http://www.isa.uniovi.es/index.html

Abstract. A non-uniform distribution of rolling pressure during steel lamination may produce flatness asymmetries in the steel sheets, causing a certain curvature on its edges. This deformation may cause stoppages in the rolling process, and damages in the machinery. A computer vision system for measuring this straightness defect is presented. This system shows the adaptation of well-known computer vision techniques to fit precision and real-time constraints. Some problems that arise during the implementation phase are also described, and the correspondent solutions outlined.

Keywords: computer vision system, straightness defect measurment

1 Introduction

The system described in this paper is designed for the hot rolling facilities that ACERALIA has in Avilés. Due to a non-uniform distribution of pressure along the cross-section of the sheet, a flatness asymmetry may appear. It causes a bigger elongation of one side of the sheet with respect to the other, presenting a curved shape. Technicians call it "camber" defect and this is the term that will be used along this paper.

When "camber" exceeds a certain limit, the steel sheet gets stuck in the rolling mill. This causes the rolling process to stop, reducing the production level. In addition, machinery may be damaged increasing the economic losses. The method employed to minimize the "camber" consists in a manual correction of the pressure put on the sheets by the rollers. At present, that correction is based on human visual inspection. The correction to be applied is decided by the operator based on subjective appreciations of the reality. The aim of the system presented in this paper is to quantify the "camber" effect. The knowledge of that measurement will help the operator to correct the rolling pressure in both sides of the sheet. Besides, it provides a uniform way to diagnose whether a sheet is valid to continue in the manufacturing process or not.

The first step in hot rolling of steel products consists in a thickness reduction from 25-30cm to approximately 3-5cm depending on the thickness of the final product. This reduction is carried out in a progressive way. When the slab moves

W. Skarbek (Ed.): CAIP 2001, LNCS 2124, pp. 427–434, 2001.

forward into the roller, it is rolled. Once it leaves the rollers, it stops and goes backwards, entering the roller for second time. As the slab moves forward again, a new rolling stage is performed. The process is repeated from five to nine steps. In the odd steps the sheets move forward while suffering a thickness reduction, until desired thickness is achieved. Pressured water is used to remove the scrap from the slab, before it enters the roller, producing a great amount of steam. Water is projected from the roller to the sheet, in the opposite direction of the sheet movement.

2   Measurement System Description

The kind of measurement proposed show some specific features that are listed below:

1. Sheet lengths range from 40 to 80m.
2. The visible area of the rolling mill is only 20m long.
3. The sheet speed is variable, ranging from -5m/s to 5m/s. The measurement must be accomplished with no stop nor slowdown of the rolling process.
4. The steel sheets temperature ranges from 800°C to 1000°C.

Due to restrictions for camera placement, and looking for a cheap solution using of the self products, a configuration based on three CCD monochrome video cameras has been chosen. All of the cameras must be synchronized to allow a simultaneous acquisition of images through a RGB framegrabber. To coordinate the system with the production line, the host PC has a multi-function data acquisition board and a RS-485 communication port. A schematic view of the arrangement is shown in Fig. 1.

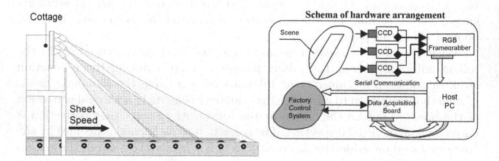

Fig. 1. Positioning of the cameras

The ideal position for the cameras would consist is placing their optical axis normal to the sheet upper surface, minimizing perspective distortion. Neverthe-less, the features of the plant do not allow it, due to cranes movements. Their trajectory would interfere with structure which should hold the cameras. Besides, the high temperature and humidity in that position would constitute a serious drawback. The chosen placement for the cameras was the front wall of a small cottage that is placed over the rolling mill as shown in Fig. 1.

The method employed in the measurement system combines image processing and some mathematical calculus. Besides, two simple assumptions are considered. In first place, the upper surface of the sheets is supposed to be flat. In second place, derivative of the curve described by the edges of the sheets is continuous. Both assumptions have proved to be reasonable.

The images of the sheet are processed to extract its edge points. From a proper camera calibration, and based on flatness assumption, real positions of edge points in a world reference frame are obtained. That information permits to define the geometry of the section of sheet captured by the camera. A mathematical algorithm, based on the second assumption, joins the different sections together in order to get the shape of the whole sheet.

3 Calibration

The aim of the camera calibrations is to be able to compute the position of points belonging to the sheet plane from their perspective images.

The well-known pinhole camera model is used. Points on a world plane are mapped to points on the image plane by a plane to plane homography, also known as a plane projective transformation [1], [2]. A homography is described by a 3×3 matrix H. Once this matrix is determined the back projection of an image point to a point on the world plane is straightforward (eq. 1). Homogeneous coordinates are used to represent world and image points, which are denoted P and p respectively.

$$\mathbf{P} \cong \mathbf{H} \cdot \mathbf{p} \tag{1}$$

In the preceding equation, $\mathbf{P}=[X, Y, W]T$, $\mathbf{p}=[x, y, 1]T$ and $\cong$ denotes equality up to scale. The scale of the matrix does not affect the equation, so only the eight degrees of freedom corresponding to the ratio of the matrix elements are significant.

H can be computed from image to world correspondences. Each correspondence between the image coordinates and the real ones of the identified calibration points, provides two linear equations in the H matrix elements. The homogeneous equations for n points are expressed in eq. 2, where the $\mathbf{H}$ matrix is expressed in vector form $\mathbf{h} = [h_{11}, h_{12}, h_{13}, h_{21}, h_{22}, h_{23}, h_{31}, h_{32}, h_{33}]$.

$$\mathbf{A} \cdot \mathbf{h} = \begin{pmatrix} \cdots\cdots\cdots\cdots\cdots\cdots & \cdots & \cdots & \cdots \\ x_i & y_i & 1 & 0 & 0 & 0 & -x_i X_i & -y_i Y_i & -X_i \\ 0 & 0 & 0 & x_i & y_i & 1 & -x_n X_i & -y_i Y_i & -Y_i \\ \cdots\cdots\cdots\cdots\cdots\cdots & \cdots & \cdots & \cdots \end{pmatrix} \cdot \mathbf{h} = \mathbf{0} \tag{2}$$

To ensure robustness of the calibration obtained, several correspondences are used, so the resultant equation system is over determined. It is a standard result

of linear algebra that the vector h that minimize the algebraic residuals —Ah—
subject to —h— =1 is given by the eigenvector of least eigenvalue of ATA. This
eigenvector is obtained directly from the SVD of A [3].

The calibration pattern and the chosen world reference frame are shown in
Fig. 2. Actually, there are three different calibration patterns, each corresponding
to one of the cameras. Two transversal stripes mark the transitions among the
three cameras.

Fig. 2. Calibration pattern and world reference frame.

Each camera pattern consists of a set of regularly sized rectangles, which
corners are used as the known world coordinates. The calibration plate must
be placed on the rolling mill and it extends along the area of inspection. It has
the same thickness as the steel sheets in the last step of the rolling process.
Due to the fact that "camber" effect is a very small deviation beside the total
length of the inspected sheet edges the construction tolerances of the calibration
plate must be very small. Special care must be paid to the alignment of the
longitudinal sides of the rectangles.

4   Detection of the Sheet Edge

By means of the camera calibrations, the position in the world reference frame
of image points belonging to the sheet plane can be computed. Therefore, the
next step is to obtain the image projection of the sheet edges. The algorithm for
detecting the sheet edges in the images is composed of three steps:

1. Conventional edge detection.
2. Removal of false edge points.
3. Subpixel edge detection.

The "camber" measurement is a real-time application. For this reason the
processing algorithms used are simplified.

Edge detection is based on Canny method [4]. First, the horizontal gradient
of the image is obtained using the corresponding Sobel operator. This operator
has the advantage of performing differentiation and at the same time providing

a smoothing effect, as mentioned in [5]. Therefore, filtering of images with that mask combines the first two steps of classical Canny edge detection. The chosen sheet edge to measure "camber" was the left one (both choices are equivalent), so possible edge image points belonging to the right edge must be removed. This is done directly assigning a null value to the points with negative gradient.[1] The image projection of the sheet edges is nearly vertical (see Fig. 3(a)). This allows the simplification of computing only the horizontal component of the intensity gradient. The processing time is substantially reduced with that simplification. Not only the spatial filtering operations are reduced to their half but, for each point in the image, computation of the modulus and the direction of the gradient is avoided. Then, the intensity gradient image is refined by looking for its local maxima.

Finally, a hysteresis thresholding is performed. This operation is controlled by two thresholds, a low one denoted t_{low} and a high one denoted t_{high}. The principle of this double thresholding is to keep pixels P that accomplish at least one of the three following constraints:

1. The horizontal gradient in P is greater than t_{high}.
2. The horizontal gradient in P is greater than t_{low} and P is 8-connected to at least one pixel whose gradient is greater than t_{high} directly.
3. The horizontal gradient in P is greater than t_{low} and P is connected to at least one pixel whose gradient is greater than t_{high} through a chain of 8-connected pixels whose gradients are all greater than t_{low}.

The choice of t_{low} and t_{high} is made automatically for each image. First, the histogram of the image with the local maxima of the gradient is made. Then t_{high} is set as the maximum value which is smaller than at least a certain number N of points in that image.[2] The value of t_{low} is a fixed percentage of t_{high}.

The shape of the image projection of sheet edges is practically straight. Besides, there is only one sheet edge on each image. Therefore, remaining false edge points can be removed through a very simple method. First, the line formed by the edge points is detected using the Hough Transform [6] with the normal equation of a straight line. Then, edge pixels whose distance from the detected line exceeds a certain value are removed from the image of edge points.

After the previous processing steps, an image with edge pixels is obtained. Each edge pixel contains the center of the true edge. In fact, the center can be anywhere within the pixel, so that the average accuracy is 0.5 pixel.

In the employed method [7], the position of the center of the true edge is estimated as the maximum of a parabola interpolating the values at the edge pixels and its two neighbours along the edge normal, namely its two row neighbours.[3] The precision achieved is less than 0.1 pixel.

---

[1] The left edge is a dark to clear transition, so the corresponding horizontal gradient is positive. The opposite occurs with the right edge.

[2] N is a preset constant. It is a fixed percentage of the expected edge points in the image.

[3] The gradient, which is normal to the edges, is considered to be horizontal.

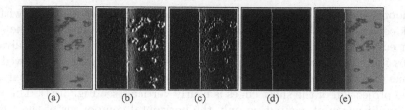

(a)   (b)   (c)   (d)   (e)

Fig. 3. Image region in different stages of the processing. (a) Intensity image. (b) Gradient image. (c) Local maxima image. (d) Edges image. (e) Overlapped intensity and edges image.

5 Concatenation of Sheet Edges

The employed method is based on the continuity of the derivative of the sheet edge shapes (second assumption of section 2). The algorithm starts with the world coordinates (X, Y) of the first set of sheet edge points (Fig. 4(a)), obtained by means of the techniques described in sections 3 and 4 .

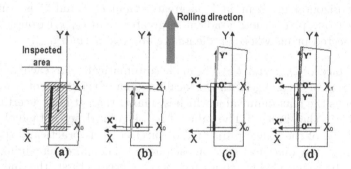

Fig. 4. World and sheet reference frames used to concatenate edges.

Then, a sheet reference frame (X', Y') is defined such that its x axis is tangent to the edge of the sheet at the point in $X = X_0$ which is also the origin of the new frame (Fig. 4(b)). The sheet frame coordinates of the edge points are obtained and those coordinates do not change with the sheet movement. When the steel sheet reaches the position in which the abscissa of O' exceeds X_1 (Fig. 4(c)), three new images are acquired from which the world coordinates of a new set of sheet edge points are obtained. In that moment, a second sheet reference frame (X'', Y'') is defined in an analogous manner as it was done with (X', Y'). This is shown in Fig. 4(d). Then, the coordinates (X'', Y'') of the world points are obtained. Now, half of the sheet edge points are known in (X', Y') coordinates, while the other half are expressed in (X'', Y'') coordinates. The (X', Y') reference frame position and orientation are determined and then a transformation between this two frames is performed so that all the known points are expressed in (X'', Y'') coordinates. At this moment, the situation is similar to that of the second step, so the operations continue the same way.

To determine the instants when images must be acquired, a concurrent worker thread in the program is receiving the instantaneous speed and computing the current advance of the sheet. Once the linkage process finishes, the "camber" is quantified as the maximum deviation of the edge from the straight line that links its ends.

6   Implementation Problems

One of the main problems is that images are composed of two interlaced fields with a time delay between their acquisition. In the current application, in which sheets move at relatively high speeds, this time delay is not acceptable so one of the fields is discarded. The vertical resolution is reduced, but due to the fact that "camber" is measured in the cross-sectional direction of the sheets, that is nearly horizontal in the images, this is not a critical problem. The vertical resolution is more important in the calibration process, in which the two fields are considered. A better, but more expensive solution would have been to use progressive scan cameras.

Other problem caused by the fast movement of the sheets is the blurring effect it has in the images. To minimize this effect, a short exposure time is used (1 ms). Due to the great amount of light emitted by the sheets, even with that exposure time it is not necessary to provide extra illumination. Furthermore, with usual exposure times the sheet images are saturated, which would have been a problem. In the calibration process the setting of the exposure time is a difficulty. The calibration pattern must be strongly illuminated or it must emit light to achieve good quality images.

Finally, a last implementation problem arises from the fact that the sheet speed is known through the rotation speed of the rollers. When the sheet leaves the rollers its speed does not coincide with the signal being received by the PC. To notify the measurement program of that event, a digital signal is provided by the factory control system. That signal is in high state if the sheet is in contact with the rollers and else it is in low state. The measurement process stops to acquire images when that digital input changes to low. Therefore there is a section of the sheets that is ignored in the measurement (the section between the rollers and the start of the inspected area).

7   Results

The described method is being tested in the ACERALIA semi-continuous rolling mill, placed in Avilés (Spain). Some results are shown in Fig. 5.

The given measure of the "camber" effect is the length of the vertical stripe. As it can be seen, the scale of the x and y axis is very different. The length of

434   R.C. González, R. Valdés, and J.A. Cancelas

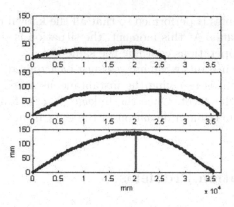

Fig. 5. Results in three inspected sheets.

the inspected edges is around 35m while the absolute deviation ranges between 25mm and 150mm.

There is no verification of this measures, but the magnitude and kind of deformations observed coincide with that described by the plant engineer.

8   Conclusion

A vision based system for measuring "camber" effect in steel sheets was proposed. The theoretical principles of the method employed and the practical difficulties were also described. One of the major characteristics of this system is its low cost. No changes in the rolling process have been made. No special illumination is needed and the installation is very simple. It can be said that the system adapts to the plant but the plant does not adapt to the system.

References

1. Semple, J., Kneebone, G.: Algebraic Projective Geometry. Oxford University Press (1979)
2. Faugeras, O. D.: Three-Dimensional Computer Vision: A Geometric Viewpoint. MIT Press, Cambridge (MA) (1993)
3. Press, W. H., Teukolsky, S. A., Vetterling, W. T., Flannery, B. P.: Numerical Recipes in C: The Art of Scientific Computing, Second Edition. Cambridge University Press.
4. Canny, J.: A Computational Approach to Edge Detection. IEEE Transactions on Pattern Analysis and Machine Intelligence, Vol. PAMI-8 (1986) 679-698
5. Gonzalez, R. C., Woods, R. E.: Digital Image Processing. Addison-Wesley, Reading (MA) (1992)
6. Pitas, I.: Digital Image Processing Algorithms. Prentice Hall (1992)
7. Devernay, F.: A Non-Maxima Suppression Method for Edge Detection with Sub-Pixel Accuracy, Programme 4-Robotique, image et vision, Project Robotvis. Rapport de recherche n°2724, INRIA (1995)

Positioning of Flexible Boom Structure Using Neural Networks

Jarno Mielikäinen[1], Ilkka Koskinen[2], Heikki Handroos[2], Pekka Toivanen[1], and Heikki Kälviäinen[1]

[1] Laboratory of Information Processing
Department of Information Technology
Lappeenranta University of Technology
P.O. BOX 20, FIN-53851 Lappeenranta
Finland
{jarno.mielikainen, pekka.toivanen, heikki.kalviainen}@lut.fi
[2] Laboratory of Machine Automation
Department of Mechanical Engineering
Lappeenranta University of Technology
P.O. BOX 20, FIN-53851 Lappeenranta
Finland
{ilkka.t.koskinen, heikki.handroos}@lut.fi

Abstract. Deflection compensation of flexible boom structures in robot positioning is becoming an important part of machine automation. Positioning is usually done using tables containing the magnitude of the deflection with inverse kinematics solutions of a rigid manipulator. In this paper, a method for locating the tip of a flexible manipulator using machine vision and a method for positioning the tip of a flexible manipulator using neural networks are proposed. A machine vision system was used in the data collection phase to locate the boom tip and the collected data was used to train MLP-networks. The developed methods improve the accuracy of manipulator positioning, and it can be integrated in the control system of the manipulator. The methods have been tested in real-time laboratory environment, and the results were promising. During the testing, the locating and the positioning were noticed to function as required, yielding reliable results with sufficient computation times.

Keywords: robot positioning, neural networks, boom structure

1 Introduction

This paper considers a concept of compensating deflections in positioning of a large-scale flexible hydraulic manipulator by integrating a machine vision system to the manipulator. Deflection compensation of flexible boom structures in robot positioning is becoming an important part of machine automation. Positioning is usually done using tables containing the magnitude of the deflection with inverse kinematics solutions of a rigid structure. There are several studies of

W. Skarbek (Ed.): CAIP 2001, LNCS 2124, pp. 435–442, 2001.

flexible manipulators, for example in [4], [11], [12]. Recently, the use of sensors in robot guidance has received a lot of attention [1], [3], [9]. The use of visual sensors with robotic manipulators has also been under research [5], [7], [8]. In this paper, a method for locating a flexible manipulator tip using machine vision and a method for positioning a flexible manipulator tip using neural networks are proposed. The work is connected to [6] and [10].

The manipulator, also called a boom, is shown in Fig. 1. It is a long and slender log-lifting crane. Its weight is minimized to allow easy manoeuvrability in varied terrain and forest work. As fully extended, the crane can lift logs from a distance of 5 meters. The boom actuator strokes are servo controlled. A simple PD-control method with low amplification is used as a control method, and thus, a fairly imprecise actuator control is achieved. The mechanical amplification of the system is also high. The control hardware is PC-based running on a dSpace DSP-card at 1kHz.

Fig. 1. Test setup of the manipulator: Boom structure and a laser cut plate field

For the user it is convenient to use the global coordinates in positioning the boom tip. However, the location of the end tip is in reality defined by cylinder strokes and boom dimensions. Therefore, a user interface that converts the global coordinates to the actuator strokes is made with a simple rigid inverse kinematics model. The model is integrated into the control system and it is invisible to the user. As the model is rigid and based on geometry it is fast to construct and implement, and it requires little calculation capacity from the control processor. However, the simplicity of the model causes very large deviations, mainly because it does not take joint wear and structural flexibility into account. Applying analytic solutions to compensate deviations is cumbersome, time consuming, and sometimes introduces no applicable results. For this reason a research project was started to find an easy and fast solution to compensate deflections.

Neural networks were thought to be suitable for automating the positioning process [10]. However, obtaining a required amount of location data is problematic. By constructing an automatic machine vision measuring cycle to measure deviations, it is possible to obtain a vast amount of data.

The proposed machine vision system consists of a camera, a laser cut plate field with colored location markers, a diode-laser pointer, and necessary software packages. The camera was fixed on the boom near the boom tip. The camera views a plate field nearby. From an obtained color picture it is possible to calculate the location of the laser pointer on the plate field. The machine vision and control systems reside on the same PC and they exchange data. Therefore, the coordinate system of the picture window is transformed to global coordinates by using a rough estimate of the real location from the inverse kinematics model. Thus, the true location of the boom tip could be solved in the global coordinates at a given time, actuator stroke, and a desired position of the boom tip known as a set point value.

2 Algorithm for Locating the Boom Tip

Two coordinate systems are used in this system. The world-coordinate system is used in a laser cut plate field and the camera coordinate system for digital image analysis. Therefore a transformation between these two systems has to be done. The general idea is that the location is known approximately and the exact location can be determined by calculating from the image the distance from the laser pointer to the nearest feature in the plate field, transforming the distance to the world coordinate system, and determining which feature would make the real point nearest to the estimate. Blue filled circles that are distributed uniformly over a rectangular grid on the backboard and a point from laser, called later on as a laser point, are used as features.

Algorithm 1 calculates the location on the boom tip. It assumes that error in estimate from the control system is smaller that the distance between blue circles and that the change in the angle of the camera is less than 90 degrees.

Algorithm 1: Locating Algorithm

1. Find the laser point L shown from the boom tip.
2. Calculate the distance D from the L to the nearest blue circle B1.
3. Find another blue circle B2 that is closest to B1.
4. Using the angle between B1 and B2 calculate a transformation which rotates the grid to the correct position according to the world coordinate system.
5. Change the coordinate values from pixels to millimeters using the known distance between two blue circles in millimeters and calculated distance between B1 and B2 in pixels.
6. Calculate an error function for each blue circle using the location estimate L from the manipulator control system and P that is laser points location with respect to the blue circle. The error function is an Euclidean distance between P and L.
7. The minimum of the error function gives the values that are used is translating the grid to the correct position according to the world coordinate system.

8. Using the rotation, scaling and translation transformations to D gives current location of the boom tip.

More detailed description of the algorithms can be found in [6].

3   Solving the Inverse Kinematics

The end tip position of the manipulator is in this case defined by the mechanical structure of the boom and actuator strokes. Thus, in the inverse kinematic problem a solution for the actuator strokes is sought after at a given end tip value. This is due to the fact, that the manipulator is controlled by setting actuator strokes while movement is monitored by the user in Cartesian coordinates.

Due to the flexibility of the manipulator (Fig. 1) the inverse kinematics cannot be solved with classical matrix-based methods.

Thus, a simplified model was introduced (Fig. 2). With the simplification a solution to the inverse kinematic problem can be produced fast and efficiently. First the angle α and the length l is solved with Pythagorean theorem. With l known the angles β and φ are solved with the cosine theorem. Thus, the lift and jib angles are known. The result are obtained from the simplified model and are then applied onto the true model and actuator strokes for jib and lift can be calculated. Thus, the inverse kinematic problem is solved.

By solving the problem, we have now obtained all lengths for the given structure. As the manipulator's end position is now defined by actuator strokes, it is now possible to define the actuator strokes (named s_1 and s_2) so that they can be used as set point values in the control loop of the manipulator. The user just feeds set point values in Cartesian coordinates that are processed by the inverse kinematic solver and transformed to actuator strokes (s_1 and s_2). The stroke values are used as set point values by the control loop of the manipulator.

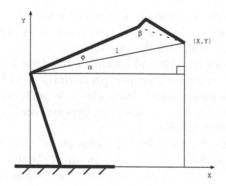

Fig. 2. Geometrical model of the inverse kinematic problem

4   Results of Using Neural Network to Control the Boom Structure

There are differences, or errors, between calculated stroke values and real end tip coordinates. Thus, a method to correct these errors is needed. A neural network is used for this purpose in this paper.

In order to correct errors in the inverse kinematic model, internal parameters of the model which are strokes s_1 and s_2 must be found out. In the position (x, y) real s_1 and s_2 were received from the control system. After that the parameters s_1^i and s_2^i were acquired from the inverse kinematic model by giving x and y as an input. Difference, i.e., error functions, between measured values and values given by the model were defined as

$$\delta s_1 = s_1 - s_1^i \tag{1}$$

$$\delta s_2 = s_2 - s_2^i. \tag{2}$$

Both δs_1 and δs_2 from Equations 1 and 2 were modelled using two separate MLP-networks which consisted of two and five neurons in one hidden layer, correspondingly. The number of neurons was determined by experiments and since the functions were so smooth only few neurons were required. The input for the neural networks were coordinate values x and y, error functions δs_1 and δs_2 were output values.

Also the used training method did not make much difference in such a simple networks, so the Levenberg-Marquardt [2] method was used.

Polynomial functions were also used for the estimation of the error functions. In case of δs_1 estimate was good with third degree polynomial, but with δs_2 the estimate was not good enough even when used polynomial was of a much higher degree.

It can be seen from the results that neural networks, which were trained by data collected with a machine vision system, achieves better results as can be seen from Figs. 4 and 6, compared to traditional methods which results can be seen in Figs. 3 and 5.

5   Conclusions

In this paper, methods for locating a flexible manipulator tip using machine vision and positioning a flexible manipulator tip using neural networks were presented. The locating and positioning systems were developed in co-operation with the Laboratory of Information Processing and the Laboratory of Machine Automation at Lappeenranta University of Technology. Blue filled circles on the backboard and a point from a laser were used as visual information. The machine vision system was supposed to locate the tip as fast and accurately as possible in order to collect data to train neural networks. Neural networks were used in positioning. The used method increases accuracy of the manipulator positioning, and it makes possible to do controlled changes during further development.

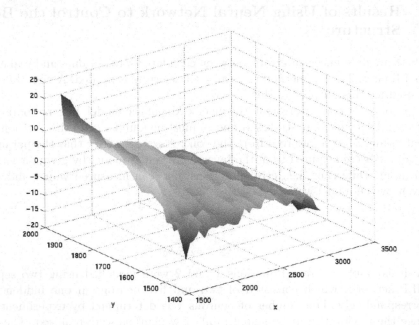

Fig. 3. Measured error in x-coordinate [mm] before correction

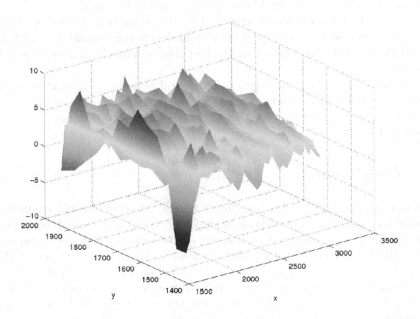

Fig. 4. Measured error in x-coordinate [mm] after correction

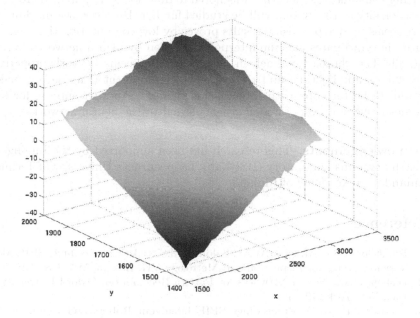

Fig. 5. Measured error in y-coordinate [mm] before correction

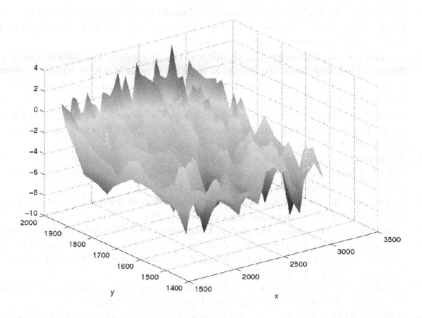

Fig. 6. Measured error in y-coordinate [mm]after correction

During real-time tests locating was noted to function as required. Increasing the functionality of the system will be studied further. However, the machine vision part seems to give promising results proven by low computation times and small errors in coordinates and the interpolation given by neural networks were good enough. The chosen approach in data collection seems to work properly only under constant lighting, and thus, further development is needed to make the system to better cope with illumination changes. Another subject for further studies is the addition of a load-sensing signal to the neural networks.

Acknowledgements. This research has been supported by the Intelligent Industrial Systems Laboratory (IISt-Lab), Lappeenranta University of Technology, Finland, and Academy of Finland.

References

1. Batchelor, B.G., Miller, J.W.V., Solomon, S.S. (eds.): Proceedings SPIE Machine Vision Systems for Inspection and Metrology VII. Boston, MA, USA (1998)
2. Bishop, C.M.: Neural Networks for Pattern Recognition. Oxford University Press Inc., New York (1995)
3. Casasent, D. (ed.): Proceedings SPIE Intelligent Robots and Computer Vision XVII: Algorithms, Techniques, and Active Vision. Boston, MA, USA (1998)
4. De Luca, A., Panzieri, S.: End-effector Regulation of Robotics with Elastic Elements by an Interactive Scheme. International Journal of Adaptive Control and Signal Processing Systems. (1996) 379–393
5. Hashimoto, H., Kubota, T., Kuou, M., Harashima, F.: Visual Control of a Robotic Manipulator Using Neural Networks, Decision and Control. Proceedings of the 29th IEEE Conference (1990) 3295–3302
6. Mielikäinen, J., Koskinen, I., Handroos, H., Toivanen, P., Kälviäinen, H.: Positioning of a Large Scale Hydraulic Manipulator by a Machine Vision System. Proceedings of Workshop on Real-Time Image Sequence Analysis, RISA 2000. Oulu, Finland (2000) 122–129
7. Mulligan, I., Mackworth, Lawrence, P.: A Model-Based Vision System for Manipulator Position Sensing, Interpretation of 3D scenes. Proceedings, Workshop on (1989) 186–193
8. Nakazawa, K.: Calibration of manipulator using vision sensor on hand unit. Industrial Electronics, Control, and Instrumentation Proceedings of the IECON '93, International Conference on (1993) 1386–1390
9. Newman, T.S., Jain, A.K.: A Survey of Automated Visual Inspection. Computer Vision and Image Understanding (1995) 231–262
10. Rouvinen, A., Handroos, H.: Robot Positioning of a Flexible Hydraulic Manipulator Utilizing Genetic Algorithms and Neural Networks. Proceedings of the Fourth Annual Conference on Mechatronics and Machine Vision in Practice. Toowoomba, Australia (Sep 1997) 379–393
11. Surdilovic, D., Vukobratovic, M.: Deflection Compensation for Large Flexible Manipulators. Mechanism and Machine Theory (1996) 317–329
12. Williams, D.W., Turcic, D.A.: An Inverses Kinematics Analysis Procedure for Flexible Open-Loop Mechanisms. Mechanism and Machine Theory (1992) 701–714

Material Identification Using Laser Spectroscopy and Pattern Recognition Algorithms

Ota Samek[1], Vladislav Krzyžánek[1], David C.S. Beddows[2], Helmut H. Telle[2], Josef Kaiser[1], and Miroslav Liška[1]

[1]Institute of Physical Engineering, University of Technology, Technicka 2, Brno, 616 69, Czech Republic
ota@fyzika.fme.vutbr.cz
[2]Department of Physics, University of Wales Swansea, Singleton Park, Swansea, UK
h.h.telle@swansea.ac.uk

Abstract. We report on pattern recognition algorithms in discriminant analysis, which were used on Laser Induced Breakdown Spectroscopy (LIBS) spectra (intensity of signal against wavelength) for metal identification and sorting purposes. In instances where accurate elemental concentrations are not needed, discriminant analysis can be applied, to compare and match spectra of "unknown" samples to library spectra of calibration samples. This type of "qualitative" pattern recognition analysis has been used here for material identification and sorting. Materials of different matrix materials (e.g. Al, Cu, Pb, Zn, vitrification glass, steels, etc.) could be identified with 100% certainty, using Principle Component Analysis and the Mahalanobis Distance algorithms. The limits within which the Mahalanobis Distance indicate a match status of *Yes, Possible* or *No* were investigated. The factors, which dictate these limits in LIBS analysis, were identified - (i) spectrum reproducibility and (ii) the sample-to-sample homogeneity. If correctly applied the combination of pattern recognition algorithms and LIBS provide a useful tool for remote and *in-situ* material identification problems, which are of a more "identify-and-sort" nature (for example those in the nuclear industry).

Keywords: pattern recognition, material identyfication, laser spectroscopy

1 Introduction

One of the spectroscopic technique based on Laser Ablation (LA) spectroscopy is Laser-Induced Breakdown Spectroscopy (LIBS). This technique offers simple, fast and real-time spectrochemical analysis, with little need for sample preparation. In the technique one utilizes the high power densities obtained by focusing the radiation from a pulsed, fixed frequency laser, to generate a luminous micro plasma from an analyte (solid, liquid and gaseous samples). To a good approximation, the plasma composition is representative of the analyte's elemental composition. In the thirty years or so since its inception the potential of LIBS as an analytical tool has been realized, leading to an ever increasing list of applications, both for analysis in the laboratory and industrial environments [1,2].

W. Skarbek (Ed.): CAIP 2001, LNCS 2124, pp. 443–450, 2001.
© Springer-Verlag Berlin Heidelberg 2001

Mostly qualitative and quantitative LIBS analysis has been applied to the analysis of solid samples. Extensive studies have been carried out for a wide range of solid samples under different operating conditions to determine parameters such as the electron density, the plasma temperature and spectral line shapes, and their relationship to the validity of analytical outcomes has been studied [3]. Detection limits for solid samples typically are in the range of a few hundreds parts per million, less in a few specific cases. When deciding on a method for elemental analysis, major advantages of LIBS over the more conventional methods are – (i) no or very little sample preparation; (ii) analysis can be carried out equally on all three physical states of matter (solids, liquids and gases); (iii) the analysis is performed in real time (approximately a few seconds when using lasers with a 10-20 Hz repetition rate); and (iv) only a small amount in the order of a few µg is ablated from the surface of solid samples, and hence the method is virtually non-destructive.

By using Discriminant Analysis, the LIBS system can be trained to recognise spectra from different samples, regardless of spectrum quality and reproducibility. Ironically, instead of collecting spectra under set ablation parameters the spectra required for the generation of these Discriminant Analysis models need to reflect all measurement conditions. This then implies that the precision and accuracy of the LIBS measurement is no longer a major issue. This is because the system is trained to evaluate the similarity of unknown spectra to its bank of data relating to the samples it is sorting. With diminished importance of spectrum quality, the ablation parameters (e.g. lens-to-sample separation) become less critical thus making a commercial LIBS system that can "point-shoot-and-identify" feasible.

2   Experimental

LIBS systems have become quite common in recent years, and full descriptions of typical systems have been given elsewhere, see e.g. in [4]. Here we only summarise the characteristic features of the laboratory system used in this study.

The laser system - a standard Nd:YAG laser was used to generate the LIBS plasma probe beam (at wavelength of 1064nm, at a repetition rate of 10Hz). Individual laser pulses had a pulse length of about 10ns; these could be adjusted for pulse energies of 10-100mJ, using a Glan polariser.

The light delivery system - mostly, a laser beam delivery system based on lens / mirror optics was used. Light from the plasma is collected by a lens/lenses, which focused the plasma light emission onto an optical fibre bundle, connected to the analysis spectrometer or in some arrangements directly to the spectrometer.

The system for spectral analysis - the system used for spectral analysis consisted of a standard spectrograph (ACR500, Acton) with a gateable, intensified photodiode array detector (IPDA, Princeton Instruments) attached to it. The gating of the detector and the timing for spectral data accumulation are controlled by a PC via a pulse delay generator (PG200, Princeton Instruments).

3   Discriminant Analysis

Each spectrum collected using a LIBS instrument is a "finger-print" of the material being analysed and the conditions under which it was collected. Most of the efforts in quantitative LIBS research have been aimed at normalising the spectrum collection conditions and procedures so that the spectrum only characterises the material. If no normalisation routines are carried out, and it is assumed that the sample is homogeneous, then the transient conditions of the LIBS measurement provide the main source of irreproducibility.

More commonly known as Discriminant Analysis in spectroscopy, the aim of any pattern recognition algorithm is to unambiguously determine the identity or quality of an unknown sample using a spectrum obtained from the sample. There are two basic applications for spectroscopic discriminant analysis: (i) material purity/quality and (ii) material identification/screening.

Material Quality Control - in its capacity for sample checking, discriminant analysis models could in principle replace many quantitative methods currently used. In effect, the algorithm gives an indication, "YES" or "NO", of whether the spectrum of the "unknown" sample matches the spectra taken from samples that were known to be of "good" quality.

Material Identification - when discriminant analysis is used in a product-identification, or a product-screening mode, the spectrum of the "unknown" is compared against multiple discriminate models. Each model is constructed from the spectra collected from samples representative of various material groups defined by the grade, purity or quality of the sample. An indication of the likelihood of the spectrum matching one of these groups is then made, and the material is therefore classified as the closest match, or no match at all.

There are many pattern recognition algorithms that can be used to assess the similarity of a measured spectrum with the training set. Here, the description is restricted to the algorithm used in this study, namely the *Mahalanobis Distance Method*.

4   Application of Mahalanobis Distance Method of Spectrum Matching to LIBS Spectra

In order to calculate the Mahalanobis Distance (M.Dist), Principle Component Analysis (PCA) is used. This decomposes the training set spectra into a series of mathematical spectra called factors which, when added together, reconstruct the original spectrum. The contribution any factor makes to each spectrum is represented by a scaling coefficient (score) which is calculated for all factors identified from the training set. Thus, by knowing the set of factors for the whole training set, the scores will represent the spectra as accurately as the original responses at all wavelengths. For a detailed description of the Mahalanobis Distance algorithm see [5].

4.1   Measurements Using the Mahalanobis Distance

The method of measurement and evaluation using the Mahalanobis Distance method will be exemplified for the distinction of three marginally different steels, encountered in the quality control for an assembly of boiler tubes. LIBS spectra were recorded in the range 320-350nm.

To fully appreciate the "mechanism" behind the M.Dist measurement, consider first the application of the M.Dist directly to the full spectral data set without using PCA, and in parallel, consider the response of just two major spectral peaks (e.g. Cu (λ=344.15nm) and Fe (λ=324.84nm)) in multiple spectra collected from a single sample. If the intensities of these two peaks are plotted against each other for numerous measurements, then an elliptical scatter of points would be expected due to the fluctuations in the measurement conditions caused by factors such as spectrometer response, sample handling and sample preparation. It is the scatter of these points which defines the M.Dist about a mean centre, in the same way that the standard deviation σ of a one-dimensional measurement x, defines the scatter about the mean measurement. However, the M.Dist applies to all pixels in the spectrum, not just to one or two pixels of a peak, and therefore can be considered to be a multi-dimensional standard deviation that is applied to the whole spectrum.

When creating the Discriminant Analysis models a list of the training set spectra was simply entered into the "in-house" / GRAMS PLSplus programs. Choosing the Discriminant Analysis option, the program generated a Discriminant Analysis model for each sample, against which test spectra were matched. The related data for the aforementioned boiler tube samples were all saved into a single calibration file, which - when loaded into the Prediction Code - performed the matching routines for each sample.

When checking the identity of the test spectra collected from each sample all were either identified as *definite* or *possible* matches to one of the Discriminant Analysis models.

Tests carried out on all recorded spectra (in total 48) showed conclusive evidence that by using the M.Dist the spectra could be correctly re-categorised into the three sample groups. However, a few of the spectra did register a *NO-NO-POSSIBLE* combination when compared with the Discriminant Analysis Models for the three steels. Even though some of the spectra returned only a *POSSIBLE* match, nevertheless a positive identification could be claimed. This was because of the relatively large M.Dist values calculated for the failing models. For a model returning a "YES" or a "POSSIBLE", the M.Dist values were between 0 and 3. For a "NO" match result the M.Dist were much greater than 3, normally at least an order of magnitude greater than the M.Dist calculated for a "YES" or POSSIBLE" result. This point is re-emphasised when using LIBS to grade steels and identify different matrix elements, as outlined in the sections below.

From the repeat analysis of the spectra collected from the samples of the three boiler tube steels, it could safely be said that Discriminant Analysis had the potential of providing a superior tool for matching LIBS spectra and identifying "unknown" materials, when compared to standard semi-quantitative analysis.

To emphasise this point Discriminant Analysis has been applied to the identification of materials with different and similar matrices.

4.2   Identification of Materials of Different Matrices

The materials used to test the ability of the LIBS system to identify different matrices were Al, Cu, Pb, Zn, mild steel, stainless steel and simulated vitrification glass. For each material, 50 spectra were collected, plus an additional 10 spectra which were treated as "unknowns".

Each spectrum was collected using a 150 lines/mm grating (centred at $\lambda = 450$ nm, and accumulated for only 50 laser induced plasma events). Note that for this part of the study, no optimisation of the plasma generation and recording was afforded but rather a "point-and-shoot" approach was taken.

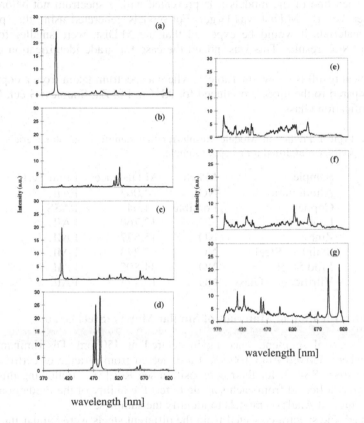

Fig. 1. LIBS spectra collected for samples of (a) aluminium, (b) copper, (c) lead, (d) zinc, (e) mild steel, (f) stainless steel and (g) vitrification glass.

As can be seen from Figure 1, the differences between the spectra collected from each of the samples reflect the completely different matrices. Only the spectra from the mild and stainless steel samples show any similarity, and the ability of LIBS Discriminant Analysis (LIBSDA) to discriminate between these samples was explored further for the identification of the grade of five ferritic steels. To make the test realistic the test spectra were coded, mixed and then sorted using the prediction module. Notice that even though the prediction made on the copper sample resulted in a POSSIBLE verdict, because the M.Dist value was greater than 1, that the spectrum had to be from copper because the other Discriminant Analysis models all gave a M.Dist value of the order 10,000. Clearly, the spectrum was not from any one of these materials. Sorting the spectra in this way a 100% identification result was obtained when using the Discriminant Analysis models to identify the 10 "unknown" spectra collected from each material.

Remembering that the M.Dist is effectively a measure of the similarity of an "unknown" spectrum to a group of training spectra, it can be expected that the M.Dist for the Discriminant Analysis models reporting a "NO" result to be high. The reasoning behind this is that spectra collected from each element are very different. Hence, when one of the models was presented with a spectrum not belonging to its native set then the M.Dist was large. For models generated using the spectra from similar materials it would be expected that the M.Dist. were smaller for a model giving a "NO" result. This was indeed the case for grade identification of stainless steels.

A typical result is shown in Table 1, where a spectrum taken from a copper sample was compared to the models produced for Al, Cu, Pb, Zn, Stainless Steel, Mild Steel, and Vitrification Glass.

Table 1. Typical Prediction Module Result for the identification of materials with totally different spectra, exemplified for a copper sample.

Sample	Match	M Distance	Limit Test
Aluminium	NO	4,063	FAIL
Copper	Possible	1.04	PASS
Lead	NO	12,760	FAIL
Zinc	NO	15,527	FAIL
Stainless Steel	NO	3,273	FAIL
Mild Steel	NO	44,805	FAIL
Vitrification Glass	NO	19,473	FAIL

4.3   Identification of Materials of Similar Matrix (Steel Grade)

Using a spectral segment, once again centred at 450 nm, Discriminant Analysis models where derived from 100 spectra collected from a range of certified stainless steels (SS469 - SS473, for their composition see Ref [5]). Following this, 10 extra spectra were collected from each sample to test the ability of the Prediction Code and the Discriminant Analysis models to identify the samples.

Although the spectra collected from the different steels were similar the Prediction Code was still capable of distinguishing between them. The M.Dist values of the (successful) identification procedure for all test-spectra collected from the five steels are given in Table 2. In general we noticed that the M.Dist of the failed models was about one order of magnitude larger compared to the worst-case average value of 1.4 for the SS469 model. This trend was observed for all the test spectra collected for this sample, and was repeatedly seen when sorting other steels. Overall, all "positive" identification results gave an M.Dist value well below 3. Comparing this with the fact that the M.Dist calculated for the models giving a *NO* result (i.e. 1 to 2 orders of magnitude greater), a 100% identification result is obtained. The higher than expected M.Dist values for the successful model were more than likely due to the poor reproducibility of the spectra. On this basis the fluctuations in the spectra may be accounted for in the Prediction Module by changing the M.Dist PASS/FAIL limits.

These limits are somewhat arbitrary anyway, reflecting the tolerance s provided by the operator. Analysts often use values greater than 1-2 and 2-3 for the YES and POSSIBLE results, e.g. 1-5 and 5-15, respectively. Therefore, these limits have to be determined. For example, for our specific case, the limit values for the identification modules could be changed from

$$0 < \text{MDist} < 1 \quad \text{PASS}; \ 1 < \text{MDist} < 3 \quad \text{POSSIBLE}; \ \text{and MDist} > 3 \quad \text{FAIL},$$

to

$$0 < \text{MDist} < 3 \quad \text{PASS}; \ 3 < \text{MDist} < 6 \quad \text{POSSIBLE}; \ \text{and MDist} > 6 \quad \text{FAIL},$$

and thus return a PASS for all the successful matches. The factors which dictate these limits in LIBS analysis are (i) spectrum reproducibility and (ii) the sample-to-sample homogeneity.

Table 2. M.Dist identification results from the „successful" models, out of the test against training reference spectra from all five steels.

Sample No	Mahalanobis Distance				
	SS469	SS470	SS471	SS472	SS473
1	1.63	0.72	0.71	0.85	0.75
2	1.61	0.84	0.68	0.99	0.46
3	2.12	0.95	0.58	1.92	0.53
4	1.33	1.36	0.93	1.13	1.06
5	1.50	0.59	2.02	0.82	0.76
6	1.30	1.72	0.73	0.99	0.69
7	1.38	0.57	0.81	0.94	0.83
8	1.03	0.86	0.86	1.13	0.91
9	0.87	0.58	0.90	0.96	0.90
10	1.29	0.81	0.75	1.13	0.60
Average MD	1.4	0.9	0.9	1.1	0.7
YES	1	8	9	6	9
POSSIBLE	9	2	1	4	1
NO	0	0	0	0	0

5 Summary

In summary the pattern recognition algorithm in discriminant analysis can be used on LIBS spectra for metal identification and sorting purposes. It is easy to apply, and the results obtained indicate a 100% identification rate for materials both of common and non-common matrices.

However, as with all Multivariate Quantitative Analysis methods, careful application is required if the technique is to be applied both correctly and successfully. For example, the limits within which the M.Dist indicate a match status of YES, POSSIBLE or NO can be changed (see section 4.3 above). By testing the models produced with randomly collected spectra from different samples of the material it represents, the range of M.Dist values which give a positive identification needs to be found. If this is not done then the model might incorrectly miss-identify materials. Furthermore, by adjusting the M.Dist limits, the poor reproducibility can in principle be accounted for, provided there are significant elemental differences in the samples being sorted, such that clear changes in the spectral responses can be observed. Therefore, if correctly applied the combination of discriminant analysis and LIBS provides a useful tool for remote and *in-situ* material identification problems.

We would like to note that the examples reported here represent only a few of the material identification applications identified by LIBS researchers. With the correct marketing, further industrially motivated applications may soon appear which require remote and *in-situ* analysis. Remote-LIBS technology exists and has been demonstrated in numerous applications (see e.g. [6]). In this work an attempt was made in an application which exploits LIBS to satisfy the commercial niche identified in the introduction, i.e. that of remote and *in-situ* material identification.

Acknowledgement. O. Samek gratefully acknowledges the financial support by Grants GACR 101/98/P282 and CEZ:J22/98:262100002. D.C.S. Beddows acknowledges the sponsorship for his Ph.D. research by BNFL plc, Sellafield.

References

1. Majidi, V., Joseph, R.: Spectroscopic applications of laser-induced plasmas. Crit. Rev. Anal. Chem. **23** (1992) 143-162.
2. Radziemski, L.: Review of selected analytical applications of laser plasmas and laser ablation 1987-1994. Microchem. J. **50** (1994) 218-243.
3. Leis, F., Sdorra, W., Ko, J.B., Niemax, K.: Basic investigation for laser microanalysis: I. Optical emission spectroscopy of laser produced sample plumes. Mikrochim. Acta II (1989) 185-199.
4. Samek, O., Beddows, D.C.S., Kaiser, J., Kukhlevsky, S., Liška, M., Telle, H.H., Young, J.: The application of laser induced breakdown spectroscopy to in situ analysis of liquid samples. Opt. Eng. **39** (2000) 2248-2262.
5. Beddows, D.C.S.: Industrial application of remote and in situ laser induced breakdown spectroscopy. Ph.D. Thesis, University of Wales Swansea (2000).
6. Davies, C.M., Telle, H.H., Montgomery, D.J., Corbett, R.E.: Quantitative analysis using remote laser-induced breakdown spectroscopy (LIBS). Spectrochim. Acta B **50** (1995) 1059-1075.

Scanner Sequence Compensation

Tomasz Toczyski and Sławomir Skoneczny

Warsaw University of Technology, Institute of Control and Industrial Electronics
ul. Koszykowa 75, 00-622 Warszawa, Poland
Tomasz.Toczyski@isep.pw.edu.pl

Abstract. Professional film scanners acting in real time (24 frames per second) are still very expensive. In most cases using a slide scanner of medium resolution equipped with additional device for transporting a film reel would be a reasonable solution. The main problem, however is a lack of accurate positioning mechanism in such sort of scanners. Therefore the position of each frame could be to some extent accidental. If frames are scanned separately from each other and this process is performed for all the frames of a movie there is usually a significant jitter in this sequence. This paper presents an efficient and simple method of obtaining jitter–free sequence i.e. finding the precise cinematic frame location in a picture that is the output of the scanning process. The procedure consists of two steps: rough estimation and the fine one. During the rough step the borders of the frame can be detected based on finding area of maximal brightness. In the second step the displacements among frame backgrounds are calculated. Additionally in order to avoid the fixed background problem the local constant component is eliminated in the postprocessing phase. As a final result a jitter is removed almost completely.

Keywords: image processing, jitter removal, slide scanner

1   Introduction

Professional film scanners acting in real time (24 frames per second) are still very expensive and quite often using a professional slide scanners (significantly cheaper) with additional device enabling celluloid tape transportation would be a reasonable solution. The only requirement is to have a precise mechanism for transporting film reel in such a manner as to assure accurate positioning of each frame, otherwise the position of each frame would have been to some extend accidental. It is not a serious problem if we scan a single frame only (a picture or a slide). The problem arises if we scan subsequent pictures separately and after that we want to build the whole sequence of them. There are usually significant displacements in the subsequent pictures among the consecutive frames. This paper presents a method of finding the precise cinematic frame location in a picture being the output of the slide scanner. After that, it is possible to extract precisely the frames' areas only, which enables creating a jitter–free sequence.

W. Skarbek (Ed.): CAIP 2001, LNCS 2124, pp. 451–456, 2001.

2 The Algorithm

Let us assume that we have a sequence I_k, $k = 0, \ldots, n-1$ of pictures from the slide scanner that includes subsequent frames from the film. Each picture, being the scanner output, includes exactly one film frame (see example in Fig. 1). All

Fig. 1. An example of the picture that includes one frame of the film.

frames are rectangles of size $w_f \times h_f$, having their sides parallel to the sides of pictures I_k (scanner outputs). The method consists of the two following steps:

– For each picture I_k obtained from the scanner an initial, rough location of the frame included in this picture should be found.
– Having found initial locations of the frames in pictures a final, accurate estimation of their locations should be performed by matching the neighboring frames to each other.

2.1 Initial Estimation of the Frame Location

Let us assume that each picture I_k is of the size $w_{sk} \times h_{sk}$. The picture I_k consists of the frame to be found and the surrounding area (see Fig. 1). The exact borders of the frame can be detected by finding the brightest rectangular block of size $w_f \times h_f$. The horizontal and vertical locations of this block can be found independent of each other. Let x_{fk} denote an initially found horizontal coordinate of the left upper corner of the frame in picture I_k. It can be calculated as follows:

$$x_{fk} = \text{argmax}_{x \in \{0,\dots,w_{sk}-w_f-1\}} \sum_{i=0}^{w_f-1} \sum_{j=0}^{h_{sk}-1} I_k(i+x,j) \tag{1}$$

where $I_k(i,j)$ is a luminance value of a pixel of coordinates (i,j) in picture I_k, and w_{sk}, h_{sk} are the width and the height of picture I_k respectively. A similar procedure is applied in order to find the vertical coordinate of the left upper corner of the picture. Namely:

$$y_{fk} = \text{argmax}_{y \in \{0,\dots,h_{sk}-h_f-1\}} \sum_{j=0}^{h_f-1} \sum_{i=0}^{w_{sk}-1} I_k(i,j+y) \tag{2}$$

The point $(0,0)$ is assumed to be in the left upper corner of the picture.

2.2   Final Estimation of Frame Location

It may seem that after applying the methodology described in the previous subsection it is possible to collect all the frames and to build the whole sequence–with no jitter. Unfortunately this method is too rough to give satisfying results. Jitter effect is still visible, which is caused by inaccuracy of the algorithm and it should be compensated. The method of compensation that will be presented further is based on the idea of detection of displacements between backgrounds of the neighboring frames.

In the most of the scenes the areas near the borders of the frame belong to the background of the scene. Objects moving with respect to the background usually lie in the middle of the frame. The displacement between the backgrounds of consecutive frames can be estimated by finding the best matching of dashed region in Fig. 2 defined as:

$$B = \{(x,y) : ((t_O \leq x < t_I) \vee (w_f - t_I \leq x < w_f - t_O)) \wedge \tag{3}$$
$$((t_O \leq y < t_I) \vee (h_f - t_I \leq y < h_f - t_O))\}$$

We compute displacement between frames backgrounds. Those frames have been extracted from images I_k in the previous section. Let now d_{xk} and d_{yk} respectively denote horizontal and vertical displacements between two backgrounds taken from frames extracted from images I_k and I_{k-1} i.e. :

$$(d_{xk}, d_{yk}) = \text{argmin}_{d_{xf}, d_{yf} \in \{-m,\dots,m\}}$$
$$\sum_{(x,y) \in B} \text{FDP}(|I_k(x_{fk-1} + x, y_{fk-1} + y) -$$
$$-I_{k-1}(x_{fk} + x + d_{xf}, y_{fk} + y + d_{yf})|) \tag{4}$$

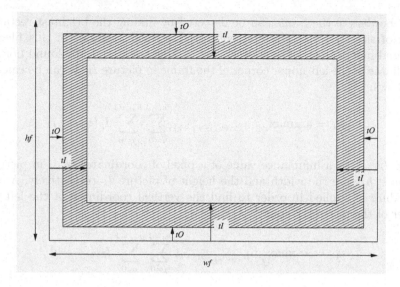

Fig. 2. B area.

where m is a parameter chosen heuristically. It is usually a number equal to a few percent of the height or the width of the frame. FDP is the function of the difference of pixel's luminances. It represents a "nonmatch" of pixels. This function should be a nonincreasing one within its domain $(0, +\infty)$. Let $d_{\mathrm{xf}k}$ and $d_{\mathrm{yf}k}$ denote displacements between background of frame from picture I_k and the background of frame from picture I_0. It is obvious that: $d_{\mathrm{xf}0} = 0$ and $d_{\mathrm{yf}0} = 0$ The displacements for the next frames can be calculated incrementally:

$$d_{\mathrm{xf}k} = d_{\mathrm{xf}k-1} + d_{\mathrm{x}k}$$
$$d_{\mathrm{yf}k} = d_{\mathrm{yf}k-1} + d_{\mathrm{y}k} \tag{5}$$

Having computed the values of $d_{\mathrm{xf}0}$ and $d_{\mathrm{yf}0}$ it is possible to obtain a jitter–compensated sequence by extracting from each frame I_k a rectangle of size $w_\mathrm{f} \times h_\mathrm{f}$. The left upper coordinates of this rectangle are

$$(x_{\mathrm{f}k} + d_{\mathrm{xf}k}, y_{\mathrm{f}k} + d_{\mathrm{yf}k}) \tag{6}$$

Such compensation causes the background to be fixed and this method could be applied if the recording camera was not moving with respect to the background. Otherwise an unnatural effect of "escaping frames" would happen. In order to remove this disadvantage it is necessary to subtract from signals $d_{\mathrm{xf}k}, d_{\mathrm{yf}k}$ their "local constant components" and then after such filtering it is possible to apply the compensation procedure according to formula (6) There are many ways of computing this "local constant component". In this paper the following method is proposed. First a median filter taken from the window that includes the six nearest neighbors of k–th frame is applied. Later, the signal filtered in this manner is smoothed by a linear filter that calculates the mean value of the three samples: previous, current and next ones. The equation of the filter can be expressed as follows:

$$d'_{xfk} = \text{median}(d_{xfk-3}, d_{xfk-2}, d_{xfk-1}, d_{xfk+1}, d_{xfk+2}, d_{xfk+3}) \qquad (7)$$
$$d'_{yfk} = \text{median}(d_{yfk-3}, d_{yfk-2}, d_{yfk-1}, d_{yfk+1}, d_{yfk+2}, d_{yfk+3})$$
$$\overline{d_{xfk}} = \frac{1}{3}(d'_{xfk-1} + d'_{xfk} + d'_{xfk+1})$$
$$\overline{d_{yfk}} = \frac{1}{3}(d'_{yfk-1} + d'_{yfk} + d'_{yfk+1})$$

Hence, a well compensated sequence will be constructed by using rectangles of size $w_f \times h_f$ and left upper corners coordinates in

$$(x_{fk} + d_{xfk} - \overline{d_{xfk}}, y_{fk} + d_{yfk} - \overline{d_{yfk}}) \qquad (8)$$

However there is still one additional problem. The rectangles may slightly overlap on black areas of pictures taken from the scanner. Sometimes they may even exceed the borders of I_k. In order to limit this effect as not to be visible in the sequence (or at least not significantly visible) instead of extracting a rectangle of size $w_f \times h_f$ from each picture, we should cut a little bit smaller ones. The final output sequence will consist of blocks of size $(w_f - k_L - k_R) \times (h_f - k_U - k_D)$ and of left upper corners in: $(x_{fk} + d_{xfk} - \overline{d_{xfk}} + k_L, y_{fk} + d_{yfk} - \overline{d_{yfk}} + k_U)$ where $k_L + k_R + k_U + k_D$ parameters are numbers of rows and columns that should be cut of from the left, the right, the upper and the lower sides of the frame. They should be equal to $1 - 2\%$ of the width or height of a frame.

3   Final Results

For testing purposes a Kodak RFS 3570 Plus scanner was used to scan separate pictures of a dancing couple directly from the film reel. The sequences are available in mpeg and avi formats on
http://www.isep.pw.edu.pl/~toczyskt/scaner.html. 93 frames (gray level images, 256 levels) were scanned in and build to make a file input. For each picture its width was $w_f = 451$ pixels and its heigh $h_f = 333$ pixels.
Next the method described in 2.1 was used in order to find a frame in each picture (in a rough manner). The sequence of such frames is in file step1 . There is a significant jitter effect that must be compensated. At this step the method described in 2.2 was used with the parameters $t_O = 10, t_I = 20, m = 10$
The FDP function was chosen as:

$$\text{FDP}(x) = \max(x, 40) \qquad (9)$$

The limit of 40 was set because it was assumed that the difference of brightness of corresponding pixels is lower than 40, and if the differences of pixels brightness were greater than 40 there would be a totally bad match.

In order to minimize the expression (4) a full search block matching method was applied [1],[2] i.e. the value of the sum from equation (4) was investigated for each possible (d_{xf}, d_{yf}). It is one of many possible methods, certainly of significant computational burden but the simplest one to implement.

The file step2 includes the sequence with static background. It is a significant visible effect of "escaping frame".

The file step3 includes the sequence after removing this effect.

The file step4 includes final sequence for the following parameters: $k_L = k_R = k_U = k_D = 5$. This algorithm works as it was expected before. The frames are quite well match to each other. However a slightly visible "rotating jitter" may be observed that which is caused by the "angle displacements" of frames in pictures being the scanner outputs. This kind of degradations are not taken into account due to their relative low influence of the visual quality of the sequence. The elaborated algorithm was implemented and tested on PII/PIII computers and it took a few minutes to compensate the jitter effect for 93 frames. This time would be significantly shorter if another method of matching the areas from Fig. 2 was applied. Anyway this time was significantly shorter then the time of frames' scanning on a Kodak slide scanner.

References

1. Sezan, M., Legandijk, R.: Motion Analysis and Image Sequence Processing. Kluwer Academic, London (1993)
2. Tekalp, M.: Digital Video Processing. Prentice Hall, New York (1995)

The Industrial Application of the Irregular 3D-Objects Image Processing in the Compact Reverse Engineering System

Dmitry Svirsky[1] and Yuri Polozkov[2]

[1]Dr. Ing. Science leader of the Computer Aided Design Centre
at Vitebsk State Technological University, 72 Moskovsky ave., 210035 Vitebsk, Belarus
Phone: 375 212 25 74 01
svirsky@vstu.unibel.by

[2]Post-grad. of Vitebsk State Technological University, 72 Moskovsky ave., 210035 Vitebsk,
Belarus, Phone: 375 212 25 76 45

Abstract. The problems of irregular 3D-objects manufacturing preparation are considered. The irregular surfaces digitizing process by video system is offered. The mathematical models of video digitizing process and software basic stages for computer 3D-models creation are shown. The forming of the computer 3D-models from video image as one from main stages of irregular 3D-objects compact reverse manufacturing is considered. The video system assubsystem of Compact Reverse Engineering System (CRES) is offered.

Keywords: image processing, reverse engineering, compact system

1 Introduction

Many industrial goods and produce equipment objects of modern industry have irregular (spatially complicated) surface form. In order to provide such products manufacturing effective, the Reverse Engineering and Rapid Prototyping technologies are widely used [1]. The Reverse Engineering technologies are based on a CAD-models construction from physical analogue 3D-objects, which were made earlier. So that to receive a number of the basic 3D-models, the analogue irregular 3D-objects surfaces must be digitized. Thus the CAD-models constructed and transformed to the needs of the production processes being used. To make a new material object from its CAD-model the Rapid Prototyping system is used. Both the Reverse Engineering system and Rapid Prototyping system integration allows to realize Compact Reverse Engineering System (CRES). Such CRES combine curtailing in space and time with a minimum level of functional and resource redundancy [2]. The application of the compact video system as CRES digitizing subsystem shall be allows small and media enterprises to react to changes of a market conjuncture more faster and to make irregular 3D-objects most effectively.

W. Skarbek (Ed.): CAIP 2001, LNCS 2124, pp. 457–464, 2001.

2   The Realization of the Irregular 3D-Object Surface Video Digitizing

The surface digitizing is the most difficult process of irregular 3D-objects manufacturing preparation. The geometrical form of the irregular 3D-object surface digitizing process consists in the measuring information reception and its processing. This process is traditionally realized by contact co-ordinate measuring machines [3]. They digitize an existing surface point by point. The contact co-ordinate measuring machines using is low effective temporary because of point physical contact with the object and necessity of many different set-ups execution for complicated 3D-objects digitizing. These measuring machines have high cost too. Therefore, the special video system was developed for irregular 3D-objects digitizing [4]. This one combine low cost with sufficient accuracy and provide high efficiency and information transference speed of digitizing process.

2.1   The Video System Configuration and Main Steps of Video Digitizing Process

The authors had carried out the quantitative analysis of functional-cost properties of different (both contact and non-contact) digitizing systems with the aim their as the subsystem for irregular 3D-objects manufacturing preparation on small and media enterprises. This analysis results are shown on the figure 1.

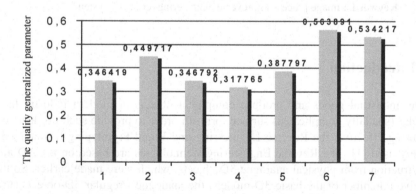

Fig. 1. The diagram of quality generalized parameters of digitizing systems

1. The contact mechanical devices; 2. The "measuring hands"; 3. The specialized automated systems; 4. The contact outliners; 5. The co-ordinate measuring machines; 6. The video systems; 7. The laser scanners.

The digitizing video systems are most effective. In the digitizing video system configuration working out the principle of optimal parity of expenses on CRES functional invariant and variable adapter in accordance with their functional importance was used [2]. The relative ease program-mathematical and technical realization was used too.

The configuration of offered video system includes with a video camera 1, a projector 4, a rotary table 6 and a computer (fig. 2). The projector is equipped with a slide 3 with the image of a co-ordinate grid with the units, which are located on equal distance from each other. The application of other digitizing video system configuration increases digitizing process complexity, CRES cost and reduces irregular 3D-objects manufacturing efficiency.

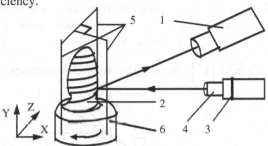

Fig. 2. The principal scheme of 3D-object digitizing by video system

The digitizing process includes an object 2 video shooting, on which surface the light strips of the slide inserted in the horizontal located projector 4 are imaging. The video camera 1 is established on a certain angle to a horizontal plane. The 3D-stage information is imported from the video camera to the computer. There the information is organized in object digital model by special software. The obtained object digital model contains the captured points co-ordinates of object surface in the chosen spatial co-ordinate system.

The object image can be registered by the video camera have following specifications: Power Source: DC 12.0 V; Power Consumption: 17.9 W; Lens: 8:1 2-Speed Power Zoom Lens; focal Length – 5 ? 40 mm; F 1.4; Resolution: more than 230 lines; Weight: 2.6 kg.; Dimensions: 130 (W) x 245 (H) x 459 (D) mm.

The projector specifications: Power Source: DC 220 V; Power Consumption: 225 W; Lens: focal Length – 2.8 ? 100 mm; Relative orifice: 1:2.8; Weight: 5.0 kg.; Dimensions: 265 (W) x 272 (H) x 125 (D) mm.

The equipment, which has other specifications, can be applied also.

2.2 The Video Digitizing Process Mathematical Modeling

The video digitizing process mathematical modeling is carry out on the basis of dependence between digitizing 3D-object points positions and their central projections positions (fig. 3) [5]. It is described following function:

$$\overline{R}_{ij} = f(x_{ij}{}'', y_{ij}{}'', \mu, \omega, R, \varphi, \eta, \nu, x_S, y_S, z_S), \tag{1}$$

where $\overline{R}_{ij} = \left(X_{ij};\ Y_{ij};\ Z_{ij} \right)^T$ - a vector, which determines a position of the i-point of j-level of an object surface in spatial co-ordinate system; $x_{ij}{}'', y_{ij}{}''$ - the i-point of j-level co-ordinates of an object surface in image co-ordinate system; ?, ?, R - the parameters, which determine the projections center position; ?, ?, ? - the orientation parameters of the image co-ordinate system relatively the object co-ordinate system; x_s, y_s, z_s - the main point co-ordinates of the image in O' x' y'.

The video camera forward central point orientation is determined as $\overline{R}_x$ vector:

$$\overline{R}s = \begin{matrix} Xs \\ Ys \\ Zs \end{matrix} = A_1(\mu)A_2(\omega)\overline{R}x = \begin{matrix} \cos\mu & 0 & \sin\mu \\ 0 & 1 & 0 \\ -\sin\mu & 0 & \cos\mu \end{matrix} \begin{matrix} \cos\omega & -\sin\omega & 0 & R \\ \sin\omega & \cos\omega & 0 & 0 \\ 0 & 0 & 1 & 0 \end{matrix}, \tag{2}$$

where $\overline{R}_x = (R; 0; 0)^T$; $|\overline{R}| = |\overline{R}_S|$; A_1, A_2 - the orientation matrixes.

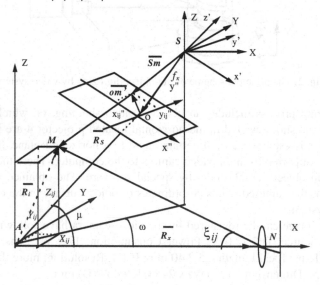

Fig. 3. The determination scheme of object points spatial co-ordinates by their images co-ordinates

The $\overline{Sm}_{ij}$ - vector determine the positions of the object points central projections in spatial co-ordinate system:

$$\overline{Sm}_{ij} = A_3(\varphi)A_4(\eta)A_5(\nu)\overline{Sm}_{ij}' = \tag{3}$$

$$= \begin{matrix} 1 & 0 & 0 \\ 0 & \cos\varphi & \sin\varphi \\ 0 & \sin\varphi & \cos\varphi \end{matrix} \begin{matrix} \cos\eta & 0 & \sin\eta \\ 0 & 1 & 0 \\ \sin\eta & 0 & \cos\eta \end{matrix} \begin{matrix} \cos\nu & \sin\nu & 0 \\ \sin\nu & \cos\nu & 0 \\ 0 & 0 & 1 \end{matrix} \begin{matrix} x''_{ij} \\ y''_{ij} \\ -f_{\mathrm{h}} \end{matrix} = (x_{ij}; \; y_{ij}; \; z_{ij})^T$$

where A3, A4, A5 - the orientation matrixes; f_k – the video camera focal distant.

The $\overline{R}_{ij}$ is determinate through $\overline{R}_S$ and $\overline{Sm}_{ij}$:

$$\overline{R}_{ij} = A_1(\mu)A_2(\omega)\overline{R}x + kA_3(\varphi)A_4(\eta)A_5(\nu)\overline{Sm}_{ij}, \tag{4}$$

where k – the scaling factor.

The light strips image on object surface are used in calculations. It allows to transform 2D-co-ordinates of the object points video images in their 3D-co-ordinates by following formulas:

$$X_{ij} = \frac{X_S z_{ij} + x_{ij}(Htg\xi_{ij} - Z_S)}{z_{ij} + x_{ij} tg\xi_{ij}}, \tag{5}$$

$$Z_{ij} = (H - X_{ij})tg\xi_{ij} = \frac{\delta_{ij}(H - X_{ij})}{f_r}, \tag{6}$$

$$Y_{ij} = Y_S + \frac{Z_{ij} - Z_S}{z_{ij}} y_{ij}, \tag{7}$$

where X_s, Y_s, Z_s – the co-ordinates of a video camera forward central point in 3D-co-ordinate system; H – the distance of digitizing object up to a projection's centre of the projector; $?_{ij}$ – the angle of a direction of the projector central ray; $?_{ij}$ - the distance from slide grid j-line to a projector main optical axis; fr – the projector focal distance.

2.3 The Image Software Processing for Irregular Surfaces Video Digitizing

The digitizing software allows to make main step for digital 3D-models creating (fig. 4) [6].

The digitizing 3D-object side images are imported from the video camera to the computer. The 3D-object side video images are vectored in computer. The vectoring process is carried out by standard software with picture recognition functions. The raster image is filtered previously. The obtained image raster elements are averaged. Then this raster image is vectored by a file – template. This file contains vectoring specific parameters: polyline width, ruptures magnitude and other. The raster and vector images should coincide after a vectoring. The irregular surface parts are removed separately in insufficient quality case. The vector images is saved as the DXF file. The 2D-co-ordinates list is formed of this file. These 2D-co-ordinates describe polylines nodes of a vector image. The 2D-co-ordinates are basic data for object 3D-co-ordinates account. In order to transform 2D-co-ordinates into 3D-co-ordinates, the special programming is carried out on known computer algebra software base. The algorithm of numerical parameters recalculation includes the operations on formation of the initial data list, numerical parameters elements input from the received file and direct recording of mathematical calculations teams. The received results are organized in the files, which describe the points spatial positions for the object separate parts. The object volumetric digital model is received by given files on boundary points synthesis. Then it is exported in graphic software through the LSP file. So the surface is fixed. This irregular digital 3D-model is used in CAD for irregular 3D-objects manufacturing.

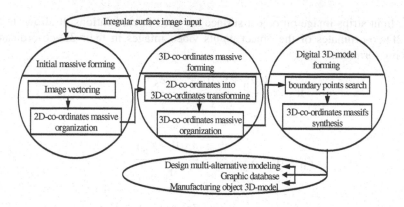

Fig. 4. The video images processing main steps

2.4 The Example of Computer 3D-Model Surface Creating by Irregular Object Video Digitizing

The irregular object (human head) was digitized by offered video system (fig. 5).

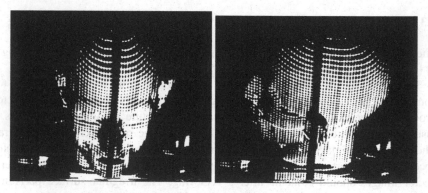

Fig. 5. The irregular 3D-object sides images by its video digitizing

The following parameters were used: ? - 45°; ?, - 0°; R – 1000 mm; ? - 0°, ? - 45°; ? - 0°; the distance of digitizing object up to a projections centre of the projector – 1336 mm; the video camera focal distance – 44 mm; the projector focal distance – 60mm. The object space model was created. This digital 3D-model fragment is shown on figure 6. It is described by 1035 points co-ordinates.

3   The Application of the Video System as CRES Subsystem

The forming of computer 3D-models is one of the main stages of irregular 3D-objects manufacturing [7]. A storage and search of the information; computer 3D- modeling;

Fig. 6. The point's digital 3D-model of the irregular object fragment

physical object making on the basis of computer model (Rapid Prototyping) are stages of this process too. The information-input module 1 - offered video system, information processing and control module 2, and industrial module 3 are included in CRES configuration for realization of irregular 3D-objects reverse creation process (fig. 7).

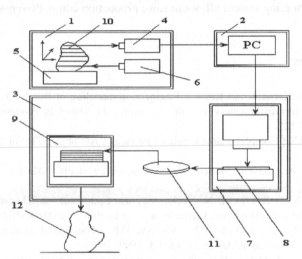

Fig. 7. The CRES configuration. 1. The information-input module; 2. The information processing module; 3. The industrial module; 4. The video camera; 5. A co-ordinate table; 6. A projector; 7. A laser cutting installation; 8. A cutting material; 9. An assembly unit; 10. Initial (analogue) object; 11. A cutout layer; 12. A ready object.

The received by the video system the digital 3D-models form the graphic database for storage and fast search of necessary information during computer aided design. The development of computer models of future products is carried out in a mode of computer design multi-alternative modeling. The fragments of digital 3D-models of the objects from the graphic database are exposed to selective transformations by means of computer modeling. This process is carried out in the information processing and control module. As a result of designing the computer model turns out. It carries full design information for future physical object manufacturing. This information will be transformed to the program, which operate of the cutting tool movement in the industrial module. The industrial module realizes the technology of level-by-level object manufacturing. It includes: a CO_2-laser; the optical channel; a control system; a coordinate table. The layers parallel connection allows monolithic products making. The method of flat elements cross connection to assemble allows 3D-object skeleton.

The industrial module specifications: Positioning Exactitude – 0.05 mm; working area – 1700 x 1200 mm; CO_2-laser; Cutting Speed – 24 m/min for unmetal 2D-bars by a thickness up to 20 mm; Power Consumption: 800 W.

4 Conclusion

The video system using in CRES allows to solve a problem of 3D-objects manufacturing process information support. The irregular 3D-objects video digitizing development and its software allows to create volumetric computer models easily. It allows also to speed up a manufacture preparation stage. The offered variant of the compact reverse engineering system allows to raise production competitiveness.

References

1. Kruth, J.P., Kerstens, A.: Reverse Engineering modeling of free-form surfaces from point clouds subject to boundary conditions. Journal of Materials Processing Technology **76** (1997) 120-127
2. Svirsky, D.N.: Compact manufacturing system as CAD object. ITC of NAS Belarus Minsk (2000) 48.
3. Makachev, A., Chaykin: Modeler 2000: 3D-scaning systems. CAD & Graphics **1** (2000) 84-86
4. Polozkov, Y.V., Svirsky, D.N.: Technology and equipment for 3D-scaning in the compact Rapid Prototiping. Materials, Technologies, Tools. Vol. 5 (2000) 97-102
5. Zavatsky, Y.A., Polozkov, Y.V., Svirsky, D.N.: The spatial objects digitizing process mathematical modeling. Annals of VSU **3** (1999) 49-53
6. Svirsky, D.N., Polozkov, Y.V.: The video images processing algorithmic support in compact Reverse Engineering system irregular surfaces. Annals of VSU **4** (2000) 90-93
7. Polozkov, Y.V., Svirsky, D.N.: 3D-image processing in the compact Reverse Engineering system. Annals of DAAM for 2000. Viena (2000) 379-380

A Local Algorithm for Real-Time Junction Detection in Contour Images

Andrzej Śluzek

Nanyang Technological University,
School of Computer Engineering and Robotics Research Centre
Singapore 639897
assluzek@ntu.edu.sg

Abstract. The paper reports development of an efficient algorithm identifying junctions in contour images. The theoretical model and selected results are presented. The algorithm combines local operators and an inexpensive matching. The algorithm scans incoming contour images with a circular window, and two levels of detection are used: structural identification and geometrical localisation. At the first level, the prospective junctions are identified based on topological properties of the window's content. At the second level, the prospective location of the junction's centre is determined, and a projection operator calculates its 1D profile. The junction is eventually accepted if the profile matches the corresponding template profile. The software implementation has been intensively tested, but the hardware implementation is the ultimate objective. Currently, a development system that directly maps C programs into FPGA structures can be used for this purpose.

Keywords: image analysis, junction detection

1 Introduction

Edges are important features in image analysis. For example, corners, T-junctions and other line junctions play an important role in visual perception and provide useful clues to 3D interpretation of images (e.g. in autonomous navigation). However, detection of such features is usually only a first step in a complex process of image interpretation. Therefore, algorithms detecting junctions should be fast (real-time), robust and reliable.

In general, real-time image processing algorithms are based on local operator (hardware-implemented in many frame grabbers) performing rather simple tasks (e.g. edge detection, image smoothing or sharpening, calculation of moments, morphological operations, etc.).

The purpose of this paper is to present a local algorithm that performs (eventually in real time) a more complex task – detection of line junctions in contour images. Initially, the algorithm was developed for a specialised robotic system within a joint research project of the Robotics Research Centre (NTU) and

W. Skarbek (Ed.): CAIP 2001, LNCS 2124, pp. 465–472, 2001.

British Gas Asia Pacific. In the system, contour images were produced by projecting a pattern of structured light on 3D objects. Later, the algorithm has been developed for more diversified applications involving contour images.

Corners and other line junctions are detected either from grey-level images or from contours. In the first group (e.g. [1], [2]) the usual approach is template matching or parameter-fitting. In contour images, junctions are typically found by grouping pixels and/or curvature analysis (e.g. [3], [4]). Such operations usually require directional analysis of contours. Sometimes, a combination of grey-level and contour techniques is used (e.g. [5]). The proposed algorithms, however, are usually too complex for real-time systems of limited capabilities (e.g. AGV on-board computers). Moreover, most of the existing real-time algorithms detecting corner-like features are using small windows (e.g. [6]) but the results may strongly depend on the scale at which the features are analysed (e.g. [7]).

For the algorithm presented in the paper, it has been assumed that:

- Junctions are detected in contour images (this is partly because of the original application).
- No directional analysis of contours is performed.
- Only junctions of significant size are detected.
- Junctions are well positioned and oriented.
- Detection is possible within contour images of any complexity and size.

The algorithm combines local operators, structural analysis and matching. Although contour extraction is not a part of the algorithm, Canny edge detector is recommended. Most of the results have been obtained by using the software implementation of the algorithm, but the hardware implementation has been developed simultaneously. Originally, a standard FPGA design approach was used (see [8] and [9] for more details). Recently, C programs are directly mapped into FPGAs using a design suite DK1 developed by Celoxica (see [10]). It allows rapid hardware implementation with virtually no circuit design. The tests, both for previous designs (see [9]) and for DK1-produced circuits, show that FPGA implementations are able to process 256x256 contour images typically in less than 40ms. This speed is satisfactory for any currently envisaged application.

2   Geometrical Model of Junctions

For a contour feature of any shape, we define its *focal point* [9]. Although for some features the focal point can be selected somehow arbitrarily, for the features of interest (i.e. junctions of lines) the selection is straightforward.

In the proposed algorithm, we assume the minimum size of junctions. Therefore, so-called junctions of magnitude **R** are introduced (see Fig. 1). In practice, the magnitude indicates the minimum size of significant junctions. Junctions of lower magnitudes should be ignored by a vision system. Although the magnitude may vary for various applications, it was found that for the majority of typical problems, the magnitude of 15–30 pixels is suitable.

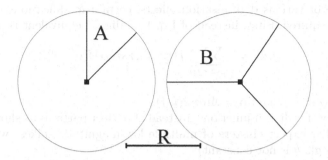

Fig. 1. Examples of junctions of magnitude **R**

Junctions of magnitude **R** can be mapped into 1D functions using the following projection operator:

$$g(\theta) = \int_{-\pi}^{+\pi} \int_{0}^{+\infty} f(r\cos\alpha, r\sin\alpha)\delta(\alpha - \theta)dr d\alpha. \tag{1}$$

Eq. 1 represents the summation of an image function $f(x,y)$ along the ray of θ direction. If the function $f(x,y)$ represents a junction of magnitude **R** (with the origin of co-ordinates shifted to the focal point) the projection operator will produce a pattern consisting of several spikes (see Fig. 2) of magnitude **R**, where the number of spikes corresponds to the number of arms and the position of spikes indicates the geometry of the junction. Therefore, the existence of such a projection pattern for a contour fragment not only indicates the existence of a junction but also reveals geometry of this junction.

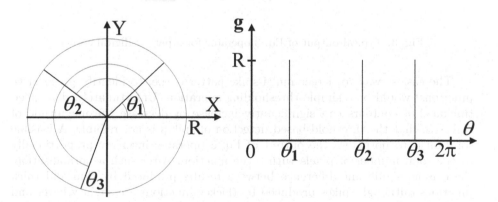

Fig. 2. Typical results of Eq. 1 projection for a junction of magnitude **R**

Because of various digitalisation effects, certain modifications of the above model are required. First, instead of Eq. 1 its digital equivalent is used:

$$g(\theta) = \sum_{i^2+j^2 \leq R^2} f(r cos\alpha, r sin\alpha)\delta(\alpha - \theta). \qquad (2)$$

where $r = \sqrt{i^2 + j^2}$ and $\alpha = Atan2(i, j)$.

Moreover, for digital junctions, instead of perfect patterns as shown in Fig. 2, we can rather expect clusters of multiple low-magnitude spikes (within digital lines, the angle θ is not constant).

Therefore, in Eq. 2 the function δ is replaced by a parameter-dependent family of normalised functions d_r:

$$g(\theta) = \sum_{i^2+j^2 \leq R^2} f(r cos\alpha, r sin\alpha)d_r(\alpha - \theta). \qquad (3)$$

where $r = \sqrt{i^2 + j^2}$ and $\alpha = Atan2(i, j)$.

The main purpose of d_r functions is to "fuzzify" each pixel so that it will contribute to a certain angular area around its actual direction. Since r respesents the distance from the focal point, the "fuzzification" should be stronger stronger for smaller values of r and weaker for distant pixels. For example, d_r could be normalised triangular functions of gradually changing width ([8]). After these modifications, Eq. 3 produces regular multi-spike patterns for digital junctions, although the spikes are diluted. An examples (for a perfect quality digital corner) is shown in Fig. 3.

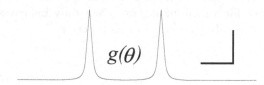

Fig. 3. Typical output of Eq. 3 operator for a perfect digital corner

The easiest way to detect multi-spike patterns (i.e. to identify prospective junctions) would be a simple thresholding operation. Unfortunately, diversified thickness of contours can significantly increase or decrease the magnitude of spikes so that the threshold-based detection of spikes is not reliable. A feasible solution is to normalise the output of Eq. 3 operator inversely proportionally to the total number of pixels within the junction. After such a normalisation, there is no significant difference between results produced by thin and thick junctions (although spikes produced by thicker junctions are slightly lower and more diluted, see examples in [12]). Normalisation, however, may affect the spikes in a different way. With the increased number of arms, the magnitude of spikes gradually decreases (Fig. 4). The proposed solution is to re-scale the profiles proportionally to the number of arms so that any junction of magnitude **R** would produce a multi-spike profile with similar heights of all spikes. Then, thresholding operation may be applied to detect junctions and to determine their geometry ([8] and [12]).

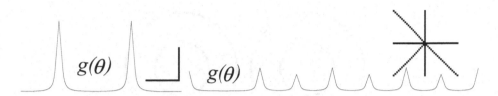

Fig. 4. Normalised profiles for junctions with different number of arms

3   Structural Model of Junctions

In digital images, certain local properties can be defined/measured by using a set of concentric circles (e.g. [13]). Based on this idea, *structural junctions* of magnitude **R** are defined as illustrated in Fig. 5.

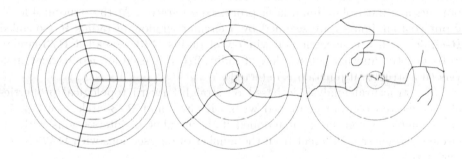

Fig. 5. Examples of structural junctions of magnitude **R**

Informally, it can be said that a structural junctions with n–arms is an intersection of a contour fragment with a set several concentric circles satisfying the following conditions:

− Each circle intersects the contour n–times.
− The intersections with the circles are properly connected (see Fig. 6).

It should be noted that if the number of concentric circles increases (and their radiuses are evenly distributed between **0** and **R**) the corresponding structural junctions converge to "geometric junctions". With only few circles available, however, there may be significant inconsistencies between structural junctions and the actual ones. First, contours not resembling actual junctions could be still accepted as structural junctions (as shown in Fig. 5). Secondly, even if a structural junction is the actual one, its focal point can be located with limited accuracy only (i.e. it can be placed anywhere within the most inner circle).

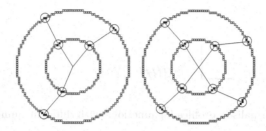

Fig. 6. Digital examples of a 3–arm and a 4–arm structural junction obtained by using two concentric circles

4   Principles of the Algorithm

The proposed algorithm detecting junctions in digital contour images makes us of both models of junction. Since in actual contour images the window may contain several unrelated contour fragments (see examples in [11], [12]), connected components are first isolated and later processed separately. At the structural level, input contour images are scanned by a circular window of radius **R** (so-called **R**-circle) with a concentric sub-window of radius **R/3** (so-called **R/3**-circle). Subsequently, another sub-window of radius **2R/3** is used for geometrical analysis (if a structural junction is detected).

The structural analysis is based on straightforward topological properties of a connected contour contained within **R**-circle and **R/3**-circle (see Fig. 7). For example, a contour is abandoned if the number of intersections with **R**-circle's perimeter is different from the number of intersections with **R/3**-circle's perimeter.

In order to apply the geometrical analysis, the junction's focal point should be localised first (the structural analysis locates it somewhere within **R/3**-circle). The assumed position of the focal point is the midpoint intersections of virtual lines joining the corresponding perimeter pixels of **R**-circle and **R/3**-circle (see Fig. 7). If the actual perfect were there, that would be the location of its focal point. When the estimated position of the focal point falls outside **R/3**-circle, the junction is abandoned (should eventually a junction be in this area, it will be detected for another location of the scanning **R**-circle). Then, a circle of radius **2R/3** is placed in the assumed focal point (since its is within **R/3**-circle, the new circle is always within the scanning **R**-circle!). Examples are shown in Fig. 7.

Finally, the Eq. 3 projection operator (including normalisation and re-scaling) is applied to the content of **2R/3**-circle (if it exhibits the same topologigal properties as the original structural junction). The profile produced by the

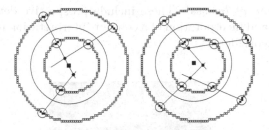

Fig. 7. Estimated location of focal points for the structural junctions from Fig. 6

operator is matched to the model junction profile. The integrated difference (over the whole profile) is a measure of disparity between the actual junction and the model. Depending on the application, various levels of disparity can be accepted. Fig. 8 illustrates the process for a poor quality junction. It explains how a model junction is created, and secondly shows differences between the profiles for high quality junctions and poor ones is shown.

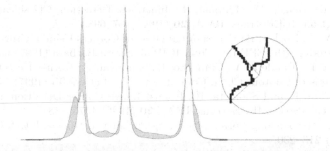

Fig. 8. Example of profile matching for a poor quality junction

In the algorithm, some operations are computationally intensive. Therefore, hardware-supported implementation has been proposed. Initially, several modules (e.g. profile extractor, perimeter analyser, etc.) have been FPGA-designed and some details can be found in [9] and [8]. Recently, DK1 design suite allows almost arbitrary decomposition of the algorithm into software and hardware components.

5   Concluding Remarks

The algorithm has been originally proposed for the navigation system of an underwater ROV (collaboration with British Gas Asia Pacific) and for that application the software implementation was able to cope with the existing time constraints ([14]). An example of the results produced there is shown in Fig. 9.

Currently, the algorithm is being developed for more demanding applications within AcRF project **RG 3/99** *"New Algorithms for Contour-Based Processing* *and Interpretation of Images"*. These include prospective development of specialised sensors, and incorporation of a contour-interpretation module (a solution based on [15] is considered).

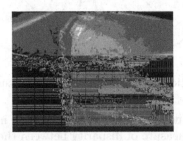

Fig. 9. The best-fit corner found in weld-seam localisation

References

1. Parida, L., Geiger, D., Hummel, R.: Junctions: Detection, Classification, and Reconstruction. IEEE Trans. PAMI **20** (1998) 687–698.
2. Chen, W.C., Rocket, P.: Bayesian Labelling of Corners Using a Grey-Level Corner Image Model. 4th IEEE Int. Conf. ICIP'97, Santa Barbara (1997) 687–690.
3. Arrebola, F., Bandera, A., Camacho, P., Sandoval, F.: Corner Detection by Local Histograms of Contour Chain Code. Electronics Letters **33** (1997) 1769–1771.
4. Mokhtarian, F. and Suometa, R.: Robust Image Corner Detection Through Curvature Scale Space. IEEE Trans. PAMI **20** (1998) 1376–1381.
5. Rohr, K.: Recognizing Corners by fitting Parametric Models. Int. J. Comp. Vision **9** (1992) 213–230.
6. Wang, H., Brady, M.: Real-Time Corner Detection Algorithm for Motion Estimation. Image & Vision Comp. **13** (1995) 695–703.
7. Lindeberg, T.: Junction Detection with Automatic Selection of Detection Scales and Localization Scales. 1st IEEE Int. Conf. ICIP'94, Austin (1994), 924–928.
8. Śluzek, A.: Hardware Supported Technique for Detecting Multi-Corners in Digital Contours. 5th IEEE Int. Workshop CAMP'2000, Padova (2000), 320–328.
9. Lai, C.H.K.: Detector of Corners for Contour Images. FYP AS00-CE-S036, School of Comp. Eng. (NTU), Singapore (2000).
10. Celoxica Ltd., http://www.celoxica.com/home.htm.
11. Śluzek, A.: Moment-based Contour Segmentation Using Multiple Segmentation Primitives. Machine Graphics & Vision **7** (1998) 269–279.
12. Śluzek, A.: A Local Algorithm for Detecting Multi-Junction Primitives in Digital Contours. 6 Int. Conf. ICARCV 2000, Singapore (2000), CD publication.
13. Śluzek, A.: Identification and inspection of 2-D objects using new moment-based shape descriptors. Pattern Rec. Lett. **16** (1995) 687–697.
14. Czajewski, W., Śluzek, A.: A Laser-Based Vision System for Robotic Weld Seam Inspection", Proc. 6 Int. Conf. ICARCV 2000, Singapore (2000), CD publication.
15. Guy, G., Medioni, G.: Inferring Global Perceptual Contours from Local Features. Int. J. Comp. Vision **20** (1996) 113–133.

A New Algorithm for Super-Resolution from Image Sequences

Fabien Dekeyser, Patrick Bouthemy, and Patrick Pérez *

IRISA/INRIA
Campus de Beaulieu, 35042 Rennes Cedex, France
firstname.lastname@irisa.fr

Abstract. This paper deals with super-resolution, i.e. the reconstruction of a high resolution image from a sequence of low resolution noisy and possibly blurred images. We have developed an iterative procedure for minimizing a measure of discrepancy based on the Csiszàr's I-divergence. One advantage of this algorithm is to provide a natural positivity constraint on the solution. We consider a block-based version to speed up convergence and we propose a computationally efficient implementation. We also introduce a temporal multiscale version of the algorithm, which proves to be more robust and stable. The algorithm requires the computation of the apparent motion in the image sequence. We use a robust multiresolution estimator of a 2D parametric motion model in order to keep computational efficiency.

Keywords: super-resolution, parametric motion estimation, block-based iterative algorithm

1   Introduction

When acquiring an image with a camera, the optical system behaves more or less as a low-pass filter. High spatial frequencies of the image are therefore lost or at least smoothed. However, some residual high frequencies (edges) can cause the sampling rate not to meet the Nyquist criterion. Therefore, digital images are generally aliased.

In this work, the term *resolution* defines the fidelity of a sampled image to the details (or high frequencies) of an original scene. The term *super-resolution* refers to the construction of an image whose resolution is higher than the resolution provided by the sensor used in the imaging system. To achieve such a goal, we exploit the information given by the sub-pixel motion between the frames of the original low resolution image sequence.

We can classify methods for solving this problem into three broad classes. Techniques from the first class solve this issue in the frequency domain. Their main drawback is that they are effective only when the inter-frame motion is a

* This work is partially supported by DGA/DSP/STTC (contract and student grant). It is carried out in cooperation with Thales-Optronique.

W. Skarbek (Ed.): CAIP 2001, LNCS 2124, pp. 473–481, 2001.

global translation. Concerning the second class of methods, the images of the sequence are first registered to get an image composed of samples on a non-uniform grid. These samples can then be interpolated and resampled on a high resolution grid. However, in that case, aliased samples are used to interpolate a high resolution image. True high frequencies cannot be reconstructed in that way. The third class of methods aims at modeling the degradation process and at stating the super-resolution issue as an inverse problem. A review of existing methods can be found in [1].

The method we propose belongs to the third class. It is based on a computationally efficient iterative algorithm for solving large linear systems and on a robust method for estimating the 2D motion parameters. It also introduces the use of a cross-entropy measure in the context of super-resolution.

The paper is organized as follows. Section 2 presents the problem modeling. Section 3 introduces the proposed super-resolution method. Section 4 describes the temporal multi-scale version of the algorithm. Section 5 outlines the motion estimation module. Experimental results are reported in Section 6.

2   Modeling the Problem

We formulate the super-resolution problem as proposed in [4]. The sequence of N observed images of M pixels is denoted $\{Y_k\}_{k=1}^{N}$. Let us assume that this sequence comes from a high resolution image X that we want to reconstruct. Each observed image k can be viewed as the result of a warping W_k (related to image motion) of the image X, followed by the convolution with a Point Spread Function (we denote B_k the matrix equivalent to this operation), downsampling and noise corruption. The downsampling operator is denoted by D_k. It performs downsampling by averaging the pixels in a block. The blur and downsampling are assumed identical for all images so that we can write $B_k = B$ and $D_k = D$. By ordering lexicographically X and Y_k to write them as vectors, we get the following matrix equation:

$$
\begin{bmatrix} Y_1 \\ \vdots \\ Y_N \end{bmatrix} = \begin{bmatrix} DBW_1 \\ \vdots \\ DBW_N \end{bmatrix} X + \begin{bmatrix} \eta_1 \\ \vdots \\ \eta_N \end{bmatrix} = \begin{bmatrix} H_1 \\ \vdots \\ H_N \end{bmatrix} X + \begin{bmatrix} \eta_1 \\ \vdots \\ \eta_N \end{bmatrix} = HX + \eta \qquad (1)
$$

where H is a matrix made of N blocks $H_k = DBW_k$, and $[\eta_1, \ldots, \eta_N]$ designates the additive noise.

3   The Super-Resolution Algorithm

The considered problem comes to the inversion of the linear system (1) where H is a huge sparse and ill-conditioned matrix. The method to solve this equation must be chosen carefully.

Some methods involve least-squares regularized solutions of the problem [4], sometimes using a Bayesian-Markovian formalism [5]. Other methods use projections onto convex sets to find a solution with bounds on the residual [8]. We consider the Csiszàr I-divergence measure and we seek the solution minimizing the following expression:

$$I(HX, Y) = \sum_{k}^{N} \sum_{i}^{M} [(H_k X)_i - Y_{k_i}] + \sum_{k}^{N} \sum_{i}^{M} Y_{k_i} \ln \frac{Y_{k_i}}{(H_k X)_i} \qquad (2)$$

where k indexes the bloc and i indexes the components of the vectors. In other words, Y_{k_i} is the intensity of the ith pixel of the kth image. This measure generalizes the Kullback divergence or cross-entropy measure. Csiszàr introduced a collection of axioms that a discrepancy measure should reasonably be required to possess [3]. He concluded that if the functions involved are all real valued, having both positive and negative values, then minimizing the least-squares measure is the only consistent choice; whereas if all the functions are required to be nonnegative, Csiszàr I-divergence is the only consistent choice. The sequence:

$$X_i^{n+1} = X_i^n \left[H^T \frac{Y}{HX^n} \right]_i \qquad (3)$$

converges towards the solution minimizing (2) when $||H|| = 1$ (see [10]). H^T is the transposed matrix of H. The ratio $\frac{Y}{HX^n}$ is a vector computed by considering each component. Its ith component is given by $\frac{y_i}{(HX)_i}$, where y_i and $(HX)_i$ are the ith pixels of vectors Y and HX respectively. Since all the matrices and vectors involved in our application are nonnegative, a crucial advantage of our algorithm is that it includes a positivity constraint on the solution. Similar multiplicative schemes were proposed by Richardson and Lucy for restoration of astronomical images [6] and by Shepp and Vardi for emission tomography [9].

To speed up convergence, we consider a block-iterative version of the algorithm [2]. Instead of using the whole matrix H at each iteration, we only use one of its blocks H_k (see (1)). We then have:

$$X_i^{n+1} = X_i^n \left[W_k^T B^T D^T \left(\frac{Y_k}{DBW_k X^n} \right)^{\gamma} \right]_i \qquad (4)$$

where $k = n \bmod N + 1$ with n denoting the iteration number, and k indexing the bloc. The estimate is thus updated with each available image Y_k and this process can be iterated until convergence or until we reach the limit number of iterations. We have added a positive exponent γ to control the rate of convergence. This parameter does not appear in the original algorithm.

In practice, the involved matrices are very sparse, but we can avoid to build them. Indeed, the iteration expressed by relation (3) is equivalent to the scheme given in Figure 1. Each module of this diagram can be computed efficiently without building any matrix.

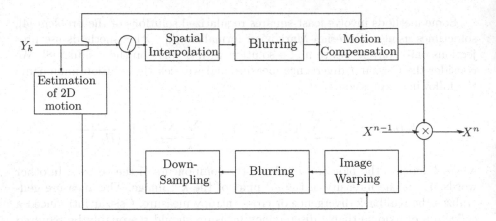

Fig. 1. Block-diagram of the super-resolution algorithm

Since D is a downsampling operator averaging the pixels in a block, D^T is a zero-order interpolation matrix (interpolation by duplication). If we assume that the convolution kernel is symmetrical, $B^T = B$. Concerning W, if it accounts for a uniform translation (respectively rotation), W^T holds for the opposite translation (respectively rotation). However, in the general case, there is no simple relationship between the direct and inverse motion compensation operator, and convergence is not ensured for complex motions. Nevertheless, we have successfully introduced 2D global, affine and quadratic, motion models. We have also considered dense motion fields. In all these cases, we explicitly compute the inverse motion and approximate W^T by the motion compensation operator corresponding to this estimated motion.

4   Temporal Multi-scale Version

In the previous section, we selected the image to be super-resolved from the observed image sequence, and we estimated the motion between this reference frame and the other images of the sequence to apply the iterative process of relation (4). We now process the original sequence in groups of three successive frames and compute a super-resolved estimate of the central frame of each group. We thus get a new sequence of improved images, but temporally subsampled by a factor of three. The process is iterated on this new reconstructed image sequence. We thus make a trade-off between spatial and temporal resolution. We call this temporal multi-scale scheme a pyramidal scheme, since it comes to build a pyramid with levels corresponding to different temporal sampling (Fig. 2).

At each level of the pyramid, motion estimation is more accurate since images have better resolution. Therefore this scheme can be viewed as a joint estimation process of the motion information and of the super-resolved image.

This pyramidal scheme is especially efficient when processing a large number of images or when motions of large magnitude are involved. As a matter of

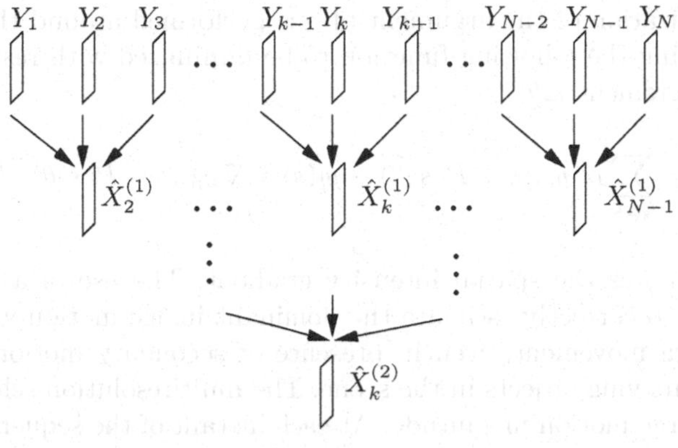

Fig. 2. The temporal pyramidal super-resolution scheme

fact, only the images in the temporal neighborhood of an image are significant, since these images have the largest common areas. We can stop the process when the images of the reconstructed sequence have too few common areas. Let us point out that the single scale version may diverge when motion estimation errors accumulate, which is not the case with the multi-scale version. There is no explicit regularization in our algorithm. Regularization nevertheless results from stopping iterations before convergence (a well known procedure in iterative image restoration). Let us also note that the main source of numerical instabilities is not image noise but errors in the motion estimates. In that context, the temporal pyramidal scheme shows slower convergence, but is numerically far more stable than the single scale version.

5   Motion Estimation

This section briefly describes the motion estimator used in our super-resolution algorithm. The global displacement between two images is modeled as a polynomial of the image point coordinates: $w_\theta(s) = P(s)\theta$, where parameter vector θ is formed by the coefficients of the two involved polynomials, s denotes pixel location, $w_\theta(s)$ stands for the displacement vector at point s supplied by the parametric motion model, and P is a matrix whose elements are monomials of x and y depending on the considered motion model. Its estimation between the lth and m^{th} image is stated as the minimization of the following robust function as introduced and described in [7]:

$$F(\theta) = \sum_{s \in S} \rho \left[y_m(s + P(s)\theta) - y_l(s) \right] \tag{5}$$

where S is the pixel grid, $y_l(s)$ and $y_m(s)$ respectively denote the intensity at pixel location s in images l and m, and ρ is a non-quadratic robust penalty function. This minimization is achieved within an incremental Gauss-Newton-type multiresolution procedure [7], able to handle large motion magnitude. At each step, a linearization of the argument of ρ is performed around the current estimate $\widehat{\theta}$, yielding the following function to be minimized with respect to motion parameter increment $\Delta\theta$:

$$G(\Delta\theta; \widehat{\theta}) = \sum_{s \in S} \rho \left[y_m(s + P(s)\widehat{\theta}) - y_l(s) + \nabla y_m(s + P(s)\widehat{\theta})^T P(s)\Delta\theta \right] \qquad (6)$$

where ∇y denotes the spatial intensity gradient. The use of a robust estimator allows us to correctly estimate the dominant image motion which is usually due to camera movement, even in presence of secondary motions arising from independent moving objects in the scene. The multiresolution scheme solves the problem of large motion magnitude. At each instant of the sequence, a Gaussian pyramid of images is built from original images y_l. At each pyramid level, parameter estimate $\widehat{\theta}$ is provided by the projection of the total parameter estimate obtained at previous level, and a new increment $\Delta\theta$ is obtained using an iterated reweighted least-square technique.

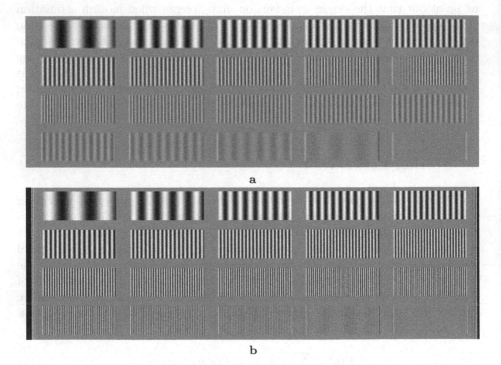

Fig. 3. (a) one image of the test chart sequence spatially interpolated, (b) super-resolved image obtained from the test chart sequence.

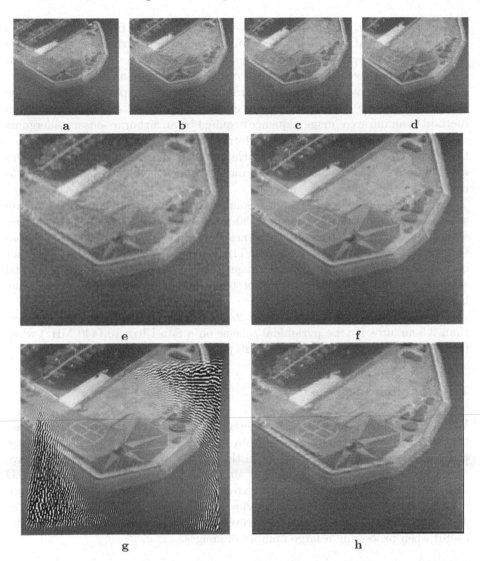

Fig. 4. (a, b, c, d) four images from the original low resolution sequence of 400 images, (e) image 3.b spatially interpolated, (f) image 3.b super-resolved using the first scheme with 50 images, (g) image 3.b super-resolved using the first scheme with 200 images, (h) image 3.b super-resolved using the temporal pyramidal scheme with the 400 images of the sequence.

6   Results

To validate our method, we have first tested it on a synthetical sequence of 100 frames that we constructed by translating and downsampling a test chart (Fig. 3). Each rectangle is formed by a sinusoidal test pattern. The first rectangle on the top left corresponds to the lowest frequency. The frequency is doubled in each subsequent rectangle. We compare on Figure 3 the results of super-resolution with results of a bicubic spatial interpolation. It is clear that our super-resolution method allows the recovery of previously aliased frequencies.

Figure 4 presents results of the pyramidal scheme compared to those of the single level scheme for a real image sequence. We applied our super-resolution method to an infrared image sequence acquired by an airborne sensor undergoing rather strong vibrations. We have performed only one pass on the sequence (i.e., the number of iterations is equal to the number of images). Parameter γ was set to 0.25. We estimated a quadratic motion model (involving 8 parameters) to account for the global image motion.

We can observe that the high resolution estimate given by the single-level scheme is correct when we use only 50 frames, however numerical errors arise when dealing with a batch of 200 frames, while the pyramidal scheme still converge when considering 400 frames. The super-resolution algorithm does not only improve the resolution of the images, it also denoises them. The pyramidal scheme does not necessarily supply better results than the single-level scheme, but it is more robust and stable.

The processing time is less than 2 minutes for the first scheme and less than 3.5 minutes for the pyramidal scheme on a Sun Ultra 10 (440 MHz) when processing 150 images of size 180×120 pixels.

7   Conclusion

We have proposed a new iterative method for the super-resolution reconstruction of an image from a low resolution image sequence. It introduces the use of the Csiszàr I-divergence measure in the context of super-resolution. We have developed a computationally efficient algorithm involving the estimation of 2D parametric motion models. We have extended this algorithm to design a temporal multi-scale version which can be viewed as a joint estimation process of the motion information and the super-resolved image reconstruction. It is especially useful when processing a large number of images.

References

1. Borman, S., Stevenson, R.L.: Super-resolution from image sequences - a review. Midwest Symp. on Circ. and Syst. (1998)
2. Browne, J., DePierro, A.: A row-action alternative to the EM algorithm for maximizing likelihoods in emission tomography. IEEE Transactions on Medical Imaging **15(5)** (October 1996) 687–699
3. Csiszàr, I.: Why least squares and maximum entropy? - an axiomatic approach to inverse problems. Annals of Statistics **19** (1991) 2033–2066
4. Elad, M., Feuer, A.: Restoration of a single superresolution image from several blurred, noisy and undersampled measured images. IEEE Transactions on Image Processing **6(12)** (December 1997) 1646–1658

5. Lorette, A., Shekarforoush, H., Zerubia, J.: Super-resolution with adaptive regularization. IEEE Int. Conf. on Im. Proc., Santa Barbara (1997)
6. Molina, R., Nuñez, J., Cortijo, F.J., Mateos, J.: Image restoration in astronomy: A Bayesian perspective. IEEE Sig. Proc. Mag. (March 2001) 11–29
7. Odobez, J.M., Bouthemy, P.: Robust multiresolution estimation of parametric motion models. Journal of Vis. Com. and Im. Rep. **6(4)** (December 1995) 348–365
8. Patti, A.J., Altunbasak, Y.: Artifact reduction for set theoretic super-resolution image reconstruction with edge adaptive constraints and higher-order interpolants. IEEE Transactions on Image Processing **10(1)** (January 2001) 179–186
9. Shepp, L.A., Vardi, Y.: Maximum-likelihood reconstruction for emission tomography. IEEE Transactions on Medical Imaging **1** (1982) 113–121
10. Snyder, L.S., Schulz, T.J., O'Sullivan, J.A.: Deblurring subject to nonnegativity constraints. IEEE Trans. on Sig. Proc. **40(5)** (May 1992) 1143–1150

Flatness Analysis of Three-Dimensional Images for Global Polyhedrization

Yukiko Kenmochi[12*], Li Chunyan[2], and Kazunori Kotani[2]

[1] Laboratoire A2SI, ESIEE, France
[2] School of Information Science, JAIST, Japan
y.kenmochi@esiee.fr, {sri,ikko}@jaist.ac.jp

Abstract. We give an overview of the problems for global polyhedrization in 3D digital images and present a solution by using our results of flatness analysis. We take both analytical and combinatorial topological approaches to define our flatness measurement which enable us to measure the degree of flatness for each point on a discretized object surface.

Keywords: 3D image analysis, global polyhedrization, flatness measurement

1 Introduction

Polyhedral representation is widely used for shape analysis and visualization in 3D computer imagery. In this paper, we focus on global polyhedrization [1, 2] rather than local polyhedrization such as the marching cubes [3]. In global polyhedrization, each plane of an polyhedron is decided by observing the configuration of object points in a larger region than local polyhedrization. Global polyhedrization has the advantage such that the number of polygons in a global polyhedron is smaller than a local polyhedron and it can consume less time for the postprocessings such as object recognition and visualization. It is also reported that global polyhedrization may have the multigrid convergence of surface area calculation which is not maintained by local polyhedrization [2,4].

However, it has been pointed out that there are many problems for global polyhedrization such as adaptation of various parameters which appear in the algorithm [1,2]. In this paper, we first give an overview of such problems. Since global polyhedrization contains a digital planar recognition, many studies have been made based on the analytical approach such as using inequation systems [2, 5]. The combinatorial topological approach is also considered because the topological structures of a 2D surface enable us to have helpful constraints for global polyhedrization [1]. In this paper, we take both analytical and combinatorial

---

* The current address of the first author is Laboratoire A2SI, ESIEE, thanks to JSPS Postdoctoral Fellowships for Research Abroad from 2000. A part of this work was supported by JSPS Grant-in-Aid for Encouragement of Young Scientists (12780207).

W. Skarbek (Ed.): CAIP 2001, LNCS 2124, pp. 482–492, 2001.

topological approaches. We first clarify the relations between the analytical representation of global polyhedra and the combinatorial topological representation. We then define and calculate a geometric feature which is for measuring the flatness for each point of an object boundary in a 3D digital image by using the topological structures of polyhedra. From the results of flatness analysis, we improve the algorithm for global polyhedrization.

2 An Overview of Global Polyhedrization Problem

Setting $\mathbf{Z}$ to be the set of all integers, $\mathbf{Z}^3$ becomes the set of all lattice points whose coordinates are integers in the three-dimensional Euclidean space $\mathbf{R}^3$. We consider the space of a three-dimensional digital image to be a finite subset of $\mathbf{Z}^3$. In this section, we introduce an analytical representation of global polyhedra in $\mathbf{Z}^3$ and give an overview of the algorithm for construction of a global polyhedron and its problems [1,2].

2.1 Analytical Definition of Global Polyhedra

A polyhedron in $\mathbf{R}^3$ is represented by a finite set of planes in $\mathbf{R}^3$. Each plane is given by

$$\mathbf{P} = \{(x, y, z) \in \mathbf{R}^3 : ax + by + cz + d = 0\} \qquad (1)$$

where a, b, c, d are real numbers. Corresponding to each $\mathbf{P}$ in $\mathbf{R}^3$, we consider discretization in $\mathbf{Z}^3$ which is called a naive plane such as

$$\mathbf{NP} = \{(x, y, z) \in \mathbf{Z}^3 : 0 \le ax + by + cz + d < w\} \qquad (2)$$

where $w = \max(|a|, |b|, |c|)$ [6]. A global polyhedron in $\mathbf{Z}^3$ is given by a finite set of pieces of naive planes such as $\cup_{i=1}^{k} \Delta\mathbf{NP}_i$ where each $\Delta\mathbf{NP}_i$ is a piece of a naive plane.

2.2 Global Polyhedrization Algorithm

Assume that a finite subset $\mathbf{V}$ of $\mathbf{Z}^3$ is given as a volumetric data which is acquired for digitization of an three-dimensional object by using three-dimensional digital imaging techniques. The algorithm for construction of a global polyhedron $\cup_{i=1}^{k} \Delta\mathbf{NP}_i$ from a given $\mathbf{V}$ is described as follow.

Algorithm 1.
Input: A volumetric data $\mathbf{V}$.
Output: A set of naive plane pieces $\Delta\mathbf{NP}_i$ for $i = 1, 2, \ldots, k$.
Begin
    1. extract the subset $\mathbf{G} \subseteq \mathbf{V}$ which consists of the border points of $\mathbf{V}$;
    2. set $i = 1$ for a first naive plane piece $\Delta\mathbf{NP}_i$;
    3. while $\mathbf{G} \ne \emptyset$ do {
        3.1. select an initial subset $\Delta\mathbf{NP}_i \subseteq \mathbf{G}$;

(a)   (b)

Fig. 1. Examples of point selection from a set of border points for the planar check of global polyhedrization; (a) is an improper selection while (b) is proper.

*3.2. while planarity(Δ**NP**$_i$) = 1 do {*
    *3.2.1. select a new point $v \in$ **G** \ Δ**NP**$_i$;*
    *3.2.2. modify Δ**NP**$_i$ such that Δ**NP**$_i$ = Δ**NP**$_i \cup \{v\}$;*
}
*3.3. set Δ**NP**$_i$ = Δ**NP**$_i$ \ $\{v\}$ and then **G** = **G** \ Δ**NP**$_i$*
*3.4. increment i such that i = i + 1 for a next naive plane piece Δ**NP**$_i$;*
}
end

In step 1, we extract **G** which is a set of the border points of **V**. The explanation of this step will be given in the next section. The following steps are devoted to decompose **G** into Δ**NP**$_i$s for $i = 1, 2, \ldots, k$ by using the planarity function such as

$$planarity(\mathbf{A}) = \begin{cases} 1, & \text{if all points in } \mathbf{A} \text{ lie on an } \mathbf{NP}; \\ 0, & \text{otherwise} \end{cases} \qquad (3)$$

in step 3.2. There are several algorithms for the planarity function [2,5,6] and we apply the Fourier elimination algorithm introduced in [5] because the algorithm fits in the analytical form (2) of global polyhedra.

2.3 Problems of Global Polyhedrization

From Algorithm 1, we list up the problems of global polyhedrization:

1. how to select an initial subset Δ**NP**$_i \subset$ **G** in step 3.1;
2. how to select a new point $v \in$ **G** \ Δ**NP**$_i$ in step 3.2.1;
3. to find an algorithm for effective calculation of the planarity in step 3.2.

Among them, we focus on the problems 1 and 2 rather than the problem 3 in this paper; for the consideration on the problem 3, see [2] for example. It is obvious that the results of global polyhedrization will be changed if we choose different Δ**NP**$_i$ and v in steps 3.1 and 3.2.1 of Algorithm 1. Following the references [2, 5], we set the constraint A for the choice of Δ**NP**$_i$ and v;

A. if we consider the triangulation of Δ**NP**$_i$ or Δ**NP**$_i \cup v$ such that the vertices of triangles are points of Δ**NP**$_i$ or Δ**NP**$_i \cup v$, then the triangulation of Δ**NP**$_i$ or Δ**NP**$_i \cup v$ is topologically equivalent to a two-dimensional disc.

This constraint enables us to avoid selecting points as shown in Figure 1 (a) and to select a new point for $\Delta\mathbf{NP}_i$ so that a triangulation of $\Delta\mathbf{NP}_i$ is topologically equivalent to a two-dimensional disc as shown in Figure 1 (b). If we consider object recognition, visualization and extraction of geometric features for the postprocessing of global polyhedrization, the following constraints will be helpful for reducing the calculation times:

B. we set the size of each naive plane piece $\Delta\mathbf{NP}_i$ to be as large as possible;
C. we set the number of naive plane pieces $\Delta\mathbf{NP}_i$ in a global polyhedron $\cup_{i=1}^k \Delta\mathbf{NP}_i$ to be as small as possible.

The constraints B and C are not independent; if the size of each naive plane piece becomes small, the number of naive plane pieces becomes large.

   If our final goal is to obtain a good approximation of a three-dimensional object surface by global polyhedron, then we may have the constraint such that

D. an obtained polyhedron gives a good approximation of our object of interest.

To discuss on a good approximation, we need, for instance, the Hausdorff distance or geometric features such as volumes, surface areas, etc. for the evaluation of approximation schemes [4]. Because it is important but also difficult to discuss what the "good approximation" is, in this paper we step away from this evaluation problem. In order to evaluate our global polyhedrization algorithm without discussion on the evaluation for approximation schemes, we deal with only polyhedral objects for our experiment. We simply compare our result of global polyhedrization with an original polyhedron and check if they are coincident or not in section 6.

3   Extraction and Triangulation of Border Points

For the constraints A and B (or C) of global polyhedrization, we need the combinatorial topological structures of the set $\mathbf{G}$ of border points, which is also called the triangulation of $\mathbf{G}$. In order to obtain the triangulation of $\mathbf{G}$ directly from a given volumetric data $\mathbf{V}$, we apply the algorithm presented in [7].

   Briefly explaining the algorithm in [7], we refer triangles for each unit cubic region in $\mathbf{V}$ to the look-up table as shown in Figure 2 and then combine these triangles to obtain a triangulation of $\mathbf{G}$. More precisely, the vertices, edges and triangles in Figure 2 are called simplexes in combinatorial topology and the set $\mathbf{G}$ of border points is triangulated to be a simplicial complex $\mathbf{B}$ formed by these simplexes [8].

Definition 1. *An n-simplex is a set whose interior is homeomorphic to the n-dimensional disc $\mathbf{D}(x) = \{y \in \mathbf{R}^n : \|x - y\| < 1\}$ with the additional property that its boundary must be divided into a finite number of lower dimensional simplexes, called the faces of the n-simplex. We write $\sigma < \tau$ if σ is a face of τ.*

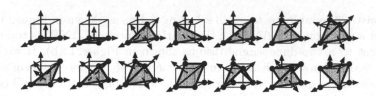

Fig. 2. The look-up table for triangles (i.e. 2-simplexes and their faces) which may constitute an object surface **B** (i.e. a 2-complex) for each unit cubic region. In the table, black points correspond to the points in a volumetric data **V**. We omit the configurations of black points in a unit cube which differ from those in the table by rotations. The arrow of each triangle is oriented to the exterior of an object.

0. A 0-simplex is a point A.

1. A 1-simplex is an open line segment $a = AB$, and $A < a$, $B < a$.

2. A 2-simplex is a triangle which is open, i.e., without the boundary points, such as $\sigma = \triangle ABC$, and then $AB, BC, AC < \sigma$. Note that $A < AB < \sigma$, so $A < \sigma$.

Definition 2. *A complex* **K** *is a finite set of simplexes,*

$$\mathbf{K} = \cup\{\sigma : \sigma \text{ is a simplex}\}$$

such that

*1. if $\sigma \in \mathbf{K}$, then all faces of σ are elements of **K**;*

*2. if σ and τ are simplexes in **K**, then $\sigma \cap \tau = \emptyset$.*

The dimension of **K** *is the dimension of its highest-dimensional simplex.*

In the process of combining triangles (i.e. 2-simplexes and the faces) at all unit cubic regions in **V**, we may have a pair of triangles such that all the vertices are coincident and the orientations are opposite. If such pair appear, we remove them from the simplicial complex **B**. The following theorem is proved in [7].

Theorem 1. *We uniquely obtain the triangulation* **B** *of the border points of* $\mathbf{V} \subset \mathbf{Z}^3$ *by referring to the look-up table in Figure 2.*

If we apply the algorithm to the subset

$$\mathbf{I}^+ = \{(x, y, z) \in \mathbf{Z}^3 : ax + by + cz + d \geq 0\}$$

with respect to a plane **P** of (1), then we obtain a 2-complex **B'**. Let us define the skeleton $Sk(\sigma)$ of a simplex σ such as the set of the vertices of σ or the union of 0-simplexes which are the faces of σ [8]. We also call the union of the skeletons of all simplexes of a complex **K** the skeleton of **K** and it is denoted by $Sk(\mathbf{K})$. We then have the following theorem in [9].

Fig. 3. Examples of (a) cyclic and (b) non-cyclic stars.

Theorem 2. *For any plane* **P** *of (1), we have the equality relations*

$$\mathbf{NP} = Sk(\mathbf{B'}).$$

From Theorem 2, we can set

$$\mathbf{G} = Sk(\mathbf{B})$$

for Algorithm 1 after construction of a 2-complex **B** as the triangulation of the border points of a volumetric data **V** by applying the algorithm in [7].

We need the following notions [8] for the subsequent sections. Let **K** be a complex and $\sigma \in \mathbf{K}$. The subcomplex consisting of σ and of all elements of **K** which follow (i.e. is greater than) σ is called the star $St_{\mathbf{K}}(\sigma)$ of σ of **K**. If $\mathbf{K}_0$ is any subcomplex of **K**, the complex consisting of all the elements of $\mathbf{K}_0$ and of all the elements of **K** which precedes (i.e. is less than) at least one element of $\mathbf{K}_0$ is called the combinatorial closure $|\mathbf{K}_0|$ of $\mathbf{K}_0$ in **K**. Then, the complex $B_{\mathbf{K}}(\sigma) = |St_{\mathbf{K}}(\sigma)| \setminus St_{\mathbf{K}}(\sigma)$ is called the outer boundary of $St_{\mathbf{K}}(\sigma)$. Let **K** be a 2-complex and σ_0 be a 0-simplex in **K**. If the outer boundary $B_{\mathbf{K}}(\sigma_0)$ is a simple closed broken line (i.e. if its elements are disposed in cyclic order, like the elements of a circle split up into sectors), the star $St_{\mathbf{K}}(\sigma_0)$ is said to be cyclic.

By using the notion of the outer boundary, we can classify all vertices or 0-simplexes of the 2-complex **B** of border points of a volumetric data **V**. Figure 3 shows examples of vertices whose stars are cyclic (a) and non-cyclic (b). We see, however, that every vertex in **B** may have a cyclic star if we assume that the boundary of a 3-dimensional object consists of a simple closed surface in $\mathbf{R}^3$ and the grid resolution of image discretization is high enough. In this paper, thus, we only consider vertices whose stars are only cyclic.

4   Spiral Ordering of Border Points

For the constraint A of global polyhedrization, we consider a spiral ordering of border points. For making the spiral order of points around a vertex v of the 2-complex **B**, we make use of the topological structures of **B**. The similar algorithm is presented in [10] for boundary traversal.

There are some functions which are used in the algorithm. Since we assume that the star $St_{\mathbf{B}}(\sigma_0)$ of every 0-simplex σ_0 in **B** is cyclic, the outer boundary of $B_{\mathbf{B}}(\sigma_0)$ constitutes a simple closed broken line. Thus, we can make a counterclockwise order of the points on $B_{\mathbf{B}}(\sigma_0)$ from v such as

$$around(v, v_1) = < v_1, v_2, \ldots, v_k > .$$

For two sequence of $b_1 = < v_1, v_2, \ldots, v_k >$ and $b_2 = < v_k, v_{k+1}, \ldots, v_l >$, the addition is obtained by

$$b_1 + b_2 = < v_1, v_2, \ldots, v_k, v_{k+1}, \ldots, v_l > .$$

For an sequence $b = < v_1, v_2, \ldots, v_l >$, the last point is obtained by

$$last(b) = < v_l >$$

and all points after v_k where $k \leq l$ are cut such as

$$cut(b, v_k) = < v_1, v_2, \ldots, v_{k-1} > .$$

Using these functions, the algorithm for ordering j points from and around a vertex v on $\mathbf{B}$ is shown. Note that $\mathbf{O}_j(v)$ is a sequence of ordering $j + 1$ points such that $< v, v_1, \ldots, v_j >$ and $\#(\mathbf{A})$ is the number of elements in a set $\mathbf{A}$.

Algorithm 2.

Input: A 2-complex $\mathbf{B}$, a starting point v and a sequence length $j + 1$.
Output: A point sequence $\mathbf{O}_j(v)$.
Begin
   1. set $\mathbf{O}_j(v) = < v > + around(v, v_1)$;
   2. set $i=1$;
   3. while $\#(\mathbf{O}_j(v)) \leq j + 1$ do {
      3.1. $\mathbf{O}_j(v)$ = $cut(\mathbf{O}_j(v), last(\mathbf{O}_j(v)))$ +
      $cut(aournd(v_i, last(\mathbf{O}_j(v))), v_{i+1})$;
      3.2. set $i = i+1$;
   }
   4. obtain $\mathbf{O}_j(v) = cut(\mathbf{O}_j(v), v_{j+1})$.
end

An example of $\mathbf{O}_j(v)$ is illustrated in Figure 4 (a). Due to the cyclic property of every point in $\mathbf{B}$, we see that the points are spirally ordered around v by using the outer boundary of the star of each point.

5   Flatness Analysis of Object Surfaces

5.1   Basic Idea of Measuring Flatness

For the constraint B (or C) of global polyhedrization, we would like to distinguish between two points x and y on a ellipsoid as illustrated in Figure 4 (b) so that a larger polygon is made around x and a smaller polygon is made around y. In order to distinguish these points, we need measuring the degree of flatness for

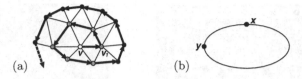

Fig. 4. (a) The order of points around v from v_1 on a 2-complex **B** by Algorithm 2. The gray points are the outer boundary of $B_{\mathbf{B}}(\{v\})$ (b) two points x and y on an ellipsoid which are expected to have the different degrees of flatness.

each point; the degree of flatness at x is expected to be larger than that at y in inverse proportion to the curvatures which are geometric features in $\mathbf{R}^3$.

We define the degree of flatness of a point $v \in Sk(\mathbf{B})$ in $\mathbf{Z}^3$ by using the planarity function of (3). Let $\mathbf{O}(v)$ be a set of as many lattice points as possible around v in $Sk(\mathbf{B})$ such that $planarity(\mathbf{O}(v)) = 1$. The degree of flatness of v is then given by

$$flatness(v) = \#(\mathbf{O}(v))$$

where $\#(\mathbf{A})$ is the number of elements in $\mathbf{A}$. For obtaining $\mathbf{O}(v)$, we use the similar algorithm to Algorithm 1. In steps 3.1 and 3.2.1 of Algorithm 1, we choose a set $\Delta\mathbf{NP}_i$ and a new point v to $\Delta\mathbf{NP}_i$, respectively. If we replace $\Delta\mathbf{NP}_i$ by $\mathbf{O}(v)$ in Algorithm 1, it is easy to see that $flatness(v)$ depends on the order of points chosen for $\mathbf{O}(v)$ around v. In this paper we use Algorithm 2 for spirally ordering points which spread around v in all directions.

5.2   Calculation of Flatness Degree

For each point v in $Sk(\mathbf{B})$, we obtain the degree of flatness by Algorithm 3. Note that $\mathbf{O}_j(v)$ in the algorithm is obtained by Algorithm 2.

Algorithm 3.
 Input: a point $v \in Sk(\mathbf{B})$.
 Output: $flatness(v)$.
 Begin
   1. set $j = 2$;
   2. while $planarity(\mathbf{O}_j(v)) = 1$ do $j = j + 1$;
   3. set $j = j - 1$;
   4. obtain $flatness(v) = \#(\mathbf{O}_j(v))$.
 end

Since we need at least three points for the planarity check, we set $j = 2$ in step 1. Applying Algorithm 3, we made the experiments for flatness analysis of two volumetric data of different objects such as a rectangular parallelepiped and an ellipsoid. For a rectangular parallelepiped, we obtained the volumetric data such that

$$\mathbf{V} = \{(x, y, z) \in \mathbf{Z}^3 : 1 \leq x \leq 3, 1 \leq y \leq 4, 1 \leq z \leq 5\} \tag{4}$$

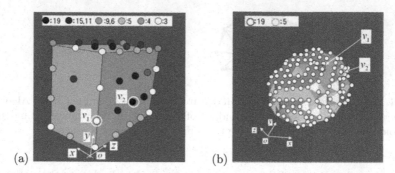

Fig. 5. The degrees of flatness for the boundary points on (a) a rectangular parallelepiped and (b) an ellipsoid.

and refer the triangles to the look-up table in Figure 2 and obtain a 2-complex **B** from **V**. We then apply Algorithm 3 for each point $v \in Sk(\mathbf{B})$. The number of points in $Sk(\mathbf{B})$ is 345 and the maximum and minimum degrees of flatness are 19 and 3 at v_1 and v_2, respectively, as illustrated in Figure 5 (a). Figure 5 (a) also shows the distribution of degrees of flatness on $Sk(\mathbf{B})$ and we see that points around the centers of rectangles have larger degrees than points near the corners. For global polyhedrization we use these flatness degrees so that we have a larger naive plane piece around a point such as v_2 and a smaller piece around a point such as v_1.

Similarly, for an ellipsoid, we make the volumetric data such that

$$\mathbf{V} = \{(x, y, z) \in \mathbf{Z}^3 : (x - 7)^2 + (y - 7)^2 + (z - 5)^2 \leq 6^2\}$$

and obtain a 2-complex **B** of the border points of **V**. In this case, the number of points in $Sk(\mathbf{B})$ is 664 and the degrees of flatness at v_1 and v_2 in Figure 5 (b) are 19 and 5, for example.

6   Global Polyhedrization by Flatness Analysis

In this section, we make use of the spiral point ordering and the results of flatness analysis presented in the previous sections for global polyhedrization under the constraints A and B. We first try the simple idea such that we choose a set $\Delta \mathbf{NP}_i$ for each i in step 3.1 of Algorithm 1 by using the degree of flatness. We thus replace steps 1 and 3.1 in Algorithm 1 by the following steps 1' and 3.1'.

1'. obtain $\mathbf{G} = Sk(\mathbf{B})$;
3.1'. select $\mathbf{O}(v)$ *such that* $flatness(v)$ *is the maximum among all* $v \in \mathbf{G}$ *and set* $\Delta \mathbf{NP}_i = \mathbf{O}(v)$;

Note that $\mathbf{O}(v)$ is obtained as $\mathbf{O}_j(v)$ in step 4 of Algorithm 3. We also considered how to choose a point v in step 3.2.1 of Algorithm 1. In Algorithm 3, we keep the boundary of $\mathbf{O}_j(v)$ or $\Delta \mathbf{NP}_i$ forming a simple closed curve and choose a new point in the counterclockwise order at the outer boundary of a boundary

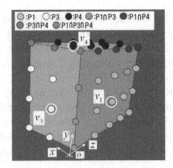

Fig. 6. A global polyhedron constructed by using the results of flatness analysis of Figure 5 (a).

point. For global polyhedrization, we also allow to choose a new point in the clockwise order at the outer boundary of a boundary point if we cannot find a new point which satisfies the planarity condition in step 3.2 of Algorithm 1.

After modifying steps 3.1 and 3.2.1 in Algorithm 1 as above, we made an experiment of global polyhedrization for a volumetric data of (4). The result is shown in Figure 6. The six points, $v_1 = (1, 2, 3)$, $v_2 = (3, 2, 2)$, $v_3 = (2, 2, 1)$, $v_4 = (2, 4, 2)$, $v_5 = (2, 1, 2)$, $v_6 = (2, 2, 5)$, whose degrees of flatness are 19, 15, 11, 11, 9 and 9, respectively, are chosen in order for v in step 3.1' and the six naive plane pieces, ΔNP_i for $i = 1, 2, \ldots, 6$ are constructed. There are some points which are shared with several naive plane pieces as shown in Figure 6. We see that our global polyhedron in Figure 6 is coincident to the original polyhedron.

7   Conclusions

In this paper, we first gave an overview of the problems for global polyhedrization and focused on some of them such as how to choose an initial set or a point in steps 3.1 and 3.2.1 of Algorithm 1 for each naive plane piece of a global polyhedron. To solve the problems, we used the relation between a naive plane which is given by an analytical form and our boundary representation which has the topological structure of a 2D manifold. By using the topological structure of an object boundary, we defined the flatness degree for each point on the boundary. Finally, we used the results of the flatness analysis and improved Algorithm 1 to solve the problems.

References

1. Françon, J., Papier, L.: Polyhedrization of the boundary of a voxel object. Discrete Geometry for Computer Imagery. LNCS **1568** Springer-Verlag, Berlin Heidelberg (1999) 425–434
2. Klette, R., Hao, J. S.: A new global polyhedrization algorithm for digital regular solids. Proceedings of IVCNZ'00 (2000) 150–155
3. Lorensen, W. E., Cline, H. E.: Marching cubes: a high-resolution 3d surface construction algorithm. Computer Graphics (SIGGRAPH '87) **21**(4) (1987) 163–169

4. Kenmochi, Y., Klette, R.: Surface area estimation for digitized regular solids. Vision Geometry IX. Proceedings of SPIE **4117** (2000) 100–111
5. Françon, J., Schramm, J. M., Tajine, M.: Recognizing arithmetic straight lines and planes. Discrete Geometry for Computer Imagery. LNCS **1176** Springer-Verlag, Berlin Heidelberg (1996) 141–150
6. Debled-Rennesson, I., Reveillès, J. P.: Incremental algorithm for recognizing pieces of digital planes. Vision Geometry V. Proceedings of SPIE **2826** (1996) 140–151
7. Kenmochi, Y., Imiya, A., Ichikawa, A.: Boundary extraction of discrete objects. Computer Vision and Image Understanding **71**(3) (1998) 281–293
8. Aleksandrov, P. S.: Combinatorial topology I. Graylock Press, Rochester N.Y. (1956)
9. Kenmochi, Y., Imiya, A.: Naive planes as discrete combinatorial surfaces. Discrete Geometry for Computer Imagery. LNCS **1953** Springer-Verlag, Berlin Heidelberg (2000) 249–261
10. Kovalevsky, V.: A topological method of surface representation: Discrete Geometry for Computer Imagery. LNCS **1568** Springer-Verlag, Berlin Heidelberg (1999) 118–135

Generalized Morphological Mosaic Interpolation and Its Application to Computer-Aided Animations

Marcin Iwanowski

Warsaw University of Technology
Institute of Control and Industrial Electronics
ul. Koszykowa 75, 00-662 Warszawa POLAND
iwanowski@isep.pw.edu.pl

Abstract. The paper describes an application of morphological mosaic interpolation based on distance function calculation to computer-aided animations. The existing method was extended and generalised in such a way that it can interpolate between any two mosaics - not only between two mosaics with non-empty intersection as the original method does. The problem for the proper generation of interpolation function - disappearing and moving particles, has been solved in the proposed method. An example of animation is also presented. It shows the change of the borders in Central Europe after the World War II.

Keywords: morphological image analysis, mosaic interpolation

1 Introduction

This paper describes a new method of computer-aided animations which makes use of morphological interpolation based on distance function calculation. Interpolation process consist in the creation of intermediate images between two given, input ones. The content of the first input (*initial*) image is evolving into the content of the second (*final*) one. The proposed method interpolates between two mosaic images consisting of clearly-defined regions marked by homogeneous values - their labels. Images of this kind are often created by computer graphics tools. The method makes use of the interpolation function. The existing method of interpolation proposed by Meyer in [4] is extended in such a way that it can interpolate between any two mosaics - not only between two mosaics with non-empty intersection as the original method does. The method analyses the input images in order to find the mosaic particles which are the obstacles for the correct generation of interpolation function - disappearing and moving particles. To solve the problem of disappearing particles, the input mosaics are modified. The moving particles are extracted and treated separately by using the morphological-affine interpolation.

Section 2 describes already existing methods of morphological interpolation based on the interpolation function. Section 3 presents proposed generalised

W. Skarbek (Ed.): CAIP 2001, LNCS 2124, pp. 493–501, 2001.
© Springer-Verlag Berlin Heidelberg 2001

method of mosaic interpolation. Section 4 is devoted to the application of the proposed method to the automated computer-aided animations. It contains an example of animation. Section 5 contains concluding remarks.

2   Interpolation Function

Morphological interpolation ([1,4,7]) is a relatively new area of mathematical morphology ([5,6]). This section recalls briefly the methods of binary and mosaic interpolation ([4]) and the morphological-affine interpolation ([2,3]) .

2.1   Binary Image Interpolation

In order to generate an interpolation function in case of two sets $U \subset V$ one should start with two auxiliary geodesic distances ([8]) defined on $V \setminus U$. A distance from the point $x \in V \setminus U$ to the set U is expressed by using the geodesic distance function d_1 and is obtained by successive geodesic dilations of U with mask V. The second geodesic distance d_2 describes the distance from point x to the border of V and is computed by geodesic dilations of V^C in U^C. The *relative* distance function is used as *the interpolation function* and is defined as:

$$int_U^V(p) = \begin{cases} 0 & in \ V^C \\ \frac{d_1(p)}{d_1(p)+d_2(p)} & in \ V \setminus U \\ 1 & in \ U \end{cases} \tag{1}$$

Interpolated set between U and V is obtained by tresholding of function (1) at a given level α:

$$Int_U^V(\alpha) = \{p : int_U^V(p) \leq \alpha\} \ ; \ \alpha \in [0,1] \tag{2}$$

In case of two sets X, Y with non-empty intersection , the interpolated set is

Fig. 1. An example of two interpolation functions (on the left) and interpolated sets at various levels.

obtained from (see also an example on fig. 1):

$$Int_X^Y(\alpha) = Int_{(X \cap Y)}^X(\alpha) \cup Int_{(X \cap Y)}^Y(1-\alpha) \ ; \ \alpha \in [0,1] \tag{3}$$

2.2   Interpolation of Mosaic Images

An important extension of the method of set interpolation by distance function has been proposed in [4]. Let A and B be the initial mosaics; $(A \cap B)$ be the their intersection; $(A \cap B)_\lambda$ be the set of pixels with label λ on both mosaics; A_λ and B_λ be partitions with label λ on mosaic A and B respectively. Similarly to the case of sets, the intersection of both mosaics must be non-empty which now means that: $\forall \lambda \, (A_\lambda \cap B_\lambda) \neq \emptyset$.

To obtain the interpolation function the geodesic distances are used. The interpolation of the mosaic can be split into a certain number of interpolations of sets. Each partition A_λ contains the partition $(A \cap B)_\lambda$. To interpolate the shape of partition λ between A_λ and $(A \cap B)_\lambda$ the interpolation function $int^{A_\lambda}_{(A \cap B)_\lambda}$ must be constructed. The interpolation functions $int^{A_\lambda}_{(A \cap B)_\lambda}$ for all the particles (all λ) is obtained by the parallel computation of the auxiliary distance functions d_1 and d_2 for all particles at once. The common interpolation function for all of the particles is denoted as: $int^A_{(A \cap B)}$. In the same way the second interpolation function: $int^B_{(A \cap B)}$ is computed. Both function are combined in order to get a single interpolation function:

$$int^B_A = 0.5 \left[int^A_{(A \cap B)} + (1 - int^B_{(A \cap B)}) \right] = 0.5(1 + int^A_{(A \cap B)} - int^B_{(A \cap B)}) \qquad (4)$$

In order to produce an interpolated mosaic at given level α, function (4) is thresholded at that level:

$$Int'_{A \to B}(\alpha)[p] = Int'_{B \to A}(1 - \alpha)[p] = \begin{cases} A(p) \; if \; int^B_A(p) \leq \alpha \\ B(p) \; if \; int^B_A(p) > \alpha \end{cases} \qquad (5)$$

2.3   Morphological-Affine Interpolation

The main problem of binary interpolation using the interpolation function is the condition of non-empty intersection of input images. Therefore it cannot be used to interpolate between two distant objects (compact binary sets). Another disadvantage is that the results of morphological interpolation are sometimes not realistic ([3]). To solve both problems *the morphological-affine interpolation* was proposed in [3,2]. The method allows to deform the shape of an object by using the morphological interpolation and to transform it using an affine transformation. It is applied to the interpolate between two objects P and Q. The method consist of three steps: moving the objects to the central, fix position using the affine transformations, morphological interpolation in the central position and the affine transformation of morphologically interpolated object to its final position.

The first step of treatment is performed separately for each of both input objects using the following auxiliary data: the *middle point* of the object (for the object P: (x_{P0}, y_{P0}), and (x_{Q0}, y_{Q0}) for Q), and the *proper angle* (η_P and η_Q). The first data is used twice: for the translation to the central position (x_C, y_C), and as a center of the rotation. The proper angle, describes the orientation of the set. There exists also the third input data, which depend on the relation between the sizes of both objects: the *scaling coefficients* s_x, s_y. All the data are computed automatically ([3,2]). The transformations which transformes the initial sets P and Q to its central positions are following:

$$A_P = T(-x_{P0}, -y_{P0}) \cdot R(\eta_P) \cdot T(x_C, y_C)$$
$$A_Q = T(-x_{Q0}, -y_{Q0}) \cdot R(\eta_Q) \cdot S(s_x, s_y) \cdot T(x_C, y_C) \tag{6}$$

where T, R, S stand for translation, rotation and scaling respectively. In the central position the interpolation functions: $int_{P' \cap Q'}^{P'}$ and $int_{P' \cap Q'}^{Q'}$ (where P' and Q' represent the transformed sets) are computed.

The interpolation functions calculated in the previous step are thresholded at given level $0 \leq \alpha \leq 1$ and in such a way the morphologically interpolated set is obtained. In order to place the it in the final position the affine transformation is applied once again. This time, however, new transformation parameters are calculated (all of them depend on α): final middle point $(x_F(\alpha), y_F(\alpha))$, rotation angle $\beta(\alpha)$ and scaling coefficient $s'_x(\alpha), s'_y(\alpha)$. The appropriate matrix of the affine transformation is following (for details - see [3,2]):

$$A_{int}(\alpha) = T(-x_C, -y_C) \cdot S(s'_x(\alpha), s'_y(\alpha)) \cdot R(-\beta(\alpha)) \cdot T(x_F(\alpha), y_F(\alpha)) \tag{7}$$

3   Generalized Method of Mosaic Interpolation

The method of mosaic interpolation, described in section 2.2, has one important disadvantage. It requires non-empty intersection of both input mosaic. The method proposed deals also with mosaics with an empty intersection.

If the intersection of two particles on both mosaics is empty, there exists at least one particle which either disappears on one of the input mosaics (*disappearing particle*) or which has been moved in such a way that it has no intersecting part on both mosaics (*moving particle*). Both situations results in the impossibility of the computation of the interpolation function. Disappearing and moving particles are shown on the examples of mosaics on fig. 2.

3.1   Disappearing Particles

Let A and B be the input mosaics ($\exists \lambda : A_\lambda \cap B_\lambda = \emptyset$). Let A' and B' be the modified mosaics ($A' \cap B' \neq \emptyset \Leftrightarrow \forall \lambda : A'_\lambda \cap B'_\lambda \neq \emptyset$). The necessary modification is done in such a way that all the particles disappearing on one of the input mosaics was reduced to one point or to a small region and added to the second initial mosaic with the appropriate label. In this case the interpolation sequence is obtained by the interpolation between modified mosaics. The input ones are however added at the beginning and at the end of the sequence:

$$A \, ; \, A' \, ; \, Int'_{A' \to B'}(\alpha_1) \, ; \, Int'_{A' \to B'}(\alpha_2) \, ; \, ... \, ; \, Int'_{A' \to B'}(\alpha_k) \, ; \, B' \, ; \, B \tag{8}$$

where α are the interpolation levels: $0 < \alpha_1 < \alpha_2 < ... < \alpha_k < 1$. Int' is the symmetric interpolator defined by (5). Interpolation sequence consist, in this case, of $k + 4$ frames. Modified mosaics are shown on fig. 2 (d) and (e), the distance function - on pos.(c). The reduction of disapering particles can be done

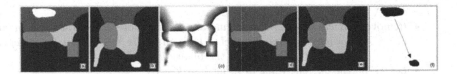

Fig. 2. Disapearing and moving mosaic particles; (a),(b) - input mosaics, (c) - interpolation function, (d),(e) - input mosaics after removal of the disapearing particles, (f) - moving particle.

in two ways. The first one is based on the ultimate erosion, the second one - on the center of gravity. In the first case the disapearing particle (treated as a binary image) is eroded by using the ultimate erosion. If necessary one can additionally perform an additional operation of thinning with the structuring element of type D in order to obtain a single pixel. In the second method disapearing particles are reduced to their centers of gravity. In case when the center of gravity does not belong to the particle, the *closest* to the gravity center pixel belonging to the particle is taken. Disapearing particles are visible on mosaics presented on fig.2 (a) and (b); modified mosaics are shown on pos.(d),(e) respectively.

3.2 Moving Particles

The second case happens when the particles of the same label exist on both mosaics but have an empty intersection. It means that this particle has been moved. The solution consist in automatic finding and removal of all the moving regions from the initial mosaics. Those particles are interpolated separately by using the morphological-affine interpolation (section 2.3). The initial mosaics with removed moving particles are interpolated in the traditional way (section 2.2), considering however the disapearing particles as described previously (section 3.1). The interpolated set at a given level is obtained by the superposition of separately interpolated particles with the modified mosaic interpolated at that level. An example of the moving particle is given on fig. 2. The moving particle itself is shown on pos.(f). The method is divided into following 3 phases:

Preparatory phase - 'find and remove'. During the preparatory phase the mosaics are modified in such a way that all the moving particles are found and removed. The algorithm is following (A' and B' are the input mosaics, such that $\forall \lambda : A'_\lambda \neq \emptyset \Leftrightarrow B'_\lambda \neq \emptyset$) is true[1]:

---

[1] This means that every label must be present on both input mosaics - disappearing particles doesn't exist.

– *Calculate the set R of labels corresponding to moving regions ('find' step):*

$$R = \left\{ \lambda : (A'_\lambda \cap B'_\lambda = \emptyset) \wedge (A'_\lambda \neq \emptyset) \wedge (B'_\lambda \neq \emptyset) \right\} \tag{9}$$

– *Let $k := 1$*
– *For each $\lambda \in R$ do: ('remove' step)*
  • *Remove particle of label λ from mosaic A' and store it separately in binary image X_k, assign temporarily label (-1) to all pixels from A' of label λ; repeat it for the mosaic B', storing it in the binary image Y_k.*
  • *Let λ_k be the current value of λ, let $k := k + 1$*

Filling the empty space. After removing the moving particles the empty space must be filled by the labels of neighboring particles. Propagation is performed by using a FIFO queue.

The algorithm (for the mosaic A') is the following:

– *For all pixels p of A' do:*
  • *If $A'(p) = -1$ and $\exists p' \in N(p) : A'(p') \neq -1$ do Put-to-queue(p)*
– *While Queue-not-empty do:*
  • *$p := $ Get-from-queue*
  • *For all $p' \in N(p)$ such that $A'(p') \neq -1$ find the dominating label e_d (label, which is assigned to the majority of neighbors)[2].*
  • *Let: $A'(p) = e_d$*
  • *For all $p' \in N(p)$ such that $A'(p') = -1$ do: Put-to-queue(p')*

where 1-pixel wide neighborhood of p is denoted by $N(p)$. The same algorthm is applied to the mosaic B'.

Generating the interpolated mosaic. In order to obtain the interpolated mosaic at given level α one uses the equation (5). The interpolated mosaic $Int'_{A' \to B'}(\alpha)$ does not include the moving particles. They are interpolated separately by using the morphological-affine interpolation. The interpolated moving particles are then added to the final, interpolated mosaic:

– *Produce interpolated mosaic: $I_\alpha = Int'_{A' \to B'}(\alpha)$*
– *For $i = 1, ..., k$ do:*
  • *Generate the interpolated set Z_i between X_i and Y_i at level α by using the morphological-affine method*
  • *Add Z_i, with label λ_i to I_α:*

$$I_\alpha(p) = \begin{cases} \lambda_i & if\ Z_i(p) = 1 \\ I_\alpha(p) & if\ Z_i(p) = 0 \end{cases} \tag{10}$$

As a result one obtains the interpolated mosaic[3] at level α.

---

[2] If there exist more majority labels assigned to the same number of pixels select e_d randomly from among them.

[3] Due to the fact that in the proposed method the separately interpolated moving particles are superimposed to the interpolated mosaic, the problem of the order of superposition can occur. We leave this topic open. It seems to be not possible to propose the order of superposition without an additional information.

3.3   Generalized Method

The generalized method considers all the remarks presented previously (A and B are the input mosaics: initial and final):

- *Find and remove disapearing particles on input mosaics A and B. Modify mosaics where necessary and store them in A' and B'.*
- *Remove the moving particles from A' and B' and store them in separate images.*
- *Interpolate between modified mosaics, interpolate moving particles separately and superimpose the results of both interpolations.*

The interpolation sequence is obtained by successive generation of interpolated mosaics for increasing values of α.

4   Animation

The example shows the animation presenting the change of shape of borders in Central Europe which took place after World War II. Initial mosaic image presents the shape of the borders in 1938, the final one - the shape in 1945. The input images are presented in fig.3. Initial mosaic images contain moving

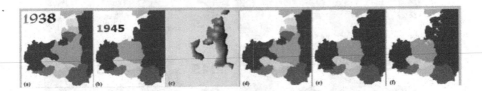

Fig. 3. Input images initial (a) and final (b); interpolation (distance) function (c); images without moving particles (d),(e); modified final image (f).

(inscriptions showing the years '1938' and '1945') and disapering (corresponding to the Baltic states) partitions. The input (final) image modified according to the rule described in section 3.1 is shown on pos.(f) (the modification of the initial image was not necessary). The modification of initial mosaic consist in the superposition of the gravity centers of disappearing particles (indicated by arrows). Some of the interpolated frames are presented on fig.4. The numbers in lower-left corner shows the appropriate interpolation levels.

5   Conclusions

The proposed method faciliates the process of computer-aided animations. It allows to produce the animation sequence in an automated way. Only the initial and final images make the input. The animation sequence is obtained as the interpolation sequence between the initial image and final one. Both images are mosaic ones. It means that they consist of clearly-defined regions characterized by their labels. They both are prepared by the operator. The animation is powered by the morphological interpolation. Simple morphological mosaic interpolation can be applied in the animation area in a very restricted way. The condition of non-empty intersection of input mosaics is a big obstacle, which make impossible to apply that method for any input mosaic images. The proposed method generalises the former one - it allows to interpolate between any input mosaics. An example of animation proves the applicability of the pro-

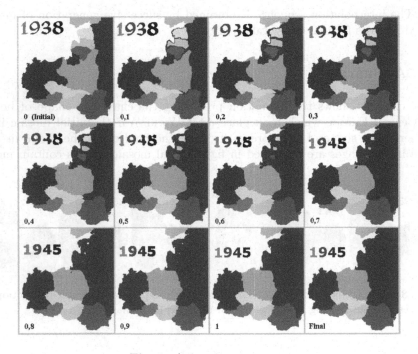

Fig. 4. Animation sequence.

posed method to the presentation graphics. The animation shows the process of change of the borders in Central Europe beeing the result of the World War II. The method can be succesfully applied in a film and TV industry as well as in the multimedia and internet branches to produce the animation sequences in fast, robust and automated way.

References

1. Beucher, S.: Intepolation of sets, of partitions and of functions. In: Heijmans, H., Roerdink, J.: Mathematical morphology and its application to image and signal processing, Kluwer (1998)

2. Iwanowski, M., Serra, J.: Morphological-affine object deformation. In: Vincent, L., Bloomberg, D.: Mathematical morphology and its application to image and signal processing, Kluwer (2000)
3. Iwanowski, M.: Application of mathematical morphology to image interpolation. Ph.D. thesis, Warsaw University of Technology, School of Mines of Paris, Warsaw-Fontainebleau (2000)
4. Meyer, F.: Morphological interpolation method for mosaic images In: Maragos, P., Schafer, R.W., Butt, M.A.: Mathematical morphology and its application to image and signal processing, Kluwer (1996)
5. Serra, J.: Image Analysis and Mathematical Morphology vol.1. Academic Press (1983)
6. Serra, J.: Image Analysis and Mathematical Morphology vol.2. Academic Press (1988)
7. Serra, J.: Hausdorff distance and interpolations. In: Heijmans, H., Roerdink, J.: Mathematical morphology and its application to image and signal processing, Kluwer (1998)
8. Soille, P.: Morphological Image Analysis. Springer Verlag (1999)

Openings and Closings by Reconstruction Using Propagation Criteria

Iván R. Terol-Villalobos[1] and Damián Vargas-Vázquez[2]

[1] CIDETEQ,S.C., Parque Tecnológico Querétaro,S/N, SanFandila-Pedro Escobedo, 76700, Querétaro Mexico,
famter@ciateq.mx

[2] Maestría en Instrumentación y Control Automático, Universidad Autónoma de Querétaro, 76000, Querétaro, México

Abstract. In this paper, a class of openings and closings is investigated using the notion of propagation criteria. The main goal in studying these transformations consists in eliminating some inconveniences of the morphological opening (closing) and the opening (closing) by reconstruction. The idea in building these new openings and closings comes from the notions of morphological filters by recontruction and levelings. Morphological filters by reconstruction extract, from an image, the connected components that are marked. The reconstruction process of the input image is achieved using geodesic dilation (or erosion) until stability. However, since thin connections exist, these filters reconstructs too much and sometimes it is impossible to eliminate some undesirable regions. Because of this inconvenience, propagation criteria must be introduced.

Keywords: mathematical morphology, propagation criteria

1 Introduction

Morphological filtering is one of the most interesting subjects of research in mathematical morphology (MM). The basic morphological filters are the morphological opening and the morphological closing. These filters present several inconveniences. In general, if the undesirable features are eliminated, the remaining structures will be changed. Recently, filters by reconstruction have become powerful tools that enable us to eliminate undesirable features without affecting desirable ones. These transformations by reconstruction, that form a class of connected filters, involve not only opening and closing, but also alternated filters, alternating sequential filters, and even new transformations called levelings (see Meyer [1]). In particular, the morphological connected filters have been studied and characterized in the binary case (see Serra and Salambier [2], Crespo et al. [3], Serra [4] among others). Filters by reconstruction are built by means of a reference image and a marker image included in the reference (see Vincent [5]). The marker image grows by iterative geodesic operations while staying always inside the reference image. This can be seen as a propagation process. The

W. Skarbek (Ed.): CAIP 2001, LNCS 2124, pp. 502–509, 2001.
© Springer-Verlag Berlin Heidelberg 2001

propagation in space is a common term in various fields of physics. In MM, the propagation algorithms are based on the geodesic dilation. The geodesic dilation allows the definition of a mapping that transforms a point of the geodesic space into another point. Thus, the marker points can be transformed by geodesic dilations. The transformed points are situated at a geodesic distance given by the size of the geodesic dilation. In this work, we will use the notion of propagation by using some criteria that restrict the propagation to some region of the mask.

Due to the interesting characteristics of filters by reconstruction, another approach for characterizing connected filters was presented by Meyer [1],[6]. By defining monotone planings and flattenings and by combining both notions, the levelings have been introduced. In the present work, the concept of leveling is used to build a new opening (closing) that enables us to avoid the inconveniences of the morphological opening (closing) and the opening (closing) by reconstruction. While the morphological opening modifies the remaining structures, the opening by reconstruction sometimes reconstructs undesirable regions. This paper is organized as follows. In section 2, the concepts of filters by reconstruction and levelings are presented. In section 3, we modify some criteria for constructing levelings in order to study two operators from a propagation point of view. An opening and a closing are proposed in this section. In section 4, an application that illustrates the interest of these transformations is presented.

2   Some Basic Concepts of Morphological Filtering

2.1   Basic Notions of Morphological Filtering

The basic morphological filters are the morphological opening $\gamma_{\mu B}$ and the morphological closing $\varphi_{\mu B}$ with a given structuring element ; where, in this work, B is an elementary structuring element (3x3 pixels) that contains its origin. $\check{B}$ is the transposed set ($\check{B} = \{-x : x \in B\}$) and μ is an homothetic parameter. The morphological opening and closing are given, respectively, by:

$$\gamma_{\mu B}(f)(x) = \delta_{\mu \check{B}}(\varepsilon_{\mu B}(f))(x) \quad \text{and} \quad \varphi_{\mu B}(f)(x) = \varepsilon_{\mu \check{B}}(\delta_{\mu B}(f))(x) \quad (1)$$

where the morphological erosion $\varepsilon_{\mu B}$ and dilation $\delta_{\mu B}$ are expressed by:

$$\varepsilon_{\mu B}(f)(x) = \wedge\{f(y) : y \in \mu \check{B}_x\} \quad \text{and} \quad \delta_{\mu B}(f)(x) = \vee\{f(y) : y \in \mu \check{B}_x\}$$

$\wedge$ is the inf operator and $\vee$ is the sup operator. In the following, we will avoid the elementary structuring element B. The expressions $\gamma_\mu, \gamma_{\mu B}$ are equivalent (i.e. $\gamma_\mu = \gamma_{\mu B}$). When the homothetic parameter is $\mu = 1$, the structuring element B will also be avoided (i.e. $\delta_B = \delta$). When $\mu = 0$, the structuring element is a set made up of one point (the origin).

2.2 Opening (Closing) by Reconstruction and Levelings

Geodesic transformations are used to build the reconstruction transformations. In the binary case, the geodesic dilation of size 1 of a set Y inside the set X is defined as $\delta_X^1(Y) = \delta(Y) \cap X$. To build a geodesic dilation of size m, the geodesic dilation of size 1 is iterated m times. Similarly, a geodesic erosion $\varepsilon_X^m(Y)$ is computed by iterating m times the geodesic erosion of size 1: $\varepsilon_X^1(Y) = \varepsilon(Y) \cup X$. When filters by reconstruction are built, the geodesic transformations are iterated until idempotence is reached (see Vincent [5]). Reconstruction transformations are frequently presented using the notion of geodesic distance. Given a set X (the mask), the geodesic distance between two pixels x and y inside X $(d_X(x,y))$ is the length of the shortest paths joining x and y which are included in X. The geodesic dilation of size $m \geq 0$ of Y within X $(Y \subset X)$ is the set of pixels of X whose distance to Y is smaller or equal to m: $\delta_X^m(Y) = \{x \in X : d_X(x,y) \leq m\}$. As expressed above, the growing process by geodesic dilations can be seen as a propagation process. In MM, the propagation in Euclidean space has been defined by means of dilations. The dilation is used to define a mapping that transforms a point x into another point y in the space. The boundaries of $\delta_m(\{x\})$ describe an isotropic propagation. Similarly, the geodesic dilations can be used to define a propagation in a geodesic space. In this case, the propagation coming from a marker stays always inside the reference (mask). The set formed by the points $\{x \in X : d_X(x,Y) = m\}$ is the propagation front, while the set of points $\{x \in X : d_X(x,Y) \leq m\}$ is composed by the regions reached by the propagation under successive iterations of the geodesic dilation of size 1 until m. The reconstruction transformation in the gray-level case is a direct extension of the binary one. In this case, the geodesic dilation $\delta_f^1(g)$ (resp. the geodesic erosion $\varepsilon_f^1(g)$) with $g \leq f$ (resp. $g \geq f$) of size 1 given by: $\delta_f^1(g) = f \wedge \delta_B(g)$ (resp. $\varepsilon_f^1(g) = f \vee \varepsilon_B(g)$) is iterated until idempotence. When the function g is equal to the erosion or the dilation of the original function, we obtain the opening and the closing by reconstruction:

$$\tilde{\gamma}_\mu(f) = \lim_{n \to \infty} \delta_f^n(\varepsilon_\lambda(f)) \quad \tilde{\varphi}_\mu(f) = \lim_{n \to \infty} \varepsilon_f^n(\delta_\lambda(f)) \tag{2}$$

The interesting filtering characteristics of filters by reconstruction, motivate the characterization by means of levelings (see Meyer [6]). In this work, we study the lower and upper levelings, defined below.

Definition 1. *A function g is a lower-leveling (resp. upper-leveling) of a function f if and only if $\forall(p,q)$ neighbors: $g(p) > g(q) \Rightarrow g(q) \geq f(q)$ (resp. $g(p) > g(q) \Rightarrow g(p) \leq f(p)$)*

In particular, the opening and the closing by reconstruction are lower and upper levelings, respectively. The criteria for building lower and upper levelings are:

Criterion 1 *A function g is a lower-leveling (resp. an upper-leveling) of a function f if and only if: $[g \geq f \wedge \delta(g)]$ (resp. $[g \leq f \vee \varepsilon(g)]$).*

Principally, we will focus on the criteria for building extended lower and upper levelings given by:

Criterion 2 *A function g is a lower-leveling (resp. an upper-leveling) of a function f if and only if:* $[g \geq f \wedge (g \vee \gamma_\lambda \delta(g))]$ *(resp.* $[g \leq f \vee (g \wedge \varphi_\lambda \varepsilon(g))]$ *)*

3 Reconstruction Transformations Using Propagation Criteria

The main drawbacks of the morphological opening (closing) and the opening (closing) by reconstruction were described above. Remember that with the opening by reconstruction the elimination of some structures can not be achieved, because the propagation process reconstructs all connected regions. The goal in this section is to introduce an opening (and a closing) that enables us to obtain intermediate results between the morphological opening (closing) and the opening (closing) by reconstruction. The process to build these new transformations involves the use of a reference image and a marker image as in the reconstruction case. Thus, a propagation process of the marker using geodesic dilation (or geodesic erosion) is used, but a propagation criterion is taken into account. We only present the case of the new opening, but a similar procedure can be made for the closing.

3.1 Opening (Closing) by Reconstruction Using a Propagation Criterion

By applying criterion (1) it is possible to say that g is an opening by reconstruction of a function f if and only if $g = f \wedge \delta(g)$. In fact, an opening by reconstruction is obtained by iterating, $f \wedge \delta(g)$ with $g = \varepsilon_\mu(f)$, until stability (the geodesic dilation). Now, let us analyze criterion (2). As for the opening by reconstruction, we will iterate the relationship $f \wedge (g \vee \gamma_\lambda \delta(g))$ until stability. However, this expression will be simplified by some condition imposed to the function g (similar for $f \vee (g \wedge \varphi_\lambda \varepsilon(g))$). We will see, that the opening γ_λ will play a main role, by restricting the propagation process to some regions of the reference image. Let f and g be the reference and the marker images, respectively. The following inequality can be established $\forall g$: $\gamma_\lambda \delta(g) \leq \delta(g)$, but nothing can be expressed for $\gamma_\lambda \delta(g)$ and g. Only, for $\lambda = 0$, the following relationship is verified: $g \leq \gamma_\lambda \delta(g) = \delta(g)$.

For $\lambda > 1$, even if the inequality $\gamma_\lambda \delta(g) \leq \delta(g)$ (due to the anti-extensivity of γ_λ) is satisfied, the equation $g \leq \gamma_\lambda(\delta(g))$ is not necessarily true. Moreover, it would be interesting to know how the propagation of successive iterations of $\gamma_\lambda \delta(g)$ is when compared to the successive iterations of $\delta(g)$. To better understand the behavior of the $\gamma_\lambda \delta(g)$, let us give some conditions for the marker image g. First, an opening by reconstruction can be obtained using a marker image given by the erosion or the morphological opening. This means:

$$\widetilde{\gamma}_\mu(f) = \underbrace{\delta_f^1 \delta_f^1 \cdots \delta_f^1}_{Until\ stability} (\varepsilon_\mu(f)) = \underbrace{\delta_f^1 \delta_f^1 \cdots \delta_f^1}_{Until\ stability} (\gamma_\mu(f))$$

On the other hand, the morphological opening satisfies the following property: *for all* λ_1, λ_2 *with* $\lambda_1 \leq \lambda_2$, $\delta_{\lambda_2}(g) = \gamma_{\lambda_1}(\delta_{\lambda_2}(g))$. By applying these properties to $\gamma_\lambda \delta(g)$ with $g = \gamma_\mu(f)$, we get for all $\lambda \leq \mu + 1$:

$$\gamma_\lambda \delta(g) = \gamma_\lambda \delta(\gamma_\mu(f)) = \gamma_\lambda \delta_{\mu+1} \varepsilon_\mu(f) = \delta_{\mu+1} \varepsilon_\mu(f) = \delta \gamma_\mu(f)$$

and $$\gamma_\mu(f) \leq \gamma_\lambda \delta(\gamma_\mu(f)) = \delta \gamma_\mu(f)$$

Thus, the term $g \vee \gamma_\lambda \delta(g)$ of $f \wedge (g \vee \gamma_\lambda \delta(g))$ is simplified to: $\gamma_\lambda \delta(g)$ and we will work with: $\omega_{\lambda,f}^1(g) = f \wedge \gamma_\lambda \delta(g)$ and $\alpha_{\lambda,f}^1(g) = f \vee \varphi_\lambda \varepsilon(g)$, with the conditions $g = \gamma_\mu(f)$ for $\omega_{\lambda,f}^1(g)$ and $g = \varphi_\mu(f)$ for $\alpha_{\lambda,f}^1(g)$. Specifically, the equation $\gamma_\lambda \delta(\gamma_\mu(f)) = \delta \gamma_\mu(f)$ expresses that when the marker image is given by $g = \gamma_\mu(f)$ for $\lambda \leq \mu + 1$, the propagation process of successive iterations of $\gamma_\lambda \delta(g)$ is similar to that generated by the succesive iterations of $\delta(g)$ and we have $g \leq \gamma_\lambda \delta(\gamma_\mu(f)) = \delta \gamma_\mu(f)$. However, this propagation process changes when the reference image f is used. The role of the propagation criterion given by γ_λ can be observed by iterating the operator $\omega_{\lambda,f}^1(g)$ until stability is reached, i.e.:

$$\lim_{n \to \infty} \omega_{\lambda,f}^n(g) = \underbrace{\omega_{\lambda,f}^1 \omega_{\lambda,f}^1 \cdots \omega_{\lambda,f}^1}_{Until\ stability}(g)$$

A dual transformation is obtained by using $\alpha_{\lambda,f}^1(g)$. Then, when the reference image modifies the propagation process of succesive iterations of $\gamma_\lambda \delta(g)$, the opening takes the role of a propagation criterion by restricting the propagation to some regions of the reference image. Figure 1 illustrates the criterion $f \wedge \gamma_\lambda \delta(g)$ seen as a propagation process. In Fig.1(a), the marker g, in gray color, is an invariant of the morphological opening γ_μ for $\mu = 14$, and in black color, we have six particles that play the role of obstacles to the propagation process coming from the marker into the reference image, in white. As expressed in section 2, the geodesic dilation of size m is the set of points whose distance to the marker set is less than or equal to m. Then, the image in Fig.1(b) illustrates the propagation process seen as a distance function of succesive iterations of $f \wedge \delta(g)$. To better illustrate the propagation, the gray level of the pixels of the distance function was multiplied by 5 and the module 255 was computed. The same procedure was computed for the propagation process using $f \wedge \gamma_\lambda \delta(g)$ with $\lambda = 6$ and 13, respectively. (see Figs. 1(c) and 1(d)).

Proposition 1. *Let* γ_μ *and* φ_μ *be the morphological opening and closing of parameter* μ, *respectively. The transformations given by the following relations*

$$\widehat{\gamma}_{\lambda,\mu}(f) = \lim_{n \to \infty} \omega_{\lambda,f}^n(\gamma_\mu(f)) \qquad \widehat{\varphi}_{\lambda,\mu}(f) = \lim_{n \to \infty} \alpha_{\lambda,f}^n(\varphi_\mu(f))$$

are an opening and a closing of size μ *with* $\lambda \leq \mu + 1$, *respectively. Where* $\omega_{\lambda,f}^1 = f \wedge \gamma_\lambda \delta$ *and* $\alpha_{\lambda,f}^1 = f \vee \varphi_\lambda \varepsilon$.

It is interesting to observe that there exits an inclusion relation between the three openings γ_μ, $\widehat{\gamma}_{\lambda,\mu}$ and $\widetilde{\gamma}_\mu$: $\gamma_\mu(f) \leq \widehat{\gamma}_{\lambda,\mu}(f) \leq \widetilde{\gamma}_\mu(f)$ with $\lambda \leq \mu + 1$. In fact, the main goal in this study is to obtain intermediate results between γ_μ and $\widetilde{\gamma}_\mu$, to reduce the problems of the morphological opening and the opening by reconstruction. Now, let us use the expression $\omega_{\lambda,f}^1 = f \wedge \delta\gamma_\lambda$ instead of $\omega_{\lambda,f}^1 = f \wedge \gamma_\lambda\delta$ to define an opening (and a closing) as the one given by proposition (1). It is possible to show that the propagation process of the successive iterations of $\delta\gamma_\lambda(g)$, using $g = \gamma_\mu(f)$, is similar to that of $\gamma_\lambda\delta(g)$. Since $\gamma_\lambda\gamma_\mu(f) = \gamma_\mu\gamma_\lambda(f) = \gamma_\mu(f)$ $\forall\lambda \leq \mu$; we have $\delta\gamma_\lambda(g) = \delta\gamma_\lambda(\gamma_\mu(f)) = \delta(\gamma_\mu(f))$, $\forall\lambda \leq \mu$. Thus, the propagations are similar with the conditions $\forall\lambda \leq \mu$ for $\delta\gamma_\lambda(g)$ and $\forall\lambda \leq \mu + 1$ for $\gamma_\lambda\delta(g)$. However, by iterating both expressions $\omega_{\lambda,f}^1(g) = f \wedge \delta\gamma_\lambda(g)$ and $\omega_{\lambda,f}^1(g) = f \wedge \gamma_\lambda\delta(g)$ until stability, different output images are obtained. The interest of the use of expression $\omega_{\lambda,f}^1(g) = f \wedge \gamma_\lambda\delta(g)$ is illustrated in section 4.

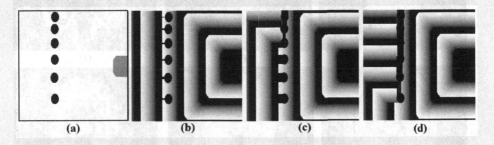

Fig. 1. Propagation process seen as a distance function

4 Magnetic Resonance Imaging (MRI)

In order to show the interest of the opening by reconstruction with a propagation criterion, we applied it to perform the segmentation of magnetic resonance imaging (MRI) of brain. MRI is characterized for its high soft tissue contrast and high spatial resolution. These two properties make MRI one of the most important and useful imaging modalities in diagnosing brain related pathologies. These transformations are applied in a 2-dimensional case (2-D). The purpose of this procedure was to segment, as accurately as possible, the skull and the brain, as well as the background. Figures 2(a) to 2(i) illustrate the procedure. To eliminate the skull from the image in Fig.2(a), an opening by reconstruction $\widetilde{\gamma}_\mu$ of size 12 was applied (Fig.2(b)). In this stage, the skull gray-level is attenuated and a simple threshold enables us to obtain the brain shown in Fig.2(c). Now, by applying the same procedure using the opening $\widehat{\gamma}_{\lambda,\mu}$ with $\mu = 12$ and $\lambda = 4$ to the image in Fig.2(a), a similar result has been obtained in Fig.2(e) by thresholding the image in Fig.2(d). However, because in some 2-D images thin connections exist between skull and brain, in some cases, it is impossible to separate both regions as it is illustrated in Fig.2(g). Using the opening by reconstruction, the operator processes the skull and the brain on the image in

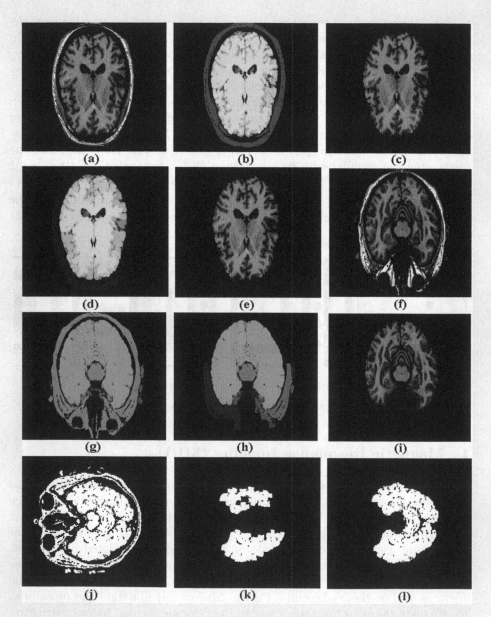

Fig. 2. (a) and (b) Original image and its opening $\widetilde{\gamma}_\mu$, (c) Brain detection from image in (b), (d) Opening $\widehat{\gamma}_{\lambda,\mu}$ of image in (a), (e) Brain detection using image in (d), (f) Original image, (g) and (h) Openings $\widetilde{\gamma}_\mu$ and $\widehat{\gamma}_{\lambda,\mu}$ of image in (f), (i) Brain detection using image in (h), (j) Original image, (k) and (l) Openings $\widehat{\gamma}_{\lambda,\mu}$ using $f \wedge \delta\gamma_\lambda$ and $f \wedge \gamma_\lambda\delta$, respectively

Fig.2(f) as a single object and this operator reconstructs "too much". By applying $\widehat{\gamma}_{\lambda,\mu}$ to image in Fig.2(f) this drawback is eliminated. Figure 2(h) illustrates the transformation with $\mu = 12$ and $\lambda = 4$. In this case, it is possible to separate the brain from the skull without considerably affecting the structure of the brain (Fig.2(i)). The transformation $\widehat{\gamma}_{\lambda,\mu}$ has been successfully tested with fifteen MR 2-D images. This set of experiments have especially shown a robust behavior to eliminate the skull. A slight difficulty has been found in a few cases when the size of the brain (2-D images) is smaller than the skull. A 3-D approach will overcome this problem. Finally, let us illustrate the use of expression $f \wedge \gamma_\lambda \delta(g)$ to build the opening $\widehat{\gamma}_{\lambda,\mu}$ instead of $f \wedge \delta\gamma_\lambda(g)$. Figure 2(j) shows a binary image, while Figs. 2(k) and 2(l) illustrate the openings $\widehat{\gamma}_{\lambda,\mu}$ with $\mu = 7$ and $\lambda = 3$, using expressions $f \wedge \delta\gamma_\lambda(g)$ and $f \wedge \gamma_\lambda \delta(g)$, respectively. Notice that when expression $f \wedge \gamma_\lambda \delta(g)$ is used, the small holes do not affect considerably the propagation process. Intuitively, since at each iteration of $f \wedge \gamma_\lambda \delta(g)$, a dilation is applied to g before the opening γ_λ, some small holes are filled during the propagation process, before the opening restricts this propagation.

5   Conclusion

In the present work, a study of a class of transformations called extended lower and upper levelings has been made. Here, this class of levelings has been studied to build new openings and closings. We have shown that these openings and closings with propagation criteria enable us to obtain intermediate results between the morphological openings (closings) and the openings (closings) by reconstruction. The opening and closing with propagation criteria do not reconstruct some connected components linked by thin connections as is the case of the opening and closing by reconstruction.

Acknowledgements. We would like to thank Marcela Sanchez Alvarez for her careful revision of the English version. The author I. Terol would like to thank Diego Rodrigo and Dario T.G. for their great encouragement. This work was partially funded by the government agency CONACyT (Mexico) under the grant 25641-A.

References

1. Meyer, F.: From connected operators to levelings. In: Heijmans, H., Roerdink, J. (eds.): Mathematical Morphology and Its Applications to Image and Signal Processing. Kluwer (1998) 191–198
2. Serra, J., Salembier, Ph.: Connected operators and pyramids. Proc. SPIE Image Algebra Math. Morphology, San Diego, CA, SPIE, **2030** (1993) 65–76
3. Crespo, J., Serra,J, Schafer, R.: Theoretical aspects of morphological filters by reconstruction. Signal Process. **47** (1995) 201–225
4. Serra, J.: Connectivity on Complete Lattices. J. of Mathematical Imaging and Vision **9** (1998) 231–251
5. Vincent, L.: Morphological grayscale reconstruction in image analysis: Applications and efficient algorithms. IEEE Transactions on Image Processing **2** (1993) 176-201
6. Meyer, F.: Levelings. In: Heijmans, H., Roerdink, J. (eds.): Mathematical Morphology and Its Applications to Image and Signal Processing. Kluwer (1998) 199–206

Multiscale Segmentation of Document Images Using *M*-Band Wavelets

Mausumi Acharyya and Malay K. Kundu

Machine Intelligence Unit, Indian Statistical Institute
203, B. T. Road, Calcutta - 700 035, INDIA
{res9522, malay}@isical.ac.in

Abstract. In this work we propose an algorithm for segmentation of
the text and non-text parts of document image using multiscale feature
vectors. We assume that the text and non-text parts have different
textural properties. *M*-band wavelets are used as the feature extractors
and the features give measures of local energies at different scales and
orientations around each pixel of the $M \times M$ bandpass channel outputs.
The resulting multiscale feature vectors are classified by an unsupervised
clustering algorithm to achieve the required segmentation, assuming no
a priori information regarding the font size, scanning resolution, type
layout etc. of the document.

Keywords: multiscale segmentation, document analysis

1 Introduction

In the present world, the advances in communication and information technology, have made it an imperative need for automated processing and reading of
documents. Recently use of papers for printed materials like newspaper, books
and other documents have been on the rise, so it has become a necessity to
digitize documents. Since digitized documents have obvious advantages in terms
of storage and transmission as well as archiving, searching, retrieval and updating, research in the area of document image analysis has become an important
issue to many researchers. For easy access of paper documents electronically,
it requires efficient extraction of information from the documents. An effective
representation of these images is to separate out the text from the non-text parts
and store the text as ASCII (character) set and the non-text parts as bit maps.
Several useful techniques for text and non-text segmentation are given in [1].
The most popular amongst these being the *bottom-up* and *top-down* approaches
[2] [3]. *Bottom-up* techniques which use connected component aggregation [2]
iteratively group together components of the same type starting from the pixel
level and form higher level descriptions of the printed regions of the document
(words, text lines, paragraphs etc.) [4]. *Top-down* approaches first split the document into blocks which are then identified and subdivided appropriately, in
terms of columns first and then split them into paragraphs, text lines and may

W. Skarbek (Ed.): CAIP 2001, LNCS 2124, pp. 510–517, 2001.

be words also [5]. Some assume these blocks to be only rectangular. The *top-down* methods are not suitable for skewed texts as these methods are restricted to rectangular blocks. Whereas the *bottom-up* methods are sensitive to character size, scanning resolution, inter-line and inter-character spacings. Several other approaches use the contours of the white space to delineate the text and non-text regions [6]. These methods can only be applied to low noise document images which are highly structured, that is all objects are separated by white background and objects do not touch each other.

Each of the above methods relies to an extent on *a priori* knowledge about the rectangularity of major blocks, consistency in horizontal and vertical spacings, independence of text, graphic and image blocks and/or assumptions about textual and graphical attributes like font size, text line orientation etc. So these methods can not work in a generic environment.

It is desirable to have segmentation techniques which do not require any *a priori* knowledge about the content and attributes of the document image or rather any such knowledge might not be available, in some applications. Jain and Bhattacharjee's [7] method has been able to overcome these restrictions and does not require an *a priori* knowledge of the document to be processed. In this work the document segmentation has been achieved by a texture segmentation scheme using Gabor wavelets as the feature extractors. One major drawback of this approach is that the use of Gabor filter makes it computationally very inefficient. Their scheme uses a subset of twenty fixed multi-channel Gabor filters consisting of four orientations and five spatial frequencies, as input features for the classifier. Randen and Husϕy [8] proposed a method using critically sampled infinite impulse response (IIR) quadrature mirror filter (QMF) banks for extracting features and thus saving a considerable computational time than that required in [7]. But both the aforementioned methods do not take into consideration the possibility of overlapped/mixed classes. Etemad *ct. al* [9] have developed an algorithm for document segmentation based on document texture using multiscale feature vectors and fuzzy local decision information.

In the present work the underlying assumption of our approach is that the text and non-text parts are considered as two different textured regions. A composite texture can be discriminated if it is possible to obtain information about the texture signal energies. The basic idea is to decompose the composite image into different frequency bands at different scales. We conjecture that a decomposition scheme yielding a large number of sub bands would definitely improve segmentation results.

One of the drawbacks of standard wavelets ($M = 2$) is that they are not suitable for the analysis of high frequency signals with relatively narrow bandwidth. So the main motivation of the present work is to use the decomposition scheme based on M-band wavelets (where $M > 2$), which yield improved segmentation accuracies. Unlike the standard wavelet decomposition which gives a logarithmic frequency resolution, the M-band decomposition gives a mixture of a logarithmic and linear frequency resolution. Further, as an additional advantage, M-band wavelet decompositions yield a large number of subbands which are required for good quality segmentation. The M-band wavelet transform performs a multiscale, multidirectional filtering of the images. It is a tool to view any signal at different scales and decomposes a signal by projecting it onto a family of functions generated from a single wavelet basis via its dilations and translations. Various combinations of the M-band wavelet filter decompose the image at different scales and orientations in the frequency plane. The filter extracts local frequencies of the image which in essence gives a measure of local energies of the image over small windows around each pixel. These energies are characteristics of a texture and give the features required for classification of the various textured regions in an image.

Some of the common difficulties that occur in documents are,

★ Text regions touching or overlapping with non-text regions.

★ Combinations of varying text and background gray level.

★ Page skew and text regions with different orientations.

We develop a texture based document image segmentation which takes care of all the above observations.

In section 2 we describe the general texture segmentation setup, and in section 3 we present the results of our experiment on different images under varying conditions.

2 Multiscale Representation

2.1 Feature Extraction

The basic system set up of our feature extraction scheme consists of a filter followed by local energy estimator and a smoothing filter. The filter bank in essence is a set of bandpass filters with frequency and orientation selective properties. In the filtering stage we make use of an eight-tap, four-band, orthogonal and linear phase wavelet transform following [10], to decompose the textured images into $M \times M$ (4×4) channels, corresponding to different direction and scales.

The 1-D M-band wavelet filters, are ψ_i, and H_i are the corresponding transfer functions, for $i=1,2,3,4$ with $M=4$. We extend the decomposition to the 2-D case by successively applying the 1-D M-band filters in a separable manner in the horizontal and vertical directions without downsampling and is denoted by ψ_{ij} for $i, j = 1, \cdots, 4$. The decomposition into 4×4 (=16) channels (resolution cells) is given in fig. 1a. and the ij^{th} resolution cell is achieved via the filter $H_{ij} = \psi_{ij}\psi_{ij}^*$, f_y and f_x correspond to the horizontal and vertical frequencies respectively. The objective of the wavelet filtering is to find out the detectable discontinuities (edges) that exist within the image. Note that the spectral response to edges of an image is strongest in direction perpendicular to the edge, while it decreases as the look direction of the filter approaches that of the edge. Therefore we can perform edge detection by using 2-D filtering of the image as,

wh : horizontal edges obtained by highpass filtering (HPF) on columns and lowpass filtering (LPF) on rows.

wv : vertical edges obtained by LPF on columns and HPF on rows.

wd : diagonal edges obtained by HPF on columns and LPF on rows.

Likewise *whd* (horizontal diagonal) and *wvd* (vertical diagonal) edges can be detected.

The frequency response of a typical edge detection filter covers a sector in the 2-D spatial frequency domain and all frequencies . Based on this concept and several wavelet decomposition filters are designed which are given by the summations $\sum_{Reg} H_{i,j}$, where Reg denotes the frequency sector of a certain direction and scale. Since the filter system we are using is orthogonal and has perfect reconstruction, quadrature mirror filter (PR-QMF) structure, that is $\sum_{i=1}^{M} \sum_{j=1}^{M} \psi_{i,j} \psi_{i,j}^{*} = 1$, the resulting filters treat all frequencies in each scale-space cell as equally possible. The number of channels and, therefore the number of possible filter combinations depend on the value of M. The wavelet filters denoted by $\sum_{Reg} H_{i,j}$ for different directions with increasing scales can be designed as follows,

– Horizontal direction filter:

$$filt_{hor1} = H_{12}$$
$$filt_{hor2} = H_{12} + H_{13}$$
$$filt_{hor3} = H_{12} + H_{13} + H_{14} + H_{24}$$

– Vertical direction filter:

$$filt_{ver1} = H_{21}$$
$$filt_{ver2} = H_{21} + H_{31}$$
$$filt_{ver3} = H_{21} + H_{31} + H_{41} + H_{42}$$

Similarly, decomposition filters corresponding to diagonal ($filt_{diag_j}$), horizontal diagonal ($filt_{hdiag_j}$) and vertical diagonal ($filt_{vdiag_j}$), (where j= 1,2,3) directions can be derived. These filter outputs basically give a measure of signal energies at different directions and scales. Frequency response of the decomposition filters are shown in fig. 1b.

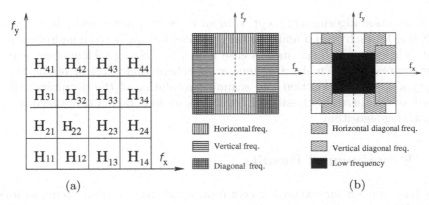

(a)   (b)

Fig. 1. (a)Frequency bands (b) Frequency response corresponding to decomposition filters

The next step is to estimate the energy of the filter responses in a local region around each pixel. The local energy estimate is utilized for the purpose of identifying areas in each channel where the bandpass frequency components are strong resulting in a high energy value and the areas where it is weak into a low energy value. We compute the standard deviation in small overlapping windows $W \times W$ around each pixel, which gives a measure of local energy. The nonlinear transform is succeeded by a Gaussian low pass (smoothing) filter. Formally, the feature image $f_k(i, j)$ corresponding to filtered image $h_k(i, j)$ is given by,

$$f_k(x, y) = \frac{1}{G^2} \sum_{(a,b) \in G_{ij}} |\Psi(h_k(a, b))| \tag{1}$$

where $\Psi(\cdot)$ is the nonlinear function and G_{ij} is a $G \times G$ window centered at pixel with coordinates (i, j).

The size G of the smoothing or the averaging window in equation (1) is an important parameter. More reliable measurement of texture feature calls for larger window sizes. On the other hand, more accurate localization of region boundaries calls for smaller windows. This is because averaging blurs the boundaries between textured regions. Another important aspect is that, Gaussian weighted windows are naturally preferable over unweighted windows, because, the former are likely to result in more accurate localization of texture boundaries. The images resulting from these operations are the feature maps $F_k(i, j)$ at different scales.

2.2 Unsupervised Classifier

Having obtained the feature images, the main task is to integrate these feature images to produce a segmentation. We define a scale - space signature as the vector of features at different scales taken at a single pixel in an image,

$$\overline{F}(i, j) = [F_0(i, j), F_1(i, j), \ldots, F_N(i, j)]$$

Segmentation algorithms accept as input a set of features and put a consistent labeling for each pixel. Fundamentally this can be considered a multidimensional data clustering problem, in this case it is a two class problem. One goal with our approach is to make a segmentation scheme independent of the font size or type layout, orientation and scanning resolution of the document, thus we need an unsupervised classifier. In this work we have used a traditional K-means clustering algorithm.

3 Experimental Results

We have applied our texture segmentation algorithm to several document images, in order to demonstrate the performance of our algorithm. These documents are scanned from parts of pages of the "Times of India" (TOI) issues in November

1999 and from pages of "Hindustan Times" issues in Febrary 2001.

Out of the total 16 features possible in our decomposition scheme we have experimentally found that only 5 features are sufficient for the required segmentation. The number of features could even be reduced without any perceptible degradation of our results.

Fig. 2a.i shows part of a page in TOI, and the successful two-class segmentation of the image. Fig. 2b.i also shows the same image fig. 2a.i oriented by an angle 55^0 and the output after image segmentation using simple histogram thresholding are shown in figures 2a.ii and 2b.ii. Fig. 2c.i shows a portion of TOI skewed by an angle 15^0 and the corresponding output after image segmentation using simple histogram thresholding is given in fig. 2c.ii.

Figures. 3a.i and 3b.i shows two other test images which have highly unstructured data. In fig. 3a.i, the text regions overlap with non-text regions, combinations of varying text and background gray level and text regions have different orientations and widely varying font size. The two class segmentation is given in fig. 3a.ii and fig. 3a.iii shows the text portion which is segmented out. While in fig. 3b.i the text portions are skewed and also have different orientations as well as gray values and have different font sizes. The corresponding two class and text segmentation results are shown in fig. 3b.ii and fig. 3b.iii respectively.

From the segmented results that we have obtained we can readily infer that our algorithm is capable of delineating the text and the non-text regions quite appreciably, while the computation is very simple and feature dimensionality is also greatly reduced. Since our technique uses overcomplete representation of the image meaning we have decomposed the image without downsampling, we get more or less accurate edge localization.

Throughout the experiment our effort has been to segment out the text part from the graphics part as accurately as possible. To compare our method with other methods we have used the same data that has been used by Randen and Hosϕy [8] (fig. 4). Using our algorithm we find that although some of the graphics part of fig 4 are misclassified as text data, we on the contrary get excellent results as far as text identification is concerned. The headings of two different font sizes could not be identified very accurately by Randen's method [8], but have been possible by our method.

4 Conclusion

It is very important to separate out the text part from the graphics part in paper documents, since it is very difficult to store or retrieve the whole digitized document even if it is in compressed form. Because if graphics are present in a data it is stored in a bit map representation and it is practically impossible to search text data in a bit map file.

In this paper we have presented a new technique for document segmentation using M-band wavelet filters. The decomposition gives a multiscale, multidirectional representation of the image and yields a large number of subbands, which ensures good quality segmentation of the images. In contrast to most traditional

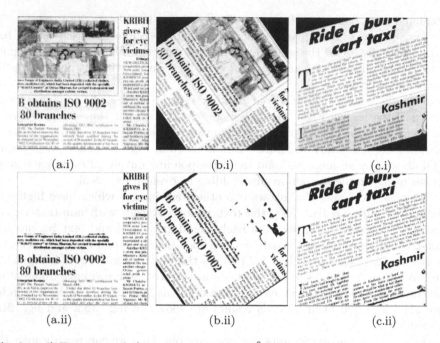

Fig. 2. a.i) Test image b.i) Image rotated by 55^0 c.i) Image skewed by an angle 15^0 a.ii) b.ii) and c.ii) segmented results

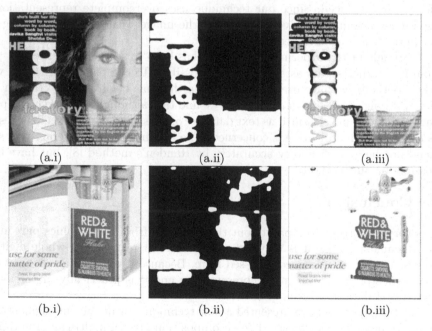

Fig. 3. a.i -b.i) Test images a.ii- b.ii) two class segmentation a.iii-b.iii) segmented result of the documents

(a)   (b)   (c)

Fig. 4. (a) Test image used in [8] and segmentation of the image using (b) our algorithm (c) Randen and Husøy [8]

methods for text-graphics segmentation we do not assume any *a priori* knowledge about the font size, column layout, orientation etc. of the input. That is our approach is purely unsupervised. Also it is to be noted that we have not done any post processing over the segmented images while simple post processing operations like median filtering or morphological dilations would have definitely improved the results.

References

1. Srihari, S.N.: Document Image Understanding. Proc. IEEE Computer Society Fall Joint Computer Conf. Nov. (1986) 87–96
2. Fletcher, L.A., Kasturi, R.: A Robust Algorithm for Text String Separation from Mixed Text/graphics Images. IEEE Trans. on Pattern Anal. and Mach. Intell. Vol. 10, Nov. (1988)
3. Chauvet, P., Lopez-Krahe, J., Taflin, E., H. Maitre, H.: System for An Intelligent Office Document Analysis, Recognition and Description. Signal Processing. Vol. 32, no. 1-2, (1993) 161–190
4. Tan, C.L., Yuan, B., Huang, W., Zang, Z.: Text/graphics Separation using Pyramid Operations. Proc. Int. Conf. on Document analysis and Recognition. (1999) 169–172
5. Nagy, G., Seth, S., Viswanathan, M.: A Prototype Document Image Analysis for Technical Journals. Computer. Vol. 25, no. 7, (1992) 10–22
6. Antaonacopulos, A.: Page Segmentation using the Description of the Background. Computer Vision and Image Understanding. Vol. 70, no. 3, (1998) 350–369
7. Jain, A.K., Bhattacharjee, S.: Text Segmentation using Gabor Filters for Automatic Document Processing. Mach. Vision and Appl. Vol. 5, no. 3, (1992) 169–184
8. Randen, T., Husøy, J.H.: Segmentation of Text/image Documents using Texture Approaches. Proc. NOBIM-konferansen-94, June (1994) 60–67
9. Etemad, K., Doermann, D., Chellappa, R.: Multiscale Segmentation of Unstructured Document Pages using Soft Decision Integration. IEEE Trans. on Pattern Anal. and Mach. Intell. Vol. 19, no. 1, (1997) 92–96
10. Alkin, O., Caglar, H.: Design of Efficient M-band Coders with Linear Phase and Perfect Reconstruction Properties. IEEE Trans. Signal Processing. Vol. 43, no. 7, (1995) 1579–1590

Length Estimation for Curves with ε-Uniform Sampling

Lyle Noakes[1], Ryszard Kozera[2], and Reinhard Klette[3]

[1] Department of Mathematics & Statistics,
The University of Western Australia, 35 Stirling Highway,
Crawley 6009 WA, Australia
lyle@maths.uwa.edu.au
http://www.maths.uwa.edu.au/~lyle/
[2] Department of Computer Science & Software Engineering,
The University of Western Australia, 35 Stirling Highway,
Crawley 6009 WA, Australia
ryszard@cs.uwa.edu.au
http://www.cs.uwa.edu.au/~ryszard/
[3] Centre for Image Technology & Robotics,
The University of Auckland,Tamaki Campus, Building 731,
Auckland, New Zealand
r.klette@auckland.ac.nz
http://www.tcs.auckland.ac.nz/~rklette/

Abstract. This paper[*] discusses the problem of how to approximate
the length of a parametric curve $\gamma : [0, T] \to \mathbb{R}^n$ from points $q_i = \gamma(t_i)$,
where the parameters t_i are not given. Of course, it is necessary to make
some assumptions about the distribution of the t_i: in the present paper
ε-uniformity. Our theoretical result concerns an algorithm which uses
piecewise-quadratic interpolants. Experiments are conducted to show
that our theoretical estimates are sharp, and that the assumption of
ε-uniformity is needed. This work may be of interest in computer graphics, approximation and complexity theory, digital and computational
geometry, and digital image processing.

Keywords: discrete curves, quadratic interpolants, length estimation

1 Introduction

In this paper we present a simple algorithm for estimating the length of a smooth
regular parametric curve $\gamma : [0, T] \to \mathbb{R}^n$ (where $0 < T < \infty$) from $m + 1$-tuples
$\mathcal{Q} = (q_0, q_1, \ldots, q_m)$ of points on the curve $q_i = \gamma(t_i)$. The t_i are not assumed to
be given, but some assumptions are needed to make our problem solvable. For
example, if none of the parameter values lie in $(0, \frac{T}{2})$ the task is intractable.

[*] This research was performed at the University of Western Australia, while the third
author was visiting under the UWA Gledden Visiting Fellowship scheme.[1,2,3] Additional support was received under an Australian Research Council Small Grant[1,2]
and under an Alexander von Humboldt Research Fellowship.[2]

W. Skarbek (Ed.): CAIP 2001, LNCS 2124, pp. 518–526, 2001.
© Springer-Verlag Berlin Heidelberg 2001

In the present paper we assume that the t_i are chosen in a fashion that is ε-*uniform*, in the sense of the following definition.

Definition 1. *For $\varepsilon \geq 0$, sampling is said to be ε-uniform when there is a reparameterization $\phi : [0, T] \to [0, T]$ of class C^k (where $k \geq 1$), and constants $0 < K_l < K_u$ such that, for any $m \geq 1$, ordered samples of size $m + 1$ satisfy*

$$\frac{K_l}{m^{1+\varepsilon}} \leq t_i - \phi\left(\frac{iT}{m}\right) \leq \frac{K_u}{m^{1+\varepsilon}}$$

(we may also write $t_i = \phi(\frac{iT}{m}) + O(\frac{1}{m^{1+\varepsilon}})$).

Note that ε-uniform sampling arises from two types of perturbations of uniform sampling: first via a diffeomorphism $\phi : [0, T] \to [0, T]$ combined subsequently with added extra distortion term $O(\frac{1}{m^{1+\varepsilon}})$. In particular, for $\phi = id$ and $\varepsilon = 0$ ($\varepsilon = 1$) the perturabtion is *linear* i.e. of uniform sampling order (*quadratic*), which constitutes asymptotically a big (small) distortion of uniform partition of $[0, T]$. For curves sampled in an ε-uniform way it is straightforward to achieve $O(\frac{1}{m^2})$ accuracy for length estimates, using piecewise-linear interpolation through the given q_i (see [14] and [15]).

In the special case where the t_i are chosen perfectly uniformly and $\gamma \in C^{r+2}$, for $r \geq 1$ integer, it is also possible to achieve at least $O(\frac{1}{m^{r+1}})$-accurate length estimates, using piecewise-polynomial interpolants of degree r (see [15]). In the present paper we show that rather similar techniques can be used, at least for $r = 2$, with ε-uniform samplings, where $\varepsilon > 0$ (the case $\varepsilon = 0$ is the subject of [14], where a more sophisticated algorithm is used).

There are other possible applications outside the scope of computer graphics and approximation theory. A recent surge of research within digital geometry and digital image processing (see e.g. [17]) constitutes further potential applications. In fact there is some analogous work for estimating lengths of digitized curves; indeed the analysis of digitized curves in $\mathbb{R}^2$ is one of the most intensively studied subjects in image data analysis. A digitized curve is the result of a process (such as contour tracing, 2D skeleton extraction, or 2D thinning) which maps a curve-like object (such as the boundary of a region) onto a computer-representable curve. An analytical description of $\gamma : [0, T] \to \mathbb{R}^2$ is not given, and numerical measurements of points on γ are corrupted by a process of *digitization*: γ is digitized within an orthogonal grid of points $(\frac{iT}{m}, \frac{jT}{m})$, where i, j are permitted to range over integer values, and m is a fixed positive integer called *the grid resolution*. Depending on the digitization model [9], γ is mapped onto a digital curve and approximated by a polygon $\hat{\gamma}_m$ whose length is an estimator for $d(\gamma)$. Approximating polygons $\hat{\gamma}_m$ based on local configurations of digital curves do not ensure multigrid length convergence, but global approximation techniques yield *linearly* convergent estimates, namely

$$d(\gamma) - d(\hat{\gamma}_m) = O\left(\frac{1}{m}\right)$$

[11], [12] or [20]. Recently, experimentally based results reported in [3] and [10] confirm a similar rate of convergence for $\gamma \subset \mathbb{R}^3$. In the special case of discrete

straight line segments in $\mathbb{R}^2$ a stronger result is proved; see e.g. [6], where a *superlinear* convergence, i.e. $O(\frac{1}{m^{1.5}})$, of asymptotic length estimation accuracy is established.

In Theorem 1 of this paper convergence is $O(\frac{1}{m^{4\varepsilon}})$, when $0 < \varepsilon \leq 1$. For these results Q arises from ε-uniform sampling as opposed to digitization. So strict comparisons cannot yet be made.

Related work can also be found in [1], [2], [4], [7], [8], [13], [16], [19], and [21]. There is also some interesting work on complexity [5], [18], [22], and [23].

2   Sampling

We are going to consider different ways of forming ordered samples

$$0 = t_0 < t_1 < t_2 < \ldots < t_m = T$$

of variable size $m + 1$ from a given interval $[0, T]$. The simplest procedure is *uniform sampling*, where $t_i = \frac{iT}{m}$. Uniform sampling is not invariant with respect to *reparameterizations*, namely order-preserving C^k diffeomorphisms $\phi : [0, S] \to [0, T]$, where $k \geq 1$. A small perturbation of uniform sampling is no longer uniform, but may approach uniformity in some asymptotic sense, at least after some suitable reparameterization. More precisely, we consider here ε-uniform samplings which are defined accordingly by Definition 1.

Note that ϕ, K_l and K_u appearing in Definition 1 are chosen independently of $m \geq 1$, and that ε-uniformity implies δ-uniformity, for all $0 \leq \delta < \varepsilon$. Uniform sampling is ε-uniform for any $\varepsilon \geq 0$. Indeed, setting $\phi = id$ and constant K appearing in the expression $O(1/m^{1+\varepsilon})$ to zero fulfills the claim. At the other extreme there are examples where sampling increments $t_i - t_{i-1}$ are neither large nor small, considering m, and yet sampling is not ε-uniform for any $\varepsilon > 0$:

Example 1. Set $\frac{iT}{m}$ or $\frac{(2i-1)T}{2m}$ according as i is even or odd. Then sampling is not ε-uniform for any $\varepsilon > 0$. However it is *more-or-less uniform*, i.e. there are constants $0 < K_l < K_u$ such that

$$\frac{K_l}{m} \leq t_i - t_{i-1} \leq \frac{K_u}{m} , \tag{1}$$

for all $1 \leq i \leq m$ and all $m \geq 1$, where here $K_l = \frac{T}{2}$ and $K_u = \frac{3T}{2}$.

To see that this sampling is not ε-uniform for $\varepsilon > 0$ assume the opposite. Then, for some at least C^1 class $\phi : [0, T] \to [0, T]$ we have

$$t_{i+1} - t_i = \frac{T}{2m} = \phi\left(\frac{(i+1)T}{m}\right) - \phi\left(\frac{iT}{m}\right) + O(\frac{1}{m^{1+\varepsilon}}) , \tag{2}$$

$$t_{i+2} - t_{i+1} = \frac{3T}{2m} = \phi\left(\frac{(i+2)T}{m}\right) - \phi\left(\frac{(i+1)T}{m}\right) + O(\frac{1}{m^{1+\varepsilon}}) . \tag{3}$$

Furthermore, by the Mean Value Theorem,

$$\frac{T}{2m} = \frac{T\phi'(\xi_{1i}^{(m)})}{m} + O(\frac{1}{m^{1+\varepsilon}}) \quad \text{and} \quad \frac{3T}{2m} = \frac{T\phi'(\xi_{2i}^{(m)})}{m} + O(\frac{1}{m^{1+\varepsilon}}) , \tag{4}$$

for some $\xi_{1i}^{(m)} \in (\frac{iT}{m}, \frac{(i+1)T}{m})$ and $\xi_{2i}^{(m)} \in (\frac{(i+1)T}{m}, \frac{(i+2)T}{m})$. By fixing i and increasing m we have

$$\phi'(\xi_{1i}^{(m)}) \to \phi'(0) \quad \text{and} \quad \phi'(\xi_{2i}^{(m)}) \to \phi'(0) .$$

On the other hand by (4) the following holds

$$\phi'(\xi_{1i}^{(m)}) \to 1/2 \quad \text{and} \quad \phi'(\xi_{2i}^{(m)}) \to 3/2 .$$

A contradiction. This proof clearly fails at $\varepsilon = 0$, for which in fact we deal with 0-uniform sampling. Formula (1) is an immediate consequence of (2) and (3).

3   Samples and Curves

For $n \geq 1$ let $< \cdot, \cdot >$ be the Euclidean inner product, and $\| \cdot \|$ the associated norm. The *length* $d(\gamma)$ of a C^k parametric curve ($k \geq 1$) $\gamma : [0, T] \to \mathbb{R}^n$ is defined as

$$d(\gamma) = \int_0^T \|\dot{\gamma}(t)\| dt ,$$

where $\dot{\gamma}(t) \in \mathbb{R}^n$ is the derivative of γ at $t \in [0, T]$. The curve γ is said to be *regular* when $\dot{\gamma}(t) \neq \mathbf{0}$, for all $t \in [0, T]$. A *reparameterization* of γ is a parametric curve of the form $\gamma \circ \phi : [0, S] \to \mathbb{R}^n$, where $\phi : [0, S] \to [0, T]$ is a C^k diffeomorphism. The reparameterization $\gamma \circ \phi$ has the same image and length as γ. Let γ be regular: then so is any reparameterization $\gamma \circ \phi$.

Definition 2. γ *is parameterized by arc-length when* $\|\dot{\gamma}(t)\| = 1$ *for all* $t \in [0, T]$.

Notice that if γ is parameterized by arc-length then $d(\gamma) = T$, and $< \dot{\gamma}, \ddot{\gamma} >$ is identically zero (provided $k \geq 2$). We want to estimate $d(\gamma)$ from ordered $m + 1$-tuples

$$\mathcal{Q} = (q_0, q_1, q_2, \ldots, q_m) \in (\mathbb{R}^n)^{m+1} ,$$

where $q_i = \gamma(t_i)$, whose parameter values $t_i \in [0, T]$ are unknown but sampled in some reasonably regular way: sampling might be ε-uniform, for some $\varepsilon \geq 0$. Since these conditions are invariant with respect to reparameterization (see [15]), suppose without loss that γ is parameterized by arc-length.

4   Quadratics and ε-Uniform Sampling

Before stating the main convergence result we recall briefly the construction of a quadratic sampler $\widetilde{\gamma}_m$ forming the so-called G^0 interpolant of the unknown curve γ (see e.g. [1] or [16]; Section 9.3).

Assume $\mathcal{Q} = (q_0, q_1, q_2, \ldots, q_{m-1}, q_m)$ are sampled from curve $\gamma \subset \mathbb{R}^n$ and, without loss of generality, assume that m is even. The Quadratic Sampler (QS) Algorithm is:

1. For each triplet of sampled points $\mathcal{Q}_i = (q_i, q_{i+1}, q_{i+2})$, where $0 \le i \le m-2$, let $Q^i : [0,2] \to \mathbb{R}^n$ be the quadratic curve satisfying

$$Q^i(0) = q_i , \qquad Q^i(1) = q_{i+1} , \qquad \text{and} \qquad Q^i(2) = q_{i+2} .$$

Then $Q^i(s) = a_0 + a_1 s + a_2 s^2$, where

$$a_0 = q_i , \qquad a_1 = \frac{4q_{i+1} - 3q_i - q_{i+2}}{2} , \qquad \text{and} \qquad a_2 = \frac{q_{i+2} - 2q_{i+1} + q_i}{2} .$$

2. Calculate the length of each Q^i according to the formula

$$d(Q^i) = \int_0^2 \sqrt{\|a_1\|^2 + 4 < a_1 | a_2 > s + 4\|a_2\|^2 s^2} ds .$$

3. Form the length estimate

$$\tilde{d}_m = \sum_{i=0}^{m/2-1} d(Q^{2i}) .$$

The proof of the following convergence result uses a C^0 piecewise-quadratic curve $\tilde{\gamma}_m$ interpolating the q_i, where $0 \le i \le m$. Let $k = 4$, so that $\gamma : [0,T] \to \mathbb{R}^n$ and its reparameterizations are at least C^4:

Theorem 1. *If sampling is ε-uniform, where $0 < \varepsilon \le 1$, then*

$$\tilde{d}_m = d(\tilde{\gamma}_m) = d(\gamma) + O(\frac{1}{m^{4\varepsilon}}) . \tag{5}$$

The proof shows also that $\tilde{\gamma}_m$ approximates γ uniformly with $O(\frac{1}{m^{1+2\varepsilon}})$ error, where $0 \le \varepsilon \le 1$. Consequently, for 1-uniform sampling the corresponding rates of convergence for

$$\lim_{m \to \infty} d(\tilde{\gamma}_m) = d(\gamma) \qquad \text{and} \qquad \lim_{m \to \infty} \|\tilde{\gamma}_m - \gamma\|_\infty = 0$$

are at least *quartic* and *cubic*, respectively. When $\varepsilon = 0$ linear convergence is guaranteed for $\|\tilde{\gamma}_m - \gamma\|_\infty$, but not for length: a more sophisticated algorithm for this case gives *quartic* convergence for length estimates [14]. In the next section we verify (for $n = 2$) that the convergence rates in Theorem 1 are sharp. Then we briefly examine the case when ε vanishes.

5   Experimentation

We test the QS Algorithm on a *semicircle* and on a *cubic* curve $\gamma_s, \gamma_c : [0, \pi] \to \mathbb{R}^2$ defined as:

$$\gamma_s(t) = (\cos(\pi - t), \sin(\pi - t)) \qquad \text{and} \qquad \gamma_c(t) = (t, (\frac{t+1}{\pi+1})^3) .$$

Note that, for each $1 \leq l \leq 3$, the following functions

$$\phi_1(t) = t, \quad \phi_2(t) = \frac{1}{\pi + 1}t(t + 1), \quad \text{and} \quad \phi_3(t) = \pi \frac{\exp(t) - 1}{\exp(\pi) - 1}$$

satisfy $\phi_l : [0, \pi] \to [0, \pi]$ such that $\dot{\phi}_l(t) \neq 0$ (for $t \in [0, \pi]$) and $\phi_l(0) = 0$ and $\phi_l(\pi) = \pi$, and therefore define three diffeomorphisms over $[0, \pi]$.

The implementation of QS was performed on Pentium III, 700MHz with 192 MB RAM using Mathematica. The sampling points are generated for fixed $\varepsilon_1 = 1$, $\varepsilon_{1/2} = 1/2$ and $\varepsilon_{1/3} = 1/3$, respectively. A numerical integration yields true lengths $d(\gamma_s) = \pi$ and $d(\gamma_c) = 3.3452$. All programs are tested for m even running from $min = 6$ up to $max = 100$ ($max = 200$). We shall only report here (see Tables 1 and 2, with $max = 100$) on some specific values obtained from the set of absolute errors

$$E_m(\gamma) = |d(\gamma) - d(\tilde{\gamma}_m)|$$

(where m is even and $min \leq m \leq max$), namely:

$$E_{min}^{max}(\gamma) = \max_{min \leq m \leq max} E_m(\gamma) \quad \text{and} \quad E_{max}(\gamma).$$

Moreover, in searching for the estimate of correct convergence rate $O(\frac{1}{m^\alpha})$ a *linear regression* is applied to pairs of points $(\log(m), -\log(E_m(\gamma)))$, where m even and $min \leq m \leq max$; for $max = 100$ see Tables 1 and 2 and for $max = 200$ see Table 3. For both curves γ_s and γ_c and all ϕ_l (where $1 \leq l \leq 3$) synthetic data $\{\gamma(t_i)\}_{i=0}^m$ (where $1 \leq i \leq m$) are derived according to ε-uniform sampling rules. Note that there is no perturbation of boundary points: $q_0 = \gamma(0)$ and $q_m = \gamma(\pi)$.

5.1 ε-Uniform Random Sampling

In this subsection we report on the QS Algorithm for the following ε-uniform sampling procedure:

$$t_i = \phi_l(\frac{i\pi}{m}) + (Random[\ ] - 0.5)\frac{1}{m^{1+\varepsilon}},$$

where $Random[\ ]$ takes the pseudorandom values from the interval $[0, 1]$. Illustrations (for $m = 6$) are available at

http://www.cs.uwa.edu.au/~ryszard/caip2001/pics/.

The following results (see Tables 1 and 2) have been recorded for both curves in question: Both experiments suggest very fast convergence. As proved in Theorem 1 the corresponding values α for each ε_1, $\varepsilon_{1/2}$ and $\varepsilon_{1/3}$ are at least of order $\alpha_1 = 4$, $\alpha_{1/2} = 2$ and $\alpha_{1/3} = 4/3$, respectively. A natural question arises however as to whether these estimates are indeed sharp. Clearly the latter experiments support this claim only for ε_1. For the remaining cases the rate of

Table 1. Results for length estimation $d(\gamma_s)$ of the semicircle γ_s

α (for $E_m \propto m^{-\alpha}$)			$E_{min}^{max}(\gamma_s) \times 10^{-2}$			$E_{max} \times 10^{-8}$		
ε_1	$\varepsilon_{1/2}$	$\varepsilon_{1/3}$	ε_1	$\varepsilon_{1/2}$	$\varepsilon_{1/3}$	ε_1	$\varepsilon_{1/2}$	$\varepsilon_{1/3}$
ϕ_1 3.986	4.284	4.207	0.36190	0.36176	0.36625	5.0750	2.5144	0.5624
ϕ_2 3.956	4.038	4.690	0.92007	0.97634	0.98982	15.486	13.179	0.1608
ϕ_3 3.867	3.879	3.831	4.19549	4.22734	4.32888	104.48	103.02	95.331

Table 2. Results for length estimation $d(\gamma_c)$ of the cubic curve γ_c

α (for $E_m \propto m^{-\alpha}$)			$E_{min}^{max}(\gamma_c) \times 10^{-4}$			$E_{max} \times 10^{-8}$		
ε_1	$\varepsilon_{1/2}$	$\varepsilon_{1/3}$	ε_1	$\varepsilon_{1/2}$	$\varepsilon_{1/3}$	ε_1	$\varepsilon_{1/2}$	$\varepsilon_{1/3}$
ϕ_1 4.272	3.190	2.451	0.02483	0.23390	0.00068	5.0750	0.0725	0.3953
ϕ_2 4.094	3.879	3.508	0.71288	1.01073	0.89811	0.1616	0.1714	0.4306
ϕ_3 4.150	4.013	3.871	1.12414	13.8679	11.1654	0.1628	2.7308	2.8737

convergence, though proved theoretically to be less than quartic, can still be faster than those established in (5). A close look at the ε-uniform random data selection shows that for each quadratic component Q^i we undershoot or over-shoot the correct length of the corresponding segment of $\gamma_{|[t_i,t_{i+2}]}$. The random choice of data provides a cancelling effect once the lengths of all Q^i forming a piecewise quadratic interpolant $\widetilde{\gamma}_m$ are summed up. To show that Theorem 1 provides sharp, estimates we look at another example.

5.2 ε-Uniform Skew Sampling

In order to cancel the effect of more or less equally distributed under and over-shooting $d(\gamma_{|[t_i,t_{i+2}]})$ with $d(Q^i)$ we test QS on the following two families of ε-uniform skew samplings (with $\phi = id$):

$$(i)\ t_i = \frac{i\pi}{m} + (-1)^{i+1}\frac{\pi}{2m^{1+\varepsilon}}\ , \qquad (ii)\ t_i = \begin{cases} \frac{i\pi}{m} & \text{if } i \text{ even}, \\ \frac{i\pi}{m} + \frac{\pi}{2m^{1+\varepsilon}} & \text{if } i \text{ odd \& } i = 4k+1, \\ \frac{i\pi}{m} - \frac{\pi}{2m^{1+\varepsilon}} & \text{if } i \text{ odd \& } i = 4k+3. \end{cases}$$

To improve robustness of estimation of α set $m = 200$. The following esti-mates (see Table 3) for α are obtained: For these examples, the convergence rates provided by Theorem 1 for $0 < \varepsilon \leq 1$ seem very sharp. On the other hand, when $\varepsilon = 0$, Table 3 suggests that $d(\gamma_c)$ and possibly $d(\gamma_s)$ are estimated in-correctly. These indications are supported by the illustrations on the web which address is given in the Subsection 5.1.

Table 3. Results for length estimation $d(\gamma_s)$ and $d(\gamma_c)$ of γ_s and γ_c

	α for semicircle γ_s				α for cubic curve γ_c			
	ε_1	$\varepsilon_{1/2}$	$\varepsilon_{1/3}$	ε_0	ε_1	$\varepsilon_{1/2}$	$\varepsilon_{1/3}$	ε_0
Sampling (i)	4.052	2.940	2.870	0.127**	4.122	3.280	3.309	-0.085**
Sampling (ii)	3.995	2.751	2.400	1.800	1.800	4.030	2.774	2.046

** Illustrations are available at the web address given previously.

6 Conclusions

We tested the theoretical results for the length estimation of an arbitrary regular curve $\gamma \subset \mathbb{R}^n$ sampled from non-uniformly distributed $m + 1$ interpolation points. The experiments (carried out for $n = 2$) confirm the sharpness or near sharpness of (5) for the class of samplings and curves studied in this paper. The case $\varepsilon = 0$ needs separate treatment [14]. An interesting question can be raised of other classes of admissible samplings for which QS or another proposed algorithm yield $d(\widetilde{\gamma}_m) \to d(\gamma)$, and if so what the corresponding convergence rates are.

References

1. Barsky, B.A., DeRose, T.D.: Geometric Continuity of Parametric Curves: Three Equivalent Characterizations. IEEE. Comp. Graph. Appl. **9**:6 (1989) 60–68
2. Boehm, W., Farin, G., Kahmann, J.: A Survey of Curve and Surface Methods in CAGD. Comput. Aid. Geom. Des. **1** (1988) 1–60
3. Bülow, T., Klette, R.: Rubber Band Algorithm for Estimating the Length of Digitized Space-Curves. In: Sneliu, A., Villanva, V.V., Vanrell, M., Alquézar, R., Crowley. J., Shirai, Y. (eds): Proceedings of 15th International Conference on Pattern Recognition. Barcelona, Spain. IEEE, Vol. III. (2000) 551-555
4. Davis, P.J.: Interpolation and Approximation. Dover Pub. Inc., New York (1975)
5. Dąbrowska, D., Kowalski, M.A.: Approximating Band- and Energy-Limited Signals in the Presence of Noise. J. Complexity **14** (1998) 557–570
6. Dorst, L., Smeulders, A.W.M.: Discrete Straight Line Segments: Parameters, Primitives and Properties. In: Melter, R., Bhattacharya, P., Rosenfeld, A. (eds): Ser. Contemp. Maths, Vol. 119. Amer. Math. Soc. (1991) 45–62
7. Epstein, M.P.: On the Influence of Parametrization in Parametric Interpolation. SIAM. J. Numer. Anal. **13**:2 (1976) 261–268
8. Hoschek, J.: Intrinsic Parametrization for Approximation. Comput. Aid. Geom. Des. **5** (1988) 27–31
9. Klette, R.: Approximation and Representation of 3D Objects. In: Klette, R., Rosenfeld, A., Sloboda, F. (eds): Advances in Digital and Computational Geometry. Springer, Singapore (1998) 161–194
10. Klette, R., Bülow, T.: Critical Edges in Simple Cube-Curves. In: Borgefors, G., Nyström, I., Sanniti di Baja, G. (eds): Proceedings of 9th Conference on Discrete Geometry for Computer Imagery. Uppsala, Sweden. Lecture Notes in Computer Science, Vol. 1953. Springer-Verlag, Berlin Heidelberg (2000) 467-478

11. Klette, R., Kovalevsky, V., Yip, B.: On the Length Estimation of Digital Curves. In: Latecki, L.J., Melter, R.A., Mount, D.A., Wu, A.Y. (eds): Proceedings of SPIE Conference, Vision Geometry VIII, Vol. 3811. Denver, USA. The International Society for Optical Engineering (1999) 52–63

12. Klette, R., Yip, B.: The Length of Digital Curves. Machine Graphics and Vision 9 (2000) 673–703

13. Moran, P.A.P.: Measuring the Length of a Curve. Biometrika 53:3/4 (1966) 359–364

14. Noakes, L., Kozera, R.: More-or-Less Uniform Sampling and Lengths of Curves. Quart. Appl. Maths. In press

15. Noakes, L., Kozera, R., and Klette R.: Length Estimation for Curves with Different Samplings. In: Bertrand, G., Imiya, A., Klette, R. (eds): Digital and Image Geometry. Submitted

16. Piegl, L., Tiller, W.: The NURBS Book. 2nd edn Springer-Verlag, Berlin Heidelberg (1997)

17. Pitas, I.: Digital Image Processing Algorithms and Applications. John Wiley & Sons Inc., New York Chichester Weinheim Brisbane Singapore Toronto (2000)

18. Plaskota, L.: Noisy Information and Computational Complexity. Cambridge Uni. Press, Cambridge (1996)

19. Sederberg, T.W., Zhao, J., Zundel, A.K.: Approximate Parametrization of Algebraic Curves. In: Strasser, W., Seidel, H.P. (eds): Theory and Practice in Geometric Modelling. Springer-Verlag, Berlin (1989) 33–54

20. Sloboda, F., Zaťko, B., Stör, J.: On approximation of Planar One-Dimensional Continua. In: Klette, R., Rosenfeld, A., Sloboda, F. (eds): Advances in Digital and Computational Geometry. Springer, Singapore (1998) 113–160

21. Steinhaus, H.: Praxis der Rektifikation und zur Längenbegriff. (in German) Akad. Wiss. Leipzig Ber. 82 (1930) 120–130

22. Traub, J.F., Werschulz, A.G.: Complexity and Information. Cambridge Uni. Press, Cambridge (1998)

23. Werschulz, A.G., Woźniakowski, H.: What is the Complexity of Surface Integration? J. Complexity. In press

Random Walk Approach to Noise Reduction in Color Images

B. Smolka[1]*, M. Szczepanski[1]*, K.N. Plataniotis[2], and A.N. Venetsanopoulos[2]

[1] Silesian University of Technology
Department of Automatic Control
Akademicka 16 Str, 44-101 Gliwice, Poland
bsmolka@ia.polsl.gliwice.pl
[2] Edward S. Rogers Sr. Department of
Electrical and Computer Engineering
University of Toronto
10 King's College Road, Toronto, Canada
kostas@dsp.toronto.edu

Abstract. In this paper we propose a new algorithm of noise reduction in color images. The new technique of image enhancement is capable of reducing impulsive and Gaussian noise and it outperforms the standard methods of noise reduction. In the paper a smoothing operator, based on a random walk model and on a fuzzy similarity measure between pixels connected by a digital self avoiding path is introduced. The efficiency of the proposed method was tested on the test color image using the objective image quality measures and the results show that the new method outperforms standard noise reduction algorithms.

Keywords: noise reduction, random walk

1   Standard Noise Reduction Filters

A number of nonlinear, multichannel filters, which utilize correlation among multivariate vectors using various distance measures, have been proposed [1,2]. The most popular nonlinear, multichannel filters are based on the ordering of vectors in a predefined moving window. The output of these filters is defined as the lowest ranked vector according to a specific vector ordering technique.

Let $\mathbf{F}(x)$ represents a multichannel image and let W be a window of finite size n (filter length). The noisy image vectors inside the filtering window W are denoted as $\mathbf{F}_j$, $j = 0, 1, ..., n - 1$. If the distance between two vectors $\mathbf{F}_i, \mathbf{F}_j$ is denoted as $\rho(\mathbf{F}_i, \mathbf{F}_j)$ then the scalar quantity $R_i = \sum_{j=0}^{n-1} \rho(\mathbf{F}_i, \mathbf{F}_j)$, is the distance associated with the noisy vector $\mathbf{F}_i$. The ordering of the R_i's: $R_{(0)} \leq R_{(1)} \leq ... \leq R_{(n-1)}$, implies the same ordering to the corresponding vectors $\mathbf{F}_i : \mathbf{F}_{(0)} \leq \mathbf{F}_{(1)} \leq ... \leq \mathbf{F}_{(n-1)}$. Nonlinear ranked type multichannel estimators define the vector $\mathbf{F}_{(0)}$ as the filter output.

---

* This work was partially supported by KBN grant 8-T11E-013-19

W. Skarbek (Ed.): CAIP 2001, LNCS 2124, pp. 527–536, 2001.

However, the concept of input ordering, initially applied to scalar quantities is not easily extended to multichannel data, since there is no universal way to define ordering in vector spaces.

To overcome this problem, distance functions are often utilized to order vectors. As an example, the *Vector Median Filter* (VMF) uses the L_1 or L_2 norm to order vectors according to their relative magnitude differences [3].

The orientation difference between two vectors can also be used as their distance measure. This so-called *vector angle criterion* is used by the *Vector Directional Filters* to remove vectors with atypical directions [4].

The *Basic Vector Directional Filter* (BVDF) is a ranked-order, nonlinear filter which parallelizes the VMF operation. However, a distance criterion, different from the L_1 norm used in VMF, is utilized to rank the input vectors. The output of the BVDF is that vector from the input set, which minimizes the sum of the angles with the other vectors. In other words, the BVDF chooses the vector most centrally located without considering the magnitudes of the input vectors.

To improve the efficiency of the directional filters, a new method called *Directional-Distance Filter* (DDF) was proposed [10]. This filter retains the structure of the BVDF but utilizes a new distance criterion to order the vectors inside the processing window.

Another efficient rank-ordered a technique called Hybrid Directional Filter was presented in [11]. This filter operates on the directional and the magnitude of the color vectors independently and then combines them to produce a unique final output. Another more complex hybrid filter, which involves the utilization of an Arithmetic Mean Filter (AMF), has also been proposed [11].

The reduction of image noise without major degradation of the image structure is one of the most important problems of the low-level image processing. A whole variety of algorithms has been developed, however none of them can be seen as a final solution of the noise problem and therefore a new filtering technique is proposed in this paper.

2 New Algorithm of Noise Reduction

Let us assume, that R^2 is the Euclidean space, W is a planar subset of R^2 and x, y are points of of the set W. A path from x to y is a continuous mapping $\mathcal{P}$: $[a, b] \rightarrow X$, such that $\mathcal{P}(a) = x$ and $\mathcal{P}(b) = y$. Point x is the starting point and y is the end point of the path $\mathcal{P}$ [5].

The notion of the path can be extended to a lattice, which is a set of discrete points, in our case image pixels. Let a digital lattice $\mathcal{H} = (\mathbf{F}, \mathcal{N})$ be defined by $\mathbf{F}$, which is the set of all points of the plane (pixels of a color image) and the neighborhood relation $\mathcal{N}$ between the lattice points.

A digital path $P = \{p_i\}_{i=0}^{n}$ on the lattice $\mathcal{H}$ is a sequence of neighboring points $(p_{i-1}, p_i) \in \mathcal{N}$. The length $L(P)$ of digital path $P\{p_i\}_{i=0}^{n}$ is simply $\sum_{i=1}^{n} \rho^{\mathcal{H}}(p_{i-1}, p_i)$, where $\rho^{\mathcal{H}}$ denotes the distance between two neighboring points (Fig. 1).

In this paper we will assign to the distance of neighboring points the value 1 and will be working with the 8-neighborhood system.

Let the pixels (i, j) and (k, l) be called connected (denoted as $(i, j) \leftrightarrow (k, l)$), if there exists a geodesic path $P^W\{(i, j), (k, l)\}$ contained in the set W starting from (i, j) and ending at (k, l).

If two pixels (x_0, y_0) and (x_n, y_n) are connected by a geodesic path of length n $P^W\{(x_0, y_0), (x_1, y_1), \ldots, (x_n, y_n)\}$ then let $\chi^{W,n}$

$$\chi^{W,n}\{(x_0, y_0), (x_n, y_n)\} = \sum_{k=0}^{n-1} \|\mathbf{F}(x_{k+1}, y_{k+1}) - \mathbf{F}(x_k, y_k)\|, \qquad (1)$$

be a measure of dissimilarity between pixels (x_0, y_0) and (x_n, y_n), along a specific geodesic path P^W joining (x_0, y_0) and (x_n, y_n) [6,7]. If a path joining two distinct points x, y, such that $\mathbf{F}(x) = \mathbf{F}(y)$ consists of lattice points of the same values, then $\chi^{W,n}(x, y) = 0$ otherwise $\chi^{W,n}(x, y) > 0$.

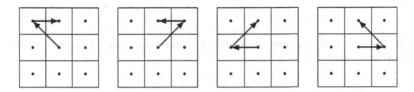

Fig. 1. There are four geodesic paths of length 2 connecting two neighboring points contained in the 3×3 window W when the 8-neighborhood system is applied.

Self avoiding random walk (SAW) is a special walk along the image lattice such that adjacent pairs of edges in the sequence share a common vertex of the lattice, but no vertex is visited more than once and in this way the trajectory never intersects itself. In other words SAW is a path on a lattice that does not pass through the same point twice.

On the two-dimensional lattice SAW is a finite sequence of distinct lattice points $(x_0, y_0), (x_1, y_1), (x_2, y_2), \ldots, (x_n, y_n)$, which are in neighborhood relation and $(x_i, y_i) \neq (x_j, y_j)$ for all $i \neq j$ [8]. Paths presented in Fig. 2 are examples of self avoiding walks.

Let us now define a fuzzy similarity function between two pixels connected along all geodesic digital paths leading from (i, j) and (k, l)

$$\mu^{W,n}\{(i, j), (k, l)\} = \frac{1}{\omega} \sum_{l=1}^{\omega} \exp\left[-\beta \cdot \chi_l^{W,n}\{(i, j), (k, l)\}\right], \qquad (2)$$

where ω is the number of all self avoiding paths connecting (i, j) and (k, l), β is a parameter and $\chi_l^{W,n}\{(i, j), (k, l)\}$ is a dissimilarity value along a specific path from a set of all ω possible paths leading from (i, j) to (k, l).

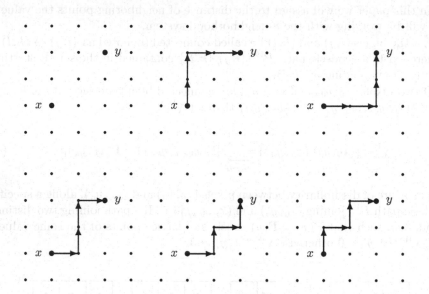

Fig. 2. There are five paths of length 4 connecting point x and y when the 4-neighborhood system is used.

For $n = 1$ and W a square mask of the size 3×3 (Fig. 1), we have

$$\mu^{W,1}\{(i,j),(k,l)\} = \exp\{-\beta\|\mathbf{F}(i,j) - \mathbf{F}(k,l)\|\}\,, \qquad (3)$$

and when $\mathbf{F}(i,j) = \mathbf{F}(k,l)$ then $\chi^{W,n}\{(i,j),(k,l)\} = 0$, $\mu\{(i,j),(k,l)\} = 1$, and for $\|\mathbf{F}(i,j) - \mathbf{F}(k,l)\| \to \infty$ then $\mu \to 0$.

The normalized similarity function takes the form

$$\psi^{W,n}\{(i,j),(k,l)\} = \frac{\mu^{W,n}\{(i,j),(k,l)\}}{\displaystyle\sum_{(l,m)\Leftrightarrow(i,j)} \mu^{W,n}\{(i,j),(l,m)\}}\,, \qquad (4)$$

and has the property that

$$\sum_{(k,l)\Leftrightarrow(i,j)} \psi^{W,n}\{(i,j),(k,l)\} = 1\,. \qquad (5)$$

Now we are in a position to define a smoothing transformation $\mathbf{T}$

$$\mathbf{T}(i,j) = \sum_{(k,l)\Leftrightarrow(i,j)} \psi^{W,n}\{(i,j),(k,l)\} \cdot \mathbf{F}(k,l)\,, \qquad (6)$$

where (k,l) are points which are connected with (i,j) by self avoiding digital paths of length n included in W.

3   Results

The effectiveness of the new smoothing operator defined by (6) was tested on the *LENA* image contaminated by Gaussian noise of $\sigma = 30$ and on the same original image contaminated by 4% impulsive noise (salt & pepper added on each channel) mixed with Gaussian noise ($\sigma = 30$). The performance of the presented method was evaluated by means of the objective image quality measures RMSE, PSNR, NMSE and NCD [2].

Tables 2 and 3 show the obtained results for $n = 2$ and $n = 3$ in comparison with the standard noise reduction algorithms shown in Tab. 1. Additionally Figs. 3 and 4 show the comparison of the new filtering technique with the standard vector median.

Table 1. Filters taken for comparison with the proposed noise reduction technique.

Notation	METHOD	REF.
AMF	Arithmetic Mean Filter	[2]
VMF	Vector Median Filter	[3]
BVDF	Basic Vector Directional Filter	[4]
GVDF	Generalized Vector Directional Filter	[9]
DDF	Directional-Distance Filter	[10]
HDF	Hybrid Directional Filter	[11]
AHDF	Adaptive Hybrid Directional Filter	[11]
FVDF	Fuzzy Vector Directional Filter	[12]
ANNF	Adaptive Nearest Neighbor Filter	[13]
ANP-EF	Adaptive Non Parametric (Exponential) Filter	[14]
ANP-GF	Adaptive Non Parametric (Gaussian) Filter	[14]
ANP-DF	Adaptive Non Parametric (Directional) Filter	[14]
VBAMMF	Vector Bayesian Adaptive Median/Mean Filter	[14]

The smoothing operator $\mathbf{T}$ in (6) has to be applied in an iterative way. Starting with low value of β enables the smoothing of the image noise components. At each iteration steps the parameter β has been increased, like in simulated annealing (we have used $\beta(k) = \beta(k - 1) \cdot \delta$). For all experiments $\beta(0) = 10$, $\delta = 1.25$ were used and five iterations were performed.

For the calculation of the similarity function we used the L_1 metric and an exponential function, however we have obtained good results using other convex functions and different vector metrics like L_2 or L_∞.

The assumption that the random walk is self avoiding leads to several interesting features of the new smoothing technique. It enables the elimination of small image objects, which consist of fewer than n pixels, when n is the number of steps of a *SAW*. This feature makes the new filtering technique similar to the mathematical morphology filters. It does not have however the drawbacks of dilation or erosion as the larger objects are not changed and the connectivity of

Table 2. Comparison of the new algorithms with the standard techniques (Tab. 1) using the *LENA* standard image corrupted by Gaussian noise $\sigma = 30$. SAW-2, SAW-3 denote the self avoiding walk with 2 and 3 steps respectively. The subscripts denote the iteration number ($\beta_0 = 20, \delta = 1.25$).

METHOD iter. $[N]$	$NMSE[10^{-3}]$	$RMSE$	SNR [dB]	$PSNR$ [dB]	NCD $[10^{-4}]$
NONE	420.550	29.075	13.762	18.860	250.090
AMF_1	66.452	11.558	21.775	26.873	95.347
AMF_3	69.307	11.803	21.592	26.691	76.286
AMF_5	91.911	13.592	20.366	25.465	75.566
VMF_1	136.560	16.568	18.647	23.745	153.330
VMF_3	93.440	13.705	20.295	25.393	123.500
VMF_5	87.314	13.248	20.589	25.688	117.170
$BVDF_1$	289.620	24.128	15.382	20.480	143.470
$BVDF_3$	279.540	23.705	15.536	20.634	117.400
$BVDF_5$	281.120	23.772	15.511	20.610	114.290
$GVDF_1$	112.450	15.035	19.490	24.589	119.890
$GVDF_3$	76.988	12.440	21.136	26.234	89.846
$GVDF_5$	76.713	12.418	21.151	26.250	84.876
DDF_1	150.830	17.412	18.215	23.314	143.530
DDF_3	106.900	14.659	19.710	24.809	114.770
DDF_5	100.500	14.213	19.979	25.077	108.960
HDF_1	119.100	15.473	19.241	24.339	131.190
HDF_3	72.515	12.073	21.396	26.494	99.236
HDF_5	66.584	11.569	21.766	26.865	92.769
$AHDF_1$	105.480	14.561	19.768	24.867	129.710
$AHDF_3$	64.519	11.388	21.903	27.002	97.873
$AHDF_5$	60.166	10.997	22.206	27.305	91.369
$FVDF_1$	78.927	12.596	21.028	26.126	101.950
$FVDF_3$	57.466	10.748	22.406	27.504	77.111
$FVDF_5$	62.269	11.188	22.057	27.156	74.235
$ANNF_1$	86.497	13.186	20.630	25.729	107.130
$ANNF_3$	63.341	11.284	21.983	27.082	82.587
$ANNF_5$	66.054	11.523	21.801	26.900	78.677
$ANP\text{-}E_1$	66.082	11.525	21.799	26.898	95.237
$ANP\text{-}E_3$	60.396	11.018	22.190	27.288	76.896
$ANP\text{-}E_5$	73.416	12.148	21.342	26.441	75.456
$ANP\text{-}G_1$	66.095	11.526	21.798	26.897	95.244
$ANP\text{-}G_3$	60.443	11.023	22.187	27.285	76.890
$ANP\text{-}G_5$	73.497	12.155	21.337	26.436	75.458
$ANP\text{-}D_1$	81.306	12.784	20.899	25.997	104.980
$ANP\text{-}D_3$	58.389	10.834	22.337	27.435	78.486
$ANP\text{-}D_5$	63.136	11.265	21.997	27.096	75.442
$SAW\text{-}2_1$	**53.904**	**10.409**	**22.684**	**27.782**	**81.871**
$SAW\text{-}2_3$	**46.717**	**9.691**	**23.305**	**28.404**	**67.476**
$SAW\text{-}2_5$	**51.782**	**10.202**	**22.858**	**27.957**	**67.420**
$SAW\text{-}3_1$	**51.203**	**10.145**	**22.907**	**28.006**	**74.855**
$SAW\text{-}3_3$	**54.012**	**10.420**	**22.675**	**27.774**	**68.139**
$SAW\text{-}3_5$	**61.901**	**11.155**	**22.083**	**27.182**	**69.735**

objects is always preserved. Unlike in mathematical morphology no structuring element is needed. Instead, only a number of steps is given as input.

This efficiency of the new algorithm as compared with the vector median filter is shown in Fig. 3 and 4. After the application of the new filter impulse pixels introduced by noise process are removed, the contrast is improved, the image is smoothed and the edges are well preserved.

Table 3. Comparison of the new algorithms with the standard techniques (Tab. 1) using the *LENA* standard image corrupted by 4% Impulse and Gaussian noise $\sigma = 30$. The subscripts denote the iteration number. SAW-2, SAW-3 denote the self avoiding with 2 and 3 steps respectively ($\beta_0 = 20, \delta = 1.25$).

METHOD iter. $_{[N]}$	$NMSE[10^{-3}]$	$RMSE$	SNR [dB]	$PSNR$ [dB]	NCD $[10^{-4}]$
NONE	905.930	42.674	10.429	15.528	305.550
AMF_1	128.940	16.099	18.896	23.995	122.880
AMF_3	97.444	13.996	20.112	25.211	95.800
AMF_5	113.760	15.122	19.440	24.539	92.312
VMF_1	161.420	18.013	17.920	23.019	161.700
VMF_3	104.280	14.478	19.818	24.916	128.620
VMF_5	96.464	13.925	20.156	25.255	121.790
$BVDF_1$	354.450	26.692	14.504	19.603	152.490
$BVDF_3$	336.460	26.006	14.731	19.829	123.930
$BVDF_5$	338.940	26.102	14.699	19.797	118.500
$GVDF_1$	140.970	16.833	18.509	23.607	126.820
$GVDF_3$	93.444	13.705	20.294	25.393	94.627
$GVDF_5$	91.118	13.534	20.404	25.503	89.277
DDF_1	176.670	18.845	17.528	22.627	152.050
DDF_3	119.330	15.488	19.232	24.331	119.940
DDF_5	110.620	14.912	19.561	24.660	113.390
HDF_1	143.190	16.966	18.441	23.539	139.360
HDF_3	82.413	12.871	20.840	25.939	104.620
HDF_5	74.487	12.236	21.279	26.378	97.596
$AHDF_1$	132.710	16.333	18.771	23.869	138.180
$AHDF_3$	75.236	12.298	21.236	26.334	103.410
$AHDF_5$	68.563	11.740	21.639	26.738	96.327
$FVDF_1$	108.760	14.786	19.635	24.734	111.220
$FVDF_3$	73.796	12.179	21.320	26.418	83.629
$FVDF_5$	76.274	12.382	21.176	26.275	80.081
$ANNF_1$	110.720	14.919	19.558	24.656	113.560
$ANNF_3$	75.652	12.332	21.212	26.310	86.836
$ANNF_5$	76.757	12.421	21.149	26.247	82.825
$ANP\text{-}E_1$	128.590	16.077	18.908	24.007	122.890
$ANP\text{-}E_3$	90.509	13.488	20.433	25.532	97.621
$ANP\text{-}E_5$	96.930	13.959	20.135	25.234	94.131
$ANP\text{-}G_1$	128.600	16.078	18.908	24.006	122.900
$ANP\text{-}G_3$	90.523	13.489	20.432	25.531	97.603
$ANP\text{-}G_5$	96.990	13.963	20.133	25.231	94.134
$ANP\text{-}D_1$	113.900	15.131	19.435	24.533	115.230
$ANP\text{-}D_3$	74.203	12.213	21.296	26.394	85.026
$ANP\text{-}D_5$	76.265	12.381	21.177	26.275	81.202
SAW-2$_1$	**70.728**	**11.924**	**21.504**	**26.603**	**88.788**
SAW-2$_3$	**51.547**	**10.179**	**22.878**	**27.976**	**70.566**
SAW-2$_5$	**55.945**	**10.605**	**22.522**	**27.621**	**70.012**
SAW-3$_1$	**60.367**	**11.016**	**22.192**	**27.291**	**79.741**
SAW-3$_3$	**58.222**	**10.818**	**22.349**	**27.448**	**70.729**
SAW-3$_5$	**66.308**	**11.545**	**21.784**	**26.883**	**72.013**

Fig. 3. Comparison of the efficiency of the vector median with the proposed noise reduction technique. **a, d)** images contaminated by Gaussian Noise ($\sigma = 30$) mixed with impulsive salt & pepper noise (4% on each R,G,B channel), **g, j)** images corrupted by strong Gaussian noise ($\sigma = 100$), **b, e, h, k)** output of the vector median filter, **c, f, i, l)** results of the filtration with the new filter ($\beta_0 = 10, \delta = 1.25, n = 3$, five iterations).

a)   b)   c)

Fig. 4. Comparison of the efficiency of the vector median with the new filter proposed in this paper. **a)** test images (a part of a scanned map and of an old manuscript), **b)** results of the standard vector median filtration (3×3 mask, 5 iterations), **c)** results of the filtration with the new filter ($\beta_0 = 20, \delta = 1.25, n = 2$, three iterations)

4   Conclusions

In this paper, a new filter for noise reduction in color images has been presented. Experimental results indicate that the new filtering technique outperforms standard procedures used to reduce mixed impulsive and Gaussian noise in color images. The efficiency of the new filtering technique is shown in Tables 2 and 3 and in Figs. 3,4.

References

1. Pitas, I., Venetsanopoulos, A. N.: Nonlinear Digital Filters : Principles and Applications. Kluwer Academic Publishers, Boston, MA (1990)
2. Plataniotis, K.N., Venetsanopoulos, A.N.: Color Image Processing and Applications. Springer Verlag (June 2000)
3. Astola, J., Haavisto, P., Neuovo, Y.: Vector median filters. IEEE Proceedings **78** (1990) 678-689
4. Trahanias, P.E., Venetsanopoulos, A.N.: Vector directional filters: A new class of multichannel image processing filters. IEEE Trans. on Image Processing. **2**(4) (1993) 528-534
5. Borgefors, G.: Distances transformations in digital images. Computer Vision, Graphics and Image Processing **34** (1986) 334-371
6. Toivanen, P.J.: New geodesic distance transforms for gray scale images. Pattern Recognition Letters **17** (1996) 437-450
7. Cuisenaire, O.: Distance transformations: fast algorithms and applications to medical image processing. PhD Thesis, Universite Catholique de Louvain (Oct. 1999)
8. Madras, N., Slade, G.: The Self-Avoiding Walk. Birkhauser, Boston (1993)
9. Trahanias, P.E., Karakos, D., Venetsanopoulos, A.N.: Directional processing of color images : theory and experimental results. IEEE Trans. on Image Processing **5**(6) (1996) 868-880
10. Karakos, D., Trahanias, P.E.: Generalized multichannel image filtering structures. IEEE Trans. on Image Processing **6**(7) (1997) 1038-1045
11. Gabbouj, M., Cheickh, F.A.: Vector median - vector directional hybrid filter for colour image restoration. Proceedings of EUSIPCO (1996) 879-881
12. Plataniotis, K.N., Androutsos, D., Venetsanopoulos, A.N.V.: Fuzzy adaptive filters for multichannel image processing Signal Processing Journal **55**(1) (1996) 93-106
13. Plataniotis, K.N., Androutsos, D., Sri, V., Venetsanopoulos, A.N.V.: A nearest neighbor multichannel filter. Electronic Letters (1995) 1910-1911
14. Plataniotis, K.N., Androutsos, D., Vinayagamoorthy, S., Venetsanopoulos, A.N.V.: Color image processing using adaptive multichannel filters. IEEE Trans. on Image Processing **6**(7) (1997) 933-950
15. Smolka, B., Wojciechowski, K.: Random walk approach to image enhancement. (to appear in:) Signal Processing, Vol. 81. No. 4 (2001)

Wigner Distributions and Ambiguity Functions in Image Analysis

Stefan L. Hahn and Kajetana M. Snopek

Institute of Radioelectronics, Warsaw University of Technology, Nowowiejska 15/19,
00-665 Warsaw, Poland
{hahn, snopek}@ire.pw.edu.pl

Abstract. The Wigner distribution of a two-dimensional image function has the form of a four-dimensional Fourier transform of a correlation product $r(x_1, x_2, \chi_1, \chi_2)$ with respect to the spatial-shift variables χ_1 and χ_2. The corresponding ambiguity function has the form of the inverse Fourier transform of $r(x_1, x_2, \chi_1, \chi_2)$ with respect to spatial variables x_1 and x_2. There exist dual definitions in the frequency domain (f_1, f_2, μ_1, μ_2), where μ_1, μ_2 are frequency-shift variables. The paper presents the properties of these distributions and describes applications for image analysis.

Keywords: image analysis, Wigner distributions

1 Introduction

Images are described by two-dimensional functions $u(x)$, where $x = (x_1, x_2)$ are Cartesian spatial coordinates of a plane. In several applications it is convenient to replace the real function $u(x)$ with the corresponding analytic signal $\psi(x)$ [9], [10]. For example, in an earlier paper the authors described the decomposition of images into amplitude and phase patterns in the spatial domain [13]. In image analysis spectral methods are frequently applied. Usually the image function is truncated by a rectangular boundary $\Pi(x)$. Therefore the spectrum of such a signal is given by the Fourier integral

$$\Gamma(f) = \int \Pi(x) u(x) e^{-j2\pi(f_1 x_1 + f_2 x_2)} dx_1 dx_2, \, x = (x_1, x_2), f = (f_1, f_2). \tag{1}$$

This spectrum yields no information about the local spectral properties of the image function. For this reason four-dimensional distributions $C(x, f)$ are applied to get information about local spectral properties of image functions. The above distributions are four-dimensional extensions of well-known time-frequency distributions. Many distributions are members of the Cohen's class [6]. The extension of the Cohen's class for 2-dimensional signals is given by the integral

W. Skarbek (Ed.): CAIP 2001, LNCS 2124, pp. 537–546, 2001.

$$C(x,f) = \qquad \phi_{fx}(\mu,\chi)\psi(u+\chi/2)\psi^*(u-\chi/2)e^{j2\pi\,\mu_1(u_1-x_1)+\mu_2(u_2-x_2)} \qquad (2)$$

$$e^{-j2\pi(f_1\chi_1+f_2\chi_2)}du_1du_2d\chi_1d\chi_2d\mu_1d\mu_2,$$

where $\chi=(\chi_1,\chi_2)$, $\mu=(\mu_1,\mu_2)$. Details are described in the Section 2 of this paper. All members of the Cohen's class are bilinear transformations producing unwanted cross-terms, which make the analysis difficult. Several distributions with attenuated or reduced cross-terms have been described [3], [7], [14], including so-called pseudo-Wigner distributions [4], [11]. The 4-D extension of pseudo-Wigner distributions [16] is described in the Section 5.

2 Selected Properties of the Cohen's Class Distributions of 2-D Signals

The integration of the six-fold integral (2) with respect to the variable $u=(u_1,u_2)$ yields the following definition of the Cohen's class distribution of 2-D signals

$$C(x,f) = \qquad \phi_{fx}(\mu,\chi)A(\mu,\chi)e^{-j2\pi(f_1\chi_1+f_2\chi_2+\mu_1x_1+\mu_2x_2)}d\mu_1d\mu_2d\chi_1d\chi_2, \qquad (3)$$

where

$$A(\mu,\chi) = \qquad \psi(x+\chi/2)\,\psi^*(x-\chi/2)e^{j2\pi(\mu_1x_1+\mu_2x_2)}dx_1dx_2, \qquad (4)$$

is the 4-D extension of the Woodward ambiguity function (AF) [19] (Remark: We use the so-called symmetrical form of the ambiguity function [7]), and $\phi_{fx}(\mu,\chi)$ is a 4-dimensional kernel function defining a particular distribution. The Eq. (3) may be rewritten in next three equivalent forms with different kernels related by Fourier transforms. In this paper we use the notation of variables presented in Table 1.

Table 1. The notation of variables in the spatial- and frequency-domains.

Spatial domain	Spatial-shift	Frequency domain	Frequency-shift
$x=(x_1, x_2)$	$\chi=(\chi_1, \chi_2)$	$f=(f_1, f_2)$	$\mu=(\mu_1, \mu_2)$

Following the classification proposed by Flandrin [7] for time-frequency kernels ("frequency-time", "time-time", "time-frequency" and "frequency-frequency"), we proposed to apply the following notations of four 2-D kernels: ϕ_{ff}, ϕ_{tt}, ϕ_{tf} and ϕ_{ff}. Let us mention that in existing literature about 2-D distributions of 1-D signals, e.g. [3], [7], these kernels are described by four different Greek letters, usually another set by particular authors. The extension of this notation for 4-D kernels is presented in Table 2.

Table 2. The 4-D kernels of the Cohen's class.

Frequency-spatial	Spatial-spatial	Spatial-frequency	Frequency-frequency
$\phi_{fx}(\mu, \chi)$	$\phi_{xx}(x, \chi)$	$\phi_{xf}(x, f)$	$\phi_{ff}(\mu, f)$

The kernel ϕ_{fx} is used in the first form denoted *"frequency-spatial"* (see Eq. (3)). The next three forms are denoted:

"Spatial-spatial" form

$$C(x, f) = \qquad \phi_{xx}\left(x' - x, \chi\right)\psi\left(x' + \chi/2\right)\psi^*\left(x' - \chi/2\right) \tag{5}$$
$$e^{-j2\pi(f_1\chi_1 + f_2\chi_2)} dx_1' dx_2' d\chi_1 d\chi_2,$$

"Spatial-frequency" form

$$C(x, f) = \qquad \phi_{xf}\left(x' - x, \mu - f\right) W\left(x', f\right) dx_1' dx_2' d\mu_1 d\mu_2, \tag{6}$$

"Frequency-frequency" form

$$C(x, f) = \qquad \phi_{ff}\left(\mu, f' - f\right)\Gamma\left(f' + \mu/2\right)\Gamma^*\left(f' - \mu/2\right) \tag{7}$$
$$e^{j2\pi(\mu_1 x_1 + \mu_2 x_2)} d\mu_1 d\mu_2 df_1' df_2'$$

The graphical representation of the above 4-D distributions using the three-dimensional coordinates system, for example $\{x_1, x_2, C(x, f)\}$ is possible using the notion of a cross-section. Let us define the following selected cross-sections

$$C(x_{10}, x_{20}, f_1, f_2) \text{ with fixed variables } x_1 = x_{10}, x_2 = x_{20}, \tag{8}$$

$$C(x_1, x_2, f_{10}, f_{20}) \text{ with fixed variables } f_1 = f_{10}, f_2 = f_{20}, \tag{9}$$

$$C(x_{10}, x_2, f_{10}, f_2) \text{ with fixed variables } x_1 = x_{10}, f_1 = f_{10}, \tag{10}$$

$$C(x_1, x_{20}, f_1, f_{20}) \text{ with fixed variables } x_2 = x_{20}, f_2 = f_{20}. \tag{11}$$

3   The 4-D Wigner and Ambiguity Functions

The 4-D Wigner distribution (WD) [18] of a complex signal $\psi(x) \overset{2F}{\Gamma} (f)$ is defined by the Fourier transform of the 4-D signal-domain auto-correlation function

$$r(x, \chi) = \psi(x_1 + \chi_1/2, x_2 + \chi_2/2) \, \psi^*(x_1 - \chi_1/2, x_2 - \chi_2/2), \tag{12}$$

with respect to the shift variables, that is,

$$W(x,f) = \quad r(x,\chi) e^{-j2\pi(f_1\chi_1 + f_2\chi_2)} d\chi_1 d\chi_2 , \tag{13}$$

or alternatively by the inverse Fourier transform of the 4-D frequency-domain auto-correlation function

$$R(\mu,f) = \Gamma\left(f_1 + \mu_1/2, f_2 + \mu_2/2\right) \Gamma^*\left(f_1 - \mu_1/2, f_2 - \mu_2/2\right), \tag{14}$$

with respect to the shift variables, that is,

$$W(x,f) = \quad R(\mu,f) e^{j2\pi(\mu_1 x_1 + \mu_2 x_2)} d\mu_1 d\mu_2 . \tag{15}$$

Notice, that r and R form a pair of Fourier transforms. In consequence, (13) and (15) define *exactly the same* function. The kernels of the Wigner distribution using the notations of Table 2 are written in Table 3.

Table 3. The kernels of the 4-D Wigner distribution

$\phi_{fx}(\mu,\chi)$				$\phi_{xx}(x,\chi)$		$\phi_{xf}(x,f)$		$\phi_{ff}(\mu,f)$		
1_{μ_1}	1_{μ_2}	1_{χ_1}	1_{χ_2}	$\delta(x_1)$	$\delta(x_2)$	$\delta(x_1)$	$\delta(x_2)$	1_{μ_1}	1_{μ_2}	$\delta(f_1)$
				1_{χ_1}	1_{χ_2}	$\delta(f_1)$	$\delta(f_2)$	$\delta(f_2)$		

$1_{\mu_1}, 1_{\mu_2}, 1_{\chi_1}, 1_{\chi_2}$ are unit distributions equal 1 for all values of the corresponding variable. We applied the symbol   of a tensor product of distributions. Usually authors in notations delete the unit distribution and the symbol   . However, our notation displays better the Fourier relations between kernels. For example: $F_{\tau\ f}(1_{\tau}\ 1_{t} \neq \delta(f)\ 1_{t}$. The 4-D ambiguity function [15] is defined by the inverse Fourier transform of $r(x,\chi)$ given by Eq.(12) with respect to spatial variables, differently to the use in the Eq.(13) of shift variables, that is,

$$A(\mu,\chi) = \quad r(x,\chi) e^{j2\pi(\mu_1 x_1 + \mu_2 x_2)} dx_1 dx_2 \tag{16}$$

or alternatively by the Fourier transform of $R(\mu,f)$ given by the Eq.(14) with respect to the frequency variables, that is,

$$A(\mu,\chi) = \quad R(\mu,f) e^{-j2\pi(\chi_1 f_1 + \chi_2 f_2)} df_1 df_2 . \tag{17}$$

Again, (16) and (17) define *exactly the same* function. The Wigner distribution $W(x,f)$ and the ambiguity function form a pair of 4-D Fourier transforms [5], [17]

$$A(\mu,\chi) = \quad W(x,f) e^{-j2\pi(\chi_1 f_1 + \mu_1 x_1 + \chi_2 f_2 + \mu_2 x_2)} dx_1 dx_2 df_1 df_2 . \tag{18}$$

$$W(x,f) = \quad A(\mu,\chi)\, e^{j2\pi\left(\chi_1 f_1 + \mu_1 x_1 + \chi_2 f_2 + \mu_2 x_2\right)} d\chi_1 d\chi_2 d\mu_1 d\mu_2. \tag{19}$$

Notice that the Wigner distribution is a real function for any signal, real or complex. Differently, the ambiguity function may be complex. The ambiguity function is real only if $W(x,f) = W_{eeee}(x,f)$ or $W(x,f) = W_{oooo}(x,f)$. In a special case, that the Wigner distribution is "causal", that is, has the support limited to the orthant $^+$ ($x_1 > 0$, $x_2 > 0$, $f_1 > 0$, $f_2 > 0$), the ambiguity function is a 4-D analytic function [9].

4   The Cross-Terms of 4-D Wigner Distribution

Consider a sum of N signals

$$\psi_N(x) = \sum_{i=1}^{N} \psi_i(x). \tag{20}$$

The correlation product defined by the Eq. (12) takes the form

$$r(x,\chi) = \sum_{i=1}^{N} \psi_i(x^+) \sum_{j=1}^{N} \psi_j(x^-), \tag{21}$$

where $x^+ = (x_1 + \chi_1/2, x_2 + \chi_2/2)$ and $x^- = (x_1 - \chi_1/2, x_2 - \chi_2/2)$.   The corresponding WD is given by

$$W_N(x,f) = \underbrace{\sum_{i=1}^{N} W_i(x,f)}_{\text{auto-terms}} + 2 \underbrace{\sum_{i=1}^{N-1} \sum_{j=i+1}^{N} W_{ij}(x,f)}_{\text{cross-terms}}. \tag{22}$$

The number of cross-terms equals $N(N-1)/2$. In order to reduce the number of cross-terms of 2-D WD's of 1-D signals, it is favourable to apply analytic signals [2], [7]. This property can be extended for 4-D WD's of 2-D analytic signals with single quadrant spectra [9], [10]. However, the spectral information is contained in four quadrants of the Fourier frequency plane. Notice that due to Hermitian symmetry, the spectral information in a half-plane is redundant. Since the spectral information contained in two quadrants may differ, in general we have to define two different analytic signals with single-quadrant spectra [10]. Using these signals we avoid cross-terms due to the interaction of signals defined by the inverse Fourier transform of single-quadrant spectra from all four quadrants. For example, a real signal may be written in the form

$$u(x) = \frac{\psi_1(x) + \psi_2(x) + \psi_3(x) + \psi_4(x)}{4}. \tag{23}$$

The number of cross-terms equal here $N(N-1)/2=6$, while using only two signals given by the formula

$$u(x) = \frac{\psi_1(x) + \psi_1^*(x) + \psi_3(x) + \psi_3^*(x)}{4}, \tag{24}$$

we get only one single cross-term.

5   Pseudo-Wigner Distribution of 2-D Signals (Images)

Claasen and Mecklenbräuker introduced the concept of the time-domain pseudo-Wigner distribution [4]. The extension of this notion for 2-D signals, called *spatial-domain pseudo-Wigner distribution* is given by the integral

$$\mathrm{PW}_x(x,f) = \quad h(\chi)\, r(x,\chi) e^{-j2\pi(f_1\chi_1 + f_2\chi_2)} d\chi_1 d\chi_2, \tag{25}$$

where $h(\chi)$ is a 2-D window function, which in general may be non-separable. Let us have an example of a separable window in the form of a product of Gaussian functions: $h(\chi_1, \chi_2) = e^{-a_1\chi_1^2} e^{-a_2\chi_2^2}$. Any other products of well-known windows (cosine-roll-off, Hanning, Hamming etc.) can be applied. Let us introduce the notion of a *frequency-domain pseudo-Wigner distribution* [11], [16] given by the integral

$$\mathrm{PW}_f(x,f) = \quad G(\mu)\, R(\mu,f) e^{j2\pi(\mu_1 x_1 + \mu_2 x_2)} d\mu_1 d\mu_2, \tag{26}$$

where $G(\mu)$ is a frequency-domain window, for example $G(\mu_1, \mu_2) = e^{-b_1\mu_1^2} e^{-b_2\mu_2^2}$. Notice that differently to the spatial- and frequency-domain WD's given by the Eqs. (13) and (15) which represent exactly the same function, the pseudo-Wigner distributions (25) and (26) *are different*. As in the case of 2-D PWD's of 1-D signals, the 4-D PW_x and PW_f serve to reduce or cancel the unwanted cross-terms of the WD. However, each one cancels another set of cross-terms. The PWD's defined above can be written in the form of the convolutions with the WD. Consider the window functions and their Fourier transforms

$$h(x) \overset{2F}{} H(f);\ G(f) \overset{2F^{-1}}{} g(x). \tag{27}$$

The convolution form of the spatial-domain PWD is

$$\mathrm{PW}_x(x,f) = \quad H(f-\mu)\, \mathrm{W}_\psi(x,\mu)\, d\mu_1 d\mu_2 = H(f) * \mathrm{W}_\psi(x,f). \tag{28}$$

Similarly, the frequency-domain PWD is

$$\mathrm{PW}_f(x,f) = \int g(x-\chi)\, \mathrm{W}_\psi(\chi,f)\, \mathrm{d}\chi_1 \mathrm{d}\chi_2 = g(x)*\mathrm{W}_\psi(x,f). \tag{29}$$

Since the spatial- and frequency-domain distributions cancel each another set of cross-terms, it is possible to derive a distribution with a four-dimensional separable window. Let us insert into the Eq. (28) instead of W_ψ the PW_f given by the Eq. (29). We get

$$\mathrm{PW}_{xf}(x,f) = \int H(f-\mu) \int g(x-\chi)\mathrm{W}_\psi(\chi,\mu)\, \mathrm{d}\chi_1 \mathrm{d}\chi_2\ \mathrm{d}\mu_1 \mathrm{d}\mu_2. \tag{30}$$

The same result can be obtained inserting into the Eq. (29) instead of W_ψ the PW_x given by the Eq. (28). Let us call the distribution defined by the Eq. (30) a 4-D *dual-dual window pseudo-Wigner distribution*. Notice that the 2-D version is known as the *smoothed pseudo-Wigner distribution* [7] or a *dual-window PWD* [11], [16]. However, to our knowledge, it has never been derived in the manner above presented.

6   Applications in Image Analysis

In the past the Wigner distribution has been applied in image analysis. For example [1] describes applications of the 4-D WD of real signals using optical methods. It seems, that the authors of the paper [8] describing applications in the analysis of ocean surfaces rediscovered the 4-D WD. As well, no reference is given to the use of analytic signals. We studied some properties of the 4-D WD using following test images:

1. An image produced by a Gaussian crater modulated by a 2-D harmonic signal (see Fig.1a) and a 2-D random field (see Fig.1b). Next figures show that the cross-sections of the WD of the form $\mathrm{W}(x_1, x_2, f_{10}, f_{20})$ (see Eq. (9)) of the sum of above signals may serve to extract the crater embedded in the random field (see Fig. 1c). Fig.1d shows the cross-section with $f_{10}=f_{20}=0$. The carter is not visible. Differently, at the cross-section at the point $f_{10}=f_{20}=2.5$ the crater is dominant. For convenience, Fig.1f shows the cross-section of the WD of the crater (no random field).

Gaussian crater

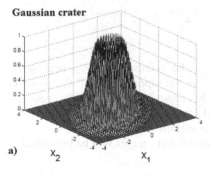

a) x_2 x_1

Random field

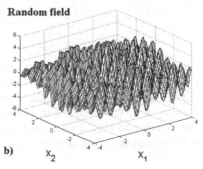

b) x_2 x_1

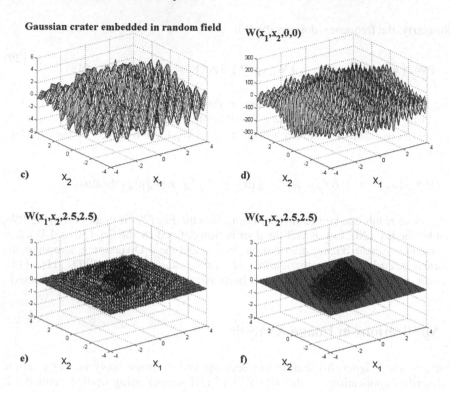

Fig. 1. a) A Gaussian crater modulated with a 2-D harmonic signal, b) a 2-D random field, c) the Gaussian crater embedded in the random field, d) the cross-section of the 4-D WD of the crater embedded in the random field, $f_{10}=f_{20}=0$, e) the cross-section of the 4-D WD of the crater embedded in the random field, $f_{10}=f_{20}=2.5$, f) the cross-section of the 4-D WD of the crater (no random field), $f_{10}=f_{20}=2.5$.

Fig.1e shows the cross-section with the crater extracted from the random field (compare with Fig.1f).

2. We studied the cross-sections of the 4-D WD of a signal in the form of a product of two Gaussian signals

$$\psi(x_1,x_2) = \tag{31}$$

$$e^{-\pi(x_1-a_{11})^2}e^{j2\pi f_{11}x_1} + e^{-\pi(x_1-a_{12})^2}e^{j2\pi f_{12}x_1} \quad e^{-\pi(x_2-a_{21})^2}e^{j2\pi f_{21}x_2} + e^{-\pi(x_2-a_{22})^2}e^{j2\pi f_{22}x_2} \quad .$$

Remarkably, we found that a proper choice of the points f_{10} and f_{20} may serve to cancel the cross-terms in complete analogy to the application of the windowed pseudo-Wigner distributions. Fig.2a shows the cross-section with all cross-terms. Fig.2b shows a cross-section with eliminated cross-terms along the x_1-axis, Fig.2c along the x_2-axis and Fig.2d with all cross-terms eliminated.

3. At the conference cross-sections of selected natural images will be presented.

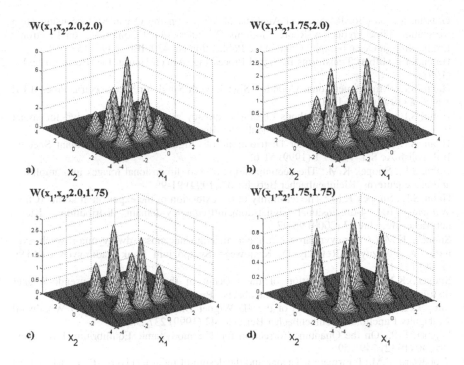

Fig. 2. The selected cross-sections of the WD of the signal given by the Eq. (31)

Acknowledgment. The research has been supported by the grant No. 8T11D 018 16 of the Committee of Scientific Research of Poland

References

1. Bamler, R., Glünder, H.: The Wigner distribution function of two-dimensional signals. Coherent-optical generation and display. Optica Acta. Vol.30 **12** (1983) 1789-1803
2. Boashash, B.: Note on the Use of the Wigner Distribution for Time-Frequency Signal Analysis. IEEE Trans. Acoust. Speech and Signal Processing. Vol.36 **9** (1988) 1518-1521
3. Boudreaux-Bartels, G.F.: Mixed Time-Frequency Signal Transformations. The Transforms and Applications Handbook. CRC Press, Inc., Boca Raton, Fl. (1996)
4. Claasen, T.C.A.M., Mecklenbräuker, W.F.G.: The Wigner Distribution-a Tool for Time-Frequency Signal Analysis, Part I – Continuous-Time Signals. Philips J. Res. Vol.35 (1980) 217-250
5. Claasen, T.C.A.M., Mecklenbräuker, W.F.G.: The Wigner Distribution-a Tool for Time-Frequency Signal Analysis, Part III – Relations with other Time-Frequency Signal Transformations. Philips J. Res. Vol.35 (1980) 372-389
6. Cohen, L.: Time-Frequency distributions – A Review. Proc. IEEE Vol. 77 **7** (1989) 941-981
7. Flandrin, P.: Time-frequency/time-scale analysis. Academic Press (1999)

8. Grassin, S., Garello, R.: Spectral Analysis of Waves on the Ocean Surface using High Resolution Time Frequency Representations. 7th International Conference on Electronic Engineering in Oceanography Conference Publication No.439, IEE (1997) 153-159
9. Hahn, S.L.: Hilbert Transforms in Signal Processing. Artech House, Inc. Boston, London (1997)
10. Hahn, S.L.: Multidimensional Complex Signals with Single-orthant Spectra. Proc. IEEE Vol.80 **8** (1992) 1287-1300
11. Hahn, S.L.: A review of methods of time-frequency analysis with extension for signal plane-frequency plane analysis. Kleinheubacher Berichte **44** (2001) (in print)
12. Hahn, S.L.: The Szilard-Wigner Distributions of Signals with Single-Orthant Spectra. Bull.Polish Ac.Sci. Vol.47 **1** (1999) 51-65
13. Hahn, S.L., Snopek K.M.: The decomposition of two-dimensional images into amplitude and phase patterns. Kleinheubacher Berichte **37** (1994) 91-99
14. Hahn, S.L., Snopek K.M.: The feasibility of the extension of the exponential kernel (Choi-Williams) f04 4-dimensional signal-domain/frequency-domain distributions. Kleinheubacher Berichte **43** (2000) 90-97
15. Snopek, K.M.: Czterowymiarowa funkcja niejednoznaczności dwuwymia-rowych sygnałów analitycznych. Referaty IX Krajowego Sympozjum Nauk Radiowych URSI'99 (1999) 125-130
16. Snopek, K.M.: The Application of the concept of the dual-window pseudo-Wigner distribution in 4-D distributions. Kleinheubacher Berichte **44** (2001) (in print)
17. Snopek, K.M.: The Comparison of the 4D Wigner Distributions and the 4D Woodward Ambiguity Functions. Kleinheubacher Berichte **42** (1999) 237-246
18. Wigner, E.P.: On the Quantum Correction for Thermodynamic Equilibrium. Phys. Rev. Vol. 40 (1932) 749-759
19. Woodward, P.M.: Information Theory and the design of radar receivers. Proc. IRE Vol.39 (1951) 1521-1524

A Markov Random Field Image Segmentation Model Using Combined Color and Texture Features

Zoltan Kato[1] and Ting-Chuen Pong[2]

[1] National University of Singapore, School of Computing, 3 Science Drive 2,
Singapore 117543, Tel: +65 874 8923, Fax: +65 779 4580, kato@comp.nus.edu.sg
[2] Hong Kong University of Science and Technology, Computer Science Dept., Clear
Water Bay, Kowloon, Hong Kong, China. Tel: +852 2358 7000, Fax: +852 2358 1477,
tcpong@cs.ust.hk

Abstract. In this paper, we propose a Markov random field (MRF) image segmentation model which aims at combining color and texture features. The theoretical framework relies on Bayesian estimation associated with combinatorial optimization (Simulated Annealing). The segmentation is obtained by classifying the pixels into different pixel classes. These classes are represented by multi-variate Gaussian distributions. Thus, the only hypothesis about the nature of the features is that an additive white noise model is suitable to describe the feature values belonging to a given class. Herein, we use the perceptually uniform CIE-$L^*u^*v^*$ color values as color features and a set of Gabor filters as texture features. We provide experimental results that illustrate the performance of our method on both synthetic and natural color images. Due to the local nature of our MRF model, the algorithm can be highly parallelized.

Keywords: image segmentation, Markov random field model

1 Introduction

Image segmentation is an important early vision task where pixels with similar features are grouped into homogeneous regions. There are many features that one can take into account during the segmentation process: gray-level, color, motion, different texture features, etc. However, most of the segmentation algorithms presented in the literature are based on only one of the above features. Recently, the segmentation of color images (textured or not) received more attention [3,6,7]. In this paper, we are interested in the segmentation of color textured images. This problem has been addressed in [7], where an unsupervised segmentation algorithm is proposed which uses Gaussian MRF models for *color textures*. These models are defined in each color plane with interactions between different color planes. The segmentation algorithm is based on agglomerative hierarchical clustering. Our approach is different in two major points. First, we use a stochastic model based segmentation framework instead of clustering. Second, we use a combination of classical, gray-level based, texture features and

W. Skarbek (Ed.): CAIP 2001, LNCS 2124, pp. 547–554, 2001.

color instead of a direct modelization of color textures. The segmentation model consists of a MRF defined over a nearest neighborhood system. The image features are represented by multi-variate Gaussian distributions (basically a noise model). Since the different texture-types are described by a set of Gaussian parameters, it is possible to classify or recognize textures based on a prior learning of the possible parameters. Of course, parameter estimation can be a difficult task, if we do not have training data. Herein, we do not address this problem but we note that the estimation task can be solved using an adaptive segmentation technique [5,10].

We use the perceptually uniform CIE-L*u*v* color values and texture features derived from the Gabor filtered gray-level image. Of course, the nature of the texture features is not crucial to the algorithm from the segmentation point of view. The only requirement in the current model is that an additive white noise model should be suitable to describe the texture features. Most of the filtering approaches (see [8] for a comparative study of different filtering techniques) fall into this category but stochastic texture models (such as Gaussian Markov random fields [7,10]) are also suitable for our segmentation model. Herein, we use a real-valued even-symmetric Gabor filter bank. Segmentation requires simultaneous measurements in both spatial and frequency domain. However, spatial localization of boundaries requires larger bandwidths whereas smaller bandwidths give better texture measurements. Fortunately, Gabor filters have optimal joint localization in both domains [4]. In addition, when we are combining texture features with color, the spatial resolution is considerably increased. We have tested the algorithm on a set of both natural and synthetic color textured images.

2   MRF Segmentation Model

We assume that images are defined over a finite lattice $S = \{s_1, s_2, \ldots, s_N\}$, where $s = (i, j)$ denotes lattice sites (or pixels). For each pixel s, the region-type (or pixel class) that the pixel belongs to is specified by a class label, ω_s, which is modeled as a discrete random variable taking values in $\Lambda = \{1, 2, \ldots, L\}$. The set of these labels $\omega = \{\omega_s, s \in S\}$ is a random field, called the *label process*. Furthermore, the observed image features (color and texture) are supposed to be a realization $\mathcal{F} = \{f_s | s \in S\}$ from another random field, which is a function of the label process ω. Basically, the *image process* $\mathcal{F}$ represents the deviation from the underlying label process. Thus, the overall segmentation model is composed of the hidden label process ω and the observable noisy image process $\mathcal{F}$. Our goal is to find an optimal labeling $\hat{\omega}$ which maximizes the a posteriori probability $P(\omega \mid \mathcal{F})$, that is the *maximum a posteriori* (MAP) estimate [2]: $\arg\max_{\omega \in \Omega} \prod_{s \in S} P(f_s \mid \omega_s))P(\omega)$, where Ω denotes the set of all possible labelings.

Herein, the image process is modeled by an additive white noise. Thus, we suppose that $P(f_s \mid \omega_s)$ follows a Gaussian distribution and pixel classes $\lambda \in \Lambda = \{1, 2, \ldots, L\}$ are represented by the mean vectors μ_λ and the covariance matrices Σ_λ. Furthermore, ω is supposed to be a MRF with respect to a first order neighborhood system. Thus, according to the *Hammersley-Clifford theorem* [2], $P(\omega)$ follows a Gibbs distribution:

$$P(\omega) = \frac{\exp(-U(\omega))}{Z(\beta)} = \frac{\prod_{C \in \mathcal{C}} \exp(-V_C(\omega_C))}{Z(\beta)} \, , \tag{1}$$

where $U(\omega)$ is called an *energy function*, $Z(\beta) = \sum_{\omega \in \Omega} \exp(-U(\omega))$ is the normalizing constant (or *partition function*) and V_C denotes the *clique potential* of clique $C \in \mathcal{C}$ having the label configuration ω_C. $\mathcal{C}$ is the set of spatial second order cliques (ie. *doubletons*). Note that the energies of *singletons* (ie. pixel sites $s \in \mathcal{S}$) directly reflect the probabilistic modeling of labels without context, while doubleton clique potentials express relationship between neighboring pixel labels. In our model, these potentials favor similar classes in neighboring pixels. Thus the energy function of the so defined MRF image segmentation model has the following form:

$$U(\omega, \mathcal{F}) = \sum_{s \in \mathcal{S}} \left(\ln(\sqrt{(2\pi)^3 \mid \Sigma_{\omega_s} \mid}) + \frac{1}{2}(\boldsymbol{f}_s - \boldsymbol{\mu}_{\omega_s}) \Sigma_{\omega_s}^{-1} (\boldsymbol{f}_s - \boldsymbol{\mu}_{\omega_s})^T \right)$$

$$+ \beta \sum_{\{s,r\} \in \mathcal{C}} \delta(\omega_s, \omega_r), \tag{2}$$

where $\delta(\omega_s, \omega_r) = 1$ if ω_s and ω_r are different and 0 otherwise. $\beta > 0$ is a parameter controlling the homogeneity of the regions. As β increases, the resulting regions become more homogeneous. Now, the segmentation problem is reduced to the minimization of the above energy function. Since it is a nonconvex function, some combinatorial optimization technique is needed to tackle the problem. Our experiments used Simulated Annealing (Gibbs sampler [2]) and Iterated Conditional Modes (ICM) [1].

2.1   Color Features

The first question, when dealing with color images, is how to measure quantitatively color difference between any two arbitrary colors. Experimental evidence suggests that the RGB tristimulus color space may be considered as a Riemannian space [9]. Due to the complexity of determining color distance in such spaces, several simple formulas have been proposed. These formulas approximate the Riemannian space by a Euclidean color space yielding a perceptually uniform spacing of colors. One of these formulas is the L*u*v* [9] color space that we use herein.

2.2   Texture Features

Many different techniques have been proposed for analyzing image texture. Herein, we will focus on the multi-channel filtering approach where the channels are represented by a bank of real-valued, even-symmetric Gabor filters. The

Fourier domain representation of the basic even-symmetric Gabor filter oriented at 0° is given by [4]:

$$H(u,v) = A \left(\exp \left(-\frac{1}{2} \left(\frac{(u-u_0)^2}{\sigma_u^2} + \frac{v^2}{\sigma_v^2} \right) \right) + \exp \left(-\frac{1}{2} \left(\frac{(u+u_0)^2}{\sigma_u^2} + \frac{v^2}{\sigma_v^2} \right) \right) \right),$$
(3)

where $\sigma_u = 1/2\pi\sigma_x$, $\sigma_v = 1/2\pi\sigma_y$, $A = 2\sigma_x\sigma_y$, u_0 is the frequency of a sinusoidal plane wave along the x-axis, and σ_x and σ_y are the deviations of the Gaussian envelope along the x and y axes. Filters with other orientations can be obtained by rotating the coordinate system. In our tests, we used four orientations: $0^o, 45^o, 90^o, 135^o$ and the radial frequencies u_0 are 1 octave apart: $\sqrt{2}, \sqrt{2}/2, \sqrt{2}/4, \sqrt{2}/8, \ldots$ For an image with a width of 2^W pixels, the highest radial frequency falling inside the image array is $\sqrt{2}/2^{W-2}$. From each filtered image g, we compute a *feature image* using the nonlinear transformation $|\tanh(\alpha g_s)|, s \in \mathcal{S}$; followed by a Gaussian blurring with a deviation proportional to the center frequency of the Gabor filter: $\sigma = k/u_0$. In our experiments, the Gabor filtered images are scaled to the interval $[-1, 1]$ and we set $\alpha = 40$ and $k = 1$.

Table 1. Computing times and segmentation error on the synthetic images.

Image:	Fig. 2				Fig. 3			
feature	Gibbs	error	ICM	error	Gibbs	error	ICM	error
texture	185 sec.	13.5%	3.4 sec.	16.6%	1349 sec.	19.0%	13 sec.	20.6%
color	105 sec.	2.5%	5.5 sec.	8.8%	732 sec.	19.7%	20 sec.	23.1%
combined	319 sec.	1.2%	16 sec.	2.7%	2581 sec.	8.9%	73 sec.	11.4%

3   Experimental Results

The proposed algorithm has been tested on a variety of color images including synthetic images (Fig. 1, Fig. 2, Fig. 3), outdoor (Fig. 4) and indoor (Fig. 5) scenes. We have used MIT's VisTex database to compose the synthetic color textured images. The test program has been implemented in C and run on a UltraSparc1 workstation. Herein, we present a few examples of these results and compare segmentation results using color only, texture only and combined features. Furthermore, we also compare the results obtained via a deterministic (ICM [1]) and stochastic (Simulated Annealing using the Gibbs sampler [2]) relaxation. The mean vectors and covariance matrices were computed over representative regions selected by the user and we set $\beta = 2.5$ in all cases. This value has been found to provide satisfactory results in all cases. An optimal Gabor filter set containing 2–4 filters has been picked manually for each image. We remark, however, that it is also possible to automatically select these filters (see [8] for a collection of methods). We found in all cases that segmentation based purely on texture gives fuzzy boundaries but usually homogeneous regions. Whereas segmentation based on color is more sensitive to local variations in color but provides sharp boundaries. As for the combined features, the advantages of both color and texture based segmentation are quite well preserved: we obtained sharp boundaries and homogeneous regions (see Fig. 1 as a good example). For example, the ball on the *tennis* image (Fig. 5) is much better detected than on either the texture or color segmentations. In terms of sharpness and homogeneity, the combined segmentation clearly outperforms the others. The power of combined features is well demonstrated by Fig. 3. Three regions contain a wooden texture with nearly matching colors and a small difference in the direction (right and lower part) or scale (middle part) in texture. The two other regions have similar texture but completely different color. Comparing color and texture only segmentations, the latter two regions are perfectly separated in the color segmentation but they are mixed in the texture one. On the other hand, the former three regions are much better separated in the texture segmentation than in the color one. As for the combined segmentation, the five regions are well separated, the error rate is half of the separate segmentation's rates. Regarding the different optimization techniques, we can see that the suboptimal ICM provides somewhat lower quality compared to the optimal Gibbs sampler but it converges much faster (see Table 1). However, this difference is less important in the case of combined features and for the real images it is nearly invisible. We also run a test on one of the images published in [7] (see Fig. 4) and we found that our results are close to the ones in [7].

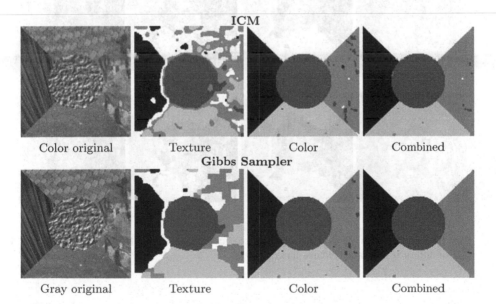

Fig. 1. Segmentation of a 128 × 128 synthetic image with 5 classes. We have used 2 Gabor filters to extract texture features.

ICM

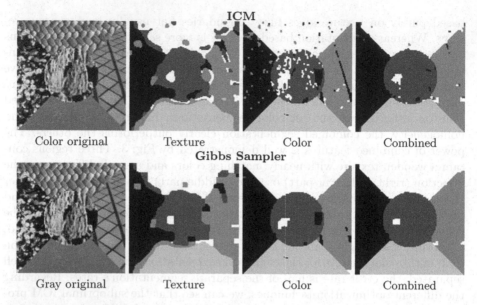

Fig. 2. Segmentation of a 128 × 128 synthetic image with 5 classes. We have used 3 Gabor filters to extract texture features.

ICM

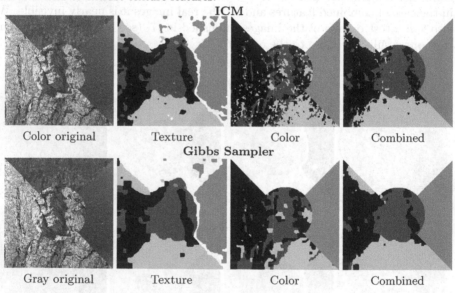

Fig. 3. Segmentation of a 256 × 256 synthetic image with 5 classes. We have used 4 Gabor filters to extract texture features.

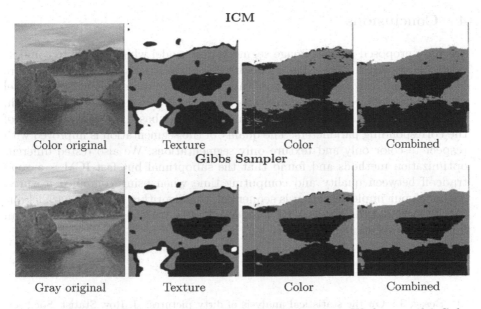

Fig. 4. Segmentation of a 384 × 384 real image with 3 classes. We have used 3 Gabor filters to extract texture features.

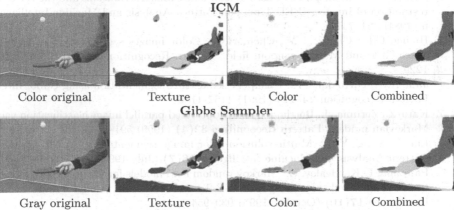

Fig. 5. Segmentation of a 360 × 240 real image with 6 classes. We have used 3 Gabor filters to extract texture features.

4   Conclusions

We have proposed a MRF image segmentation model which is able to combine color and texture features. The model itself is not restricted to a specific texture feature. In fact, any feature is suitable as far as feature values belonging to a pixel class can be modeled by a random white noise. Due to our model based approach, it is also possible to classify different kind of textures based on a prior training of the corresponding parameters. The quality of the segmentation is improved with respect to color only and texture only segmentations. We also tested different optimization methods and found that the suboptimal but fast ICM is a good tradeoff between quality and computing time when using combined features. Although our implementation is sequential, the algorithm is highly parallel due to the local nature of the MRF model. Thus, a parallel implementation can further improve the computing speed.

References

1. Besag, J.: On the statistical analysis of dirty pictures. J. Roy. Statist. Soc., ser. B, 1986
2. Geman, S., Geman, D.: Stochastic relaxation, Gibbs distributions and the Bayesian restoration of images. IEEE Trans. on Pattern Analysis and Machine Intelligence **6** (1984) 721–741
3. Huang, C.L., Cheng, T.Y., Chen, C.C.: Color images segmentation using scale space filter and Markov random field. Pattern Recognition **25(10)** (1992) 1217–1229
4. Jain, A.K., Farrokhnia, F.: Unsupervised texture segmentation using Gabor filters. Pattern Recognition **24(12)** (1991) 1167–1186
5. Kato, Z., Zerubia, J., Berthod, M.: Unsupervised parallel image classification using Markovian models. Pattern Recognition **32(4)** (1999) 591–604
6. Liu, J., Yang, Y.H.: Multiresolution color image segmentation. IEEE Trans. on Pattern Analysis and Machine Intelligence **16(7)** (July 1994) 689–700
7. Panjwani, D.K., Healey, G.: Markov random field models for unsupervised segmentation of textured color images. IEEE Trans. on Pattern Analysis and Machine Intelligence **17(10)** (October 1995) 939–954
8. Randen, T., Husoy, J.H.: Filtering for texture classification: A comparative study. IEEE Trans. on Pattern Analysis and Machine Intelligence **21(4)** (April 1999) 291–310
9. Sangwine, S.J., Horne, R.E.N. (eds): The Colour Image Processing Handbook. Chapman & Hall (1998)
10. Won, C.S., Derin, H.: Unsupervised segmentation of noisy and textured images using Markov random fields. Computer Graphics and Image Processing: Graphical Models and Image Processing **54(4)** (July 1992) 208–328

Application of Adaptive Hypergraph Model to Impulsive Noise Detection*

Soufiane Rital[1], Alain Bretto[2], Driss Aboutajdine[1], and Hocine Cherifi[3]

[1] Université Mohammed V -Agdal- GSC-LEESA B.P. 1014, Faculté des
Sciences,Rabat - Maroc `aboutaj@fsr.ac.ma`
[2] Université Jean Monnet LIGIV (E. A. 3070), Site G.I.A.T Industries, rue Javelin
Pagnon BP 505, 42007 Saint-Etienne Cedex 1 France `bretto@univ-st-etienne.fr` ,
[3] Université de Bourgogne LIRSIA,Faculté des Sciences Mirande - France
`cherifi@crid.u-bourgogne.fr`

Abstract. In this paper, using hypergraph theory, we introduce an image model called Adaptive Image Neighborhood Hypergraph (AINH). From this model we propose a combinatorial definition of noisy data. A detection procedure is used to classify the hyperedges either as noisy or clean data. Similar to other techniques, the proposed algorithm uses an estimation procedure to remove the effects of the noise. Extensive simulations show that the proposed scheme consistently works well in suppressing of impulsive noise.

Keywords: hypergraph theory, combinatorial model, coise detection

1 Introduction

Combinatorial field models have been successfully applied to many fundamental problems of image analysis and computer vision [12], [13] such as edge detection, noise cancellation and image segmentation [9],[16]. Perhaps the most popular mathematical framework is graph theory, which models binary relations. Graphs provide an efficient and powerful tool for specifying the relations between similar features in an image. The vertices correspond to objects and the edges represent the relations between objects.

However, it is difficult for the geometry and the topology of an image to be represented by a graph, because both the geometry and the topology of an image can not necessarily be expressed by binary relations.

The notion of a hypergraph has been developed both by the mathematical Hungarian school and the mathematical French school simultaneously around 1960 [2]. This concept which represents more general relations than binary ones can be used in many areas of mathematics and computer sciences such as topology and geometry, data structures, networks, and many other domains where relations between objects are fundamental [8]. Hypergraph theory was introduced in computer vision in 1997 [3]. By defining a comprehensive global model,

---

* This work has been supported by the project 'Pars_CNR n° 36' and the 'comité mixte inter_universitaire franco_marocain under grant AI n° 166/SI/98'

W. Skarbek (Ed.): CAIP 2001, LNCS 2124, pp. 555–562, 2001.

this theory provides significant improvement in low level image processing [3], [4], [5], [11]. The corruption of images by noise is a frequently encountered problem in many image processing fields. The observed noise can be often modeled as impulsive and smoothing techniques that can remove this kind of noise while preserving the edges are required.

In this paper, we present an adaptive model which is based on image neighborhood hypergraph model (INH) [3]. The adaptive image neighborhood model (AINH) captures the local properties of the image. This leads to a new noise definition which is more pertinant than our previous work [4], [5] to preserve edges.

This paper is organized as follows. In section II, we briefly introduce the necessary terminology and notations related to graphs and hypergraphs. Image adaptive hypergraph model is described in section III. Section IV introduces the concept of a noisy hyperedge and shows how this model can be exploited to develop a noise detection algorithm. Here we also examine some basic properties of this algorithm. To illustrate the effectiveness of our method we present some experimental results in section V. In the last section we make some concluding remarks.

2   Graph and Hypergraph Terminology

2.1   Graph

The general terminology concerning graphs and hypergraphs is similar to [1], [2]. All graphs in this paper are finite, undirected, with neither loops nor multiple edges. All graphs will be considered connected with no isolated vertex. Let $G = (X; E)$ be a graph, we denote by $\Gamma(x)$ the *neighborhood* of a vertex x, i. e. the set consisting of all vertices adjacent to x which is defined by: $\Gamma(x) = \{y \in X, \{x, y\} \in E\}$. In the same way, we define the *neighborhood of* $A \subseteq X$ as $\Gamma(A) = \bigcup_{x \in A} \Gamma(x)$. The *open neighborhood* of A is $\Gamma^\circ(A) = \Gamma(A) \setminus A$.

2.2   Hypergraph

A *hypergraph* H on a set X is a family $(E_i)_{i \in I}$ of non-empty subsets of X called *hyperedges* with: $\bigcup_{i \in I} E_i = X$, $I = \{1, 2, \ldots, n\}$, $n \in \mathbb{N}$. Let us note: $H = (X; (E_i)_{i \in I})$. An hyperedge E_i is *isolated* if and only if $\forall j \in I, j \neq i$ if $E_i \cap E_j \neq \emptyset$ then $E_j \subseteq E_i$.

Neighborhood Hypergraph : Given a graph G, the hypergraph having the vertices of G as vertices and the neighborhood of these vertices as hyperedges (including these vertices) is called the *neighborhood hypergraph* of G. So, to each graph we can associate a neighborhood hypergraph:

$$H_G = (X, (E_x = \{x\} \cup \Gamma(x))) \tag{1}$$

3   Image Adaptive Hypergraph Model

This section is devoted to the definition of our model. First we recall some definitions about digital images.

A distance d' on X defines a *grid* (a graph connected, regular, without both loop and multi-edge).

A *digital image* (on a grid) is a two-dimensional discrete function that has been digitized both in spatial coordinates and in magnitude feature value. Throughout this paper a digital image will be represented by the application :

$$I : X \subseteq \mathbb{Z}^2 \to \mathcal{C} \subseteq \mathbb{Z}^n \ \text{ with } \ n \geq 1 \tag{2}$$

where $\mathcal{C}$ identifies the *feature intensity level* and X identifies a set of points called *image points*. The couple $(x, I(x))$ is called a *pixel*.

Let d be a distance on $\mathcal{C}$, we have a neighborhood relation on an image defined by:

$$\forall x \in X, \ \Gamma_{\alpha,\beta}(x) =$$
$$\{x' \in X, x' \neq x \mid d(\mathcal{I}(x), \mathcal{I}(x')) < \alpha \text{ and } d'(x, x') \leq \beta\} \tag{3}$$

The neighborhood of x on the grid will be denoted by $\Gamma_\beta(x)$. So to each image we can associate a hypergraph called *Image Adaptive Neighborhood Hypergraph* (IANH):

$$\mathcal{H}_{\alpha,\beta} = \big(X, (\{x\} \cup \Gamma_{\alpha,\beta}(x))_{x \in X}\big). \tag{4}$$

The attribute α can be computed in an adaptive way depending on local properties of the image. Throughout this paper α will be estimated by the standard deviation of the pixels $\{x\} \cup \Gamma_\beta(x)$ This image representation is more relevant than the previous one introduced in [3], because it takes in account both local and global aspects of the image. Hence IANH offers new facilities for handling topology and geometry of the image. Consequently it gives more information about the nature of the image to analyze. From this model, we will now develop a noise detection application.

4   Detection of Impulsive Noise

A common type of corruption that occurs in image data is corruption by an impulsive noise process. Attenuation of noise and preservation of details are usually two contradictory aspects of image. Various noise reduction algorithms make various assumptions, depending on the type of imagery and the goals of the restoration [6],[7].

In this section, we present a noise cancellation algorithm that exploits a lack of homogeneity criterion. We consider that the global homogeneity characterizes regions, local homogeneity characterizes edges, no homogeneity characterizes a noise. A noise reduction algorithm is based on the following criterion : binary classification of hyperedge of image (H_0 noisy hyperedge and H_1 no noisy hyperedge) and filtering the noisy parts.

4.1 Noise Definition

The impulsive noise detection based on the hypergraph model as well as its modifications and generalizations [4][5][11], has been efficient in the removal of impulse noise. However, because these approaches are typically implemented invariantly across an image, they also tend to alter pixels which are undisturbed by noise. Additionally, they are prone to edge jitter in cases where the noise ratio is high. One way to avoid this situation is to improve the definition of noisy hyperedge in the noise detection framework.

Definition: We will call *disjoined chain* a succession of hyperedges disconnected two by two. A disjoined chain is *thin* if the cardinality of each hyperedge is equal to 1. To model a noise we propose the following definition:
We say that $E_{\alpha,\beta}(x)$ is a *noise hyperedge* if it verifies one of the two conditions:

- the cardinality of $E_{\alpha,\beta}(x)$ is equal to 1: $E_{\alpha,\beta}(x)$ is not contained in disjoined thin chain having five elements at least.
- $E_{\alpha,\beta}(x)$ is an isolated hyperedge and there exists an element y belonging to the open neighborhood of $E_{\alpha,\beta}(x)$ on the grid, such that $E_{\alpha,\beta}(y)$ is isolated. (i.e. $E_{\alpha,\beta}$ is isolated and it has an isolated hyperedge in its neighborhood on the grid).

This new definition allows a better discrimination between edge pixels and noisy pixels than our previous definition [5].

Property. The lemma below shows that a noisy hyperedge must be isolated.

Lemma 1. *If the cardinality of an hyperedge is equal to one, then this hyperedge is isolated.*

Proof. Given an hyperedge $E_{\alpha,\beta}(x)$ such that its cardinality is equal to 1. We suppose that it is not isolated. It is included in the hyperedge $E_{\alpha,\beta}(y)$, so $x \in \Gamma_{\alpha,\beta}(y)$ but $x \in \Gamma_{\alpha,\beta}(y)$ and the cardinality of $E_{\alpha,\beta}(x)$ is greater than 1. This brings us to a contradiction.

4.2 Noise Cancellation Algorithm

Step 1 Data :
    Image I, and neighborhood order β
Step 2 Construction of the adaptive neighborhood hypergraph $H_{\alpha,\beta}$ on I
    – a. $X \longleftarrow \emptyset$
    – b. For each pixel $(x, I(x))$ of I
    –– i. $\alpha = \sigma$, with σ being the standard deviation estimate of the pixels of $\{x\} \cup \Gamma_\beta(x)$, ($\Gamma_\beta(x)$ stands for neighborhood of x on the grid).
    –– ii. $\Gamma_{\alpha,\beta}(x) \longleftarrow \emptyset$
    –– iii. for each pixel $(y, I(y))$ of $\Gamma_\beta(x)$, if $d(I(x), I(y)) \leq \alpha$ then $\Gamma_{\alpha,\beta}(x) \longleftarrow \Gamma_{\alpha,\beta}(x) \cup \{y\}$

      -- iv. $X \longleftarrow X \cup \{x\}$
      -- v. $E_{\alpha,\beta}(x) \longleftarrow \{\Gamma_{\alpha,\beta}(x) \cup \{x\}\}$
      - c. $H_{\alpha,\beta}(x) \longleftarrow (X, (E_{\alpha,\beta}(x))_{x \in X})$
Step 3   Determination of isolated hyperedges of $H_{\alpha,\beta}$
      For each vertex x of X
      -- i. $E' \longleftarrow \emptyset$
      -- ii. $E' \longleftarrow \bigcup_{y \in E_{\alpha,\beta}(x)} E_{\alpha,\beta}(y)$
      -- iii. if $E' = E_{\alpha,\beta}(x)$, ($E_{\alpha,\beta}(x)$ is an isolated hyperedge) then;
$E_{\alpha,\beta}^{is}(x) = E_{\alpha,\beta}(x)$.
Step 4   Detection of noise hyperedges
      For every $E_{\alpha,\beta}^{is}(x)$
      -- i. If the cardinality of $E_{\alpha,\beta}^{is}(x)$ is equal to 1, and if $E_{\alpha,\beta}^{is}(x)$ is not contained
in a disjoined thin chain having five elements at least, then $E_{\alpha,\beta}^{h}(x) = E_{\alpha,\beta}^{is}$;
      -- ii. If there exists y of $\Gamma^{o}(E_{\alpha,\beta}^{is}(x))$ such as $E_{\alpha,\beta}(y)$ is isolated, then classify
$E_{\alpha,\beta}(x)$ as a noisy hyperedge : $E_{\alpha,\beta}^{b}(x) = E_{\alpha,\beta}$.
Step 5   Estimation
      For each noisy hyperedge $E_{\alpha,\beta}^{b}(x)$, replace the intensity of $E_{\alpha,\beta}^{b}(x)$ by the
value of a functional dependent of intensity of $\Gamma^{o}(E_{\alpha,\beta}^{b}(x))$.

Properties of the Algorithm.

Correctness and Complexity Proposition 1. *For β given, a unique solution exists. The algorithm complexity is in $O(n)$ (n standing for the pixels number of the image).*

Proof. The complexity of each step of the algorithm is raised by the number of pixels of the image up to a multiplicative coefficient. This coefficient is the number of restrained pixels in hyperedge, raised itself by the cardinality maximum of hyperedges. Therefore, the complexity of the algorithm is in $O(n)$. For $\alpha = \sigma$ and β fixed, there exists a single hypergraph $H_{\alpha,\beta}$ associated to an image data, so from the definition of noisy hyperedges, the existence of a unique classification and a unique solution is proven.

5 Experimental Results

The performance of the proposed impulse noise detection has been evaluated and compared with those of some existing detection algorithms, including the Stevenson strategy [14] and Tovar filter[15].

In our simulations, a group of 256×256 gray-scale images corrupted by fixed-valued impulses with various noise ratios are used. The noise intensity of fixed-valued impulses corresponds to 0 or 255 with equal probability. The peak signal-to-noise ratio (**PSNR**) is used to measure the image quality.

The first experiment is conducted to gauge the efficacy of the proposed technique for filtering images corrupted with different noise ratios.

The results of filtering the peppers image are given in Figure 1. The noise free image in figure (1.a) was corrupted by a 5% impulsive noise in figure (1.b). The output image using the median filter 3×3 on the noisy hyepredges by our algorithm is shown in figure (1.c).

Figures (1.d ,1.e) presents the superposition of the false alarm noise image and the original one. It is interesting to note that, the false alarm noise is not localized on the edges and the detected noise image (1.f) doesn't contain any information about edges. That means that our algorithm preserves edges. In addition, the AINH model yields better subjective quality with respect to noise suppression and detail preservation.

As any classical method the performances of our model decrease when the percentage of noisy pixels increases. That is because the spatial homogeneity of noisy data grows and the model has difficulty to discern noisy hyperedges from noise free ones; see figure 2.

To assess the effectiveness of the proposed technique with other methods. Table 1 presents the comparative PSNR, detection and false alarm probabilities **pd and pf** [10] results of removing the two ratios 5% and 10% of impulsive noise.

It is obviously seen in table 1 that the proposed algorithm provides superior results to the other methods in removing impulsive noise.

Table 1. Comparative filtering results in PSNR, pd and pf

Image 256x256	Alg.	% injected	PSNR in dB of I_b	PSNR in dB of I_f	pd	pf
Peppers	Our algo	5	18.08	32.32	0.872	0.007
−	Our algo.	10	15.06	27.00	0.82	0.005
−	Stv. algo. order 1	5	18.08	23.45	0.73	0.01
−	− − order 2	5	18.08	25.08	0.85	0.03
−	− − order 3	5	18.08	25.95	0.86	0.042
−	− − order 1	10	15.06	19.09	0.574	0.09
−	− − order 2	10	15.06	22.86	0.80	0.01
−	− − order 3	10	15.06	24.24	0.84	0.02
−	Tovar algo.	5	18.08	25.88	0.96	0.55
−	−	10	15.06	25.55	0.97	0.52

I_b : Corrupted image;
I_f : The filtering results by median 3x3 filter.

6   Conclusion and Perspectives

A key problem that arises in many recent approaches of noise detection and cancellation methods is that these approaches are location-invariant in nature, and thus they also tend to alter the pixels not corrupted by noise. In this paper a new algorithm for image noise reduction based on properties of hypergraph theory has been proposed. When compared with other detection based filtering schemes, the proposed technique consistently yields satisfactory results in suppressing impulse noise while still possessing a simple computational structure. Noticeable gains have been achieved over other methods in terms of PSNR as well as the perceived image quality.

Further works will be focused on :

– Definition of edges, and development of an algorithm to detect these edges from the adaptive image neighborhood hypergraph model;
– High level image processing algorithms using IANH.

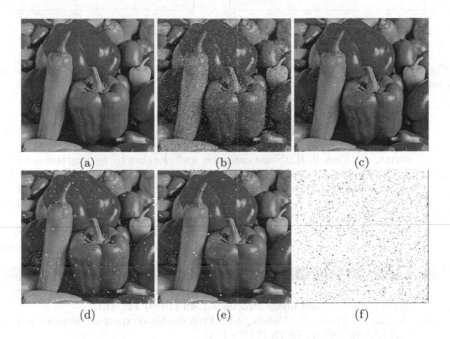

Fig. 1. (*a*) The original peppers image, (*b*) peppers image corrupted by 5% of impulsive noise, $PSNR = 18.08dB$, (*c*) image obtained by noisy hyperedges filtering using median 3x3 filter resulting; $PSNR = 32.32dB$. (d,e) false alarm noise for the 5% and 10 % of impulse noise percentage, (f) the detected noise for the 5%.

References

1. Berge, C.: Graphs North Holland (1987).
2. Berge, C.: Hypergraphs. North-Holland, Amsterdam (1987)
3. Bretto, A., Azema, Cherifi, H., Laget, B.: Combinatorics and image processing. Computer Vision Graphic in Image Processing, **59** (5), September (1997) 256–277

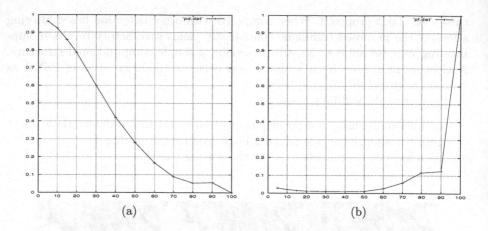

Fig. 2. Evolution of the (a) Pd and (b) Pf according to injected impulsive noise percentage in the image

4. Bretto, A., Cherifi, H.: A noise cancellation algorithm based on hypergraph modeling. IEEE Workshop on Digital Signal Processing (1996) 1–4.
5. Bretto, A., Cherifi, H.: Noise detection and cleaning by hypergraph model. In: IEEE Computer Sciences (ed.): International Symposium on Information Technology: Coding and computing. IEEE Computer Sciences (2000) 416–419
6. Chen, T., Wu, H.R.: Adaptive Impulse Detection Using Center Weighted Median Filters. IEEE Signal Processing Letters **8** (2001) 1–3
7. Chen, T., Wu, H.R.: Application of Partition-Based Median Type Filters for Suppressing Noise in Images. (accepted) IEEE Transactions on Image Processing (2001)
8. Gondran, M., Minoux: Graphs and Algorithms. Wiley, Chichester (1984)
9. Kovalevsky, V.A.: Finite topology as applied to image processing. Computer vision Graphics, and Image Processing. **46** (1989) 141–161
10. Macchi, O., Faure, C., Caliste, J.P.: Probabilités d'erreur de détecteurs. Revue CETHEDEC, Cah. NS **75** (1975) 1–51
11. Rital, S., Bretto, A., Cherifi, H., Aboutajdine, D.: Modélisation d'images par hypergraphe application au Débruitage. ISIVC Rabat (2000)
12. Rosenfeld, A.: Digital Image Processing. Academic Press, San Diego (1982)
13. Russ, J.C.: The image processing handbook. CRC Press, IEEE press. Berlin, Springer (1999)
14. Stevenson, R.L., Schweizer, S.M.: Nonlinear Filtering Structure for Image Smoothing in Mixed-Noise Environments. Journal of Mathematical Imaging and Vision, **2** (1992) 137–154
15. Tovar, R., Esleves, C., Rustarnante, R., Psenicka, B., Mitra, S.K.: Implementation of an Impulse Noise Removal Algorithm from Images. (2000) preprint.
16. Voss, K.: Discrete Images, Object and Functions in $\mathbb{Z}^n$. Algorithms and Combinatorics. Springer Verlag (1990)

Estimation of Fusarium Head Blight of Triticale Using Digital Image Analysis of Grain

Marian Wiwart, Irena Koczowska, and Andrzej Borusiewicz

University of Warmia and Mazury, Department of Plant Breeding and Seed Science
10-724 Olsztyn, pl. Lodzki 3, Poland
wiwart@moskit.uwm.edu.pl

Abstract. The response of spring triticale to the infection of heads with mixture of *Fusarium culmorum* and *F. avenaceum* isolates was investigated with the application of colour image analysis of grains. The results seem to suggest that there is a strong relationship between declining values of the two yield components: kernels weight per spike and one thousand kernels weight (TKW), and the values of colour components H S and I (hue, saturation, intensity); the presence of the relationship is confirmed by high values of simple correlation coefficients. The technique applied in the experiment makes it possible to diagnose a case of *Fusarium* infestation at a nearly 90% probability, the result that creates the basis for further research towards the elaboration of a completely automatic system of colour image analysis of grain.

Keywords: image analysis, Fusarium infestation, colour model

1 Introduction

Fusarium head blight (FHB), generally caused by *Fusarium culmorum* and *F. avenaceum,* occurs in humid and semi humid cereal-growing areas throughout the world [16]. This very important disease may results in significant yield reduction and low quality of seed material. Isolates of *Fusarium* sp. obtained from small grain cereals often produce mycotoxins, e.g. deoxynivalenol (DON) and other trichothecenes – they have been shown to be harmful to human and animal health [5]. A rapid method of analysis of head blight, might be very useful in resistance breeding research. Evaluation of health status of an individual sample based on the calculation of the percentage of *Fusarium* infected kernels is not reliable unless symptoms of fusariosis are evident. However, it happens that symptoms of the disease are hardly noticeable; on the other hand, sometimes it is possible to detect the presence of fungi in kernels which visually differ only insignificant from healthy ones especially when infection proceeds at the latter stages of kernels development [9]. Germination of such kernels may be abnormal or else they may become a source of primary inoculum, thus leading to further spread of the infection [14]. Digital image analysis, gaining more and more popularity in biological research, has supplied the investigator with tools of

W. Skarbek (Ed.): CAIP 2001, LNCS 2124, pp. 563–569, 2001.

a new, higher quality, with which analysis of shape, intensity or colour of various plant structures and organs, including seeds, is more precise [8, 20]. The studies, carried out in this area, focus mainly on identification of cultivars as well as determination of certain relationships between geometrical parameters of seeds and some yield components [1,11]. Analysis of colour image exploits a full range of possibilities created by 24-bit computer systems, commonly available today. Irrespective of numerous technical and computer related problems, caused by huge amounts of information to be processed [7, 13, 18], there are already some reports in the literature suggesting a considerable progress in the experiments conducted on the application of this type of analysis to the evaluation of health status and purity of seeds as well as some of its commercial properties [3,4, 21,17].

The study presented in this paper focused on the assessment of the applicability of computer colour image analysis of spring triticale grains to the determination of a degree of their infestation with *Fusarium* sp. after artificial inoculation of heads.

2 Material and Methods

The material consisted of 620 breeding strains (F_4 generations) of spring triticale (own breeding materials of Department of Plant Breeding and Seed Production, University of Warmia and Mazury in Olsztyn, Poland) obtained by crossing four Polish cultivars: Migo, Maja, Jago and Gabo; as well as 2 isolates of *Fusarium culmorum* (W.G. Smith)Sacc. and 2 isolates of *F. avenaceum* (Fr.)Sacc. [15], obtained from naturally infected kernels of spring triticale.

Inoculation was performed at the full flowering stage. Thirty heads of each strain, chosen randomly, were treated by aqueous suspension of conidia of booth *Fusarium* species (concentration of 500 000 spores/cm^3). Spores were obtained from 14-day old cultures grown on a PDA medium (Merck®). Inoculated heads were immediately covered with polyethylene bags for 48 hours. A reduction of the kernels weight per spike and TKW calculated for infected heads versus non-inoculated heads (control) was used as a measure of the response to infection.

Digital image analysis was conducted on a Pentium III PC (equipped with a SPEA ShowTime Plus® card and measuring software MultiScan®6.0) attached to a colour CCD camera Sony EVI-1011P. MS Windows® compatible software was applied: MS Excel® 97 and Corel® 7.0. Two samples of kernels were taken from each combination and placed on a Ø94 mm polystyrene Petri's dishes. The samples were illuminated with 4 halogen bulbs of 3500°K colour temperature. Each image was entered into the MultiScan®, where a stereologic grid (4x4 square gridding meshes, each of which was a separate replication) was applied on it. For each mesh, average values of the three primaries: red (R), green (G) and blue (B), were determined to be used subsequently for the calculation of the three colour components: H, S, I (hue, saturation, intensity) [7, 12, 13]:

$$\text{if } 0 \leq R, G, B \leq 1 \text{ then}$$

$$H = \text{acos} \frac{\frac{1}{2}\left[(R-G)+(R-B)\right]}{\left[(R-G)^2 + (R-B)(G-B)\right]^{\frac{1}{2}}} \tag{1}$$

$$S = 1 - \min(R,G,B) \tag{2}$$

$$I = R + G + B \tag{3}$$

The resulting values of the H, S, I were included in the analysis of simple correlation, by referring them to a decrease in the kernels weight per spike and the weight of one thousand kernels. The results of the image analysis were presented graphically and, by making use of the relationships exist in the colour circle [6, 7], areas typical of the grains obtained from a control and inoculated heads were established.

3 Results

The inoculation of heads led to a considerable reduction of the two major yield components of spring triticale: weight of kernels per spike and TKW. After infection, the weight of kernels per spike dropped by an average 26.36% in comparison to the control, while TKW declined by 23.28% (range of differences for both characters was 38.02% ±13.29 and 32.89%±11.46, respectively). Obtained result corresponding with the data cited in the literature on various species of small grain cereals [5, 16] and indicating quite strong pathogenicity of the isolates applied as well as a wide range of the response of the strains studied

The digital analysis of the colour images of kernels revealed great variation in the values of the three colour components, H S I, between the control and inoculated objects (tab. 1).

Table 1. Mean values of H S I (bold) for the colour of grains after artificial inoculation of heads (min. max.) in comparison to non-inoculated control (** differences between control and inoculation significant at p=0.01)

	H	S	I
Control	**53.6** (43.9÷63.2)	**0.58** (0.517÷0.642)	**1.75** (1.66÷1.84)
Inoculated	**42.8**[**] (31.0÷54.6)	**0.56** (0.429÷0.619)	**1.81**[**] (1.70÷1.92)

The colour of all the inoculated objects, compared with the control, was lower in hue (nearer the red). A similar tendency was observed in the case of the S component – lower values here are indicative of lower saturation of the colour of kernels. Values of I (intensity) were higher for the samples obtained from the inoculated heads, which means that this grain was lighter than from the control objects.

The calculated values of Pearson's correlation coefficients showed a strong relationship between the per cent decline in the two yield components and the values of the all three colour components; the relationship was more evident in the case of

TKW than in the kernels weight per spike (tab. 2). Higher absolute values of the correlation coefficient calculated for saturation and intensity than for hue may suggest that a change in the colour of the seed coat means lighter and paler colour rather than a shift towards the red.

Table 2. Values of Pearson's correlation coefficients calculated for a decline (after infection) in the values of booth yield components and H S I (**significant at p=0.01)

Decline in per cent of the control of:	H	S	I
kernels weight per spike	0.08	0.51**	-0.45**
TKW	0.30**	0.77**	-0.66**

Mutual relationships between H and S may be presented graphically, by referring them to the regularities which occur in the colour circle; this is the best way to visualise the range of changes in the colour of seed coats (fig. 3). Based on the boundary empirical values of H and S for all 620 strains investigated, the following areas were set: A – for the control kernels, B – for kernels from inoculated heads and C (shared area), in which the values for H and S can be identical regardless the origin of the grain. Percentage of each area for all the results obtained was as follows: 27.89% of A, 61.91% for B and 10.20% for C. It means that with the application of the method discussed hereby it is possible to discriminate clearly between infected and health grain (the probability nearing 90%).

4 Discussion

Presented results show potential applications of computer aided image analysis to the evaluation of health status of small grain cereals. A change of the seed coat colour observed after infection by *Fusarium* sp. is, next to the shape deformation, one of the major indicators of infestation. With these symptoms, it is possible not only to perform organoleptic analysis of the health status of a sample [16, 9, 14] but also to assess the content of some mycotoxins (DON) on the basis of the frequency of the FDK (*Fusarium* damaged kernels) fraction [5]. However, the latter is possible only for a very serious case of infection, when kernels reveal clear symptoms of disease. Such evaluation can also be burdened by some error due to the wear of the observer's sight, who has to analyse numerous samples of grain; another source of error is the fact that poor formation or malformation of kernels need not be indicative of fusariosis [10]. Therefore, it seems recommendable to search for new techniques, with which health screening of kernels will be faster and, most importantly, more objective. Implementation of such methods, especially in terms of individual kernels, will perhaps play a significant role in the research on seed production, cereal grain processing and resistance breeding. It would also be important to design a machine for sorting kernels automatically. The existing solutions make it possible to separate damaged kernels out. In 1995 a team of German authors from Hohenheim University published a design of a machine, equipped with a CCD camera and PC, used for automatic removal of mechanically damaged kernels [2].

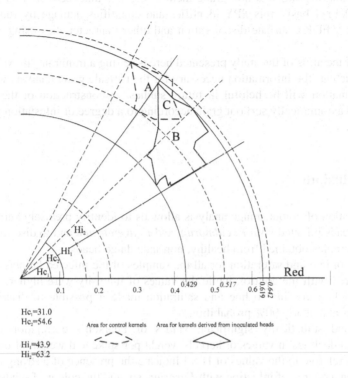

Fig. 1. Areas determined for the control (A) and inoculated (B) kernels. C is the shared area. H-indexed values refer to Hue (H_i – inoculated objects, H_c – control objects); the scale on the red primary axis refers to the values of saturation. The graphic presentation is executed according to the pattern of the colour circle

Although highly effective (83.7 to 98.4% damaged kernels sorted out), the machine has a disadvantage – the analysis of kernels is time consuming (app. 3s per kernel). On the other hand, specialists from the company Foss Tecator have designed an apparatus called 310 GrainChec®, in which digital image analysis is performed with the use of a CCD camera mounted and computer system equipped with a Pentium processor as well as the software based on the technique of Artificial Neural Networks. This precalibrated instrument performs analysis of purity, colour and size distribution as well as end-user characteristics. The purity test identifies the sample composition and the result is presented as classes. Capacity of this apparatus is equal to 500 objects/min (Foss Tecator, patent no 470 465). Very interesting results present Ruan et al [17]. Those authors applied of four-layer backpropagation artificial neural networks architecture to estimate of *Fusarium* scabby wheat kernels. The trained neural network produced excellent estimation of % FDK and is much more accurate then human expert panel. Using this technique maximum and absolute errors, by correlation coefficient equal to 0.97, are 5.14% and 1.93%, respectively. Canadian company Maztech Micro Vision Ltd [19] offers full-automatically system for objective determination of grade and non-grade factors in grains and seeds SPY Grain Grader. On a per kernel basis, this SPY identifies and quantifies damage by fungal disease (especially %FDK), surface discoloration and other characteristics using 36 bit RGB colour.

The one of the aims of the study presented here is setting a minimum-maximum range, within which all the information necessary for the triticale grain analysis will be held. This information will be helpful in future studies on construction of the equipment which will automatically sort out grain according to a degree of infestation with fungal pathogens.

5   Conclusions

1. Application of colour image analysis allow us to identify precisely kernel samples from heads infected with *F. culmorum* and *F. avenaceum* and to discriminate them from samples obtained from healthy, non-inoculated heads.
2. Values of hue and saturation for all the samples of the infected kernels were lower compared with the control, while the values of intensity were higher; distribution of kernels according to hue and saturation made it possible to discriminate the samples at a nearly 90% probability.
3. High and statistically significant values of Pearson's correlation coefficients between declines in values of kernels weight per spike and weight of one thousand kernels relative to the values of H S I indicate the presence of a strong relationship between a degree of infection with *Fusarium* sp. and the colour of seeds.
4. Colour image analysis is unlikely to replace precise instrument methods or standard mycological analyses; however, it may greatly assist the investigations, by making it possible to design some equipment in the future which will automatically sort the seeds out.

References

1. Berman, M., Bason, L., Ellison, F., Peden, G., Wrigley, C.: Image Analysis of Whole Grains to Screen for Flour – Milling Yield in Wheat Breeding. Cereal Chem. **73**(3) (1996) 323-327
2. Büerman, M., Schmidt, R., Reitz, P.: Bruchkornbestimmung. Automatiesiertes Verfahren mit digitaler Blidanalyse. Landtechnik **50**(4) (1995) 200-201
3. Chtioui, Y., Bertrand, D., Dattee, Y., Devaux, M.F.: Identification of Seeds by Colour Imaging: Comparison of Discriminant Analysis and Artificial Neural Network. J.Sci.Food. Agric. **71** (1996) 433-441
4. Coles, G.: An Objective Dry Pea Colour Scoring System for Commercial and Plant Breeding Applications. J. Sci. Food Agric. **74** (1997) 435-440

5. Chełkowski, J.: Formation of mycotoxins produced by Fusarium in heads of wheat, triticale and rye. In: J. Chełkowski (eds.): Fusarium – Mycotoxins, Taxonomy and Pathogenicity. Elsevier (1989) 63-84
6. Foley, J.D., Van Damm, A., Feiner, S.K., Huges, J.E.: Computer Graphics, Principles and Practice. 2nd edn. Reading, Massachusetts: Addison Wesley (1990)
7. Fornter, B., Meyer, T.E.: Number by Colors. Springer-Verlag, Telos (1997)
8. Glasbey, C.A., Horgan, G.W.: Image Analysis for the Biological Sciences. John Willey & Sons Ltd. (1995)
9. Häni, F.: Zur Biologie und Bekämpfung von Fusariosen bei Weizen und Roggen. Phytopath. Z. **100** (1981) 44-87
10. Jones, D.G., Clifford, B.C.: Cereal Diseases. Their Pathology and Control. John Willey & Sons Ltd. (1983)
11. Kubiak, A., Fornal, L.: Interaction between geometrical features and technological quality of polish wheat grains. Pol. J. Food Nutr. Sci. **3** (44) 4 (1994) 75-85
12. Luong, Q.T.: Color in computer vision. In: Chen, C.H., Pau, L.F. and Wang, P.S.P (eds.) Handbook of Pattern Recognition and Computer Vision (1993) 311-368
13. Mailutha, J.T.: Application of Image Analysis to Agricultural and non-Agricultural Problems. An Overview. Engineering the Economy, KSAE Conf. (1995) 30-32
14. Nelson, P.E., Toussoun, T.A., Cook, R.J.: Fusarium, Biology and Taxonomy. The Pennsylvania State University (1981)
15. Nelson, P.E., Toussoun, T.A., Marasas, W.F.O.: Fusarium – An Illustrated Manual for Identification. The Pennsylvania State University (1983)
16. Parry, D.W., Jenkinson, P., McLeod, L.: Fusarium ear blight (scab) in small grain cereals – a review. Plant Path. **44** (1995) 207-238
17. Ruan, R., Ning, S., Song, A., Ning, A., Jones, R., Chen, P.: Estimation of Fusarium Scab in Wheat Using Machine Vision and a Neural Network. Cereal Chem. **75** (4) (1998) 455-459.
18. Sonka, M., Hlavac, V., Boyle, R.: Image Processing, Analysis, and Machine Vision. PWS Publishing (1999)
19. The new standard for objective quality analysis. Maztech Micro Vision Ltd., Canada (1999)
20. Wiwart, M.: Computer image analysis – a new tool in agricultural researches. PNR, Polish Academy of Sciences **5** (1999) 3-15 (eng. summary.)
21. Wiwart, M., Korona, A.: Application of a colour image analysis of kernels in evaluation of the infection of triticale grown in different cultivation systems. Pl. Breed. Seed Sci. 42, **1** (1998) 69-79

Fast Modified Vector Median Filter

B. Smolka[1], M. Szczepanski[1]*, K.N. Plataniotis[2], and A.N. Venetsanopoulos[2]

[1] Silesian University of Technology
Department of Automatic Control
Akademicka 16 Str, 44-101 Gliwice, Poland
bsmolka@ia.polsl.gliwice.pl

[2] Edward S. Rogers Sr. Department of
Electrical and Computer Engineering
University of Toronto
10 King's College Road, Toronto, Canada
kostas@dsp.toronto.edu

Abstract. A new filtering approach designed to eliminate impulsive noise in color images, while preserving fine image details is presented in this paper. The computational complexity of the new filter is significantly lower than that of the Vector Median Filter. The comparison shows that the new filter outperforms the VMF, as well as other standard procedures used in color image processing, when the impulse noise is to be eliminated.

Keywords: image processing, vector median filter

1 Introduction

The processing of color image data has received much attention in the last years. The amount of research published to date indicates an increasing interest in the area of color image processing [1-6]. It is widely accepted that color conveys very important information about the scene objects and this information can be used to further refine the performance of an imaging system.

The most common image processing tasks are noise filtering and image enhancement. These tasks are an essential part of any image processor whether the final image is utilized for visual interpretation or for automatic analysis [1, 2].

As it has been generally recognized, that the nonlinear vector processing of color images is the most effective way to filter out noise and to enhance color images, the new filtering technique presented in the paper is also nonlinear and utilizes the correlation among the channels of a color image.

---

* This work was partially supported by KBN grant 8-T11E-013-19

W. Skarbek (Ed.): CAIP 2001, LNCS 2124, pp. 570–580, 2001.

This paper is divided into three parts. In the first section, a brief overview of the standard noise reduction operations for color images based on the concept of vector median is presented. The second part shows the construction of the new algorithm of image enhancement and the last part depicts the results of noise attenuation achieved using the proposed algorithm in comparison with the standard noise suppression techniques described at the beginning of this paper.

2   Standard Noise Reduction Filters

A number of nonlinear, multichannel filters, which utilize correlation among multivariate vectors using various distance measures, have been proposed [1,2]. The most popular nonlinear, multichannel filters are based on the ordering of vectors in a predefined moving window. The output of these filters is defined as the lowest ranked vector according to a specific ordering technique.

Let $\mathbf{F}(x)$ represents a multichannel image and let W be a window of finite size n (filter length). The noisy image vectors inside the window W are denoted as $\mathbf{F}_j$, $j = 0, 1, \ldots, n - 1$. If the distance between two vectors $\mathbf{F}_i, \mathbf{F}_j$ is denoted as $\rho(\mathbf{F}_i, \mathbf{F}_j)$ then the scalar quantity

$$R_i = \sum_{j=1}^{n} \rho(\mathbf{F}_i, \mathbf{F}_j), \tag{1}$$

is the distance associated with the noisy vector $\mathbf{F}_i$ inside the processing window. An ordering of the R_i's

$$R_{(0)} \leq R_{(1)} \leq \ldots \leq R_{(n-1)}, \tag{2}$$

implies the same ordering to the corresponding vectors $\mathbf{F}_j$

$$\mathbf{F}_{(0)} \leq \mathbf{F}_{(1)} \leq \ldots \leq \mathbf{F}_{(n-1)}. \tag{3}$$

Nonlinear ranked type multichannel estimators define the vector $\mathbf{F}_{(1)}$ as the filter output. This selection is due to the fact that vectors that diverge greatly from the data population usually appear in higher indexed locations in the ordered sequence. However, the concept of input ordering, initially applied in scalar quantities is not easily extended to multichannel data, since there is no universal way to define ordering in vector spaces.

To overcome the problem, distance functions are often utilized to order vectors. As an example, the *Vector Median Filter* (VMF) uses the L_1 norm in (1) to order vectors according to their relative magnitude differences [7].

The output of the VMF is the pixel $\mathbf{F}_k \in W$ for which the following condition is satisfied:

$$\sum_{j=0}^{n-1} \rho(\mathbf{F}_k, \mathbf{F}_j) < \sum_{j=0}^{n-1} \rho(\mathbf{F}_i, \mathbf{F}_j), \; k \neq i. \tag{4}$$

In this way the VMF consists of computing and comparing the values of R_i and the output is the vector $\mathbf{F}_k$ for which R reaches its minimum. In other words if for some k the value

$$R_k = \sum_{j=0}^{n-1} \rho(\mathbf{F}_k, \mathbf{F}_j), \tag{5}$$

is smaller than R_0:

$$R_0 = \sum_{j=0}^{n-1} \rho(\mathbf{F}_0, \mathbf{F}_j), \quad (\rho(\mathbf{F}_k, \mathbf{F}_k) = 0), \tag{6}$$

then the original value of the pixel $\mathbf{F}_0$ in the filter window W is being replaced by $\mathbf{F}_k$ which satisfies the condition (4), which means that $k = \arg\min_i R_i$.

Figure 1 a, b) shows the situation where the pixel $\mathbf{F}_2$ is moved to the center of the filter window and the original pixel $\mathbf{F}_0$ is replaced by $\mathbf{F}_2$ if the distance R_2 associated with $\mathbf{F}_2$ is smaller than R_0 and it is the minimal distance associated with the vectors belonging to the filtering window W.

The construction of the VMF is shown in Fig. 2, where the Euclidean distance is used. However different norms can be applied for noise suppression using the VMF concept. Most commonly the L_p metric is used for the construction of the Vector Median Filter, then

$$\rho(\mathbf{F}_i, \mathbf{F}_j) = \left\{ \sum_{k=1}^{l} (F_i^k - F_j^k)^p \right\}^{\frac{1}{p}}. \tag{7}$$

Choosing a specific value of p we can obtain the following popular distance measures :

$$\rho(\mathbf{F}_i, \mathbf{F}_j) = \sum_{k=1}^{l} |F_i^k - F_j^k|, \quad p = 1 \tag{8}$$

$$\rho(\mathbf{F}_i, \mathbf{F}_j) = \sqrt{\sum_{k=1}^{l} (F_i^k - F_j^k)^2}, \quad p = 2 \tag{9}$$

$$\rho(\mathbf{F}_i, \mathbf{F}_j) = \max_k |F_i^k - F_j^k|, \quad p = \infty \tag{10}$$

where l is the vector dimension. In this paper the L_1, L_2 and L_∞ was used.

The *Basic Vector Directional Filter* (BVDF) is a ranked-order, nonlinear filter which parallelizes the VMF operation [8]. However, a distance criterion, different from the L_1 norm used in VMF, is utilized to rank the input vectors. The angular distance distance criterion used in BVDF is defined as a scalar measure

$$R_i = \sum_{j=1}^{n} r(\mathbf{F}_i, \mathbf{F}_j), \quad \text{with} \quad r(\mathbf{F}_i, \mathbf{F}_j) = \cos^{-1}\left(\frac{\mathbf{F}_i \, \mathbf{F}_j^T}{|\mathbf{F}_i| \, |\mathbf{F}_j|} \right), \tag{11}$$

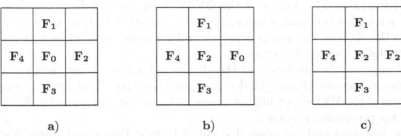

a)   b)   c)

Fig. 1. Illustration of the construction of the new filtering technique for the 4-neighborhood system, as compared with the vector median. In case of the vector median the center pixel is replaced by its neighbor and the sum of distances between the center and neighbor pixel is being calculated a) and b). So the sum of distances is $R_0 = \rho\{F_0, F_1\} + \rho\{F_0, F_2\} + \rho\{F_0, F_3\} + \rho\{F_0, F_4\}$ a) and $R_2 = \rho\{F_2, F_1\} + \rho\{F_2, F_3\} + \rho\{F_2, F_4\} + \rho\{F_2, F_4\}$ b). If $R_2 < R_0$ then the temporary output of the vector median operator is F_2, then next pixels are being checked, whether they minimize the sum of distances. In the new method, if the center pixel $\mathbf{F_0}$ is to be replaced by its neighbor F_2, then the pixel $\mathbf{F_0}$ is removed from the filter window (**c**)) and the total distance $R_2 = \rho\{F_2, F_1\} + \rho\{F_2, F_3\} + \rho\{F_2, F_4\}$ between $\mathbf{F_2}$ (new center pixel) is calculated. If the total distance R_2 is greater than $R_0 = -\beta + \rho\{F_0, F_1\} + \rho\{F_0, F_2\} + \rho\{F_0, F_3\} + \rho\{F_0, F_4\}$ then the center pixel is replaced by $\mathbf{F_2}$, otherwise it is retained.

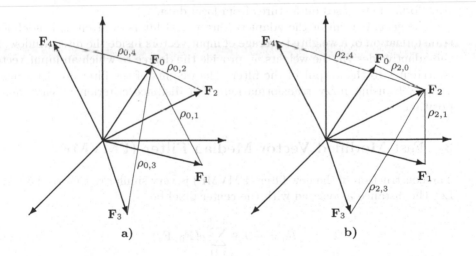

a)   b)

Fig. 2. In the vector median filter, the distance R_0 associated with the vector $\mathbf{F_0}$ equals $R_0 = \rho(0,1) + \rho(0,2) + \rho(0,3) + \rho(0,4)$ (**a**) and the distance R_2 associated with the vector $\mathbf{F_2}$ equals $R_2 = \rho(2,0) + \rho(2,1) + \rho(2,3) + \rho(2,4)$ (**b**).

as the distance associated with the noisy vector $\mathbf{F}_i$ inside the processing window of length n. The output of the BVDF is that vector from the input set, which minimizes the sum of the angles with the other vectors. In other words, the BVDF chooses the vector most centrally located without considering the magnitudes of the input vectors.

To improve the efficiency of the directional filters, a new method called *Directional-Distance Filter* (DDF) was proposed [10]. This filter retains the structure of the BVDF but utilizes a new distance criterion to order the vectors inside the processing window.

The new rank-ordered operation called Hybrid Directional Filter was proposed in [11]. This filter operates on the directional and the magnitude of the color vectors independently and then combines them to produce a unique final output. This hybrid filter, which can be viewed as a nonlinear combination of the VMF and BVDF filters, produces an output according to the following rule:

$$
\mathbf{F}_{HyF} = \begin{cases} \mathbf{F}_{VMF} & \text{if } \mathbf{F}_{VMF} = \mathbf{F}_{BVDF} \\ \left(\dfrac{\|\mathbf{F}_{VMF}\|}{\|\mathbf{F}_{BVDF}\|} \right) \mathbf{F}_{BVDF} & \text{otherwise} \end{cases}, \tag{12}
$$

where $\mathbf{F}_{BVDF}$ is the output of the BVDF filter, $\mathbf{F}_{VMF}$ is the output of the VMF and $\|\cdot\|$ denotes the magnitude of the vector.

A different approach in the development of directional filters was taken with the introduction of a new class of adaptive directional filters [14,15]. The adaptive filters proposed there, utilize data-dependent coefficients to adapt to local image characteristics. The weights of the adaptive filters are determined by fuzzy transformations based on features from local data.

The general form of the adaptive directional filters is given as a nonlinear transformation of a weighted average of input vectors inside the filter window. In the adaptive design, the weights w_i provide the degree to which an input vector contributes to the output of the filter. The weights of the filter are determined adaptively using fuzzy transformations of a distance criterion at each image position.

3   Fast Modified Vector Median Filter (FMVMF)

The construction of the new filter (FMVMF) is very similar to that of the VMF. Let the distance associated with the center pixel be

$$
R_0 = -\beta + \sum_{j=1}^{n-1} \rho(\mathbf{F}_0, \mathbf{F}_j), \tag{13}
$$

where β is a threshold parameter and let the distance associated with the neighbors of $\mathbf{F}_0$ be

$$
R_i = \sum_{j=1}^{n-1} \rho(\mathbf{F}_i, \mathbf{F}_j), i = 1, \ldots, n - 1. \tag{14}
$$

Then, if for some k, R_k is smaller than R_0

$$R_k = \sum_{j=1}^{n-1} \rho(\mathbf{F}_k, \mathbf{F}_j) \ < R_0 \,, \tag{15}$$

then $\mathbf{F}_0$ is being replaced by $\mathbf{F}_k$. It happens when

$$\sum_{j=1}^{n-1} \rho(\mathbf{F}_k, \mathbf{F}_j) < -\beta + \sum_{j=1}^{n-1} \rho(\mathbf{F}_0, \mathbf{F}_j) \,, \tag{16}$$

so the condition is

$$\beta < \sum_{j=1}^{n-1} \{\rho(\mathbf{F}_0, \mathbf{F}_j) - \rho(\mathbf{F}_k, \mathbf{F}_j)\} \,. \tag{17}$$

The construction of the new filter is very similar to that of VMF, however the major difference is presented in 1c) and Fig. 5. The pixel $\mathbf{F}_2$ is put to the center of W and the center pixel $\mathbf{F}_0$ is removed from the window. The center pixel $\mathbf{F}_0$ will be replaced by $\mathbf{F}_2$ if the distance R_2 associated with $\mathbf{F}_2$ is smaller than R_0 and is the minimal distance associated with the vectors belonging to W.

The rejection of the center pixel $\mathbf{F}_0$ is the most important feature of the new algorithm. As the center pixel is suspected to be noisy, it is not taken into account when calculating the distances associated with the neighbors of $\mathbf{F}_0$. This is illustrated in Fig. 1 c) and Fig. 5.

Table 1. Filters taken for comparison with the new modified Vector median filter.

Notation	Filter	Ref.
AMF	Arithmetic Mean	[2]
VMF	Vector Median	[7]
BVDF	Basic Vector Directional	[8]
GVDF	Generalized Vector Directional	[8]
DDF	Directional-Distance	[5]
HDF	Hybrid Directional	[9]
AHDF	Adapt. Hybrid Directional	[9]
FVDF	Fuzzy Vector Directional	[10]
ANNF	Adapt. Nearest Neighbor	[11]
ANP-EF	Adapt. Non-Param. (Exponential)	[14]
ANP-GF	Adapt. Non-Param. (Gaussian)	[14]
ANP-DF	Adapt. Non-Param. (Directional)	[14]

In finding the maximum in (5), we obtain $n - 1$ nonzero components in R_0. If we replace the central pixel by one of its neighborhood (by $\mathbf{F}_2$ in Figs. 1 and 4), then we obtain only $n - 2$ nonzero components in R (14), as the center pixel disappears from the filter window. In this way the filter is faster than the VMF and replaces the central pixel only when it is really noisy while preserving the original undisturbed image structures.

Fig. 3. Comparison of the efficiency of the vector median and the proposed filter: **a)** parts of the *BARBARA, GOLDHIL* and *BOATS* color images, **b)** images contaminated by 4% impulsive noise, **c)** images filtered using the proposed technique, **d)** the result of the filtering with the vector median.

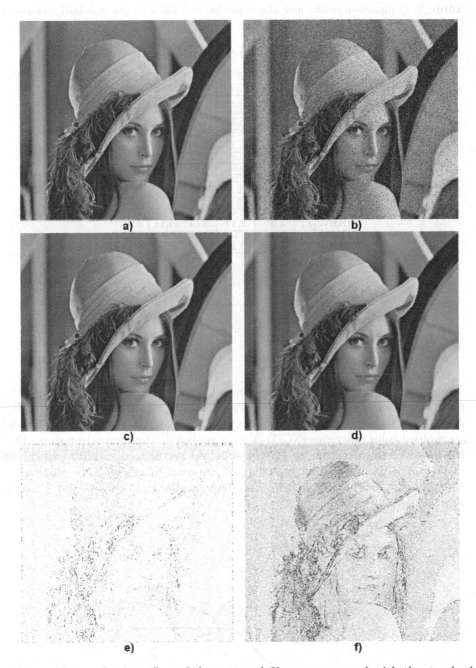

Fig. 4. Noise reduction effect of the proposed filter as compared with the standard VMF: a) colour test image *LENA*, b) image distorted by 4 % impulsive noise, c) new method (3×3 window, $\beta = 0.5$), d) VMF, e) and f) the absolute difference between the original and filtered image (the RGB values were multiplied by factor 10).

Table 2. Comparison of the new algorithm ($\rho_0 = 0.75$) with the standard techniques (Tab. 1) using the *LENA* standard image.

FILTER *LENA*	NMSE $[10^{-3}]$	RMSE	SNR [dB]	PSNR [dB]	NCD $[10^{-4}]$
NONE	514.72	32.17	12.88	17.98	79.17
AMF	79.32	12.63	21.01	26.11	82.75
VMF	18.77	6.14	27.27	32.375	40.47
BVDF	24.59	7.03	26.09	31.19	41.15
GVDF	19.47	6.26	27.11	32.20	41.77
DDF	18.87	6.16	27.24	32.34	40.24
HDF	18.61	6.12	27.30	32.40	41.28
AHDF	18.31	6.07	27.37	32.47	41.17
FVDF	22.25	6.69	26.53	31.63	44.69
ANNF	26.80	7.34	25.72	30.82	48.01
ANP-E	78.60	12.57	21.05	26.14	82.46
ANP-G	78.62	12.57	21.05	26.14	82.48
ANP-D	24.18	6.97	26.17	31.26	46.07
NEW-L_1	**4.99**	**3.17**	**33.01**	**38.11**	**6.62**
NEW-L_2	**5.24**	**3.256**	**32.80**	**37.90**	**6.51**
NEW-L_∞	**5.24**	**3.25**	**32.80**	**37.90**	**6.51**

4   Results

The color images *Lena* and *Peppers* have been contaminated by 4% impulsive noise. The root of the mean squared error (RMSE), signal to noise ratio (SNR), peak signal to noise ratio (PSNR), normalized mean square error (NMSE) and normalized color difference (NCD) [3] were analyzed. The comparison shows that the new filter outperforms by far the standard vector median filter (which can be treated as a reference filter), when the impulsive noise has to be eliminated. The efficiency of the new filtering technique is shown in Tabs. 2 and 3 and in Figs. 3 and 4.

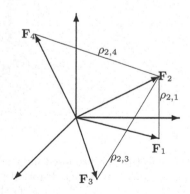

Fig. 5. In the new method the distance R_2 equals $R_2 = \rho(2,1) + \rho(2,3) + \rho(2,4)$.

The threshold β in (13) is a design parameter of the new filter. For $\beta = 0$ similar results to that obtained with the standard VMF have been achieved. However if $\beta > 0$ then the new filter has the ability of noise removal while preserving fine image details (lines, edges, corners, texture). If the RGB values of the color image are normalized to the range $[0, 1]$, then good filtering results have been achieved for β from the $[0.5, 0.75]$ interval.

Table 3. Comparison of the new algorithm ($\rho_0 = 0.75$) with the standard techniques (Tab. 1) using the *PEPPERS* standard image.

FILTER *PEPPERS*	NMSE $[10^{-3}]$	RMSE	SNR [dB]	PSNR [dB]	NCD $[10^{-4}]$
NONE:	650.33	32.63	11.87	17.86	75.23
AMF	108.65	13.34	19.64	25.63	100.47
VMF	27.57	6.72	25.60	31.59	51.64
BVDF	47.94	8.86	23.19	29.18	54.67
GVDF	31.86	7.22	24.97	30.96	52.70
DDF	28.18	6.79	25.50	31.49	51.14
HDF	26.82	6.63	25.72	31.71	51.42
AHDF	26.43	6.58	25.78	31.77	51.32
FVDF	33.34	7.39	24.77	30.76	54.07
ANNF	45.12	8.59	23.46	29.45	65.89
ANP-DF	37.24	7.81	24.29	30.28	56.39
ANP-EF	106.70	13.22	19.72	25.71	99.76
ANP-GF	106.69	13.22	19.72	25.71	99.75
NEW-L_1	**8.93**	**3.82**	**30.49**	**36.48**	**9.81**
NEW-L_2	**8.51**	**3.73**	**30.70**	**36.69**	**8.71**
NEW-L_∞	**8.51**	**3.73**	**30.70**	**36.69**	**8.71**

5   Conclusions

The new algorithm presented in this paper can be seen as a modification and improvement of the commonly used vector median filter. The computational complexity of the new filter is significantly lower than that of the Vector Median. The comparison shows that the new filter outperforms the VMF, as well as other standard procedures used in color image processing, when the impulse noise should be eliminated.

References

1. Pitas, I., Venetsanopoulos, A. N.: Nonlinear Digital Filters: Principles and Applications. Kluwer Academic Publishers, Boston, MA (1990)
2. Plataniotis, K.N., Venetsanopoulos, A.N.: Color Image Processing and Applications. Springer Verlag (June 2000)
3. Venetsanopoulos, A.N., Plataniotis, K.N.: Multichannel image processing. Proceedings of the IEEE Workshop on Nonlinear Signal/Image Processing **2-6** (1995)
4. Pitas, I., Tsakalides, P.: Multivariate ordering in color image processing. IEEE Trans. on Circuits and Systems for Video Technology **1, 3** (1991) 247-256

5. Karakos, D.G., Trahanias, P.E.: Combining vector median and vector directional filters: The directional-distance filters. Proceedings of the IEEE Conf. on Image Processing, ICIP-95 (Oct 1995) 171-174

6. Pitas, I., Venetsanopoulos, A.N.: Order statistics in digital image processing. Proceedings of IEEE **80**(12) (1992) 1893-1923

7. Astola, J., Haavisto, P., Neuovo, Y.: Vector median filters. IEEE Proceedings **78** (1990) 678-689

8. Trahanias, P.E., Venetsanopoulos, A.N.: Vector directional filters: A new class of multichannel image processing filters. IEEE Trans. on Image Processing **2**(4) (1993) 528-534

9. Trahanias, P.E., Karakos, D., Venetsanopoulos, A.N.: Directional processing of color images : theory and experimental results. IEEE Trans. on Image Processing **5**(6) (1996) 868-880

10. Karakos, D., Trahanias, P.E.: Generalized multichannel image filtering structures. IEEE Trans. on Image Processing **6**(7) (1997) 1038-1045

11. Gabbouj, M., Cheickh, F.A.: Vector median - vector directional hybrid filter for colour image restoration. Proceedings of EUSIPCO (1996) 879-881

12. Plataniotis, K. N., Androutsos, D., Venetsanopoulos, A.N.V.: Fuzzy adaptive filters for multichannel image processing. Signal Processing Journal **55**(1) (1996) 93-106

13. Plataniotis, K. N., Androutsos, D., Sri, V., Venetsanopoulos, A.N.V.: A nearest neighbour multichannel filter. Electronic Letters (1995) 1910-1911

14. Plataniotis, K.N., Androutsos, D., Vinayagamoorthy, S., Venetsanopoulos, A.N.V.: Color image processing using adaptive multichannel filters. IEEE Trans. on Image Processing **6**(7) (1997) 933-950

15. Plataniotis, K.N., Androutsos, D., Venetsanopoulos, A.N.V.: Colour Image Processing Using Fuzzy Vector Directional Filters. Proc. of the IEEE Workshop on Nonlinear Signal/Image Processing. Greece (1995) 535-538

Hierarchical Method of Digital Image Segmentation Using Multidimensional Mathematical Morphology

Grzegorz Kukiełka and Jerzy Woźnicki

Institute of Microelectronics and Optoelectronics, Warsaw University of Technology
ul. Koszykowa 75, 00-663 Warsaw, POLAND
{J.Woznicki, G.Kukielka }@imio.pw.edu.pl

Abstract. We consider the problem of image segmentation as the general image processing task. The proposed algorithm is modification of the approach described in [14]. Based on the multidimensional morphological filter theory a universal segmentation algorithm is developed. We also present the results of the described segmentation method on several examples containing grayscale images of different objects.

Keywords: mathemathical morphology, image segmentation, image sequences

1. Introduction

Segmentation is the process that subdivides an image into its constituent parts or objects. The main goal of the segmentation task is to find a partition for a given image. Each segment created during the segmentation process should be as close as possible to the real objects perceived by the human visual system. The following criteria of homogeneity are usually used during segmentation process: colour, texture and gray level. Mathematically speaking segmentation is a computational process (natural or artificial) that translates the original image representation into the form of described segments (regions). After segmentation properties of the regions as well as their locations are known.

As a fundamental step of many image processing system segmentation is very active research area. In the literature many techniques are described [2][7][12][14]. Some of them are considered as general approaches while others are designed for specific applications. In this work general approach is described which is based on the methods of mathematical morphology.

Using a set theoretical methodology for image analysis, mathematical morphology can estimate many features of the geometrical structure in the image such as size, connectivity, shape, and others. This corresponds with nature of human visual system. For this reason we use mathematical morphology methods to solve the image segmentation problem.

The rest of the paper is organized as follows. Section 2 provides an overview of the proposed algorithm of segmentation. Section 3 discusses possible applications and presents examples of segmentation. Section 4 concludes the paper and presents possible future work.

W. Skarbek (Ed.): CAIP 2001, LNCS 2124, pp. 581–588, 2001.

2. Proposed Method of Image Segmentation

In this section a description of the proposed method of image segmentation is presented. Based on the m-valued morphological filter theory a universal segmentation algorithm is developed. The proposed algorithm is modification of the approach described in [14]. Because we propose general segmentation method we use n-dimensional in space and m-dimensional in value domain images (definition 1) as inputs.

Definition 1. n-*dimensional, m-valued image is a transformation* $\mathbf{f}$: $\mathbf{E}$ $\mathbf{T}$, *where* $\mathbf{E}$ *is* n-*dimensional digital space* ($\mathbf{E}$ $\mathbf{Z}^n$) *and* $\mathbf{T}$ *is the cartesian product of* m *totally ordered complete lattices* $T_1, T_2, \ldots, T_m$:

$$\mathbf{T} = T_1 \ T_2 \ \ldots \ T_m. \tag{1}$$

Some examples of n-dimensional (n>1) and m-valued (m>1) types of images are presented in table 1.

Table 1. Examples of n-dimensional, m-valued images.

Mapping type	Image sizes	Type of image
$\mathbf{F}$: (x, y) Y	n = 2, m = 1	Gray level
$\mathbf{F}$: (x, y, z) Y	n = 3, m = 1	3D gray level
$\mathbf{F}$: (x, y) (r, g, b)	n = 2, m = 3	Color image
$\mathbf{F}$: (x, y, z) (r, g, b)	n = 3, m = 3	3D color image
$\mathbf{F}$: (x, y, t) (r, g, b)	n = 3, m = 3	Color image sequence
$\mathbf{F}$: (x, y) (v_x, v_y)	n = 2, m = 2	Estimated motion field
$\mathbf{F}$: (x, y, z, t) (r, g, b)	n = 4, m = 3	3D color moving image

Mathematically speaking the problem of hierarchical segmentation is to find a set of image partition R_i, i 0 $[1, \ldots, N]$ for $\mathbf{E}$ which satisfy the condition:

$$\forall x \ \mathbf{E} \ ; \quad R_1 \ R_2 \ \ldots \ R_N \tag{2}$$

These partition images create a hierarchy for the segmented input image. From relation (2) we know that the hierarchical segmentation propagate the boundary of partition image precisely from iteration to iteration of the algorithm.

In the literature various segmentation algorithms are described. Although these algorithms appear in different forms depends on the application, usually they are composed of four main steps: preprocessing, feature extraction, decision and post-processing. Using mathematical morphology methods these steps can be implemented as shown in the figure 1.

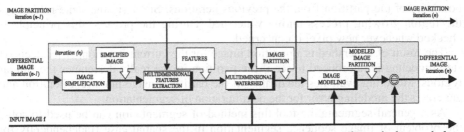

Fig. 1. Block diagram of the hierarchical segmentation using mathematical morphology methods.

Simplification filter should control the type and amount of data removed from the image before the next step of the segmentation. We used following mathematical morphology filters during this phase: opening (or closing) by reconstruction, and area opening (or closing). Opening by reconstruction preserves all connected components that are not totally removed by erosion by a structuring element of size r. This opening has a size oriented simplification effect for example in the case of gray level images it removes bright elements that are smaller than the structuring element. By duality closing by reconstruction has the same effect but on the dark components. Area opening (by duality closing) is similar to the previous one except that it preserves the connected components that have a number of pixels larger than a limit 8. Difference between the reconstruction and area filters is following: the area filters preserves thin but large (>8(size limit) r) objects, reconstruction filters removes components if the width is greater than the size of the structuring element. Results of filtration by area filters are located between original data and results of filtration by reconstruction filters.

Feature extraction is the next step of the algorithm. During these phase the regions of interest are extracted. Extraction procedure depends on the definition of similarity. The similarity criterion used in our algorithm is a planar one. The similarity criterion is also called the distance function. During tests we used city block distance because of the maximal difference in all channels, but there are several definitions of distance functions which can be used. The result of the feature extraction step is the partition image where zero values represents uncertain areas. It is important that any feature extracted may not cross the boundary of the partition from the previous iterations.

The goal of the decision step is to determine the remaining uncertain regions after feature extraction procedure. During these phase we used modified version of the watershed algorithm. The idea is to rely on classical binary segmentation tools.

One of the simplest approaches consists in computing the distance function l_f on the binary image f and in computing the watershed of $(-l_f)$. The watershed transform associates to each minima of $(-l_f)$ a region called a catchment basin. Note that minima of $(-l_f)$ are the maxima of the distance function. Is starts from the extracted features and extent them until they occupy all the available space E (see definition 1). In modified version of the watershed algorithm a pixel is chosen because it is the neighborhood of the feature (marker) and the similarity between them is the highest at that time than any other pair of pixel and neighborhood feature. The same similarity measurement called distance function is used as in the previous step (feature extraction). Constraint from the previous level is important and should be taken into account during current iteration of algorithm. Any feature extracted may not cross the boundary of the partition from the previous iterations. Similarly the same rule obeys the region growing process during watershed. During the process, this principle is checked whenever new pixel is concerned.

The result of watershed is a partition image for the current iteration. After decision step modeling is introduced to see if the partition image really designates the similar regions. Modeling evaluates the quality of the partition image in current iteration and generates the modeling error for the next iteration of segmentation.

As a general segmentation tool this method of segmentation can be used to solve the problem of image sequence segmentation in the spatial case (independently for each image frame) as well as in time case (problem of the motion field segmentation). The basic structure of the motion segmentation system using described method of segmentation is shown on the figure 2.

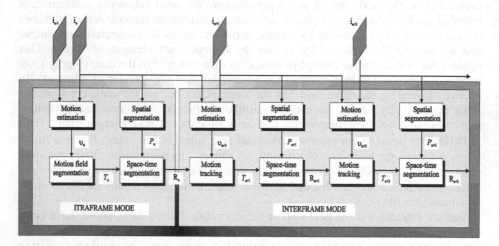

Fig. 2. Block diagram of the spatio-temporal segmentation system with motion tracking function. i_n – original image frame number n, υ_n – estimated motion field for frame number n, T_n – image partition (result of segmentation of υ_n), S_n – image partition (result of segmentation of i_n), R_n – partition image; result of spatio-temporal segmentation of i_n and υ_n.

During experiments the best block matching technique and spatio-temporal differential algorithm are used as the motion estimation methods. Obviously the quality of estimated motion field is not a problem because we test the performance of the segmentation algorithm. Segmentation process is relatively easy in the case of good quality estimated motion field.

3. Experimental Results

The performance of the proposed segmentation algorithm will be illustrated for real images (spatial case) and image sequences (spatio-temporal case).

3.1 Spatial Segmentation

Attempt to use image analysis for the assessment of tumour-induced angiogenesis on an animal model.

Angiogenesis is the formation of new blood vessels. We aimed our study to assess the effects of various stimuli on the development of new blood vessels using the digital image processing system. The purpose of using image segmentation for the evaluation of the test was to introduce automatic quantification with a simultaneous definition of the surface, shape and size of new blood vessels.

In this paper we describe the first step of our studies, which concerns the ability of spatial image segmentation of microscopic image to a form of known number of regions representing blood vessels differentiated from the background.

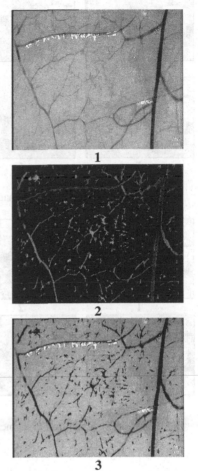

Fig. 3. The goal of the segmentation is to find regions of blood vessels. Original image is shown in (1), result of image segmentation is shown in (2). Superposition of (1) and (2) after thresholding is shown in (3).

3.2 Spatio-Temporal (Motion) Segmentation

Motion segmentation can provide the basis for content based video accessibility because we know where the real moving objects are in the dynamic scene and where they will go. Of course it is good example of spatio-temporal segmentation using segmentation model shown in the figure 2. In the figure 4 the original image sequence is shown. The results of the spatial segmentation for images from figure 4 are shown in the figure 5.

Using the motion estimation algorithm described in [11] the computed motion fields images are shown in the figure 6. These motion fields was segmented by the hierarchical segmentation algorithm described in section 2.

Fig. 4. Original image sequence of three moving objects.

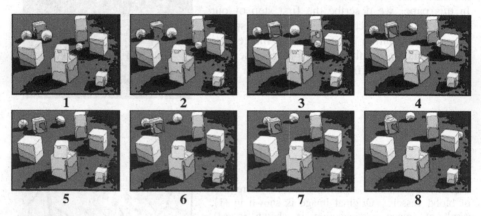

Fig. 5. Results of spatial frame segmentation (inter mode) of image sequence from figure 4.

Finally the spatio-temporal segmentation combines spatial partitions with motion partitions into a precise partitions shown in the figure (7). One example of content based image data manipulation is shown in the figure (8). The manipulated image sequence is shown in the figure (4) where the foreground moving objects are and the background is from the sequence presenting rotating chair.

4. Conclusions

Being independent of any specific application, this mathematical morphology segmentation method can be used in many fields. The described segmentation method does not destroy the contours of the extracted regions because of the properties of morphological filters used during simplification phase of the algorithm. Because all morphological filters and operators are based on very simple operations so can be easy implemented both in software and hardware.

Fig. 6. Estimated motion fields for the pairs of images shown in the figure (4). Spatio-temporal differential algorithm [11] is used as the motion estimation method with the following parameters: $\nabla^2 = 600$, number of iterations 500. Estimated motion fields are shown on (a) for images 1 i 2, on (b) for images 2 and 3, on (c) for images 3 and 4, on (d) for images 4 and 5 respectively.

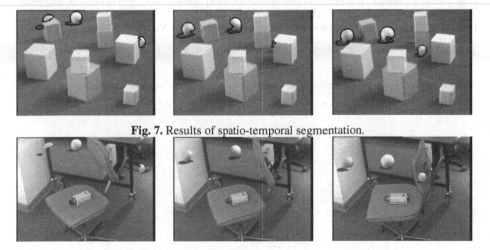

Fig. 7. Results of spatio-temporal segmentation.

Fig. 8. Example of content based image data manipulation.

Acknowledgements. This work was partially supported by KBN grant 7-T08A-050-16.

References

[1] Aggarwal, J.K., Nandhakumar, N.: On the computation of motion from sequences of images – a review. Proc. IEEE, Vol.76, No. 8, August (1988) 917-935

[2] Ayer, S., Schroeter, P., Bigun, J.: Segmentation of moving objects by robust motion parameter estimation over multiple frames. Third European Conference on Computer Vision, ECCV'94, Vol.2. Stockholm, May (1994) 316-327

[3] Bar-Shalam Y., Xiao-Rong, L.: Estimation and tracking: principles, techniques, and software. Artech House, Boston, MA, (1993)

[4] Bar-Shalom, Y., Fortmann, T.E.: Tracking and Data Association. Academic Press (1988)

[5] Csinger, A.: The psychology of Visualization. University of British Columbia (1992)

[6] Dougherty, E.: Mathematical Morphology in Image Processing. Optical Engineering Ser. Vol. 34. Dekker Inc. U.S. (1992)

[7] Haralick, R.M.: Mathematical Morphology: Theory & Hardware. Oxford Series on Optical & Imaging Sciences No.12. Oxford University Press.

[8] Haralick, R.M., Shapiro, L.G.: Computer and Robot Vision. Vol 2, Addison-Wesley (1993)

[9] Haralick, R.M., Sternberg, S.R., Zhuang, X.: Image analysis using mathematical morphology. IEEE Transaction on Pattern and Machine Intelligence 9 (1987) 532-550

[10] Heijmans, H.J.A.M.: Morphological Image Operators. Academic Press, Boston (1994)

[11] Horn, B.K.P., Schunck, B.G.: Determining Optical Flow, Artificial Intelligence, Vol. 17 (1981)

[12] Meyer, F., Beucher, S.: Morphological segmentation. Journal of Visual Communication and Image Processing, Vol. 1, No. 1 (1990) 21-46

[13] Nieniewski, M.: Morfologia matematyczna w przetwarzaniu obrazów. Akademicka Oficyna Wydawnicza PLJ, Warszawa (1998)

[14] Salembier, P.: Morphological multiscale segmentation for image coding. Signal Processing, Vol. 38, No. 3 (1994) 359-386

[15] Serra, J.: Image Analysis and Mathematical Morphology, Vol. 2: Theoretical Advances. Academic Press Inc., U.S. (1988)

[16] Serra, J.: Image Analysis and Mathematical Morphology. Vol. 1, Academic Press (1982)

Images of Imperfectly Ordered Motifs: Properties, Analysis, and Reconstruction

Vladislav Krzyžánek[1,2*], Ota Samek[1], and Rudolf Reichelt[2]

[1] Institute of Physical Engineering, Brno University of Technology,
Technická 2, CZ-61669 Brno, Czech Republic
[2] Institute for Medical Physics and Biophysics, University of Münster,
Robert-Koch-Str. 31, D-48149 Münster, Germany
krzyzane@uni-muenster.de

Abstract. We report on a study of images with imperfectly ordered motifs (distorted 2D lattices). Our aim was to study the Fourier transform properties of the images according to the Fourier transform of perfect lattices and to improve methods for analysis of continuously distorted 2D lattices and techniques for 2D reconstruction. For this purpose, the locally normalized correlation is used in the method of correlation averaging. In order to reduce the image distortion the function of the distortion is approximated step by step by the Lagrange interpolation polynomials. The modification of the methods is demonstrated on electron micrographs of the S-layers of differently prepared *Cyanobacteria*.

Keywords: image analysis, imperfectly ordered motifs, Lagrange interpolation

1 Introduction

Images which consist of imperfectly ordered motifs are often dealt with in many scientific areas. Natural 2D crystals[1] are typical representatives of such images. However, perfect crystals are a prerequisite for electron crystallography, but are rarely perfectly ordered in nature (see for example Fig. 1). Distortion of an ideal lattice often limits the resolution that can be obtained from electron micrographs by methods of image analysis (e.g. [6], for overview of computational tools for microscopy-based structural biology see [3]). The image analysis is mostly needed on one hand to characterize the repeating motifs (or more precisely the unit cells), especially their lattice parameters, the type of symmetry and on the other hand for reconstruction. Two classical methods are often used for this analysis: the method of spatial-frequency spectrum modification (SFSM) [1] and the method of correlation averaging (CA) [4]. In our study we will use the

---

* Present address in Münster.
[1] In our case of 2D crystals, surface layers (S-layers) of *Cyanobacteria* which represent a noisy continuously distorted lattices were studied.

W. Skarbek (Ed.): CAIP 2001, LNCS 2124, pp. 589–600, 2001.

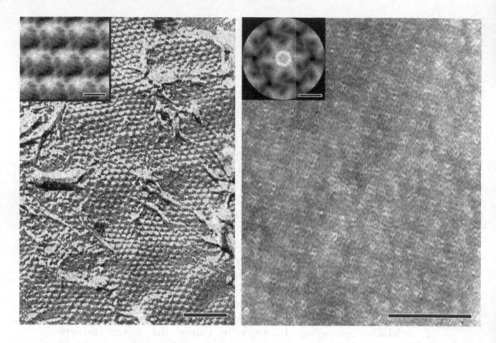

Fig. 1. Electron micrographs represent continuously distorted lattices, and the related 2D reconstructions. Left: S-layer of *Microcystis* cf. *wesenbergii* Bítov 1994; the sample was freeze-dried and subsequently uni-directionally shadowed with metal. Right: S-layer of negatively stained *Synechocystis aquatilis* Gromov 1965/428. (Bars: 100 nm for micrographs, 10 nm for reconstructions.) The transmission electron micrographs were kindly provided by J. Šmarda and D. Šmajs

SFSM method because of higher precision, elegancy, and simplicity. Thus, the Fourier transform (FT) of distorted lattices is important for our study, mainly to outline the effect of lattice distortions on the accuracy of the SFSM method.

2   Perfect 2D Crystal Lattices

2D crystal lattices and their FT are the basis of our image analysis. The notation used for the following mathematical considerations is according to [7].

2.1   Fourier Transform of Perfect Lattices

Following rather general definitions of the FT and the inverse FT are used:

$$F(\boldsymbol{X}) = A^2 \iint\limits_{-\infty}^{\infty} f(\boldsymbol{x}) \exp(-\mathrm{i}k\boldsymbol{X} \cdot \boldsymbol{x}) \, \mathrm{d}^2 \boldsymbol{x}$$

and

$$f(x) = B^2 \int\!\!\int\limits_{-\infty}^{\infty} F(X) \exp(ikX \cdot x)\, \mathrm{d}^2 X \; ,$$

where both functions f and F have to satisfy the well known conditions for FT. A, B, and k are real constants[2] satisfying the condition

$$AB = \frac{|k|}{2\pi} \; .$$

The finite 2D crystal lattice can be described by two forms:

1. as a general lattice detached by a shape function from the infinite lattice

$$f_{\text{perfect}}(x) = f_U(x) * \sum_{n \in \inf} \delta(x - x_n)s(x) \; , \tag{1}$$

2. and as a lattice consisting only of entire unit cells

$$f_{\text{perfect}}(x) = f_U(x) * \sum_{n \in V} \delta(x - x_n) \; , \tag{2}$$

where the function $f_U(x)$ characterizes the unit cell (an ideal motif), a_1 and a_2 are the basis lattice vectors, and $x_n = n_1 a_1 + n_2 a_2$ is a lattice vector. The finiteness of the lattices is defined by the shape function $s(x)$ expressed by the unit-step function in Eq. (1) and by V in Eq. (2). The FTs of Eqs. (1) and (2) are given as

$$F_{\text{perfect}}(X) = \frac{1}{A^2 V_U} F_U(X) \sum_{h \in \inf} \delta\left(X - \frac{2\pi}{k} X_h\right) * S(X) \; , \tag{3}$$

and

$$F_{\text{perfect}}(X) = F_U(X) \sum_{n \in V} \exp(-ikx_n \cdot X) \; , \tag{4}$$

where $V_U = |a_1 \times a_2|$ is the volume of the unit cell, a_1^{-1} and a_2^{-1} are the basis vectors of the reciprocal lattice, $X_h = h_1 a_1^{-1} + h_2 a_2^{-1}$ is a lattice vector of the reciprocal lattice and $S(X)$ is the shape amplitude (the FT of the shape function $s(x)$). Because of the position of diffraction maxima of the order h lies on the position $\frac{2\pi}{k} X_h$, in the next, the positions of the maxima will be chosen as positions of the reciprocal lattice points and are given by vector $\tilde{X}_h$, i.e. $\tilde{X}_h = \frac{2\pi}{k} X_h$.

Hence, the FT of a perfect infinite lattice consists of the reciprocal lattice points and, in the case of perfect finite lattice, these reciprocal lattice points are superimposed by the shape amplitude $S(X)$.

---

[2] Generally $A, B \in \mathbb{C}$ but we confine there to $\mathbb{R}$ for simplicity. The use of the 3 constants were introduced [7] because of a generalization; the constants $[A, B, k]$ have often different values, such as $[1, 1, 2\pi]$, $[1, 1, -2\pi]$, $[1/2\pi, 1, -1]$.

2.2 Method of Spatial-Frequency Spectrum Modification

The SFSM method [1] is based on the fact that the FT of crystal lattices is a reciprocal lattice (Sect. 2.1). However, Aebi et al. [1] discuss reconstruction of crystal lattices only with additive noise and without any space distortion. The FT of distorted lattices gives a "reciprocal lattice" only at a vicinity of the origin. The distortion influences not only the FT functional values at all reciprocal lattice points but also the area in the vicinity of well recognized reciprocal lattice points (Sect. 3). Hence, the method is ideal only for images with added noise.

The 2D reconstruction of the lattice consisting of many unit cells in each direction can be performed just from the FT values $F_{\text{perfect}}(\tilde{\boldsymbol{X}}_h)$ at lattice points of the reciprocal lattice. It can be done either by the Fourier series

$$f(\boldsymbol{x}) \sim \sum_h F_{\text{perfect}}(\tilde{\boldsymbol{X}}_h) \exp(\mathrm{i}2\pi \boldsymbol{X}_h \cdot \boldsymbol{x}) \tag{5}$$

or by using the inverse FT of filtered function $F_{\text{perfect}}(\boldsymbol{X})$

$$f(\boldsymbol{x}) \sim \mathrm{FT}^{-1}\left\{ F_{\text{perfect}}(\boldsymbol{X}); \quad F_{\text{perfect}}(\boldsymbol{X}) = 0 \text{ for } \forall \ \boldsymbol{X} \neq \tilde{\boldsymbol{X}}_h \right\} . \tag{6}$$

It is the special case of the SFSM method where a set of δ-functions defines the filter.

2.3 Method of Correlation Averaging

The CA method [4] performs calculation of 2D reconstruction by arithmetic averaging of all unit cells of the image. Their exact superposition is determined by the correlation. Provided that the local maxima of the correlation of the whole image $f(\boldsymbol{x})$ and a selected small region (reference region or reference function) $f_{\text{ref}}(\boldsymbol{x})$ of the image, which contains at least one unit cell, describes the positions of the lattice points. The correlation is defined by

$$f_{\text{ref}}(\boldsymbol{x}) \star f(\boldsymbol{x}) = \iint\limits_{V_{\text{ref}}} f_{\text{ref}}(\boldsymbol{y}) f(\boldsymbol{x}+\boldsymbol{y}) \, \mathrm{d}^2\boldsymbol{y} , \tag{7}$$

where the integral is taken over the area V_{ref} of the reference region. Also this method is useful for noisy images without distortion, but in contradiction to the SFSM method, it is also useful for images consisting of randomly deviated motifs. In the case of other distortions, undesirable artefacts can occur (cf. Sect. 4.1). A partial elimination of the artefacts can be done using an averaged motif obtained by the CA method as the reference function [10].

3   Fourier Transform of Distorted Lattices

The influence of a distortion of lattices to their FT considerably affects the accuracy and is of major interest. As already mentioned above, continuous distortion

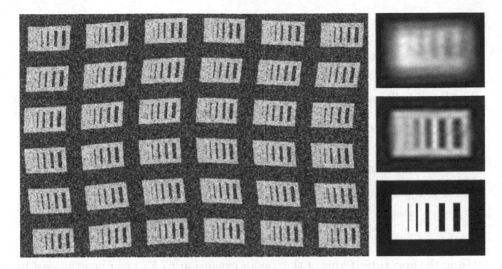

Fig. 2. Model of a continuously distorted lattice and its 2D reconstructions obtained by the SFSM method (on the right from top to bottom) without and with reduction of the distortion. The original nondistorted image where the smallest black line has a width of 1 pixel is shown at the bottom right

is the main type of distortion of our images (Fig. 1, 2). However, to determine the FT of such lattices is not an easy task. The model of a general continuously distorted lattice

$$f(\boldsymbol{x}) = f_{\text{perfect}} \{\boldsymbol{f}_{\text{dist}}(\boldsymbol{x})\} \ ,$$

where $\boldsymbol{f}_{\text{dist}}(\boldsymbol{x})$ describes a continuous transformation of the coordinates $\boldsymbol{x}$, has not been solved as yet. That is why we introduce another, simpler type of distortion: an image consisting of randomly displaced, differently oriented, and noisy motifs having different size. The FT of this image is similar to a continuously distorted lattice.

Images consisting of randomly displaced, differently oriented, and noisy motifs having different size can be described by

$$f(\boldsymbol{x}) = \sum_{n \in V} f_{\text{U}n}(\beta_n \mathcal{A}_n \boldsymbol{x}) * \delta(\boldsymbol{x} - \boldsymbol{\varepsilon}_n) \ , \tag{8}$$

where V specifies the finite range of the lattice, $\boldsymbol{\varepsilon}_n = n_1 \boldsymbol{a}_1 + n_2 \boldsymbol{a}_2 + \boldsymbol{\xi}_n$ is lattice vector with random deviation $\boldsymbol{\xi}_n$ from the periodic lattice points, and the function $f_{\text{U}n}(\boldsymbol{x})$ characterizes the individual noisy unit cell (motif), which can be expressed by the sum of the ideal motif $f_{\text{U}}(\boldsymbol{x})$ and a random noise $g_n(\boldsymbol{x})$, i.e. $f_{\text{U}n}(\boldsymbol{x}) = f_{\text{U}}(\boldsymbol{x}) + g_n(\boldsymbol{x})$. Coefficients β_n describe the individual magnification of particular motifs and the matrix $\mathcal{A}_n = \begin{pmatrix} \cos \varphi_n & -\sin \varphi_n \\ \sin \varphi_n & \cos \varphi_n \end{pmatrix}$ their rotations by the individual angle φ_n.

The FT of Eq. (8) is given by

$$F(X) = \sum_{n \in V} \frac{1}{\beta_n} F_{Un} \left(\frac{1}{\beta_n} X \mathcal{A}_n^{-1} \right) \exp(-ik\varepsilon_n \cdot X) \ . \tag{9}$$

The Eq. (9) allows to study in particular simple distortions and in particular the FT values in the original reciprocal lattice points. Equation (9) represents the basic formula for that work.

3.1   Images Consisting of Randomly Displaced Motifs

The FT in points $\tilde{X}_h$ is given by

$$F(\tilde{X}_h) = F_U(\tilde{X}_h) \sum_{n \in V} \exp(-ik\xi_n \cdot \tilde{X}_h) \ .$$

Using the first three terms of the Taylor expansion $F(\tilde{X}_h)$ can be expressed by

$$F(\tilde{X}_h) \approx F_U(\tilde{X}_h) \left[\sum_{n \in V} 1 - ik\tilde{X}_h \cdot \sum_{n \in V} \xi_n - \frac{k^2}{2} \sum_{n \in V} \left(\xi_n \cdot \tilde{X}_h \right)^2 \right] \ .$$

The first sum is equal to the number of motifs in the image (denoted as N_V). The second sum is a multiple of N_V and the mean value of the deviations ξ_n. Provided that the mean value is equal 0 (i.e. symmetric distribution of random deviations), the second sum is then also equal 0. The argument in the third sum can be rewritten as $\sum_{n \in V} (\xi_n \cdot \tilde{X}_h)^2 = \tilde{X}_h^2 \sum_{n \in V} [\xi_n \cos \angle(\xi_n, \tilde{X}_h)]^2$, i.e. the sum can be reduced to a variance of the deviations ξ_n in the direction of the vector $\tilde{X}_h$ (denoted as $\Delta_{\xi_n, \tilde{X}_h}$). Then one obtains a more simple relation

$$F(\tilde{X}_h) \approx F_U(\tilde{X}_h) \left(N_V - \frac{k^2}{2} |\tilde{X}_h|^2 \Delta_{\xi_n, \tilde{X}_h} \right) \ . \tag{10}$$

From the Eq. (10) it is obvious that the values of the function $F(\tilde{X}_h)$ (i) are noisy (the values decrease with increasing distance from the vicinity), and (ii) are dependent on the dispersion of the motif displacements.

3.2   Images Consisting of Motifs with Noise

The FT after decomposition of (9) into two components (the first one corresponds to a perfect lattice and the second one to the noise) is given as

$$F(X) = N_V F_U(X) + \sum_{n \in V} G_n(X) \exp(-ikx_n \cdot X)$$

and in the points $\tilde{X}_h$

$$F(\tilde{X}_h) = N_V F_U(\tilde{X}_h) + \sum_{n \in V} G_n(\tilde{X}_h) \ .$$

Supposing that the mean value of the noise according to particular motifs is equal to 0, the sum is also 0. In that case one obtains

$$F(\tilde{\boldsymbol{X}}_h) = N_V F_U(\tilde{\boldsymbol{X}}_h) \quad \text{for} \quad \sum_{n \in V} g_n(\boldsymbol{x}) = 0 \ . \tag{11}$$

Hence, the FT values in points $\tilde{\boldsymbol{X}}_h$ are not effected in this case. Note that Eq. (11) represents essentially the SFSM method.

3.3 Image Consisting of Motifs with Different Angular Orientation

The FT in points $\tilde{\boldsymbol{X}}_h$ is given by

$$F(\tilde{\boldsymbol{X}}_h) = \sum_{n \in V} F_U(\tilde{\boldsymbol{X}}_h \mathcal{A}_n^{-1}) \ . \tag{12}$$

The Eq. (12) shows that the FT values of this image in original lattice points $\tilde{\boldsymbol{X}}_h$ is a multiple of N_V and the arithmetic average of the FT of the particular motifs rotated by angles φ_n.

3.4 Image Consisting of Motifs with Different Size

The FT in points $\tilde{\boldsymbol{X}}_h$ is given by

$$F(\tilde{\boldsymbol{X}}_h) = \sum_{n \in V} \frac{1}{\beta_n} F_U \left(\frac{1}{\beta_n} \tilde{\boldsymbol{X}}_h \right) \ . \tag{13}$$

This case of distortion is similar to the previous case, however, in addition, weighting coefficients $1/\beta_n$ in the arithmetic averaging occur.

4   Image Analysis of Distorted Lattices

4.1   Estimation of Lattice Points

The correlation (7) is used for this purpose in the CA method. The result of this correlation at every point can be considered as a measure of squared distance $d^2(\boldsymbol{x})$ of two functions (in Euclidean space) and can be written as

$$d^2(\boldsymbol{x}) = \iint_{V_{\text{ref}}} \left[f_{\text{ref}}^2(\boldsymbol{y}) - 2f_{\text{ref}}(\boldsymbol{y})f(\boldsymbol{x} + \boldsymbol{y}) + f^2(\boldsymbol{x} + \boldsymbol{y}) \right] \ \mathrm{d}^2\boldsymbol{y} \ .$$

The first integral is a constant, the second integral is the correlation according to Eq. (7), and the third integral depends only on the image function and on the position $\boldsymbol{x}$. In the case of a perfect lattice, the third integral is a constant; in the case of distorted lattice the integral fluctuates around a constant (supposing that all the motifs have almost the same intensity). The correlation is the most effective way to estimate the lattice points of a very noisy image that, however, may be deformed only by random displacements of the motifs. For other distortions the correlation (Eq. (7)) can have following disadvantages:

1. dependence of the result of correlation on the selected reference area,
2. decreasing authenticity of estimation of lattice points with increasing distortion,
3. dependence of the function values of the correlation at the points of its local maxima on intensity of particular motifs,
4. occurrence of other local maxima,
5. difficulty to display the correlation image.

These disadvantages can be partially removed by two modifications.

The first modification is the replacement of the correlation (7) by the locally normalized correlation (LNC)[3] defined by

$$
f_{\text{ref}}(\boldsymbol{x}) \overset{N}{\star} f(\boldsymbol{x}) = \frac{\iint\limits_{V_{\text{ref}}} f_{\text{ref}}(\boldsymbol{y}) f(\boldsymbol{x}+\boldsymbol{y}) \, \mathrm{d}^2 \boldsymbol{y}}{\sqrt{\iint\limits_{V_{\text{ref}}} f_{\text{ref}}^2(\boldsymbol{y}) \, \mathrm{d}^2 \boldsymbol{y} \iint\limits_{V_{\text{ref}}} f^2(\boldsymbol{x}+\boldsymbol{y}) \, \mathrm{d}^2 \boldsymbol{y}}} \,, \tag{14}
$$

where function $f(\boldsymbol{x})$ characterizes the intensity of whole image, $f_{\text{ref}}(\boldsymbol{x})$ is the reference function. The integration is over the area V_{ref} of the reference function. The Eq. (14) was derived from (7) by normalization. According to the Schwarz inequality the LNC takes values from interval $\langle 0, 1 \rangle$ – then the LNC is independent of intensity of particular motifs and is less dependent on random deviations of the values of the correlated functions[4].

The second modification reducing considerably the dependence of the correlation (or the LNC) on the selected reference region is the use of an "averaged motif" as a more smooth reference function. In [10] the averaged motif is obtained as a result of the CA method but we prefer the motif obtained from the distorted image by the SFSM method. It is useful to use only the first (or the first and the second) diffraction orders. This motif does not contain the fine structure of the ideal motif but only its rough outer shape and the periodicity. The correlation is then less dependent on the rotation of the particular motifs caused by the continuous distortion.

4.2   Techniques for Reduction of Continuous Distortion

We have developed a technique for reduction of a continuous distortion only from the distorted image itself. The original idea was to find a conformal mapping from the analysis of functions of a complex variable (based on the assumption that the function of distortion is a continuous function) that is described only by positions of lattice points in the distorted image and corresponding lattice

---

[3] The LNC is often called as cross-correlation coefficient (from statistics) and used in reconstructions of particles (e.g. [5]).

[4] In our experience (analysis of cyanobacterial S-layers) the values of the LNC were often within the interval $\langle 0.4, 0.8 \rangle$.

points in a perfect lattice similar to the technique used in SEMPER system[5] [9]. The first result was an approximation of the distortion function step by step using the linear fractional function [8]. A generalization of the approximation is based on the Lagrange interpolation polynomials of a complex variable

$$L_n(z) = \sum_{i=1}^{n} w_i \prod_{\substack{j=1 \\ i \neq j}}^{n} \frac{z - z_j}{z_i - z_j} \,, \tag{15}$$

where $z_j, w_j \in \mathbb{C}$ and n is the degree of the polynomial. Here, we use the knowledge, that the function is uniquely determined by n pairs of corresponding points.

For $n = 4$, the approximated areas are small tetragons subdividing the whole image. According to an acceptable estimation of lattice points which divide the image into tetragons with vertices ε_{n_1,n_2}, ε_{n_1+1,n_2}, $\varepsilon_{n_1+1,n_2+1}$, ε_{n_1,n_2+1} (Fig. 3), it is helpful to use them. Then every tetragon in the distorted image corresponds to one rhomboid in the periodical image by Eq. (15). For this transformation it is necessary to change coordinates: the position of the point in Cartesian reference coordinate system $[x_1, x_2]$ is now characterized by complex number $x_1 + ix_2$. Note that the approximation of the distortion in one unit cell can be done by more neighbouring points.

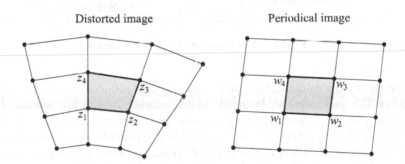

Fig. 3. Scheme of reduction of the distortion

Comparison of the 2D reconstruction of the continuously distorted image, both without reduction of the distortion and with reduction of the distortion, is illustrated in Fig. 2.

---

[5] For determination of the image distortion, the SEMPER system [9] uses the CA method for determination of lattice points, least-square method for the estimation of parameters of the perfect lattice, and subsequently computes 5 parameters of the distortion for each unit cell. This method is very precise but takes time for computation.

4.3   Algorithm for Image Analysis

The algorithm for the image analysis is shown in Fig. 4. It enables the estimation of the lattice parameters, the lattice symmetry (from both the FFT and the LNC), and 2D reconstruction.

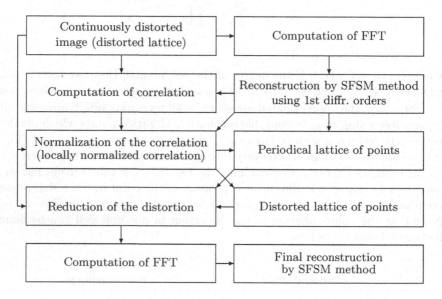

Fig. 4. Algorithm for the image analysis

For the final reconstruction of the image with reduced distortion both the SFSM or CA methods can be used. More accurate results are reached by the SFSM method, because the latter processes the complete image globally and the spatial-frequency spectrum of the analyzed image provides other information for more precise specification of the type of symmetry and the lattice parameters. In addition, the error of estimation of the lattice points which occurs in the CA method does not appear here.

The calculation was performed from digitized images using the MATLAB software package. Some functions were programmed using [2].

5   Experimental Results and Discussion

The described method of image analysis was tested using several models (e.g., see Fig. 2) and applied to analyse electron micrographs of several representatives of various strains of S-layers of *Cyanobacteria* [11,12]. The cyanobacterial S-layers present the most monotonous ultrastructure of a biologic membrane, formed by 2D crystalline arrays of identical protein (or glycoprotein) units; the center-to-center spacing of two neighbouring units (lattice parameters) ranges from 2.5 to 22 nm.

In an ideal case, the S-layers form 2D crystal lattices, however, in real samples the protein units as well as their regular arrangement may be distorted in various ways (Fig. 1). The used S-layers are situated on a non-plane surface of bacteria and may be distorted during their preparation because of different reasons. The S-layers prepared by freeze-fracturing and subsequent freeze-drying show significant distortions. In the case of S-layers prepared by the negative staining the distortions are much smaller than in the freeze-dried sample. Unfortunately, these images possess a much lower contrast.

6 Conclusion

The paper describes a method of analysis of continuously distorted quasi-periodical images applied for the analysis of transmission electron micrographs of S-layers of *Cyanobacteria*. Here, the 2D reconstruction was used, however, it may be used for both the direct and the subsequent 3D reconstruction from projections, respectively. We show that the most frequently used methods, namely the SFSM and the CA, do not provide authentic results for strongly distorted images. The distortion should be reduced before the application of these methods. In the case of 2D reconstruction we focused on a continuous distortion that we reduce using an approximation of the distortion function. For estimation of lattice points of distorted images two modifications of the CA method were used.

Distortions of images may be caused also by some effects like creeping of piezos in scanning probe microscopes and specimen drift, respectively, or by the effect of electromagnetic stray-fields on the electron beam of scanning electron microscopes. Presently, the algorithm for image reconstruction is tested for scanning electron micrographs of periodic samples where the image is distorted because of electromagnetic stray-fields.

Acknowledgments. The authors thank Prof. Jiří Komrska for his helpful discussion. Furthermore, we are grateful to Prof. Jan Šmarda and Dr. David Šmajs for providing the electron micrographs. V. Krzyžánek is indebted to the Grant Agency of the Czech Republic for financial support of the work by Grant No. 101/99/D077.

References

1. Aebi, U., Smith, P.R., Dubochet, J., Henry, C., Kellenberger, E.: A study of the structure of the T-layer of Bacillus brevis. Journal of Supramolecular Structure **1** (1973) 498–522
2. Brigham, E.O.: The Fast Fourier Transform and its Applications, Prentice-Hall, Englewood Cliffs, New Jersey (1988)
3. Carragher, B., Smith, P.R.: Advances in Computational Image Processing for Microscopy. Journal of Structural Biology **116** (1996) 2–8
4. Frank, J.: New methods for averaging non-periodic objects and distorted crystals in biologic electron microscopy. Optik **63** (1982) 67–89

5. Frank, J.: Three-Dimensional Electron Microscopy of Macromolecular Assemblies. Academic Press, San Diego (1996)
6. Hawkes, P.W., Kasper, E.: Principles of Electron Optics, Vol. 3, Part XV. Academic Press, London (1994)
7. Komrska, J.: The Fourier Transform of Lattices. In: Eckertová, L., Růžička, R. (eds.): Proceedings of the International Summer School Diagnostics and Applications of Thin Films, May 27th–June 5th 1991. IOP Publishing, Bristol (1992) 87–113
8. Krzyžánek, V.: Analysis of continuously distorted quasi-periodic images: two-dimensional reconstruction of S-layers of Cyanobacteria. Optical Engineering **39** (2000) 872–878
9. Saxton, W.O.: Semper: Distortion Compensation, Selective Averaging, 3-D Reconstruction, and Transfer Correction in a Highly Programmable System. Journal of Structural Biology **116** (1996) 230–236
10. Saxton, W.O., Baumeister, W.: The correlation averaging of a regularly arranged bacterial cell envelope protein. Journal of Microscopy **127** (1982) 127–138
11. Šmajs, D., Šmarda, J., Krzyžánek, V.: New Finding of S-layers among Cyanobacteria. Algological Studies **94** / Arch. Hydrobiol. Suppl. **129** (1999) 317–332
12. Šmarda, J., Šmajs, D., Krzyžánek, V.: S-Layers on Cell Walls of Cyanobacteria. Micron (accepted)

Implementation and Advanced Results on the Non-interrupted Skeletonization Algorithm

Khalid Saeed, Mariusz Rybnik, and Marek Tabedzki

Computer Engineering Department
Faculty of Computer Science
Bialystok University of Technology
Wiejska 45A, 15 351 Bialystok[1]
POLAND
aidabt@ii.pb.bialystok.pl

Abstract. This paper is a continuation to the work in [1], in which a new algorithm for skeletonization is introduced. The algorithm given there and implemented for script and text is applied here on images like pictures, medical organs and signatures. This is very important for a lot of applications in pattern recognition, like, for example, data compression, transmission or saving. Some interesting results have been obtained and presented in this article. Comparing our results with others we can conclude that if it comes to thinning of scripts, words or sentences our method is as good as some of the latest approaches, when considering cursive script. However, when it comes to pictures, signatures or other more complicated images, our algorithm showed better and more precise results [6].

Keywords: image skeletonization, cursive script thinning

1 Introduction

Skeletonization is a very important stage in pattern recognition when considering almost all methods and approaches of classification. Despite this fact, in most cases, the thinning is not as precise as required, although the authors of such approaches declare that their methods are almost ideal [1,2,3,4]. In fact, there exist some methods that lead to an almost one-pixel-skeletonized image [1,3,4]. These methods proved to really be good when considering scripts or words, or some special applications [2,5] but not pictures or some complicated images like medical ones [1,6]. Since not all authors reveal the details of their algorithms or the computer programs, and they usually do not mention their drawbacks as discussed in the general survey in [7], the comparison is really problematic. In many cases it was made in such a way that the authors of this paper had implemented the algorithms of others according to the basic ideas mentioned in their published proceedings and then compared the results with ours under the same conditions. The experiments showed, in almost all considered

---

[1] *The Rector of Bialystok University of Technology financially supports this manuscript. Grant no. W/II/3/01.*

W. Skarbek (Ed.): CAIP 2001, LNCS 2124, pp. 601–609, 2001.

cases, high efficiency in keeping the connectivity (non-interruption) of the image contour without missing the main features of it. This is essential, particularly in compressing, transmitting or saving data of large size, or of necessary features to keep after thinning.

2   Theoretical Considerations

The most essential considerations in this aspect are concentrated on the following algorithm and its implementation. The general criterion for thinning scripts, words and sentences of different languages is given in [1]. Here, however, we introduce the main stages of the algorithm, which lead to a recognizable contour.

2.1   Algorithm of Thinning

The algorithm of thinning to one-pixel-skeleton image presented in [2] is modified [1] to have a non-interrupted image. By non-interrupted image we mean the one with a continuous contour of one-pixel-wide skeleton. This is basic for most of the applications of the algorithm whose essential steps are given below. We are considering the Arabic letter ﻲ - pronounced *yaa*, but without dots. Arabic letters are of cursive character and hence very good examples of other handwritten languages or images like signatures or pictures. The same letter was used in the basic algorithm [2].

1. The image is bitmapped, first with its black pixels designated 1's:

```
                    1111
                    11111
        1          111111
        1      1
        1      11
        1      11111
        1       1111
        11         11
        111        11
   111111111111
    111111111
     111111
```

2.   The 1's of the contour (that sticking the 0's background) are changed into 2's; those in the elbow corners into 3's. Then we obtain the following figure:

```
                    2222
                    23112
        2          222222
        2      2
        2      22
        2      23222
        2       2222
        22         22
        232        22
   233222222222
    231111322
     222222
```

Remark 1. Note that this stage may end the algorithm indicating the final contoured shape defined by the outside and inside circumferences. If it comes to pictures or some applications of image processing, this is the main aim of skeletonization. Fig.1 shows the letter  together with its contour shape after stage 2:

Fig. 1. The letter ى without dots.

However, to show the whole method of thinning to a one-pixel-skeleton image, being the most required and practical case, consider the following additional stages [1].

3. Consider the points with 2, 3 or 4 sticking neighbors and change them into 4's:

```
                                4224
                               23112
                              222224
              2              2
              2              22
              2              43224
              2               4222
              2                 42
             22                 42
            232
         433222222224
          231111324
           422224
```

Notice that there are 8 such possibilities in each case depending on the way the pixels surrounding the under test point *x* are filled. The weight of each of these pixels is taken from the following array:

128	1	2
64	**x**	4
32	16	8

Therefore, we have 24 different possibilities given in three different arrays with two, three or four neighbors, respectively:

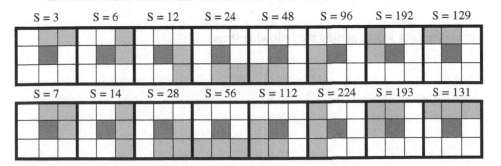

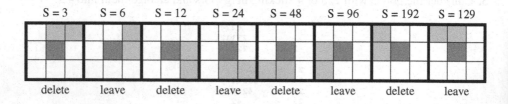

The dark points (x's) show the pixels to either be removed or left, depending on their position in the image. As an example, the following array shows the successive points of removing or leaving for the case of two-neighbour points.

The sum s of the surrounding pixels, ranging from 0 to 255, decides whether to delete the point or not:

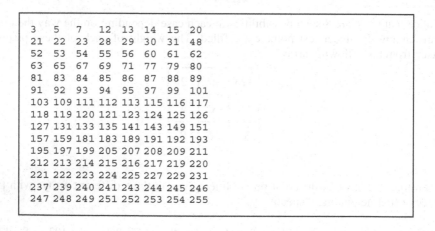

The following Deletion Array gives the sums of points to be removed:

```
  3   5   7  12  13  14  15  20
 21  22  23  28  29  30  31  48
 52  53  54  55  56  60  61  62
 63  65  67  69  71  77  79  80
 81  83  84  85  86  87  88  89
 91  92  93  94  95  97  99 101
103 109 111 112 113 115 116 117
118 119 120 121 123 124 125 126
127 131 133 135 141 143 149 151
157 159 181 183 189 191 192 193
195 197 199 205 207 208 209 211
212 213 214 215 216 217 219 220
221 222 223 224 225 227 229 231
237 239 240 241 243 244 245 246
247 248 249 251 252 253 254 255
```

4. Delete the 4's until getting the following shape:

```
                    22
                    23112
        2           22222
        2           2
        2           22
        2           322
        2           222
        22            2
        232           2
        3322222222
        23111132
            2222
```

5. Check for deleting the unnecessary 2's and 3's in the figure above without interrupting the connectivity of the image, designating the essential-to-be-left points by 1's, to get:

```
                    1311
        1           1
        1           1
        1           1
        1           311
        1             1
        1             1
        3             1
        33    111
        311113
```

This stage is sometimes repeated until reaching the following final image:

```
                    1111
        1           1
        1           1
        1           1
        1           11
        1             1
        1             1
        1             1
        1    111
        11111
```

2.2 Computer Flow Chart

The algorithm has been programmed using *MFC* C^{++} language and given the name *KMM*. The flow chart of the computer program is given in the Appendix. A number of examples are tested by this program and compared with other approaches for speed, complexity and cost.

3 Examples

The examples we consider here differ from those considered in [1], as we are interested in picture images than texts. We are considering Fourier picture, the colon as an example of medical organs, and the signature of one of the authors.

3.1 Pictures

A number of pictures have been tested according to the algorithm of this paper. Pictures and photos - color, black and white - were considered, of which Fourier photo was one. Fig.2 shows Fourier's original photo and its contour thinned form. Notice that the important features of the image are still saved.

Fig. 2. Fourier Image thinned for the purpose of compression and data saving

3.2 Signatures

Another example used to demonstrate feature vector extraction is a typical signature recognition [8]. Although the algorithm used in signature recognition does not need to thin the image of the signature before classifying it, the thinning proved to simplify the process of description and classification. Fig.3 shows a signature sample with its thinned shape, with all features kept unchanged.

Fig. 3. Signature Thinning

3.3 Medical Images

Consider the colon in Fig.4 in its three different forms, original three-dimensional colour image, black and white one and the thinned to its contour final shape. This is one of several examples tested to contour the medical images for preparation to compress before transmitting or saving as the skeletonized image.

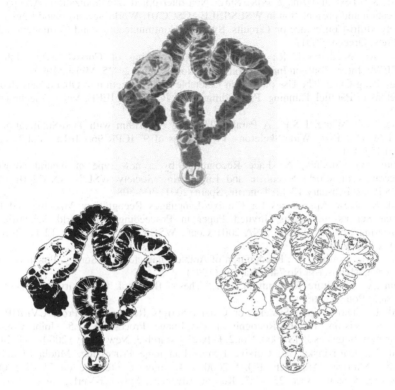

Fig. 4. Colon Thinning

4 Conclusions

For the aim of pattern recognition [9,10,11], the algorithm described in [1] is sufficient to achieve good results. The interruption in the continuity of lines forming the skeleton of an image, in many cases, showed more practical features for classification and description following the criterion in [2]. However, that method does not provide one-pixel width skeletonization. The one presented here, though, satisfies the conditions for having uninterrupted skeleton of the tested image. This condition is required before applying many algorithms in image processing. The algorithm presented in this paper showed significant improvement and in some cases higher efficiency and more practical results.

References

1. Saeed, K.: Text and Image Processing: Non-Interrupted Skeletonization. Accepted for publication and presentation in WSES/IEEE - CSCC'01, World Scientific and Engineering Society Multi-Conference on Circuits, Systems, Communications and Computers, July 8-15. Crete, Greece (2001)

2. Saeed, K., Niedzielski, R.: Experiments on Thinning of Cursive-Style Alphabets. ITESB'99, Inter. Conf. on Information Technologies, June 24-25. Mińsk (1999)

3. Yung-Sheng Chen: The Use of Hidden Deletable Pixel Detection to Obtain Bias-Reduced Skeletons in Parallel Thinning. Proceedings of 9th ICPR'96 - IEEE, Vol.1. Vienna (1996) 91-95

4. Zhang, Y.Y., Wang, P.S.P.: A Parallel Thinning Algorithm with Two-Subiteration that Generates One-Pixel-Wide Skeletons. Proceedings of 9th ICPR'96 - IEEE, Vol.2. Vienna (1996) 457-461

5. Ososkov, G., Stadnik, A.: Face Recognition by a new type of Neural Networks. Proceedings of World Scientific and Engineering Society WSES- NNA'2001 Conf., WSES Press, February 12-14. Tenerife, Spain (2001) 304-308

6. Saeed, K.: New Approaches for Cursive Languages Recognition: Machine and Hand Written Scripts and Tests. Invited Paper in Proceedings of World Scientific and Engineering Society WSES- NNA'2001 Conf., WSES Press, February 12-14. Tenerife, Spain (2001) 304-308

7. Ghuwar, M., Skarbek, W.: Recognition of Arabic Characters - A Survey. Polish Academy of Science, Manuscript No.740. Warsaw (1994)

8. Hodun, A.: Signature Recognition. B.Sc. Thesis. Bialystok University of Technology, Bialystok, Poland (2001)

9. Saeed, K.: Three-Agent System for Cursive-Scripts Recognition. Proc. CVPRIP'2000 Computer Vision, Pattern Recognition and Image Processing - 5th Joint Conf. on Information Sciences JCIS'2000, Vol.2, Feb. 27 - March 3. New Jersey (2000) 244 -247

10. Saeed, K., Dardzińska, A.: Cursive Letters Language Processing: Muqla Model and Toeplitz Matrices Approach. FQAS'2000 – 4th Inter. Conference on Flexible Query Answering Systems, Oct. 25 – 27. Recent Advances, Springer-Verlag, Warsaw (2000) 326-333

11. Saeed, K.: A Projection Approach for Arabic Handwritten Characters Recognition. Proc. of ISCI - International Symposium on Computational Intelligence, Aug. 31 – Sep. 1. New Trends and App. in Comp. Intelligence, Springer-Verlag, Kosice, Slovakia (2000) 106-111

Appendix: The Computer Flow Chart of the Program *KMM*

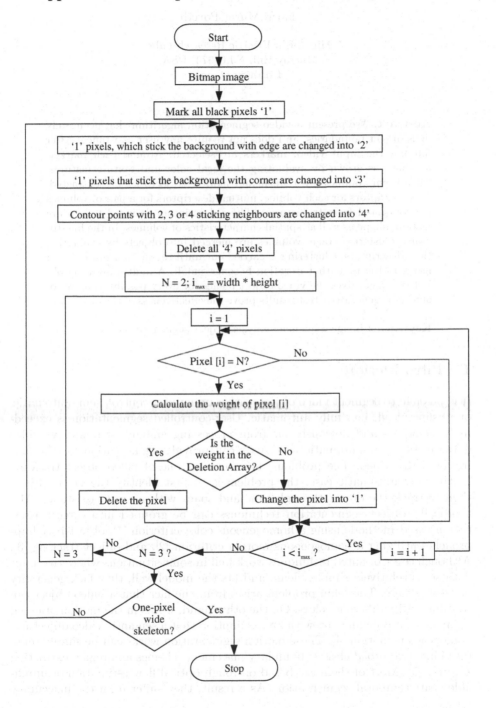

Object Segmentation of Color Video Sequences

Fatih Murat Porikli

Mitsubishi Electric Research Labs,
Murray Hill, NJ 07974, USA,
fatih@merl.com

Abstract. We present a video segmentation algorithm that accurately finds object boundaries, and does not require any user assistance. After filtering the input video, markers are selected. Around each marker, a volume is grown by evaluating the local color and texture features. The grown volumes are refined and motion trajectories are extracted. Self-descriptors for each volume, mutual-descriptors for a pair of volumes are computed from trajectories. These descriptors designed to capture motion, shape as well as spatial characteristics of volumes. In the fine-to-coarse clustering stage, volumes are merged into objects by evaluating their descriptors. Clustering is carried out until the motion similarity of merged objects at that iteration becomes small. A multi-resolution object tree that gives the video object planes for every possible number of objects is generated. Test results prove the effectiveness of the algorithm.

Keywords: image sequence analysis, object segmentation

1 Introduction

It is possible to segment video objects either under user control, semi-automatic, or unsupervised, i.e., fully automatic. User-controlled segmentation is exceedingly laborious and obviously far from processing amount of nowadays video data. In the semi-automatic case, user can provide segmentation for the first frame of the video. The problem then becomes one of video object tracking. In the fully automatic case, the problem is to first identify the video object, then to track the object through time and space with no user assistance. Methodically, object segmentation techniques can be grouped into three classes: region-based methods using a homogeneous color criterion [1], object-based approaches utilizing a homogeneous motion criterion [2], and object tracking [3]. Although color-oriented techniques work well in some situations where the input data set is relatively simple, clean, and fits the model well, they lack generality and robustness. The main problem arises from the fact that a video object can contain totally different colors. On the other hand, works in the motion oriented segmentation domain start with an assumption that a semantic video object has homogeneous motion [4]. These motion segmentation works can be simply separated into two broad classes: boundary placement schemes and region extraction schemes [5]. Most of them are based on rough optical flow estimation or unreliable spatiotemporal segmentation. As a result, they suffer from the inaccuracy

W. Skarbek (Ed.): CAIP 2001, LNCS 2124, pp. 610–619, 2001.

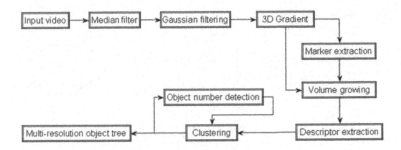

Fig. 1. Flow diagram of the video segmentation algorithm.

of motion boundaries. The last class of methods that is related to semantic video object extraction is tracking [6]. However, the tracking algorithms need user interface, and the performance of the tracking algorithms depends extensively on the initial segmentation. In general, most of the object extraction algorithms treat segmentation as a 2-D inter-or-intra frame processing problem with some additional motion model assumptions or smoothing constraints by ignoring the 3-D nature of the video data.

To develop an algorithm that blends intra-frame color and texture based spatial segmentation schemes with inter-frame motion estimation techniques, we consider video sequence as a 3-D volumetric data, that we call it as video-cube, but not a collection of 2-D images. Thus, the semantic object information can be propagated forward and as well as backward in time without saddling into initial segmentation accuracy or tracking limitations.

A general framework of the algorithm is shown in Fig. 1. In Section II, the video-cube concept is introduced. Section III describes the stages of filtering, marker selection, volume growing, refining, and clustering volumes into objects. The test results and discussion are included in the last section.

2   Formation of Video-Cube

By registering the raw and processed image frames along the time axis as shown in Fig. 2, a video-cube $V(x, y, t)$ $1 \leq x \leq x_M$, $1 \leq y \leq y_M$, and $1 \leq t \leq t_M$ is constricted. Each element of video-cube V corresponds to a vector $\mathbf{v}(x, y, t) = [y, u, v, \theta_1, \ldots, \theta_K]^T$ that consists of color and texture features of the spatiotemporal point (x, y, t). Here, y, u, and v stand for the luminance and chrominance features, $\theta_1, \ldots, \theta_K$ are the normalized texture features. For simplicity, we will denote each component as a subscript, e.g., $\mathbf{v}_y$ instead of $\mathbf{v}(x, y, t)_y$.

We preferred the YUV color space over the RGB. Most of the existing color segmentation approaches have utilized the RGB color space although the RGB space has machine oriented chromatics rather than human oriented chromatics.

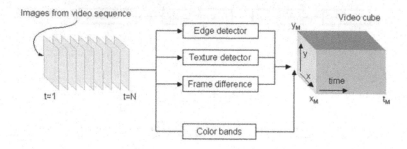

Fig. 2. Video-cube generation from the raw and processed images of the video sequence.

One other disadvantage of the RGB space is the dependency of all three parameters from the light intensity. The YUV space performs in accordance with human reception and more importantly, inter-color distances can be computed using the L^2-norm.

The texture components $\mathbf{v}_{\theta_1}, \ldots, \mathbf{v}_{\theta_K}$ are obtained by Gabor transform [7]. Gabor filters are quadrature filters and can be used to extract a certain wavelength and orientation from an image with a specified bandwidth. 2-D Gabor filters $h(x, y)$ have the functional form

$$h(x,y) = g(x,y)e^{-2\pi i(ux+vy)}, \quad g(x,y) = \frac{1}{2\pi\sigma_g^2}e^{-\frac{x^2+y^2}{2\pi\sigma_g^2}} \tag{1}$$

where σ_g^2 specifies effective width, and u, v specify modulation that has spatial frequency $f = \sqrt{u^2 + v^2}$ and direction $\theta = tan^{-1}(v/u)$. Then the texture scores are found by

$$\mathbf{v}_\theta = |h(x,y) \otimes I(x,y)|. \tag{2}$$

We chose the values for the spatial frequency $f = 2, 4, 8$ and the direction $\theta = 0, \pi/4, \pi/2, 3\pi/4$, which leads to a total of 12 features. The computed texture scores are normalized as described in [8].

3   Object Segmentation

3.1   Pre-filtering

The color channels of the input video $\mathbf{v}_y, \mathbf{v}_u, \mathbf{v}_v$ are first filtered to remove out noise. Another reason of pre-filtering is to prepare video-cube to the volume growing stage. Elimination of the image flickers prevents from excessive segmentation, and decreases computation load significantly. A fast 3×3 median filter [9] that exploits 2-D coherence is utilized together with a 5×5 Gaussian filter to remove noise and smoothen image irregularities.

3.2   Marker Assignment

A video object is assumed to be made of smaller parts. Such parts are the group of points that are spatially consistent, i.e., color and texture distributions are uniform within. The grouped points, called as volumes, are expanded from seed points, called markers, as in the watershed transform [10]. The marker points serves as "basins" in the growing process. Note that, not the spatial position of a marker but its features are intended here. For each marker, a volume is assigned and volume's attributes are initialized. Therefore, a marker point should be selected such that it can characterize its enclosing volume as relevant as possible. The points have low local diversity are good candidates to represent their local neighborhood. A marker point m_i can be selected in three ways:

- Uniformly distributed: The V is divided into identical smaller cubes and their centers are selected as markers. However, an edge point can be chosen as well, which reduces the accuracy of volume growing.
- Minimum gradient magnitude: Markers are selected if the gradient magnitude is minimum. Let S be the set of all possible spatiotemporal points, i.e., it is all the points of V initially. The gradient magnitude is computed from the color channels, and the minimum gradient magnitude point is chosen as a marker. A preset neighborhood around that marker is removed from the set S. The next minimum in the remaining set is chosen, and selection process repeated until no point remains in the video-cube.
- Minimum gradient with volume growing: The minimum m_i is chosen as above. Instead of removing a preset neighborhood around the marker, a volume W_i is grown as explained in the next section, and all the points of the volume is removed from the set S

$$m_i = \arg \min_S \nabla V(x, y, t) \ ; \ \ S = V - \bigcup_{j=1}^{i} W_j. \tag{3}$$

Finding minimum is a computationally expensive process. Rather than searching the full-resolution video-cube V, a down-converted version is used. More computational reduction is achieved by dividing down-converted video-cube V into slices in time or other axes. Minimum is found for the first slice, and a volume is grown, then the next minimum is searched in the next slice, and so forth.

3.3   Volume Growing

Volumes are enlarged as "inflating balloons" from markers by applying distance criteria. For each volume W_i, a feature vector ω^i that is similar to the video-cube point's feature vector is defined. Two distance criteria d_g, d_l are designed. The first criterion d_g measures the distance between the feature vector of the current volume and the candidate point. In the spatial sense, this criterion is a volume-wise "global" measure. The second criterion d_l determines the distance between the feature vectors of the current volume and another point that is already included in the current volume and also adjoint to the candidate. Thus, the second criterion can be viewed as a "local" measure. A global threshold ϵ_g helps limiting the range of feature distance, i.e., color variation in the volume. Local threshold ϵ_l prevents from trespassing edges even the global threshold permits. The global and local thresholds are adaptively determined from the video-cube V. Let x^- be an unmarked candidate point that is adjoint to the current volume. Let x^+ be another point adjoint to x^- but already included in the current volume W_i. Then, the first global distance d_g is defined as

$$d_g(\omega^i, \mathbf{v}^-) = \sum_k |\omega_k^i - \mathbf{v}_k^-| \quad k : y, u, v, \theta_1, ..., \theta_{12} \tag{4}$$

where $\mathbf{v}^-$ and $\mathbf{v}^+$ are the feature vectors of x^- and x^+. Similarly, the second local distance d_n is

$$d_l(\omega^i, \mathbf{v}^-) = \sum_k |\mathbf{v}_k^+ - \mathbf{v}_k^-| \quad k : y, u, v, \theta_1, ..., \theta_{12}. \tag{5}$$

If the distances d_g and d_l are smaller than ϵ_g and ϵ_l, the point x^- is included in the volume W_i. The neighboring point x^- is set as an active surface point for W_i, and the feature vector for the marker is updated accordingly. In the next iteration, the neighboring pixels of the active surface points are examined. Volume growing is repeated until no point remains in the video-cube.

The thresholds are made adaptable to input video by using the variance and dynamic range of the features. Variance of a feature gives information about the distribution of that feature in the video-cube. A small variance indicates smooth distribution, whereas, high variance refers to texture and edgeness. To The global threshold should be tuned up large if the variance is high and it should be scaled to the dynamic range. Let I represent a feature, i.e, $I \equiv Y$ for luminance. The global variance σ^2 and mean η are simply

$$\sigma^2 = \frac{1}{M} \sum_{x,y,t \in V} (I(x,y,t) - \eta)^2, \qquad \eta = \frac{1}{M} \sum_{x,y,t \in V} I(x,y,t). \tag{6}$$

where M is the total number of points. The dynamic range μ for is

$$\mu = \max I(x,y,t) - \min I(x,y,t) \ , \quad x,y,t \in V. \tag{7}$$

Then the global threshold ϵ_g is then assigned by scaling the dynamic range as

$$\epsilon_g = \kappa \frac{\mu}{\sigma + 1} \tag{8}$$

where $\kappa > 1$ is the sensitivity parameter that sets how fine the final segmentation should be. It is observed that a good choice is $\kappa \approx 2$. The local threshold $\epsilon_l(x, y, t)$ is the average of the discontinuity between the neighboring point features scaled with a relative edgeness score:

$$\epsilon_l(x,y,t) = \tilde{\eta} \cdot \tilde{\mu}(x,y,t)$$

$$\tilde{\eta} = \frac{1}{2M} \sum_{x,y,t \in V} |I(x,y,t) - I(x-1,y,t)| + |I(x,y,t) - I(x,y-1,t)|$$

$$\tilde{\mu}(x,y,t) = \max I(i,j,t) - \min I(i,j,t) \quad x-2, y-2 \le i,j \le x+2, y+2$$

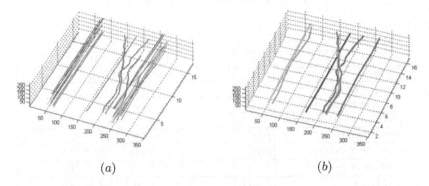

Fig. 3. Trajectories; (a) before clustering, (b) at the 7^{ht} object level of Children seq.

To fasten extraction of these statistics, only the first image of the video can be used instead of the whole video-cube V.

After volume growing, the video-cube is divided into multiple smaller parts. Some of these parts are negligible in size, however, they effect the computational load if the clustering stage. Also, some points, such as edges, are not grouped into any volume at all. To assign ungrouped points to a volume, the volume growing thresholds are relaxed iteratively. At each iteration, volumes are inflated towards the unmarked points until no more point remains. Small volumes are blended into the bordering most similar volumes that gives the best combination of the greatest mutual surface, the smallest color distance, the smallest mutual volume, and the highest compactness ratio as defined in the next section.

3.4 Self and Mutual Descriptors

Descriptors are used to understand various aspects of the volumes. The volumes W_i are represented by a set of self descriptors $f(i)$ and mutual descriptors $g(i,j)$ as summarized in Table 1. These descriptors identify motion, spatial, and color characteristics, as well as the mutual correlation. The volumes will be grouped with respect to their descriptors at the clustering stage in order to assemble the objects. For each volume W_i, a trajectory $T_i(t) = [X_i(t), Y_i(t)]^T$ is extracted by computing the frame-wise averages of volume's points coordinates

$$T_i(t) = \begin{bmatrix} X_i(t) \\ Y_i(t) \end{bmatrix} = \begin{bmatrix} \frac{1}{N_t} \sum x_t \\ \frac{1}{N_t} \sum y_t \end{bmatrix} ; \quad (x_t, y_t, t) \in W_i. \qquad (9)$$

Trajectories are the center of masses of regions in an image frame, hence they approximate the translational motion of the region. This is a nice property that can be used to initialize parameters in motion model fitting stage. Sample trajectories can be seen in Fig. 3 a-b. Then, the distance $\Delta d_{ij}(t)$ between the trajectories

Table 1. Self and Mutual descriptors

color mean	$f_1(i)$	$\frac{1}{N_i}\sum \mathbf{v}_y(x_k)$, $x_k \in W_i$
volume	$f_2(i)$	$\bigcup x_k$, $x_k \in W_i$
surface	$f_3(i)$	$\sum x_k \cap x_l$, $x_k \in W_i\ x_l \in W_j\ i \neq j$
compactness	$f_4(i)$	$f_2(i)/(f_3(i))^2$
vertical translation	$f_5(i)$	$y_{1,i} - y_{N,i}$
horizontal translation	$f_6(i)$	$x_{1,i} - x_{N,i}$
route length	$f_7(i)$	$\sum |T_i(t) - T_i(t-1)|$
average x position	$f_8(i)$	$\frac{1}{N_i}\sum X(x_k)$, $x_k \in W_i$
average y position	$f_9(i)$	$\frac{1}{N_i}\sum Y(y_k)$, $x_k \in W_i$
existence	$f_{10}(i)$	$\sum i_t$; $i_t = 1 \leftarrow T_i(t) \neq 0$
mean of distance	$g_1(i,j)$	$\frac{1}{N_i \cap N_j}\sum \Delta d_{ij}(t)$
variance of distance	$g_2(i,j)$	$\frac{1}{N_i \cap N_j}\sum (\Delta d_{ij}(t) - g_1(i,j))^2$
maximum distance	$g_3(i,j)$	$max\,\Delta d_{ij}(t)$
directional difference	$g_4(i,j)$	$\sum |T_i(t) - T_i(t-1) - T_j(t) + T_j(t-1)|$
compactness ratio	$g_5(i,j)$	$\frac{f_4(W_i \cup W_j)}{f_4(W_i) + f_4(W_j)}$
mutual boundary ratio	$g_6(i,j)$	$f_3(i) + f_3(j) - f_3(W_i \cup W_j)/f_3(i)$
color difference	$g_7(i,j)$	$|f_1(i) - f_1(j)|$
coexistence	$g_8(i,j)$	$\sum i_t \wedge j_t$; $i_t = 1 \leftarrow T_i(t) \neq 0$

$T_i(t)$ and $T_j(t)$ at time t is calculated to determine motion based descriptors

$$\Delta d_{ij}(t) = \sqrt{(X_i(t) - X_j(t))^2 + (Y_i(t) - Y_j(t))^2}. \qquad (10)$$

Motion characteristics such as vertical and horizontal motion, route length, mean and variance of distance, direction difference, and average change in the distance are derived from the trajectories. Therefore, without a computationally expensive method, the motion information is blended into segmentation efficiently. Each descriptor is linearly normalized to $[0, 1]$ by using its maximum and minimum at last.

3.5   Clustering

A fine-to-coarse type hierarchical merging method is chosen since the video-cube already divided into small parts before the clustering. The most similar volume pairs are merged to decrease the number of the volumes at each iteration. We define "similarity" as the degree of relevance of volumes in motion and shape. Two volumes are similar if their motion trajectories are consistent and they built a compact shape when combined together. Color aspects are omitted; partly because it was already included in volume growing, and also portions of a semantically meaningful object do not have to possess the same color aspects, i.e., a human face made up from different color regions, mouth, skin, hair, etc.

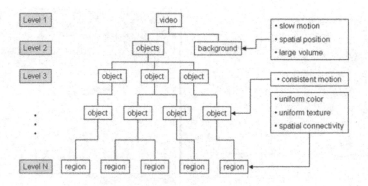

Fig. 4. Multi-resolution object tree.

The most suitable descriptors are found to be variance of trajectory distance $g_2(i,j)$ for motion, and the compactness $g_4(i,j)$ and mutual boundary ratio $g_6(i,j)$ for shape relevance. Each volume W_i is compared to its neighboring volumes W_j, and a similarity score $S(i,j)$ for the pair is determined if the pair satisfies a set of constraints. The constraint set is embedded to prevent from degenerate cases. An example of such a case is that a volume encircling a smaller volume and causing the highest compactness ratio. Although such volumes motions may be inconsistent, still they will get a high similarity score. Another example of degenerate case is parts of the two objects that are not adjoint but perfectly aligned in motion. The shape relevance will be very low; although the motion is consistent, the similarity score could be small. An effective similarity score is formulated as

$$S_{ij} = -\log(g_2(i,j)) + g_4(i,j) + g_6(i,j). \tag{11}$$

After all similarity scores are computed for the possible volume pairs, the volume pair (i', j') that gives the maximum similarity score are merged together, the descriptors are updated and normalized accordingly.

3.6 Object Number Estimation

Clustering is performed until the candidate volumes to be merged become inconsistent in motion. Motion inconsistency is measured by the variance of the trajectory distance descriptor $g_2(i,j)$. If the trajectory distance $g_2(i'_m, j'_m)$ of the most consistent pair in the current object level m is considerably higher than the $g_2(i'_{m-1}, j'_{m-1})$ at the previous level, the last merge is assumed to be a violation of the motion similarity. Thus, segmentation is halted:

$$\frac{\partial g_2(i'_m, j'_m)}{\partial m} \gg 1 \; ; \quad i'_m, j'_m = \arg\max(S_{ij}) \quad at\ level\ m \tag{12}$$

A multi-resolution object tree as illustrated in Fig. 4 is generated from the clustering results. By multi-resolution tree, segmentation is not repeated in case the number of objects is changed later. Also, this structure enables imposing relational constraints and analyzing object properties by graph theory methods.

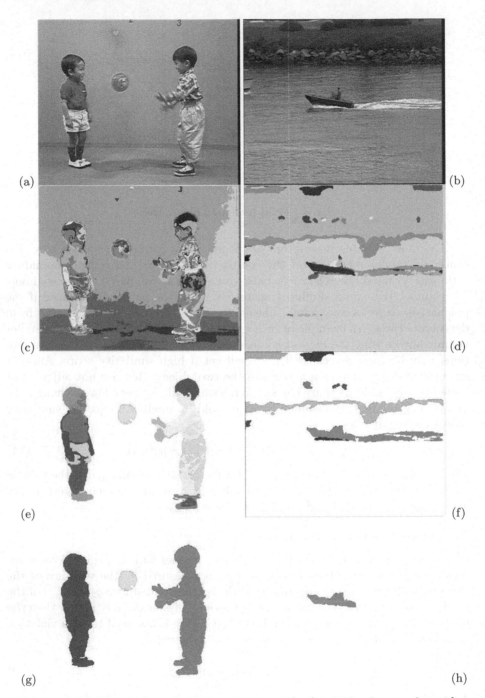

Fig. 5. (a-b) Frames form the input sequences, (c-d) initial volumes after volume growing, (e-f) intermediate clustering results at object levels $m = 20 - 12$ respectively, and (g-h) the final segmentation results.

4 Test Results and Conclusion

The proposed algorithm has been tested with standard MPEG sequences. Fig. 5 presents sample results. The first row (Fig. 5 a-b) shows original frames from two test sequences. The initial volumes after volume growing stage are given in the second row (Fig. 5 c-d). Here, volumes are color coded for illustration purposes. The initial number of volumes are 42 and 27 respectively. We noticed that the adaptive threshold estimation method enables us to confine the initial number of volumes to 20 $\sim$ 50 which is a very reasonable initial range for most sequences. Figures 5 e-f are the intermediate clustering results. Segmentation stopped at $m = 4$ and $m = 2$ object levels respectively without any user interface. The backgrounds and object boundaries were detected accurately.

The test results confirmed that the object extraction is robust even when the motion is large. Because no separate motion computation is involved in segmentation, our algorithm is computationally faster than any optical flow based or motion field modeling method that usually need some initial segmentation. Having an obvious advantage over the stochastic segmentation techniques, an object-wise multi-resolution representation is generated after clustering; therefore the extraction objects is not repeated in case the number of objects is changed. No user segmentation of object regions is involved, which is mostly required by object tracking based methods.

References

1. Kunt, M., Ikonomopoulos, A., Kocher, M.: Second generation image coding. Proceedings of IEEE, no. 73 (1985) 549–574
2. Bouthemy, P., Francois, E.: Motion segmentation and qualitative dynamic scene analysis from an image sequence. Int. J. Comput. Vision, no. 10 (1993) 157–187
3. Meyer, F., Bouthemy, P.: Region-based tracking using affine motion models in long image sequences. CVGIP-Image Understanding no. 60 (1994) 119–140
4. Duc, B., Schtoeter, P., Bigun, J.: Spatio-temporal robust motion estimation and segmentation. Proc. Comput. Anall. Images and Patterns (1995) 238–245
5. Wang, J., Adelson, E.: Representing moving images with layers. IEEE Transaction on Image Processing, no.3 (1994)
6. Aggarwal, J.K., Davis, L.S., Martin, W.N.: Corresponding processes in dynamic scene analysis. Proceedings of IEEE, no.69 (1981) 562–572
7. Jain, A.K., Farrokhnia, F.: Unsupervised texture segmentation using Gabor filters. Pattern Recognition, vol. 24 (1991) 1167–1186
8. Pichler, O., Teuner, A., Hosticka, B.: An unsupervised texture segmentation algorithm with feature space reduction and motion feedback. IEEE Transaction on Image Processing (1998) 53–61
9. Kopp, M., Purgathofer, W.: Efficient 3x3 median filter computations. Technical University, Vienna (1994)
10. Serra, J., Soille, P.: Watershed, hierarchical segmentation and waterfall algorithm. Proc. of Mathematical Morphology Appl. Image Process. (1994) 69–76
11. Jain, A., Murty, M., Flynn, J.: Data Clustering: A Review. ACM Computing Surveys, Vol. 31 (1999) 264–323

Thresholding Image Segmentation Based on the Volume Analysis of Spatial Regions

Dominik Sankowski[1,2] and Volodymyr Mosorov[1,2]

[1] Computer Engineering Department of Technical University of Lodz, Poland,
90-924, Al. Politechniki 11
{dsan, mosorow}@kis.p.lodz.pl
[2] The Academy of Humanities & Economics in Lodz, Poland, 90-922,
Rewolucji 1905 Str. 52

Abstract. In the first part of the paper a new theoretical approach to the problem of image segmentation is described. A method for automatic segmenting of an unknown number and unknown location of objects in an image has been proposed. This method is based on both local properties of neighbouring pixels and global image features. To allow for automated segmentation, slices are formed at different values of the threshold level, which contain spatial uniformity regions. In the second part, the image segmentation is considered as a problem of selection of slices, which should comprise regions with features satisfying the requirements desired. The selection is based on the proposed minima criterion including a volume analysis of neighbouring slices. An important characteristic of the approach is that it reflects object shapes devoid of noise, and does not use heuristic parameters such as an edge value. The results of this method are presented on several examples containing greyscale images of objects of different brightness.

Keywords: - image segmentation, spatial uniformity regions, a volume analysis of neighbouring slices.

W. Skarbek (Ed.): CAIP 2001, LNCS 2124, p. 620, 2001.
© Springer-Verlag Berlin Heidelberg 2001

Topographic Feature Identification Based on Triangular Meshes

Hélio Pedrini and William Robson Schwartz

Federal University of Paraná, Curitiba-PR 81531-990, Brazil

Abstract. A new method for extracting topographic features from images approximated by triangular meshes is presented. Peaks, pits, passes, ridges, valleys, and flat regions are defined by considering the topological and geometric relationship between the triangular elements. The approach is suitable for several computer-based recognition tasks, such as navigation of autonomous vehicles, planetary exploration, and reverse engineering. The method has been applied to a wide range of images, producing very promising results.

Keywords: image analysis, topographic feature identification

1 Introduction

The extraction of topographic features in digital images is a primary problem encountered in any general computer vision system. Several computer-based recognition tasks such as navigation of autonomous vehicles, planetary exploration, reverse engineering, and medical image analysis require the construction of accurate models based on shape descriptors in order to represent surface information in an efficient and consistent way.

Peaks, pits, ridges, valleys, passes, and flat regions are some useful topographic features used in image analysis. A peak is a point such that in some neighborhood of it, there is no higher point. Similarly, in some neighborhood of a pit, there is no lower point. A ridge correspond to a long, narrow chain of higher points or crest, while a valley correspond to a chain of points with lower elevations. A pass is a low point on a ridge or between adjacent peaks.

Several methods have been proposed to identify topographic features in digital images. The vast majority of the methods are defined in terms of cell-neighbor comparisons within a local window over the image [6,9,16] or derived from contour lines of the image [2,10,17]. Concepts from differential geometry are often used in surface feature recognition [1]. For instance, basic surface types can be determined by estimating directional derivatives of the intensity image. Since the computation of surface derivatives is extremely sensitive to noise or other small fluctuations, image smoothing is usually required to reduce the effects of noise. However, such smoothing can also suppress relevant information or cause edge displacement.

Peucker and Johnston [15] characterize the surface shape by the sequence of positive and negative differences as surrounding points are compared to the central point. Peucker and Douglas [14] describe several variations of this method for detecting surface specific points and lines in terrain data.

W. Skarbek (Ed.): CAIP 2001, LNCS 2124, pp. 621–629, 2001.

The method proposed by Johnston and Rosenfeld [9] detects peaks (pits) by finding all points P such that no points in an n by n neighborhood surrounding P have higher (lower) elevation than that of P. To find ridges (valleys), their method identifies points that are either east-west or north-south elevation maxima (minima) through a "smoothed" array in which each point is given the highest elevation in a 2×2 square containing it.

Paton [12] uses a six-term quadratic expansion in Legendre polynomials fitted to a small disk around each pixel. The most significant coefficients of the second-order polynomial are used to classify each pixel into a descriptive label. Grender [5] compares the grey level elevation of a central point with surrounding elevations at a given distance around the perimeter of a circular window and the radius of the window may be increased in successive passes through the image.

Hsu, Mundy, and Beaudet [8] use a quadratic surface approximation at every pixel on the image surface. Lines emanating from the central point in the principal axes of this approximation provide natural boundaries of patches representing the surface. The principal axes from some critical points distributed over the image are selectively chosen and interconnected into a network to produce an approximation of the image data. Mask matching and state transition rules are used to extract a set of primitive features from this network.

Toriwaki and Fukumura [18] use two local measures of grey level pictures, connectivity number and coefficient of curvature for classification of each pixel into a descriptive label, which is then used to extract structural information from the image.

Watson, Laffey, and Haralick [7,19] provide a method for classifying topographic features based on the first and second directional derivatives of the surface estimated by bicubic polynomials, generalized splines, or the discrete cosine transformation. A technique proposed by Gauch and Pizer [13] locates regions where the intensity changes sharply in two opposite directions. The curvature calculation is based on level curves of the image, requiring the evaluation of a large polynomial in the first-, second-, and third-order partial derivatives.

A more recent evaluation of some methods for ridge and valley detection is presented by López *et al.* [11]. A survey describing efficient data structures and geometric algorithms for extracting topographic features in terrain models is given in [4].

The method proposed in this paper differs from the majority of the feature detection algorithms found in literature, which are generally based on regular grid models. A disadvantage of regular grids is their inherent spatial invariability, since the structure is not adaptive to the irregularity of the object surface. This may produce a large amount of data redundancy, especially where the topographic information is minimal. The method proposed by Falcidieno and Spagnuolo [3] is the most similar to the one presented here.

In our method, triangulated irregular networks represent the object surface as a mesh of adjacent triangles, whose vertices are the data points. The points need not lie in any particular pattern and the density may vary over space. There are many advantages associated with triangulated irregular networks. First, complex data are commonly irregularly distributed in space, therefore, the structure of the triangulation can be adjusted to reflect the density of the data. Consequently, cells become larger where data are sparse, and smaller where data are dense. Second, topographic features can be incorporated into the model. For instance, vertices in a triangulation can describe nodal features such as peaks, pits or passes, while edges can represent linear features such as break, ridge or channel lines. Finally, triangles are simple geometric objects which can be easily manipulated and rendered.

An improved triangular model is presented in Section 2. The definition of topographic features as descriptive elements of the surface is given in Section 3. Section 4 presents some experimental results and implementation issues. Section 5 concludes with some final remarks and directions for future research.

2   Improved Triangular Model

The construction of our triangular meshes is performed by a hybrid refinement and decimation approach, incrementally determining a better distribution of the data points.

An approximation formed by two triangles is initially constructed. This mesh is then incrementally refined until either a specified error is achieved or a given number of points is reached. Once the desired level of accuracy has been satisfied, the approximation is simplified by eliminating a small number of points based on a vertex removal criterion. Finally, the approximation is again refined to the given error tolerance and partially resimplified. This alternate refinement and decimation process is repeated until either no further improvement in the accuracy of the approximation can be achieved or a given number of vertices is reached.

A constrained Delaunay triangulation is used to maintain the topology of the data points, whose vertices lie at a subset of the input data. While most of the traditional triangulation techniques are based on the approximation error as a criterion for the mesh quality, the objective of our approach is to construct triangular meshes that preserve important topographic features in the approximated surface.

A new local error metric is used to select points to be inserted into the triangulation, which is based on the maximum vertical error weighted by the standard deviation calculated in a neighborhood of the candidate point, given by

$$C = \frac{|h(p) - z(p)|}{\sigma(p)} \tag{1}$$

where $h(p)$ is the height value of point p, $z(p)$ is the height value of the interpolated surface at point p, and $\sigma(p)$ is the standard deviation calculated in a 3×3 neighborhood of the candidate point p.

The idea of the above metric is to associate greater importance to the points in regions where the local variability of the data is high, allowing the surface to conform to the local trends in the data. In flat regions, where $\sigma(p) = 0$, the algorithm uses only the numerator $|h(p) - z(p)|$ to select new points.

A priority queue stores the sequence of vertices used to refine the triangulation, ordered by increasing approximation error. For each refinement step, only those vertices affected by the insertion process need to have their approximation error recalculated.

Conversely, a measure of angle between surface normals is used to determine whether a vertex should be removed from the triangulation. The criterion for removing a vertex v is computed by averaging the surface normals n_i of the triangles surrounding v weighted with their areas A_i and taking the maximum angle, α_{max}, between the averaged normal, n_{av} and the surrounding triangles, that is

$$\alpha_{max} = \max\left(\arccos\frac{\vec{n}_{av}\cdot\vec{n}_i}{|\vec{n}_{av}|\cdot|\vec{n}_i|}\right) \qquad (2)$$

where $\vec{n}_{av} = \dfrac{\sum\vec{n}_i.A_i}{\sum A_i}$ and $0 \leq \vec{n}_i \leq \pi$.

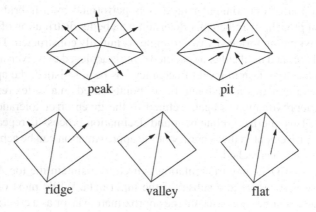

peak   pit

ridge   valley   flat

Fig. 1. Topographic features in triangular meshes.

This criterion describes a measure of roughness, indicating a local variability of the data. Each vertex having a value less than the specified maximum angle will be removed during the decimation process. The area around the removed point is retriangulated, which requires careful consideration, in particular when such area is not a convex polygon. The edges that form the triangulation of the polygon surrounding the vertex v must be checked to determine if they do not intersect one another. The iterative application of this decimation process reduces the number of points and edges that are not representative of the scene features.

The sequence of local modifications generated during the refinement and decimation steps is applied to a triangulation until either the desired accuracy of the approximation is achieved or a given number of points is reached. The extraction of a representation of the terrain at a given tolerance level is obtained by using a coarse triangulation and iteratively inserting vertices into the triangulation until the desired precision is satisfied. If a given triangulation already guarantees a smaller error tolerance, then vertices are removed from the triangulation, starting with the vertex with the smallest error.

Fig. 2. Application of the proposed method to (a) Lena image; (b) triangular mesh obtained by our hybrid triangulation technique. The mesh contains only 2.8% of the original points; (c) peaks (dark points) and pits (grey points) extracted by using the triangular mesh; (d) the extracted ridges (dark lines) and valleys (grey lines).

3   Topographic Feature Extraction

The triangulated surface is defined as graph consisting of vertices, directed edges, and triangles. In our method, the identification of topographic features is derived from the triangular models. This can be achieved by analyzing the structure of the triangles in the mesh.

Figure 1 illustrates some of the most common topographic features, which are classified according to the topological and geometric relationship between the triangulation elements. The angle between the normals of adjacent triangles indicates concave, convex, and planar shapes along the surface.

For each edge e not belonging to the surface boundary, it is assigned a label according to the information about the angle between the two triangles t_1 and t_2 sharing the edge. The angle between t_1 and t_2 is concave (convex) if for any points $p_1 \in t_1$ and $p_2 \in t_2$, the straight line segment $p_1 p_2$ is located completely above (below) the surface identified by t_1 and t_2. Otherwise, it is plane. A surface characteristic point P is classified as peak (pit) if its z value is greater (smaller) than the z value of each point belonging to the lines with intersect in P.

These extracted feature elements, describing nodal features (such as peaks, pits, or passes) and linear features (such as ridges, rivers, roads, channels, or cliffs), are incorporated into the triangulation as constrained vertices and edges, respectively, in a such way that subsequent operations will preserve them.

4   Experimental Results

Our method has been tested on a number of data sets in order to illustrate its performance. Due to space limitations, only three data sets are presented here. Figure 2(a) shows the classic *Lena* image (512×512), Figure 3(a) shows a digital terrain model consisting of 1201×1201 elevation points and 3- by 3-arc-second data spacing (90 meters), and Figure 4(a) shows a high altitude aerial image (512×512). For each one of these three images, the corresponding triangular meshes obtained by our hybrid triangulation method (Figures 2-4(b)), the extracted peaks and pits (Figures 2-4(c)), and the extracted ridges and valleys (Figures 2-4(d)) are illustrated.

Although the meshes shown above have only a small percentage of the original number of points, the models still capture the main features of the images.

The algorithms were implemented in C++ programming language on Unix/Linux platform. The triangulation algorithm is able to select 55,000 points in approximately 60 seconds on an SGI O2 workstation (IRIX 6.5, R5000 with a 200MHz MIPS processor and 64 Mbytes of main memory).

5   Conclusions

Unlike many other approaches which are based on regular grid models, we presented a method for extracting topographic features from images approximated by triangular meshes. Characteristic points, lines, and regions are defined by considering the topological and geometric relationship between the triangular elements.

This technique provides an effective compromise between fidelity and time requirements, producing approximations with great flexibility while retaining the most relevant surface features. The method has been applied to a wide range of images containing different properties, achieving encouraging results despite the fact that no additional refinement technique has been performed on the extracted features.

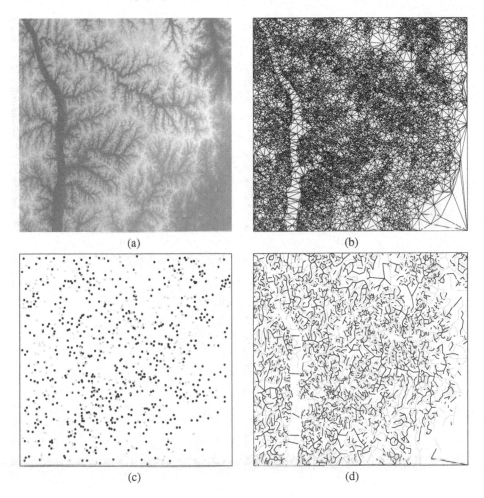

Fig. 3. Application of the proposed method to (a) terrain image; (b) triangular mesh obtained by our hybrid triangulation technique. The mesh contains only 4.5% of the original points; (c) peaks (dark points) and pits (grey points) extracted by using the triangular mesh; (d) the extracted ridges (dark lines) and valleys (grey lines).

References

1. Besl, P.J., Jain, R.C.: Segmentation through variable-order surface fitting. IEEE Transactions on Pattern Analysis and Machine Intelligence **10** (1988) 167–192
2. Christensen, A. H.J.: Fitting a triangulation to contour lines. Proceedings of the Eighth International Symposium on Computer-Assisted Cartography, Baltimore, Maryland, USA (1987) 57–67
3. Falcidieno, B., Spagnuolo, M.: A new method for the characterization of topographic surfaces. International Journal of Geographical Information Systems **5** (1991) 397–412
4. Floriani, L. D., Puppo, E., Magillo, P.: Applications of computational geometry to geographic information systems. In: Sack, J., Urrutia, J. (eds.): Handbook of Computational Geometry. Elsevier Science (1999) 333–388

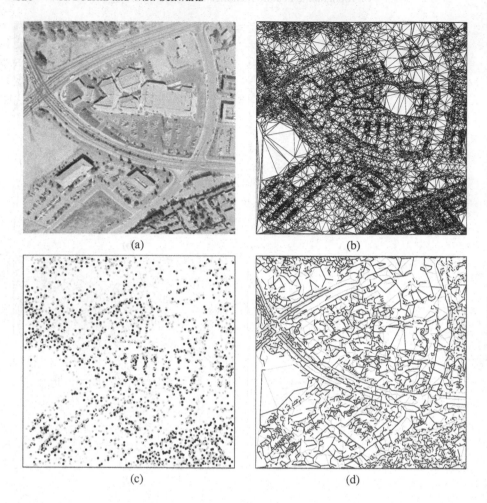

(a) (b)

(c) (d)

Fig. 4. Application of the proposed method to (a) aerial image; (b) triangular mesh obtained by our hybrid triangulation technique. The mesh contains only 5% of the original points; (c) peaks (dark points) and pits (grey points) extracted by using the triangular mesh; (d) the extracted ridges (dark lines) and valleys (grey lines).

5. Grender, G.C.: TOPO III: A Fortran program for terrain analysis. Computers & Geosciences **2** (1976) 195–209

6. Haralick, R.M.: Ridges and valleys on digital images. Computer Vision, Graphics, and Image Processing **22** (1983) 28–38

7. Haralick, R.M., Watson, L.T., and Laffey, T.J.: The topographic primal sketch. The International Journal for Robotics Research **2** (1983) 50–72

8. Hsu, S., Mundy, J.L., Beaudet, P.R.: WEB representation of image data. Proceedings of the Fourth International Joint Conference on Pattern Recognition, Kyoto, Japan (1978) 675–680

9. Johnston, E.G., Rosenfeld, A.: Digital detection of pits, peaks, ridges, and ravines. IEEE Transactions on Systems, Man, and Cybernetics **5** (1975) 472–480

10. Lo, S.H.: Automatic mesh generation and adaptation by using contours. International Journal for Numerical Methods in Engineering **31** (1991) 689–707
11. López, A.M., Lumbreras, F., S., J., Villanueva, J.J.: Evaluation of methods for ridge and valley detection. IEEE Transactions on Pattern Analysis and Machine Intelligence **21** (1999) 327–335
12. Paton, K.: Picture description using legendre polynomials. Computer Graphics and Image Processing **4** (1975) 40–54
13. Paul, J.G., and Pizer, S.: Multiresolution analysis of ridges and valleys in grey-scale images. IEEE Transactions on Pattern Analysis and Machine Intelligence **15** (1993) 635–646
14. Peucker, T.K., Douglas, D.H.: Detection of surface-specific points by local parallel processing of discrete terrain elevation data. Computer Graphics and Image Processing **4** (1975) 375–387
15. Peucker, T.K., Johnston, E.G.: Detection of surface-specific points by local parallel processing of discrete terrain elevation data. Technical Report **206**, University of Maryland (1972)
16. Skidmore, A.K.: Terrain position as mapped from gridded digital elevation model. International Journal of Geographical Information Systems **4**(1) (1990) 33–49
17. Tang, L.: Automatic extraction of specific geomorphological elements from contours. Proceedings 5th International Symposium on Spatial Data Handling, IGU Commission on GIS, Charleston, South Carolina, USA (1992) 554–566
18. Toriwaki, J., Fukumura, T.: Extraction of structural information from grey pictures. Computer Graphics and Image Processing **7** (1978) 30–51
19. Watson, L.T., Laffey, T.J., Haralick, R.M.: Topographic classification of digital image intensity surfaces using generalized splines and the discrete cosine transformation. Computer Vision, Graphics, and Image Processing **29** (1985) 143–167

Visual Attention Guided Seed Selection for Color Image Segmentation

Nabil Ouerhani[1], Neculai Archip[2], Heinz Hügli[1], and Pierre-Jean Erard[2]

[1] Institute of Microtechnology, University of Neuchâtel
Rue A.-L. Breguet 2, CH-2000 Neuchâtel, Switzerland
{Nabil.Ouerhani, Heinz.Hugli}@unine.ch
[2] Institute of Computer Sciences, University of Neuchâtel
Emile-Argand 11, CH-2000 Neuchâtel, Switzerland
{Neculai.Archip, Pierre-Jean.Erard}@unine.ch

Abstract. The "seeded region growing" (SRG) is a segmentation technique which performs an image segmentation with respect to a set of initial points, known as seeds. Given a set of seeds, SRG then grows the regions around each seed, based on the conventional region growing postulate of similarity of pixels within regions. The choice of the seeds is considered as one of the key steps on which the performance of the SRG technique depends. Thus, numerous knowledge-based and pure data-driven techniques have been already proposed to select these seeds. This paper studies the usefulness of visual attention in the seed selection process for performing color image segmentation. The purely data-driven visual attention model, considered in this paper, provides the required points of attention which are then used as seeds in a SRG segmentation algorithm using a color homogeneity criterion. A first part of this paper is devoted to the presentation of the multicue saliency-based visual attention model, which detects the most salient parts of a given scene. A second part discusses the possibility of using the so far detected regions as seeds to achieve the region growing task. The last part is dedicated to experiments involving a variety of color images.

Keywords: color image segmentation, visual attention, seed selection

1 Introduction

Visual attention is the ability to rapidly detect interesting parts of a given scene. Using visual attention in a computer vision system permits a rapid selection of a subset of the available sensory information before further processing. The selected locations are supposed to represent the conspicuous parts of the scene. Higher level computer vision tasks can then focus on these locations.

Various computational models of visual attention have been presented in previous works [1,2,3]. These models are, in general, data-driven and based on the feature integration principle [4]. Known as saliency-based, the model presented in [1] considers a variety of scene features (intensity, orientation and color) to compute a set of conspicuity maps which are then combined into the final saliency

W. Skarbek (Ed.): CAIP 2001, LNCS 2124, pp. 630–637, 2001.

map. The conspicuity operator is a kind of "contrast detector" which, applied on a feature map, detects the regions of the scene containing relevant information. Visual attention processes have been used to speed up some tasks, for instance object recognition [5], landmarks detection for robot navigation [6] and 3D scene analysis [7]. The image segmentation task considered further should also benefit from visual attention.

The "seeded region growing" (SRG) presented in [8] is a segmentation technique which performs a segmentation of an image with respect to a set of points, known as seeds. SRG is based on the conventional region growing postulate of similarity of pixels within regions. Given a set of seeds, SRG then finds a tessellation of the image into homogeneous regions. Each of which is grown around one of the seeds.

It is obvious that the performance of SRG technique depends strongly on the choice of the seeds. Some previous works have dealt with the seed selection problem [9]. Knowledge-based as well as pure data-driven solutions have been proposed. The first class of methods is usually used in specific contexts where information about the regions of interest is available. Automatic knowledge based methods as well as pure interactive seed selection belong to this class. The data-driven methods are, however, more general and can be applied on scene images without any a priori knowledge. Consequently, a wider range of images can be processed using the latter class of techniques. Numerous data-driven seed selection methods are based on histogram analysis [10]. Using either original images or even gradient images, the technique aims to find peaks on the histogram. These peaks are supposed to constitute homogeneous regions on the image. A suitable thresholding permits the selection of these regions. The selected locations of the image are then used as seed regions to achieve the SRG task. This seed selection method is straightforward for gray level images and can be extended to deal with color images. A seed selection based on intensity and color needs, however, a mechanism which combines both features.

In this work we study the possibility to use visual attention as a method for seed selection from color images. This idea is motivated by the performance of the bottom-up saliency-based model of visual attention to detect interesting locations of an image, taking into account a variety of scene features. The automatically detected salient regions are used as seeds to apply the SRG technique on color images. A first part of this paper is devoted to the presentation of the visual attention model used in this work. The SRG algorithm is then described. The last part of the paper is devoted to experimental results carried out on various color images.

2   Visual Attention Model

2.1   Saliency-Based Model

According to a generally admitted model of visual perception [2], a visual attention task can be achieved in three main steps (see Fig. 1).

1) First, a number (n) of features are extracted from the scene by computing the

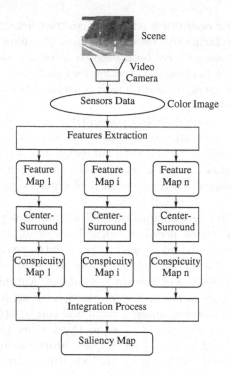

Fig. 1. Scheme of a computational model of attention.

so-called feature maps. Such a map represents the image of the scene, based on a well-defined feature. This leads to a multi-feature representation of the scene. The features used in this work are intensity, color components, and gradient orientation.

2) In a second step, each feature map is transformed in its conspicuity map. Each conspicuity map highlights the parts of the scene that strongly differ, according to a specific feature, from its surrounding. In biologically plausible models, this is usually achieved by using a *center-surround*-mechanism. Practically, this mechanism can be implemented with a *difference-of-Gaussians*-filter, which can be applied on feature maps to extract local activities for each feature type.

3) In the last stage of the attention model, the n conspicuity maps are integrated together, in a competitive way, into a *saliency map S* in accordance with equation 1.

$$S = \sum_{i=1}^{n} w_i C_i \tag{1}$$

The competition between conspicuity maps is usually established by selecting weights w_i according to a weighting function w, like the one presented in [1]: $w = (M - \overline{m})^2$, where M is the maximum activity of the conspicuity map and $\overline{m}$ is the average of all its local maxima. w measures how the most active locations differ from the average. Thus, this weighting function promotes conspicuity maps in which a small number of strong peaks of activity is present. Maps that contain numerous comparable peak responses are demoted. It is obvious that this competitive mechanism is purely data-driven and does not require any a priori knowledge about the analyzed scene.

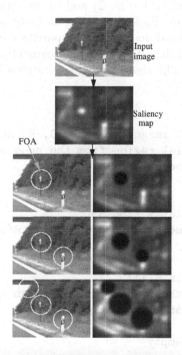

Fig. 2. Selection of salient locations. Applying a WTA mechanism to a saliency map permits the selection of the most salient locations of the image.

2.2   Selection of Salient Locations

At any given time, the maximum of the saliency map defines the most salient location, to which the focus of attention (FOA) should be directed. A "winner-take-all" (WTA) mechanism [1] is used to detect, successively, the significant regions. Given a saliency map computed by the saliency-based model of visual attention, the WTA mechanism starts with selecting the location with the maximum value of the map. This selected region is considered as the most salient part of the image (winner). The FOA is then shifted to this location. Local inhibition is activated in the saliency map, in an area around the actual FOA. This yields dynamical shifts of the FOA by allowing the next most salient location to subsequently become the winner. Besides, the inhibition mechanism prevents the FOA from returning to a previously attended locations. An example of salient regions selection based on the WTA mechanism is given in Figure 2.

3   Seeded Region Growing

Given a number of seeds, the SRG algorithm finds homogeneous regions around these points [8]. Originally, this method was applied to gray-scale images. We extend this algorithm to deal with color images. On the original image, we compute the FOA points (see Fig. 2) and we use them as seeds input for the region growing algorithm. Thus, for each spot we obtain a segmented region. It is obvious that this method does not find a partition of the input image, as it is the case in the classical definition of image segmentation. Only visually salient regions are segmented. The algorithm we use to grow a region is:

Seeded algorithm

```
decompose original image on the R, G, B channels
create R(i), initial region from the seed point
add the neighborhoods  of R(i) in SSL
begin
repeat
  remove first point x from SSL
  if(x satisfy a membership criteria in R(i))
    add x to R(i)
  end if
  if(x was added in R(i))
    add into SSL the neighbors of x which are not in SSL
    update the mean of the region R(i)
  end if
until SSL is not empty
```

By SSL we denote a list with candidate points called sequentially sorted list. Initially it contains the neighborhoods of the seed pixel. To decide the criterion homogeneity of the regions, initially we compute the $Tolerance(I_R)$, $Tolerance(I_G)$, $Tolerance(I_B)$ the tolerance rates for the three channels, using the cluster decomposition of histograms. Let f be the function which denotes the image, and f_R, f_G, f_B the image functions for the three channels. We express:

$$\delta_R(x) = |f_R(x) - mean_{y \in R_i}[f_R(y)]|$$
$$\delta_G(x) = |f_G(x) - mean_{y \in R_i}[f_G(y)]|$$
$$\delta_B(x) = |f_B(x) - mean_{y \in R_i}[f_B(y)]|$$

Thus, the criteria that must be accomplished by the candidate point x is:

$\delta_R(x) < Tolerance(I_R)$ and
$\delta_G(x) < Tolerance(I_G)$ and
$\delta_B(x) < Tolerance(I_B)$

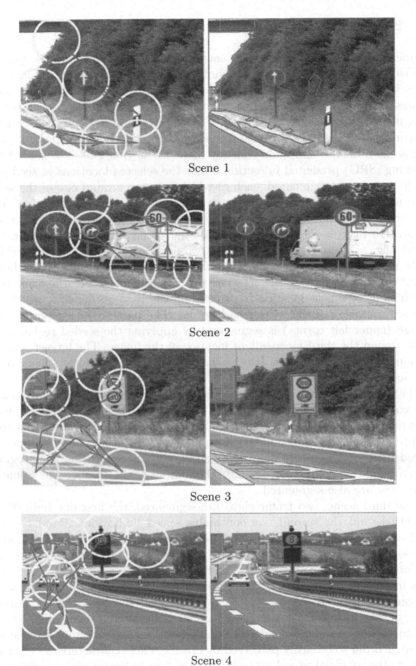

Scene 1

Scene 2

Scene 3

Scene 4

Fig. 3. Experimental results. Left: the color images with the eight most salient locations. Right: segmented salient regions.

4   Experiments

Numerous experiments have been carried out in order to study the usefulness of visual attention for the seeded region growing algorithm. Four outdoor scenes have been considered in the experiments presented in Figure 3. Each scene is represented by its color image. A saliency map is computed for each color image, using the saliency-based model of visual attention presented in section 2.1. A winner-take-all (WTA) mechanism selects the eight most conspicuous parts of the scene from the computed saliency map (see section 2.2). The seeded region growing (SRG) presented in section 3 uses the selected locations as seed points. Eight regions are segmented, each of which is grown around one of the seeds.

The most salient point of the first scene is located on the traffic sign. This is due to color and intensity contrast in this part of the image. Starting the segmentation task at this location permits the segmentation of the arrow of the traffic sign. Due to intensity contrast two parts of the signpost are within the eight most salient locations. Through a targeted region growing around these two points, the main part of the signpost can be segmented. For the same reason, the fifth most salient location is situated on the road border line, which allows the segmentation of the whole road border. The part of the sky visible on the image (upper left corner) is segmented by applying the seeded region growing task around the third most salient location of the image. The largest segmented region represents the forest. It is grown around the fourth most salient point of the scene.

Eight salient locations are also selected from the saliency map computed from the color image of the second scene. Consequently, eight regions are segmented around these seeds. For instance, the arrows of the two blue traffic signs are segmented around the third and the sixth most salient positions of the image. The speed limitation sign contains the second most salient location. The segmentation around this point easily delimits the contour of the number '60'. Some parts of the car are also segmented.

In the third scene, two traffic signs are segmented, the first one indicating the directions and the second one containing speed limitations. Road borders as well as a part of the forest are also segmented. Three important traffic signs are segmented in the fourth scene. A part of the white car and a part of the discontinuous line separating two road lanes are also segmented.

It is important to notice that neither the visual attention model nor the segmentation algorithm are adapted to road scenes analysis. Nevertheless, important objects such as traffic signs or road border lines are often segmented. This kind of objects contain relevant information and often stand out from the rest of the scene, in order to be easily perceived by drivers. These characteristics are natural help to the artificial visual attention mechanism to detect these relevant scene elements.

The presented experiments clearly show the usefulness of a visual attention mechanism for color image segmentation by means of seeded region growing. The salient locations of the image are natural candidates of seeds since they are often surrounded by relevant information. An additional benefit of the combination of visual attention and the SRG algorithm is the speed up of the segmentation task.

5   Conclusion

This work studies the usefulness of visual attention in the seed selection process for color image segmentation. The considered purely data-driven visual attention process, built around the concepts of conspicuity and saliency maps, provides the required points of attention which are then used as seeds in a SRG segmentation algorithm using a color homogeneity criterion. The experiments presented in this paper clearly validate the idea of using salient locations as start points for the SRG algorithm. The conducted experiments concern outdoor road traffic scenes. Despite the unavailability, to the segmenter, of any a priori knowledge about the analyzed scenes, the segmentation performs well, as the segmented objects include, in all cases, the most conspicuous road signs of the scene. The results speak for the good performance of the attention guided segmentation in similar scenes. Due to its bottom-up character, the reported segmentation method is expected to have similar performance in different scenes.

Acknowledgment. Part of this work was supported by the Swiss National Science Foundation under project number 54124.

References

1. Itti, L., Koch, Ch., Niebur, E.: A model of saliency-based visual attention for rapid scene analysis. IEEE Transactions on Pattern Analysis and Machine Intelligence (PAMI), Vol. 20(11) (1998) 1254-1259
2. Milanese, R.: Detecting salient regions in an image: from biological evidence to computer implementation. Ph.D. Thesis, Dept. of Computer Science, University of Geneva, Switzerland (Dec 1993)
3. Culhane, S.M., Tsotsos, J.K.: A prototype for data-driven visual attention. ICPR92, Vol. 1. Hague, Netherland (1992) 36-40
4. Treisman, A.M., Gelade, G.: A feature-integration theory of attention. Cognitive Psychology **12** (1980) 97-136
5. Ratan, A.L.: The role of fixation and visual attention in object recognition. MIT AI-Technical Report **1529** (1995)
6. Todt, E., Torras, C.: Detection of natural landmarks through multi-scale opponent features. ICPR 2000, Vol. 3. Barcelona (Sep 2000) 988-1001
7. Ouerhani, N., Hugli, H.: Computing visual attention from scene depth. ICPR 2000, Vol. 1. Barcelona (Sep 2000) 375-378
8. Adams, R., Bischof, L.: Seeded region growing. IEEE Trans. on Pattern Analysis and Maschine Intelligence, vol. 16, no. 6 (1994)
9. Koethe, U.: Primary image segmentation. 17. DAGM-Symposium. Springer (1995)
10. O'Gorman, L., Sanderson, A.C.: The converging squares algorithm: An efficient method for locating peaks in multidimensions. IEEE Trans. Pattern Analysis and Machine Intelligence, PAMI, Vol. 6 (1984) 280-288

Theoretical Analysis of Finite Difference Algorithms for Linear Shape from Shading

Tiangong Wei and Reinhard Klette

CITR, University of Auckland, Tamaki Campus
Building 731, Auckland, New Zealand
{tiangong, rklette}@citr.auckland.ac.nz

Abstract. This paper analyzes four explicit, two implicit and four semi-implicit finite difference algorithms for the linear shape from shading problem. Comparisons of accuracy, solvability, stability and convergence of these schemes indicate that the weighted semi-implicit scheme and the box scheme are better than the other ones because they can be calculated more easily, they are more accurate, faster in convergence and unconditionally stable.

Keywords: shape from shading, finite difference scheme, stability

1 Introduction

The basic problem in shape from shading is to recover surface values $Z(x, y)$ of an object surface from its variation in brightness. The surface function Z is assumed to be defined in image coordinates (x, y). It is identical to the *depth* of surface points visualized in an image, i.e., to the Euclidean distance between image plane and surface point.

We assume parallel illumination of a Lambertian surface. The illumination is characterized by an orientation $(p_s, q_s, -1)$ and its intensity E_0. For the surface we assume that it may be modeled by a function $Z(x, y)$, and $\rho(x, y)$, with $0 \leq \rho(x, y) \leq 1$, denotes the *albedo* (i.e. the reflectance constant) at point (x, y). The surface function $Z(x, y)$ satisfies the following *image irradiance equation*

$$\frac{1 + p_s p + q_s q}{\sqrt{1 + p_s^2 + q_s^2}\sqrt{1 + p^2 + q^2}} = \rho(x, y) \cdot E_0 \cdot E(x, y) \tag{1}$$

over a compact image domain Ω, where $(p, q) = (p(x, y), q(x, y))$ is the surface gradient with $p = \partial Z / \partial x$ and $q = \partial Z / \partial y$ at point $(x, y) \in \Omega$, and $E(x, y)$ is the image brightness at this point formed by an orthographic (parallel) projection of reflected light onto the xy-image plane. Throughout this paper we assume that $E_0 \cdot \rho(x, y) = 1$, for all points $(x, y) \in \Omega$, i.e., we assume a constant albedo for all projected surface points. This approach is called *albedo-dependent shape recovery*, see Klette et al. [5].

The above nonlinear, first-order partial differential equation has been studied with a variety of different techniques (see, e.g., Horn [1,3]; Horn and Brooks

W. Skarbek (Ed.): CAIP 2001, LNCS 2124, pp. 638–645, 2001.
© Springer-Verlag Berlin Heidelberg 2001

[2]; Tsai and Shah [13]; Lee and Kuo [9]; Kimmel and Bruckstein [4]). The traditional approaches employ regularization techniques. However, Oliensis [10] discovered that in general, shape from shading not be assumed to be ill-posed, and regularization techniques should be used with caution. Furthermore, Zhang et al. [16] pointed out that all shape from shading algorithms produce generally poor results. Therefore, new shape from shading methods should be developed to provide more accurate, and realistic results. Pentland [11] proposed a method based on the linearity of the reflectance map in the surface gradient (p, q), which greatly simplifies the shape from shading problem. This leads to the following *linear image irradiance equation:*

$$\frac{1 + p_s p + q_s q}{\sqrt{1 + p_s^2 + q_s^2}} = E(x, y). \tag{2}$$

As an example, such a special case arises, e.g., in recovering the shape of parts of the lunar surface ("Maria of the moon"). we can rewrite (2) as

$$p_s \frac{\partial Z}{\partial x}(x, y) + q_s \frac{\partial Z}{\partial y}(x, y) = F(x, y). \tag{3}$$

where $F(x, y) = E(x, y)\sqrt{1 + p_s^2 + q_s^2} - 1$. For the sufficient conditions assuring the well-posedness of the problem (3) we refer the reader to Kozera [6]. Horn [1] first proposed a method for recovery of shapes described by (3). Kozera and Klette [7,8] presented four algorithms based on explicit finite difference methods. Ulich [14] also discussed two explicit and one implicit finite difference algorithms for (3). So far it has not yet been theoretically studied which finite difference algorithms for (3) are better. The method used for the proof of stability and convergence are relatively complicated.In this paper, we consider (3) over a rectangle domain

$$\Omega = \{(x, y) \in \mathbb{R}^2 : \quad 0 \leq x \leq a, \quad 0 \leq y \leq b\}$$

with the following initial condition $Z(x, 0) = \phi(x)$ $(0 \leq x \leq a)$ and boundary conditions $Z(0, y) = \psi_0(y)$ and $Z(a, y) = \psi_1(y)$ $(0 \leq y \leq b)$, where the given functions $\phi(x), \psi_0(y)$ and $\psi_1(y)$ satisfy $\phi \in C([0, a]) \cap C^2((0, a)), \psi_0, \psi_1 \in C([0, b]) \cap C^2((0, b)), \phi(0) = \psi_0(0), \phi(a) = \psi_1(a)$, and $(p_s, q_s) \neq (0, 0)$. Throughout this paper we assume that the above Cauchy problem is *well-posed* over a rectangle Ω, that is, there exists a unique solution $Z(x, y)$ to the corresponding partial differential equation satisfying the boundary conditions and depending continuously on the given initial condition, and we also suppose that the solution $Z(x, y)$ is sufficiently smooth, at least $Z(x, y) \in C^2(\bar{\Omega})$, see Kozera [6].

The organization of the rest of the paper is as follows. In Section 2 we present ten different discretizations of equation (3): four explicit, two implicit and four semi-implicit schemes. The initial condition $Z(x, 0)$ is used for all these methods, but different boundary conditions $Z(0, y)$ and/or $Z(a, y)$ are required. In Section 3 we discuss the accuracy, solvability, consistency, stability and convergence of these methods. The conclusions are given in Section 4.

2   Finite Difference Algorithms

Suppose that the rectangular domain Ω is divided into small grids by parallel lines $x = x_i$ ($i = 0, 1, \ldots, M$) and $y = y_j$ ($j = 0, 1, \ldots, N$), where $x_i = ih$, $y_j = jk$ and $Mh = a$, $Nk = b$, M and N are integers, h is the grid constant in x-direction (i.e., distance between neighboring grid lines) and k is the grid constant in y-direction. For convenience, we shall denote by $Z(i, j)$ the value $Z(x_i, y_j)$ of solution $Z(x, y)$ on the grid point (x_i, y_j), and by $Z_{i,j}$ an approximation of $Z(i, j)$.

2.1   Explicit Schemes

Forward-Forward (FF) Scheme: Approximating $\partial Z / \partial x$ and $\partial Z / \partial y$ with the forward difference quotient gives the following discretization for (3):

$$p_s \frac{Z(i+1,j) - Z(i,j)}{h} + q_s \frac{Z(i,j+1) - Z(i,j)}{k} + O(h+k) = F(i,j) \,,$$

where $O(h+k) = -hp_s Z_{xx}(\theta_1, y_j)/2 - kq_s Z_{yy}(x_i, \theta_2)/2$, $x_i \leq \theta_1 \leq x_{i+1}$, $y_j \leq \theta_2 \leq y_{j+1}$. Dropping the truncation error $O(h+k)$, and rearranging the above equation gives

$$Z_{i,j+1} = (1+c)Z_{i,j} - cZ_{i+1,j} + \frac{k}{q_s}F_{i,j}, \tag{4}$$

$$i = 0, 1, \ldots, M-1; j = 0, 1, \ldots, N-1,$$

where the corresponding finite difference initial conditions $Z_{i,0}(i = 0, 1 \ldots, M)$ and boundary conditions $Z_{M,j}(j = 0, 1, \ldots, N)$ are given, $c = \frac{p_s k}{q_s h}$, $q_s \neq 0$. The truncation error of the FF scheme is in the order of $O(h+k)$. Given an linear shape from shading problem (3), the FF scheme with the above boundary condition recovers the unknown shape over a *domain of influence* of which coincides with the entire Ω. But if we only give the following boundary conditions $Z_{0,j}(j = 0, 1, \ldots, N)$, then the domain of influence of the FF scheme is as follows

$$D_{FF_0} = \{(x,y) \in \mathbb{R}^2 : \ 0 \leq x \leq a, \ 0 \leq y \leq (-b/a)(x-a)\}.$$

The same scheme with different boundary conditions coincide with different domain of influence. Therefore, boundary conditions are very important to the finite difference algorithms for the linear shape from shading (3).

Backward-Forward (BF) Scheme: If we use the above techniques, we can get the following two-level explicit scheme

$$Z_{i,j+1} = cZ_{i-1,j} + (1-c)Z_{i,j} + \frac{k}{q_s}F_{i,j}, \tag{5}$$

$$i = 1, \ldots, M; j = 0, \ldots, N-1,$$

where initial conditions $Z_{i,0}(i = 0, 1, \ldots, M)$ and boundary conditions $Z_{0,j}(j = 0, 1, \ldots, N)$ are given, $c = \frac{p_s k}{q_s h}$, $q_s \neq 0$. The truncation error of the BF scheme is $O(h+k)$, and the domain of influence of the BF scheme with the above boundary condition is entire Ω.

Lax-Friedrichs (LF) Scheme: Approximating $\partial Z/\partial x$ with the central difference quotient and $\partial Z/\partial y$ with the forward difference approximation, and then replacing $Z_{i,j}$ by its average at the $(i+1)$th and $(i-1)$th levels.

$$Z_{i,j+1} = \frac{1-c}{2}Z_{i+1,j} + \frac{1+c}{2}Z_{i-1,j} + \frac{k}{q_s}F_{i,j}, \tag{6}$$
$$i = 1, \ldots, M-1; j = 0, 1, \ldots, N-1,$$

where initial conditions $Z_{i,0}(i = 0, 1, \ldots, M)$, boundary conditions $Z_{0,j}$ and $Z_{M,j}(j = 0, 1, \ldots, N)$ are given, $c = \frac{p_s k}{q_s h}$, $q_s \neq 0$. It holds that the truncation error of the LF scheme is $O\left(h^2 + k + h^2/k\right)$, and the domain of influence of the LF scheme is entire Ω.

Leapfrog Scheme: Approximating both $\partial Z/\partial x$ and $\partial Z/\partial y$ with the central difference quotient, this leads to the following three-level explicit scheme

$$Z_{i,j+1} = Z_{i,j-1} + c(Z_{i-1,j} - Z_{i+1,j}) + \frac{2k}{q_s}F_{i,j}, \tag{7}$$
$$i = 1, \ldots, M-1; j = 1, \ldots, N-1,$$

where $Z_{i,0}(i = 0, \ldots, M), Z_{0,j}$ and $Z_{M,j}(j = 0, \ldots, N)$ are given, $c = \frac{p_s k}{q_s h}$, $q_s \neq 0$. The truncation error of the leapfrog scheme is $O(h + k)$, and the domain of influence of the leapfrog scheme is entire Ω. To start the computations of the leapfrog scheme we must specify the values of $Z_{i,0}$ and $Z_{i,1}$ for all i, usually $Z_{i,1}$ can be calculated by another scheme, e.g., the FF scheme, the BF scheme or others.

2.2 Implicit Schemes

The finite difference schemes described previously are all explicit. They are easy to be calculated, but the accuracy of the explicit schemes is usually lower. The following *implicit* schemes will overcome this drawback.

Central-Backward (CB) Scheme: As before, approximating $\partial Z/\partial x$ with the central difference quotient and $\partial Z/\partial y$ with the backward difference quotient, we get the following two-level implicit scheme.

$$-\frac{c}{2}Z_{i-1,j+1} + Z_{i,j+1} + \frac{c}{2}Z_{i+1,j+1} = Z_{i,j} + \frac{k}{q_s}F_{i,j}, \tag{8}$$
$$i = 1, \ldots, M-1; j = 0, \ldots, N-1,$$

where the initial conditions $Z_{i,0}(i = 0, \ldots, M)$, the boundary conditions $Z_{0,j}$ and $Z_{M,j}(j = 0, \ldots, N)$ are given, $c = \frac{p_s k}{q_s h}$, $q_s \neq 0$. The truncation error of the CB scheme is $O\left(h^2 + k\right)$. The computations of the CB scheme will take much more time because it requires to solve a linear algebraic systems at each j level.

Crank-Nicolson (CN) Scheme: Another implicit finite difference algorithm used to solve the linear shape from shading problem is the CN scheme:

$$-\frac{c}{4}Z_{i-1,j+1} + Z_{i,j+1} + \frac{c}{4}Z_{i+1,j+1} = \frac{c}{4}Z_{i-1,j} + Z_{i,j} - \frac{c}{4}Z_{i+1,j} + \frac{k}{q_s}F_{i,j} \,, \quad (9)$$

$$i = 1, \ldots, M-1; j = 0, 1, \ldots, N-1,$$

where $Z_{i,0}(i = 0, 1, \ldots, M), Z_{0,j}$ and $Z_{M,j}(j = 0, \ldots, N)$ are given, $c = \frac{p_s k}{q_s h}$, $q_s \neq 0$. The truncation error of the CN scheme is $O\left(h^2 + k^2\right)$.

2.3   Semi-implicit Schemes

An *explicit* finite difference scheme for (3) contains only one unknown value of Z at each j level. The unknown value is calculated directly from the known values of Z at the previous levels. Therefore, explicit schemes are easy to be computed. The disadvantage of explicit schemes is that their accuracy is lower since the order of their truncation errors is usually lower. For an *implicit* scheme, there are three unknown values of Z at each j level. Implicit schemes are more accurate than explicit schemes since the order of the truncation errors of implicit schemes is higher than that of explicit schemes. However, the computation of implicit schemes takes much more time than that of explicit schemes because implicit schemes require to solve a linear algebraic systems for each j. In order to overcome the drawbacks and take the advantages of explicit and implicit schemes, we consider the following *semi-implicit* schemes which contain two unknown values of Z at each j level.

Forward-Backward (FB) Scheme: Approximating $\partial Z/\partial x$ with the forward difference quotient and $\partial Z/\partial y$ with the backward difference quotient gives

$$Z_{i+1,j} = (1-d)Z_{i,j} + dZ_{i,j-1} + \frac{h}{p_s}F_{i,j}, \quad (10)$$

$$i = 0, 1, \ldots, M-1; j = 1, 2, \ldots, N,$$

where $Z_{i,0}(i = 0, \ldots, M)$ and $Z_{0,j}(j = 0, \ldots, N)$ are given, $p_s \neq 0, d = 1/c$. The truncation error of the FB scheme is $O(h+k)$.

Backward-Backward (BB) Scheme: Approximating both $\partial Z/\partial x$ and $\partial Z/\partial y$ with the backward difference scheme yields

$$Z_{i,j} = \frac{1}{1+c}Z_{i,j-1} + \frac{c}{1+c}Z_{i-1,j} + \frac{k}{q_s(1+c)}F_{i,j}, \quad (11)$$

$$i = 1, \ldots, M; j = 1, \ldots, N,$$

where $Z_{i,0}(i = 0, 1, \ldots, M)$ and $Z_{0,j}(j = 0, 1, \ldots, N)$ are given, $c = \frac{p_s k}{q_s h}, c \neq -1$, $q_s \neq 0$. The truncation error of the BB scheme is $O(h+k)$.

Weighted Semi-implicit (WS) Scheme:

$$Z_{i,j+1} = Z_{i,j} + \frac{c}{2+c}(Z_{i-1,j+1} - Z_{i+1,j}) + \frac{k}{q_s(2+c)}F_{i,j}, \qquad (12)$$
$$i = 1, \ldots, M-1; j = 0, 1, \ldots, N-1,$$

where $Z_{i,0}(i = 0, 1, \ldots, M), Z_{0,j}$ and $Z_{M,j}(j = 0, \ldots, N)$ are given, $c = \frac{p_s k}{q_s h}$, $c \neq -2$, $q_s \neq 0$. The truncation error of the WS scheme is $O\left(h + k^2\right)$.

Box Scheme: The box scheme for solving (3) is as follows:

$$Z_{i+1,j+1} = Z_{i,j} + \frac{1-c}{1+c}(Z_{i+1,j} - Z_{i,j+1}) + \frac{2k}{q_s(1+c)}F_{i,j}, \qquad (13)$$
$$i = 0, 1, \ldots, M-1; j = 0, 1, \ldots, N-1,$$

where $Z_{i,0}(i = 0, 1, \ldots, M)$ and $Z_{0,j}(j = 0, 1, \ldots, N)$ are given, $c = \frac{p_s k}{q_s h}$, $c \neq -1$, $q_s \neq 0$. The truncation error of the box scheme is $O\left(h^2 + k^2\right)$.

Given a linear shape from shading problem (3), the *domain of influence* of each scheme above coincides with the entire domain Ω.

3   Analysis of Finite Difference Algorithms

Given the above list of schemes we are naturally led to the question of which of them are useful and which are not. In this section we firstly determine which schemes have solutions that approximate solutions of the shape from shading problem (3). Later on we determine which schemes are more accurate than others and also investigate the efficiency of the various schemes.

3.1   Consistency

Definition 1. *A finite difference scheme is said to be* consistent *with a partial differential equation iff as the grid constants tend to zero, the difference scheme becomes in the limit the same as the partial differential equation at each point in the solution domain.*

Theorem 1. *All the above finite difference schemes are consistent with (3), the LF scheme is consistent if k/h is constant.*

3.2   Solvability

Theorem 2. *Let $c = \frac{p_s k}{q_s h}$ be a fixed constant, $q_s \neq 0$. Then,*
(a) *all the above explicit schemes are solvable;*
(b) *the CB scheme is solvable if $|c| < 1$, the CN scheme is solvable if $|c| < 2$;*
(c) *the FB scheme is solvable if $p_s \neq 0$, the BB and box schemes are solvable if $c \neq -1$, the WS scheme is solvable if $c \neq -2$.*

3.3   Stability

Definition 2. *A finite difference scheme is said to be* stable *iff the difference between the numerical solution and the exact solution of the difference scheme does not increase as the number of rows of calculation at successive j levels in the solution domain is increased.*

Theorem 3. *Let $c = \frac{p_s k}{q_s h}$ be a fixed constant, $q_s \neq 0$. Then,*
(a) *the FF scheme is stable iff $-1 \leq c \leq 0$;*
(b) *the BF scheme is stable iff $0 \leq c \leq 1$;*
(c) *the LF and the leapfrog schemes are stable iff $|c| \leq 1$;*
(d) *the CB and the CN schemes are unconditionally stable;*
(e) *the FB scheme is stable iff $d \leq 1$, where $d = 1/c$;*
(f) *the BB scheme is stable iff $c \geq 0$ or $c < -1$;*
(g) *the WS and the box schemes are unconditionally stable.*

3.4   Convergence

Definition 3. *A solution to a finite difference scheme which approximates a given partial differential equation is said to be* convergent *iff at each grid-point in the solution domain, the solution of the difference scheme approaches the solution of the corresponding partial differential equation as the grid constants tend to zero.*

Theorem 4. *Let $c = \frac{p_s k}{q_s h}$ be a fixed constant, $q_s \neq 0$. Then,*
(a) *the FF scheme is convergent if $-1 \leq c \leq 0$;*
(b) *the BF scheme is convergent if $0 \leq c \leq 1$;*
(c) *the LF and the leapfrog schemes are convergent if $|c| \leq 1$;*
(d) *the CB and the CN schemes are convergent for all $c \in \mathbb{R}$;*
(e) *the FB scheme is convergent if $d \leq 1$, where $d = 1/c$;*
(f) *the BB scheme is convergent if $c \geq 0$ or $c < -1$;*
(g) *the WS scheme is convergent if $c \neq -2$; the box scheme is convergent if $c \neq -1$.*

For the proof of the theorem 1, 2, 3 and 4 see [15].

4   Conclusions

The semi-implicit finite difference algorithms for linear shape from shading are discussed in this paper for the first time. Comparisons of accuracy, solvability, stability and convergence of each scheme indicate that the weighted semi-implicit scheme and the box scheme are more useful than the others.

All schemes presented in this paper are supplemented by a specification of the domain of influence, truncation error, consistency, solvability, stability and convergence analysis. The domain of influence of each scheme in this paper coincides with the entire domain Ω.

In comparison with results obtained by Kozera and Klette [7,8] and Ulich [14], the ranges of the stability and convergence of the FB and BB schemes are identified as being larger.

Acknowledgment. The authors thank Dr Ryszard Kozera (UWA Nedlands, Australia) for helpful comments on an initial draft of this paper.

References

1. Horn, B.K.P.: Robot Vision. McGraw-Hill, New York, Cambridge M.A. (1986)
2. Horn, B.K.P., Brooks, M.J.: The variational approach to shape from shading. Computer Vision, Graphics, and Image Processing **33** (1986) 174–208
3. Horn, B.K.P.: Height and gradient from shading. International Journal of Computer Vision **5** (1990) 37–75
4. Kimmel, R., Bruckstein, A.M.: Tracking level sets by level sets: a method for solving the shape from shading problem. Computer Vision and Image Understanding **62** (1995) 47–58
5. Klette, R., K. Schlüns, K., Koschan, A.: Computer Vision - Three-dimensional Data from Images. Springer, Singapore (1998)
6. Kozera, R.: Existence and uniqueness on photometric stereo. Applied Mathematics and Computation **44** (1991) 1–104
7. Kozera, R., Klette, R.: Finite difference based algorithms in linear shape from shading. Machine Graphics and Vision, **2** (1997) 157–201
8. Kozera, R., Klette, R.: Criteria for differential equations in computer vision. CITR-TR-27, The University of Auckland, Tamaki Campus (Aug 1998)
9. Lee, K.M., Kuo, C.J.: Shape from shading with a linear triangular element surface model. IEEE Transactions on Pattern Analysis and Machine Intelligence **15** (1993) 815–822
10. Oliensis, J.: Uniqueness in shape from shading. International Journal of Computer Vision **6** (1991) 75–104
11. Pentland, A.P.: Linear shape from shading. International Journal of Computer Vision **4** (1991) 153–162
12. Strikwerda, J.C.: Finite Difference Schemes and Partial Differential Equations. Wordsworth & Brooks/Cole Advanced Books & Software. Pacific Grove, California (1989)
13. Tsai, P.S., Shah, M.: Shape from shading using linear approximation. Image and Vision Computing **12** (1994) 487–498
14. Ulich, G.: Provably convergent methods for the linear and nonlinear shape from shading problem. Journal of Mathematical Imaging and Vision **9** (1998) 69–82
15. T. Wei and R. Klette: *Analysis of finite difference algorithms for linear shape from shading.* CITR-TR-70, Tamaki Campus, The University of Auckland (Oct 2000)
16. Zhang, R., Tsai, P.S., Cryer, J.E., Shah, M.: Shape from shading: a survey. IEEE Trans. Pattern Analysis and Machine Intelligence **21** (1999) 690–706

Relational Constraints for Point Distribution Models

Bin Luo[1,2] and Edwin R. Hancock[1]

[1] Department of Computer Science,
University of York, York YO1 5DD, UK.
[2] Anhui University, P.R. China

Abstract. In this paper we present a new method for aligning point distribution models to noisy and unlabelled image data. The aim is to construct an enhanced version of the point distribution model of Cootes and Taylor in which the point-position information is augmented with a neighbourhood graph which represents the relational arrangement of the landmark points. We show how this augmented point distribution model can be matched to unlabelled point-sets which are subject to both additional clutter and point drop-out. The statistical framework adopted for this study interleaves the processes of finding point correspondences and estimating the alignment parameters of the point distribution model. The utility measure underpinning the work is the cross entropy between two probability distributions which respectively model alignment errors and correspondence errors. In the case of the point alignment process, we assume that the registration errors follow a Gaussian distribution. The correspondence errors are modelled using probability distribution which has been used for symbolic graph-matching. Experimental results are presented using medical image sequences.

Keywords: point distribution models, relational constraints, landmark points

1 Introduction

Point pattern matching is a problem of pivotal importance in computer vision that continues to attract considerable interest. The problem may be abstracted as either alignment or correspondence. Alignment involves explicitly transforming the point positions either under a predefined rigid geometry or under a non-rigid deformation so as to maximise a measure of correlation. Examples here include Procrustes normalisation [6], affine template matching [15] and deformable point models [2]. Correspondence, on the other hand, involves recovering a consistent arrangement of point assignment labels. The correspondence problem can be solved using a variety of eigendecomposition [9] and graph matching [4,1,16] algorithms. The problem of point pattern matching has attracted sustained interest in both the vision and statistics communities for several decades. For instance, Kendall [6] has generalised the process to projective manifolds using the concept of Procrustes distance. Ullman [11] was one of the first to recognise the importance of exploiting rigidity constraints in the correspondence matching of point-sets. Recently, several authors have drawn inspiration from Ullman's ideas in developing general purpose correspondence matching algorithms using the Gaussian weighted proximity matrix. As a concrete example, Scott and Longuet-Higgins [9] locate correspondences by

W. Skarbek (Ed.): CAIP 2001, LNCS 2124, pp. 646–656, 2001.

finding a singular value decomposition of the inter-image proximity matrix. Shapiro and Brady [10], on the other hand, match by comparing the modal eigenstructure of the intra-image proximity matrix. In fact these two ideas provide some of the basic groundwork on which the deformable shape models of Cootes *et al* [2] and Sclaroff and Pentland [8] build. This work on the co-ordinate proximity matrix is closely akin to that of Umeyama [12] who shows how point-sets abstracted in a structural manner using weighted adjacency graphs can be matched using an eigen-decomposition method. These ideas have been extended to accommodate parametererised transformations [13] which can be applied to the matching of articulated objects [14]. More recently, there have been several attempts at modelling the structural deformation of point-sets. For instance, Amit and Kong [1] have used a graph-based representation (graphical templates) to model deforming two-dimensional shapes in medical images. Lades *et al* [7] have used a dynamic mesh to model intensity-based appearance in images.

The motivation for the work reported in this paper is that the dichotomy normally drawn between the two processes overlooks considerable scope for synergistic interchange of information. In other words, there must always be bounds on alignment before correspondence analysis can be attempted, and vice versa. To this end, we develop a new point-pattern matching method in which we exploit constraints on the spatial arrangement of correspondences to augment the recovery of alignment parameters. Although there have been recent attempts at realising this goal using relaxation algorithms [5] and the EM algorithm [3], in this paper we put the alignment and correspondence processes on a symmetric footing. The two processes communicate in a symmetric manner via an integrated utility measure. The utility measure is the cross-entropy between the probability distributions for alignment and correspondence.

We apply the new matching framework to the non-rigid alignment of point-sets which deform according to a point distribution model. This is a class of deformable shape model recently developed by Cootes and Taylor [2]. The idea underpinning point distribution models is to learn the modes of variation of point-patterns by computing the eigenmodes of the co-variance matrix for a set of training examples, The eigenvectors of the co-variance matrix define directions in which the points can move with respect to the mean-pattern. Once trained in this way, a point distribution model can be fitted to data by estimating the proportions of each eigen-mode that minimise the distance between the data and the aligned model. When carefully trained, the method can be used to model quite complex point deformations. Moreover, the modal deformations defined by the eigenvectors can be reconciled with plausible natural modes of shape variation. While much effort has been expended in improving the training of PDM's, there are a number of shortcomings that have limited their effective matching to noisy point-data. Firstly, in order to estimate the modal displacements necessary to align a point-distribution model with data, the data points must be labelled. In other words, the correspondences between points in the model and the data must be known *a priori*. Secondly, the point-sets must be of the same size. As a result, point distribution models can not be aligned reliably to point-sets which are subject to contamination or dropout. Finally, information concerning the relational arrangement of the points is overlooked. Our new matching framework provides a natural way of overcoming these shortcomings.

Our aim in this paper is to present a statistical framework for aligning point distribution models. We depart from the conventional treatment in three ways. First, we deal with unlabelled point-sets. That is to say we commence without knowledge of the correspondences between the model and the data to be matched. Our second contribution is to deal with the case in which the model and the data contain different numbers of points. This might be due to poor feature location or contamination by noise. Finally, we aim to exploit constraints on the relational arrangement of the points to improve the fitting process.

2 Point Distribution Models

The point distribution model of Cootes and Taylor commences from a set training patterns. Each training pattern is a configuration of labelled point co-ordinates or landmarks. The patterns of landmark points are collected as the the object in question undergoes representative changes in shape. To be more formal, each pattern of landmark points consists of L labelled points whose co-ordinates are represented by the set of position co-ordinates $\{x_1, x_2, \ldots, x_L\} =$
$\{(x_1, y_1), \ldots\ldots(x_L, y_L)\}$. Suppose that there are N patterns of landmark points. The tth training pattern is represented using the long-vector of landmark co-ordinates $X_t = (x_1, y_1, x_2, y_2, \cdots, x_L, y_L)^T$, where the subscripts of the co-ordinates are the landmark labels. For each training pattern the labelled landmarks are identically ordered. The mean landmark pattern is represented by the average long-vector of co-ordinates $\hat{X} = \frac{1}{N} \sum_{t=1}^{N} X_t$. The covariance matrix for the landmark positions is

$$U = \frac{1}{N} \sum_{t=1}^{N} (X_t - \hat{X})(X_t - \hat{X})^T \tag{1}$$

The eigenmodes of the landmark covariance matrix are used to construct the point-distribution model. First, the eigenvalues λ of the landmark covariance matrix are found by solving the eigenvalue equation $|U - \lambda I| = 0$ where I is the $2L \times 2L$ identity matrix. The eigen-vector ϕ^{λ_i} corresponding to the eigenvalue λ_i is found by solving the eigenvector equation $U\phi^{\lambda_i} = \lambda_i \phi^{\lambda_i}$. According to Cootes and Taylor, the landmark points are allowed to undergo displacements relative to the mean-shape in directions defined by the eigenvectors of the covariance matrix U. To compute the set of possible displacement directions, the K most significant eigenvectors are ordered according to the magnitudes of their corresponding eigenvalues to form the matrix of column-vectors $\Phi = (\phi^{\lambda_1} | \phi^{\lambda_2} | \ldots | \phi^{\lambda_K})$, where $\lambda_1, \lambda_2, \ldots, \lambda_K$ is the order of the magnitudes of the eigenvectors. The landmark points are allowed to move in a direction which is a linear combination of the eigenvectors. The updated landmark positions are given by $X^{(n)} = \hat{X} + \Phi r$, where r is a vector of modal co-efficients. This vector represents the free-parameters of the global shape-model.

2.1 Landmark Displacements

The matrix formulation of the point-distribution model allows the global shape deformation to be computed. However, in order to develop our matching method we will be interested in individual point displacements. We will focus our attention on the displacement vector for the landmark point indexed i produced by the eigenmode indexed λ. The two components of displacement are the elements long-vector ϕ_{λ_i} indexed $2i - 1$ and $2i$. We denote the displacement vector by $v_i^{\lambda_i} = (\phi_{2i-1}^{\lambda_i}, \phi_{2i}^{\lambda_i})^T$. For each landmark point the set of displacement vectors associated with the individual eigenmodes are concatenated to form a $2 \times K$ displacement matrix. For the jth landmark, the displacement matrix is $\Delta_j = (v_j^{\lambda_1}|v_j^{\lambda_2}|...|v_j^{\lambda_K})$. The point-distribution model allows the landmark points to be displaced by a vector amount which is equal to a linear superposition of the displacement-vectors associated with the individual eigenmodes. To this end let $r = (r_1, r_2, ..., r_K)^T$ represent a vector of modal superposition co-efficients for the different eigenmodes. With the modal superposition co-efficients to hand, the position of the landmark j is displaced by an amount $\Delta_j r$ from its mean position $\hat{x}_j = \frac{1}{N}\sum_{t=1}^{N} x_i$. The aim in this paper is to develop an iterative method for aligning the point distribution model. At iteration n of the algorithm we denote the aligned position of the landmark point j by the vector $x_j^{(n)} = \hat{x}_j + \Delta_j r^{(n)}$.

We wish to align the point distribution model represented in this way to a set of observed data-points. The data-points to be fitted are represented by an unlabelled set of D point position vectors $\mathbf{w} = \{w_1, w_2, ..., w_D\}$. This size of this point set may be different to the number of landmark points L used in the training of the point-distribution model. The free parameters that must be adjusted to align the landmark points with $\mathbf{w}$ are the modal co-efficients r.

2.2 Relational Constraints

One of our goals in this paper is to exploit structural constraints to improve the recovery of alignment parameters from sets of feature points. To this end we represent point adjacency using a neighbourhood graph. There are many alternatives including the N-nearest neighbour graph, the Delaunay graph, the Gabriel graph and the relative neighbourhood graph. Because of its well documented robustness to noise and change of viewpoint, we adopt the Delaunay triangulation as our basic representation of image structure. We establish Delaunay triangulations on the data and the model, by seeding Voronoi tessellations from the feature-points.

The process of Delaunay triangulation generates relational graphs from the two sets of point-features. More formally, the point-sets are the nodes of a data graph $G_D = \{D, E_D\}$ and a model graph $G_M = \{M, E_M\}$, where $M = \{1, ..., L\}$ is the index-set of the landmark points. Here $E_D \subseteq D \times D$ and $E_M \subseteq M \times M$ are the edge-sets of the data and model graphs. Later on we will cast our optimisation process into a matrix representation. Here we use the notation $\hat{E}_D(i, i')$ to represent the elements of the adjacency matrix for the data graph; the elements are unity if $i = i'$ or if (i, i') is an edge and are zero otherwise. We represent the state of correspondence between the two graph using the function $f : D \to M$ from the nodes of the model graph onto the nodes of the data-graph.

3   Dual Step Matching Algorithm

We characterise the matching problem in terms of separate probability distributions for alignment and correspondence. In the case of alignment, the distribution models the registration errors between the landmarks of the point distribution model and the observed data-points under the vector of model displacements r. The correspondence process on the other hand captures the consistency of the pattern of matching assignments to the graph representing the point-sets. The set of assignments is represented by the function $f : D \rightarrow M$. Suppose that $P_{i,j}^{(n)}$ is the probability that observed data-point indexed i is in alignment with the landmark point indexed j at iteration n. Similarly, $Q_{i,j}^{(n)}$ is the probability that data point i is in correspondence with landmark point j. Further suppose that $p_{i,j}^{(n)} = p(w_i | x_j, \Phi, r^{(n)})$ is the probability distribution for the alignment error between the dat-point i and the landmark point j under the vector of modal coefficients parameters $r^{(n)}$ recovered at iteration n. The distribution of the correspondence errors associated with the assignment function $f^{(n)}$ at iteration n is $q_{i,j}^{(n)}$. With these ingredients the utility measure which we aim to maximise in the dual alignment and correspondence steps is

$$\mathcal{E} = \sum_{i=1}^{D} \sum_{j=1}^{L} \left[Q_{i,j}^{(n)} \ln p_{i,j}^{(n+1)} + P_{i,j}^{(n)} \ln q_{i,j}^{(n+1)} \right] \tag{2}$$

In other words, the two processes interact via a symmetric expected log-likelihood function. The correspondence probabilities weight contributions to the expected log-likelihood function for the alignment errors, and vice-versa. Cross and Hancock showed how the first term arises through the gating of the log-likelihood function of the EM algorithm [3].

The alignment parameters and correspondence matches are recovered via the dual maximisation equations

$$r^{(n+1)} = \arg\max_{\Phi} \sum_{i \in D} \sum_{j \in M} Q_{i,j}^{(n)} \ln p_{i,j}^{(n+1)}, \quad f^{(n+1)} = \arg\max_{f} \sum_{i \in D} \sum_{j \in M} P_{i,j}^{(n)} \ln q_{i,j}^{(n+1)} \tag{3}$$

3.1   Alignment

To develop a useful alignment algorithm we require a model for the measurement process. Here we assume that the observed position vectors, i.e. w_i are derived from the model points through a Gaussian error process. According to our Gaussian model of the alignment errors,

$$p(w_i | x_j, \Phi, r^{(n)}) = \frac{1}{2\pi\sqrt{|\Sigma|}} \exp\left[-\frac{1}{2}(w_i - x_j^{(n)})^T \Sigma^{-1} (w_i - x_j^{(n)}) \right] \tag{4}$$

where Σ is the variance-covariance matrix for the point measurement errors. Here we assume that the position errors are isotropic, in other words the errors in the x and y

directions are identical and uncorrelated. As a result we write $\Sigma = \sigma^2 I_2$ where I_2 is the 2x2 identity matrix and σ^2 is the isotropic noise variance for the point positions. With this model, the alignment step is concerned with minimising the weighted square error measure

$$\mathcal{E}_c = \sum_{i=1}^{D} \sum_{j=1}^{L} Q_{i,j}^{(n)} (\boldsymbol{w}_i - \boldsymbol{x}_j^{(n+1)})^T (\boldsymbol{w}_i - \boldsymbol{x}_j^{(n+1)}) \tag{5}$$

With the point-distribution model displacement process detailed earlier, the weighted squared-error becomes

$$\mathcal{E}_c = \sum_{i=1}^{D} \sum_{j=1}^{L} Q_{ij}^{(n)} (\boldsymbol{w}_i - \hat{x}_j - \Delta_j \boldsymbol{r}^{(n+1)})^T (\boldsymbol{w}_i - \hat{x}_j - \Delta_j \boldsymbol{r}^{(n+1)}) \tag{6}$$

Our aim is to recover the vector of modal co-efficients which minimize this weighted squared error. To do this we solve the system of saddle-point equations which results by setting $\partial \mathcal{E}_c / \partial \boldsymbol{r}^{(n+1)} = 0$. After applying the rules of matrix differentiation and simplifying the resulting saddle-point equations, the solution vector is

$$\underline{\boldsymbol{r}}^{(n+1)} = (\sum_{j=1}^{L} \Delta_j^T \Delta_j)^{-1} \{ \sum_{i=1}^{D} \sum_{j=1}^{L} Q_{ij}^{(n)} \boldsymbol{w}_i^T \Delta_j - \sum_{j=1}^{L} \hat{x}_j^T \Delta_j \} \tag{7}$$

Further simplification results if we note that the landmark covariance matrix U is symmetric. As a result its individual eigenvectors are orthogonal to one another, i.e.. $\phi^{\alpha T} \phi^\beta = 0$ if $\alpha \neq \beta$. As a result the modal co-efficient for the eigenmode indexed λ_k is

$$r_k^{(n+1)} = \frac{\sum_{i=1}^{D} \sum_{j=1}^{L} Q_{ij}^{(n)} \sum_{k=1}^{K} \boldsymbol{w}_i^T \boldsymbol{v}_i^{\lambda_k} - \sum_{j=1}^{L} \sum_{k=1}^{K} \hat{x}_j^T \boldsymbol{v}_j^{\lambda_k}}{\phi^{\lambda_k T} \phi^{\lambda_k}} \tag{8}$$

Finally, we update the *a posteriori* alignment probabilities. This is done by substituting the revised parameter vector into the conditional measurement distribution. Using the Bayes rule, we can re-write the *a posteriori* alignment probabilities using the measurement density function

$$P_{ij}^{(n+1)} = \frac{Q_{ij}^{(n)} p(\boldsymbol{w}_i | \boldsymbol{x}_j, \Phi, \boldsymbol{r}^{(n)})}{\sum_{j=1}^{L} Q_{ij}^{(n)} p(\boldsymbol{w}_i | \boldsymbol{x}_j, \Phi, \boldsymbol{r}^{(n)})} \tag{9}$$

Upon substituting the Gaussian distribution appearing in Equation (15), the revised alignment probabilities are related to the updated point positions in the following manner

$$P_{ij}^{(n+1)} = \frac{Q_{ij}^{(n)} \exp(-\frac{1}{2\sigma^2} (\boldsymbol{w}_i - \boldsymbol{x}_j^{(n+1)})^T (\boldsymbol{w}_i - \boldsymbol{x}_j^{(n+1)}))}{\sum_{j=1}^{L} Q_{ij}^{(n)} \exp(-\frac{1}{2\sigma^2} (\boldsymbol{w}_i - \boldsymbol{x}_j^{(n+1)})^T (\boldsymbol{w}_i - \boldsymbol{x}_j^{(n+1)}))} \tag{10}$$

3.2   Correspondences

The correspondences are recovered via maximisation of the quantity

$$\mathcal{E}_c = \sum_{i \in \mathcal{D}} \sum_{j \in \mathcal{M}} P_{i,j}^{(n)} \ln q_{i,j}^{(n+1)} \tag{11}$$

Suppose that $V_D(i) = \{i'|(i,i') \in E_D\}$ represents the set of nodes connected to the node i by an edge in the graph with edge-set E_D. Furthermore, let us introduce a $D \times L$ matrix of assignment variables S whose elements that convey the following meaning

$$s_{i,j}^{(n)} = \begin{cases} 1 & \text{if } f^{(n)}(i) = j \\ 0 & \text{otherwise} \end{cases} \tag{12}$$

In a recent study [16], we have shown that the probability distribution for the assignment variables is

$$q_{i,j}^{(n)} = K \exp\left[-k_e \sum_{i' \in V_D(i)} \sum_{j' \in V_M(j)} (1 - s_{i',j'}^{(n+1)})\right] \tag{13}$$

where K and k_e are constants. With this distribution to hand, the correspondence assignment step reduces to one of maximising the quantity

$$\mathcal{F}_c = \sum_{i \in D} \sum_{j \in M} \sum_{i' \in D} \sum_{j' \in M} \hat{E}_D(i,i') \hat{E}_M(j,j') P_{i,j}^{(n)} s_{i',j'}^{(n+1)} \tag{14}$$

where $\hat{E}_D(i,i')$ and $\hat{E}_M(j,j')$ are the elements of the adjacency matrices for the data and model graphs. In more compact notation, the updated matrix of correspondence indicators $S^{(n+1)}$ satisfies the condition

$$S^{(n+1)} = \arg \max_S Tr[\hat{E}_D P^{(n)} \hat{E}_M S^{(n)}] \tag{15}$$

where $P^{(n)}$ is a matrix whose elements are the alignment probability $P_{i,j}^{(n)}$. In other words, the utility measure gauges the degree of correlation between the edge-sets of the two graphs under the permutation structure induced by the alignment and correspondence probabilities. Following Scott and Longuet-Higgins [9] we recover the matrix of assignment variables that maximises $\mathcal{F}_c$ by performing the singular value decomposition $\hat{E}_D P^{(n)} \hat{E}_M = V \Delta U^T$, where Δ is again a diagonal matrix and U and V are orthogonal matrices. The matrices U and V are used to compute an assignment matrix $R^{(n+1)} = V U^T$. To compute the associated matrix of correspondence probabilities, $Q^{(n+1)}$, we perform row normalisation on $R^{(n+1)}$. As a result

$$Q_{i,j}^{(n)} = \frac{R_{i,j}^{(n+1)} \cdot}{\sum_{j \in M} R_{i,j}^{(n+1)}} \tag{16}$$

This is clearly simplistic and violates symmetry. In our future work we plan to improve the algorithm to include Sinckhorn normalisation and slack variables for unmatchable nodes along the lines of Gold and Rangarajan [4].

4 Experiments

We have experimented with our new alignment method on an X-ray angiogram image sequence of a beating heart. Here the feature points are hand-labelled locations of maximum curvature on the outline of the heart. There are 16 feature points in each image.

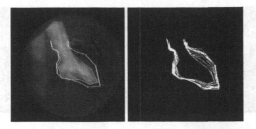

Fig. 1. Overlaped mean shape and training images

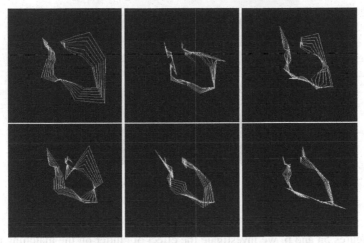

Fig. 2. Eigenmodes of the training images

In total we have used 19 frames to train the point-distribution model. The mean shape is shown in Figure 1a superimposed on one of the images from the sequence; the different frames used for training are shown in Figure 1b. In Figure 2 we show the modal displacements corresponding to the first 6 eigenmodes.

An example of the alignment of the PDM which was not part of the training set is shown in Figure 3. The sequence shows the PDM iterating from the mean shape to the final alignment. The different panels in the figure show different iterations of the algorithm. The process converges in 10 iterations. In this example there is no initial correspondence assignment. Each point in the model is assigned a probability which is evenly distributed. The final set of correspondences obtained are shown in Figure 4. These are all correct.

We now illustrate the effect of removing the adjacency graph from the point representation and the correspondence step from the matching process. To meet this goal we simply fit the point distribution model so as to minimise the quantity

$$\mathcal{E}_{EM} = \sum_{i=1}^{D} \sum_{j=1}^{L} \left[P_{i,j}^{(n)} \ln p_{i,j}^{(n+1)} \right] \tag{17}$$

This means that we use the EM algorithm to estimate the vector of modal parameters r.

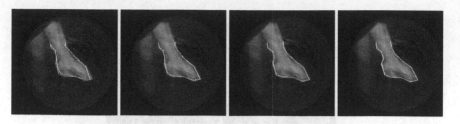

Fig. 3. Alignment results

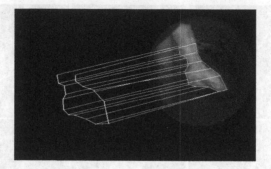

Fig. 4. Correspondence

In Figures 5a and b, we investigate the effect of clutter on the matching process. In Figure 5 a we show the evolution of one of the fitted PDM parameters with iteration number. The upper curve is the result obtained with the EM algorithm, while the lower curve is the result obtained when we use the new method reported in this paper. The main feature to notice is that both methods converge to the same parameter values, but that the convergence of the graph-based method is faster. In Figure 5 b we repeat this experiment when random noise points are added to the data. Here we should recover the same parameter values as in Figure 5 a above. Both algorithms result in a significant parameter error. However, the graph-based method gives a final result which is closer to the correct answer. Moreover, its convergence is again much faster than the EM method.

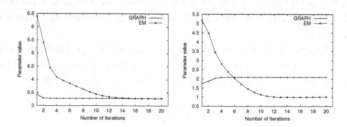

Fig. 5. Comparison of PDM parameter value (a)Without clutter (b)With 18.75% clutter

Next, we investigate the effect of added clutter on the alignment error. Figure 6 shows the alignment error as a function of iteration number for the EM algorithm and the graph-based method. In Figure 6 a the fraction of added clutter is 6.25%. Here both methods converge to a result in which the alignment error is consistent with zero. However, the graph-based method converges at a much faster rate. Figure 6 b repeats this experiment when the fraction of added clutter is 18.75%. Now both methods are subject to a substantial alignment error. However, in the case of the graph-based method this is smaller than that incurred by the EM method.

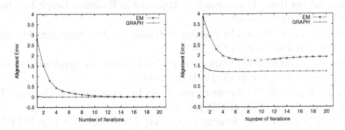

Fig. 6. Comparison of the alignment error (a) 6.25% clutter (b) 18.75% clutter

5   Conclusions

Our main contribution in this paper are two-fold. First, we show how the point distribution models can be augmented with point adjacency information. Second, we show how to fit the resulting model to noisy and unlabelled point-sets using a unified approach to correspondence and alignment. The method is shown to operate effectively when the landmark points are both unlabelled and subject to structural corruption. The method is both rapid to converge and robust to point-set contamination.

References

1. Amit, Y., Kong, A.: Graphical Templates for Model Registration. IEEE PAMI **18** (1996) 225–236
2. Cootes, T.F., Taylor, C.J., Cooper, D.H., Graham, J.: Active Shape Models - Their Training and Application. CVIU **61** (1995) 38–59
3. Cross, A.D.J., Hancock, E.R.: Graph Matching with a Dual-Step EM Algorithm. IEEE PAMI **20**(11) (1998) 1236–1253
4. Gold, S., Rangarajan, S.: A graduated assignment algorithm for graph matching. IEEE PAMI **18**(4) (1996) 377–388
5. Chui, H., Rangarajan, A.: A new algorithm for non-rigid point matching. Proceedings of IEEE Conference, CVPR 2000, vol. 2 (2000) 44–51
6. Kendall, D.G.: Shape Manifolds: Procrustean metrics and complex projective spaces. Bulletin of the London Mathematical Society **16** (1984) 81–121

7. Lades, M., Vorbruggen, J.C., Buhmann, J., Lange, J., von der Maalsburg, C., Wurtz, R.P., Konen, W.: Distortion-invariant object-recognition in a dynamic link architecture. IEEE Transactions on Computers **42** (1993) 300–311
8. Sclaroff, S., Pentland, A.P.: Modal Matching for Correspondence and Recognition. IEEE PAMI **17** (1995) 545–661
9. Scott, G.L., Longuet-Higgins, H.C.: An Algorithm for Associating the Features of 2 Images. Proceedings of the Royal Society of London Series B - Biological **244**(1309) (1991) 21–26
10. Shapiro, L.S., Brady, J.M.: Feature-based Correspondence - An Eigenvector Approach. Image and Vision Computing **10** (1992) 283–288
11. Ullman, S.: The Interpretation of Visual Motion. MIT Press (1979)
12. Umeyama, S.: An Eigen Decomposition Approach to Weighted Graph Matching Problems. IEEE PAMI **10** (1988) 695–703
13. Umeyama, S.: Least Squares Estimation of Transformation Parameters between Two Point sets. IEEE PAMI **13**(4) (1991) 376–380
14. Umeyama, S.: Parameterised Point Pattern Matching and its Application to Recognition of Object Families. IEEE PAMI **15** (1993) 136–144
15. Werman, M., Weinshall, D.: Similarity and Affine Invariant Distances between 2D Point Sets. IEEE Trans. on PAMI, **17**(8) (1995) 810–814
16. Wilson, R.C., Hancock, E.R.: Structural matching by discrete relaxation. IEEE T-PAMI **19**(6) (June 1997) 634–648

Shape-from-Shading Using Darboux Smoothing

Hossein Ragheb and Edwin R. Hancock

Department of Computer Science, University of York, York YO1 5DD, UK.
{hossein,erh}@minster.cs.york.ac.uk

Abstract. This paper describes a new surface normal smoothing process which can be used in conjunction with shape-from-shading. Rather than directly smoothing the surface normal vectors, we exert control over their directions by smoothing the field of principal curvature vectors. To do this we develop a topography sensitive smoothing process which overcomes the problems of singularities in the field of principal curvature directions at the locations of umbilics and saddles. The method is evaluated on both synthetic and real world images.

Keywords: shape-from-shading, Darboux smoothing

1 Introduction

Shape-from-shading (SFS) is concerned with recovering surface orientation from local variations in measured brightness. There is strong psychophysical evidence for its role in surface perception and recognition [1,9,10]. However, despite considerable effort over the past two decades, reliable SFS recovery has proved to be an elusive goal [6,3]. The reasons for this are two-fold. Firstly, the recovery of surface orientation from the image irradiance equation is an under-constrained process which requires the provision of boundary conditions and constraints on surface smoothness to be rendered tractable. Secondly, real-world imagery rarely satisfies these constraints. Several authors have attempted to develop shape-from-shading methods which overcome these shortcomings. For instance, Oliensis and Dupuis [11], and Bichsel and Pentland [2] have developed solutions for which SFS is not under-constrained, but which require prior knowledge of the heights of singular points of the surface. Meanwhile, Kimmel and Brookstein have shown how the apparatus of level-set theory can be used to solve the image irradiance equation as a boundary value problem [4,7].

However, in general SFS has suffered from the dual problems of model dominance (which results in oversmoothing of the recovered surface) and poor data-closeness. Recently, Worthington and Hancock have developed a new SFS scheme which has gone some way to overcoming some of the shortcomings of existing, iterative SFS algorithms [14,15]. Specifically, they have shown that the image irradiance equation can be treated as a hard constraint [14]. The idea is a simple one. For a Lambertian surface, the image irradiance equation constrains the local surface normal to fall on a cone. The axis of this cone points in the light source direction while the cosine of the apex angle of the cone is proportional to

W. Skarbek (Ed.): CAIP 2001, LNCS 2124, pp. 657–667, 2001.

the local image brightness. Worthington and Hancock exploit this property to develop a geometric method for iteratively updating the set of local surface normal directions. At each image location the surface normal is initially positioned on the irradiance cone so that its projection onto the image plane points in the direction of the local image gradient. The field of surface normal directions is then subjected to a smoothing process which is sensitive to the local topographic surface structure. Smoothing is effected using robust error kernels whose width is governed by the variance of the local shape-index (an angular measure of local surface topography first suggested by Koenderink and Van Doorn [8]). This smoothing process results in surface normals which violate the image irradiance equation, i.e. do not fall on the irradiance cone. To restore data-closeness, the smoothed surface normals are projected onto the closest position on the cone. This process is iterated to convergence.

The observation underpinning this paper is that although the use of topographic information improves the pattern of surface normals, it overlooks a considerable body of information concerning the differential structure of the underlying surface residing in the local Darboux frame. At each point on the surface the z-axis of the Darboux frame points in the direction of the local surface normal, while the x and y axes are aligned in the directions of the local maximum and minimum surface curvatures. There have been several attempts to improve the consistency of recovered surface information by smoothing Darboux frames. For instance Sander and Zucker [13] have incorporated a least-squares process for smoothing the principal curvature directions in the inference of surfaces from 3D position data. This smoothing process has been applied as a post-processing step in shape-from-shading by Lagarde and Ferrie [5]. However, there are limitations to the use of least-squares smoothing of the Darboux frames. These relate to fact that the principal curvature direction is undefined at umbilic surface points and that the direction is discontinuous at saddles.

Our aim in this paper is to refine the information returned by shape-from-shading, not by smoothing the field of surface normal directions, but by smoothing the field of maximum curvature directions. There are two novel contributions: First, to overcome the problems encountered at umbilics (the centres of cups and domes) and saddles, we develop a topography sensitive weighting process. Second, we incorporate the smoothing process into the Worthington and Hancock [14] framework which ensures that the recovered surface normals satisfy the data-closeness constraints imposed by the image irradiance equation. By smoothing in this way we avoid using surface normal directions in parabolic surface regions, i.e. at the locations of ridges and ravines. These topographic features are associated with rapid rates of change in surface normal direction and, hence, rapid variations in the local image intensity (i.e. they are locations where the image gradient is large). Since the change in principal curvature direction is smaller than the change in surface normal direction at such features, Darboux smoothing may moderate the over-smoothing of ridge and ravine topography. Experiments on synthetic and real-world data reveal that the method delivers surface orientation information that is both accurate and stable.

2   Shape-from-Shading

The shape-from-shading algorithm of Worthington and Hancock has been demonstrated to deliver needle-maps which preserve fine surface detail [14]. The observation underpinning the method is that for Lambertian reflectance from a matte surface, the image irradiance equation defines a cone of possible surface normal directions. The axis of this cone points in the light-source direction and the opening angle is determined by the measured brightness. If the recovered needle-map is to satisfy the image irradiance equation as a hard constraint, then the surface normals must each fall on their respective reflectance cones. Initially, the surface normals are positioned so that their projections onto the image plane point in the direction of the image gradient. Subsequently there is iterative adjustment of the surface normal directions so as to improve the consistency of the needle-map. In other words, each surface normal is free to rotate about its reflectance cone in such a way as to improve its consistency with its neighbours. This rotation is a two-step process. First, we apply a smoothing process to the current surface normal estimates. This may be done in a number of ways. The simplest is local averaging. More sophisticated alternatives include robust smoothing with outlier reject and, smoothing with curvature or image gradient consistency constraints. This results in an off-cone direction for the surface normal. The hard data-closeness constraint of the image irradiance equation is restored by projecting the smoothed off-cone surface normal back onto the nearest position on the reflectance cone.

To be more formal let L be a unit vector in the light source direction and let $E(i, j)$ be the brightness at the image location (i, j). Further, suppose that $N^k(i, j)$ is the corresponding estimate of the surface normal at iteration k of the algorithm. The image irradiance equation is $E(i, j) = N^k(i, j) \cdot L$. As a result, the reflectance cone has opening angle $\cos^{-1}(E(i, j))$. After local smoothing, the off-cone surface normal is $N_S^k(i, j)$. The updated on-cone surface normal which satisfies the image irradiance equation as a hard constraint is obtained via the rotation $N^{k+1}(i, j) = \Phi N_S^k(i, j)$. The matrix Φ rotates the smoothed off-cone surface normal estimate by the angle difference between the apex angle of the cone, and the angle subtended between the off-cone normal and the light source direction. This angle is $\theta = \cos^{-1}(E(i, j)) - \cos^{-1}(N_S^k(i, j) \cdot L)$. This rotation takes place about the axis whose direction is given by the vector $(u, v, w)^T = N_S^k(i, j) \times L$. This rotation axis is perpendicular to both the light source direction and the off-cone normal. Hence, the rotation matrix is

$$
\Phi = \begin{pmatrix} c + u^2 c' & -ws + uvc' & vs + uwc' \\ ws + uvc' & c + v^2 c' & -us + vwc' \\ -vs + uwc' & us + vwc' & c + w^2 c' \end{pmatrix}
$$

where $c = \cos\theta$, $c' = 1 - c$, and $s = \sin\theta$.

3   Local Darboux Frames

Once local estimates of the local surface normal directions are to hand, we can represent the local surface structure using a Darboux frame. We commence by using the local surface normals to compute the Hessian matrix

$$H = \begin{bmatrix} \frac{\partial}{\partial x}(N^k)_x & \frac{\partial}{\partial x}(N^k)_y \\ \frac{\partial}{\partial y}(N^k)_x & \frac{\partial}{\partial y}(N^k)_y \end{bmatrix} = \begin{bmatrix} h_{11} & h_{12} \\ h_{21} & h_{22} \end{bmatrix} \tag{1}$$

where for simplicity we have suppressed the co-ordinate suffixes on the surface normals. The two eigenvalues of the Hessian matrix are the maximum and minimum curvatures:

$$\lambda_M^k = -\frac{1}{2}(h_{11} + h_{22} - S) \quad , \quad \lambda_m^k = -\frac{1}{2}(h_{11} + h_{22} + S) \tag{2}$$

where $S = \sqrt{(h_{11} - h_{22})^2 + 4(h_{21}h_{12})}$. The eigenvector associated with the maximum curvature λ_M^k is the principal curvature direction. On the tangent-plane to the surface, the principal curvature direction is given by the 2-component vector

$$M_2^k = \begin{bmatrix} (h_{12}, -\frac{1}{2}(h_{11} - h_{22} + S))^T & h_{11} \geq h_{22} \\ (\frac{1}{2}(h_{11} - h_{22} - S), h_{21})^T & h_{11} < h_{22} \end{bmatrix} \tag{3}$$

From the known direction of the surface normal and the direction of the principal curvature direction on the tangent-plane, we compute the 3 component unit-vector M^k in the image co-ordinate system. The direction of the surface normal together with the principal curvature direction define a local frame of reference referred to as a Darboux frame. From the eigenvalues of the Hessian, we can also compute the shape-index of Koenderink and Van Doorn [8]

$$\phi^k = \frac{2}{\pi} \arctan \left(\frac{\lambda_M^k + \lambda_m^k}{\lambda_M^k - \lambda_m^k} \right) \quad \lambda_M^k \geq \lambda_m^k \tag{4}$$

The relationship between the shape-index and the topographic class of the underlying surface are summarised in Table 1. The table lists the topographic classes (i.e. dome, ridge, saddle ridge etc.) and the corresponding shape-index interval.

4   Smoothing Principal Curvature Directions

We would like to smooth surface normal directions using the field of principal curvature directions. However, this is not a straightforward task. The reason for this is that certain classes of local surface topography are characterised by rapid changes in principal curvature direction. There are two important cases. Firstly, for elliptic structures, i.e. domes and cups, there is a singularity since the principal curvature is not defined at the associated umbilic or spherical point, and the surrounding vector field is radial in structure. Secondly, for hyperbolic structures, i.e. saddles, there is a discontinuity in the principal curvature direction.

Table 1. Topographic classes.

Class	Region-type	Shape index
Dome	Elliptic	$\left[\frac{5}{8}, 1\right)$
Ridge	Parabolic	$\left[\frac{3}{8}, \frac{5}{8}\right)$
Saddle ridge	Hyperbolic	$\left[\frac{1}{8}, \frac{3}{8}\right)$
Plane	Hyperbolic	Undefined
Saddle-point	Hyperbolic	$\left[-\frac{1}{8}, \frac{1}{8}\right)$
Cup	Elliptic	$\left[-\frac{5}{8}, -1\right)$
Rut	Parabolic	$\left[-\frac{5}{8}, -\frac{3}{8}\right)$
Saddle-rut	Hyperbolic	$\left[-\frac{3}{8}, -\frac{1}{8}\right)$

This is due to the fact that the directions of maximum and minimum curvature flip as the saddle is travered.

To overcome these problems we weight against domes and saddles in the smoothing process. To do this we note that the shape-index associated with parabolic structures is $\pm\frac{1}{2}$. We therefore use the topographic weight

$$w_{i,j}^{(k)} = \exp\left(-\mu\left(|\phi_{i,j}^k| - 0.5\right)^2\right) \tag{5}$$

where μ is a constant to compute the smoothed principal curvature direction

$$M_S^k(i,j) = \sum_{(l,m)\in R_{i,j}} w_{l,m}^{(k)} M^k(l,m) \tag{6}$$

over the neighbourhood $R_{i,j}$ of the pixel indexed (i,j). Our aim is to use the smoothed principal curvature direction to compute an updated Darboux frame.

5   Surface Normal Adjustment

Once the smoothing of the field of principal curvature directions is complete, then we can use their direction vectors (M^k) to compute revised estimates of the surface normal directions (N_D^k). This is a two-step process. First, rotate the unsmoothed surface normals around the irradiance cone in a manner which is consistent with the rotation of the principal curvature directions which results from the Darboux smoothing process. To do this we project the smoothed and unsmoothed principal curvature directions onto the plane which is the perpendicular bisector of irradiance cone. Once these projections are to hand, then we compute their angle-difference on the plane. We rotate the on-cone surface normal around the cone by an amount equal to this angle. The second step is to restore the orthogonality of the Darboux frame by rotating the surface normal away from the irradiance cone so that it is perpendicular to the smoothed principal curvature direction. The procedure adopted is illustrated in Figures 1 and 2.

5.1   Rotation about the Irradiance Cone

To be more formal, the projection of the unsmoothed principal curvature direction onto the plane perpendicular to the irradiance cone is $M_P^k = L \times M^k \times L$ while the projection of the smoothed principal curvature direction is $M_{SP}^k = L \times M_S^k \times L$. On the plane perpendicular to the irradiance cone, the unsigned angle between the projected principal curvature directions is $|\alpha^k| = \cos^{-1}(M_P^k \cdot M_{SP}^k)$. To perform the rotation of the on-cone surface normal in a consistent manner, we require the sign or sense of this angle. Because the vectors M_P^k and M_{SP}^k are both perpendicular to the light source direction L, the cross product $M_P^k \times M_{SP}^k$ is either in the direction of the light source direction L or in the opposite direction. So, we can compute the signed rotation angle α^k using the following equation : $\alpha^k = (L \cdot (M_P^k \times M_{SP}^k))|\alpha^k|$. We rotate the current surface normal around the irradiance cone defined by the light source direction L and the apex angle $\cos^{-1}(E)$ by the angle α^k. The updated surface normal direction is given by the formula

$$N_R^k = N^k \cos(\alpha^k) + (N^k \cdot L)L(1 - \cos(\alpha^k)) + (L \times N^k)\sin(\alpha^k) \qquad (7)$$

The rotation is illustrated in Figure 1. The surface normals N_R^k computed in this way are not necessarily perpendicular to the corresponding smoothed principal curvature directions.

5.2   Computing the Surface Normal Direction

We restore the geometry of the Darboux frame, we rotate the surface normals off the irradiance cone so that they are perpendicular to the smoothed principal curvature directions. To do this we compute the plane P_D which contains the vectors N_R^k and M_S^k (Figure 2). The cross product $M_S^k \times N_R^k$ identifies a vector perpendicular to the plane P_D. So, the cross product of this vector with the smoothed principal curvature direction M_S^k gives a vector which belongs to the plane P_D and is also perpendicular to M_S^k. We take the direction of this vector to be our new surface normal $N_D^k = M_S^k \times N_R^k \times M_S^k$. These surface normals do not fall on the irradiance cone.

5.3   Imposing the Irradiance Constraint

To restore the irradiance constraint we can clearly apply the procedure of Worthington and Hancock and project the normals onto the nearest position on the irradiance cone. However, there are situations where the principal curvature directions are unavailable. For instance, the principal curvature direction is not defined at the location of umbilics. Also, it can not be computed when the Hessian matrix is singular.

There is a further reason to use smoothed surface normals N_S^k in conjunction with Darboux surface normals N_D^k. The reason for this is that there are certain surface regions for which the Darboux surface normals N_D^k may be more accurate and may be closer to the irradiance cone than smoothed surface normals N_S^k.

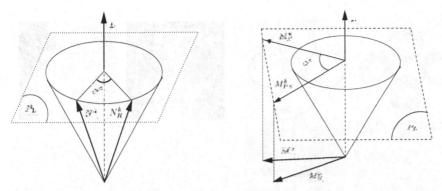

Fig. 1. Rotating the Darboux frame around the light-source direction a) rotating the current surface normal around the cone ; b) finding the rotation angle using the principal curvature directions.

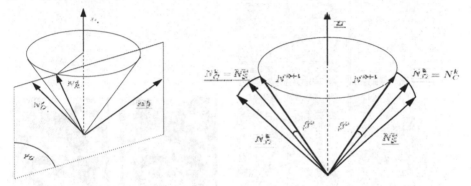

Fig. 2. a) Finding the direction of the local surface normal to the rotated Darboux frame; b) rotating the closest surface normal onto the irradiance cone

The cases in question are a) locations where image intensity gradients are large; b) at the locations of ridges and ruts. On the other hand, the smoothed surface normals N_S^k may be more accurate when a) where the intensity gradients are small and b) at the locations of domes.

To overcome these problems, we have used a switching process. For each location in the image we have computed two surface normal directions. The first of these is the surface normal N_D^k computed from the principal curvature direction. The second is the surface normals N_S^k found by applying the robust smoothing method of Worthington and Hancock to the field of current surface normal estimates (N^k). From these two alternatives, we select the surface normal which is closest to the irradiance cone. In other words, the surface normal used for projection onto the irradiance cone is

$$N_C^k = \begin{cases} N_D^k & \text{if } |\cos^{-1}(N_D^k \cdot L) - \cos^{-1}(E)| \leq |\cos^{-1}(N_S^k \cdot L) - \cos^{-1}(E)| \\ N_S^k & \text{otherwise} \end{cases}$$

$$(8)$$

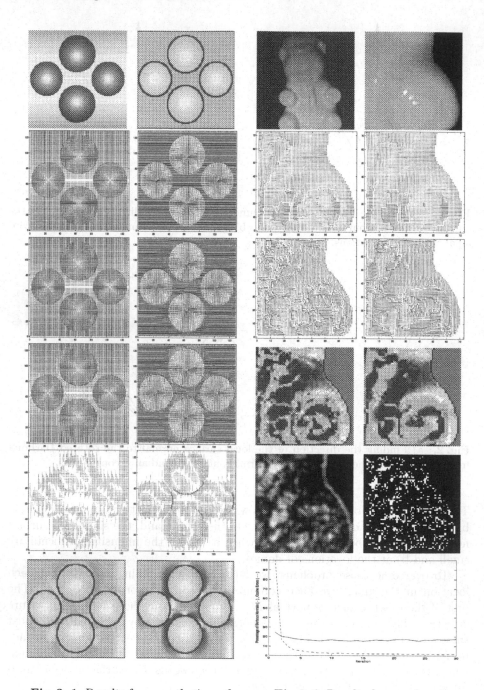

Fig. 3. 1. Results for a synthetic surface   **Fig.3.** 2. Results for a real surface

With the closest surface normal to hand, we can effect projection onto the irradiance cone. The rotation angle required for the projection is

$$\beta^k = \cos^{-1}(N_C^k \cdot L) - \cos^{-1}(E) \tag{9}$$

and after projection the updated surface normal is

$$N^{k+1} = N_C^k \cos(\beta^k) + \left(\left(\frac{N_C^k \times L}{\|N_C^k \times L\|} \right) \times N_C^k \right) \sin(\beta^k) \tag{10}$$

The geometry of this procedure is described in Figure 2.

6 Experiments

We have experimented with our new shape-from-shading method on synthetic and real-world imagery. The experimentation with synthetic data is designed to measure the effectiveness of the method on surface which contain a variety of topographic structure. This aspect of the work also allows us to measure some of the quantitive properties of the method. The real world examples allow us to demonstrate the utility of method on complex surfaces, which do not necessarily satisfy our Lambertian reflectance model.

6.1 Synthetic Data

Figure 3.1 shows an example of the results obtained with synthetic surfaces used for our experiments. The original object consists of four ellipsoids placed on a paraboloid surface. In the top row, the left-hand image shows the intensity image obtained when the surface is illuminated under Lambertian reflectance, while the right-hand image shows the shape-index for the ground truth data. The second, third and fourth rows show the surface normal directions (left-hand column) and the principal curvature directions (right-hand column) extracted from the surface. In the second row we show the ground-truth data obtained from the elevation data. The third row shows the results obtained using our Darboux smoothing process. The fourth row shows the results obtained using the algorithm of Worthington and Hancock. In the fifth row of the figure, we show the the difference between the ground-truth surface normals and the normals delivered by the Darboux smoothing method (left) and the Worthington and Hancock method (right). The errors delivered by the Darboux smoothing method are considerably smaller than those delivered by the Worthington and Hancock method especially at high curvature locations. The bottom row of the figure shows the shape-index from Darboux smoothing (left) and Worthington and Hancock method (right). The main feature to note from these examples is that the new Darboux smoothing method gives a cleaner definition of both the paraboloid and its intersections with the ellipsoids. Moreover, the agreement between the principal curvature directions is also better. Finally, the best agreement between the ground-truth shape-index is obtained with Darboux smoothing method.

6.2   Real World Experiments

In Figure 3.2, we provide some detailed experimental results for a real world image. The full image is shown in the top left-hand panel and the detail used in our analysis (the bottom right-hand leg) is shown in the top right-hand panel. The object is shiny (and hence stretches our Lambertian model) and as a result the image contains specularities. The detail under study is approximately cylindrical and the end of the cylinder contains a depression or dimple. The second, third and fourth rows of the figure respectively show the surface normals, the principal curvature directions and the shape-index. The left-hand column shows the results obtained with Darboux smoothing, while the right-hand column shows the results obtained using the method of Worthington and Hancock. The main feature to note is that when Darboux smoothing is used, then the cylindrical structure and the dimple stand out better in both the surface normal and the shape-index plots. In the fifth row we show the final weights used in Darboux smoothing process (left) and the location of the final Darboux surface normals on the surface (right). It can be seen that the weights are weaker at hyperbolic and elliptic locations of the surface. Finally, in the bottom row , we show two curves to demonstrate the iterative characteristics of our new Darboux smoothing method. The dotted curve shows the average angle difference between the chosen off-cone surface normals (N_C^k) and the on-cone surface normals (N^{k+1}). As the algorithm iterates then this angle decreases rapidly. The solid curve shows the fraction of pixels for which the Darboux surface normal is the closest to the irradiance cone. This varies slowly with iteration number. Hence, the amount of iterative adjustment of the surface normal directions decreases rapidly with iteration number, while the fraction of Darboux normals closest to the cone remains relatively constant.

7   Conclusions

To conclude, we have presented a new smoothing method for use in conjunction with shape-from-shading. The method smooths the field of principal curvature directions in a topography sensitive way which preserves discontinuities at umbilics and saddles. The method offers enhanced performance over smoothing the field of surface normal directions.

References

1. Belhumeur, P.N., Kriegman, D.J.: What is the Set of Images of an Object Under All Possible Lighting Conditions? CVPR (1996) 270-277
2. Bichsel, M., Pentland, A.P.: A Simple Algorithm for Shape from Shading. CVPR (1992) 459-465
3. Brooks, M.J., Horn, B.K.P.: Shape and Source from Shading. IJCAI (1986) 932-936
4. Bruckstein, A.M.: On Shape from Shading, CVGIP, Vol. 44 (1988) 139-154
5. Ferrie, F.P., Lagarde, J.: Curvature Consistency Improves Local Shading Analysis. ICPR, Vol. 1 (1990) 70-76

6. Horn, B.K.P., Brooks, M.J.: The Variational Approach to Shape from Shading. CVGIP, Vol. 33, No. 2 (1986) 174-208
7. Kimmel, R., Bruckstein, A.M.: Tracking Level-sets by Level-sets: A Method for Solving the Shape from Shading Problem. CVIU, Vol. 62, No. 1 (1995) 47-58
8. Koenderink, J.J., van Doorn, A.J.: Surface Shape and Curvature Scales. IVC, Vol. 10 (1992) 557-565
9. Koenderink, J.J., van Doorn, A.J., Kappers, A.M.L.: Surface Perception in Pictures. Perception and Psychophysics, Vol. 52, No. 5 (1992) 487-496
10. Koenderink, J.J. van Doorn, A.J.: The Internal Representation of Solid Shape with Respect to Vision. Biological Cybernetics, Vol. 32 (1979) 211-216
11. Oliensis, J., Dupuis, P.: A Global Algorithm for Shape from Shading. CVPR (1993) 692-701
12. Samaras, D., Metaxas, D.: Variable Albedo Surface Reconstruction from Stereo and Shape from Shading. CVPR, Vol. 1 (2000) 480-487
13. Sander, P.T., Zucker, S.W.: Inferring Surface Structure and Differential Structure from 3D Images. PAMI, Vol. 12, No. 9 (1990) 833-854
14. Worthington, P.L. and Hancock, E.R.: New Constraints on Data-closeness and Consistency for SFS. PAMI, Vol. 21 (1999) 1250-1267
15. Worthington, P.L., Hancock, E.R.: Needle Map Recovery Using Robust Regularizers. IVC, Vol. 17 (1999) 545-557

A Novel Robust Statistical Design of the Repeated Genetic Algorithm

Shiu Yin Yuen, Hoi Shan Lam, and Chun Ki Fong

Department of Electronic Engineering,
City University of Hong Kong,
Tat Chee Avenue, Hong Kong
eekelvin@cityu.edu.hk

Abstract. The genetic algorithm is a simple optimization method for a wide variety of computer vision problems. However, its performance is often brittle and degrades drastically with increasing input problem complexity. While this problem is difficult to overcome due to the stochastic nature of the algorithm, this paper shows that a robust statistical design using repeated independent trials and hypothesis testing can be used to greatly alleviate the degradation. The working principle is as follows: The probability of success P of a stochastic algorithm A (genetic algorithm) can be estimated by running N copies of A simultaneously or running A repeatedly N times. By hypothesis testing, it is shown that P can be estimated to a required figure of merit (i.e. the level of significance). Knowing P, the overall probability of success $P_{repeated}$ for N applications of A can be computed. This is used in turn to adjust N in an iterative scheme to maintain a constant $P_{repeated}$, achieving a robust feedback loop. Experimental results are reported on the application of this novel algorithm to an affine object detection problem.

Keywords: genetic algorithm, statistical design, object location problem

1 Introduction

Genetic Algorithm (GA) is an interesting stochastic optimization technique that is directly inspired by analogy with genetics and evolution theories. In marked contrast to serial optimization techniques such as the gradient ascent, the GA employs a population of potential solutions, known as chromosomes. A set of M chromosomes form the current generation. The next generation of chromosomes are obtained by modifying the chromosomes by three operators, selection, followed by crossover and mutation. The procedure of a GA can be summarized below:

1. Randomly initialize a population of M chromosomes.
2. Select two chromosomes for mating from the population.
3. Perform crossover with probability P_c to obtain two offspring chromosomes.
4. Perform mutation with probability P_m to the offspring chromosomes, calculate their fitness and then add them to a new population.

W. Skarbek (Ed.): CAIP 2001, LNCS 2124, pp. 668–675, 2001.
© Springer-Verlag Berlin Heidelberg 2001

5.   Repeat step 2-4 until the new population reaches the same size as the current population.

In spite of promising successes in the application of the GA to computer vision (e.g. [1-4]), its performance is often brittle and degrades with increasing problem complexity. Owing to the fact that it is merely an intelligent random search algorithm, there can be no guarantee that it has found the correct solution (i.e. the global maximum) given an unknown problem landscape, and theoretical work on the GA's probability of success [5] has not yielded sufficient insight for practical application. The GA's probability of success P for a particular landmark test image *supplied by the researcher* can be obtained by running it on the same image N times, then takes the number of times $N_{success}$ of which the GA successfully finds the correct solution. In this case $N_{success}/N$ approaches P as N approaches infinity by the law of large numbers. This is usually the way to justify that the GA works well in a particular application in previous researches. However, this figure of merit P has no bearing on the actual probability of success of a totally new and unknown image. In fact, the GA's performance is often brittle, failing completely when given a new, unknown image of higher complexity than the landmark test image.

Whilst this problem cannot be completely overcome due to the stochastic nature of the algorithm, this paper shows that a statistical design using repeated independent trials and hypothesis testing can be used to design a more robust algorithm to input images of varying complexities. The idea is to test online the probability of success P, then modify N to meet the requirement on the overall probability of success $P_{repeated}$ for N applications of the GA in an iterative feedback loop.

Section 2 introduces the repeated GA (RGA) [6] and our iterative hypothesis testing (IHT) method for increasing the robustness of the GA. Section 3 reports the experimental results on applying the IHT-GA to an affine object detection problem. Section 4 gives a conclusion.

2 Repeated Genetic Algorithm

Suppose instead of a single run of the GA, the algorithm runs the GA N ($N > 1$) times on the same image. Each run is kept independent, with no information passed between each run. At the end of each run, the best solution found is recorded. Then the chromosomes are randomly initialized to begin the next run. Such an arrangement is known as the repeated GA (RGA). It is a special case of the parallel GA in the literature [7].

Let the probability of success of a single run be P_{SGA}. It is the probability that the best solution found in a single run is the correct (i.e. globally optimal) solution. Let the probability of success after running N times be P_{RGA}. It is the probability that any one of the N best solutions is the correct solution. Since the N runs are independent,

$$P_{RGA} = 1 - (1 - P_{SGA})^N \qquad (1)$$

Given a user specified figure of merit P_{RGA} (eg. P_{RGA} 0.95), the required number of runs N is simply

$$N \ge \frac{\ln(1 - P_{RGA})}{\ln(1 - P_{SGA})} \tag{2}$$

Thus provided that P_{SGA} is known, one can guarantee the RGA's overall probability of success. In [6], we report a hypothesis testing method to estimate P_{SGA} from a set of training images. In this paper, we explore another route, which is to estimate P_{SGA} from N runs on a *single* input image. The reasoning is as follows:

Suppose s is the best solution amongst the N solutions. If it is assumed that it is the correct solution, then it must appear around $N \cdot P_{SGA}$ times amongst the N solutions. Thus if the assumption is correct, then from the actual number of times that s appears, we can say something about P_{SGA}.

Let c_1 be the number of times s appears in the N runs. Let $c_2 = N$. To prove that P_{SGA} is larger than some lower bound P_{est}, the following hypothesis testing may be used:

$$H_0 : P_{SGA}(I) = P_{est}$$
$$H_1 : P_{SGA}(I) > P_{est} \tag{3}$$

In this case, the binomial experiment has been conducted and it is found that c_1 out of c_2 trials are successful. If the type I error (also called the level of significance) α is user specified, then P_{est} is the probability that the null hypothesis H_0 is just accepted.

Let X be the number of successes in c_2 trials. Then $c_1 + 0.5$ is the critical value, and

$$\alpha = P(\text{Type I error}) = P(X \ge c_1 + 1 \,|\, P_{SGA}(I) = P_{est}) = \sum_{x=c_1+1}^{c_2} b(x; c_2, P_{est}) \tag{4}$$

where $b()$ is the binomial distribution. Since α, c_1 and c_2 are given, P_{est} can be found by a table lookup.

The physical meaning is as follows: We are confident that if s is indeed the correct solution, then P_{SGA} has a lower bound of P_{est}, and the level of significance of this statement is α. The lower the level of significance, the higher is our confidence that the statement is correct.

The value of P_{est} can now be used to give a strong indication of whether s is indeed the correct solution. If P_{est} is exceedingly small, there would be cause for caution, and more runs should be made by increasing N. The equation which governs P_{SGA} and the overall probability of success P_{RGA} is equation (2), which constitutes a necessary condition. Substitute P_{est} into P_{SGA}, equation (2) becomes

$$N_{required} \ge \frac{\ln(1 - P_{RGA})}{\ln(1 - P_{est})} \tag{5}$$

These insights can be put together to design a novel repeated genetic algorithm. Initially, the GA is repeated $N = N_{init}$ times. The best solution s is found amongst the $c_2 = N_{init}$ runs. The number of runs c_1 which finds the same solution s is also found. Then P_{est} is calculated (equation (4)) which in turn is used to calculate $N_{required}$ (equation (5)). If $N_{required} \le N$, then the algorithm exits with success. Otherwise, N is increased by

1 and the algorithm iterates unless $N > N_{limit}$, in which case the algorithm exits with failure.

The complete algorithm is summarized below:

Iterative Hypothesis Testing (IHT) GA
Input: a single image I, α, P_{RGA}, N_{init}, N_{limit}.
1.   Set $N = N_{init}$. Run RGA.
2.   Find the best solution s amongst the N runs and the number of times c_1 that it occurs.
3.   Use equation (4) to calculate P_{est}.
4.   Check whether equation (5) is satisfied. If yes, exit with success.
5.   Otherwise run the GA a single more time ($N \leftarrow N+1$).
6.   If N N_{limit}, goto step 2, otherwise exit with failure.
Output: Success/failure flag, best solution s.

Let give an illustrative example. Let set N_{init} to 5, i.e. initially the GA is run five times. Suppose the solutions found in the five runs are s_1, s_2, s_3, s_4 and s_5. Suppose the fitness values are $f(s_1) = f(s_4) > f(s_2) > f(s_3) > f(s_5)$ and s_1 and s_4 are identical solutions (e.g. an identical pose in a matching problem). Then $s = s_1 = s_4$ is the best solution amongst the five runs, and s occurs in $c_1 = 2$ out of $c_2 = N = 5$ runs. Using equation (4), it is found that P_{SGA} should be at least 0.1056 with a level of significance of $\alpha = 0.01$. However, when $P_{est} = 0.1056$ is put into equation (5), it is found that $N_{required} = 27 \gg 5$. This gives a clear signal that s may not be the correct (i.e. globally best) solution. Thus the GA is run one more time and the situation is evaluated again. The process is repeated until a solution is found that satisfies the necessary condition (equation (5)) or N is larger than N_{limit}.

Note that N_{init} cannot be 1, otherwise the IHT algorithm will exit with success trivially. In practice, N_{init} may be set to any reasonable value larger than 1, or better still, from a priori knowledge about the expected complexity of the input image [6].

On the other hand, N_{limit} is a threshold which is determined by the maximum amount of computational resources which the algorithm is prepared to expend on finding the solution. A clear physical meaning of N_{limit} can be seen from the following equation, obtained by rearranging equation (1):

$$P_{SGAlimit} \quad 1 - (1 - P_{RGA})^{\frac{1}{N_{limit}}} \tag{6}$$

Thus N_{limit} gives the minimum probability of success for a single run of GA on an input image. In other words, it represents a threshold probability below which the engineer prefers not to determine the best solution, for example, because of lack of time or computing resources. Another reason may be because finding the best solution with such a low probability has little importance in a particular application.

A major advantage of this RGA algorithm is that it is able to automatically adjust the number of runs N. For easy images, N shall be small whereas N shall be large for complex images. This automatic adjustment is done *without* knowing a priori the complexity of the image. Thus the algorithm is more robust. This desirable characteristic is lacking in many existing GA implementations. These implementations are typically brittle, working well with some fairly complicated images but may not work as well with other images.

The algorithm exploits the necessary condition that the probability of success of a single run P_{SGA} must satisfy. We hasten to remark that the condition is not sufficient, since it is impossible to guarantee that the best solution s found from the N runs is indeed the correct (i.e. globally best) solution. In fact, no stochastic algorithm can guarantee that such is the case unless the search is so exhaustive that it tests the entire search space once.

Thus the algorithm cannot guarantee that it has found the correct solution if it exits with success. Paradoxically, if it exits with failure, then it can guarantee that the best solution found has less than a probability of P_{RGA} of being the correct solution, and the level of significance of this statement is α. Thus if it fails to find the correct solution, it can detect the failure confidently.

3   Application to an Affine Object Location Problem

The novel method above is tested on an affine object location problem [3]. Given a template $T = \{(x, y)\}$ and an image $I = \{(x', y')\}$ containing a single instance of an affine transformed copy of T, the problem is to recover the unknown affine transformation (a, b, c, d, e, f), where

$$\begin{matrix} x' \\ y' \end{matrix} = \begin{matrix} a & b \\ c & d \end{matrix} \begin{matrix} x \\ y \end{matrix} + \begin{matrix} e \\ f \end{matrix} \tag{7}$$

Instead of (a, b, c, d, e, f), the chromosome is alternatively coded as $(x_1', y_1', x_2', y_2', x_3', y_3')$, where (x_i, y_i), i = 1, 2, 3 are three fixed template points, since the three point mapping uniquely determines an affine transformation [3]. The following settings are used for a single GA run: $M = 100$, $P_c = 0.65$, $P_m = 0.025$. Uniform crossover of x and y are used. The number of generations N_g is 150. In each generation, M new chromosomes are generated. The $2M$ chromosomes are ranked by fitness and the M least fit chromosomes are discarded. The remaining M chromosomes form a new generation. During the initialization, (x_i', y_i') are randomly selected from edge points.

The following settings are used for the IHT: $P_{RGA} = 0.95$, $N_{init} = 4$, $N_{limit} =$. The input images are 128 x 128 and a Pentium 550 MHz computer is used. The following parameter restrictions are imposed on the input images:

$$0.8 \quad abs(\begin{vmatrix} a & b \\ c & d \end{vmatrix}) \quad 1.0 \qquad -1 \quad a, b, c, d \quad 1 \tag{8}$$

Chromosomes whose affine transform does not satisfy equation (8) will be assigned a fitness of 0.

To investigate the performance of the IHT RGA on images of varying complexities, 100 images of a pair of scissors generated by random affine transformations with 0%, 2%, and 5% (of the whole image) added random noise are used as input (fig. 1).

Let digress briefly to discuss the issue of the amount of noise. When we say "2%" noise, we mean 2% of the 128x128 images are noise points. The presence of a noise point will flip the image point value from 1 to 0 or vice versa. For the scissors template, there are 367 data points. For a scissors image with 367 data points and "2%" noise added, it corresponds to 47% noise. If it were "5%" noise added, it corresponds to 69% noise. Typically, the scissors image will have less than this amount of points due to the scaling down effect (equation (8)). Thus for a typical "2% noise image", the amount of noise is somewhat larger than 47%; for a typical "5% noise image", the amount of noise is somewhat larger than 69%.

The results for α =0.1 is shown in table 1. The average N for the 0%, 2% and 5% noise levels are 6.14, 48 and 67.87. The actual success rates are 97%, 89% and 80%. The Q factor [6] measures the speedup over an exhaustive search. The average Q factors are 120238, 15486 and 10956.

Fig. 1. Random transformations image with 0%, 2%, and 5% noise of the whole image

Table 1. The result for α =0.1 and α =0.01

	α =0.1			α =0.01		
Noise	0%	2%	5%	0%	2%	5%
Max N	19	231	342	35	298	1011
Min N	4	4	4	4	4	4
Average N	6.14	48	67.87	9.68	66.39	122.58
Max time (second)	41	422	765	84	575	1826
Min time (second)	7	7	7	8	7	8
Average time (second)	13.9	88.1	134.5	21.58	127.63	223.2
Actual success rate	97%	89%	80%	99%	92%	90%

The results for α =0.01 is shown in table 1. The average N for the 0%, 2% and 5% noise levels are 9.68, 66.39 and 122.58. The actual success rate are 99%, 92% and 90%. The average Q factors are 76486, 11200 and 6068.

Clearly, a smaller α i.e. a larger confidence, will give a larger success rate. However, the average N required is also larger, which gives a smaller average Q factor.

It is clearly shown that the algorithm can automatically adjust the number of N in proportion to the input complexity without knowing the input complexity a priori. The algorithm increases the number of runs *automatically* for difficult scenes.

Though the hypothesis testing is only a necessary but not sufficient condition for finding the solution with P_{RGA} = 95%+, the average probability of success is 88.7% (for α =0.1) and 93.7% (for α =0.01).

This indicates that the algorithm is working robustly against scenes of various complexities, *without knowing the complexity a priori*. It is interesting since it automatically adjusts itself when presented with difficult scenes to maintain a fairly constant level of probability of success. It is also clear that a smaller α will give a better average probability of success.

As control, When α =0.1, the average N for 0% noise image is 6.14. If a fixed N = 7 is used, then for 2% noise images, the success rate is 30%, for 5% noise images, the success rate is 20%. When α = 0.01, the average N for 0% noise images is 9.68. If a fixed N = 10 is used, then for 2% noise images, the success rate is 43%, for 5% noise images, the success rate is 28%. This shows clearly that both the GA and the RGA designs with a fixed N are a lot more brittle. From the Q factors, it is shown that the algorithm takes more time to analyze complex scenes. The same phenomenon is displayed by human beings (eg. the Dalmatian dog image).

Refer to table 1. For α =0.01 and 5% noise, the maximum N is 1011, which corresponds to a Q factor of 736. This shows that the IHT RGA has a hard time working with *that particular* 5% noise level image. To check whether this image is indeed a difficult image, the IHT algorithm is run 100 times on this image and it is found that the mean N is 747. In contrast, the average N for the set of 5% images is 122.58. Using t-test with a level of significance of 0.001, it is also found that the two means are different statistically. Hence this image is indeed a more difficult image than an average 5% image. In this situation, if we have had set the N_{limit} to some smaller limiting value, the IHT would know the image is beyond its ability to handle. This control signal may be used to trigger a more sophisticated vision algorithm. Thus this robust algorithm also has the ability to detect that it is not working properly. Note that this is a very desirable characteristic.

4   Conclusions

The genetic algorithm is an interesting optimization method for a wide variety of engineering problems. However, its performance is often brittle and degrades drastically with increasing input complexity. Since the complexity of an input image is not known a priori, this seriously limits the applicability of the GA as a general tool. While this problem is difficult to overcome due to the stochastic nature of the algorithm, this paper shows that a statistical design using repeated GA trials and hypothesis testing can be used to design an iterative (feedback) algorithm which is adaptive and substantially more robust. Experimental results are reported on the application of the algorithm to an object detection problem. The results confirm its prowess and demonstrates some interesting characteristics. Finally, note that the statistical design can equally be applied to other stochastic algorithms (e.g. simulated annealing) to increase their robustness.

References

1. Hill, A. and Taylor, C.J.: Model-based image interpretation using genetic algorithms. Image and Vision Computing **10**(5) (1992) 295-300
2. Roth, G. and Levine, M.D.: Geometric primitive extraction using a genetic algorithm. IEEE Trans. on Pattern Analysis and Machine Intelligence **16**(9) (1994) 901-905

3. Tsang, W.M.: A genetic algorithm for aligning object shapes. Image and Vision Computing **15** (1997) 819-831
4. Yuen, S.Y. and Ma, C.H.: Genetic algorithm with competitive image labelling and least square. Pattern Recognition **33**(12) (2000) 1949-1966
5. Suzuki, J.: A Markov chain analysis on simple genetic algorithms. IEEE Trans. On Systems, Man, and Cybernetics **25**(4) (1995) 655-659
6. Yuen, S.Y., Fong, C.K. and Lam, H.S.: Guaranteeing the probability of success using repeated runs of genetic algorithm, to appear in Image and Vision Computing.
7. Cantu-Paz, E. and Goldberg, D.E.: Efficient parallel genetic algorithms: theory and practice, Comput. Methods Appl. Mech. Engrg. **186** (2000) 211-238

Binocular Stereo Matching by Local Attraction

Herbert Jahn

Deutsches Zentrum für Luft und Raumfahrt e. V. (DLR)
Institut für Weltraumsensorik und Planetenerkundung
Rutherfordstrasse 2, 12489 Berlin, Germany
Phone: +49-30-67055-510
Fax: +49-30-67055-512
Herbert.Jahn@dlr.de

Abstract. A new approach to binocular stereo matching for epipolar geometry is presented. It is based on the idea that some features (edges) in the left image exert forces on similar features in the right image in order to attract them. Each feature point (i,j) of the right image is described by a coordinate x(i,j). The coordinates obey a system of time discrete Newtonian equations, which allow the recursive updating of the coordinates until they match the corresponding points in the left image. That model is very flexible. It allows shift, expansion and compression of image regions of the right image, and it takes into account occlusion to a certain amount. Furthermore, it can be implemented in parallel-sequential network structures allowing future real-time stereo processing (when corresponding hardware is available). The algorithm, which is confined here as a first step only to image points along edges, was applied to some stereo image pairs with a certain success, which gives hope for further improvements.

Keywords: matching, parallel processing, Newton's equations of motion

1   Introduction

Real-time stereo processing which is necessary in many applications needs very fast algorithms and processing hardware. The stereo processing capability of the human visual system together with the parallel-sequential neural network structures [1] of the brain lead to the conjecture that there exist parallel-sequential algorithms which do the job very efficiently. Therefore, it seems to be natural to concentrate effort to the development of such algorithms.

In prior attempts to develop parallel-sequential matching algorithms [2, 3] some promising results have been obtained. But in some image regions serious errors occurred which have led to a new attempt to be presented here.

If one de-aligns both our eyes by pressing one eye with the thumb then one has the impression, as if one of the images is pulled to the other until matching is achieved. This has led to the idea that prominent features (especially edge elements) of one image exert forces to corresponding features in the other image in order to attract them. A (homogeneous) region between such features is shifted together with the region bounding features whereas it can be compressed or stretched, because corresponding regions may have different extensions. Therefore, an adequate model

W. Skarbek (Ed.): CAIP 2001, LNCS 2124, pp. 676–683, 2001.

for the matching process seems to be a system of Newtonian equations of motion governing the shift of the pixels of one image. Assuming epipolar geometry a pixel (i,j) of the left image corresponds to a pixel (i',j) of the right image of the same image row. If a mass point with coordinate $x(i',j)$ and mass m is assigned to that pixel then with appropriate forces of various origins acting on that point it can be shifted to match the corresponding point (i,j). To match points inside homogeneous regions, the idea is to couple neighbored points by springs in order to shift these points together with the edge points. The model then resembles a little bit the old model of Julesz which he proposed in [4] for stereo matching.

In chapter 2 the method is explained in detail. Some results are shown in chapter 3. Finally, in the conclusions some ideas for future research are presented.

2   The Equations of Motion

We consider a left image $g_L(i,j)$ and a right image $g_R(i,j)$ $(i,j = 0,...,N\text{-}1)$. Corresponding points (i,j) of g_L and (i',j) of g_R on epipolar lines j are connected by $i' = i+s$ where $s(i',j)$ is the disparity. Here, $s(i',j)$ is assigned to the coordinates of the right image, but it is also possible to assign it to the left image or to a centered (cyclopean) image. In corresponding points the following equation approximately holds:

$$g_L(i, j) \quad g_R(i + s, j) \tag{1}$$

In most images there are points which are absent in the other image (occluded points). Those points (for which (1) of course does not hold) must be considered carefully in order to avoid mismatch with wrong disparities.

Now, to each pixel (i',j) of the right image a coordinate $x(i',j)$, a velocity $v(i',j)$, and a mass m are assigned in order to describe the motion of such a point. Let (i_e,j) be an edge point of the left image and (i_e',j) an edge point of the right image, respectively. Then, the edge point in the left image exerts a force $K(i_e, i_e',j)$ on the edge point in the right image in order to attract that point. We consider that force as an external force. Furthermore, on mass point (i',j) there can be acting internal forces such as the spring type forces $K_{spring}(i'\text{-}1, i',j)$, $K_{spring}(i'+1, i',j)$ and a force $-\gamma\, v(i',j)$ describing the friction with the background. Other internal forces such as friction between neighboring image rows j, j 1 can also be included. More general, the forces, here denoted as $K(i',j, j\ k)$, can also depend on edges and grey values in other image rows $j\ k$ of both images which means a coupling of different image rows.

With those forces Newton's equations are:

$$m\ \ddot{x}_t(i', j) = -\gamma\ \dot{x}_t(i', j) + K_t(i', j, j\ k) \tag{2}$$

Introducing the velocities $v = \dot{x}$, the system of differential equations (2) of second order can be converted into a system of first order equations

$$\dot{x}_t(i', j) = v_t(i', j) \tag{3}$$
$$m \; \dot{v}_t(i', j) = -\gamma \; v_t(i', j) + K_t(i', j, j \;\; k)$$

Here, $z = \dfrac{x}{v}$ are the state variables of the system.

Now, approximating $\dot{z}_t$ by $\dfrac{z_{t+\Delta t} - z_t}{\Delta t}$, the system of differential equations turns into a system of difference equations or discrete time state equations:

$$x_{t+\Delta t}(i', j) = x_t(i', j) + \Delta t \; v_t(i', j) \tag{4}$$
$$m \; v_{t+\Delta t}(i', j) = (m - \gamma \; \Delta t) \; v_t(i', j) + K_t(i', j, j \;\; k)$$

That system allows to calculate the system state z_t recursively. The initial conditions are:

$$x_0(i', j) = i' \tag{5}$$
$$v_0(i', j) = 0$$

When that recursive system of equations has reached its final state $x_{t_{max}}(i', j)$, then the disparity can be calculated according to

$$s(i', j) = i' - x_{t_{max}}(i', j) \tag{6}$$

which is the final shift of $x_t(i', j)$ from its initial position $x_0(i', j) = i'$.

The recursive calculation of the disparity according to (4) allows the incorporation of some countermeasures against ambiguities. In particular, the so-called ordering constraint [5] can be included: Let

$$\Delta_t x(i', j) = \Delta t \; v_t(i', j) \tag{7}$$

be the increment of $x_t(i', j)$. Then, the initial order $x_t(i'+1, j) > x_t(i', j)$ of the pixels of the right image can be guaranteed if the limitation of $\Delta x_t(i', j)$ introduced in [2] is used:

$$\Delta x_t(i', j) = \begin{array}{ll} d_+/2 & if \quad \Delta t \ v_t(i', j) > d_+/2 \\ -d_-/2 & if \quad \Delta t \ v_t(i', j) < -d_-/2 \\ \Delta t \ v_t(i', j) & elsewhere \end{array} \qquad (8)$$

Here, $d_+ = x_t(i'+1, j) - x_t(i', j)$, $d_- = x_t(i', j) - x_t(i'-1, j)$. Conditions such as (8) can be checked easily in each step of recursion.

The algorithm (4) defines a parallel-sequential processing structure. Parallel means here, that at every discrete time t all calculations can be carried out in parallel for each (i',j). Then, all these t – layers are processed one after the other, i. e. sequentially. If one assigns a processing element or neuron to each image point (i',j) of the right image then the algorithm can be implemented with a Multi Layer Neural Network where each layer corresponds to one level t of recursion. Another possibility is a Recurrent Neural Network with one layer only where the output of the layer is fed into the layer as the new input. To calculate the forces K each neuron (i',j) must have access to a limited region of the left image around (i',j). Of course, a pure sequential computation using very fast PC's or workstations can also be very efficient.

We come now to the calculation of the forces K. First, it must be acknowledged that essential stereo information is only present in image regions with significant changes of grey level, and especially near edges. Furthermore, in the epipolar geometry assumed here only the x – dependence of the grey values, i. e. $_xg(i,j) = g(i,j) -g (i-1,j)$ is essential. Let's assume that there is a step edge between (i,j) and $(i-1,j)$ with $| \ _xg(i,j)| > threshold$. That means that (i,j) belongs to an image segment and $(i-1,j)$ to another one. When e. g. such an edge is at the border of a roof of a building then often left or right of that edge we have occlusion. Then, if the pixel (i,j) has a corresponding pixel in the other stereo image this may be not the case for pixel $(i-1,j)$ or vice versa. Therefore, both pixels (i.e. pixels left hand and right hand of an edge) must be considered separately. They can have different disparities or even worse: in one of them cannot be calculated a disparity at all. To such pixels only with prior information or by some kind of interpolation a (often inaccurate) disparity can be assigned. With respect to our attracting forces that means the following: If there is an edge in the left image between $(i-1,j)$ and (i,j) and another one between $(i'-1,j)$ and (i',j) in the right image then there is a force $K_R(i,i',j)$ originating from (i,j) and attracting (i',j) and another force $K_L(i-1,i'-1,j)$ acting from $(i-1,j)$ to $(i'-1,j)$. This is necessary for coping with occlusion.

Let's consider the external force $K_R(i,i',j)$. Then, first, that force depends on the difference $|g_L(i,j)-g_R(i',j)|$ or, more general, on a certain mean value of that difference. That mean value should be calculated only over pixels which are in the same image regions as pixels (i,j) (in left image) and (i',j) (in right image) in order to exclude problems with occlusion. To guarantee this, the averaging is performed only over image points (i_k,j_k) with $|g_L(i,j)-g_L(i_k,j_k)|$ threshold and (i'_l,j_l) with $|g_R(i',j)-g_R(i'_l,j_l)|$ threshold, respectively. We denote that mean value as

$$\Delta g(i,i'; j) = \langle |g_L(i, j) - g_R(i', j)| \rangle \qquad (9)$$

Secondly, as it was mentioned already in [2], pure radiometric criteria are not sufficient. Therefore, geometric deviations are taken into account too. To do that, we consider two region border lines (one in the left image and the other in the right image) which contain the points (i,j) and (i',j), respectively. The situation is shown in figure 1.

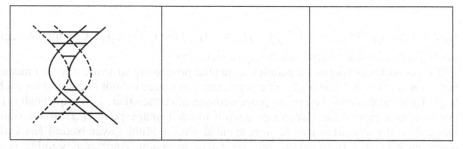

Fig. 1. Borderlines in left image, in right image, and overlaid

We see that both borderlines are different and do not match. A useful quantity for measuring that mismatch is the sum of the border point distances along the horizontal lines drawn in figure 1. Be $(i_k,j+k)$ a point on the left borderline and $(i'_k,j+k)$ a point on the right borderline, respectively. For $k = 0$ the points are identical with the points (i,j) and (i',j). Be $d_o = |i - i'|$. Then, a useful border point distance is

$$d_w(i,i';j) = \sum_{k=-w}^{w} (|i_k - i_k| - d_0) \qquad (10)$$

The distance d_w accomplishes a certain coupling between epipolar image rows which are no longer independent. This sometimes can reduce mismatches efficiently. With Δg and d_w the total distance is

$$d(i,i';j) = \alpha_1 \, d_w(i,i';j) + \alpha_2 \, \Delta g(i,i';j) \qquad (11)$$

The smaller that distance between edge points (i,j) and (i',j) is, the bigger is the force $K_R(i,i',j)$. Therefore,

$$K_R(i,i';j) \quad \exp[-d(i,i';j)] \qquad (12)$$

seems to be a good measure for the force K_R. The calculation of K_L is fulfilled analogously.

Now, the (external) force $K_{ext}(i',j)$ acting on point (i',j) can be computed as the maximum of all forces $K_R(i,i',j)$ with different i or as a weighted sum of these forces. Here, we take into account only points (i,j) with $| i - i' |$ Max_disparity. The maximum disparity used here is often known a priori. The introduction of *Max_disparity* is not necessary. One can also use distance depending weighting and calculate the resulting force as

$$K_{R,ext,t}(i',j) = \sum_i K_R(i,i',j)\ f\big(\big|i - x_t(i',j)\big|\big)\ sign\big(i - x_t(i',j)\big) \qquad (13)$$

with $f(|\ i - x_t(i',j)|)$ being a certain weighting function which decreases with increasing distance $|\ i - x_t(i',j)|$. Here, we use the special function

$$f(x) = \begin{array}{ll} 1 & if \quad |x| \quad Max\_disparity \\ 0 & elsewhere \end{array} \qquad (14)$$

Up to now we have considered only forces which act only on image points (i',j) near edges. But we must assign a disparity to each point of the right image. Therefore, the disparity information from the edges must be transferred into the image regions. Within the model presented here, it is useful to do this by means of adequate forces, which connect the edge points with interior points (i.e. points inside regions). Local forces of spring type have been studied for that purpose. Let $x_t(i',j)$ and $x_t(i'+1,j)$ be two neighbored mass points which we assume to be connected by a spring. Then, point $x_t(i'+1,j)$ exerts the following (attracting or repulsive) force on point $x_t(i',j)$:

$$K_{spring,t}(i'+1,i',j) = \kappa\ \big[x_t(i'+1,j) - x_t(i',j) - 1\big] \qquad (15)$$

The same force, but with the opposite sign, acts from $x_t(i',j)$ on $x_t(i'+1,j)$ according to Newton's law of action and reaction.

Experiments with those and other local forces (e.g. internal friction) have not brought much success up to now. Of course, the stereo information is transferred from the edges into the regions, but very slowly. One needs many recursions until convergence. Therefore, one result of these investigations is that local forces are not sufficient. We need far-field interaction between points $x_t(i',j)$ and $x_t(i'+k,j)$ which can easily be introduced into our equations of motion. First experiments with such forces have given some promising results but this must be studied more detailed in future. Therefore, we confine our studies here to the calculation of the disparities along edges without taking into account forces (15) and others.

3 Some Results

As in older papers [2], [3] the algorithm was applied to the standard Pentagon stereo pair. This is useful in order to compare results. Furthermore, that image pair is a big challenge because of the many similar structures and the many occlusions. Figure 2 shows the image pair.

Fig. 2. Stereo pair "Pentagon"

If one applies to these images the gradient operator $_x g(i,j) = g(i,j) - g(i-1,j)$ with $|\ _x g(i,j)| > threshold$ in order to find the vertical edges then one obtains a too big number of edge points if the threshold is small or too few edges if the threshold is big. To avoid that we apply a special parallel-sequential algorithm for edge preserving smoothing [6] after that the gradient operator can be applied successfully.

Several computer experiments have been carried out in order to choose proper parameters of the algorithm. It turned out that the quantity w in (10) should have a size between 5 and 10 (but of course, that value depends on the structure of the image) whereas the best values of parameters α_1 and α_2 of equation (11) seem to be in the vicinity of $\alpha_1 = \alpha_2 = 1$. But these investigations are only preliminary. It is also not clear which the best law of the external forces (11) – (13) is.

The result of the algorithm after 20 iterations (disparity image) is shown in figure 3. It shows that in most edge points good matching was achieved. But it shows also that there are left many mismatches resulting in wrong disparities.

Other image stereo pairs with completely different image structures were processed with the same parameters with qualitatively the same results. Because of limited space they cannot be presented here.

Fig. 3. Disparity image

4 Conclusions

The results show that the introduced parallel-sequential model based on Newton's equations of motion and attracting forces between edges may be a promising approach to real-time stereo processing. The algorithm gives the right disparities in most edge points but there remain errors. Therefore, new efforts are necessary to enhance the quality of the approach. Some ideas for improvement are the following: First, far - field forces should be introduced. Secondly, the assumed force law (11) – (14) must be optimized or changed. When the right position $x_{tmax}(i',j)$ is reached then the external forces should reduce to zero in order to avoid oscillations which are small but not zero now.

Finally, it must be mentioned that the model can be extended to the non-epipolar case introducing coordinates $y(i',j)$ and forces acting in y – direction.

References

1. Science 1689.Springer (1999) 568-577 Hubel, D.: Eye, Brain, and Vision. Scientific American Library; New York (1995)
2. Jahn, H.: Stereo Matching for Pushbroom Stereo Cameras. Int. Archives of Photogrammetry and Remote Sensing, Vol. XXXIII, Part B3. Amsterdam (2000) 436-443
3. Jahn, H.: Parallel Epipolar Stereo Matching. 15th Int. Conf. on Pattern Recognition, Vol. 1. Barcelona (2000) 402-405
4. Julesz, B.: Foundations of Cyclopean Perception. The University of Chicago Press, Chicago (1971) 203-215
5. Klette, R., Schlüns, K., Koschan, A.: Computer Vision. Springer, Singapore (1998)
6. Jahn, H.: Feature Grouping based on Graphs and Neural Networks. Proc. of CAIP'99, Lecture Notes in Computer

Characterizations of Image Acquisition and Epipolar Geometry of Multiple Panoramas

Shou-Kang Wei, Fay Huang, and Reinhard Klette

CITR, Computer Science Department, The University of Auckland,
Tamaki Campus, Auckland, New Zealand

Abstract. Recently multiple panoramic images have emerged and received increasingly interests in applications of 3D scene visualization and reconstruction. There is a need to characterize and clarify their common natures and differences so that a more general form/framework or a better computational model can be further discovered or developed. This paper introduces some notions at an abstract level for characterizing the essential components of panoramic image acquisition models. A general computational model is proposed to describe the family of cylindrical panoramas. The epipolar geometry of the cylindrical panoramic pairs for a general and a leveled case are particularly studied.

Keywords: panoramic image, epipolar geometry

1 Introduction

Traditionally, a 360 degree panorama can be acquired by rotating a normal camera with respect to a fixed rotation center and taking images consecutively at equidistant angles. More recently, panoramic images taken with a line camera acquired with respect to single rotation axis or multiple rotation axes have emerged and received increasingly interests in applications of 3D scene visualization and reconstruction.

Since many different panoramic image models have been proposed and used in various intended applications, there is a need to characterize them and clarify their differences so that better understanding of them and the related properties can be achieved. By observing the common characteristics among them, a more general form/framework or a better computational model may be further discovered or developed.

Geometric studies such as epipolar geometry are well established for pairs of planar images [4,8,10,12,21]. Compared to that, the computer vision literature still lacks work on pairs of panoramic images. Due to differences in geometry between the planar and the panoramic image models, geometric properties for planar images may not necessarily be true for panoramic images.

The rest of the paper is organized as follows. Brief reviews of the related literatures are provided in Section 2. In Section 3 we introduce some basic/general components/notions for design, analysis and assessment of image acquisition models. The computational model and epipolar geometry of the cylindrical panorama family are presented in Section 4. Conclusions are drawn in Section 5.

W. Skarbek (Ed.): CAIP 2001, LNCS 2124, pp. 684–691, 2001.

2   Brief Reviews

A well-known and typical example for 3D scene visualization using a single-focal-point panorama is QuickTimeVR [3] from Apple Inc.. Using multiple single-focal-point panoramas to reconstruct a 3D scene, S.B. Kang and R. Szeliski discussed different methods and their performances in [11]. Other similar works can be found in [6,13,14]. The direct merit of this approach is by-passing the complicate and erroneous[1] process of multiple depth-maps merging that has to be done in multiple planar image cases.

The family of cataoptrical panoramas [2,5,19,22] provide real-time and highly portable imaging capabilities at affordable cost. The applications include robot navigation, teleportation, 3D scene reconstructions, etc.. For the latter case, the epipolar geometry coincides with image columns if the two panoramic camcorders are specially arranged such that the optical axes are co-axis and each acquired panoramic image is warped to a cylinder. The drawbacks of this approach include low resolution; inefficient usage of images (e.g. the self-occluded and mirror-occluded area of an image); and potentially inaccurate image acquisition along the peripheral of the spherical mirror (e.g. spherical aberration or distortion).

H. Ishiguro et al. first proposed an image acquisition model that is able to produce multiple panoramas by a single swiveling of a pinhole-projection camera. The model was invented for the 3D reconstruction of an indoor environment. Their approach reported in 1992 in [9] already details essential features of the multi-perspective panoramic image acquisition model. The modifications or extensions of their model have been discussed by other works such as [15,17,18,20]. S. Peleg and M. Beh-Ezra [15] described a model using circular projections[2] for stereo panoramic image generation, which allows the left and right eye perceptions of panoramic images. H.-Y. Shum and L.-W. He [17] proposed a concentric model in which novel views within an inner circle and between the inner and outer circles were approximated by the circular projections in normal direction (the same as in [15]) and in tangential direction. With the concentric model H.-Y. Shum and R. Szeliski [18] show that epipolar geometry consists of horizontal lines if two panoramic images form a symmetric pair.

3   Image Acquisition Characterization

This section introduces some notions at an abstract level along with examples for characterizing the essential components of panoramic image acquisition models.

Definition 1. A focal set $\mathcal{F}$ is a non-empty (finite) set of focal points in 3D space. A focal point, an element of $\mathcal{F}$, can be represented as a 3-vector in $\mathbb{R}^3$.

---

[1] Various sources of the errors, such as inexact estimation of relative poses between cameras, may cause serious degradation to the resulting quality.

[2] See following sections for further explanations.

Definition 2. *A receptor set S is a non-empty infinite or finite set of receptors (photon-sensing elements) in 3D space. A receptor, an element of S, can be characterized geometrically as 3-vectors in $\mathbb{R}^3$.*

It is convenient to express a collection of points by a supporting geometric primitive such as a straight line, curve, plane, quadratic surface etc. where all of the points lie on. For examples, the single-center panoramic model (e.g. Quick-TimeVR) consists of a single focal point (i.e. $\#\mathcal{F} = 1$) and a set of receptors lie on a cylindrical or spherical surface. The multi-center image model consists of a set of focal points on various geometrical forms (such as a vertical straight line, a 2D circular path, a disk, or a cylinder etc.) and a set of receptors being on a cylindrical, cubic or spherical surface.

A single light ray with respect to a point in 3D space at one moment of time can be described by seven parameters, which is known as *plenoptical function* [1]. All possible light rays in a specified 3D space and time interval form a *light field*, denoted as $\mathcal{L}$.

The association between focal points in $\mathcal{F}$ and receptors in S determines a particular proper subset of the light field $\mathcal{L}$. For instance, a complete bipartite set of focal and receptor sets is defined as

$$\mathcal{B}_{\mathcal{F} \times S} = \{(p, q) : p \in \mathcal{F} \wedge q \in S\},$$

where each element (p, q) specifies a light ray emitting at point p and passing through point q. Note that a complete bipartite set of focal and receptor sets is a proper subset of the light field (i.e. $\mathcal{B}_{\mathcal{F} \times S} \subset \mathcal{L}$).

Definition 3. *A focal-to-receptor association rule defines an association between a focal point and a receptor, where a receptor is said to be associated with a focal point if and only if any light ray which is incident with the receptor passes through the focal point.*

Each image acquisition model has it's own association rule for the focal and receptor sets. Sometimes, a single rule is not enough to specify complicate associating conditions between the two sets, thus a list of association rules is required. A pair of elements satisfies a list of association rules if and only if the pair satisfies each individual association rule.

Definition 4. *A set of projection-rays $\mathcal{U}$ is a non-empty subset of the complete bipartite set of focal and receptor sets (i.e. $\mathcal{U} \subseteq \mathcal{B}_{\mathcal{F} \times S} \subset \mathcal{L}$), which satisfies the following conditions:*

1. *It holds $(p, q) \in \mathcal{U}$ iff (p, q) satisfies one (a list of) pre-defined association rule(s);*
2. *For every $p \in \mathcal{F}$, there is at least one $q \in S$ such that $(p, q) \in \mathcal{U}$;*
3. *For every $q \in S$, there is at least one $p \in \mathcal{F}$ such that $(p, q) \in \mathcal{U}$.*

The projection-ray set $\mathcal{U}$ of a multi-perspective panoramic image acquisition model [7,15] is a subset of the complete bipartite set of focal and receptor sets and can be characterized formally as follows. The focal points in $\mathcal{F}$ are an ordered finite sequence, $p_1, p_2, \ldots, p_n$, which all lie on a 1D circular path in 3D space. The set of receptors form a uniform (orthogonal) 2D grid and lie on a 2D cylindrical surface that is co-axial to the circular path of the focal points. The number of columns of the grid is equal to n. The association rules determining whether (p, q) belongs to the projection-ray set $\mathcal{U}$ are as follows:

1. All $q \in \mathcal{S}$ which belong to the same column must be assigned to an unique $p_i \in \mathcal{F}$.
2. There is an ordered one-to-one mapping between the focal points $p_i \in \mathcal{F}$ and the columns of the grid. In other words, the columns of the grid, either counterclockwise or clockwise, may be indexed as $c_1, c_2, \ldots, c_n$ such that every $q \in c_i$ is mapped to p_i, with $1 \leq i \leq n$.

Definition 5. *A reflector set $\mathcal{R}$ is a set of reflectors' surface equations, usually a set of first or second order continuous and differentiable surfaces in 3D space.*

A reflector set is used to characterize how light rays can be captured indirectly by the receptors. For instance, a hyperbolic mirror is used in conjunction with the pinhole projection model for acquiring a wide visual field of a scene. Such type of image acquisition model allows that all the reflected projection rays intersect at the focus of the hyperboloid [2,19], which possess a simple computational model for supporting possible applications.

Let $\mathcal{P}(\mathcal{R})$ denote the power set of the reflector set. Define a geometrical transformation T as follows:

$$T : \mathcal{U} \times \mathcal{P}(\mathcal{R}) \to \mathcal{A},$$
$$((p, q), s) \mapsto (p', q'),$$

where $\mathcal{A}$ is a non-empty subset of the light field. Any element of $\mathcal{A}$, i.e. a light ray, is represented by a pair of points, denoted as (p', q'), specifying its location and the orientation. The transformation T is a function which transforms a projection ray with respect to an element of $\mathcal{P}(\mathcal{R})$ to a reflected ray.

Definition 6. *A reflected-ray set $\mathcal{V}$ is a non-empty set of light rays, which is a subset of the light field. Formally,*

$$\mathcal{V} = \{T((p, q), s) : (p, q) \in \mathcal{U} \wedge s \in \mathcal{P}(\mathcal{R})\}.$$

Note that, when a transformation of a projection-ray set takes place, only one element of $\mathcal{P}(\mathcal{R})$ is used. In particular, if $\emptyset \in \mathcal{P}(\mathcal{R})$ is chosen, the resulting reflected-ray set is identical to the original projection-ray set. When the number of elements of the chosen s is more than one, the transformation behaves like ray-tracing.

A single projection-ray set (or a reflected-ray set) is referred to as a set of light rays defined by an image acquisition model at a moment of time and a specific location. Two more factors are regarded to characterize multiple projection-ray sets: a temporal factor describes the acquisition time, and a spatial factor

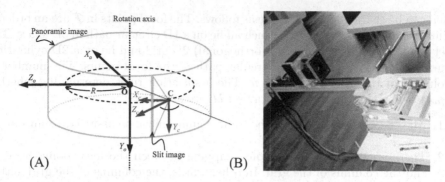

Fig. 1. A general image acquisition model of the cylindrical panorama family shown in (A) and the actual setup in (B).

describes the pose of the model. Multiple images, i.e. a collection of projection-ray sets acquired at different times or poses, $\{\mathcal{U}_{t,\rho}\}$, are a subset of the light field.

Some applications [3,17] use only a single projection-ray set to approximate a complete light field in a restricted zone and some [18] require multiples in order to perform special tasks such as depth from stereo. Usually, a few sampled projection-ray sets are acquired for approximating a complete light field of a medium-to-large scale space. The selection of sets of optimal projection-ray samples become an important factor to the quality of the intended applications.

4   Epipolar Geometry

4.1   Computational Model

A general computational model of the cylindrical panorama family is described below and depicted in Figure 1(A). A slit camera [16] with the actual setup is shown in Figure 1(B). A slit camera can be characterized geometrically by a single focal point $\mathbf{C}$ (the effective focal length is denoted as f) and a 1D linear receptors (i.e. a slit image). The distance between the slit camera's focal point and the rotation axis, denoted as R, remains constant during a single panoramic image acquisition process. The angular interval of every subsequent rotation step is assumed to be constant. Each slit image contributes to one column of a panoramic image of dimension $H \times W$.

An angle, ω, is defined by the angle between the normal vector of the focal circle at the associated focal point and the optical axis of the slit camera for more flexibility in generating different viewing-angled panoramic images, which has been reported being useful in various applications [15,17].

4.2   General and Leveled Cases

Let us consider an arbitrary pair of polycentric panoramic images, a *source image* E and a *destination image* E'. Given an image point $\mathbf{p}$ with image coordinates (x, y) on E. The possible locations of the corresponding point of $\mathbf{p}$, denoted as $\mathbf{p}'$ with the image coordinates (x', y'), on E' may be constrained by an epipolar curve. Let a 3×3 rotation matrix $[\mathbf{r}_1^T \mathbf{r}_2^T \mathbf{r}_3^T]^T$ and a 3×1 translation vector $(t_x, t_y, t_z)^T$ specify the orientation and the location of the destination turning-rig coordinate system with respect to the source turning-rig coordinate system.

Consider x and y as being given. Let $\theta = \frac{2\pi x}{W}$, $\theta = \frac{2\pi x'}{W'}$, $\delta = (\theta + \omega)$, $\delta' = (\theta' + \omega')$, and $\phi = tan^{-1}(y/f)$. The relationship between x' and y' can be described by the epipolar curve equation

$$y' = \frac{f' \mathbf{r}_2^T \cdot \mathbf{V}}{\mathbf{r}_1^T \cdot \mathbf{V} \sin \delta' + \mathbf{r}_3^T \cdot \mathbf{V} cos\delta' - R' \cos \omega'},$$

which is only valid if the value of the denominator is greater than zero. The vector $\mathbf{V}$ is defined as follows:

$$\mathbf{V} = \mathbf{A} + \frac{R' \sin \omega' + \mathbf{r}_1^T \cdot \mathbf{A} \cos \delta' - \mathbf{r}_3^T \cdot \mathbf{A} \sin \delta'}{\mathbf{r}_3^T \cdot \mathbf{B} \sin \delta' - \mathbf{r}_1^T \cdot \mathbf{B} \cos \delta'} \mathbf{B},$$

where

$$\mathbf{A} = \begin{pmatrix} R \sin \theta - t_x \\ -t_y \\ R \cos \theta - t_z \end{pmatrix} \quad \text{and} \quad \mathbf{B} = \begin{pmatrix} \sin \delta \cos \phi \\ \sin \phi \\ \cos \delta \cos \phi \end{pmatrix}.$$

In practice, the image acquisition rigs are usually leveled for allowing most scene objects to be in natural orientation in the image over completely 360° view. However, the heights can be different. Thus, the row vectors of the rotation matrix that specifies the orientation between the destination and the source turning-rig coordinate systems becomes: $\mathbf{r}_1^T = (\cos \gamma, 0, -\sin \gamma)^T$, $\mathbf{r}_2^T = (0, 1, 0)^T$, and $\mathbf{r}_3^T = (\sin \gamma, 0, \cos \gamma)^T$, where the angle γ determines the rotation with respect to the y-axis. Therefore, the epipolar curve equation, simplified from the general case, for an arbitrary leveled polycentric panorama pair is as follows:

$$y' = \frac{f' \left(\frac{y}{f} \right)(R' \sin \omega' - R \sin(\delta' - \theta) - t_x \cos \delta' + t_z \sin \delta') - t_y f' \sin(\delta' - \delta)}{R' \sin(\delta - \theta' - \gamma) - R \sin \omega - t_x \cos \delta + t_z \sin \delta},$$

where $\delta = (\theta + \omega)$ and $\delta' = (\theta' + \omega' + \gamma)$. Figure 2 illustrates the leveled (but at different heights) case. The images are taken by a high-resolution line-camera [16] with up to 5184 pixels for a scan-line. The effective focal length of the line-camera is 21.7 mm. The rotation axes of these two panoramic images are 1.4 m apart and 45 mm difference in height. The radiuses of the base circles are both equal to 100 mm. The associated ω's for these images in (A) and in (B) are equal to 205° and 155°, respectively. The line-camera takes 22000 slit images for each panoramic image.

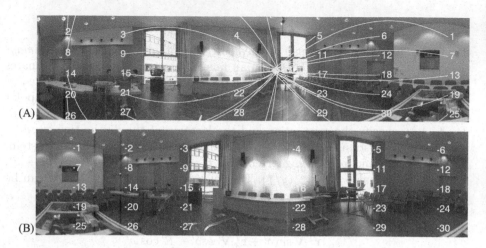

Fig. 2. A pair of leveled polycentric panoramas at different heights. (B) is the source panoramic image with 30 test points labeled by '·' and indexed by numbers. (A) is the destination panoramic image superimposed with the corresponding epipolar curves.

5 Conclusion

In this paper, we characterize the image acquisition and epipolar geometry of multiple panoramic images. The existing approaches using multiple panoramic images are briefly reviewed. The emphasis has been placed on demonstrating the flexibility and compactness in characterizing different types of acquisition models and epipolar geometry of a general and leveled polycentric panoramic pair.

A family of cylindrical panoramic images, which serves for a wide range of applications, has been discussed. A general computational model of this family is described and used in computing epipolar curve of the family for both general and leveled case.

In future, we will look further into the relationship between applications (with some knowledge from the scene of interest) and various panoramic image acquisition models. The limitations of each model and the evaluation criteria should also be analyzed.

Acknowledgments. The authors would like to thank Ralf Reulke, Anko Börner, Martin Scheele and Karsten Scheibe in Institute of Space Sensor Technology and Planetary Exploration, German Aerospace Center for their helps in acquiring and processing all the panoramic images.

References

1. Adelson, E.H., Bergen, J.R.: The plenoptic function and the elements of early vision. Computational Models of Visual Proceeding. Cambridge, Massachusetts, USA (March 1991) 3–20

2. Baker, S., Nayar, S.K.: A theory of single-viewpoint catadioptric image formation. IJCV **35**(2) (Nov 1999) 1–22
3. Chen, S.E.: QuickTimeVR - an image-based approach to virtual environment navigation. Proc. SIGGRAPH'95. Los Angeles, California, USA (Aug 1995) 29–38
4. Faugeras, O.: Three-Dimensional Computer Vision: A Geometric Viewpoint. The MIT Press, London, England (1993)
5. Gluckman, J., Nayar, S.K., Thorek, K.J.: Real-time panoramic stereo. Proc. DARPA98, Monterey, California, USA (Nov 1998) 299–303
6. Huang, H.C., Hung, Y.P.: Panoramic stereo imaging system with automatic disparity warping and seaming. GMIP **60**(3) (May 1998) 196–208
7. Huang, F., Wei, S.K., Klette, R.: Geometrical Fundamentals of Polycentric Panoramas. (to appear in:) Proc. ICCV'2001, Vancouver, British Columbia, Canada (July 2001)
8. Hartley, R., Zisserman, A.: Multiple View Geometry in Computer Vision. Cambridge Uni. Press, United Kingdom (2000)
9. Ishiguro, H., Yamamoto, M., Tsuji, S.: Omni-directional stereo. PAMI **14**(2) (Feb 1992) 257–262
10. Kanatani, K.: Geometric Computation for Machine Vision. Oxford Uni. Press, New York (1993)
11. Kang, S.B., Szeliski, R.: 3-d scene data recovery using omnidirectional multibaseline stereo. IJCV **25**(2) (Nov 1997) 167–183
12. Klette, R., Schliins, K., Koschan, A.: Computer Vision - Three-Dimensional Data from Images. Springer, Singapore (1998)
13. McMillan, L., Bishop, G.: Plenoptic modeling: an image-based rendering system. Proc. SIGGRAPH'95, Los Angeles, California, USA (Aug 1995) 39–46
14. Matsuyama, T., Wada, T.: Cooperative distributed vision - dynamic integration of visual perception, action, and communication. Proc. CDV-WS98, Kyoto, Japan (Nov 1998) 1–40
15. Peleg, S., Ben-Ezra, M.: Stereo panorama with a single camera. Proc. CVPR99, Fort Collins, Colorado, USA (Jun 1999) 395–401
16. Reulke, R., Scheele, M.: Der drei-zeilen ccd-stereoscanner waac: Grundaufbau und anwendungen in der photogrammetrie. Photogrammetrie, Fernerkundung, Geoinformation **3** (1998) 157-163
17. Shum, H.Y., He, L.W.: Rendering with concentric mosaics. Proc. SIGGRAPH'99, Los Angeles, California, USA (Aug 1999) 299–306
18. Shum, H.Y., Szeliski, R.: Stereo reconstruction from multiperspective panoramas. Proc. ICCV99, Korfu, Greece (Sep 1999) 14–21
19. Svoboda, T.: Central Panoramic Cameras Design, Geometry, Egomotion. PhD Thesis, Czech Technical University, Prague, Czech Republic (1999)
20. Wei, S.K., Huang, F., Klette, R.: Three-dimensional scene navigation through anaglyphic panorama visualization. Proc. CAIP99, Ljubljana, Slovenia (Sep 1999) 542–549
21. Xu, G., Zhang, Z.: Epipolar Geometry in Stereo, Motion and Object Recognition - A Unified Approach. Kluwer Academic Publishers, Netherlands (1996)
22. Zheng, J.Y., Tsuji, S.: Panoramic representation for route recognition by a mobile robot. IJCV **9**(1) (Oct 1992) 55–76

Interclass Fuzzy Rule Generation for Road Scene Recognition from Colour Images

Malcolm Wilson

Electronics and Mathematics
University of Luton
Park Square,
Luton, LU1 3JU,
Bedforddfordshire
ENGLAND
Malcolm.Wilson@Luton.ac.uk

Abstract. In many image classification problems the extent of usefulness of any variable for the purposes of discrimination apriori is unknown. This paper describes a unique fuzzy rule generation system developed to overcome this problem. By investigating interclass relationships very compact rule sets are produced with redundant variables removed. This approach to fuzzy system development is applied to two problems. The first is the classification of the Fisher Iris data [4] and the second is a road scene classification problem, based on features extracted from video images taken by a camera mounted in a motor vehicle.

Keywords: feature redundancy, fuzzy training, image classification.

1 Introduction

This paper examines the production of a fuzzy logic based environment classification system for automotive applications. As cars become more advanced it will be desirable to incorporate some "knowledge" of the driving environment into their decision making systems. For example, in the past it was possible to select "reverse" on cars with automatic gears while traveling forward, with the predictable consequences of a non-intelligent "automatic" system. This would not be possible today. It would be possible, however, to select "cruise" when the vehicle was just entering a build up area, which is an unsafe action. The fuzzy system presented here is designed to classify "Urban", "Suburban", "Highway" and "Country" driving environments. This information could then be used to modulate the behavior of the vehicle's systems and facilitate automatic driving.

In this safety critical application a fuzzy logic classifier is preferred to neural network classification, because the fuzzy rules may be examined and modified by experts to ensure that hazardous outputs are recognised and eliminated. The "black box" nature of neural network classifiers prohibit such intervention [8]. The descriptions of the driving environment are also linguistic in nature, which can be readily derived from a fuzzy rule system.

W. Skarbek (Ed.): CAIP 2001, LNCS 2124, pp. 692–699, 2001.

The classification of the driving environment is made using features extracted from a moving picture sequence captured by a camera in the car, but it is intended that the system is extendable. Then, if new inputs (say from radar or even further pre processing of the visual data) become available in the future, they may be easily incorporated into the system.

In complex problems (in terms of number of input variables) it is difficult to anticipate or visualise the effect or usefulness of any one variable on classification. It is possible that some variables are completely redundant or that some subset of variables necessary for the classification of one class is a different subset necessary for the classification of another. The inclusion of redundant variables, however, can lead to weakening of the rules and a poor linguistic interpretation of the classification. The method of fuzzy rule generation presented here, was developed specifically to overcome the above problem, by taking an interclass approach, to eliminate the redundant variables when classifying each class. This in turn may highlight any overall redundant variables in the entire data set. Although developed for this particular classification problem, the techniques used are applicable to other multi variable classification problems; particularly where the usefulness of each variable for the separation of each class is unknown. The system is demonstrated on the Fisher Iris data [4] and on image data from road scenes.

2   Redundant Variables

Imagine the classification of the fruit set {apples pears oranges tangerines lemons limes}. To classify the whole set, variables:

- *size*
- *colour*
- *texture*
- *aspect ratio*

may be needed. So if we look across all classes we would see no redundant variables. However, looking at the class of oranges, only 2 variables (*colour* and *size*) are necessary to uniquely identify oranges from the total set. Looking more closely the variable size is only necessary due to the inclusion of tangerines in the fruit set. A similar argument can be used for all of the other classes in this fruit set. If an extra variable *hardness* were to be included in the variable set, then *hardness* would be redundant across all classes. The main point to be noted here is that looking for redundancy across the total data set may reveal none, but the inter class investigation can reveal many redundant variables.

3   The Road Scene Classification Problem

The problem is the classification of colour images taken from a vehicle in Britain into the categories "Country", "Urban", "Highway" and "Suburban". The colour images were taken with a hand held professional camcorder which recorded colour pictures onto tape in DVCAM format. The DVCAM tape was transferred onto hard disk in "bmp" format. The bitmap images were processed to give the input variables. While driving on the left of the road, the most useful discriminating information would be contained by the area bounded by the road side, the skyline and full left of the pictures. Areas outside this region of interest would generally contain similar information in all classification situations. In addition colours of other vehicles (outside of the region of interest) would distort colour histograms. The isolation of a region of interest also reduces the computationally intensive RGB $\rightarrow$ HSL conversion. The region of interest is determined using a road edge detector [6].The input variables extracted from the images were:

- 1. Redbrick colour hue by normalised pixel count.
- 2. Grass colour hue by normalised pixel count.
- 3. Colour Saturation (mean).
- 4. Long vertical edge count.
- 5. Road edge disturbance.
- 6. Road edge curvature.

4   Development of the Fuzzy Logic Classification System

Initially, a pilot study into the viability of the project resulted in a simple "hand crafted" system, being produced based upon two simple fuzzy rules:

- If grass colour is high and redbrick colour is low, Then environment is country.
- If grass colour is low and redbrick colour is high, Then environment is urban.

The high and low membership functions were also "hand crafted". These simple rules gave correct outcomes in clear situations. When additional variables were used to cope with the anomalies found, it became impossible to produce sensible "hand crafted" rules and membership functions. A supervised learning system had to be developed.

Training data was generated by extracting the variables described above from 40 hand labelled images to produce 40 training feature vectors. The problem is extremely non- linear and the data is not linearly separable. The use of clustering techniques and principal component analysis failed to separate the road scene data into the "urban", "suburban", "country" and "highway classes.

Some of the 6 variables available to the system may be redundant, either completely or for the discrimination of particular classes. A method of redundant variable identification is used to evaluate the usefulness of any variable. Many techniques exist for the training of fuzzy logic systems from numerical data [1] [5] [2]. The two which showed most promise for this application were Wang and Mendel [1] and Abe and Lan's [5] approaches. Neither of these techniques deal in a satisfactory way with the problem of redundant variables. It is shown in [7] that the Wang and Mendel rule set [1] can be reduced in terms of numbers of rules and variables used, if a suitable variable reduction algorithm is applied. This is possible because of the arbitrary selection of the membership functions. The Abe and Lan's [5] rule generation system does not lend itself so easily to variable reduction and certainly not on a class by class basis, because of the inhibition rectangle structure and the generation of "NOT" clauses in the rule set, (see Interclass examination). A variation of Abe and Lan's hyperbox rule generation system is developed here which is designed to support variable reduction. The identification of the redundant variable is useful to the designer of the classification system, not only so they may discard the variable, but also so they can identify which variables are used for each classification.

4.1 Interclass Examination

The Wang and Mendel rule generation system [1] does not take account of the interaction of other class' data in the production of the rule set. As shown in [7] a Wang Mendel rule generation system [1] need only generate one rule (with one variable) for the identification of the "setosa" Iris type. Abe and Lan's rule generation algorithm [5] describes a system which compares pairs of classes in terms of its data's maximum and minimum values, and generates hyperboxes spanning these values. When intersection is found between the hyperboxes the area is marked as an "inhibition region". The inhibition region is reconsidered in terms of maximum and minimum values and the process is recursively repeated until any inhibition region is broken down into smaller non-overlapping hyperboxes are produced. A consequence of the inhibition regions is the generation of linguistically complex "NOT" clauses in the fuzzy rule set. The inhibition rectangles were Abe and Lan's approach [5] to the problem of overlapping rectangles proposed by Simpson [3]. Simpson considered interclass separation of any class with any other and did not allow overlapping of hyperboxes from different classes. This was based on the assumption that data from two or more classes would exist in any overlapping regions of the different class' hyperboxes. The system described here is based on hyperboxes, but differs from Abe and Lan's [5] approach in 4 important ways:

- 1. It compares each class with all other classes in the initial generation of rules, rather than comparing class pairs.
- 2. It does not specifically consider the minimum and maximum of each class' data set in each dimension. It, instead, expands the size of the classes' hyperboxes by the addition of the individual class' data points. The hyperboxes are never allowed to grow such that they encapsulate another class' data point, but subject to the above, hyperboxes of different classes are allowed to intersect(if no data points exist in the intersection).
- 3. As a consequence of item 2 inhibition hyperboxes are not created. That is, each class is separated into its own distinct set of hyperboxes.
- 4. As a product of the above, variable redundancy is easily identified from the outset on a class by class basis. See [7].

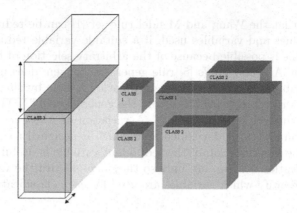

Fig. 1.

4.2   Identification and Removal of Redundant Variables from the Rule Set

The identification of variable redundancy, in this system is based upon the fact that removal of variables from rules of the form shown below,

If variable 1 is X1 AND variable 2 is X2 AND variable 3 is X3 AND variable 4 is X4: then class is Y

is equivalent to setting the removed variables' membership function to unity for all of the domain of the removed variables.

The above rules in this rule generation system are represented by hyperboxes. Note that within the boundary of the hyperbox a membership function of value one is represented. Therefore, if we take a rule generation hyperbox and extend it along one or more variables to the extents of the domain of those variables, we are effectively removing the representation of that variable from the rule. If we can extend a hyperbox in this way without intersection with a hyperbox of another class then the variable is redundant, because that variable cannot be necessary to differentiate it from another class. Fig. 1 below indicates the principle. Class 3 may be infinitely extended in the Y and Z dimensions without intersecting another hyperbox (as indicated by the arrows in Fig. 1). Therefore it may be determined by the variable X alone.

The identification and reduction algorithm follows the above principle:

Generate all combinations of 1,2,3,4,5 variables from 6 variables
    For classes 1 → 4
        For all hyperboxes in that class
            For each combination
                Produce reduced dimension hyperbox by discarding variables
                    in the current combination
                Check for overlap of the reduced variable hyperboxes
                If no overlap
                    Mark combination as redundant

Table 1. Classification results for road scene data

Category	no in sequence	Misclassifications reduced var	Misclassifications full var
Country	91	11	9
Highway	75	7	21
Suburban	28	7	7
Urban	136	5	1
Country	6	0	0
Highway	64	56	63
Country	41	12	11
Suburban	21	15	15
Country	15	2	1
Total	480	115	133
Percentage correct		76	72
Percentage correct ignoring second highway		88	85

5   Results

5.1   Fisher Iris Data

The approach was initially applied to the classification of the Fisher Iris data [4] by training with the first 25 data from each class. This resulted in the production of just 5 rules with only 2 errors in the entire set. The above result cannot be repeated if the order of the input data is changed. Data pre ordered by minimum distance or minimum sum of squares distance, gives 3 errors and 5 rules, while random ordering increases the maximum number of errors to 6 (for 5 rules). This compares favourably with Abe and Lan's [5] system which produced 17 rules in order to reduce the total number of errors to two (and 5 rules to reduce the number of errors to 6). By removing the rules generated by all of the outlying points the number of errors totals 4, while the entire Iris set is now classified using only 2 variables and 3 rules.

5.2   The Road Scene Classification Data

The road scene data was gathered by recording a journey of about 32 minutes duration. The journey encompassed country, highway, urban and suburban environments. A frame in every 4 seconds were processed to extract the following: redbrick colour, grass colour, long horizontal line count, mean colour saturation, curve and curve disturbance. This resulted in 480 frame description vectors. Ten pictures were selected from each classification and hand labelled. The system was then trained using the associated training vectors and target classifications. Then the trained system produced classifications based upon the 480 sample frames. The classifications were associated with the original 480 pictures and an animated sequence was generated. The 480 frame "journey" travelled through

91 frames of "country", 75 frames of "highway", 28 frames of "suburban", 136 of "urban", 6 frames of "country", 64 frames of "highway", 41 frames of "country",21 frames of "suburban", 18 frames of "country". The same sequence was run on the fuzzy rule set produced without variable suppression and the resulting outputs are compared in Table 1.

The results were generally reliable except at road junctions, roundabouts and other points of lost road edge. Examples are shown in Fig.2. The second highway sequence produced poor results compared with the other sequences. This was due to a rapid adverse change in lighting conditions.

Fig. 2.

6   Conclusions

This paper has shown that a compact fuzzy rule set may be produced by considering each classification with respect to all other classifications. Comparisons between the variable reduced road scene recognition and the full variable version show slighly less errors generally in the reduced variable version. However, in the system described, ordering of the input data has an influential effect on the results. The technique is scalable, but complexity increases with the production of $1 \rightarrow$ n-1 subset combinations of n variables. Pre-ordering of the data by minimum sum of squares distance does not ensure the smallest number of errors, and this would be a basis for further work. The advantages of identifying redundant variables on an interclass basis are:

- 1. A better linguistic interpretation of the system.
- 2. Rules gain confidence as unnecessary dependencies on the redundant variables membership function are removed.
- 3. It identifies which variables are necessary for each class individually.
- 4. Sometimes variables are offered to intelligent systems without real knowledge of their worth. Worthless variables are identified.

Voting techniques have been used to resolve the problem of variable subsets. An example is [9] which allows different subsets of variables to vote for the candidate classification. The reduced variable subsets, produced by the algorithm presented in this paper, could be similarly post-processed. This will be the subject of further work.

References

1. Wang, L.X., Mendel J.M.: Generating fuzzy rules by learning from examples. IEEE Transactions on Systems, Man, and Cybernetics, vol. 28 **6** (1992) 1414–1427
2. Onisawa, T., Anzai, T.: Acquistion of Intelligible Fuzzy Rules Systems. Man, and Cybernetics, 1999. IEEE SMC '99 Conference Proceedings. 1999 IEEE International Conference on, vol. 5 (1999) 268–273
3. Simpson, P.K.: Fuzzy Min-Max Neural Networks-Part 1: Classification. Neural Networks, IEEE Transactions, Vol. 3, **5** (1992) 776–786
4. Fisher, R.A.: The use of multiple measurements in taxonomic problems. Annals of Eugenics **7** (1936) 179–188
5. Abe, S., Lan M S: A method for fuzzy Rules extraction directly from numerical data and its application to pattern classification. IEEE Transactions on fuzzy systems, vol. 1, **1** (1995) 18–28
6. Wilson, M., Dickson, S.: Poppet: A Robust Road Boundary Detection and Tracking Algorithm. Proceedings of the 10th British Machine Vision Conference (1999) 352–361
7. Wilson, M.: Exploitation of interclass variable redundancy for compact fuzzy rulesets. Proceedings of 2001 WSES International Conference on: Fuzzy Sets & Fuzzy Systems (FSFS '01), Tenerife (2001)
8. Andrews, R., Diederich, J., Tickle, A. B.: Survey and critique of techniques for extracting rules from trained artificial neural networks. Knowledge-Based Systems, Elsevier Science, vol. 1, **6** (1995) 373–389
9. Carpenter, G., Gjaja, M. N.,Gopal, S., Woodcock, C. E.: ART Neural Networks for Remote Sensing:Vegatation Classification from Lansat TM and Terrain Data. IEEE Transactions on Geoscience and Remote Sensing, vol.35, **2** (1997) 308–325

Unsupervised Learning of Part-Based Representations*

David Guillamet and Jordi Vitrià

Centre de Visió per Computador-Dept. Informàtica, Universitat Autònoma de
Barcelona, 08193 Bellaterra, Barcelona, Spain
Tel. +34 93-581 30 73 Fax. +34 93-581 16 70
{davidg, jordi}@cvc.uab.es,
http://www.cvc.uab.es/~davidg

Abstract. This article introduces a segmentation method to automatically extract object parts from a reduced set of images. Given a
database of objects and dividing all of them using local color histograms,
we obtain an object part as the conjunction of the most similar ones.
The similarity measure is obtained analyzing the behaviour of a local
vector with respect to the whole object database. Furthermore, the
proposed technique is able to associate an energy to each object part
being possible to find the most discriminant object parts. We present
the non-negative matrix factorization (NMF) technique to improve the
internal data representation by compacting the original local histograms
($50D$ instead of $512D$). Moreover, the NMF based projected histograms
only contain a few activated components and this fact improves the
clustering results with respect to the use of the original local color
histograms. We present a set of experimental results validating the use
of the NMF in conjunction with the clustering technique.

Keywords: object segmentation, part representation, non-negative matrix factorization

1   Introduction

It is not clear how the brain stores all the information about the objects that can
appear in an image (or scene). But, it is clear that this information is represented
using an intermediate structure between local features and whole objects [5].

In computer vision, several approaches have focused on the local extraction
of features. For example, Pope [4] models an object with a series of views, representing each view with a large variety of local features, and describing each
feature with a probabilistic model. Weber [9] uses the assumption that an object
class is a collection of objects which share visually similar features that occur in
similar spatial configurations. Ohba and Ikeuchi [3] extract some local features
from a set of representative keypoints of an object and uses all of them to create

---

* This work is supported by Comissionat per a Universitats i Recerca de la Generalitat
de Catalunya and Ministerio de Ciencia y Tecnología grant TIC2000-0399-C02-01.

W. Skarbek (Ed.): CAIP 2001, LNCS 2124, pp. 700–708, 2001.

an eigenspace where they are rejected according to their relative distances. These local approaches are used instead of global representations to overcome important problems like partial occlusions, local illumination changes, rotations and different scales. The most important global method is the Principal Component Analysis (PCA) [8].

In an intermediate level between local features and global object representations, Bierderman [1] introduced the Recognition By Components (RBC) model that postulates a few dozen generic shape parts, called *geons*, joined by categorical spatial relationships chosen from an equally small fixed set. Biederman claims that the only computational approach that can cope with all the problems of object processing is structural decomposition. The work of Shams [5] is based on learning the *geons* that compose a whole object without any previous knowledge about them. These intermediate techniques are based on decomposing an object in local structural shapes.

Our technique is based on the definition of object class mentioned by Weber [9], which defines an object class as a collection of objects which share characteristics features or parts that are visually similar and occur in similar spatial configurations. That is, an object is composed of different *parts* and from the conjunction of all the parts emerges the object entity. Weber [9] uses a local feature detector and by applying a clustering algorithm, he finds the most distinctive parts of objects. Our approach is based on the same definition of part but we use the behaviour of each local vector of an object with respect to the other local vectors of the database trying to find an object part containing similar behaviours. Furthermore, assuming this definition of part, we are able to classify the object parts according to its discriminant information with respect to the other parts of the object.

2   Object Clustering

Assuming that we have a database of different images that contain similar structures (i.e. newspapers), our main goal is to obtain the regions of all the images where the objects have a certain behaviour. Assuming that each instance of our database has a set of representative local vectors $V = \{v_i\}$, that in our particular case is a set of local color histograms, we define a measure of similarity between two normalized color histograms (v_i and v_j) [7] as $s(v_i, v_j) = \sum_1^k min(v_i^k, v_j^k)$.

Our object segmentation algorithm is based on finding a region of an object that has a similar behaviour with all the other objects of the database. This goal can be achieved by defining a similarity measure M_i^{kl} that reflects the highest similarity between the local vector i from object k and all the local vectors from object l.

$$M_i^{kl} = argmax\{s(v_i^k, v_j^l)\} \quad \forall v_j^l \in O_l \tag{1}$$

Thus, it has to be carried out an exhaustive search for the most similar local vector in object O_l to obtain this similarity measure. This expression was also used by Shams [5] to partition an object in local parts according to its similarity to an object database of similar objects. Denoting as $\mathbf{M_i^k}$ the vector that contains

Fig. 1. 3 different objects belonging to the same object family. It can be seen that these objects contain similar and different regions according to their color information.

all the l similarity measures M_i^{kl}, we obtain a vector that reflects how a local vector of an object i can be found in the rest of the database. It is expected that if the database is composed of objects that contain similar regions, this similarity vector will have a certain behaviour that would be different from the other zones of the image. On this expectation, we calculate the correlations:

$$R_{ij}^k = \frac{\mathbf{M_i^k}^\mathbf{T}\mathbf{M_j^k}}{\| \mathbf{M_i^k} \| \cdot \| \mathbf{M_j^k} \|} \qquad (2)$$

This matrix reflects how all the local responses of an object k are correlated between them. If we assume that our object database is composed of objects containing similar regions, we clearly assume that this matrix will contain two different clusters: the similar regions of an object and the other regions. Given this correlation matrix, we use the algorithm developed by Shi and Malik [6] to obtain two or more clusters.

As example, let us assume that we have a database of 3 objects, as shown in figure (1), of the same object family but containing different color regions. Applying the explained clustering algorithm by using local color histograms, we obtain 2 clusters (common parts and different parts) as shown in figure (2).

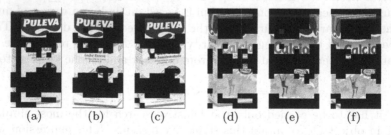

Fig. 2. (a), (b) and (c) are the common regions extracted from the three objects of figure (1). Different regions can be seen in (d), (e) and (f).

As seen in figure (2), the object parts are clearly separated according to their behaviour with respect to the other local object vectors. It has to be mentioned that the two obtained clusters are calculated in a totally unsupervised way given that the clustering algorithm finds the optimal partition. Thus, this clustering algorithm is useful when we are working in unsupervised environments.

Once we have obtained a partition of the original object into two clusters, we can iterate this algorithm until a measure of similarity between the local vectors of a given part is activated (the NCut value in the case of the algorithm of Shi and Malik [6]). Assuming that we have divided an object into N different parts, we can know the discrimation of each object part with respect to the whole object database. Using expression (2) we have partitioned an object into N parts by examining the correlation between all the local vectors defined in expression (1) that codify the similarities between all the object database. Thus, we can extract a measure of discrimination of each object part by considering the energy of its vectors M as:

$$E^k_{\text{part}} = \frac{1}{P} \sum_{i=1}^{P} \| M^k_i \| \quad \forall M^k_i \in \text{Part}(O_k) \tag{3}$$

and P indicates the number of local vectors in object part $\text{Part}(O_k)$. By sorting the local regions according to its energy (E^k_{part}), we can obtain the most discriminant local part of an object with respect to the whole object database.

3   Non-negative Matrix Factorization

The previous clustering algorithm is based on partitioning the local vectors of an object in two clusters according to their similarities with the other local object vectors of the database. An iterative version of the previous algorithm would obtain a hierarchy of local parts that we can sort according to the energy of vectors M (see expression (3)). Although this clustering algorithm works, it can fail to segment very similar regions given that is unsupervised. A sparse coding of the input vectors would improve some ambiguities of the actual results (similar regions can be expressed by different basis). A non-negative matrix factorization of the whole original space (color histograms) can solve this problem because it will try to represent the whole original space with a few basis and the projections of the original local vectors with respect to this set of basis will configure an approximated sparse code that will be more easy to clusterize.

3.1   Non-negative Matrix Factorization

The non-negative matrix factorization (NMF) is a method to obtain a representation of a certain data only using non-negativity constraints. These constraints lead to a part-based representation because they allow only additive, not subtractive, combinations of the original data [2].

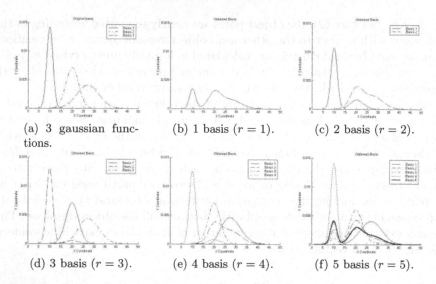

(a) 3 gaussian func-   (b) 1 basis ($r = 1$).   (c) 2 basis ($r = 2$).
tions.

(d) 3 basis ($r = 3$).   (e) 4 basis ($r = 4$).   (f) 5 basis ($r = 5$).

Fig. 3. (a) 3 original basis functions used to make up the set of 10 training vectors used in our experiment. (b) to (f) are the different basis functions obtained after applying the NMF method with different configurations of the parameter r. If we expect to find a small amount of basis ($r = 1, 2$), the obtained basis are combinations of the global behaviours of the sample data. If we only want 3 basis functions ($r = 3$), the obtained basis are exactly the same as the original ones (a). Furthermore, if we want to extract a large amount of basis functions ($r = 4, 5$), the extracted basis functions are combinations of the specific behaviours of the sample data.

The NMF method is based on representing the object database with an $n \times m$ matrix V where each column is a n non-negative local vector belonging to the original database (m local vectors). Using two new matrices (W and H) we can obtain an approximation of the whole object database (V) as $V_{i\mu} \approx (WH)_{i\mu} = \sum_{a=1}^{r} W_{ia}H_{a\mu}$. The dimensions of the matrix factors W and H are $n \times r$ and $r \times m$, respectively. Each column of matrix W contains a basis vector and each column of matrix H is an encoding that corresponds to a one-to-one correspondence with an original local vector in V. Using the PCA technique, each column of matrix W would be named as eigenimage and the matrix factors of H would be named eigenprojections. Furthermore, PCA constraints the columns of W to be orthonormal and the rows of H to be orthogonal to each other ([2]). In contrast to PCA, NMF does not allow negative entries in the matrix factors W and H. These non-negativity constraints permit the combination of multiple basis images to represent an object.

The implementation of NMF is based on an iterative update rule for matrices W and H. These update rules are given by [2]:

$$W_{ia} \leftarrow W_{ia} \sum_{\mu} \frac{V_{i\mu}}{(WH)_{i\mu}} H_{a\mu}, \quad W_{ia} \leftarrow \frac{W_{ia}}{\sum_j W_{ja}}, \quad H_{a\mu} \leftarrow H_{a\mu} \sum_i W_{ia} \frac{V_{i\mu}}{(WH)_{i\mu}} \quad (4)$$

We start with positive random initial conditions for matrices W and H and we define an objective function $F = \sum_{i=1}^{n} \sum_{\mu=1}^{m} [V_{i\mu} \log(WH)_{i\mu} - (WH)_{i\mu}]$ that must be maximised. By this way, the iterative update rules will converge to a local maximum of this objective function. As can be noted, this iterative process is easy to implement and is not time consuming given that is based on simple operations. See ([2]) for more information about NMF and its implementation.

Fig. 4. Set of 20 color newspapers (360×549) containing a large variety of regions.

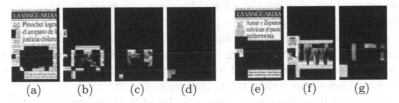

(a)   (b)   (c)   (d)   (e)   (f)   (g)

Fig. 5. The first four images are the four object parts that belong to one newspaper being sorted according to their energies: (a) Is the most common object part (newspaper title and all the text) and the other object parts corresponds to graphic zones of the object. (d) Shows that the most discriminant object part of this object is an orange region that does not appear in the rest of the object database. The last three images belong to another newspaper.

3.2   Visual Example Using NMF

We have generated a set of synthetic vectors to test the availability of NMF to obtain the basis functions used to make up the training set. Using 3 gaussian basis functions that can be seen in figure (3.a), we have created a set of 10 samples by weighting the original basis.

From each generated vector we have extracted a $1000D$ histogram by dividing the X axis in 1000 partitions. Thus, each bin of this histogram contains the value of the corresponding Y axis. Applying the NMF algorithm over this set of 10 histograms and expecting to find a controlled number of sparse basis, the results that we have found are shown in figure (3.b - 3.f). As seen in figure (3), if we extract $r = 3$ basis functions from our sample data, we obtain the original functions that were used to make up the training set. This result is interesting because in a real application, even if we do not know the number of original basis used to create the data we are dealing with, we can find an approximation of the local behaviour of the sample data.

4   Experimental Results

The main aim of this paper is to demonstrate that the above mentioned clustering algorithm extracts local parts of objects in an unsupervised way. To reflect this objective, we have selected the use of local color histograms instead of other representations (i.e. gaussian derivatives) because the results are more visual understandable. Nevertheless, other representations can be chosen depending on the problem nature.

To test our clustering algorithm, we have selected a set of 20 color newspapers containing a large variety of information (text and graphics). These newspapers can be seen in figure (4).

Applying the previous explained clustering algorithm, we have extracted a set of regions that have different behaviours according to expression (1). Sorting these regions with respect to their energy (as noted by expression (3)), we can obtain a sorted list of discriminant local regions. An example of two newspapers can be seen in figure (5).

From all the sorted regions obtained using this clustering algorithm, we only show two object divisions in figure (5) that are representative enough to demonstrate that the algorithm can find a local division of a global object and can sort all the local regions according to their energy.

Using the NMF technique, we have calculated 50 basis histograms (see matrix W in expression (4)) and we have projected all the local histograms to this reduced space (the original space is $512D$). The basis histograms obtained after 1000 iterations is shown in figure (6). It can be seen that the predominance of white regions in the input database implies a large amount of local basis containing white regions.

Using the projections of the original local histograms as the data input of the previous explained clustering algorithm, we have obtained nearly the same results, but with specific differences. For example, figure (7) shows the obtained object parts for the objects shown in figure (5).

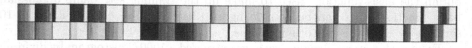

Fig. 6. 50 local basis histograms obtained after 1000 iterations of the NMF technique.

As seen in figure (7), we can check that the green region is divided into different levels of green (light, dark) and this behaviour is not manifested when

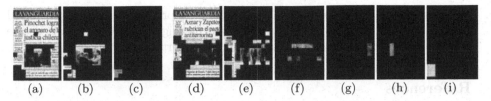

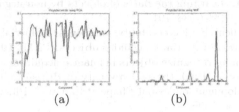

Fig. 7. Different object parts obtained by projecting the database to the basis found using the NMF technique. These parts can be compared with the ones in fig. (5).

(a)   (b)

Fig. 8. (a) is a vector projection using the well-known PCA technique, (b) is the same vector but being projected using the NMF technique.

we do not use the NMF technique. This is due because these specific tonalities are represented with different basis (see figure (6)). The two obtained results are perceptually good enough to consider this object part segmentation method as a good one. Nevertheless, using the NMF technique we can improve the clustering algorithm for two main reasons: (i) We work with $50D$ vectors instead of $512D$ local histograms; (ii) Only a reduced number of the components of each projected vector are activated. This independency between the components of the projected vectors can be seen in figure (8) where we compare a projected vector using the NMF technique and the same vector but projected using the well-known PCA technique. From figure (8) we can check that the obtained projected local vector using the NMF technique contain only a few activated components with respect to the PCA projection.

5 Conclusions

In this paper we have presented an unsupervised method to automatically extract local object parts according to the similarities of each part with respect to the whole object database. Furthermore, each obtained object part has an associated internal energy that can be used for a further discriminant analysis.

We have divided each object using local histograms and we have obtained a part based representation by means of these local vectors. Each object part is extracted by applying the explained clustering algorithm. Using the Non-negative Matrix Factorization (NMF), we have created a set of basis histograms that are used to project the original database. The projected vectors (that are positive defined) are also clusterized using the proposed clustering algorithm and the results are nearly the same. Nevertheless, we propose to use the projected vectors because they are more compact than the original ones and nearly sparse (independent components) and can be used in a further analysis.

References

1. Biederman, I.: Recognition-by-components: A theory of human understanding. Psychological Review **94**(2) (1987) 115–147
2. Lee, D., Seung, H.: Learning the parts of objects by non-negative matrix factorization. Nature **401** (1999) 788–791
3. Ohba, K.: and Ikeuchi, K.: Detectability, uniqueness and reliability of eigen windows for stable verification of partially occluded objects. PAMI **19**(9) (1997) 1043–1048
4. Pope, A.: Learning to recognize objects in images: acquiring and using probabilistic models of appearance. PhD. thesis, University of British Columbia (1995)
5. Shams, L.: Development of Visual Shape Primitives. PhD. thesis. University of Southern California (1999)
6. Shi, J., Malik, J.: Normalized cuts and image segmentation. IEEE Transaction on Pattern Analysis and Machine Intelligence **22**(8) (2000) 888–905
7. Swain, M., Ballard, D.: Color indexing. IJCV **7**(1) (1991) 11–32
8. Turk, M., Pentland, A.: Eigenfaces for recognition. Journal of Neuroscience **3**(1) (1991) 71–86
9. M. Weber, M., Welling, M., Perona, P.: Unsupervised learning of models for recognition. Proc. of 6th European Conference of Computer Vision (2000)

A Comparative Study of Performance and Implementation of Some Area-Based Stereo Algorithms

Bogusław Cyganek and Jan Borgosz

University of Mining and Metallurgy
Department of Electronics
Al. Mickiewicza 30, 30-059 Kraków, Poland
{cyganek,borgosz}@uci.agh.edu.pl

Abstract. The paper presents a comparison of a practical implementation of some area-based stereo algorithms. There are many stereo algorithms known that employ area based processing for matching of two or more images. However, much less information is available on practical implementations and applications of such methods, as well as their advantages and limitation. The work has been done to fill this gap and facilitate choice of the right stereo algorithm for machine vision applications, often using off-the-shelf cameras

Keywords: stereo matching, depth recovery, area-based stereo

1 Introduction

The purpose of this work is to report some results and conclusions on practical implementation and the use of area-based stereo processing algorithms for machine vision applications. Theoretical background of processing of two images for depth recovery can be found in [5][10]. The most troublesome part of the process is to find corresponding image points. This is mainly due to non-uniqueness of the matching problem, as well as image noise and distortions, or object occlusions. There are many variations of stereo methods based on choice of primary features for matching. One category, called area-based matching, relies on intensity comparison of some areas in local vicinity of potentially corresponding pixels.

The real problem encountered by many researchers and developers is when decision should be made on a sort of a stereo algorithm for a given application. Very little information is available on practical implementations of stereo methods with all implications [9]. Since the authors of this paper were in such a situation, the intention of this paper is to present our approach and results of software implementation of some of the area-based stereo algorithms.

W. Skarbek (Ed.): CAIP 2001, LNCS 2124, pp. 709–716, 2001.

2 Area-Based Stereo

Throughout rest of this paper we assume standard stereo setup, so epipolar lines follow horizontal scan lines of a stereo pair [5][8][10]. No other camera calibration is assumed, although if used for precise depth recovery, the cameras calibration parameters have to be known. The classical approach for area-based stereo is to choose size and shape for a region to be compared and then, using some metric, find the regions that give the closest match. If such an operation is performed for each pixel of the reference image, and potentially matching pixels from the second image, then a dense disparity map is obtained. Some of the commonly used metrics [1][3] are presented in Table 1. The two compared areas are matched in the sense of a measure if its value takes on an extreme. In all cases except, *SCP*, *SCP-N*, and *CoVar*, it is a minimum. Special precautions have to be undertaken when using *SSDN*, *ZSSD-V*, *SCP-N* and *CoVar* measures for area with zero intensity. In all such cases, denominators of related formulas take on zero, causing overflow in outcome values. This is especially frequent for black spots in an image or very small matching areas. The special attention should be devoted to the *Census* and *Rank* measures, that belong to the non-parametric methods of image processing. Introduced by Zabih and Woodfill [13], have been investigated by Woodfill and Von Herzen [12] for real time applications, as well as by Dunn and Corke [4]. They outstanding abilities for signal transformation to facilitate the input layer of a neural network have been reported by Cyganek [2].

Table 1. Most popular measures for area comparison in stereo algorithms. *SAD* – Sum of Absolute Differences, *ZSAD* – zero mean *SAD*, *SSD* – Sum of Squared Differences, *ZSSD* – zero mean *SSD*, *SSD-N* – *SSD* Normalized, *SCP* – Sum of Cross Products, *SCP-N* – Sum of Cross Products Normalized, *CoVar* – Covariance-Variance

Name	Formula
SAD	$\sum_{(i,j)\,U} \left\| I_1(x+i,y+j) - I_2(x+d_x+i,y+d_y+j) \right\|$
ZSAD	$\sum_{(i,j)\,U} \left\| \left(I_1(x+i,y+j) - \overline{I_1(x,y)}\right) - \left(I_2(x+d_x+i,y+d_y+j) - \overline{I_2(x+d_x,y+d_y)}\right) \right\|$
SSD	$\sum_{(i,j)\,U} \left(I_1(x+i,y+j) - I_2(x+d_x+i,y+d_y+j) \right)^2$
ZSSD	$\sum_{(i,j)\,U} \left[\left(I_1(x+i,y+j) - \overline{I_1(x,y)}\right) - \left(I_2(x+d_x+i,y+d_y+j) - \overline{I_2(x+d_x,y+d_y)}\right) \right]^2$
SSD-N	$\dfrac{\sum\limits_{(i,j)\,U} \left[I_1(x+i,y+j) - I_2(x+d_x+i,y+d_y+j) \right]^2}{\sqrt{\sum\limits_{(i,j)\,U} I_1(x+i,y+j)^2 \quad \sum\limits_{(i,j)\,U} I_2(x+d_x+i,y+d_y+j)^2}}$
ZSSD	$\dfrac{\sum\limits_{(i,j)\,U} \left[\left(I_1(x+i,y+j) - \overline{I_1(x,y)}\right) - \left(I_2(x+d_x+i,y+d_y+j) - \overline{I_2(x+d_x,y+d_y)}\right) \right]^2}{\sqrt{\sum\limits_{(i,j)\,U} \left(I_1(x+i,y+j) - \overline{I_1(x,y)}\right)^2 \quad \sum\limits_{(i,j)\,U} \left(I_2(x+d_x+i,y+d_y+j) - \overline{I_2(x+d_x,y+d_y)}\right)^2}}$

Name	Formula
SCP	$\displaystyle\bigotimes_{(i,j)\,U} I_1(x+i,y+j)\; I_2(x+d_x+i,y+d_y+j)$
SCP-N	$\displaystyle\frac{\bigotimes_{(i,j)\,U} I_1(x+i,y+j)\; I_2(x+d_x+i,y+d_y+j)}{\sqrt{\bigotimes_{(i,j)\,U} I_1(x+i,y+j)^2 \quad \bigotimes_{(i,j)\,U} I_2(x+d_x+i,y+d_y+j)^2}}$
CoVar	$\displaystyle\frac{\bigotimes_{(i,j)\,U}\left(I_1(x+i,y+j)-\overline{I_1(x,y)}\right)\left(I_2(x+d_x+i,y+d_y+j)-\overline{I_2(x+d_x,y+d_y)}\right)}{\sqrt{\bigotimes_{(i,j)\,U}\left(I_1(x+i,y+j)-\overline{I_1(x,y)}\right)^2 \quad \bigotimes_{(i,j)\,U}\left(I_2(x+d_x+i,y+d_y+j)-\overline{I_2(x+d_x,y+d_y)}\right)}}$
CENSUS	$\displaystyle\bigotimes_{(i,j)\,U} IC_1(x+i,y+j)\; IC_2(x+d_x+i,y+d_y+j)$

Where: $I_k(x,y)$, $\overline{I_k}(x,y)$, $IC_k(x,y)$ stand for intensity, mean intensity, and *Census* value for a k-*th* image at a point *(x,y)* in image coordinates, *i* and *j* are integer indices, d_x and d_y disparity values for *x* and *y* direction, respectively, *U* defines a set of points of a local neighborhood of a point at *(x,y)*, i.e. $U=\{(i,j){:}p(x+i,y+j)\ I_k\}$, finally, denotes a Hamming operator.

Census measure in our implementation has been computed as depicted in Fig. 1a.

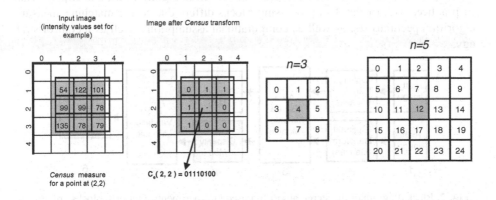

Fig. 1. a – Computation of the *Census* measure for a pixel at position (2,2). b – pixel numbering when computing *Census* measure

For a given central pixel at *(i,j)* and its *n-th* neighborhood, the corresponding *Census* measure *IC(i,j)* can be expressed as a series of bits:

$$IC(i, j) = b_{n^2-1} \ldots b_k \ldots b_3 b_2 b_1 b_0 \text{, where}$$

$$k \in \left[0, \ldots, n^2 - 1\right] / \left\{ \frac{n^2}{2} \right\}. \tag{1}$$

Bit-pixel numbering in a neighborhood depicts Fig. 1b. The b_k parameter can be expressed as follows:

$$b_k = \begin{array}{l} 1, \quad when \ \ I\left(i - \dfrac{n}{2} + \dfrac{k}{n}, j - \dfrac{n}{2} + k \bmod n\right) \ \ I(i, j) \\ 0 \quad otherwise \end{array} \tag{2}$$

Where: $I(i, j)$ stands for an intensity value of an image at a point (i,j) in image coordinates, k/n and k **mod** n denote integer and modulo divisions, respectively.

Rank measure returns number of surrounding pixels in a neighborhood that are in certain relation to the central pixel, e.g. their intensity is less than intensity of the central pixel. For both, *Rank* and *Census* measures, the outcome of their computation in an *n-th* pixel neighborhood is a stream of *n-1* bits (1). In practice, these measures become very useful for strongly noised or not equally lightened images.

3   Structure of an Area-Based Stereo Algorithm

Fig. 2 presents the overall block diagram of the stereo algorithm used in experiments. In practice, each of the data processing blocks differs due to the matching measure, algorithm peculiarities, as well as computational assumptions, such as a control strategy.

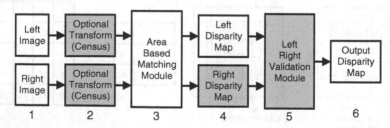

Fig. 2. Block diagram of the stereo algorithm used in experiments. Optional blocks are grayed

The purpose of the first module is to acquire a stereo pair at the same time and, possibly with the same acquisition parameters. Often used catadioptric systems can greatly help in this task. Second block is characteristic in the case of ordinal measures, such as *Census* or *Rank*, when new values have to be computed from intensity samples. The most important is the third module for the area-based matching process. Details of its implementation follows in the next section. The output of this module originates in so called left or right disparity map, based on choice of the left of right image as a reference for matching. The optional fifth step is devoted to the disparity map validation, in this case by a cross checking method [6]. Optional disparity map processing module can be placed after the fifth block for further disparity map improvements. This can be achieved e.g. by means of a non-linear filtering such as median [3].

4 Experimental Results

In our experiments most of the metrics presented in Table 1 have been employed in real algorithms. More than dozen of stereo pairs have been used for testing, four of which are presented in this paper, see Fig. 3.

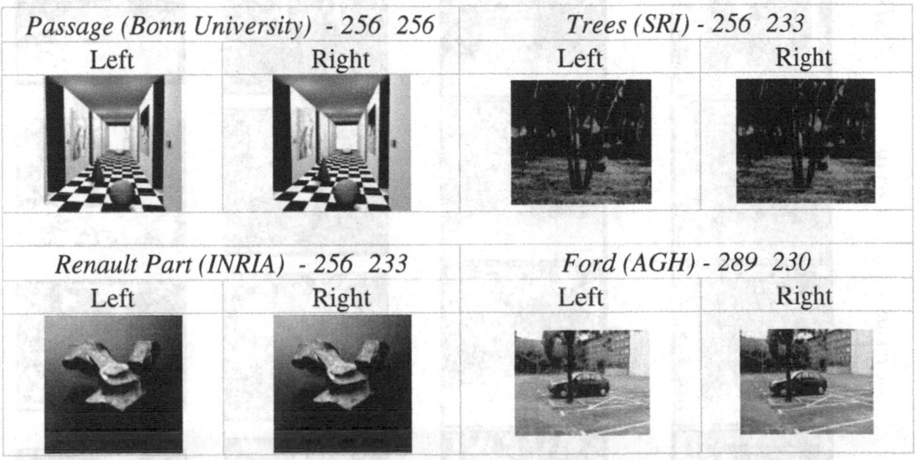

Fig. 3. Stereo pairs used in experiments

Depth maps for the aforementioned test images presents Fig. 4. All the disparity maps have been obtained by means of the methods described in Sect. 3.

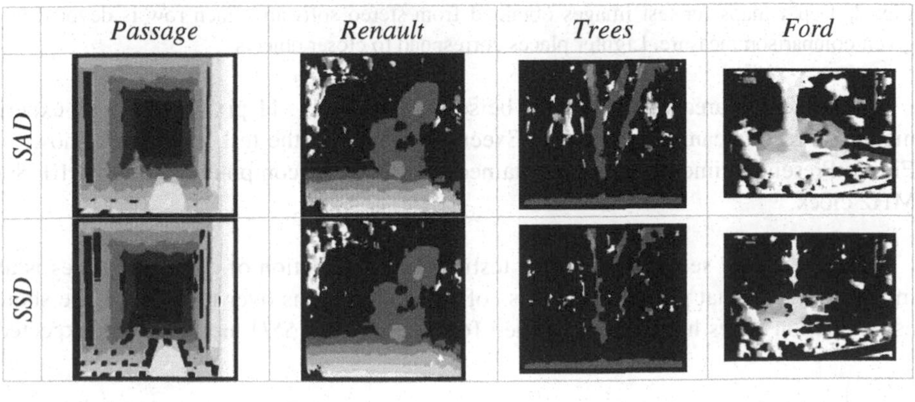

Fig. 4. Depth maps for test images obtained from stereo software. Each row is devoted to a given comparison measure. Lighter places correspond to closer objects

The matching area was chosen to be square of size 11 11 pixels during all experiments, so results can be compared. Execution times for the test images are shown in Fig. 5. Presented times have been obtained on IBM PC® computer, Pentium® III, 800 MHz clock.

It can be easily seen that for each testing pair distribution of execution times is almost identical, what proves robustness of some algorithms over the other. The shortest execution times have been obtained for the *SAD* and *SSD* measures. As expected,

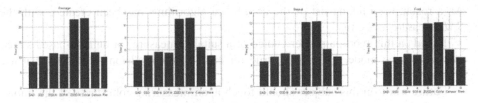

Fig. 5. Execution times for tested stereo pairs

the worst times are for the measures that for each matching area have to compute average intensity value in that area. This is a case of *ZSSD-N* and *CoVar* measures.

Based on the experimental results and another tests performed over stereo pairs the ranking list of algorithms has been created. Two factors have been taken into account: *a quality of an output disparity map* and *an execution time*. To assess the accuracy of each algorithm, subjective judgments have been made by comparing the computed disparity map with human perception. However, to gain more quantitative measure, additional disparity-checking has been performed by generating random points in an image and then comparing a computed disparity value at that place with disparity measured by a human. This method is similar to the one presented by Wei et.al. [11]. Results of this comparisons are available in Table 2.

Table 2. Area-based stereo algorithms classification

Place	Quality	Time	Place	Quality	Time
1	SAD	SAD	5	Rank	SSD-N
2	SSD-N	Rank	6	CoVar	Census
3	SCP-N	SSD	7	ZSSD-N	ZSSD-N
4	Census	SCP-N	8	SSD	CoVar

It can be surprising, but the best algorithm, due to its execution time and obtained quality, seems to be algorithm with the *SAD* measure employed. It is also very interesting that in the case of more sophisticated measures, such as *ZSSD-N* and *CoVar*, that subtract mean value for each matching area, not only performance time is very long, but also the quality of obtained disparity maps is rather questionable.

Having this in mind, it is not surprising that many commercially available stereo systems use area-based algorithm with *SAD* metrics [7].

5 Conclusions and Future Work

In this paper, we have presented our implementation and tests of the area-based stereo methods that employed broad range of image comparison measures. All of the mentioned measures and many stereo pairs have been tested to obtain reliable results. Four testing pairs and eight measures have been presented in this paper. For each test, execution time and quality of the output disparity map have been recorded and compared.

As a result, the ranking list of the area-based algorithms has been created, with the *SAD* measure at the top.

Based on this work, a proper choice of a suitable measure for an area-based stereo method can be made. This is also true for other machine vision algorithms that use such measures.

Future work on this subject should touch upon measuring robustness of the implemented algorithms against noise, image distortions and occlusions that are often detected in stereo images. This is especially important for stereo systems e.g. with simple cameras that are mounted in robots, navigating in a real environment. In such applications fast and reliable response from the stereo module is of the utmost importance.

We also plan to compare other stereo methods with area-based algorithms, presented in this paper.

References

1. Banks, J., Porter, R., Bennamoun, M., Corke, P.: A Generic Implementation Framework for Stereo Matching Algorithms. Technical Report, CSIRO Manufacturing Science and Technology, Australia (1997)
2. Cyganek, B.: Neural Networks Application to The Correlation-Based Stereo-Images Matching. Engineering Applications of Neural Networks, Proceedings of the 5th International Conference EANN '99, Warsaw, Poland (1999)
3. Cyganek, B.: Stereo images matching by means of the tensor representation of a simple neighborhood. PhD Thesis, University of Mining and Metallurgy, Poland (2000)
4. Dunn, P., Corke, P.: Real-time Stereopsis using FPGAs. CSIRO Division of Manufacturing Technology, Australia (1997)
5. Faugeras, O.: Three-dimensional computer vision. A geometric viewpoint. MIT (1993)
6. Fua, P.: A Parallel Stereo Algorithm that Produces Dense Depth Maps and Preserves Image Features. INRIA Technical Report No 1369 (1991)
7. Konolige, K.: Small Vision Systems: Hardware and Implementation. Artificial Intelligence Center, SRI International (1999)
8. Scharstein, D.: View Synthesis Using Stereo Vision. Lecture Notes in Computer Science 1583. Springer-Verlag, Berlin Heidelberg New York (1999)
9. Szeliski, R., Zabih, R.: An Experimental Comparison of Stereo Algorithms. Vision Algorithms: Theory and Practice. Lecture Notes in Computer Science 1883. Springer-Verlag, Berlin Heidelberg New York (2000)
10. Trucco, E., Verri, A.: Introductory Techniques for 3-D Computer Vision. Prentice-Hall (1998)
11. Wei, G.-Q., Brauer, W., Hirzinger, G.: Intensity- and Gradient-Based Stereo Matching Using Hierarchical Gaussian Basis Functions. IEEE Transactions on Pattern Analysis and Machine Intelligence, Vol. 20, No. 11 (1998) 1143-1160
12. Woodfill, J., Von Herzen, B.: Real-Time Stereo Vision on the PARTS Reconfigurable Computer. IEEE Symposium on FPGAs for Custom Computing Machines (1997)
13. Zabih, R., Woodfill, J.: Non-parametric Local Transforms for Computing Visual Correspondence. Computer Science Department, Cornell University, Ithaca (1998)

A New Autocalibration Algorithm: Experimental Evaluation

Andrea Fusiello*

Dipartimento di Informatica - Università degli Studi di Verona
Strada Le Grazie, 15 - 37134 Verona, Italy
fusiello@sci.univr.it

Abstract. A new autocalibration algorithm has been recently presented by Mendonça and Cipolla which is both simple and nearly globally convergent. Analysis of convergence is missing in the original article. This paper fills the gap, presenting an extensive experimental evaluation of the Mendonça and Cipolla algorithm, aimed at assessing both accuracy and sensitivity to initialization. Results show that its accuracy is fair, and – remarkably – it converges from almost everywhere. This is very significant, because most of the existing algorithms are either complicated or they need to be started very close to the solution.

Keywords: computer vision, self-calibration

The classical approach to *autocalibration* (or *self-calibration*), in the case of a single moving camera with constant but unknown intrinsic parameters and unknown motion, is based on the recovery of the intrinsic parameters by solving the Kruppa equations [1,2], which have been found to be very sensitive to noise [2]. Recently new methods based on the *stratification* approach have appeared, which upgrade a projective reconstruction to an Euclidean one without solving explicitly for the intrinsic parameters (see [3] for a review). An algorithm has been recently presented by Mendonça and Cipolla [4], which, like the Kruppa equations, is based on the direct recovery of intrinsic parameters, but it is simpler.

Apart from sensitivity to noise, the applicability of autocalibration techniques in the real world depends on the issue of initialization. Since a non-linear minimization is always required, convergence to the global minimum is guaranteed only if the algorithm is initialized in the proper basin of attraction. Unfortunately, this issue was not addressed by Mendonça and Cipolla.

This paper gives an account of the experimental evaluation of the Mendonça and Cipolla algorithm (in mine implementation), aimed at assessing its performances, especially the sensitivity to initialization. Results are quite interesting, as it turns out that the algorithm converges to the global minimum from *almost everywhere.*

* This article has been written while the author was a Visiting Research Fellow at the Department of Computing and Electrical Engineering - Heriot-Watt University, supported by EPSRC (Grant GR/M40844).

W. Skarbek (Ed.): CAIP 2001, LNCS 2124, pp. 717–724, 2001.

1 Notation and Basics

This section introduces the mathematical background on perspective projections necessary for our purposes.

A pinhole camera is modeled by its 3×4 *perspective projection matrix* (or simply *camera matrix*) $\tilde{\mathbf{P}}$, which can be decomposed into

$$\tilde{\mathbf{P}} = \mathbf{A}[\mathbf{R} \mid \mathbf{t}]. \tag{1}$$

The matrix $\mathbf{A}$ depends on the *intrinsic parameters*, and has the following form:

$$\mathbf{A} = \begin{bmatrix} \alpha_u & \gamma & u_0 \\ 0 & \alpha_v & v_0 \\ 0 & 0 & 1 \end{bmatrix}, \tag{2}$$

where α_u, α_v are the focal lengths in horizontal and vertical pixels, respectively, (u_0, v_0) are the coordinates of the *principal point*, given by the intersection of the optical axis with the retinal plane, and γ is the *skew* factor. The camera position and orientation (*extrinsic parameters*), are encoded by the 3×3 rotation matrix $\mathbf{R}$ and the translation $\mathbf{t}$.

Let $\tilde{\mathbf{w}} = [x, y, z, 1]^\top$ be the homogeneous coordinates of a 3D point in the world reference frame (fixed arbitrarily) and $\tilde{\mathbf{m}} = [u, v, 1]^\top$ the homogeneous coordinates of its projection onto the image. The transformation from $\tilde{\mathbf{w}}$ to $\tilde{\mathbf{m}}$ is given by

$$\kappa \tilde{\mathbf{m}} = \tilde{\mathbf{P}} \tilde{\mathbf{w}}, \tag{3}$$

where κ is a scale factor.

Let us consider the case of two cameras. A three-dimensional point $\mathbf{w}$ is projected onto both image planes, to points $\tilde{\mathbf{m}} = \tilde{\mathbf{P}} \tilde{\mathbf{w}}$ and $\tilde{\mathbf{m}}' = \tilde{\mathbf{P}}' \tilde{\mathbf{w}}$, which constitute a *conjugate pair*. It can be shown [5] that the following equation holds:

$$\tilde{\mathbf{m}}'^\top \mathbf{F} \tilde{\mathbf{m}} = 0, \tag{4}$$

where $\mathbf{F}$ is the *fundamental matrix*. The rank of $\mathbf{F}$ is in general two and, being defined up to a scale factor, it depends upon seven parameters. In the most general case, all the geometrical information that can be computed from pairs of images are encoded by the fundamental matrix. Its computation requires a minimum of eight conjugate points to obtain a unique solution [5]. It can be shown [5] that

$$\mathbf{F} = \mathbf{A}'^{-\top} \mathbf{E} \mathbf{A}^{-1}. \tag{5}$$

where $\mathbf{E}$ is the *essential matrix*, which can be obtained from conjugate pairs when intrinsic parameters are known. The essential matrix encodes the rigid transformation between the two cameras, and, being defined up to a scale factor, it depends upon five independent parameters: three for the rotation and two for the translation up to a scale factor. Unlike the fundamental matrix, the only property of which is being of rank two, the essential matrix is characterized by the following Theorem (see [6] for a proof).

Theorem 11 *A real matrix* $\mathbf{E}$ 3×3 *can be factorized as product of a nonzero skew-symmetric matrix and a rotation matrix if and only if* $\mathbf{E}$ *has two identical singular values and a zero singular value.*

2   Autocalibration

In many practical cases, the intrinsic parameters are unknown and the only information that can be extracted from a sequence are point correspondences, which allow to compute a set of fundamental matrices. *Autocalibration* consist in computing the intrinsic parameters, or – in general – Euclidean information, starting from fundamental matrices (or, equivalently, from point correspondences). In this section we will see which constraints are available for the autocalibration.

2.1   Two-Views Constraints

As we saw in Section 1, the epipolar geometry of two views is described by the fundamental matrix, which depends on seven parameters. Since the five parameters of the essential matrix are needed to describe the rigid displacement, at most two independent constraints are available for the computation of the intrinsic parameters from the fundamental matrix.

These two constraints come from the characterization of the essential matrix given by Theorem 11. Indeed, the condition that the matrix $\mathbf{E}$ has a zero singular value and two non-zero equal singular values is equivalent to the following conditions, found by Huang and Faugeras [7]:

$$\det(\mathbf{E}) = 0 \quad \text{and} \quad \mathbf{trace}((\mathbf{EE}^\top))^2 - 2\mathbf{trace}((\mathbf{EE}^\top)^2) = 0. \tag{6}$$

The first condition is automatically satisfied, since $\det(\mathbf{F}) = 0$, but the second condition can be decomposed [5] in two independent polynomial relations that are equivalent to the two equations found by Trivedi [8].

This is an algebraic interpretation of the so-called *rigidity constraint*, namely the fact that for any fundamental matrix $\mathbf{F}$ there exist two intrinsic parameters matrix $\mathbf{A}$ and $\mathbf{A}'$ and a rigid motion represented by $\mathbf{t}$ and $\mathbf{R}$ such that $\mathbf{F} = \mathbf{A}'^{-\top}([\mathbf{t}]_\wedge \mathbf{R})\mathbf{A}^{-1}$. By exploiting this constraint, Hartley [6] devised an algorithm to factorize the fundamental matrix that yields the five motion parameters and the two different focal lengths. He also pointed out that no more information could be extracted from the fundamental matrix without making additional assumptions (e.g. constant intrinsic parameters).

2.2   N-Views Constraints

The case of three views is not a straightforward generalization of the two-views case. The epipolar geometry can be described using the *canonical decomposition* [9] or the *trifocal tensor*, both of which use the minimal number of parameters, that turns out to be 18. The rigid displacement is described by 11 parameters:

6 for 2 rotations, 4 for two directions of translation and 1 ratio of translation norms. Therefore, in this case there are seven constraints available on the intrinsic parameters. If they are constant, three views are sufficient to recover all the five intrinsic parameters.

In the general case of n views, Luong demonstrated that at least $11n - 15$ parameters are needed to describe the epipolar geometry, using his canonical decomposition. The rigid displacement is described by $6n-7$ parameters: $3(n-1)$ for rotations, $2(n-1)$ for translations, and $n-2$ ratios of translation norms. There are, thus, $5n - 8$ constraints available for computing the intrinsic parameters. Let us suppose that n_k parameters are known and n_c parameters are constant. Every view apart from the first one introduces $5 - n_k - n_c$ unknowns; the first view introduces $5 - n_k$ unknowns, therefore the unknown intrinsic parameters can be computed provided that

$$5n - 8 \geq (n - 1)(5 - n_k - n_c) + 5 - n_k, \tag{7}$$

which is equivalent to the following equation reported in [10]:

$$nn_k + (n - 1)n_c \geq 8. \tag{8}$$

As pointed out in [9], the $n(n - 1)/2$ fundamental matrices are not independent, hence the $n(n - 1)$ constraints like (Eq. 6) that can be derived from them are not independent. Nevertheless they can be used for computing the intrinsic parameters, since redundancy improves stability, as mentioned in [4].

2.3   The Mendonça and Cipolla Algorithm

Mendonça and Cipolla method for autocalibration is based on the exploitation Theorem 11. A cost function is designed, which takes the intrinsic parameters as arguments, and the fundamental matrices as parameters, and returns a positive value proportional to the difference between the two non-zero singular value of the essential matrix. Let $\mathbf{F}_{ij}$ be the fundamental matrix relating views i and j, and let $\mathbf{A}_i$ and $\mathbf{A}_j$ be the respective intrinsic parameters matrices. Let ${}^1\sigma_{ij} > {}^2\sigma_{ij}$ be the non zero singular values of $\mathbf{E}_{ij} = \mathbf{A}_i^\top \mathbf{F}_{ij} \mathbf{A}_j$. The cost function is

$$C(\mathbf{A}_i\ i = 1\dots n) = \sum_{i=1}^{n} \sum_{j>n}^{n} w_{ij} \frac{{}^1\sigma_{ij} - {}^2\sigma_{ij}}{{}^2\sigma_{ij}}, \tag{9}$$

where w_{ij} are normalized weight factors.

3   Experiments

In these experiments, intrinsic parameters were kept constant, hence the following cost function was actually used:

$$C(\mathbf{A}) = \sum_{i=1}^{n} \sum_{j>n}^{n} w_{ij} \frac{{}^1\sigma_{ij} - {}^2\sigma_{ij}}{{}^1\sigma_{ij} + {}^2\sigma_{ij}} \tag{10}$$

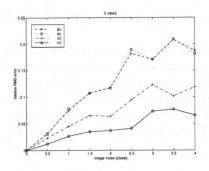 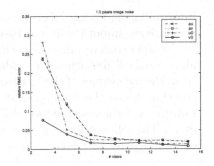

Fig. 1. Relative RMS error on intrinsic parameters versus image noise standard deviation (left) and number of views (right).

As customary it was assumed $\gamma = 0$. The weight w_{ij} was the residual of the estimation of $\mathbf{F}_{ij}$, as suggested by [4]. The minimum number of views required to achieve autocalibration in this case is three, according to (8). Fundamental matrices were computed using the linear 8-point algorithm with data normalization.

The algorithm was tested on synthetic data, which consisted of 50 points randomly scattered in a sphere of radius 1 unit, centered at the origin. Random views were generated by placing cameras at random positions, at a mean distance from the centre of 2.5 units with a standard deviation of 0.25 units. The orientations of the cameras were chosen randomly with the constraint that the optical axis should point towards the centre. The intrinsic parameters were given a known value: $\alpha_u = \alpha_v = 800, u_0 = v_0 = 256$. Image points were (roughly) contained in a 512x512 image.

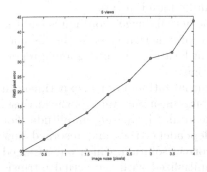 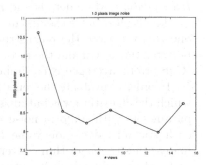

Fig. 2. Reconstruction residual RMS pixel error versus image noise standard deviation (left) and number of views (right).

I used the Nelder-Meads simplex method (implemented in the `fmins` function of MATLAB), to minimise the cost function. This methods does not use gradient information and is less efficient than Newton methods, particularly if the function is relatively smooth as in our case.

In order to determine the accuracy of the algorithm, Gaussian noise with variable standard deviation was added to image points. The algorithm was started from the true values of the intrinsic parameters and always converged to a nearby solution. Since the fundamental matrices are affected by the image noise, the minimum of the cost function does not coincide with the actual intrinsic parameters. The relative RMS error is reported in Figure 1. Each point is the average of 70 independent trials.

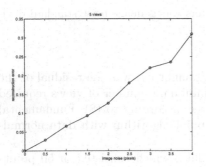 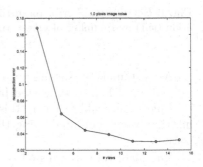

Fig. 3. Reconstruction RMS error versus image noise standard deviation (left) and number of views (right).

Using the intrinsic parameters computed by autocalibration, and the fundamental matrices, structure was recovered by first factorizing out the motion from the essential matrices, then recovering the projection matrices and finally computing 3-D structure from projection matrices and point correspondences by *triangulation*. More details and references can be found in [11].

The *pixel error* is the distance between the actual image coordinates and the ones derived from the reconstruction. The reconstruction error is the distance between the actual and the reconstructed point locations. Figures 2 and 3 report RMS errors, averaged over 70 independent trials.

In order to evaluate the sensitivity to the initialization, I ran an experiment in which the algorithm was initialized by perturbing the actual value of the intrinsic parameters with uniform noise with zero mean and increasing amplitude. For each standard deviation value I ran 70 independent trials and recorded how many times the algorithm converged to the correct solution, which was assumed to be the one to which it converged when initialized with the actual intrinsic parameters. Perturbation was obtained with the following formula (in MATLAB syntax):

```
a0 = a_true + pert * a_true.*(rand(1,4)-0.5)
```

where a0 is the initialization, a_true is a vector containing the true intrinsic parameters, and **pert** is a value ranging from 0 to 10 (corresponding to 1000%!).

Figure 4(a) shows the result with 5 views and 1.0 pixels image noise. The same experiment with 15 views yielded very similar result, not shown here. In another experiments I used only positive uniform noise:

$$a0 = a\_true + pert * a\_true.*(rand(1,4))$$

and the results are shown in Fig 4(b).

Finally, the algorithm was initialized with a random point in the 4D cube $[0, 2000] \times [0, 2000] \times [0, 2000] \times [0, 2000]$ and it converged in the 86% of cases, with 5 views and 1.0 pixels image noise.

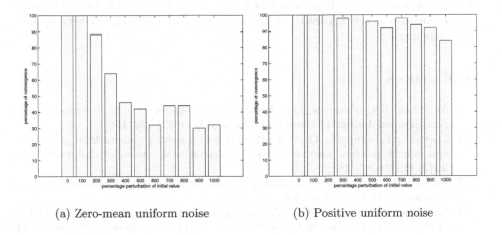

(a) Zero-mean uniform noise   (b) Positive uniform noise

Fig. 4. Percentage of convergence vs initial value perturbation (percentage) for 5 views.

On the average, it takes 5 seconds of CPU time on a Sun Ultra 10 running MATLAB to compute intrinsic parameters with 5 views.

The MATLAB code used in the experiments is available on the web from http://www.sci.univr.it/~fusiello/demo/mc.

4   Discussion

Intrinsic parameters are recovered with fair, but not excellent, accuracy. The error consistently increases with image noise and decreases with the number of views. With 1.0 pixel noise no appreciable improvement is gained by using more than seven views, but this number is expected to increase with the noise. It is not advisable to use the minimum number of views (three).

As for the reconstruction, the residual pixel error depends only on the image noise and not sensibly on the number of views (excluding the three views case). The reconstruction error, consistently decreases with the number of views. With 5 views and image noise of 1.0 pixel, the accuracy is about 30%. This figure depends only partially on the computation of the intrinsic parameters. It also depends on the recovery of motion parameters and on the triangulation. In both cases linear algorithm were used. Improvements can be expected by using non-linear refinement.

The algorithm shows excellent convergence properties. Remarkably, even when true values are perturbed with a relative error of 200% convergence is achieved in the 90% of the cases (Figure 4(a)). Results suggest that failure occurs when the sign of the parameters is changed. Indeed, figures improve dramatically when perturbation is a positive uniform random variable: in this case the algorithm converges from almost everywhere (Figure 4(b)).

In summary, the algorithm is fast and converges in a wide basin, but accuracy is not its best feature. If accuracy is a concern, it is advisable to run a bundle adjustment, which is known to be the most accurate method, but very sensitive to initialization.

References

1. Maybank, S.J., Faugeras, O.: A theory of self-calibration of a moving camera. International Journal of Computer Vision **8**(2) (1992) 123–151
2. Luong, Q.T., Faugeras, O.: Self-calibration of a moving camera from point correspondences and fundamental matrices. International Journal of Computer Vision **22**(3) (1997) 261–289
3. Fusiello, A.: Uncalibrated Euclidean reconstruction: A review. Image and Vision Computing **18**(6-7) (May 2000) 555–563
4. Mendonça, P.R.S., Cipolla, R.: A simple technique for self-calibration. Proceedings of the IEEE Conference on Computer Vision and Pattern Recognition (1999) 500–505
5. Luong, Q.T., Faugeras, O.D.: The fundamental matrix: Theory, algorithms, and stability analysis. International Journal of Computer Vision **17** (1996)
6. Hartley, R.I.: Estimation of relative camera position for uncalibrated cameras. Proceedings of the European Conference on Computer Vision. Santa Margherita L. (1992) 579–587
7. Huang, T.S., Faugeras, O.D.: Some properties of the E matrix in two-view motion estimation. IEEE Transactions on Pattern Analysis and Machine Intelligence **11**(12) (Dec 1989) 1310–1312
8. Trivedi, H.P.: Can multiple views make up for lack of camera registration? Image and Vision Computing, **6**(1) (1988) 29–32
9. Luong, Q.-T., Viéville, T.: Canonical representations for the geometries of multiple projective views. Computer Vision and Image Understanding **64**(2) (1996) 193–229
10. Pollefeys, M., Koch, R., Van Gool, L.: Self-calibration and metric reconstruction in spite of varying and unknown internal camera parameters. Proceedings of the IEEE International Conference on Computer Vision. Bombay (1998) 90–95
11. Fusiello, A.: The Mendonça and Cipolla self-calibration algorithm: Experimental evaluation. Research Memorandum RM/99/12. Department of Computing and Electrical Engineering, Heriot-Watt University, Edinburgh, UK (1999). Available at ftp://ftp.sci.univr.it/pub/Papers/Fusiello/RM-99-12.ps.gz

An Iconic Classification Scheme for Video-Based Traffic Sensor Tasks

Włodzimierz Kasprzak*

Inst. of Control and Computation Eng., Warsaw Univ. of Technology
ul. Nowowiejska 15-19, PL-00-665 WARSZAWA
W.Kasprzak@ia.pw.edu.pl

Abstract. An application-oriented vision-based traffic scene sensor system is designed. Its most important vision modules are identified and their algorithms are described in details: the on-line auto-calibration modules and three optional modules for 2–D measurement tasks (i.e. queue length detection, license plate identification and vehicle classification). It is shown that all three tasks may be regarded as applications of an iconic image classification scheme. Such a general scheme is developed and it can be applied for the above mentioned tasks by exchanging the application-dependent modules for pre-segmentation and feature extraction. The practical background of described work constitutes the IST project OMNI, dealing with the development of a network-wide intersection-driven model that can take advantage from the existence of advanced sensors, i.e. video sensors and vehicles equipped with GPS/GSM.

Keywords: computer vision, traffic sensor system

1 Introduction

The current solutions in the field of road traffic management are reaching their limits: new solutions are needed to further improvement of the performance of our road networks [2,4,9]. The current situation in traffic network technology is that it does not present standard interfacing for new applications or devices. Cities that already have traffic control systems find problems in the integration of new technology (surveillance applications, advanced sensors, real-time traffic information in WWW, fleet management) from different vendors. Therefore, the IST-project OMNI [5,3] concentrates its efforts on the development of a network-wide intersection-driven model that includes information flow from/to advanced sensors, i.e. video sensors [10,3,8] and vehicles equipped with GPS and GSM.

In the context of road traffic control, in this work a vision–based sensor system is developed (sec. 2). The main vision modules are described: the auto-calibration task (sec. 3) and three 2-D measurement tasks, which can be seen as applications of a general iconic classification scheme (sec. 4), differing only by the use of application-specific feature extraction methods (sec. 5).

* This work was supported by the EC project IST–11250: "Open Model For Network-Wide Heterogeneous Intersection-Based Transport Management (OMNI)".

W. Skarbek (Ed.): CAIP 2001, LNCS 2124, pp. 725–732, 2001.
© Springer-Verlag Berlin Heidelberg 2001

 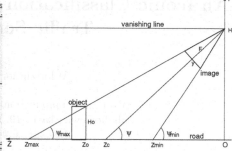

Fig. 1. Using the GUI module for user-defined road environment [3].

Fig. 2. The on-line computation of current viewing direction ((known focal length F, location over the road plane H).

2   The Vision-Based Traffic Sensor System

The proposed vision sensor system posseses a modular architecture, where each module is responsible for some (basic) control- or (optional) image analysis-functions. There are two basic modules and three (mostly obligatory) vision modules. The basic obligatory modules are: the *control module* which provides a GUI interface and control functions for the overall system (Fig. 1), and the *communication module* for the integration with the traffic information system. One particular vision module is usually obligatory - the *auto-calibration* of the vision system [12,7]. Other vision modules are optional and they can be classified by the type of delivered measurements as belonging to *punctual (1-D)*, *2-D* or *2-D&T* – measurements.

2.1   Punctual Measurements

This type of measurements means the detection of 1-D signals - due to a repeated analysis of single images. As an example the traffic flow detection module (module 2) can be defined. The task is for each predefined lane of the road to detect and to estimate the average traffic flow, i.e. the number of vehicles per minute. Usually the user pre-defines a rectangular area in the image, where vehicles should be counted. A popular method assumes, that a vehicle is detected if a histogram performed along a scan line in predefined area takes a characteristic shape [9].

2.2   Two-Dimensional Measurements

The provided information entities are related to locations in the image. Again a repeated analysis of only single images (independently of each other) is performed. Three modules of this type are considered.

Queue length detection (by space occupancy detection) – for each predefined lane of the road the current queue length and/or number of cars is detected ([3]).

Classification of vehicles – the front part of the vehicle in the lane can be detected or the overall sizes of the vehicle can be estimated and on the base of this measurements the detected vehicle can be classified into: small car, lorry, bus, van. [2,8].

Vehicle identification by license plate recognition – its task is to extract the license plate section of a previously detected vehicle and to identify the plate's symbols [2,11].

2.3 2-D and T-Measurements

This includes the average speed detection of vehicles in each lane, obtained due to a recursive analysis of a sequence of images in time. Particularly, a speed detection module can be considered Its task is for each road lane to estimate the average speed of moving vehicles [Fer94, Kas00]. In general, the shape of a vehicle may be modelled as a combination of two boxes, with equal width. For speed estimation it is sufficient to track the on-ground center point of the front/back area, which corresponds to the side face of lower model box . For the object tracking process in infinite image sequences usually a recursive estimator is required (for example an extended Kalman filter (EKF) or a tracking-error minimization filter can be used for this purpose).

3 Functional Description of the Auto-Calibration Module

Most of proposed image analysis methods depend critically on having an accurately calibrated camera. For a pin-hole camera model 6 geometric parameters (the position and orientation of the camera) and 4 intrinsic parameters (the focal length of the projection, the pixel size and the principal point in the image) are required. The intrinsic parameters (except focal length, which may be changed) are usually calibrated before the analysis system starts its operation, whereas the 6 geometric parameters should be modified on-line, in accordance with instantaneous, real position and orientation of the camera.

Our auto-calibration procedure for traffic scene analysis assumes for simplicity (Fig. 2), that the height H over the road plane is known and it remains constant. In this way the camera direction angles have to be on-line computed. The auto-calibration procedure (inspired by [12,7]) consists of the following steps:

1. It starts with edge detection, followed by linear segment detection.
2. If (nearly) vertically elongated image lines correspond to road markings then they induce the set of vanishing - lines (VP-lines), from which the road's vanishing point in the image plane is detected (Fig. 3(a)).
3. From the VP point the two camera orientation angles (alpha, beta) can be computed.
4. Then we set the focal length to some default value and made a back–projection of VP–lines onto the road's plane (Fig. 3(b)).

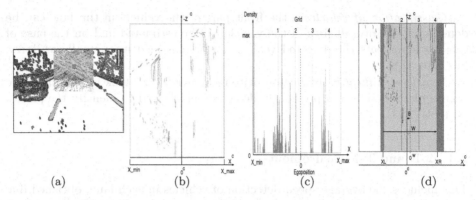

(a)   (b)   (c)   (d)

Fig. 3. Results of the auto-calibration procedure: (a) VP-detection due to density search in the VP-line image, (b) top view of VP-line segments back-projected to the road plane, (c) fitting a road grid model to density peaks along the road's cross section, (d) top view of the computed 3–lane grid (and ego-position).

5. By projecting the above lines onto the cross line (parallel to the OX axis), with weights corresponding to the distance from the observer, we obtain a histogram of road markings.
6. In the road histogram a known 1–D reference model - a grid corresponding to road lanes is placed. By changing the scale and position of this grid (expected focal length, camera position relative to road's center line) a best fit of this grid to the histogram image of VP-lines is searched for (Fig. 3(c)).
7. From the best fit of known grid with the densities in the histogram image, current focal length and camera positions are obtained immediately (Fig. 3(d)).

4 The Iconic Object Detection/Classification Scheme

In the following we concentrate on the 2–D vision modules only. For them a common general iconic object detection scheme was developed, that consists of following subsequent processing steps (Fig. 4):

1. application-dependent pre-segmentation (optional step),
2. image scanning into a frame pyramid with different window sizes,
3. frame normalization step (size- and shape-, light- and histogram-based window normalization),
4. an application-dependent feature extraction step - it performs a transformation of the normalized window to a feature image (or vector) accordingly to the specific task.
5. the window classification step - it performs object detection or classification.

The classification network may take the form of a many-layer feed-forward network, of which the input corresponds to the feature vector. The classifier is

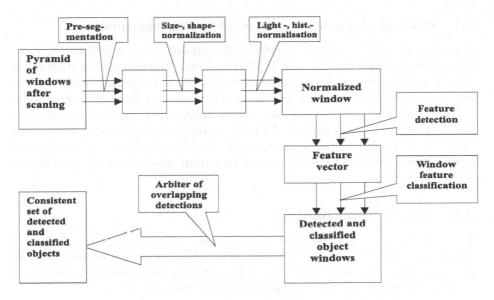

Fig. 4. The general iconic object detection/classification scheme.

(a)   (b)   (c)

Fig. 5. Motion based image window generation: (a) a normalized image, (b) the motion mask image, (c) detected candidate windows with moving objects.

trained with the inputs that have fixed size. The output indicates a class to which the input belongs. For locating an object in an input image of arbitrary size, one should scan the entire image.

As mentioned, the classification step itself is based is on ANN (artificial neural networks). Hence, two operation modes are distinguished: the learning mode and the active working mode. In the active mode following detailed steps are possible, after an object of some class has been detected inside of some window: (1) assuming the exact front view position of the object is given , it allows further detection processes (for example license plate recognition), or (2) the original size of the window gives clues about the real size (width and height) of the object.

4.1   Example of Application-Dependent Pre-segmentation

If moving objects have to be detected/classified only, then a preliminary pixel-based visual motion mask can be applied for a fast detection of important image windows during the generation of the image frame pyramid. It allows an image segmentation into moving and non-moving regions. Any visual motion detector can be applied to two or more consecutive images (Fig. 5). For example, a so called dynamic mask is generated by the method of adaptive difference image (Fig. 5(b)). The image pixels are then classified into "moving" or "stationary" pixels, which leads to the detection of important windows (with "moving objects") (Fig. 5(c)).

4.2   Window Normalization

The entire image or only selected image regions are scanned to variously located windows of different sizes. After size normalization to some predefined window size, iconic pre-processing steps, like light-correction (subtracting the estimated scene light) and histogram stretching can be performed.

4.3   Image Window Classification by ANNs

Let an image window of some normalized size is scanned to an input vector Y. A standard numeric approach to classification would be very time consuming, as the vector sizes are very large. The approach usually requires matrix inversion of matrices with sizes equal to the size of a cross-correlation matrix of input vector Y. An ANN-learning approach is preferred instead. In course of diploma works we have tested two ANNs: (a) the multi-layer perceptron with classical back-propagation algorithm and (b) a supervised LVQ method. A simple adaptive solution of the classification problem can be provided by means of the following supervised LVQ algorithm [8]:

1. Every output neuron i represents a class label C_i.
2. A learning sample p^d consists of an input vector x_p and its class label d.
3. Two winners are determined, the correct one k (which corresponds to the proper class label of the input sample) and the nearest incorrect one l (by distance measures between weights w_i of remaining i-th output neurons and the input x).
4. In the learning rule the vector w_k corresponding to the correct label is moved towards the input vector, whereas the vector w_l with the incorrect label, which may even be nearest the input, is moved away from it.

5   Some Applications of the Iconic Classification Scheme

5.1   Vehicle Detection in Many-Object Scenes

Its task is to detect the vehicles in the lane and to estimate the queue length. A particular feature detection process may even not be required in this case, as well

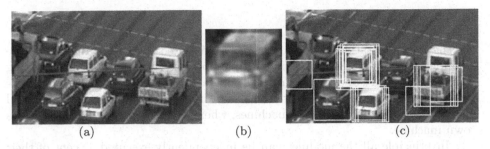

(a)   (b)   (c)

Fig. 6. White vehicle detection in a grey-scale image: (a) original image, (b) learned class pattern, (c) detected hypotheses.

as the arbitration step may be simplified. In Fig. 6 an example of objects from class "white vehicle", detected directly in the grey-scale image, is shown. The lane occupancy of even overlapping hypotheses may be relatively easy computed as the overlapping of windows from such a set.

5.2 Car Identification by License Plate Recognition

Its task is to extract the license plate section of a detected vehicle and to identify all characters, digits and eventually graphic patterns contained at the plate. For example, in [11] a Morphological Filter Neural Network is proposed for license plate character classification at the pixel level. In experimental studies, this morphological feature-based neural network (MFNN) system is applied among others to automatic recognition of vehicle license plates.

In our system the detection of license plate is fitted into the iconic classification scheme by using a specific feature extraction method. This method (inspired by [2]) consists of following subsequent steps: (1) local mean removal and binarization, (2) connected component detection, (3) filtering out small- and large-sized candidate components, (4) nearly-horizontal adjacency detection, (5) transforming the detected chain of components to a sequence of output windows, corresponding to single characters.

5.3 Vehicle Classification

Its task is to classify a detected vehicle object into: small car, lorry, bus, van, etc. [10,8]. Two general approaches are possible: a 2-D approach, assuming that only the front view is available, or a 3-D approach, assuming that the length of the vehicle is also detectable. Due to the earlier performed motion-based segmentation step, image windows with single cars are already available. In order to get the feature vector for the classifier, the 2–D or 3–D boundary boxes of the vehicle are searched for. This requires the detection and grouping of nearly horizontal and nearly vertical line segments in the image window.

6 Conclusions

In the implementation of the sensor, each presented system module should be mapped onto a single process or multiple processes of a concurrently working computer system (for example under the operating system QNX). Different hardware alternatives can be considered - starting from a single computer for all modules, and ending by many machines, where every module is executed on its own machine.

In principle all the modules can be independently executed, except of their dependence from the image acquisition and the auto-calibration modules. The image acquisition module controls a critical section - the vision sensor. Other application modules are gaining access to image data via this image acquisition module. Additionally, the sensor parameters must be on-line available and accessible by all modules.

References

1. Aggarwal, J.K., Nandhakumar, N.: On the computation of motion from sequences of images - a review. Proceedings of the IEEE **76**, no. 8 (1988) 917–935.
2. Blissett, R.: Eyes on the road. Roke Manor Research, IP Magazine, UK, May/June (1992) WWW page: www.roke.co.uk
3. India4 - video sensor. Product information. Citilog, Paris, France (1999) WWW page: www.citilog.fr
4. Ferrier, N.J., Rowe, S.M., Blake, A.: Real-Time Traffic Monitoring. Proceedings of the Second IEEE Workshop on Applications of Computer Vision. IEEE Computer Society Press (1994) 81–88
5. OMNI project: Open Model For Network–wide Heterogeneous Intersection–based Transport Management, EC IST-11250, www.cordis.lu/ist/projects/99-11250.htm, www.etra.es/english/tecnologias/curso/omni.html
6. OMNI specification: Omni network model - functional specifications, Delivery D2.1, IST-Project 11250 - OMNI, Brussels (Sept. 2000)
7. Kasprzak, W., Niemann, H.: Adaptive Road Recognition and Egostate Tracking in the Presence of Obstacles. International Journal of Computer Vision **28**, no. 1. Kluwer Academic Publ., Boston Dordrecht London (1998) 6–27
8. Kasprzak, W.: Adaptive methods of moving car detection in monocular image sequences. Machine Graphics & Vision **9** no. 1/2. ICS PAS Warsaw (2000) 167–186.
9. Takahashi, K. et al.: Traffic flow measuring system by image processing. MVA'96, IAPR Workshop on Machine Vision Applications. Tokyo, Japan (1996) 245–248
10. Tan, T.N., Sullivan, G.D., Baker, K.D.: Fast Vehicle Localisation and Recognition Without Line Extraction and Matching. BMVC94. Proceedings of the 5th British Machine Vision Conference. BMVA Press, Sheffield, UK (1994) 85–94
11. Won, Y.G., Park, Y.K.: Property of greyscale hit-or-miss transform and its applications. Machine Graphics & Vision **9**. ICS PAS Warsaw (2000) 539–547
12. Worrall, A.D., Sullivan, G.D., Baker, K.D.: A simple, intuitive camera calibration tool for natural images. BMVC94. Proceedings of the 5th British Machine Vision Conference. BMVA Press, Sheffield, UK (1994) 782–790

Matching in Catadioptric Images with Appropriate Windows, and Outliers Removal

Tomáš Svoboda[1] and Tomáš Pajdla[2]

[1] Computer Vision Group, Swiss Federal Institut of Technology, Gloriastrasse 35,
8092 Zürich, Switzerland, svoboda@vision.ee.ethz.ch
[2] Center for Machine Perception, Czech Technical University, Karlovo nám. 13,
CZ 121-35 Praha 2, Czech Republic

Abstract. Active matching windows for matching in panoramic images
taken by a catadioptric camera are proposed. The shape and the size
of the windows vary depending on the position of an interest point.
The windows size is then normalized and a standard correlation is used
for measuring similarities of the points. A semi-iterative method based
on sorting correspondences according to their similarity is suggested
to remove possible outliers. It is experimentally shown that using this
method the matching is successful for small and also big displacement
of corresponding points.

Keywords: image matching, catadioptric panoramic cameras, active
windows

1 Introduction

The process of establishing correspondences between a pair of images is called
image matching and it is a well established topic in computer vision, see [2]
or [14] for further references. The image matching consists of three main steps.
First, points of interest have to be detected in both images. The Harris corner
detector [5] is a commonly used algorithm for solving this task. Second, the
interest points are mutually compared and the most similar ones are paired into
the corresponding pairs, shortly *correspondences*. Some points are often matched
incorrectly. These false matches, called *outliers*, have to be removed, which is
the final stage of an image matching algorithm.

In this work, we are interested in the second two steps of the matching algorithm. We show that fixed, rectangular windows, usually used for cameras with
perspective projections, are no longer appropriate for catadioptric cameras with
a nonlinear projection. Similar observation is stated in [1] and homogeneous
filtering for viewsphere is proposed. We propose active windows for measuring similarities between points. Similarities between points are determined by
computing by the standard 2D correlation. For removing outliers, we suggest a
sequential method based on sorting correspondences.

The text is organized as follows. Section 2 introduces appropriate matching windows for panoramic cameras. A method for removing false matches is
explained in Section 3. Experiments with real cameras are shown in Section 4.

W. Skarbek (Ed.): CAIP 2001, LNCS 2124, pp. 733–740, 2001.

2   Windows for Panoramic Images

2.1   Why Do Not Standard Windows Work?

The crucial premise to a successful image matching is the fact that point similarities are measured on the projection of the same 3D scene planar patch. Rectangular windows commonly used for matching in perspective projections are no more adequate to the catadioptric cameras. The main reason may be a big camera rotation around its vertical axis, which however, is quite common case. The big camera rotation invoke large movement of corresponding points between consecutive frames. It is worse when the camera movement is combined with some significant translation. Let us consider a 3D point in the scene. Its projections in the cameras could not only change its azimuth but also the the distance from the image center. The standard rectangular window with fixed size and accordingly oriented w.r.t. the image axis clearly does not represent the same surroundings of the scene point. Hence, the main assumption of the successful matching is not fulfilled.

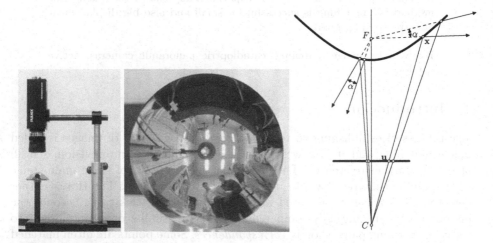

Fig. 1. Central catadioptric camera with a hyperbolic mirror (left) and a catadioptric image (middle). Size of the optimal window depends on the point's position (right). Parameters of the hyperbola are the same as for our real camera. For clarity, elevation range α is in the drawing enlarged to $12°$. F denotes the focal point of the mirror and C is the projection center of the camera.

2.2   Derivation of an Appropriate Window

Suppose that the displacement of the catadioptric camera center is significantly smaller than the depth of the scene. It means that the scale of the point's surroundings does not change as seen from these viewpoints, in other words, the surroundings is seen under the same spatial angle. This spatial angle intersects the mirror creating a small patch that, projected to the image plane, gives us the desired optimal window for matching. Note that the matching window will be bigger if the point is close to the mirror's periphery and smaller if the point is close to the image center, see Fig 1.

Statement 1 *Optimal (under assumptions stated above) window is bounded by four conics.*

Consider a surroundings of a space point bounded by lines. These lines project as conics in the image plane [3].

Practical considerations. Debate spatial cones with the rectangular cross section. The dimensions of the cones are then defined by the *elevation* and *azimuth* range. These ranges play the role of the height and the width of the usual rectangular window and are to be preset, i.e. they will not be modified by the matching algorithm. The ranges has to be set carefully. Too small cones give us low descriptive information and more points will be mutually similar. However, too large cones highly probably breaks the assumption of locally planar neighborhood of the scene point. Moreover, they are computationally expensive during the matching process. One can afford relatively large cones, since our camera moves within a man made environment with a lot of planar structures, like walls, floors, ceilings or cabinets. The computation of conics would be impractical for the implementation. Let us put a constraint on the movements of cameras that is commonly satisfied in practice. We assume that the changes in the inclination of the cameras are neglectable. Therefore, we can assume that surroundings of the points are bounded by lines that are parallel and perpendicular to the common horizon of the cameras. Horizontal lines are projected to the conics with the common main axis, moreover, they can be approximated by the circles centered at the image center, since only small parts of the conics are used. Vertical lines project to the radial lines that intersect in the image center. Using this approximation, the points within the optimal window can be addressed by polar coordinates.

Computation of the window parameters. Assume a point of interest, $\mathbf{u}$, found by, e.g., the Harris detector. This point is projected to the mirror point $\mathbf{x}$, see Fig 1. We omit the transformation formulas here in order to keep the derivation lucid. Interested readers are referred to [8,10] for details. The Cartesian coordinates, $\mathbf{x} = [x, y, z]^\top$ are transformed to the spherical coordinates $\mathbf{x} = [\theta, \varphi, r]^\top$, where θ denotes the elevation, φ the azimuth, and r the distance of the point from the origin. Recall, that the origin is placed at the effective viewpoint, the focal point of the mirror. The corner points of the area made by the intersection of the spatial cone and the mirror surface are computed using the preset ranges of the azimuth and the elevation. Their radii r are computed from

$$r = \frac{-b^2}{e \sin \varphi - a},$$
(1)

which is the equation in polar coordinates of the hyperbola centered at the focal point of the mirror, where a, b are the parameters of the hyperbolic mirror, and $e = \sqrt{a^2 + b^2}$. These corner points, after being projected back to the image plane, bound the window where the measure of similarity is to be computed. Note, that they do not bound a rectangular window but a part of an inter-circular area.

2.3 Measure of Similarity

Direct comparison of different windows is not possible because of their different size and non-integer coordinates of their boundaries. Differently sized windows would be normalized to some shape and size that enables direct measure of similarity, e.g. the correlation or sum of square differences. We re-map the irregular windows to rectangular shape, being the azimuth on the horizontal and the radius on the vertical axis. The resolution of the transformed window is determined by the largest window that surrounds a point close to the periphery of the mirror. It means that the smaller window are mapped oversampled. No interpolation is used in this backward mapping. The normalized rectangular windows are then resampled to a smaller resolution using bicubic interpolation. These windows can then be fed into a standard algorithm that measures the similarity of points. We use the standard 2D correlation

$$c = \frac{\sum_{ij}(I_{ij} - \overline{I})(J_{ij} - \overline{J})}{\sqrt{\sum_{ij}(I_{ij} - \overline{I})^2 \sum_{ij}(J_{ij} - \overline{J})^2}} , \tag{2}$$

where I and J are intensities in the image windows and $\overline{I}$ and $\overline{J}$ are average intensities in the windows.

2.4 Pairing

Mutual correlation coefficients of all points of interest in two images are computed and arranged in the matrix $C_{m,n}$, where m and n denote the number of interest points in the first and the second image respectively. Maximal correlation coefficients are found in each row and column of the matrix. Consider that a maximum has been found in row r. Its position denotes a column c. If the maximum of this column lies in the row r then the points, indexed by r in the first and by c in the second image, are considered as 1-1 corresponding pair.

3   Removing False Matches

We exploit the properties of the *essential matrix*. Each pair of corresponding points has to satisfy the following equation

$$\mathbf{x}_2^\top E \mathbf{x}_1 = 0 , \tag{3}$$

where $\mathbf{x}$ are point coordinates on the mirror and E is the essential matrix. A derivation of (3) can be found in [6] for perspective cameras, and in [11] for catadioptric cameras. Equation (3) is a homogeneous linear equation in elements of E. It can be rearranged into the form

$$Ae = 0 . \tag{4}$$

Using the Singular Value Decomposition (SVD) [4] of A, e is obtained as the right singular vector corresponding to the smallest singular value. To improve the conditionality of the estimation, all mirror points are normalized to unit length $\|x\| = 1$, see [8] for details.

Many robust methods were proposed in the last two decades for solving (4), since its form is the same for both calibrated and uncalibrated perspective cameras. Robust, RANSAC based methods are suggested in [12,13]. These methods are proved to be very robust but require to know the percentage of outliers a priori and they are computationally expensive. They are useful in the case that a lot of correspondences is provided, since their complexity does not depend on the total number of correspondences.

Here, we propose a semi-iterative method that requires no assumptions about outliers. We assume that the higher correlated points make better pair than the less correlated points. This assumption may fail when highly repetitive patterns are presented in the scene. Thus, more constraints are necessary. Ranked correspondences are used to generate sample sets of the correspondences, S_i. First set S_1 contains the 9 best correspondences[1]. Each S_{i+1} set is then constructed as a union of S_i with some (one or a few) best ranked correspondences not yet involved in S_i. Maximally $N - 9$ sample sets can be constructed from N correspondences, yet practically, if N is large, 5 sets seem to be enough. Adding more correspondences, S_{i+1} should generate better results than the set S_i (only Gaussian noise is present) until the first outlier is included into the growing set.

For each sample set of correspondences, S_i, an essential matrix E_i is computed from (4). The quality of E_i has to be measured with respect to its expected properties. The matrix E has, in the noise-free case, rank 2 and its nonzero singular values are equal [7]. We propose the difference of two largest singular values of E_i as a quality measure

$$q_E = \sigma_1 - \sigma_2 , \tag{5}$$

where σ_1 and σ_2 are the two biggest singular values of E_i. The correspondences that were used for computation the best E_i, i.e. E_i with the lowest q_E, are denoted as inliers and the rest as *outliers*. A more detailed description of this outlier removal can be, with alternative methods for sorting correspondences, found in [10].

---

[1] We do not need to compute E from the minimal set of correspondences. Instead of using 7 points, we use 9 that much is more robust in the presence of the Gaussian noise.

 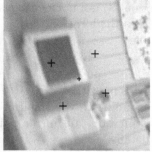 

Fig. 2. The panoramic image is shown on the left. The point of interest is on the left, approximately in the middle of the image. Cropped surrounding around the point of interest. Four crosses denote boundary of the optimal window. The re-mapped and re-sampled normalized optimal window with a resolution of 25×25 pixels is shown on the right.

4   Experiments

An example of the optimal window is shown in Fig 2. The elevation and azimuth range are both equal to $5°$. The point of interest was selected manually at the corner of one computer screen.

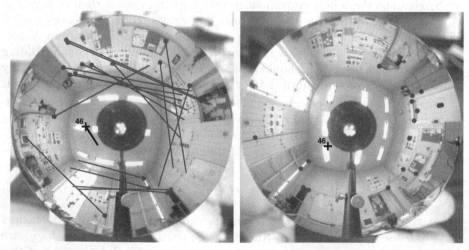

Fig. 3. The pair of the panoramic image. The movement between viewpoints includes a huge rotation around the vertical axis. The displacement of the corresponding points is larger than the half of the image. Positions of corrects pairs are denoted by the thick points. Lines show their displacement. The point No 46, denoted by a cross, is clearly an outlier because of the incoherent displacement.

In the following experiments, all interest points were located by the Harris corner detector [5]. The smoothing parameter σ for image derivatives is set to

6.5, the detector parameter $k = 0.04$ is taken from [5] and the local maxima of the filter's response are searched in the 8-neighborhood of each point. The local maximum is considered as an interest point if it is greater then the threshold 10^4. All detector parameters above follow from the experiments presented in [9]. Around 200 interest points were found in images. Many points had no counterparts in the second image. Interest points lying outside the mirror and very close to its center (camera itself reflection) as well as points on the mirror holder were removed. The parameters of the active widows were $3° \times 3°$ spatial cone and 15×15 pixels normalized window size.

Fig 3 shows the reliability of the matching when the camera undergoes substantial rotation and the displacement of the corresponding points is very large. The amount of the rotation is best seen on the position of the window (very bright, saturated area). Points of interest are shown as small crosses.

Possible 75 corresponding pairs were detected by the 1-1 pairing. The pairs were sorted according to their correlation coefficients and the essential matrix E was estimated with increasing number of pairs, see Figure 4. Together 17 points were denoted as correct pairs until the point No 46 was detected as the first outlier. The correctness of this decision can be visually verified on Figure 3.

The experiments with smaller displacement are omitted here because of space restrictions. The results are similar.

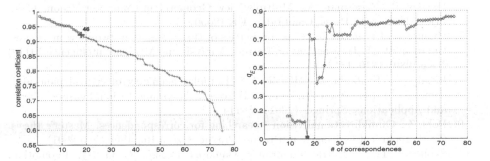

Fig. 4. Sorted correspondences on the left. Quality measure of the essential matrix on the right.

5 Conclusions

Active windows for matching in panoramic images were proposed. Active shape of the window yields, even with simple correlation, good matching results also in the case substantial displacement of the corresponding points. The performance of the active shape overcomes ones of the fixed, rectangular window mainly in the case of substantial rotation and of the changes in the distance of the point from the image center changes. A semi-iterative method based on sorting correspondences was suggested to remove outliers.

Highly repetitive patterns in the scene may spoil the matching. Many pairs exhibit good similarity however, they can be mutually commuted. Alternative method for sorting correspondences were proposed in [10].

Acknowledgment. This research was supported by the Czech Ministry of Education under the grant VS96049 and 4/11/AIP CR, by the Grant Agency of the Czech Republic under the grant GACR 102/01/0971, by the Research Program J04/98:212300013 Decision and control for industry, and by EU Fifth Framework Program project Omniviews No. 1999-29017.

References

1. Chahl, J., Srinivasan, M.: A complete panoramic vision system, incorporating imaging, ranging, and three dimensional navigation. IEEE Workshop on Omnidirectional Vision. IEEE Computer Society Press (June 2000) 104–111
2. Faugeras, O.: Three-Dimensional Computer Vision. MIT Press, 1993.
3. Geyer, C., Daniilidis, K.: A unifying theory for central panoramic systems and practical implications. In: Vernon, D. (ed.): Proceedings of the 6th European Conference on Computer Vision. Lecture Notes in Computer Science no. 1843. Springer, Berlin, Germany (June/July 2000) 445–461
4. Golub, G.H., v. Loan, C.F.: Matrix Computation. 2nd edn. The Johns Hopkins University Press (1989)
5. Harris, C., Stephen, M.: A combined corner and edge detection. In: Matthews, M.M. (ed.): Proceedings of the 4th ALVEY vision conference. University of Manchester, England (Sep 1988) 147–151
6. Hartley, R., Zisserman, A.: Multiple View Geometry in Computer Vision. Cambridge University Press, Cambridge, UK (2000)
7. Huang, T., Faugeras, O.: Some properties of the E matrix in two-view motion estimation. IEEE Transactions on Pattern Analysis and Machine Intelligence **11**(12) (Dec 1989) 1310–1312
8. Pajdla, T., Svoboda, T., Hlaváč, V.: Epipolar geometry of central panoramic cameras. In: Benosman, R., Kang, S.B. (eds.): Panoramic Vision: Sensors, Theory and Applications. Springer (2001) 85–114
9. Pohl, Z.: Omnidirectional vision – searching for correspondences. Master's thesis. Czech Technical University in Prague, Prague, Czech Republic (Feb 2001)
10. Svoboda, T.: Central Panoramic Cameras Design, Geometry, Egomotion. PhD Thesis, Center for Machine Perception. Czech Technical University, Prague, Czech Republic (Apr 2000)
11. Svoboda, T., Pajdla, T., Hlaváč, V.: Epipolar geometry for panoramic cameras. In: Burkhardt, H., Bernd, N., (eds.): the Fifth European Conference on Computer Vision, Freiburg, Germany. Lecture Notes in Computer Science, no. 1406. Springer, Berlin, Germany (Jun 1998) 218–232
12. Torr, P., Murray, D.: The development and comparison of robust methods for estimating the fundamental matrix. IJCV **24**(3) (Sep 1997) 271–300.
13. Zhang, Z.: Determining the epipolar geometry and its uncertainty: A review. International Journal of Computer Vision **27**(2) (1998) 161–195
14. Zhang, Z., Deriche, R., Faugeras, O., Luog, Q.T.: A robust technique for matching two uncalibrated images through the recovery of the unknown epipolar geometry. Artificial Intelligence **78** (Oct 1995) 87–119. Also: Research Report No.2273, INRIA Sophia-Antipolis

Author Index

Lecture Notes in Computer Science

For information about Vols. 1–2048
please contact your bookseller or Springer-Verlag